An Elsevier Civil Engineering Compendium

STRUCTURAL ENGINEERING COMPENDIUM I

An Elsevier Civil Engineering Compendium

STRUCTURAL ENGINEERING COMPENDIUM I

A collection of papers from the journals

Journal of Constructional Steel Research
Thin-Walled Structures
Engineering Structures
Computers and Structures
Construction and Building Materials
Journal of Wind Engineering and Industrial Aerodynamics
Marine Structures

2000

ELSEVIER
2002
AMSTERDAM – LONDON – NEW YORK – OXFORD – PARIS – SHANNON – TOKYO

ELSEVIER SCIENCE Ltd
The Boulevard, Langford Lane
Kidlington, Oxford OX5 1GB, UK

First edition 2002

Library of Congress Cataloging in Publication Data
A catalog record from the Library of Congress has been applied for.

British Library Cataloguing in Publication Data
A catalogue record from the British Library has been applied for.

ISBN: 0 08 044038 X

Transferred to digital printing 2005
Printed and bound by Antony Rowe Ltd, Eastbourne

PREFACE

This compendium is made up of a selection of the best and most representative papers from a group of Elsevier's structural engineering journals. Selections were made by the journals' Editorial teams.

The papers all appeared during 2000 and we hope they will provide a useful snapshot of the field for researchers, students and practitioners.

The journals from which papers have been selected and their Editors are:

Journal of Constructional Steel Research *P.J. Dowling, J.E. Harding, R. Bjorhovde*
Thin-Walled Structures *J. Loughlan, K.P. Chong*
Engineering Structures *P.L. Gould*
Computers and Structures *K.J. Bathe, B.H.V. Topping*
Construction and Building Materials *M.C. Forde*
Journal of Wind Engineering & Industrial Aerodynamics *N.P. Jones*
Marine Structures *A. Mansour, T. Moan, Tetsuya Yao*

Each paper appears in the same format as it was published in the journal; citations should be made using the original journal publication details.

It is intended that this compendium will be the first in a series of such collections. A compendium has also been published in the area of geotechnical engineering-
(ISBN: 0-08-044095-9)

For full details of Elsevier's structural engineering publications, please visit the Elsevier website at www.elsevier.com or (where access is allowed) log on to ScienceDirect at www.sciencedirect.com.

Ian Salusbury
Elsevier Science Ltd
Langford Lane
Kidlington
Oxford OX5 1GB
United Kingdom

Email: i.salusbury@elsevier.co.uk

CONTENTS

Papers from

Journal of Constructional Steel Research

Deformation limit and ultimate strength of welded T-joints in cold-formed RHS sections

Xiao-Ling Zhao *

Department of Civil Engineering, Monash University, Clayton, VIC 3168, Australia

Abstract

This paper describes the deformation limit and ultimate strength of welded T-joints in cold-formed RHS sections. Both web buckling failure mode and chord flange failure mode are investigated. The strength at a certain deformation (chord flange indentation) limit can be regarded as the ultimate strength of a T-joint. The deformation limit mainly depends on the ratio β $(=b_1/b_o)$. Based on the test results of T-joints in cold-formed RHS sections, the deformation limit is found as $3\%b_o$ for $0.6\leq\beta\leq0.8$ or $2\gamma\leq15$, and $1\%b_o$ for $0.3\leq\beta<0.6$ and $2\gamma>15$. The ultimate strength so determined is compared with the existing design formulae. Proposed formulae for ultimate strength of web buckling failure and chord flange failure are given. © 2000 Elsevier Science Ltd. All rights reserved.

Keywords: Cold-formed steel; Deformation limit; Hollow sections; Ultimate strength; Welded joints

1. Introduction

A typical welded T-joint is shown in Fig. 1 where a branch is welded to a chord. Symbols used in this paper are defined above. Post-yield response has been observed in different types of tubular joints [1–15] due to the effect of membrane forces in the chord and strain hardening of the material. The deformation limit used to determine the ultimate strength of a joint has been investigated by many researchers: Mouty [16], Yura et al. [17], Korol and Mirza [18], Zhao and Hancock [19], Lu et al. [15]. However, these deformation limits are only valid for certain cases. A more general deformation limit based on the local indentation of the chord flange face

* Tel.: +61-3-990-54972; fax: +61-3-990-54944.
E-mail address: zxl@eng.monash.edu.au (X.-L. Zhao)

Reprinted from *Journal of Constructional Steel Research* **53 (2)**, 149-165 (2000)

Nomenclature

N_s	section capacity of web buckling
P_{max}	peak load
P_{ult}	ultimate load
$P_{1\%bo}$	load at deformation of $1\%b_o$
$P_{3\%bo}$	load at deformation of $3\%b_o$
P_{CIDECT}	predicted capacity using CIDECT model
P_{kato}	predicted capacity using Kato model
P_{zh1}	predicted capacity using the modified Kato model
P_{zh2}	predicted capacity using the membrane mechanism model
b_o	chord width
b_1	branch width
f_y	yield stress of RHS
h_o	chord depth
h_1	branch width
r_{ext}	external corner radius of RHS
t_o	chord thickness
t_1	branch thickness
α_c	reduction factor for web buckling capacity
β	b_1/b_o
2γ	b_o/t_o

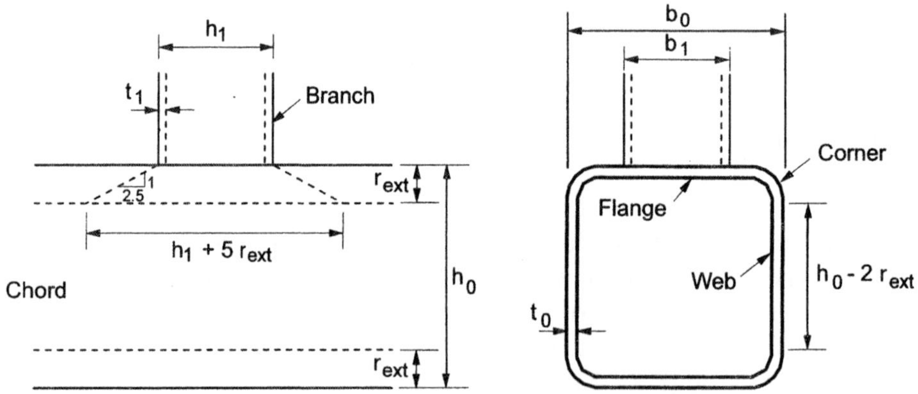

Fig. 1. Cold-formed RHS sections and web buckling model.

was proposed by Lu et al. [20] to cover all types of welded tubular joints. It can be summarised as:

- for a joint which has an obvious peak load at a deformation around $3\%b_o$, the peak load or the load at $3\%b_o$ deformation is considered to be the ultimate load, where b_o is the width of the chord member as shown in Fig. 1;

- for a joint which does not have a pronounced peak load, the ultimate deformation limit depends on the ratio of the load at $3\%b_o$ to the load at $1\%b_o$. If the ratio is greater than 1.5, the deformation limit is $1\%b_o$, i.e. serviceability is in control. The ultimate strength is taken as 1.5 times the load at $1\%b_o$. If the ratio is less than 1.5, the deformation limit is $3\%b_o$, i.e. strength is in control. The ultimate strength is taken as the load at $3\%b_o$;
- a validity range of β $(=b_1/b_o)$ and 2γ $(=b_o/t_o)$ is given to determine whether the design is governed by serviceability or by strength.

The proposal [20] was mainly based on tests of hot-rolled sections. There is a need to verify the proposed deformation limit for welded T-joints in cold-formed RHS sections.

This paper describes the verification of the deformation limit using the test results of welded T-joints in cold-formed RHS sections [2,7]. Both web buckling failure mode and chord flange failure mode are investigated. Based on the test results, the deformation limit is found to be $3\%b_o$ for $0.6\leq\beta\leq0.8$ or $2\gamma\leq15$ and $1\%b_o$ for $0.3\leq\beta<0.6$ and $2\gamma>15$. The ultimate strength of the web buckling is compared with the existing design formulae given by Packer [21], Packer et al. [22] and Zhang et al. [23]. The ultimate strength of chord flange failure is compared with the capacities of T-joints determined using CIDECT design formula [22], the Kato model [24], the modified Kato model [7] and the membrane mechanism model [19]. Proposed formulae for the ultimate strength of web buckling failure and chord flange failure are given, where the corners of cold-formed RHS sections are taken into account.

2. Experimental investigation

2.1. Failure modes

Tests on T-joints in cold-formed RHS sections were performed by Zhao and Hancock [7] in Australia and Kato and Nishiyama [2] in Japan. There are three main failure modes, namely web buckling failure, chord flange failure and branch local buckling failure. The failure mode of branch local buckling is similar to that observed in a stub column test. This failure mode is not discussed in this paper since local buckling can be prevented by using a plate slenderness which is lower than the plate yield slenderness limit of RHS sections as reported by Zhao and Hancock [25] and Hancock and Zhao [26]. A clear peak load is normally found for a web buckling failure mode. The chord flange failure usually has a post-yield response due to the effect of membrane forces in the chord and strain hardening of the material [19].

2.2. Test results

The section dimensions and material properties are summarised in Tables 1 and 2 for web buckling failure and for chord flange failure, respectively. The following

Table 1
Dimensions and material properties (specimens with web buckling failure)

Specimen label	Chord h_o (mm)	b_o (mm)	t_o (mm)	Branch h_1 (mm)	b_1 (mm)	t_1 (mm)	β (b_1/b_o)	2γ (b_o/t_o)	Yield stress (MPa)
S1B1C11	102	51	4.9	51	51	4.9	1.0	10.4	379
S1B1C12	102	51	3.2	51	51	4.9	1.0	15.9	373
S1B1C13	102	51	2	51	51	4.9	1.0	25.5	400
S1B2C21	102	102	9.5	102	102	8	1.0	10.7	421
S1B2C22	102	102	6.3	102	102	8	1.0	16.2	412
Kato1	127	127	7.9	102	102	6.4	0.80	16.1	404
Kato1′	127	127	7.9	102	102	6.4	0.80	16.1	342
Kato2	150	150	6	125	125	6	0.83	25.0	366
Kato2′	150	150	6	125	125	6	0.83	25.0	328
Kato3	200	200	6	178	178	12.7	0.89	33.3	368
Kato4	127	127	3	102	102	6.4	0.80	42.3	382
Kato6	127	127	3	152	102	6.4	0.80	42.3	382
Kato7	127	127	3	203	102	6.4	0.80	42.3	382
Kato8	203	203	4.8	178	178	12.7	0.88	42.3	348
Kato23	127	127	7.9	102	102	4.8	0.80	16.1	404
Kato24	150	150	6	127	127	6.4	0.85	25.0	366
Kato37	150	150	6	127	127	3	0.85	25.0	366
Kato44	350	350	12	300	300	6	0.86	29.2	264

Table 2
Dimensions and material properties (specimens with chord flange failure)

Specimen label	Chord h_o (mm)	b_o (mm)	t_o (mm)	Branch h_1 (mm)	b_1 (mm)	t_1 (mm)	β (b_1/b_o)	2γ (b_o/t_o)	Yield stress (MPa)
S1B1C21	102	102	9.5	51	51	4.9	0.50	10.7	421
S1B1C22	102	102	6.3	51	51	4.9	0.50	16.2	412
S1B1C23	102	102	4	51	51	4.9	0.50	25.5	417
Kato5	127	127	3	51	102	6.4	0.80	42.3	382
Kato11	150	150	6	100	100	6	0.67	25.0	366
Kato12	200	200	6	150	150	6	0.75	33.3	368
Kato13	250	250	6	200	200	9	0.80	41.7	400
Kato15	150	250	9	178	178	12.7	0.71	27.8	387
Kato16	150	150	6	75	75	3.2	0.50	25.0	366
Kato17	200	200	6	125	125	6	0.63	33.3	368
Kato19	178	229	4.6	102	102	6.4	0.45	49.8	375
Kato21	254	254	9.5	127	127	6.4	0.50	26.7	380
Kato25	250	250	6	75	75	2.3	0.30	41.7	400
Kato26	150	150	6	75	75	2.3	0.50	25.0	366
Kato27	150	150	6	102	102	3.2	0.68	25.0	366
Kato32	350	350	12	102	102	2.4	0.30	29.2	264
Kato33	254	254	9.5	127	127	3	0.50	26.7	380

information is included in Tables 1 and 2: specimen label, dimensions of the chord members and branch members, the ratio (β) of the width of branch (b_1) to that of chord (b_o), the ratio (2γ) of the width of chord (b_o) to the chord thickness (t_o) and the measured yield stress. The ratio β varies from 0.3 to 1.0. The ratio 2γ varies from 10.4 to 50. All the RHS sections are manufactured using a cold-forming process. The measured yield stress (the 0.2% proof stress for rounded stress–strain curves) varies from 264 to 421 MPa. The load versus deflection (local indentation of the chord flange face) curves are reported in Zhao [27] and Kato and Nishiyama [2], which form the basis of the verification described in Sections 3 and 4. Detailed test procedures are described in Zhao and Hancock [7] and Kato and Nishiyama [2]. Displacement transducers were used to measure the deformation (Δ) as shown in Figure 2 of Zhao and Hancock [7].

3. Deformation limit for web buckling failure mode

For a joint which has a peak load (P_{max}) at a deformation smaller than $3\%b_o$ the peak load is considered to be the ultimate load (P_{ult}), as shown in Fig. 2(a). For a joint which has a peak load (P_{max}) at a deformation larger than $3\%b_o$ the load at the deformation limit $3\%b_o$ is considered to be the ultimate load (P_{ult}), as shown in Fig. 2(b). The ultimate loads so determined are shown in Table 3 for T-joints with a web buckling failure mode. The ultimate load (P_{ult}) is compared with the peak load (P_{max}) obtained in the test. A mean ratio of 0.998 is reached with a small COV (coefficient of variation) of 0.003. It can be concluded that for the web buckling failure mode the $3\%b_o$ deformation limit for the ultimate strength proposed by Lu et al. [20] applies to T-joints in cold-formed RHS sections with $0.8 \leq \beta \leq 1.0$.

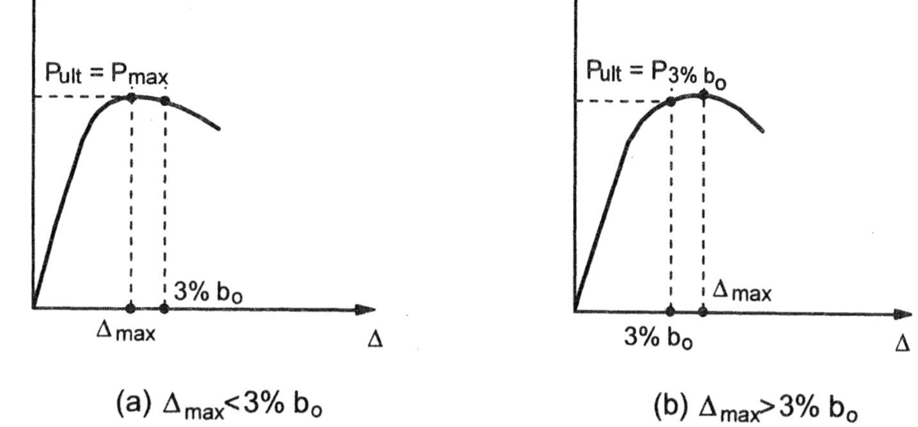

Fig. 2. Load–deformation curves (web buckling failure).

Table 3
Deformation limit and ultimate strength (web buckling failure)

Specimen label	Δ_{max} (%b_o)	P_{max} (kN)	P_{ult} (kN)	P_{ult}/P_{max}
S1B1C11	3.14	326	324	0.994
S1B1C12	2.25	163	163	1.000
S1B1C13	2.31	75.4	75.4	1.000
S1B2C21	4.08	1207	1193	0.988
S1B2C22	2.33	652	652	1.000
Kato1	3.25	565	563	0.996
Kato1′	3.44	541	539	0.996
Kato2	2.33	353	353	1.000
Kato2′	2.02	332	332	1.000
Kato3	1.31	514	514	1.000
Kato4	2.68	88	88	1.000
Kato6	2.56	111	111	1.000
Kato7	1.28	148	148	1.000
Kato8	1.58	270	270	1.000
Kato23	3.94	548	544	0.993
Kato24	2.08	430	430	1.000
Kato37	2.27	316	316	1.000
Kato44	1.92	1065	1065	1.000
MEAN				0.998
COV				0.003

4. Deformation limit for chord flange failure mode

4.1. Ultimate load (strength versus serviceability)

From Lu et al. [20], the ultimate deformation limit depends on the ratio of the ultimate load ($P_{3\%bo}$) to the serviceability load ($P_{1\%bo}$). If the ratio is less than 1.5, the ultimate deformation limit is 3%b_o, i.e. the strength is in control. The ultimate strength is taken as $P_{3\%bo}$, as shown in Fig. 3(a). If the ratio is greater than 1.5, the

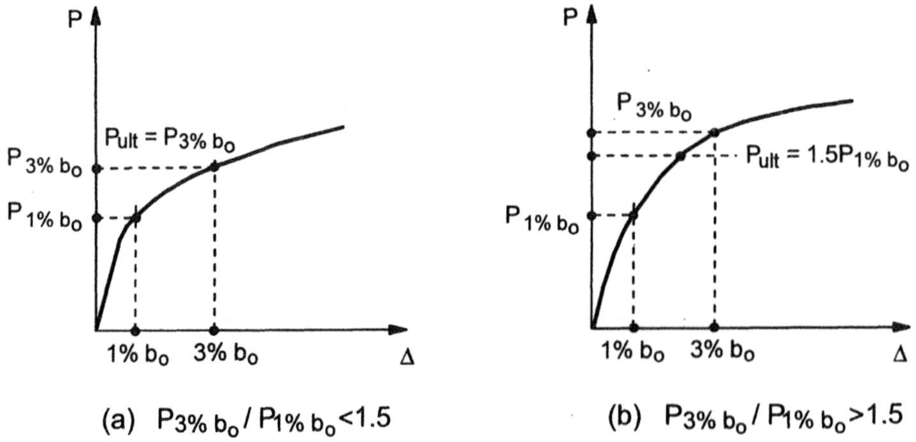

Fig. 3. Load–deformation curves (chord flange failure).

serviceability deformation limit is 1%b_o, i.e. the serviceability is in control. The ultimate strength is taken as 1.5 times $P_{1\%bo}$, as shown in Fig. 3(b). The ultimate load (P_{ult}) so determined is listed in Table 4 where the control criterion and the corresponding values of β and 2γ are also given.

4.2. Validity ranges

A validity range of β and 2γ was given in Lu et al. [20] to determine whether the design is governed by serviceability or by strength. The β values in Table 4 are plotted in Fig. 4 against the corresponding 2γ values, where the existing validity range given by Lu et al. [20] is also plotted as a dashed line. A proposed validity range is plotted in Fig. 4 as a dot–dashed line based on the values in Table 4, which has a slightly lower cut-off value of 2γ. It can be seen from Fig. 4 that the ratio β is more important than the ratio 2γ in determining the control criterion. A simple validity range can be summarised as:

- for $0.6 \leq \beta \leq 0.8$ or $2\gamma \leq 15$, strength is in control, i.e. the deformation limit is 3%b_o and $P_{ult}=P_{3\%bo}$;
- for $0.3 \leq \beta < 0.6$ and $2\gamma > 15$, serviceability is in control, i.e. the deformation limit is 1%b_o and $P_{ult}=1.5 \cdot P_{1\%bo}$.

This agrees with the experimental observation that larger deformation is obtained for joints with smaller β due to the effect of membrane forces in the chord member [2,27].

Table 4
Deformation limit and ultimate strength (chord flange failure)

Specimen label	$P_{1\%bo}$ (kN)	$P_{3\%bo}$ (kN)	$P_{3\%bo}/P_{1\%bo}$	P_{ult} (kN)	Control criterion	β	2γ
S1B1C21	299	410	1.369	410	Strength	0.50	10.7
S1B1C22	91.2	170	1.861	137	Serviceability	0.50	16.2
S1B1C23	41.9	73.7	1.759	63	Serviceability	0.50	25.5
Kato5	60.8	75.8	1.246	76	Strength	0.80	42.3
Kato11	160	195	1.219	195	Strength	0.67	25.0
Kato12	206	255	1.238	255	Strength	0.75	33.3
Kato13	245	297	1.209	297	Strength	0.80	41.7
Kato15	465	569	1.224	569	Strength	0.71	27.8
Kato16	74.6	113	1.520	112	Serviceability	0.50	25.0
Kato17	129	158	1.229	158	Strength	0.63	33.3
Kato19	39	59.3	1.520	59	Serviceability	0.45	49.8
Kato21	203	306	1.512	305	Serviceability	0.50	26.7
Kato25	55	86.9	1.580	83	Serviceability	0.30	41.7
Kato26	64.3	104	1.621	97	Serviceability	0.50	25.0
Kato27	160	193	1.206	193	Strength	0.68	25.0
Kato32	144	224	1.560	216	Serviceability	0.30	29.2
Kato33	176	279	1.586	264	Serviceability	0.50	26.7

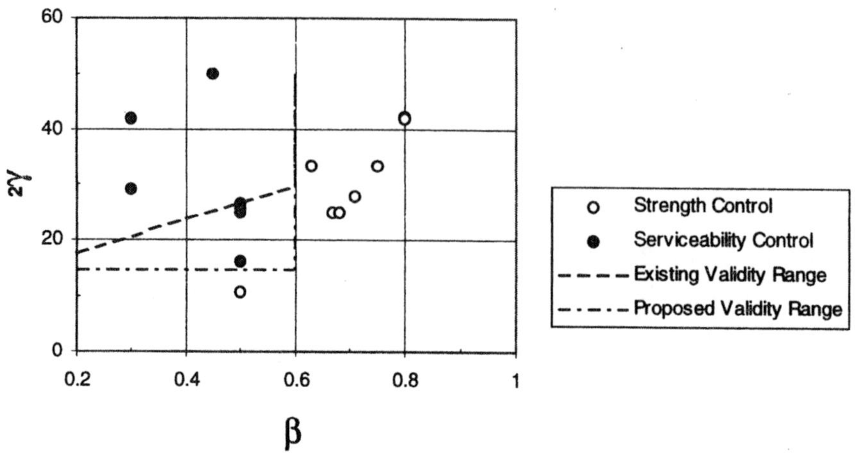

Fig. 4. Validity range (chord flange failure).

5. Ultimate strength of web buckling

5.1. Comparison with existing formulae for web buckling

The existing formulae for web buckling of RHS T-joints are summarised in Appendix A. The details can be found in Packer et al. [22], Packer [21] and Zhang et al. [23]. The experimental ultimate strength (P_{ult}) determined in Section 3 is compared with predicted ultimate strength using the existing formulae. The comparison is presented in Table 5. When the ratio is less than 1.0, it means that the formula underestimates the capacity. When the ratio is larger than 1.0, it means that the formula overestimates the capacity.

It can be seen that the CIDECT formula [22] underestimates the web buckling strength. The formula by Zhang et al. [23] underestimates the web buckling strength for β=1.0 and overestimates the web buckling strength for β<1.0. The formula by Packer [21] gives the best prediction with a mean ratio of 1.056 and a COV of 0.181. The following aspects are not considered in the existing formulae:

- rounded corners of cold-formed RHS sections in calculating the flat web depth;
- effect of β ratio which represents, to some extent, the influence of load eccentricity; or
- effect of $(h_o-2r_{ext})/t_o$ which represents, to some extent, the influence of column slenderness.

5.2. Proposed formulae for web buckling

In this paper the web buckling of RHS sections is treated as a column buckling problem. The proposed model is shown in Fig. 1. The column length is assumed to be (h_o-2r_{ext}) where h_o is the overall depth of the chord member and r_{ext} is the

Table 5
Comparison of ultimate strength (web buckling failure)

Specimen label	β	P_{ult} (kN)	P_{CIDECT} (kN)	P_{packer} (kN)	P_{zhang} (kN)	P_{CIDECT}/P_{ult}	P_{packer}/P_{ult}	P_{zhang}/P_{ult}
S1B1C11	1.0	324	198	267	270	0.611	0.824	0.833
S1B1C12	1.0	163	80	128	137	0.491	0.785	0.840
S1B1C13	1.0	75.4	14	62	60	0.186	0.822	0.796
S1B2C21	1.0	1193	1067	1127	1085	0.894	0.945	0.909
S1B2C22	1.0	652	548	549	623	0.840	0.842	0.956
Kato1	0.80	563	433	623	803	0.769	1.107	1.426
Kato1′	0.80	539	367	527	680	0.681	0.978	1.262
Kato2	0.83	353	261	390	530	0.739	1.105	1.501
Kato2′	0.83	332	234	349	475	0.705	1.051	1.431
Kato3	0.89	514	390	467	649	0.759	0.909	1.263
Kato4	0.80	88	59	114	164	0.670	1.295	1.864
Kato6	0.80	111	73	154	226	0.658	1.387	2.036
Kato7	0.80	148	87	204	285	0.588	1.378	1.926
Kato8	0.88	270	206	297	411	0.763	1.100	1.522
Kato23	0.80	544	433	623	803	0.796	1.145	1.476
Kato24	0.85	430	280	398	537	0.651	0.926	1.249
Kato37	0.85	316	280	398	537	0.886	1.259	1.699
Kato44	0.86	1065	877	1223	1704	0.823	1.148	1.600
MEAN						0.695	1.056	1.366
COV						0.238	0.181	0.283

external corner radius. This assumption is the same as that used in previous research on web buckling of RHS sections under bearing forces [28–30]. The column area is $(h_1+5r_{ext})\cdot t_o$ where h_1 is the overall depth of branch member and t_o is the web thickness. The column buckling strength can be expressed as

$$P_{web-buckling}=\alpha_c\cdot N_s \tag{1}$$

where α_c is a reduction factor and N_s is the section capacity, i.e.

$$N_s=2(h_1+5r_{ext})\cdot t_o\cdot f_y \tag{2}$$

in which f_y is the yield stress of the chord member.

The reduction factor (α_c) depends on the value of β and $(h_o-2r_{ext})/t_o$. The external radius of corners (r_{ext}) is taken as $2.5t$ when the thickness of the tube is larger than 3 mm, otherwise the external radius of corners is taken to be twice the thickness [31]. The value of β represents the effect of load eccentricity while the value of $(h_o-2r_{ext})/t_o$ represents the effect of column slenderness. The expression of α_c can be calibrated using the test results (P_{ult}) in Table 3.

The ratio P_{ult}/N_s ($=\alpha_c$) versus $(h_o-2r_{ext})/t_o$ is plotted in Fig. 5 for specimens with web buckling failure. It seems that the ratio P_{ult}/N_s ($=\alpha_c$) is about 0.7 for β=1. The ratio P_{ult}/N_s ($=\alpha_c$) decreases as $(h_o-2r_{ext})/t_o$ increases for $0.8<\beta<0.9$. The simple regression lines are plotted in Fig. 6. They are expressed as

$$\alpha_c=0.7 \text{ for } \beta=1.0 \tag{3}$$

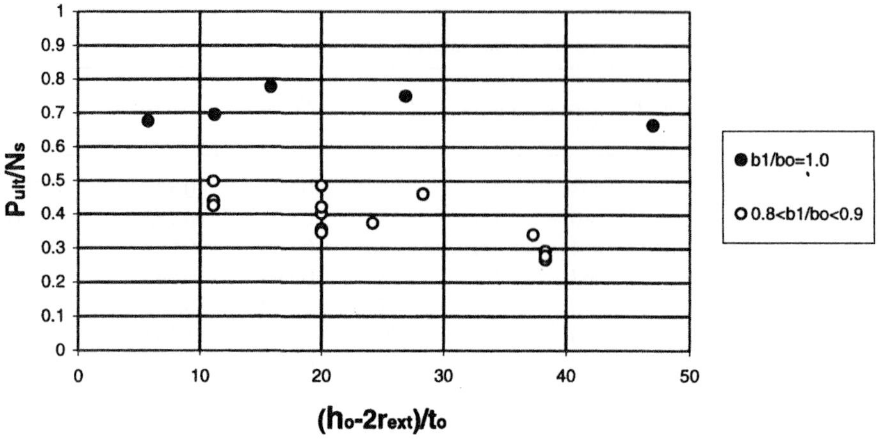

Fig. 5. P_{ult}/N_s versus $(h_o-2r_{ext})/t_o$.

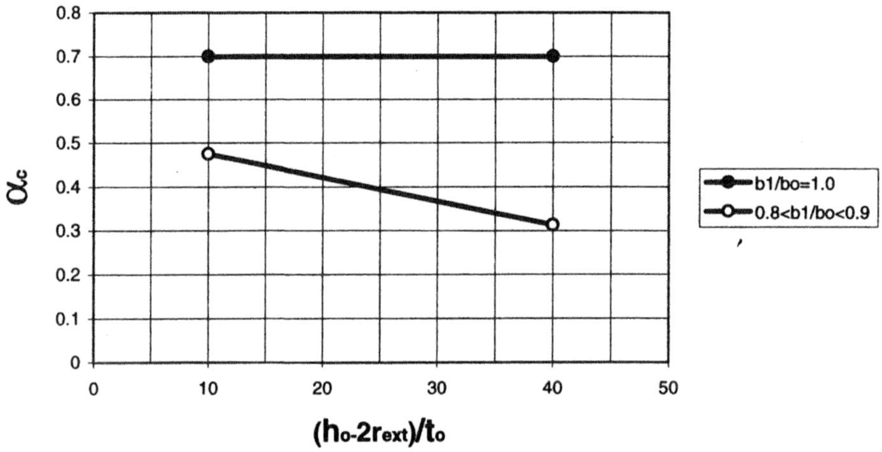

Fig. 6. α_c versus $(h_o-2r_{ext})/t_o$.

$$\alpha_c=0.529-0.0054\cdot(h_o-2r_{ext})/t_o \text{ for } 0.8<\beta<0.9 \tag{4}$$

A linear interpolation may be used for $0.9<\beta<1.0$.

The predicted web buckling strength $(\alpha_c\cdot N_s)$ is plotted in Fig. 7 against the experimental web buckling strength (P_{ult}). A good agreement is obtained with a mean ratio $(\alpha_c\cdot N_s/P_{ult})$ of 1.026 and a coefficient of variation of 0.1020.

6. Ultimate strength of chord flange failure

The existing formulae for chord flange failure of RHS T-joints are summarised in Appendix B. They are all based on yield line mechanisms. The details can be found in Packer et al. [22] for the CIDECT model, in Kato and Nishiyama [24] for

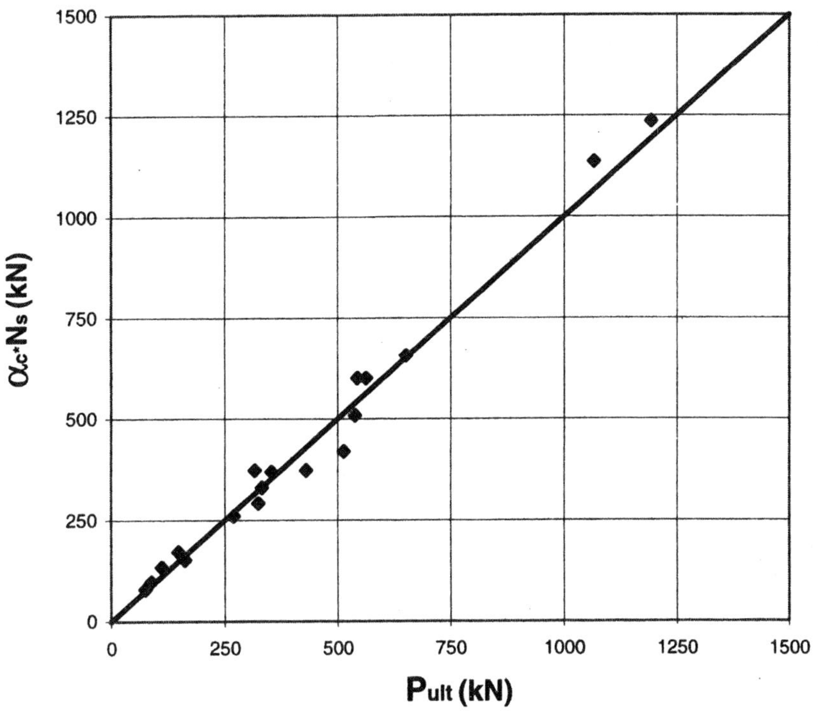

Fig. 7. Experimental ultimate strength (P_{ult}) versus proposed web buckling strength ($\alpha_c \cdot N_s$).

the Kato model, in Zhao and Hancock [7] for the modified Kato model and in Zhao and Hancock [19] for the membrane mechanism model. The experimental ultimate strength (P_{ult}) determined in Section 3 is compared with the predicted ultimate strength using the existing formulae. The comparison is presented in Table 6.

It can be seen that the CIDECT model (P_{CIDECT}) underestimates the chord flange strength. The reasons for the conservatism of the CIDECT model are the effect of membrane forces in the chord and the effect of local thickening of the corners on the negative plastic moments along the corners [32]. The Kato model (P_{kato}) overestimates the chord flange strength. This is most likely due to consideration of hinges on the centre of the corners, which did not occur as a result of much higher yield stress in the corners. Better results are obtained from the membrane mechanism model (P_{zh2}). The modified Kato model (P_{zh1}) gives the best results. The predicted strength (P_{zh1}) using the modified Kato model is plotted in Fig. 8 against the experimental ultimate strength (P_{ult}). Good agreement is obtained with a mean ratio (P_{zh1}/P_{ult}) of 1.049 and a COV of 0.091. The positions of hinges in the modified Kato model are at the top of the web adjacent to the corners rather than at the centre of the corners. It seems that the corners of cold-formed RHS sections should be considered in predicting both web buckling strength and chord flange strength. The modified Kato model is proposed to predict the chord flange strength.

Table 6
Comparison of ultimate strength (chord flange failure)

Specimen label	P_{ult} (kN)	P_{CIDECT} (kN)	P_{kato} (kN)	P_{zh1} (kN)	P_{zh2} (kN)	P_{CIDECT}/P_{ult}	P_{kato}/P_{ult}	P_{zh1}/P_{ult}	P_{zh2}/P_{ult}
S1B1C21	410	291	577	386	303	0.710	1.407	0.941	0.739
S1B1C22	137	125	208	166	141	0.912	1.518	1.212	1.029
S1B1C23	63	50.6	76.4	67.1	65.5	0.803	1.213	1.065	1.040
Kato5	76	45	126	80.3	65.6	0.592	1.658	1.057	0.863
Kato11	195	144	230	182	182	0.738	1.179	0.933	0.933
Kato12	255	185	301	234	246	0.725	1.180	0.918	0.965
Kato13	297	244	499	357	336	0.822	1.680	1.202	1.131
Kato15	569	389	781	564	483	0.684	1.373	0.991	0.849
Kato16	112	101	131	118	127	0.902	1.170	1.054	1.134
Kato17	158	131	171	152	182	0.829	1.082	0.962	1.152
Kato19	59	55.4	63.1	60.9	95.8	0.939	1.069	1.032	1.624
Kato21	305	263	333	303	340	0.862	1.092	0.993	1.115
Kato25	83	81.2	88.7	86.7	128	0.978	1.069	1.045	1.542
Kato26	97	101	131	118	127	1.041	1.351	1.216	1.309
Kato27	193	149	246	191	189	0.772	1.275	0.990	0.979
Kato32	216	212	242	233	239	0.981	1.120	1.079	1.106
Kato33	264	263	333	303	340	0.996	1.261	1.148	1.288
MEAN						0.840	1.276	1.049	1.106
COV						0.151	0.154	0.091	0.211

7. Conclusions

For the web buckling failure mode the $3\%b_o$ deformation limit for the ultimate strength proposed by Lu et al. [20] applies to T-joints in cold-formed RHS sections with $0.8 \le \beta \le 1.0$.

For the chord flange failure mode:

- for $0.6 \le \beta \le 0.8$ or $2\gamma \le 15$, strength is in control, i.e. the deformation limit is $3\%b_o$ and $P_{ult} = P_{3\%bo}$;
- for $0.3 \le \beta < 0.6$ and $2\gamma > 15$, serviceability is in control, i.e. the deformation limit is $1\%b_o$ and $P_{ult} = 1.5 \cdot P_{1\%bo}$.

A proposed web buckling formula (see Eq. (1)) has been given, which considers the rounded corners of cold-formed RHS sections, the effect of β ratio and the effect of $(h_o - 2r_{ext})/t_o$. The section capacity N_s is given in Eq. (2). The reduction factor α_c is given in Fig. 6 or in Eqs. (3) and (4).

The modified Kato model has been found to give the best prediction for chord flange failure.

It can be concluded that the corners of cold-formed RHS sections should be considered in predicting both web buckling strength and chord flange strength.

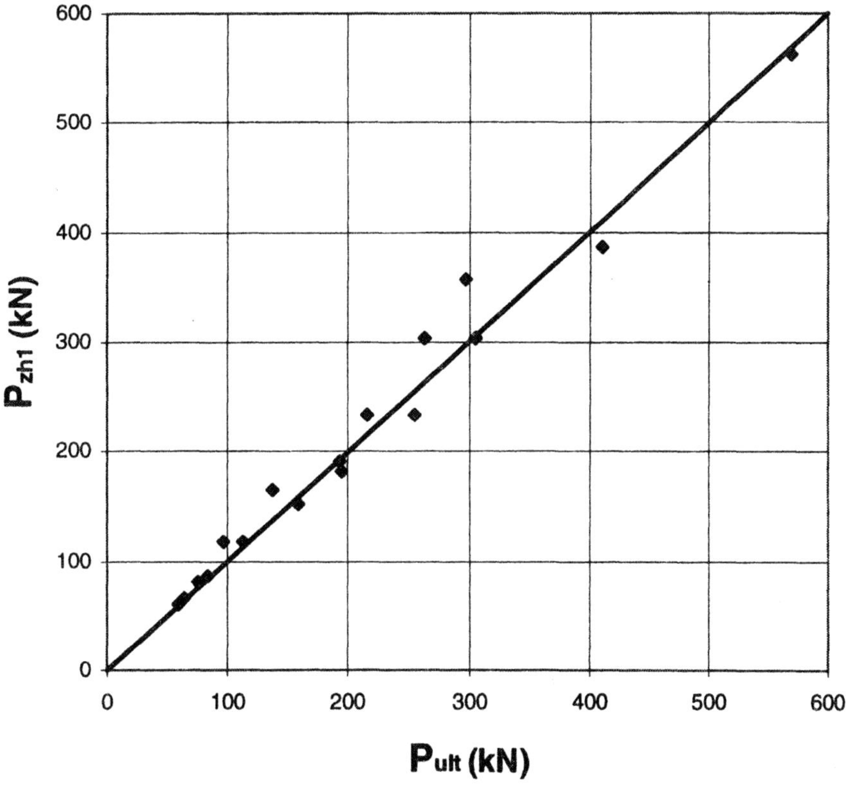

Fig. 8. Experimental ultimate strength (P_{ult}) versus proposed chord flange strength (P_{zh1}).

Appendix A. Existing formulae for web buckling

A.1. Formulae in CIDECT design guide [22]

For β=1.0

$$P_{CIDECT}=f_y.t_o\cdot(2h_1+10t_o)$$

For $\beta\leq 0.85$

$$P_{CIDECT}=\frac{f_y\cdot t_o^2}{1-\beta}\cdot\left(2\frac{h_1}{b_o}+4\sqrt{1-\beta}\right)$$

For $0.85<\beta<1.0$ use linear interpolation.

A.2. Formula by Packer [21]

$$P_{packer}=f_y\cdot b_o^{0.3}\cdot t_o^{1.7}\cdot\left[3.8+10.75\left(\frac{b_1+h_1}{2b_o}\right)^2\right]$$

A.3. Formula by Zhang et al. [23]

$$P_{zhang}=2f_y \cdot t_o \cdot h_{1e} \cdot k_1$$

in which

$$k_1=1.75-0.030 \cdot h_o/t_o \text{ when } h_o/t_o \leq 25$$

$$k_1=1.40-0.016 \cdot h_o/t_o \text{ when } h_o/t_o > 25$$

$$h_{1e}=h_1 \cdot k_2$$

$$k_2=(0.7 \cdot h_o/h_1)^{0.7} \text{ when } h_1/h_o \leq 0.7$$

$$k_2=(0.7 \cdot h_o/h_1)^{0.2} \text{ when } h_1/h_o > 0.7$$

Appendix B. Existing formulae for chord flange failure

B.1. Formula in CIDECT design guide [22] (CIDECT model)

$$P_{CIDECT}=\frac{f_y \cdot t_o^2}{1-\beta} \cdot \left(2\frac{h_1}{b_o}+4\sqrt{1-\beta}\right)$$

where $\beta=b_1/b_o$.

B.2. Formula by Kato and Nishiyama [24] (Kato model)

$$P_{kato}=\frac{f_y \cdot t_o^2}{1-\cdot\dfrac{b_1+2 \cdot t_o}{b_o-1.88 \cdot t_o}} \cdot \left(2 \cdot \frac{h_1+2 \cdot t_o}{b_o-1.88 \cdot t_o}+4\sqrt{1-\cdot\frac{b_1+2t \cdot_o}{b_o-1.88 \cdot t_o}}\right)$$

B.3. Formula by Zhao and Hancock [7] (modified Kato model)

$$P_{zh1}=\frac{f_y \cdot t_o^2}{1-\cdot\dfrac{b_1+2 \cdot t_o}{b_o}}\left(2 \cdot \frac{h_1+2 \cdot t_o}{b_o}+4\sqrt{1-\cdot\frac{b_1+2 \cdot t_o}{b_o}}\right)$$

B.4. Formula by Zhao and Hancock [19] (membrane mechanism model)

$$P_{zh2} = P_m + \sum_{i=3,5}^{9} k_i P_i$$

$$P_m = 2 \cdot S_y \cdot \sin\alpha_y$$

$$S_y = 0.5 \cdot (b_o + b_1) \cdot t_o \cdot f_y$$

$$\sin\alpha_y = \sqrt{1 - \cdot \frac{1}{(1+\varepsilon_y)^2}}$$

Table 7
Terms k_i and P_i

Yield line type no. (i)	Number of yield lines (k_i)	Contribution of each ith type yield line (P_i)
3	2	$P_3 = M_p \cdot \left(\frac{h_1}{n}\right)$
5	2	$P_5 = M_p \cdot \left(\frac{h_1}{n}\right) \cdot \left(\frac{1}{1-e}\right)$
6	4	$P_6 = M_p \cdot \left(\frac{e^2 \cdot \left(\frac{h_o}{b_o}\right)^2 + K^2}{K \cdot \left(\frac{n}{b_o}\right)}\right)$
7	4	$P_7 = M_p \cdot \left(\frac{(1-e)^2 \left(\frac{h_1}{b_o}\right)^2 + K^2}{K \cdot \left(\frac{n}{b_o}\right)}\right) \left(\frac{e}{1-e}\right)$
8	4	$P_8 = M_p \cdot \left(\frac{e}{K\left(\frac{n}{b_o}\right)}\right) \cdot \left(\frac{h_o}{b_o}\right)^2$
9	2	$P_9 = M_p \cdot \left(2K\left(\frac{b_o}{n}\right) + \frac{h_1}{n}\right)\left(\frac{e}{1-e}\right)$

$$\varepsilon_y = f_y / E$$

where E is taken as 200,000 MPa.

The terms k_i and P_i are given in Table 7 in which

$$M_p = \frac{f_y \cdot t_o^2}{4}$$

$$e = \left(\frac{n}{h_o}\right) \cdot \beta$$

$$n = \frac{b_o \cdot (1-\beta)}{2}$$

$$\beta = b_1 / b_o$$

$$K = 0.5 \cdot D_1^2 \cdot D_3 + 0.5 D_1 \cdot \sqrt{D_1^2 \cdot D_3^2 + 4 \cdot D_2}$$

$$D_1 = \sqrt{1-\beta} \cdot \sqrt{\frac{1-e}{1+e}}$$

$$D_2 = \left(\frac{e \cdot h_o}{n}\right) \cdot \left(\frac{h_o}{b_o}\right)$$

$$D_3 = \left(\frac{b_o}{t_o}\right) \cdot \left(\frac{1+\beta}{2}\right) \cdot \sqrt{(1+\varepsilon_y)^2 - 1}$$

References

[1] Mouty J. Calus des changes ultimes des assemblages soudes de profils creux carres et rectangularies. Construction Metallique 1976;2:37–58.

[2] Kato B, Nishiyama I. The static strength of R.R.-joints with large b/β ratio. CIDECT prog. 5y, Department of Architecture, Faculty of Engineering, University of Tokyo (Tokyo, Japan), 1979.

[3] Stark JWB, Soetens F. Welded connections in cold-formed sections. In: Proceedings of the 5th International Specialty Conference on Cold-Formed Steel Structures, St. Louis (MI, USA), 1980.

[4] Packer JA, Davies G, Coutie M. Ultimate strength of gapped joints in RHS trusses. J Struct Div, ASCE 1982;ST2:411–31.

[5] Panjehshahi E. The behaviour of RHS tee joints under axial load and bending moment. Master Thesis, Nottingham University, UK, 1983.

[6] Labed A. Membrane action in steel hollow section welded joints. Master Thesis, Nottingham University, UK, 1989.

[7] Zhao XL, Hancock GJ. T-joints in rectangular hollow sections subject to combined actions. J Struct Engng, ASCE 1991;117(8):2258–77.

[8] Lu LH, Puthli RS, Wardenier J. The static strength of uniplanar tubular X-joints loaded by in-plane and out-of-plane bending. Stevin Report 25.6.91.28/A1, Delft Univ. Tech., 1991.

[9] van der Vegte GJ, Lu LH, Puthli RS, Wardenier J. The ultimate strength and stiffness of uniplanar tubular steel X-joints loaded by in-plane bending. In: Proceedings of the 1st World Conference on Construction Steel Design, Mexico, 1992.

[10] de Winkel GD, Rink HD, Puthli RS, Wardenier J. In: The behaviour and the static strength of unstiffened I-beam to circular column connections under multiplanar in-plane bending moments. Proceedings of the 3rd International Offshore and Polar Engineering Conference, ISOPE-93, Singapore, 1993.
[11] de Winkel GD, Rink HD, Puthli RS, Wardenier J. In: The behaviour and static strength of plate to circular column connections under multiplanar axial loadings. Tubular structures V. London: E& FN Spon, 1993:703–11.
[12] de Winkel GD, Wardenier J. In: Parametric study on the static behaviour of I-beams to tubular column connections under in-plane bending moments. Tubular structures VI. Rotterdam: Balkema, 1994:317–24.
[13] Yu Y, Liu DK, Puthli RS, Wardenier J. In: Numerical investigation into the static behaviour of multiplanar welded T-joints. Tubular structures V. London: E&FN Spon, 1993:732–40.
[14] Yu Y, Wardenier J. In: Influence of the types of welds on the static strength of RHS T- and X-joints loaded in compression. Tubular structures VI. Rotterdam: Balkema, 1994:597–605.
[15] Lu LH, Puthli RS, Wardenier J. In: Ultimate deformation criteria for uniplanar connections between I-beams and RHS columns under in-plane bending. Proceedings of the 4th International Offshore and Polar Engineering Conference, ISOPE-94, Osaka (Japan), 1994.
[16] Mouty J. In: Theoretical prediction of welded joint strength. Proceedings of the International Symposium on Hollow Structural Sections, Toronto (Canada), 1977.
[17] Yura JA, Zettlemoyer N, Edwards IF. Ultimate capacity equations for tubular joints. OTC Proc 1980;1:3690.
[18] Korol RM, Mirza FA. Finite element analysis of RHS T-joints. J Struct Engng, ASCE 1982;108(9):2081–98.
[19] Zhao XL, Hancock GJ. Plastic mechanism analysis of T-joints in RHS under concentrated force. J Singapore Struct Steel Soc 1991;2(1):31–44.
[20] Lu LH, de Winkel GD, Yu Y, Wardenier J. In: Deformation limit for the ultimate strength of hollow section joints. Tubular structures VI. Rotterdam: Balkema, 1994:341–7.
[21] Packer JA. Web crippling of rectangular hollow sections. J Struct Engng, ASCE 1984;110(10):2357–73.
[22] Packer JA, Wardenier J, Kurobane Y, Dutta D, Yeomans N. Design guide for RHS joints under predominantly static loading. Köln (Germany): Verlag TÜV Rheinland, 1992.
[23] Zhang ZL et al. Nonlinear FEM analysis and experimental study of ultimate capacity of welded RHS joints. In: Proceedings of the 3rd International Symposium on Tubular Structures, Lappeenranta (Finland), 1989.
[24] Kato B, Nishiyama I. T-joints made of rectangular tubes. In: Proceedings of the 5th International Conference on Cold-Formed Steel Structures, St. Louis (MI, USA), 1980.
[25] Zhao XL, Hancock GJ. Tests to determine plate slenderness limits for cold-formed rectangular hollow sections of grade C450. J Australian Inst Steel Construct 1991;25(4):2–16.
[26] Hancock GJ, Zhao XL. Research into the strength of cold-formed tubular sections. J Construct Steel Res 1992;123:55–72.
[27] Zhao XL. The behaviour of cold-formed RHS beams under combined actions. Ph.D Thesis, The University of Sydney, Sydney, 1992.
[28] Zhao XL, Hancock GJ. Square and rectangular hollow sections subject to combined actions. J Struct Engng, ASCE 1992;118(3):648–68.
[29] Zhao XL, Hancock GJ. Square and rectangular hollow sections under transverse end bearing force. J Struct Engng, ASCE 1995;121(11):1565–73.
[30] Zhao XL, Hancock GJ, Sully RM. Design of tubular members and connections using amendment No. 3 to AS4100. J Australian Inst Steel Construct 1996;30(4):2–15.
[31] AISC Design capacity tables for structural steel hollow sections. Sydney (Australia): Australian Institute of Steel Construction, 1992.
[32] CIDECT The strength and behaviour of statically loaded welded connections in structural hollow sections, monograph No. 6. Corby (UK): CIDECT, 1986.

Structural design of stainless steel members — comparison between Eurocode 3, Part 1.4 and test results

B.A. Burgan *, N.R. Baddoo, K.A. Gilsenan

The Steel Construction Institute, Silwood Park, Ascot, Berkshire SL5 7QN, UK

Abstract

The paper describes the test results of an ongoing major European research project which is concerned with the further development and refinement of structural design guidance for stainless steel. The paper concentrates on the work carried out to date on the design of beams, columns and beam–columns and compares the test results with resistances predicted by the design pre-standard for structural stainless steel, ENV 1993-1-4. In general, the design guidance is conservative. The tests on CHS beams indicate that the limiting diameter-to-thickness ratios for section classification can be considerably increased. For welded I-section beams, the ENV 1993-1-4 lateral torsional buckling curve appears very conservative and the less conservative curve adopted in ENV 1993-1-1 for carbon steel appears to give a better fit to the data. © 2000 Elsevier Science Ltd. All rights reserved.

Keywords: Stainless steel; Eurocode 3 Part 1.4; Structural design

1. Introduction

The attractive appearance, corrosion resistance, ease of maintenance and low life cycle cost of stainless steels have led to their use within the construction industry for over 60 years. Typical applications include fixings, fasteners and cladding. However, structural materials with exceptionally high durability and corrosion resistance are required for certain applications within many industries such as the offshore, nuclear and paper-making industries. In many cases, stainless steel can provide a cost-effective and low maintenance structural solution to meet these demands. Austenitic

* Corresponding author. Tel.: +44-1344-623345; fax: +44-1344-622944.
E-mail address: b.burgan@steel-sci.com (B.A. Burgan).

Reprinted from *Journal of Constructional Steel Research* **54 (1)**, 51-73 (2000)

grades of stainless steel also exhibit exceptional ductility, good fire resistance and non-magnetic properties, all of which may lead to structural applications in particular circumstances.

Despite the fact that around 10% of stainless steel produced is used structurally or architecturally, comparatively little research has been carried out on structural behaviour. This has led to a lack of suitable design guidance for structural engineers, the one notable exception being the American code for cold-formed sections [1]. In 1988, a joint industry project was undertaken by the Steel Construction Institute to develop design guidance for European offshore and onshore stainless steel structural applications. The design recommendations arising from this project were published by EURO INOX in 1994 as the 'Design manual for structural stainless steel' [2].

2. European design standard for structural stainless steel

Eurocode 3 deals with the design of steel structures. Part 1.1, containing general rules and rules for buildings, was issued by CEN as ENV 1993-1-1 in 1992 [3]. Around this time, work started on preparing a Eurocode covering the design of structural stainless steel and this was later designated ENV 1993-1-4 (Part 1.4 of Eurocode 3) [4]. The 'Design manual for structural stainless steel' was used as a starting point for ENV 1993-1-4, with modifications and additions made to reflect the results of ongoing research and the new European material standard for stainless steel, EN 10088 [5]. It was published by CEN in 1996 and gives supplementary provisions for the design of buildings and civil engineering works which extend the application of ENV 1993-1-1 to austenitic and duplex stainless steels.

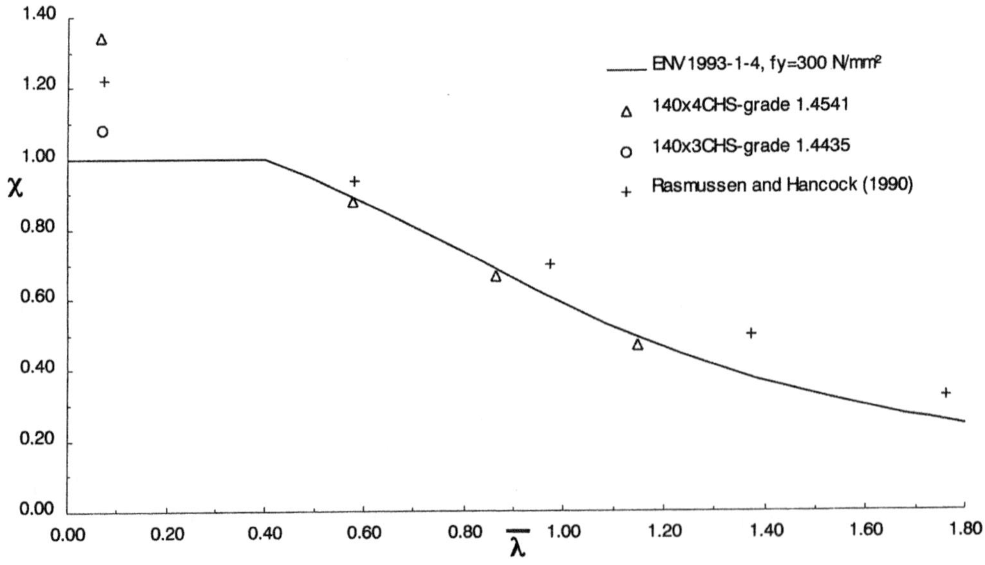

Fig. 1. Reduction factor, χ, versus generalised slenderness $\bar{\lambda}$ for CHS columns — test results and ENV 1993-1-4 design. curve

Table 1
Geometric and material properties of the CHS column specimens

Specimen reference name	140×4-SC	140×4-C1	14×4-C2	140×4-C3	140×3-SC	140×2-SC	140×2-C1	140×2-C2	140×2-C3
Steel grade to EN 10088	1.4541	1.4541	1.4541	1.4541	1.4435	1.4435	1.4435	1.4435	1.4435
Column length, L (mm)	499	2251	3350	4450	498	499	2250	3350	4449
End conditions (pinned/fixed)	F	P	P	P	F	F	P	P	P
Diameter, d (mm)	138.6	140	139.1	140.1	139.3	139.4	139.8	139.8	139.9
Thickness, t (mm)	3.98	3.99	3.99	3.98	2.87	1.97	1.95	1.95	1.97
Yield strength, f_y (N/mm^2)	294	293	293	294	352	318	319	319	318
Ultimate tensile strength, f_u (N/mm^2)	576	576	576	576	577	598	598	598	598
Modulus of elasticity, E (kN/mm^2)	193	195	195	193	190	201	197	197.5	201
Measured failure load, N_{test} (kN)	665	437	330	236	468	278	202	156	122

Table 2
Section classification and predicted buckling resistance of the CHS column specimens

Specimen reference name	140×4-SC	140×4-C1	14×4-C2	140×4-C3	140×3-SC	140×2-SC	140×2-C1	140×2-C2	140×2-C3
$(d/t)/\varepsilon^2$	47.4	47.11	46.81	47.92	80.35	100.04	103.74	103.48	100.4
Section class	1	1	1	1	3	4	4	4	4
$\bar{\lambda}$	0.065	0.577	0.865	1.148	0.071	**0.065**	**0.591**	**0.879**	**1.155**
ϕ	0.420	0.710	0.988	1.343	0.422	**0.420**	**0.722**	**1.004**	**1.352**
χ	1.000	0.890	0.682	0.491	1.000	**1.000**	**0.881**	**0.672**	**0.487**
Design buckling resistance, $N_{b.Rd}$ (kN)	494.87	444.6	338.63	245.49	433.00	**270.47**	**237.26**	**180.98**	**132.12**
$N_{test}/N_{b.Rd}$	1.344	0.983	0.975	0.961	1.081	**1.028**	**0.851**	**0.862**	**0.923**

Table 3
Geometric and material properties of the I-section column specimens — minor axis tests, grade 1.4301

Specimen reference name (minor axis tests)	I-160×80 C1	I-160×80 C2	I-160×80 C3	I-160×160 C1	I-160×160 C2	I-160×160 C3
Specimen length, L (mm)	650	1248	2046	1248	2049	3347
Web yield strength, f_{yw} (N/mm^2)	300	300	300	300	300	300
Web ultimate tensile strength, f_{uw} (N/mm^2)	624	624	624	624	624	624
Flange yield strength, f_{yf} (N/mm^2)	299	299	299	304	302	304
Flange ultimate tensile strength, f_{uf} (N/mm^2)	609	609	609	614	612	614
Modulus of elasticity, E (kN/mm^2)	200	200	200	200	199	200
Section depth, H (mm)	158	161.7	161.4	158.3	157.7	158
Section width, B (mm)	79.5	80.8	79.8	159.2	159.9	160.1
Web thickness, t_w (mm)	6	6	6	6	6	6
Flange thickness, t_f (mm)	9.8	9.8	9.8	9.8	9.9	9.8
Weld throat thickness, a (mm)	3	3	3	3	3	3
Measured failure load, N_{test} (kN)	627	420	270	1120	745	582

The basic approach followed during the preparation of the 'Design manual for structural stainless steel' was to adopt the rules for carbon steel, making modifications as necessary where stainless steel test data indicated different behaviour. In the cases where no data were available, the rules for carbon steel were generally suggested. Although this approach almost certainly led to 'safe' designs, the complex material behaviour of stainless steel was not being taken into account and its desirable properties not fully exploited. Since the material cost of stainless steel is high by normal construction material standards, economic design is of paramount importance.

3. Ongoing development of design guidance

In January 1997, a major European project started which is concerned with the further development and refinement of structural design guidance. The project is

Table 4
Geometric and material properties of the I-section column specimens — major axis tests, grade 1.4301

Specimen reference name (major axis tests)	I-160×80 C1	I-160×80 C2	I-160×80 C3	I-160×160 C1	I-160×160 C2	I-160×160 C3
Specimen length, L (mm)	2048	3343	5031	2025	3348	5145
Web yield strength, f_{yw} (N/mm^2)	300	300	300	300	300	300
Web ultimate tensile strength, f_{uw} (N/mm^2)	624	624	624	624	624	624
Flange yield strength, f_{yf} (N/mm^2)	299	299	299	300	300	299
Flange ultimate tensile strength, f_{uf} (N/mm^2)	609	609	609	610	610	610
Modulus of elasticity, E (kN/mm^2)	200	200	200	198	198	199
Section depth, H (mm)	157	157.6	158.5	158.3	158.4	158
Section width, B (mm)	79.4	78.9	80.1	160	159.8	159.2
Web thickness, t_w (mm)	6	6	6	6	6	6
Flange thickness, t_f (mm)	9.8	9.8	9.8	9.9	9.9	9.9
Weld throat thickness, a (mm)	3	3	3	3	3	3
Measured failure load, N_{test} (kN)	668	535	402	1130	860	725

being supported by the European Coal and Steel Community, the Nickel Development Institute and stainless steel producers in the UK, Sweden, Finland, France, Germany and Italy.

Both experimental and modern numerical methods (including non-linear finite element analyses) are being utilized to produce the data which is required to develop a structurally efficient design method for stainless steel structures. The tests measure actual resistances which include strain hardening and residual stresses. Numerical methods are used for describing the effects of different material stress–strain curves, for simulating the experimental tests and for analysing the effects of a wider range of parameters than those tested. Design rules are under development which will be suitable for updating ENV 1993-1-4 before it is converted to an EN (European Standard).

The scope of work covers static loading on members and connections, cyclic loading on welded connections and the behaviour of stainless steel members in fire.

Table 5
Geometric and material properties of the I-section column specimens — major axis tests, grade 1.4462

Specimen reference name (major axis tests)	I-160×160 C1	I-160×160 C2	I-160×160 C3
Specimen length, L (mm)	2050	3348	5046
Web yield strength, f_{yw} (N/mm^2)	523	523	523
Web ultimate tensile strength, f_{uw} (N/mm^2)	777	777	777
Flange yield strength, f_{yf} (N/mm^2)	522	522	522
Flange ultimate tensile strength, f_{uf} (N/mm^2)	755	755	755
Modulus of elasticity, E (kN/mm^2)	201	201	201
Section depth, H (mm)	162.7	161.4	160.4
Section width, B (mm)	159.8	159.5	161
Web thickness, t_w (mm)	6.8	6.8	6.8
Flange thickness, t_f (mm)	10.6	10.6	10.6
Weld throat thickness, a (mm)	3	3	3
Measured failure load, N_{test} (kN)	1930	1490	990

Table 6
Section classification and predicted buckling resistance of the I-section column specimens — minor axis tests, grade 1.4301

Specimen reference name (minor axis tests)	I-160×80 C1	I-160×80 C2	I-160×80 C3	I-160×160 C1	I-160×160 C2	I-160×160 C3
$d/(t_w\varepsilon)$	25.1	25.8	25.7	25.1	25.0	25.1
$c/(t_f\varepsilon)$	3.8	3.9	3.9	8.6	8.6	8.7
Section class	1	2	2	1	1	1
$\bar{\lambda}$	0.4309	0.8156	1.3562	1.0170	0.6181	1.0105
ϕ	0.6806	1.0665	1.8590	0.64	0.8499	1.3185
χ	0.8282	0.5702	0.3195	0.8654	0.6977	0.4618
Design buckling resistance, $N_{b.Rd}$ (kN)	592.2	415.9	230.9	549.4	840.3	555.6
$N_{test}/N_{b.Rd}$	1.06	1.01	1.17	1.08	0.89	1.05

The main objectives of the structural member tests are to provide test data on the effects of:

- a non-linear stress–strain curve on the cross-section resistance and buckling resistance of different members,
- higher residual stresses arising from fabricating stainless steel,
- a non-linear stress–strain curve on member deflection.

As part of this project, VTT Building Technology in Finland tested over 80 stainless

Table 7
Section classification and predicted buckling resistance of the I-section column specimens — major axis tests, grade 1.4301

Specimen reference name (major axis tests)	I-160×80 C1	I-160×80 C2	I-160×80 C3	I-160×160 C1	I-160×160 C2	I-160×160 C3
$d/(t_w\varepsilon)$	24.9	25	25.2	25.2	25.2	25.1
$c/(t_f\varepsilon)$	3.8	3.8	3.9	8.6	8.5	8.5
Section class	1	1	1	1	1	1
$\bar{\lambda}$	0.3941	0.6414	0.9584	0.366	0.6047	0.928
ϕ	0.6514	0.8734	1.2475	0.63	0.8367	1.2072
χ	0.8547	0.682	0.4888	0.875	0.7068	0.5052
Design buckling resistance, $N_{b.Rd}$ (kN)	609.1	484.8	351.6	1049.7	847.2	601.8
$N_{test}/N_{b.Rd}$	1.1	1.1	1.14	1.08	1.02	1.2

Table 8
Section classification and predicted compression resistance of the I-section column specimens — major axis tests, grade 1.4462

Specimen reference name (major axis tests)	I-160×160 C1	I-160×160 C2	I-160×160 C3
$d/(t_w\varepsilon)$	29.8	29.5	29.3
$c/(t_f\varepsilon)$	10.4	10.4	10.5
Section class	3	3	3
$\bar{\lambda}$	0.476	0.7837	1.188
ϕ	0.7182	1.0289	1.5806
χ	0.7962	0.5898	0.3812
Design buckling resistance, $N_{b.Rd}$ (kN)	1808.7	1335.1	867.8
$N_{test}/N_{b.Rd}$	1.07	1.12	1.14

steel members under compression, bending and combined compression and bending. A summary of the results of tests carried out on circular hollow sections (CHS) and welded I-section members is given below. The test results are compared with the design curves proposed in ENV 1993-1-4.

Stub-column tests were carried out on all the cross-sections to determine the effects of local buckling, strain-hardening (due to cold-forming) and residual stresses (due to welding). A stub-column test gives an average stress–strain curve for a given member.

3.1. Compression members — circular hollow sections

Nine CHS specimens, three stub columns with fixed-end boundary conditions and six columns of varying wall thickness with pin-ended boundary conditions were

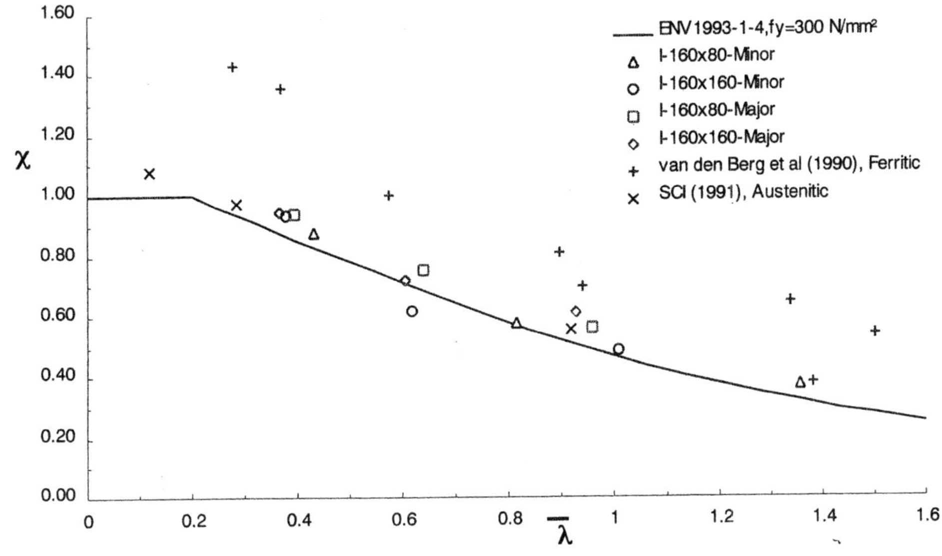

Fig. 2. Reduction factor, χ, versus generalised slenderness $\bar{\lambda}$ for I-section columns — test results and ENV 1993-1-4 design curve (grade 1.4301).

tested. A summary of the geometric and material properties of the specimens is given in Table 1. The specimens were manufactured by roll-forming stainless steel strip into tubes and seam welding along the length of the tubes. Four of the specimens were grade 1.4541 (321) and the remaining five were grade 1.4435 (316L).

Table 2 shows the section classification of all the tested specimens calculated in accordance with ENV 1993-1-4 using the measured properties. The 2-mm thick sections are Class 4 and therefore are not covered by the standard (slender CHS being beyond the scope of ENV 1993-1-4). Also in this table, the test results (characterised by the maximum applied load in the test) are compared with the resistances predicted from ENV 1993-1-4. The bold type in the table highlights the specimens for which

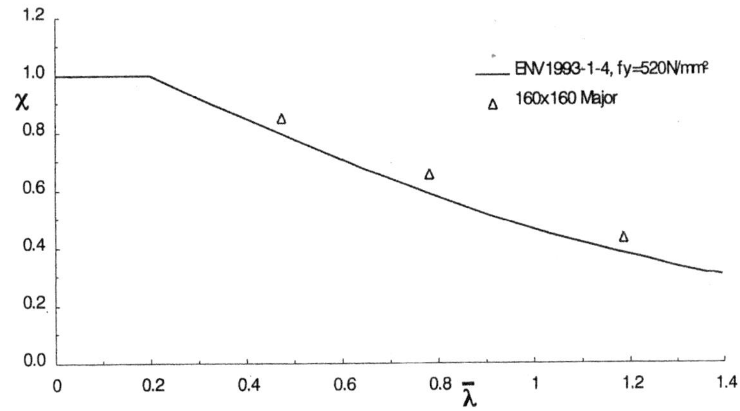

Fig. 3. Reduction factor, χ, versus generalised slenderness $\bar{\lambda}$ for I-section columns — test results and ENV 1993-1-4 design curve (grade 1.4462).

Table 9
Geometric and material properties of the CHS beam specimens

Specimen reference name	103 ×1.5	153 ×1.5	203 ×1.5	219 ×2.0	114.3 ×3.05	168.3 ×3.4	219.1 ×3.0	219.1 ×3.76	140 ×4	140 ×3	140 ×2
Steel grade to EN 10088	1.4301	1.4301	1.4301	1.4301	1.4462	1.4462	1.4462	1.4462	1.4541	1.4435	1.4435
Diameter, d (mm)	103	153	203	219	114.3	168.3	219.1	219.1	139.6	139	140.2
Thickness, t (mm)	1.3	1.3	1.3	1.8	2.7	3.7	3	3.76	4	2.87	2.04
Yield strength, f_y (N/mm^2)	461	456	370	332	643	602	598	560	292	352	313
Ultimate tensile strength, f_u (N/mm^2)	778	801	739	621	836	811	824	782	573	577	602
Modulus of elasticity, E (kN/mm^2)	200	200	200	200	200	200	200	200	198	190	195
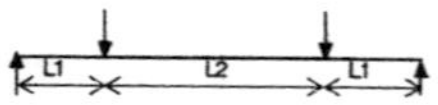	L_1=750 mm, L_2=500 mm								L_1=765 mm, L_2=500 mm		
Measured failure load, F (kN)	17.4	35.7	47.1	73.7	58.6	162	167	272	68	51.4	30.3

Table 10
Section classification and predicted moment resistance of the CHS beam specimens

Specimen reference name	103 ×1.5	153 ×1.5	203 ×1.5	219 ×2.0	114.3 ×3.05	168.3 ×3.4	219.1 ×3.0	219.1 ×3.76	140 ×4	140 ×3	140 ×2
$(d/t)/\varepsilon^2$	163.2	239.8	258.2	180.5	121.6	122.4	196.1	145.8	45.99	80.18	98.58
Section class	4	4	4	4	4	4	4	4	1	3	4
Elastic moment resistance of gross section, $M_{el.Rd}$ ($W_{el}f_y$) (kNm)	4.81	10.62	15.27	21.96	16.59	46.38	65.24	75.39	16.40	14.41	9.44
Plastic moment resistance of gross section, $M_{pl.Rd}$ ($W_{pl}f_y$) (kNm)	6.18	13.61	19.52	28.12	21.57	60.21	83.99	97.41	21.43	18.68	12.16
Test moment capacity, M_{test} (kNm)	6.53	13.39	17.66	27.64	21.98	60.75	62.63	102	26.01	19.66	11.59
$M_{test}/M_{el.Rd}$	1.357	1.260	1.157	1.258	1.325	1.310	0.965	1.353	1.586	1.365	1.228
$M_{test}/M_{pl.Rd}$	1.055	0.984	0.905	0.983	1.019	1.009	0.746	1.047	1.214	1.053	0.953
Section classification based on test	≤2	3	3	3	≤2	≤2	4	≤2	≤2	≤2	3

the standard is not applicable. It can be seen that overall the standard predicts the flexural buckling resistance very accurately. However, the following observations can be made:

- The design standard underestimates the resistance of the Class 1 stub column section by some 34% as no benefit was taken of the strain hardening of the material.
- The design standard consistently over-estimates the column resistance by a very small margin (<4%).

The results are summarised in Fig. 1 which plots the critical buckling reduction factor, χ, as a function of the generalised slenderness $\bar{\lambda}$. The figure also displays results from earlier tests by Rasmussen and Hancock [6] on seam welded CHS members in grade 1.4306 (304L). The figure shows that the recommended buckling curve (α=0.49, $\bar{\lambda}_0$=0.40) represents good agreement with the test results.

Table 11
Geometric and material properties of the I-section beam specimens — 160×80 and 160×160, grade 1.4301

Specimen reference name	I-160×80-B0	I-160×80-B1	I-160×80-B2	I-160×160-B0	I-160×160-B1	I-160×160-B2
Specimen length, L (mm)	1025	1024	2522	1025	2520	5018
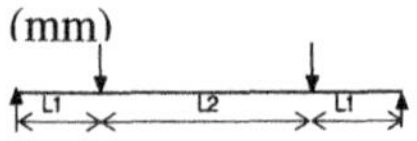 L1	266	266	266	266	266	266
L2	493	492	1990	493	1988	4486
Web yield strength, f_{yw} (N/mm^2)	300	300	300	300	300	300
Web ultimate tensile strength, f_{uw} (N/mm^2)	624	624	624	624	624	624
Flange yield strength, f_{yf} (N/mm^2)	299	299	299	302	302	300
Flange ultimate tensile strength, f_{uf} (N/mm^2)	609	609	609	612	612	610
Modulus of elasticity, E (kN/mm^2)	200	200	200	199	199	198
Section depth, H (mm)	161	158.7	158.2	158.8	158	158.5
Section width, B (mm)	80.3	80.6	80.6	159	159.4	160
Web thickness, t_w (mm)	6	6	6	6	6	6
Flange thickness, t_f (mm)	9.8	9.8	9.8	9.9	9.9	9.9
Weld throat thickness, a (mm)	3	3	3	3	3	3
Specimen failure load, F (kN)	409	366	248	687	578	396

Table 12
Geometric and material properties of the I-section beam specimens — 160×160, grade 1.4462 and 320×160, grade 1.4301

Specimen reference name	I-160×160-B0	I-160×160-B1	I-160×160-B2	I-320×160-B0	I-320×160-B1	I-320×160-B2
Steel grade to EN 10088	1.4462	1.4462	1.4462	1.4301	1.4301	1.4301
Specimen length, L (mm)	1027	2528	5025	2025	2526	5028
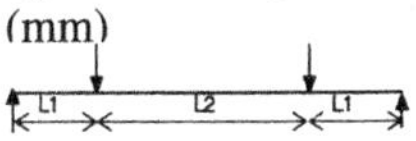 L1	266	266	266	517	517	517
L2	495	1996	4493	991	1492	3994
Web yield strength, f_{yw} (N/mm^2)	523	523	523	300	300	300
Web ultimate tensile strength, f_{uw} (N/mm^2)	777	777	777	624	624	624
Flange yield strength, f_{yf} (N/mm^2)	522	522	522	304	304	304
Flange ultimate tensile strength, f_{uf} (N/mm^2)	755	755	755	614	614	614
Modulus of elasticity, E (kN/mm^2)	201	201	201	200	200	200
Section depth, H (mm)	159.2	158.7	160.2	319.6	318.9	319.9
Section width, B (mm)	161.6	161.2	160.2	160.6	160.5	159.6
Web thickness, t_w (mm)	6.8	6.8	6.8	6	6	6
Flange thickness, t_f (mm)	10.6	10.6	10.6	9.8	9.8	9.8
Weld throat thickness, a (mm)	3	3	3	3	3	3
Specimen failure load, F (kN)	1225	955	715	835	705	444

3.2. Compression members — welded I sections

Twelve grade 1.4301 (304) column tests with three different heights were performed for two I sections. The flexural buckling tests were carried out about both the major and minor axes. A further three tests using one cross-section and three column heights were carried out on specimens fabricated from grade 1.4462 (duplex 2205) stainless steel. The tests were designed with some specimens failing about their major axis and others about their minor axis. A summary of the geometric and material properties of the specimens is given in Tables 3–5. The specimens were fabricated by continuous submerged arc welding.

Tables 6–8 show the section classification of all the tested specimens calculated in accordance with ENV 1993-1-4, using the measured properties. This shows that ten of the grade 1.4301 specimens were Class 1 and the remaining two were Class

Table 13
Section classification and predicted moment resistance of the I-section beam specimens — 160×80 and 160×160, grade 1.4301

Specimen reference name	I-160×80-B0	I-160×80-B1	I-160×80-B2	I-160×160-B0	I-160×160-B1	I-160×160-B2
$d/(t_w\varepsilon)$	25.6	25.2	25.1	25.2	25.1	25.3
$c/(t_f\varepsilon)$	3.9	3.9	3.9	8.5	8.5	8.6
Section class	1	1	1	1	1	1
Moment resistance of the cross-section, $M_{c.Rd}$ (kNm)	44.6	43.9	43.7	79.5	79.2	79.3
Elastic critical moment, M_{cr} (kNm)	1056.9	1057.0	77.6	8012.6	520.6	123.5
$\bar{\lambda}_{LT}$	0.205	0.204	0.750	0.100	0.390	0.801
ϕ_{LT}	0.52	0.52	0.99	0.47	0.65	1.05
χ_{LT}	1.00	1.00	0.61	1.00	1.00	0.58
Buckling resistance moment, $M_{b.Rd}$ (kNm)	44.6	43.9	26.7	79.5	79.2	45.9
Test moment resistance, M_{test} (kNm)	54.4	48.7	33.0	91.4	76.9	52.7
$M_{test}/M_{b.Rd}$	1.22	1.11	1.24	1.15	0.97	1.15

2. The grade 1.4462 specimens were all Class 3. The test results (characterised by the maximum applied load in the test) are also compared with the resistances predicted from ENV 1993-1-4 in these tables. It can be seen that the standard predicts the flexural buckling resistance very accurately.

The results for grade 1.4301 (304) are summarised in Fig. 2 which plots the critical buckling reduction factor, χ, as a function of the generalised slenderness $\bar{\lambda}$. The figure also displays results from earlier tests on 3CR12 steel carried out by van den Berg et al. [7] and results from tests on grade 1.4404 (316L) carried out by the Steel Construction Institute [8]. The figure shows that the selected buckling curve (α=0.76, $\bar{\lambda}_0$=0.20) represents good agreement with the test results.

The results for grade 1.4462 (duplex 2205) are summarised in Fig. 3, which also plots the critical buckling reduction factor, χ, as a function of the generalised slenderness $\bar{\lambda}$. The figure shows that the selected buckling curve (α=0.76, $\bar{\lambda}_0$=0.20) is about 10% conservative. This may be attributed to the lower residual stresses in duplex stainless steel compared to those in austenitic stainless steel.

3.3. Flexural members — circular hollow sections

A total of 11 four-point bending tests were carried out on stainless steel CHS of varying cross-sectional slenderness and material grade in order to determine their cross-sectional behaviour and moment resistance. The specimens were manufactured by roll-forming stainless steel strip into tubes and seam welding along the length of

Table 14
Section classification and predicted moment resistance of the I-section beam specimens — 160×160, grade 1.4462 and 320×160, grade 1.4301

Specimen reference name	I-160×160-B0	I-160×160-B1	I-160×160-B2	I-320×160-B0	I-320×160-B1	I-320×160-B2
Steel grade to EN 10088	1.4462	1.4462	1.4462	1.4301	1.4301	1.4301
$d/(t_w\varepsilon)$	29.0	28.9	29.3	56.3	56.1	56.3
$c/(t_f\varepsilon)$	10.5	10.5	10.4	8.7	8.7	8.6
Section class	3	3	3	2	2	2
Moment resistance of the cross-section, $M_{c.Rd}$ (kNm)	134.0	133.2	134.0	188.7	188.1	188.0
Elastic critical moment, M_{cr} (kNm)	9011.1	581.8	138.6	4231.4	1867.6	269.7
$\bar{\lambda}_{LT}$	0.122	0.478	0.983	0.211	0.317	0.835
ϕ_{LT}	0.48	0.72	1.28	0.53	0.59	1.09
χ_{LT}	1.00	0.79	0.48	1.00	1.00	0.56
Buckling resistance moment, $M_{b.Rd}$ (kNm)	134.0	105.8	63.7	188.7	188.1	105.0
Test moment resistance, M_{test} (kNm)	162.9	127.0	95.1	215.8	182.2	114.8
$M_{test}/M_{b.Rd}$	1.22	1.20	1.49	1.14	0.97	1.09

the tubes. A summary of the geometric and material properties of the specimens is given in Table 9. Three tests were carried out by VTT and the remaining eight were commissioned by SCI at Imperial College.

Table 10 shows the section classification of all the tested specimens calculated in accordance with ENV 1993-1-4, using the measured properties. This shows that all but two of the specimens are Class 4 and are therefore predicted to fail by local buckling before reaching the elastic moment of the gross section. The test results (characterised by the maximum moment reached in the tests) are compared with the resistances predicted from ENV 1993-1-4 in the table, assuming that the section was fully effective in all cases. It can be seen that all but one of the specimens with $(d/t)/\varepsilon^2<163$ did in fact achieve a resistance in excess of the theoretical plastic moment. The one exception ($(d/t)/\varepsilon^2=98.6$) warrants further investigation.

Only one specimen with $(d/t)/\varepsilon^2=196.1$ failed to reach the elastic moment resistance of the gross section by a margin of 3.5%, although two samples with higher $(d/t)/\varepsilon^2$ values (239.8 and 258.2) had resistances which exceeded the elastic moment resistance by 26% and 16%. The results therefore suggest that the limit on $(d/t)/\varepsilon^2$ between Class 3 and 4 can be increased substantially for circular hollow sections acting in bending, compared to the present limit in the standard of 90.

3.4. Flexural members — welded I sections

The cross-sectional behaviour of the I sections in bending was studied by bending tests on short beams. In addition, lateral–torsional buckling tests with two different spans were performed. Nine of the beams were grade 1.4301 (304) and three were 1.4462 (duplex 2205). A summary of the geometric and material properties of the specimens is given in Table 11 and Table 12. The specimens were fabricated by continuous submerged arc welding.

Tables 13 and 14 show the section classification of all the tested specimens calculated in accordance with ENV 1993-1-4, using the measured properties. The deepest grade 1.4301 (304) sections were found to be just outside the standard limit for Class 1 based on the web classification and the 1.4462 (duplex 2205) sections were all Class 3 based on the flange classification. The test results (characterised by the buckling reduction factor) are compared with the design curve from ENV 1993-1-4 in the tables and show that the design standard is conservative. The elastic critical buckling moment, M_{cr} was calculated in accordance with Annex F of ENV 1993-1-1.

The results for grade 1.4301 (304) are summarised in Fig. 4 which plots the lateral torsional buckling factor χ_{LT} as a function of the generalised slenderness for lateral torsional buckling $\bar{\lambda}_{LT}$. The figure also displays results from earlier tests by van Wyk et al. [9] and the Japanese Institution of Architecture [10]. The figure shows that the design curve in the standard (α_{LT}=0.76) is conservative. The carbon steel buckling curve in ENV 1993-1-1 (α_{LT}=0.49) is also shown in the figure and it can be seen

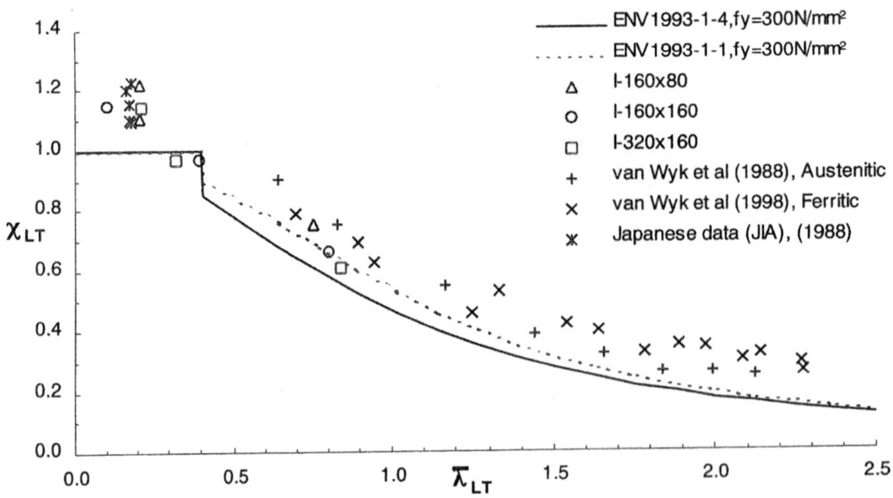

Fig. 4. Reduction factor, χ_{LT}, versus generalised slenderness $\bar{\lambda}_{LT}$ for I-section beams — test results and ENV 1993-1-1 and ENV 1993-1-4 design curves (grade 1.4301).

to be a closer fit to the data than the more conservative curve currently adopted in ENV 1993-1-4.

In a similar way, the results for grade 1.4462 are summarised in Fig. 5. Again there appears to be a closer fit to the carbon steel buckling curve.

3.5. Members subject to combined loading — circular hollow sections

A total of eight pin-ended specimens of varying wall thickness were tested with an axial load applied eccentrically through the centre of wall thickness. The specimens were manufactured by roll-forming stainless steel strip into tube and seam welding along the length of the tube. Four specimens were grade 1.4541 (321) and four were grade 1.4435 (316L). The specimen dimensions were selected so that failure would not occur by local buckling and the measured properties are presented in Table 15.

Local buckling occurred at failure in the two shorter specimens of these Class 4 members. The test results (characterised by the maximum applied load in the test) are compared with the resistances predicted from ENV 1993-1-4 in Table 16. The bold type in the table highlights the specimens for which the standard is not applicable. It can be seen that the standard is conservative in all cases, including the two specimens which buckled locally. (For the Class 4 sections, the elastic section modulus was used to calculate the moment resistance.) It can also be seen that the extent to which the standard predictions are conservative decreases as overall buckling becomes more dominant (i.e. as the specimen length increases).

Fig. 6 shows the interaction curve between flexural buckling and moment resistance predicted in ENV 1993-1-4. The test points plotted on the curve are calculated

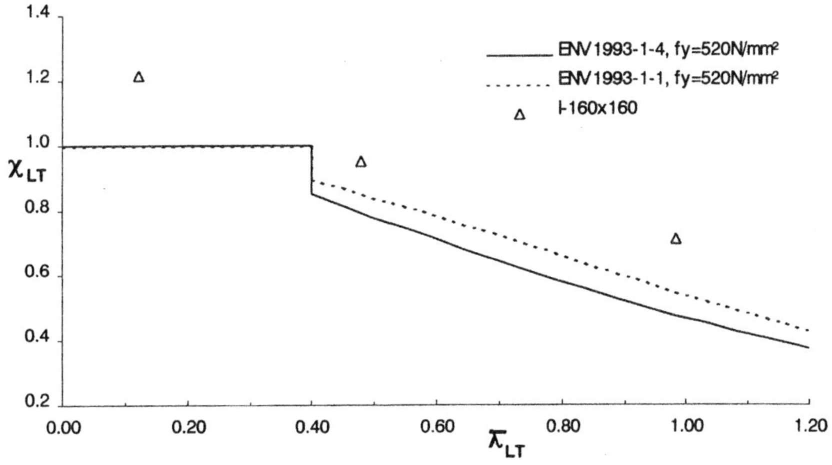

Fig. 5. Reduction factor, χ_{LT}, versus generalised slenderness $\bar{\lambda}_{LT}$ for I-section beams — test results and ENV 1993-1-4 and ENV 1993-1-1 design curves (grade 1.4462).

Table 15
Geometric and material properties of the CHS beam–column specimens

Section	140×4-EC0	140×4-EC1	140×4-EC2	140×4-EC3	140×2-EC0	140×2-EC1	140×2-EC2	140×2-EC3
Steel grade to EN 10088	1.4541	1.4541	1.4541	1.4541	1.4435	1.4435	1.4435	1.4435
Diameter, d (mm)	139	139.2	140.1	140.1	139.9	139.6	139.8	139.8
Thickness, t (mm)	3.98	3.98	3.98	3.99	1.97	1.95	1.95	1.95
Nominal yield strength, f_y (N/mm^2)	294.5	297.5	297.5	293.5	318.5	320	320	316
Ultimate tensile strength, f_u (N/mm^2)	576	573	573	572	598.5	599	599	597.5
Modulus of elasticity E (kN/mm^2)	193	194	194	196	201	195	195	199
Column length, L (mm)	550	2250	3351	4451	550	2251	3351	4451
Test failure load, N_{test} (kN)	297	202	155	121	122	89	73	58

Table 16
Section classification and predicted buckling resistance of the CHS beam–column specimens

Specimen reference name	140×4-EC0	140×4-EC1	140×4-EC2	140×4-EC3	140×2-EC0	140×2-EC1	140×2-EC2	140×2-EC3
Steel grade to EN 10088	1.4541	1.4541	1.4541	1.4541	1.4435	1.4435	1.4435	1.4435
$(d/t)/\varepsilon^2$	47.62	47.93	48.24	46.99	100.56	104.98	105.13	101.73
Section class	1	1	1	1	**4**	**4**	**4**	**4**
$M_{c.Rd}$[a]	21.3	21.6	21.9	21.6	**9.2**	**9.2**	**9.2**	**9.1**
$\bar{\lambda}$	0.143	0.586	0.868	1.139	**0.143**	**0.596**	**0.886**	**1.158**
ϕ	0.447	0.718	0.991	1.329	**0.447**	**0.726**	**1.012**	**1.357**
χ	1.000	0.884	0.680	0.496	**1.000**	**0.877**	**0.666**	**0.485**
$N_{b.Rd}$ (kN)	497.18	444.64	344.53	248.46	**271.88**	**236.73**	**180.09**	**129.37**
Buckling resistance in the presence of a moment, $N_{b.M.Rd}$ (kN)	195	161	133	114	**85**	**71**	**60**	**52**
Axial load ratio, $N_{test}/N_{b.M.Rd}$	1.52	1.26	1.16	1.06	**1.43**	**1.26**	**1.23**	**1.11**

[a] For Class 4 sections, the elastic modulus was adopted in the calculations; this is not in accordance with the standard.

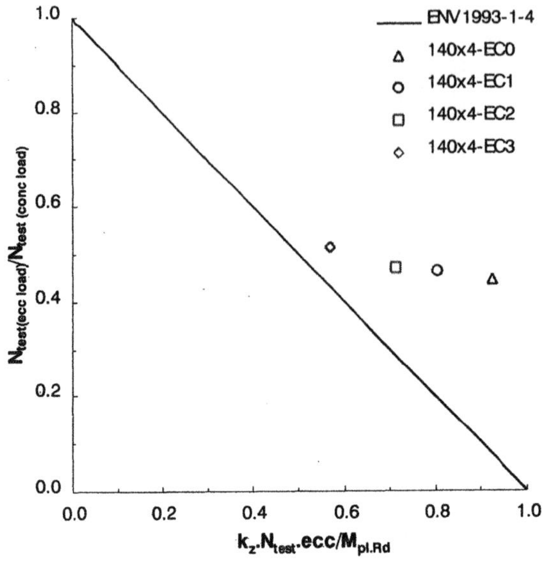

Fig. 6. CHS beam–column specimens — test results and ENV 1993-1-4 interaction curve.

on the basis of the measured axial resistance and predicted plastic moment resistance. In all cases the test points show that the standard is conservative.

3.6. Members subject to combined loading — welded I sections

A total of eight eccentric compression tests were carried out on two different I-section members of varying lengths. Bending about the major axis only was studied. The load was applied to the centreline of the flange. The specimens were fabricated by continuous submerged arc welding and were all from grade 1.4301 (304) material. The specimen dimensions were selected so that failure would not occur by local buckling and the measured properties are presented in Table 17.

Table 18 shows the section classification of all the tested specimens calculated in accordance with ENV 1993-1-4, using the measured properties. The test results (characterised by the maximum applied load in the test) are compared with the resistances predicted from ENV 1993-1-4 in the table. It can be seen that the standard is conservative in all cases. As with the CHS beam–column tests, the extent to which the standard predictions are conservative tends to decrease as overall buckling becomes more dominant (i.e. as the specimen length increases).

4. Conclusions

The results of a series of member tests forming part of an ongoing ECSC research project concerned with further development and refinement of structural design guidance are presented. Comparisons of this experimental data with the design provisions in ENV 1993-1-4 are made. In general, the guidance in ENV 1993-1-4 is conserva-

Table 17
Geometric and material properties of the I-section beam–column specimens, grade 1.4301

Specimen reference name (failure about major axis)	I-160×80-EC0	I-160×80-EC1	I-160×80-EC2	I-160×80-EC3	I-160×160-EC0	I-160×160-EC1	I-160×160-EC2	I-160×160-EC3
Specimen length, L (mm)	500	2045	3339	5041	502	2048	3345	5043
Web yield strength, f_{yw} (N/mm^2)	300	300	300	300	300	300	300	300
Web ultimate tensile strength, f_{uw} (N/mm^2)	624	624	624	624	624	624	624	624
Flange yield strength, f_{yf} (N/mm^2)	299	299	299	299	300	299	299	300
Flange ultimate tensile strength, f_{uf} (N/mm^2)	609	609	609	609	610	609	609	610
Modulus of elasticity E (kN/mm^2)	200	200	200	200	198	200	200	198
Section depth, H (mm)	158.3	160.3	158.9	158.7	162.3	159	158.5	159.5
Section width, B (mm)	82.7	79.4	79.1	80.9	159.8	160.8	160.1	160.4
Web thickness, t_w (mm)	6	6	6	6	6	6	6	6
Flange thickness, t_f (mm)	9.8	9.8	9.8	9.8	9.9	9.8	9.8	9.9
Weld throat thickness, a (mm)	3	3	3	3	3	3	3	3
Measured failure load, N_{test} (kN)	505	338	270	222	705	540	454	356

Table 18
Section classification and predicted moment resistance of the I-section beam–column specimens, grade 1.4301

Specimen reference number	I-160×80-ECO	I-160×80-EC1	I-160×80-EC2	I-160×80-EC3	I-160×160-ECO	I-160×160-EC1	I-160×160-EC2	I-160×160-EC3
$(d/t_w)(13\alpha-1)/\varepsilon$	301.5	306.1	302.9	302.4	311.9	303.1	302.0	305.4
$c/(t_f\varepsilon)$	4	3.8	3.8	3.9	8.5	8.6	8.6	8.6
Section class	1	1	1	1	2	1	1	1
$M_{c.Rd}$	44.6	43.9	43.3	44.0	81.5	79	78.4	80
$\bar{\lambda}$	0.095	0.386	0.635	0.958	0.089	0.366	0.600	0.904
ϕ	0.465	0.645	0.867	1.247	0.462	0.630	0.832	1.177
χ	1	0.861	0.686	0.489	1.000	0.875	0.710	0.518
$N_{b.Rd}$ (kN)	734.3	618.4	490.0	354.2	1205.7	1044.1	843.9	624.1
Buckling resistance in the presence of a moment, $N_{b.M.Rd}$ (kN)	329.1	268.1	220.8	186.7	560.4	469.1	391.2	333.2
Axial load ratio, $N_{test}/N_{b.M.Rd}$	1.53	1.26	1.22	1.19	1.26	1.15	1.16	1.07

tive. The tests on CHS beams indicate that the limiting diameter-to-thickness ratios for section classification can be considerably increased. For welded I-section beams, the ENV 1993-1-4 lateral torsional buckling curve appears very conservative and the less conservative curve adopted in ENV 1993-1-1 for carbon steel appears to give a better fit to the data.

Acknowledgements

The following organisations sponsored the projects described in this paper: Acciai Speciali Terni SpA, Avesta Sheffield AB Research Foundation, Avesta Sheffield Ltd, Department of the Environment, Transport and the Regions, European Coal and Steel Community, Outokumpu Polarit Oy, Studiengesellschaft Stahlanwendung eV, The Nickel Development Institute and Ugine S.A. Their support is gratefully acknowledged.

References

[1] ANSI/ASCE-8-90. Specification for the design of cold-formed stainless steel structural members. American Society of Engineers, USA, 1991.
[2] EURO INOX. Design manual for structural stainless steel, 1994.
[3] ENV 1993-1-1. Eurocode 3 Design of steel structures: Part 1.1, General rules and rules for buildings. CEN, 1992.
[4] ENV 1993-1-4. Eurocode 3 Design of steel structures: Part 1.4, General rules and supplementary rules for stainless steels. CEN, 1996.
[5] EN 10088. Stainless steels. CEN, 1995.
[6] Rasmussen KJR, Hancock GJ. Stainless steel tubular columns — test and design. Tenth International Speciality Conference on Cold Formed Steel Structures, St Louis, MO, USA, 23–24 October 1990.
[7] van den Berg GJ, van der Merwe P, Bredenkamp PJ. The strength of Type 3CR12 corrosion resisting steel built-up I section column and beams. Report MD-51, Faculty of Engineering, Rand Afrikaans University, March 1990.
[8] The Steel Construction Institute. Technical report 29: Tests on stainless steel beams and columns. SCI report no. RT/231, July 1991.
[9] van Wyk ML, van den Berg GJ, van der Merwe P. The lateral torsional buckling strength of doubly symmetric stainless steel beams. Report MD-58, Faculty of Engineering, Rand Afrikaans University, May 1990.
[10] Japanese Institution of Architecture. Strength and deformation of H-shaped stainless steel beams. Journal of the Kanto Branch, 1988 (in Japanese).

Recent research and design developments in steel and composite steel–concrete structures in USA

Theodore V. Galambos*

University of Minnesota, Minneapolis, USA

Abstract

A brief review of the status of structural steel research in the US at the end of the Twentieth Century is presented in this paper to show that while many problems are being solved, there are new and challenging problems remaining. The chief impetus for continued research is that provided by natural disasters, such as earthquakes, tropical storms, tornadoes and floods occurring in densely populated urban areas. New materials and new experimental and computational technologies also give rise to new and exciting research problems. © 2000 Elsevier Science Ltd. All rights reserved.

Keywords: Bridges; Buildings; Design; Research; Steel structures; United States of America; Seismic behavior; High-performance materials

1. Introduction

The purpose of this paper is to give a brief overview of the current developments in structural steel research in the US, and of the future directions that the structural steel engineering research may take in the coming Century. The driving forces of research in this field are the following:

- new construction methods and construction products
- new materials
- economic considerations
- natural disasters

* Corresponding author. Tel.: +1-612-625-5522 fax: +1-612-626-7750.

Reprinted from *Journal of Constructional Steel Research* **55 (1-3)**, 289-303 (2000)

Three of these motivations are common to all engineering developments, not just to structural engineering. However, the impetus due to natural disasters is unique to our field. Recent major natural disasters in the US, such as the Northridge earthquake in California and hurricane Andrew in Florida, have spurred much of the current research activity.

The presentation here is of necessity incomplete, because the author is not aware of all research going on everywhere in the country and there is not enough space in this presentation. The overview is meant to give a general flavor of the research activities, and to show that a significant effort is going on in the US. The following is a list of 10 major topics in steel research:

1. Limit States Design for bridges
2. Monitoring of structural performance in the field
3. Design of seismically resistant connections
4. Curved girder bridges
5. Composite columns with high-performance concrete
6. Building frames with semi-rigid joints
7. "Advanced Structural Analysis" for buildings
8. Repair and retrofit of structures
9. Steel structures with high-performance steels
10. Cold-formed steel structures

The next parts of this paper will give brief discussions on some of these topics. Several topics will then be elaborated in more detail. The paper will conclude with a look toward the future of structural steel research.

2. Research on steel bridges

The American Association of State Transportation and Highway Officials (AASHTO) is the authority that promulgates design standards for bridges in the US. In 1994 it has issued a new design specification which is a Limit States Design standard that is based on the principles of reliability theory. A great deal of work went into the development of this code in the past decade, especially on calibration and on the probabilistic evaluation of the previous specification. The code is now being implemented in the design office, together with the introduction of the Systeme Internationale units. Many questions remain open about the new method of design, and there are many new projects that deal with the reliability studies of the bridge as a system. One such current project is a study to develop probabilistic models, load factors, and rational load-combination rules for the combined effects of live-load and wind; live-load and earthquake; live-load, wind and ship collision; and ship collision, wind, and scour. There are also many field measurements of bridge behavior, using modern tools of inspection and monitoring such as acoustic emission techniques and other means of non-destructive evaluation. Such fieldwork necessitates parallel studies in the laboratory, and the evolution of ever more sophisticated high-technology data transmission methods.

America has an aging steel bridge population and many problems arise from fatigue and corrosion. Fatigue studies on full-scale components of the Williamsburg Bridge in New York have recently been completed at Lehigh University. A probabilistic AASHTO bridge evaluation regulation has been in effect since 1989, and it is employed to assess the future useful life of structures using rational methods that include field observation and measurement together with probabilistic analysis. Such an activity also fosters additional research because many issues are still unresolved. One such area is the study of the shakedown of shear connectors in composite bridges. This work has been recently completed at the University of Missouri.

In addition to fatigue and corrosion, the major danger to bridges is the possibility of earthquake induced damage. This also has spawned many research projects on the repair and retrofit of steel superstructures and the supporting concrete piers. Many bridges in the country are being strengthened for earthquake resistance. One area that is receiving much research attention is the strengthening of concrete piers by "jacketing" them by sheets of high-performance reinforced plastic.

The previously described research deals mainly with the behavior of existing structures and the design of new bridges. However, there is also a vigorous activity on novel bridge systems. This research is centered on the application of high-performance steels for the design of innovative plate and box-girder bridges, such as corrugated webs, combinations of open and closed shapes, and longer spans for truss bridges. It should be mentioned here that, in addition to work on steel bridges, there is also very active research going on in the study of the behavior of prestressed concrete girders made from very high strength concrete. The performance and design of smaller bridges using pultruded high-performance plastic composite members is also being studied extensively at present. New continuous bridge systems with steel-concrete composite segments in both the positive moment and the negative moment regions are being considered. Several researchers have developed strong capabilities to model the three-dimensional non-linear behavior of individual plate girders, and many studies are being performed on the buckling and post-buckling characteristics of such structures. Companion experimental studies are also made, especially on members built from high-performance steels. A full-scale bridge of such steel has been designed, and will soon be constructed and then tested under traffic loading. Research efforts are also underway on the study of the fatigue of large expansion joint elements and on the fatigue of highway sign structures.

The final subject to be mentioned is the resurgence of studies of composite steel-concrete horizontally curved steel girder bridges. A just completed project at the University of Minnesota monitored the stresses and the deflections in a skewed and curved bridge during all phases of construction, starting from the fabrication yard to the completed bridge. Excellent correlation was found to exist between the measured stresses and deformations and the calculated values. The stresses and deflections during construction were found to be relatively small, that is, the construction process did not cause severe trauma to the system. The bridge has now been tested under service loading, using fully loaded gravel trucks, for two years, and it will continue to be studied for further years to measure changes in performance under service over time. A major testing project is being conducted at the Federal Highway Administration laboratory in Washington, DC, where a half-scale curved composite girder bridge is currently being tested to determine its limit states. The test-bridge was designed to act as its own test-frame, where various portions can be replaced after testing. Multiple flexure tests, shear tests, and tests under combined bending and shear, are thus performed with realistic end-conditions and restraints. The experiments are also modeled by finite element analysis to check conformance between reality and prediction. Finally design standards will be evolved from the knowledge gained. This last project is the largest bridge research project in the USA at the present time.

From the discussion above it can be seen that even though there is no large expansion of the nation's highway and railroad system, there is extensive work going on in bridge research. The major challenge facing both the researcher and the transportation engineer is the maintenance of a healthy but aging system, seeing to its gradual replacement while keeping it safe and serviceable.

3. Research on steel members and frames

There are many research studies on the strength and behavior of steel building structures. The most important of these have to do with the behavior and design of steel structures under severe seismic events. This topic will be discussed later in this paper. The most significant trends of the non-seismic research are the following:

- "Advanced" methods of structural analysis and design are actively studied at many Universities, notably at Cornell, Purdue, Stanford, and Georgia Tech Universities. Such analysis methods are meant to determine the load-deformation behavior of frames up to and beyond failure, including inelastic behavior, force redistribution, plastic hinge formation, second-order effects and frame instability. When these methods are fully operational, the structure will not have to undergo a member check, because the finite element analysis of the frame automatically performs this job. In addition to the research on the best approaches to do this advanced analysis, there are also many studies on simplifications that can be easily utilized in the design office while still maintaining the advantages of a more complex analysis. The advanced analysis method is well developed for in-plane behavior, but much work is yet to be done on the cases where bi-axial bending or lateral-torsional buckling must be considered. Some successes have been achieved, but the research is far from complete.
- Another aspect of the frame behavior work is the study of the frames with semi-rigid joints. The American Institute of Steel construction (AISC) has published design methods for office use. Current research is concentrating on the behavior of such structures under seismic loading. It appears that it is possible to use such frames in some seismic situations, that is, frames under about 8 to 10 stories in height under moderate earthquake loads. The future of structures with semi-rigid

frames looks very promising, mainly because of the efforts of researchers such as Leon at Georgia Tech University [1], and many others.

- Research on member behavior is concerned with studying the buckling and post-buckling behavior of compact angle and wide-flange beam members by advanced commercial finite element programs. Such research is going back to examine the assumptions made in the 1950s and 1960s when the plastic design compactness and bracing requirements were first formulated on a semi-empirical basis. The non-linear finite element computations permit the "re-testing" of the old experiments and the performing of new computer experiments to study new types of members and new types of steels. White of Georgia Tech is one of the pioneers in this work. Some current research at the US military Academy and at the University of Minnesota by Earls is discussed later in this report. The significance of this type of research is that the phenomena of extreme yielding and distortion can be efficiently examined in parameter studies performed on the computer. The computer results can be verified with old experiments, or a small number of new experiments. These studies show a good prospect for new insights into old problems that heretofore were never fully solved.

4. Research on cold-formed steel structures

Next to seismic work, the most active part of research in the US is on cold-formed steel structures. The reason for this is that the supporting industry is expanding, especially in the area of individual family dwellings. As the cost of wood goes up, steel framed houses become more and more economical. The intellectual problems of thin-walled structures buckling in multiple modes under very large deformations have attracted some of the best minds in stability research. As a consequence, many new problems have been solved: complex member stiffening systems, stability and bracing of C and Z beams, composite slabs, perforated columns, standing-seam roof systems, bracing and stability of beams with very complicated shapes, cold-formed members with steels of high yield stress-to-tensile strength ratio, and many other interesting applications. The American Iron and Steel Institute (AISI) has issued a new expanded standard in 1996 that brought many of these research results into the hands of the designer [2].

5. Research on steel-concrete composite structures

Almost all structural steel bridges and buildings in the US are built with composite beams or girders. In contrast, very few columns are built as composite members. The area of composite column research is very active presently to fill up the gap of technical information on the behavior of such members. The subject of steel tubes filled with high-strength concrete is especially active. One of the aims of research performed by Hajjar at the University of Minnesota is to develop a fundamental understanding of the various interacting phenomena that occur in concrete-filled columns and beam-columns under monotonic and cyclic load. The other aim is to obtain a basic understanding of the behavior of connections of wide-flange beams to concrete filled tubes.

Other major research work concerns the behavior and design of built-up composite wide-flange bridge girders under both positive and negative bending. This work is performed by Frank at the University of Texas at Austin and by White of Georgia Tech, and it involves extensive studies of the buckling and post-buckling of thin stiffened webs. Already mentioned is the examination of the shakedown of composite bridges. The question to be answered is whether a composite bridge girder loses composite action under repeated cycles of loads which are greater than the elastic limit load and less than the plastic mechanism load. A new study has been initiated at the University of Minnesota on the interaction between a semi-rigid steel frame system and a concrete shear wall connected by stud shear connectors.

6. Research on connections

Connection research continues to interest researchers because of the great variety of joint types. The majority of the connection work is currently related to the seismic problems that will be discussed in the next section of this paper. The most interest in non-seismic connections is the characterization of the monotonic moment-rotation behavior of various types of semi-rigid joints.

7. Research on structures and connections subject to seismic forces

The most compelling driving force for the present structural steel research effort in the US was the January 17, 1994 earthquake in Northridge, California, North of Los Angeles. The major problem for steel structures was the extensive failure of prequalified welded rigid joints by brittle fracture. In over 150 buildings of one to 26 stories high there were over a thousand fractured joints. The buildings did not collapse, nor did they show any external signs of distress, and there were no human injuries or deaths. A typical joint is shown in Fig. 1.

In this connection the flanges of the beams are welded to the flanges of the column by full-penetration butt welds. The webs are bolted to the beams and welded to the columns. The characteristic features of this type of connection are the backing bars at the bottom of the beam flange, and the cope-holes left open to facilitate the field welding of the beam flanges. Fractures occurred in the welds, in the beam flanges, and/or in the column flanges, sometimes penetrating into the webs.

Once the problem was discovered several large research projects were initiated at various university laboratories, such as The University of California at San Diego, the University of Washington in Seattle, the University of Texas at Austin, Lehigh University at Bethlehem, Pennsylvania, and at other places. The US Government under the leadership of the Federal Emergency Management Agency (FEMA) insti-

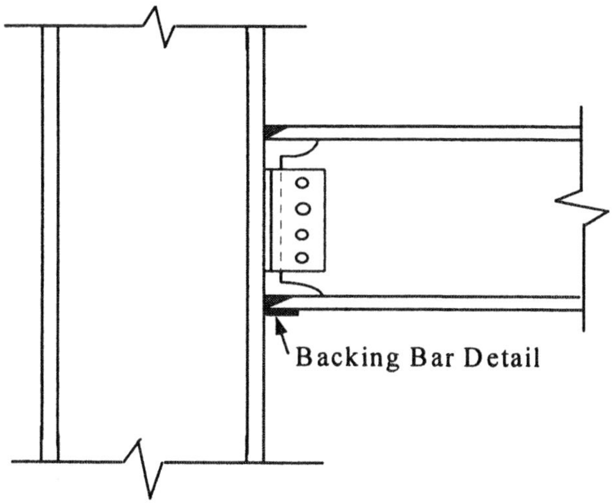

Fig. 1. Pre-Northridge connection.

tuted a major national research effort. The needed work was deemed so extensive that no single research agency could hope to cope with it. Consequently three California groups formed a consortium which manages the work:

1. **S**tructural Engineering Association of California
2. **A**pplied Technology Council
3. **C**alifornia Universities for Research in Earthquake Engineering

The first letters in the name of each agency were combined to form the acronym **SAC,** which is the name of the joint venture that manages the research. We shall read much from this agency as the results of the massive amounts of research performed under its aegis are being published in the next few years.

The goals of the program are to develop reliable, practical and cost-effective guidelines for the identification and inspection of *at-risk steel moment frame buildings*, the repair or upgrading of *damaged buildings*, the design of *new construction*, and the rehabilitation of *undamaged buildings*. As can be seen, the scope far exceeds the narrow look at the connections only.

The first phase of the research was completed at the end of 1996, and its main aim was to arrive at interim guidelines so that design work could proceed. It consisted of the following components:

- A state-of-the-art assessment of knowledge on steel connections
- A survey of building damage
- The evaluation of ground motion
- Detailed building analyses and case studies
- A preliminary experimental program
- Professional training and quality assurance programs
- Publishing of the *Interim Design Guidelines*

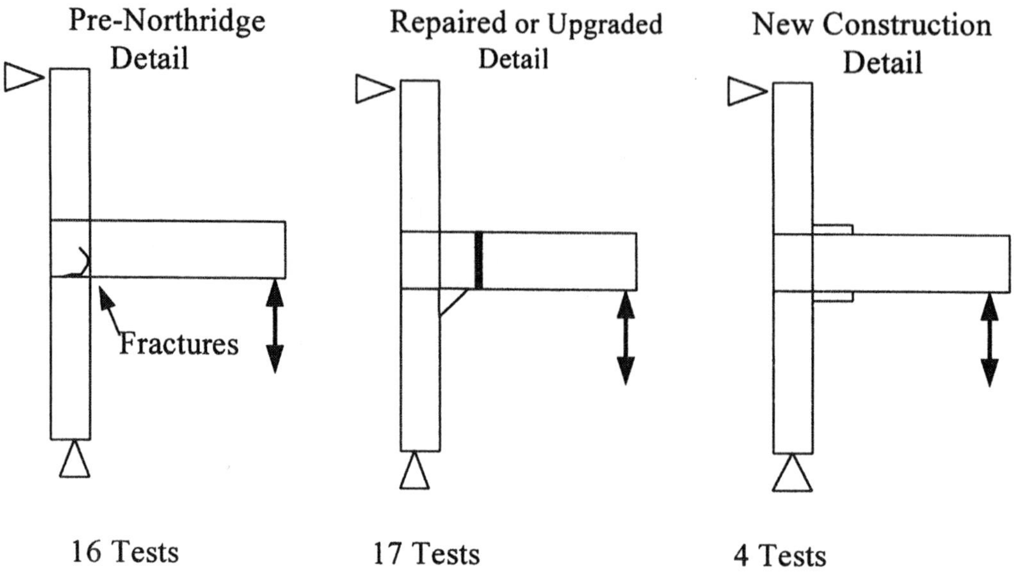

Fig. 2. Preliminary full-scale connection tests.

A number of reports were issued in this first phase of the work. A partial list of these is appended at the end of this paper.

During the first phase of the SAC project a series of full-scale connection tests under static and, occasionally, dynamic cyclic tests were performed. Tests were of pre-Northridge-type connections (that is, connections as they existed at the time of the earthquake), of repaired and upgraded details, and of new recommended connection details. A schematic view of the testing program is illustrated in Fig. 2. Some recommended strategies for new design are schematically shown in Fig. 3.

The following possible causes, and their combinations, were found to have contributed to the connection failures:

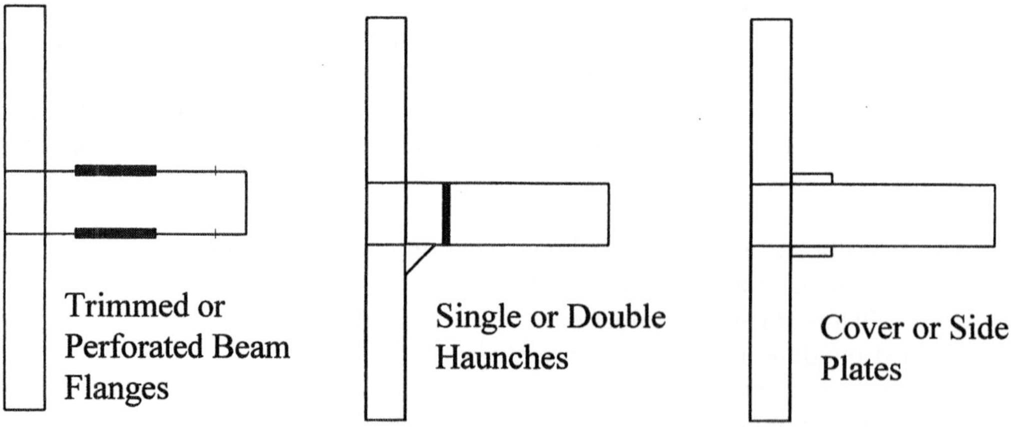

Fig. 3. Some recommended improvements in the interim guidelines.

- Inadequate workmanship in the field welds
- Insufficient notch-toughness of the weld metal
- Stress raisers caused by the backing bars
- Lack of complete fusion near the backing bar
- Weld bead sizes were too big
- Slag inclusion in the welds

While many of the failures can be directly attributed to the welding and the material of the joints, there are more serious questions relative to the structural system that had evolved over the years mainly based on economic considerations. The structural system used relatively few rigid-frames of heavy members that were designed to absorb the seismic forces for large parts of the structure. These few lateral-force resistant frames provide insufficient redundancy. More rigid-frames with smaller members could have provided a tougher and more ductile structural system. There is a question of size effect: test results from joints of smaller members were extrapolated to joints with larger members without adequate test verification. The effect of a large initial pulse may have triggered dynamic forces that could have caused brittle fracture in joints with fracture critical details and materials. Furthermore, the yield stress of the beams was about 30 to 40% larger than the minimum specified values assumed in design, and so the connection failed before the beams, which were supposed to form plastic hinges.

As can be seen, there are many possible reasons for this massive failure rate, and there is blame to go around for everyone. No doubt, the discussion about why and how the joints failed will go on for many more years. The structural system just did not measure up to demands that were more severe than expected. What should be kept in mind, however, is that no structure collapsed or caused even superficial non-structural damage, and no person was injured or killed. In the strictest sense the structure sacrificed itself so that no physical harm was done to its users. The economic harm, of course, was enormous.

Phase 2 of the SAC project started on Jan. 1, 1996 and is planned to be completed on Dec. 31, 1999. Its aims are to provide advice and guidance to code officials, designers, steel makers, welding engineers, and fabricators, in fact, to anyone connected with earthquake resistant design of steel buildings. The work includes the development of design-criteria for new buildings, and inspection, evaluation, repair and retrofit procedures for existing buildings that are at risk. A broad scope of professional issues is being examined. Ultimately, a *performance-based* methodology will be recommended to the professions dealing with seismic design problems. All types of moment-frame connections will be studied: bolted and welded connections, semi-rigid connections, connections made with special steels, energy-dissipating connections, etc. The research consists of many new experiments on joints, as well as a *systems-reliability*-based probabilistic method for optimizing the best structural design and evaluation procedures.

The research work of the Phase 2 SAC Project is essentially complete as of the date of this conference (Sep. 1999). The basic analytical and experimental work consists of the following topics:

- Materials and fracture issues
- Welding, joining and inspection
- Analysis and testing of connections
- Earthquake performance of structural systems
- Simulation of seismic response

Data and concepts from these five teams have been absorbed and utilized by the team working on the development of the reliability framework for performance prediction and evaluation. A number of extensive State-Of-The-Art (SOA) reports based on the research are now in the final stages of completion. The material from these SOA reports, as well as results from trial designs, cost analyses, loss analyses, and from an evaluation of social, economic and policy issues, will then be the basis of new seismic design criteria for use by building codes.

Phase 2 of the SAC Project is by far the largest and most expensive cooperative structural engineering effort in the history of US structural steel research. Much is expected to come of it. The way steel structures will be designed for steel structures is going to be deeply affected. The Northridge earthquake of January 17, 1994 proved a warning and a lesson, as well as a major impetus to learn more and to apply this knowledge more effectively.

8. Research on the required properties of high-performance steels

One other example will be elaborated on a research topic that is not motivated by natural disaster but by technological development, as an illustration among many which could have been presented. Steel makers have recently developed the capability to produce so-called "high-performance" steels economically, and there is a desire to use these steels in civil and military construction. Such steels are of high strength, with yield points of around 500 to 700 MPa, they can be produced to a variety of weldability, corrosion and toughness characteristics. Much work has been done on these steels in Japan (see [3,4] and many more papers) with their application in seismic structures in mind. Structures from a steel, HSLA80, have been extensively studied at Lehigh University in the US [5]. The research question to be answered is not *"Given a steel of certain properties, what are the member and structural characteristics?"* but *"Given the desired structural characteristics, what should the properties of the steel be?"*. These questions were discussed in a workshop sponsored by the US National Institute of Standards in Technology (NIST) at the University of Minnesota on July 1, 1996. The purpose of this meeting was to define the research needs to adapt the high-performance steels to the requirements of the structural design standards. Many issues were raised, but here only the subject of compactness and lateral bracing will be briefly touched. The shape of the stress–strain curve has a profound effect on the inelastic load-deformation behavior of members, as illustrated by the following example. The idealized form of a tensile stress–strain diagram is shown in Fig. 4. Data for four representative steels are given in Table 1. Steel A [4] is a new steel in Japan that has very good ductility and a low yield stress-to-

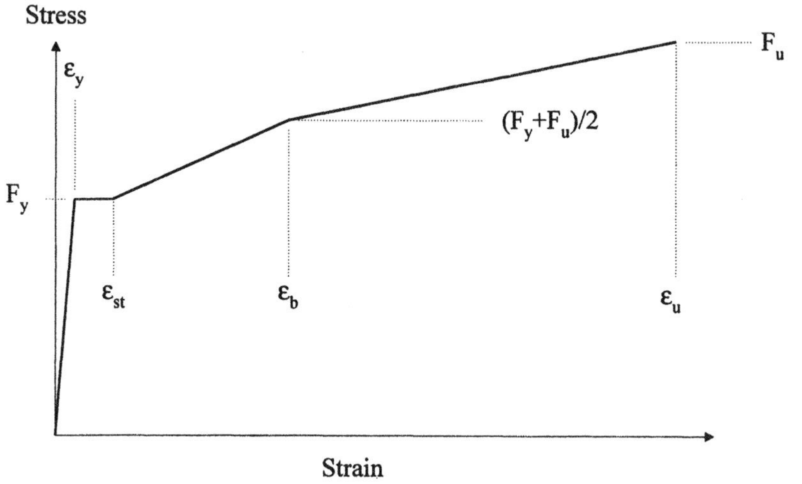

Fig. 4. Stress-strain model.

Table 1
Representative material properties for analysis

Material	F_y, *MPa*	F_y/F_u	ϵ_{st}/ϵ_y	ϵ_b/ϵ_y	ϵ_u/ϵ_y	E/E_{st}
Steel A, new Japanese	483	0.74	2	10	50	46
Steel B, ASTM A514	820	0.93	2	7	17	118
Steel C, ASTM A36	331	0.71	10	25	136	30
Steel D, HSLA80	586	0.88	2	19	29	136

tensile strength ratio (*yield ratio*), that is, it has about the same capacity to strain-harden at structural carbon steel (Steel C [6]; Steel B is a quenched and tempered steel with a very high yield stress but a high yield ratio [7]; Steel D is the steel HSLA80 from the research at Lehigh [6].

The load-deflection curves in Fig. 5 were obtained from a finite element analysis using the commercial program ABAQUS. The structure was a simply supported beam under a three-point loading. Lateral bracing was provided at the end-supports and under the central load-point. The section was a W200×46 (W8×31 in US units) profile, with a flange slenderness ratio b_f/t_f=7.8, a web slenderness ratio of h/t_w=29.9 and an unbraced length slenderness of L_b/r_y=71. As seen from Fig. 5, the shape of the stress–strain curve can have a tremendous difference on the inelastic rotation capacity of a structural member. The most important parameter appears to be the yield ratio and the ductility of the steel.

In addition to research on the high-performance steels, new work on the definition and improvement of conventional steels is also being conducted, spurred by the realization that the physical properties of steels as they are presently being produced

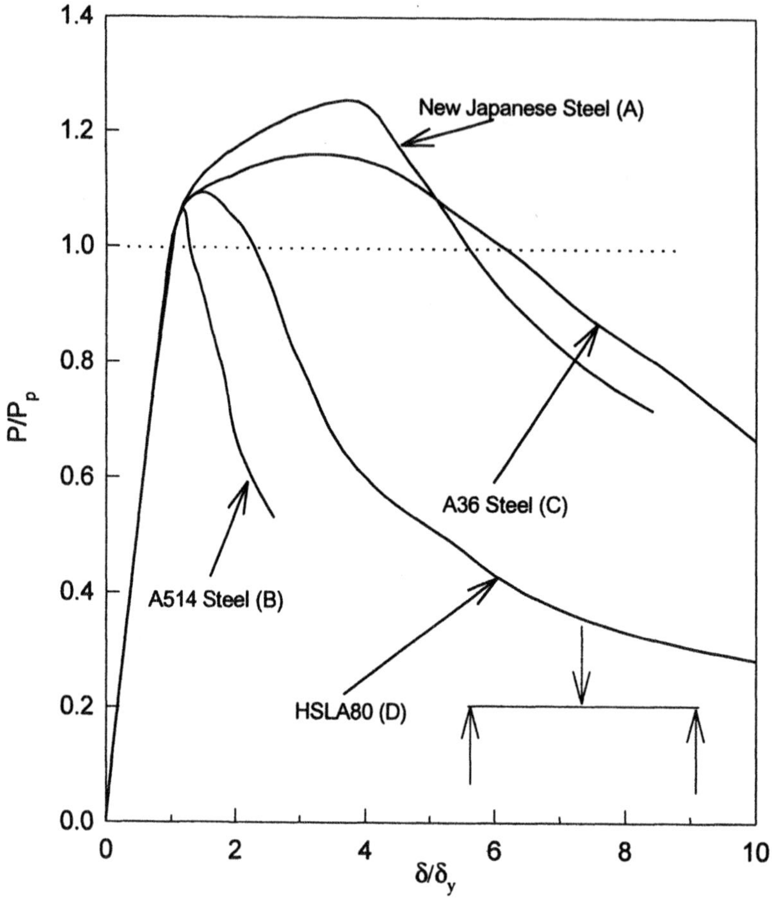

Fig. 5. Load-deflection curves for three-point loading.

are quite different from the steels for which the plastic design research was done 30 years ago. The yield stress is higher and it seems that due to the rotary straightening process the larger shapes end up with zones in their cross section where the ductility is unacceptably low.

Further work on this subject is being pursued by Earls at the US Military Academy [8,9] and by Ricles and his co-workers at Lehigh University [10]. More finite element analyses and laboratory experiments are being conducted to establish the desired stress–strain characteristics of high-performance steel to achieve optimal dimensions for compactness limits, so that this material can be effectively used in seismic design applications. Additional work is done on the design of the best shapes for bridge-girders, and a full-scale girder bridge will be fabricated and tested at the structural laboratory of the Federal Highway Administration at Washington, DC.

9. Future directions of structural steel research and conclusion

The future holds many challenges for structural steel research. The ongoing work necessitated by the two recent earthquakes that most affected conventional design methods, namely, the Northridge earthquake in the US and the Kobe earthquake in Japan, will continue well into the first decade of the next Century. It is very likely that future disasters of this type will bring yet other problems to the steel research community. There is a profound change in the philosophy of design for disasters: we can no longer be content with saving lives only, but we must also design structures which will not be so damaged as to require extensive repairs.

Another major challenge will be the emergence of many new materials such as high-performance concrete and plastic composite structures. Steel structures will continually have to face the problem of having to demonstrate viability in the marketplace. This can only be accomplished by more innovative research. Furthermore, the new comprehensive limit-states design codes which are being implemented worldwide, need research to back up the assumptions used in the theories.

Specifically, the following list highlights some of the needed research in steel structures:

- Systems reliability tools have been developed to a high degree of sophistication. These tools should be applied to the studies of bridge and building structures to define the optimal locations of monitoring instruments, to assess the condition and the remaining life of structures, and to intelligently design economic repair and retrofit operations.
- New developments in instrumentation, data transfer and large-scale computation will enable researchers to know more about the response of structures under severe actions, so that a better understanding of "real-life" behavior can be achieved.
- The state of knowledge about the strength of structures is well above the knowledge about serviceability and durability. Research is needed on detecting and preventing damage in service and from deterioration.
- The areas of fatigue and fracture mechanics on the one hand, and the fields of structural stability on the other hand, should converge into a more unified conceptual entity.
- The problems resulting from the combination of inelastic stability and low-cycle fatigue in connections subject to severe cyclic loads due to seismic action will need to be solved.
- The performance of members, connections and connectors (e.g., shear connectors) under severe cyclic and dynamic loading requires extensive new research, including shakedown behavior.

The list could go on, but one should never be too dogmatic about the future of such a highly creative activity as research. Nature, society and economics will provide sufficient challenges for the future generation of structural engineers.

10. For further reading

SAC Joint Venture Technical Reports of Phase 1 Research, 555 University Avenue, Suite 126, Sacramento, CA 95825 (these are preliminary reports which were prepared in a short time frame):

SAC 95-03 Characterization of Ground Motions During the Northridge Earthquake of January 17, 1994, Dec. 1995

SAC 95-04 Analytical and Field Investigations of Buildings Affected by the Northridge Earthquake of January 17, 1994, Dec. 1995

SAC 95-05 Parametric Analytical Investigations of Ground Motion and Structural Response, Northridge Earthquake of January 17, 1994, Dec. 1995

SAC 95-06 Surveys and Assessment of Damage to Buildings Affected by the Northridge Earthquake of January 17, 1994, Dec. 1995

SAC 95-07 Case Studies of Steel Moment Frame Building Performance in the Northridge Earthquake of January 17, 1994, Dec. 1995

SAC 95-08 Experimental Investigations of Materials, Weldments and Nondestructive Examination Techniques, Dec. 1995

SAC 96-01 Experimental Investigations of Beam-Column Subassemblages March 1996

Federal Emergency Management Agency Interim Guidelines: Evaluation, Repair, Modification and Design of Welded Steel Moment Frame Structures FEMA 267, Aug. 1965

Acknowledgements

This paper is an up-dated version of a presentation given by the author on October 19, 1996 at Osaka, Japan on the occasion of the retirement of Professor Yuhshi Fukumoto. The major topics of that presentation are retained here, but the topics were all brought up to the status of 1999.

References

[1] Leon RT, Hoffman JJ, Staeger T. Partially restrained composite connections. Chicago, IL: American Institute of Steel Construction, 1996.

[2] American Iron and Steel Institute. Specification For the Design of Cold-Formed Steel Structural Members, adopted October 25, 1996.

[3] Fukumoto Y. New constructional steels and structural stability. In: Proc. 50th Anniv. Conference, Structural Stability Research Council, June 21–22, Bethlehem, PA, 1994:211–25.

[4] Kuwamura H. Effect of yield ratio on the ductility of high-strength steels under seismic loading. In: Proc. SSRC Ann. Tech. Sess., Univ. of Minnesota, Minneapolis, MN, 1988:201–10.

[5] Green PS, Sooi TK, Ricles JM, Sause R, Kelly T. Inelastic behavior of structural members fabricated from high-performance steel. In: Proc. SSRC Ann. Tech. Sess., Lehigh Univ., Bethlehem, PA, 1994:435–56.

[6] Sooi TK, Green PS, Sause R, Ricles JM. Stress-strain properties of high-performance steel and the

implications for civil-structure design. In: Proc. Intl. Symp. on High-Performance Steel for Structural Applications, Oct. 30–Nov. 1, Cleveland, OH, 1995:35–43.
[7] McDermott JF. Local plastic buckling of A514 beams. ASCE J Struct Div 1969;95(ST9 Sep.):1837–50.
[8] Earls CJ, Galambos TV, Hajjar JF. On inelastic buckling of high strength steel beams under moment gradient. In: Proc. Struct. Stab. Res. Council, June 9–11, Toronto, Canada, 1997:553–68.
[9] Earls CJ. On the inelastic failure of high strength steel I-shaped beams. J Const Steel Res 1999;49(1):1–24.
[10] Ricles JM, Lu LW, Peng S-W. Cyclic behavior of composite CFT column-WF beam connections. In: Proc. Struct. Stab. Res. Council, June 9–11, Toronto, Canada, 1997:369–90.

Semi-compact steel plates with unilateral restraint subjected to bending, compression and shear

M.A. Bradford [a,*], S.T. Smith [b], D.J. Oehlers [c]

[a] *Professor of Civil Engineering, School of Civil and Environmental Engineering, The University of New South Wales, Sydney, NSW 2052, Australia*

[b] *Postdoctoral Research Fellow, Department of Civil and Structural Engineering, The Hong Kong Polytechnic University, Hung Hom, Kowloon, Hong Kong*

[c] *Senior Lecturer, Department of Civil and Environmental Engineering, The University of Adelaide, Adelaide, SA 5005, Australia*

Received 20 November 1997; received in revised form 24 August 1998; accepted 27 September 1999

Abstract

The buckling of thin-walled steel plates which are juxtaposed with a rigid medium must be treated as a contact problem, since the local buckles may only form away from the rigid medium in a unilateral mode. One particular example is the bolting of plates to the sides of reinforced concrete beams in order to stiffen and strengthen the original concrete beam. Plates which experience first yield before local buckling are referred to as semi-compact, and this paper uses an energy method developed elsewhere by the authors to study the unilateral local buckling of restrained plates, so that limiting depth to thickness ratios may be obtained that delineate the semi-compact section classification when the plate is subjected to a combination of bending, compressive and shearing actions. The application to retrofitting concrete beams is discussed. © 2000 Elsevier Science Ltd. All rights reserved.

Keywords: Contact problem; Elasticity; Rayleigh-Ritz method; Section classification; Unilateral buckling; Yielding

* Corresponding author. Tel.: +612-9385-5014; fax: +612-9385-6139.

Reprinted from *Journal of Constructional Steel Research* **56 (1)**, 47-67 (2000)

1. Introduction

Unilateral buckling is a buckling mode that occurs when the buckling member is restrained against movement in one direction by a rigid medium. Problems of this type which have received attention have been the buckling of pipelines in trenches or on the seabed, the buckling of floating ice sheets, and the upheaval failure of pavements. In structural engineering, the unilateral buckling problem is usually germane to thin plates that are restrained by a rigid medium that is often a concrete core.

The traditional studies of local buckling of plates with elastic or rigid restraint have been concerned with plates bonded or attached to foundations. The unilateral buckling problem is termed in the mathematical literature as a contact problem. Three general methods have been used to solve the contact problem, viz. the penalty method, the method of mathematical programming and direct substitution of a unilateral buckling mode shape, displacement function or specified contact surface into the appropriate analysis procedure. The penalty method, as described below, has been adopted in the present study.

The penalty method may take many formulations. In the formulation of Oktake [1,2] and Oden and Kikuchi [3], a finite element approximation was used which was augmented with an additional penalty term due to the foundation that adds to the total potential energy of the system. The penalty term is zero when unilateral constraint is satisfied at a node point, but takes on a positive value when it is violated. The solution of the penalty problem was found to converge to the solution of the nonlinear plate problem with unilateral constraint if a large enough foundation modulus is adopted. The method is advantageous as the number of degrees of freedom is not increased, but ill-conditioning can occur if too large a penalty parameter is adopted. The Lagrange multiplier method [4] and the augmented Lagrangian method [5] are other forms of the penalty method. In the former method, Lagrange multipliers representing contact forces enforce the contact constraint with the rigid medium. These terms are considered as additional degrees of freedom. Because Lagrange multipliers may produce zero diagonal terms in the tangent stiffness matrix, the numerical difficulties may be overcome by using a penalty parameter and Lagrange multipliers. This method, although giving satisfactory results, possesses slow convergence characteristics [6].

Other methods of solution of the unilateral contact problem have been the early finite element treatments of Cheung and his colleagues [7,8] in the late 1960's. These were refined some 20 years later in a number of studies [9,10], while the finite difference method has also been used [11] as well as the boundary element method [12].

The Rayleigh-Ritz method, as set out in Timoshenko's text [13], has been found to be a very powerful tool for implementing the penalty method into the solution of unilateral buckling problems. The plate displacement function adopted for this is a two-dimensional Ritz function, and this can handle arbitrary support conditions with bending, compression and shear. The pb-2 Rayleigh-Ritz method was developed by Liew and his colleagues [14,15], and the degree of the displacement function can be increased until the desired accuracy is reached. The method was modified and applied by the authors to handle pure shear [16], and combined bending, compression and shear [17], and these results were validated experimentally [18].

Because it is important that plates subjected to unilateral buckling do not buckle before they reach first yield, particularly if the plates are bolted to a reinforced concrete beam to increase strength and stiffness, the present paper deploys the Rayleigh-Ritz method to determine the relationships between bending, compression and shear actions that define the semi-compact limit in plate buckling [19]. This problem has been solved for bilateral buckling that is normally encountered in thin-walled sections [20,21], but it is believed that the depth to thickness ratios obtained in the unilateral study are the results of unique research. The application of these depth to thickness limits to plated reinforced concrete beams are discussed.

2. Theory

2.1. General

The method employed in this paper is the Rayleigh-Ritz method with a nonlinear elastic foundation that exhibits a sign-dependent foundation stiffness. The full theory is set out in Ref. 17, and only the pertinent points in the derivation are outlined here. The flat plate that is subjected to bending, compression and shearing actions, and to a variety of edge conditions, is shown in Fig. 1.

The plate is discretised into a rectangular grid, but the elements of the grid are segments and not to be confused with finite elements. The contact status (i.e. the propensity of the plate to buckle into the rigid medium or away from it) is ascertained at each grid coordinate (x_g, y_g).

2.2. Displacement function

The Rayleigh-Ritz method requires a displacement function that satisfies the boundary conditions a priori. The flexural displacement used here is [14,15]

$$w(x,y)=\varphi_b(x,y)\sum_{q=0}^{p}\sum_{r=0}^{q}a_m\phi_m(x,y) \tag{1}$$

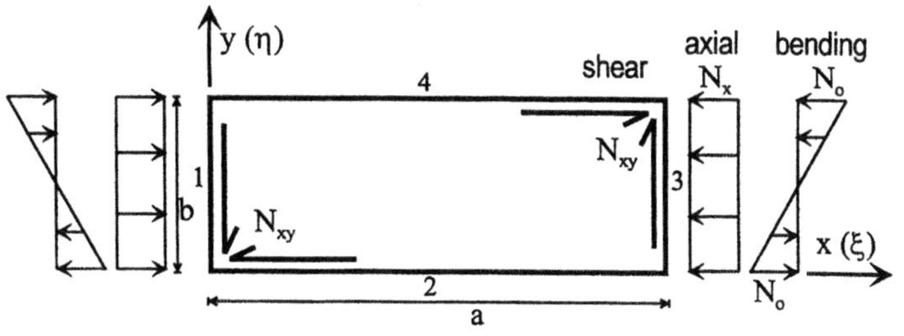

Fig. 1. Rectangular plate.

where p is the degree of a two-dimensional polynomial, a_m is the arbitrary Ritz coefficient, and $\phi_m(x,y)$ is the mth term of a complete two-dimensional polynomial given by

$$\phi_m(x,y)=x^r y^{q-r} \tag{2}$$

where $m=(q+1)(q+2)/2$. The term $\varphi_b(x,y)$ is a polynomial function expressing the boundary conditions that are functions of the geometry and restraint. These are given in Refs. 14, 15 and 16.

2.3. Strain energy stored in plate

The strain energy stored in the plate of thickness t may be written as

$$U=\frac{D}{2}\int\int_A\left[\left(\frac{\partial^2 w}{\partial x^2}+\frac{\partial^2 w}{\partial y^2}\right)^2+2(1-\nu)\left(\frac{\partial^2 w}{\partial x^2}\frac{\partial^2 w}{\partial y^2}-\left\{\frac{\partial^2 w}{\partial x\partial y}\right\}^2\right)\right]\mathrm{d}A \tag{3}$$

where, $D=Et^3/12(1-\nu^2)$, E is Young's modulus and ν is Poisson's ratio.

2.4. Work done during buckling

The work done during buckling due to longitudinal and shear forces V can be written as

$$V=\int\int_A\left[\frac{N_x}{2}\frac{\partial^2 w}{\partial x^2}+\frac{N_0}{2}\left(1-\alpha\frac{y}{b}\right)\frac{\partial^2 w}{\partial x^2}+N_{xy}\frac{\partial^2 w}{\partial x\partial y}\right]\mathrm{d}A \tag{4}$$

where A is the area of the plate, b is the plate width, N_x is the applied axial compression, N_0 is the maximum compressive force at the edge y=0, N_{xy} is the applied shear force and α specifies the bending gradient (α=0 for pure compression and α=2 for pure bending).

2.5. Strain energy stored due to rigid restraint

The condition of unilateral restraint is accounted for indirectly through the use of a nonlinear foundation stiffness with a sign dependent relationship. The contact condition X models unilateral constraint of the plate by prohibiting the formation of tensile normal contact stresses, and any penetration into the rigid medium. The relationship between X and w is shown in Fig. 2. The contact condition is chosen to model the tensionless nature of the rigid foundation, so that

$$X=\begin{cases}0 & \text{(separation)}\\ 1 & \text{(contact)}\end{cases} \tag{5}$$

An elastic spring of large foundation stiffness k_{found} is added at each grid location

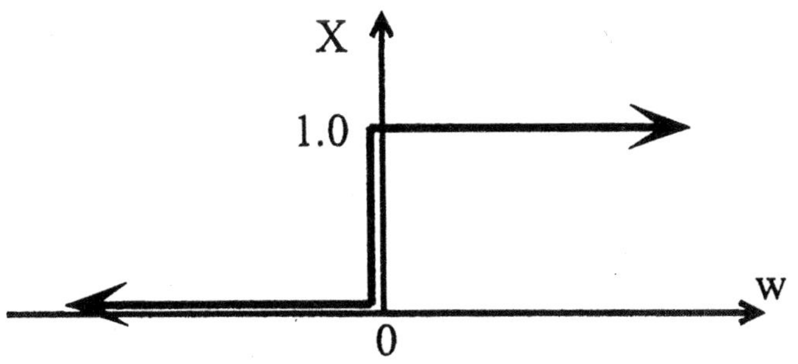

Fig. 2. Contact function.

(x_g, y_g) to model the rigid foundation. The strain energy due to the deformation of the activated springs is S is then given by

$$S = \sum_A \frac{1}{2} X k_{found} \{w(x_g, y_g)\}^2 \tag{6}$$

where $w(x_g, y_g)$ is the deformation at a particular grid point.

2.6. Solution technique

The total potential can be written as

$$\Pi = U + S - V \tag{7}$$

which can be written in matrix format as

$$\Pi = \frac{1}{2}\{a_m\}^T [K] \{a_m\} \tag{8}$$

where $\{a_m\}$ is the vector of Ritz coefficients appropriate to the degree of the polynomial adopted. The matrix $[K]$ is nonlinear owing to the unilateral contact condition, which is not known at each grid position a priori.

In accordance with the Rayleigh-Ritz procedure, the total potential is minimised with respect to $\{a_m\}$. When this minimisation is performed,

$$[K]\{a_m\} = \{0\} \tag{9}$$

which for nontrivial buckling displacements requires that $|[K]| = 0$. The procedure for the solution of this nonlinear determinantial equation, as well as explicit entries for the matrix $[K]$, is set out fully in Ref. 16. It is worth noting that numerical difficulties require Aitken's Δ^2 extrapolation [22] to be invoked in the solution of Eq. (9) for the buckling actions (eigenvalues) and buckled shape (eigenvector).

2.7. Verification of solution

References 17 and 18 present an experimental verification of the theory presented herein, in the absence of other benchmark solutions in the open literature for unilateral plate buckling under combined loading. In addition, it has been shown for bilateral buckling (when the contact condition X is relaxed) that the buckling solutions agree well with the interaction solutions for the buckling of "conventional" plates with different edge conditions. Convergence studies in the aforementioned references have indicated that a sufficiently accurate converged solution is obtainable with a 9th degree complete polynomial, using a grid size of 40×40 and a foundation modulus of k_{found} =1.0×10^5. These values were adopted in all the analyses presented in the next section. It is worth noting that a higher order polynomial and a stiffer foundation modulus were found to cause numerical instability. The solution of the buckling problem was found to be rapid on contemporary personal computers and work stations.

3. Slenderness limits for bending and shear

3.1. General

The theory presented in the previous section may be used to determine the plate slenderness limits that define the boundary between semi-compact and slender sections. These width to thickness ratios are called the yield slenderness limits λ_{ey} and are defined by

$$\lambda_{ey}=\frac{b}{t}\sqrt{\frac{f_y}{250}} \tag{10}$$

where b is the plate width, t is its thickness, and f_y is the uniaxial tensile yield strength in N/mm^2. The yield stress is normalised with respect to f_y =250 N/mm^2 in Eq. (10) so that steels with this yield stress have a b/t ratio that corresponds to λ_{ey}. The unilateral analysis allows the critical combination of the elastic bending stress $\sigma_{b,cr}$ and shear stress τ_{cr} to be obtained, where [19]

$$\sigma_{b,cr}=k_{b,s}\frac{\pi^2 E}{12(1-\nu^2)}\frac{1}{(b/t)^2} \tag{11}$$

and

$$\tau_{cr}=k_{s,b}\frac{\pi^2 E}{12(1-\nu^2)}\frac{1}{(b/t)^2} \tag{12}$$

where $k_{b,s}$ is the elastic buckling coefficient for bending in the presence of shear, and $k_{s,b}$ is the elastic buckling coefficient for shear in the presence of bending.

For an initially flat plate free from residual stresses, the relationship between $\sigma_{b,cr}$ and τ_{cr} at yield may be obtained from the Von Mises yield criterion [19]

$$\sigma_{b,cr}^2+3\tau_{cr}^2=f_y^2 \tag{13}$$

If Eqs. (11) and (12) are substituted into Eq. (13), and the material properties E =200 kN/mm^2 and ν = 0.3 are adopted, then

$$26.89(k_{b,s}^2+3k_{s,b}^2)^{1/4}=\frac{b}{t}\sqrt{\frac{f_y}{250}}=\lambda_{ey} \tag{14}$$

The computer program is set up to calculate the relationship between $k_{b,s}$ and $k_{s,b}$ at elastic buckling as a function of $r = \sigma_{b,cr}/\tau_{cr}$, so that from Eqs. (11) and (12), $r = k_{b,s}/k_{s,b}$. This relationship may be substituted into Eq. (14) to obtain the slenderness limit λ_{ey} at elastic buckling. The ratio r may be transformed conveniently into the stress ratio parameter R by

$$R=\frac{r}{r+1} \tag{15}$$

Hence $0 \leq R \leq 1$, with $R=0$ corresponding to pure shear and $R=1$ corresponding to pure bending.

3.2. *Plate with four simply supported edges*

Fig. 3 shows the relationship between the stress ratio parameter R and the yield slenderness limit λ_{ey} for a plate with four simply supported edges, denoted in the figure as S-S-S-S. The curves relating $k_{b,s}$ and $k_{s,b}$ depend on the aspect ratio $\gamma = a/b$, and remain fairly constant for values of $\gamma \geq 3$. The results in Fig. 3 for a simply supported plate were thus calculated for $\gamma=3$.

The limit for unilateral buckling is shown in Fig. 3, and also the limit for bilateral buckling obtained by relaxing the constraint condition and letting $k_{found}=0$. It is clear that the slenderness limit for which yielding and shear coincide is lowest for pure shear in both the unilateral and bilateral cases. The unilateral value for pure shear is 82, and this is the same as the value given in the Australian standard AS4100 [23]. For pure bending, the value of λ_{ey} increases to 131, which is 14% higher than the AS4100 limit of 115 that is intended to account for residual stresses and out-of straightness. The corresponding slenderness limits for unilateral buckling in pure shear and pure bending are 92 and 138 respectively. It is thus evident that the unilateral restraint increases the width to thickness ratio for which first yield will occur before buckling, although not dramatically so. It is also worth noting that a linear interaction between the value of λ_{ey} in pure shear and in pure bending is unconservative for both unilateral and bilateral buckling, owing to the convexity of the computer-generated curve.

3.3. *Plate with two clamped and two free edges*

A plate with its transverse edges loaded in bending clamped and its longitudinal edges free (denoted C-F-C-F) was analysed by the penalty method. It was found that the local buckling coefficients $k_{b,s}$ and $k_{s,b}$ tended to zero as the aspect ratio γ was increased. Fig. 4 thus plots the relationship between the stress ratio parameter R and the yield slenderness limit λ_{ey} for aspect ratios varying between 1.0 and 3.0. It is

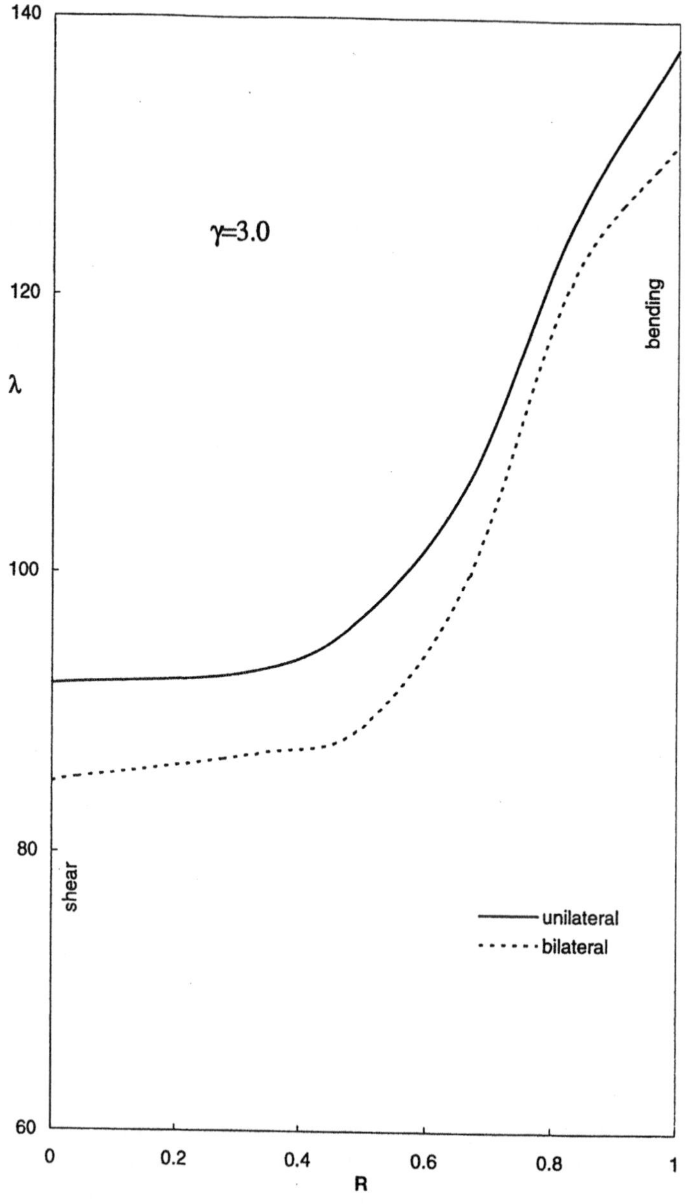

Fig. 3. Interaction of shear and bending for a S-S-S-S plate.

interesting to note that for the lower values of the aspect ratio, the critical case (for which λ_{ey} is lowest) corresponds to pure bending and not pure shear as was the case with all edges simply supported. The yield slenderness limits are also sensitive to the aspect ratio, ranging from minimum values of 72 (pure bending) with γ=1.0 to 30 (pure shear) for γ=3.0.

The results presented in Fig. 4 have practical ramifications for a steel plate that is bolted to a reinforced concrete beam in order to stiffen and/or strengthen it. If it is assumed (as has been shown by tests [18]) that the vertical or transverse lines of

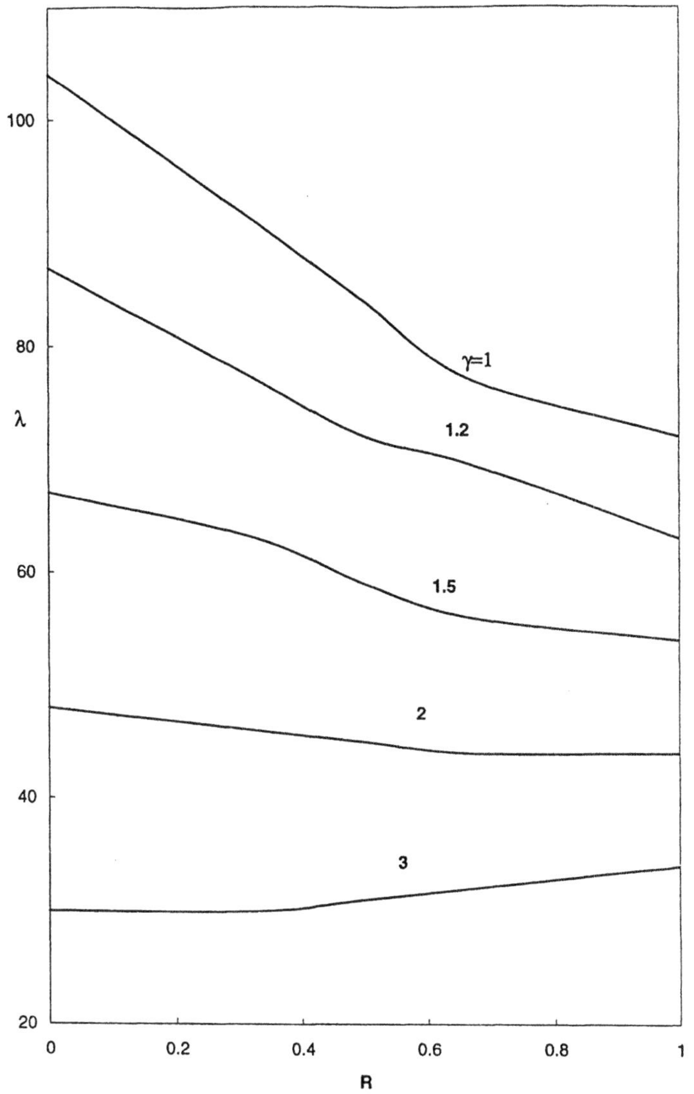

Fig. 4. Interaction of shear and bending for a C-F-C-F plate.

bolts maintain a node so that the buckling deformations are zero along the pitch line, then the plate may be thought of as restrained with clamped transverse edges and free longitudinal edges. The results in Fig. 4 may thus be used to determine the pitch of the bolting for which yielding will precede elastic local buckling. Because the plates used would be relatively slender, it is clear from Fig. 4 that the aspect ratio, which is proportional to the pitch of the bolts, must be kept to a minimum so that yielding will precede local buckling.

3.4. Plate with two clamped edges, and one edge simply supported and one edge free

The interaction between the loading parameter R for shear and bending and the yield slenderness limit λ_{ey} for unilateral buckling has been determined for a plate with clamped transverse edges, and one longitudinal edge free and the other longitudinal edge simply supported (denoted C-F-C-S). This is another boundary condition that may, in theory, be encountered in the problem of bolting steel plates to a reinforced concrete beam. Two cases were considered, these being when the free edge is subjected to the tensile portion of the bending stress and when this edge is subjected to the compressive portion of the bending stress. In both of these cases, the local buckling coefficients are dependent on the aspect ratio, and tend to zero as the aspect ratio is increased.

Fig. 5 illustrates the case when the free edge is subjected to compression. As expected, this case is somewhat similar to that with two free longitudinal edges, since the local buckling is not overly sensitive to whether the tensile edge is simply supported or free. It is thus worth noting that the limits in Figs. 4 and 5 are fairly similar for aspect ratios less than about 2. In all cases in Fig. 5, the critical condition is one of pure bending (R =1), which in all cases produces the lowest value of the yield slenderness limit. The case when the free edge is subjected to tension is shown in Fig. 6. Of course, the limits are the same in Figs. 5 and 6 for pure shear (R =0), but the behaviour is vastly different as the bending increases and $R \rightarrow 1$. Here, the unsupported region in which the plate is the less resistant to buckling is subjected to tension, and this interaction of shear and opposing tension produces large values of the yield slenderness limit as the tension is increased relative to the shear at the free edge. For this case, the most critical loading condition that produces the lowest values of λ_{ey} is pure shear.

4. Slenderness limits for compression and shear

4.1. General

A similar analysis to that undertaken for bending and shear was also undertaken for the combination of compression and shear for which first yielding and elastic buckling coincide for a unilaterally restrained plate free from residual stresses and geometric imperfections. The critical stress for a plate in uniform compression $\sigma_{c,cr}$ was obtained similarly to Eq. (11) as

$$\sigma_{c,cr}=k_{c,s}\frac{\pi^2 E}{12(1-\nu^2)}\frac{1}{(b/t)^2} \tag{16}$$

where $k_{c,s}$ is the elastic unilateral buckling coefficient for uniform compression in the presence of shear. The Von Mises yield criterion stated in Eq. (13) still applies, with $\sigma_{b,cr}$ being replaced by $\sigma_{c,cr}$. Again, the computer program was set up to calcu-

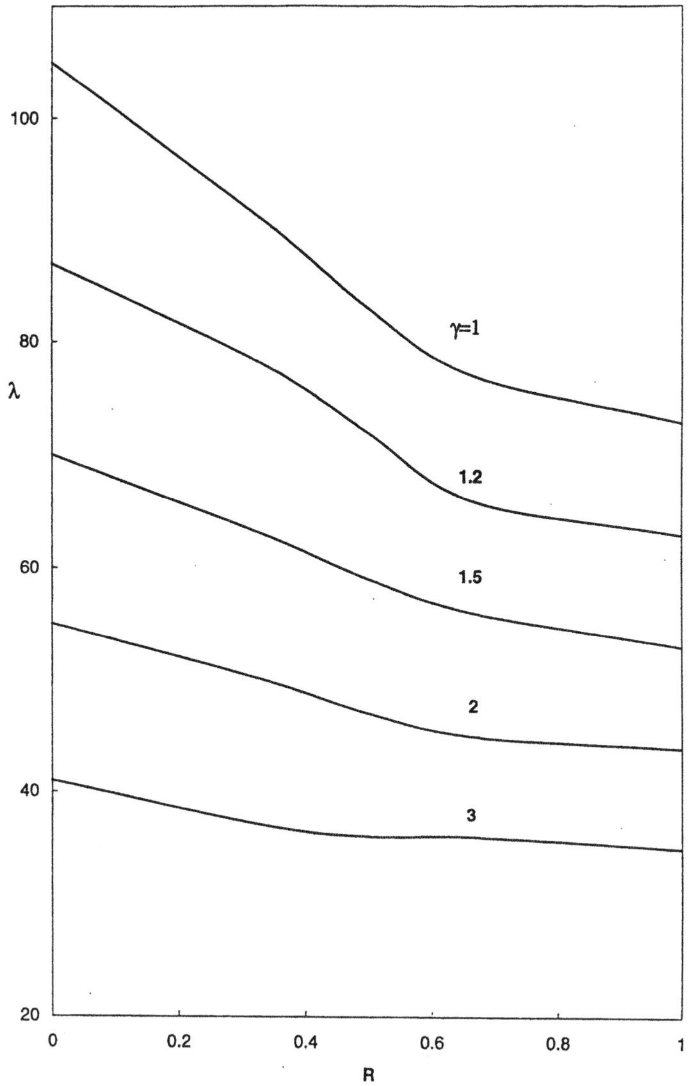

Fig. 5. Interaction of shear and bending for a C-F-C-S plate (F-edge compressed).

late the relationship between $k_{c,s}$ and $k_{s,c}$ as $s = \sigma_{c,cr}/\tau_{cr}$, which is converted into the stress ratio parameter S in a similar way to Eq. (15) as $S = s/(s+1)$. Hence $0 \leq S \leq 1$, with $S = 0$ corresponding to pure shear and $S = 1$ corresponding to pure compression.

4.2. Plate with four simply supported edges

Fig. 7 shows the relationship between the stress ratio parameter S and the yield slenderness limit λ_{ey} for a plate with four simply supported edges. As for the case of bending and shear, the curves depend on the aspect ratio γ, and are identical for unilateral buckling for values of $\gamma \geq 2$. However, unlike the shear and bending case,

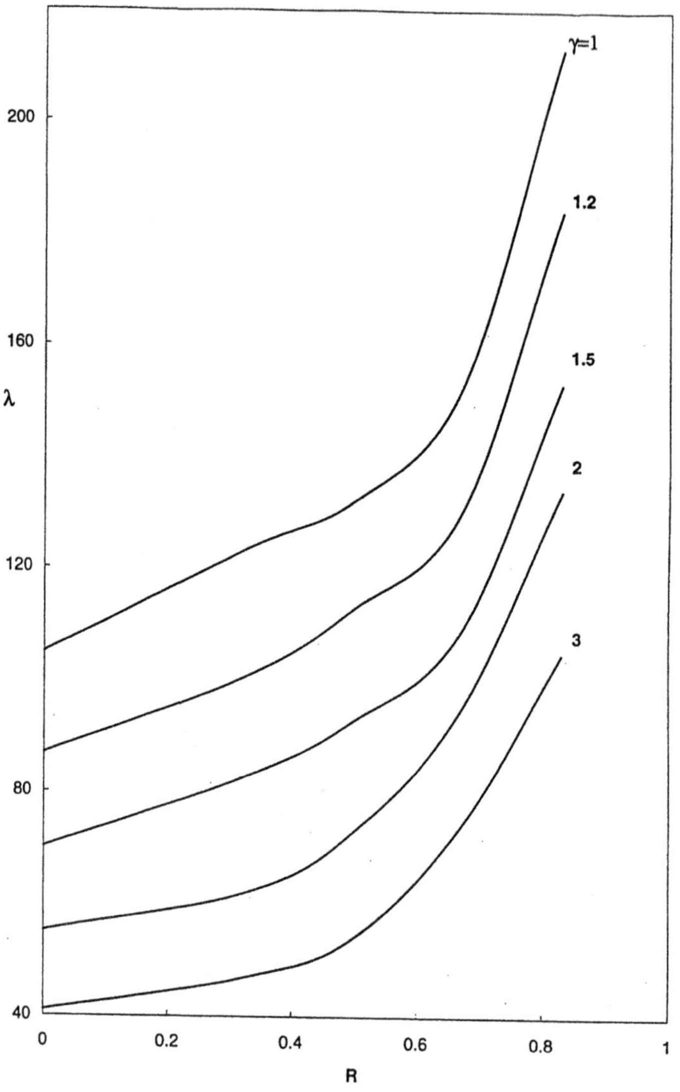

Fig. 6. Interaction of shear and bending for a C-S-C-F plate (S-edge compressed).

the critical load combination (that determines the lowest slenderness limit) is for pure compression.

For an aspect ratio of 3.0, the bilateral and unilateral limits in pure compression are λ_{ey} =54. This compares with the values given in the AS4100 [23], that range from 45 for a stress relieved simply supported plate to 35 for a heavily welded plate. The 20% difference between the values of 45 and 54 reflects the empirical inclusion of the effects of initial out-of-straightness in the code limit. It is interesting to note that λ_{ey} increases to 57 for the case of unilateral buckling with aspect ratios greater than 2. This is a consequence of the change of mode shape associated with the unilateral buckling. It is also interesting that the yield slenderness limit for bilateral

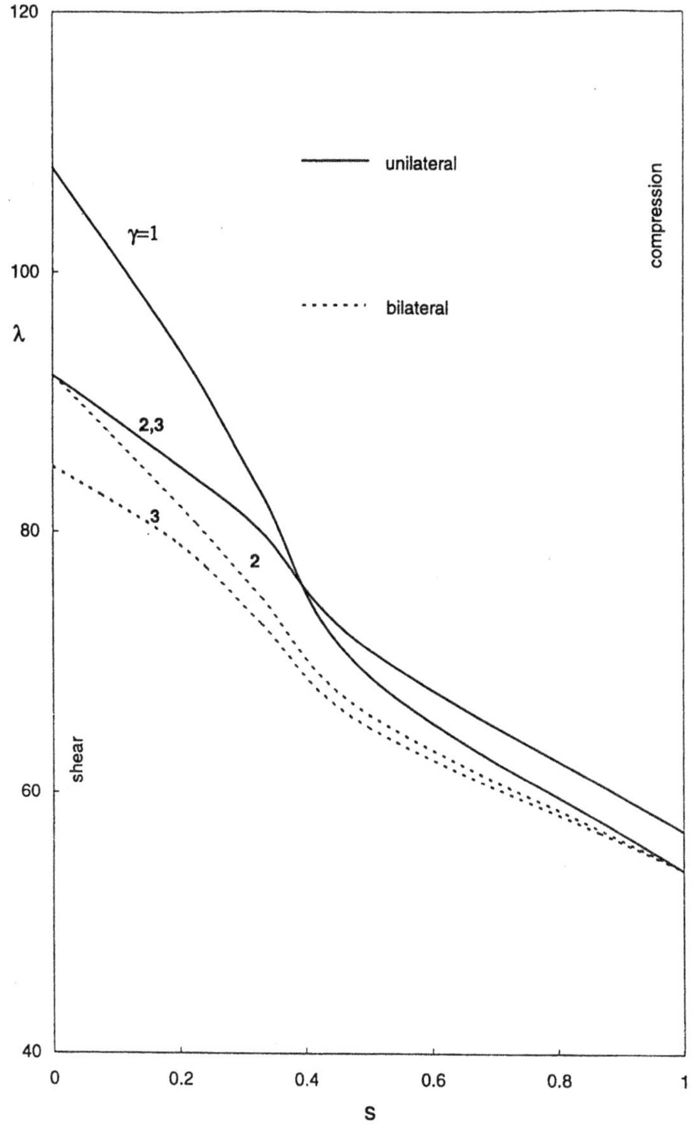

Fig. 7. Interaction of shear and compression for a S-S-S-S plate.

buckling cases with $\gamma=2$ corresponds exactly to the unilateral limit for $\gamma=1$ in pure compression, and to the unilateral limit for $\gamma=2$ for pure shear.

4.3. Plate with two clamped and two free edges

The numerical procedure has been used to analyse a plate with its two compressed edges clamped and the two uncompressed (longitudinal) edges free. The yield slenderness limit is shown in Fig. 8 as a function of the stress ratio parameter *S*, where it can be seen that again the limit is a function of the aspect ratio. The behaviour shown in Fig. 8 is similar to that shown in Fig. 4, where generally the critical load

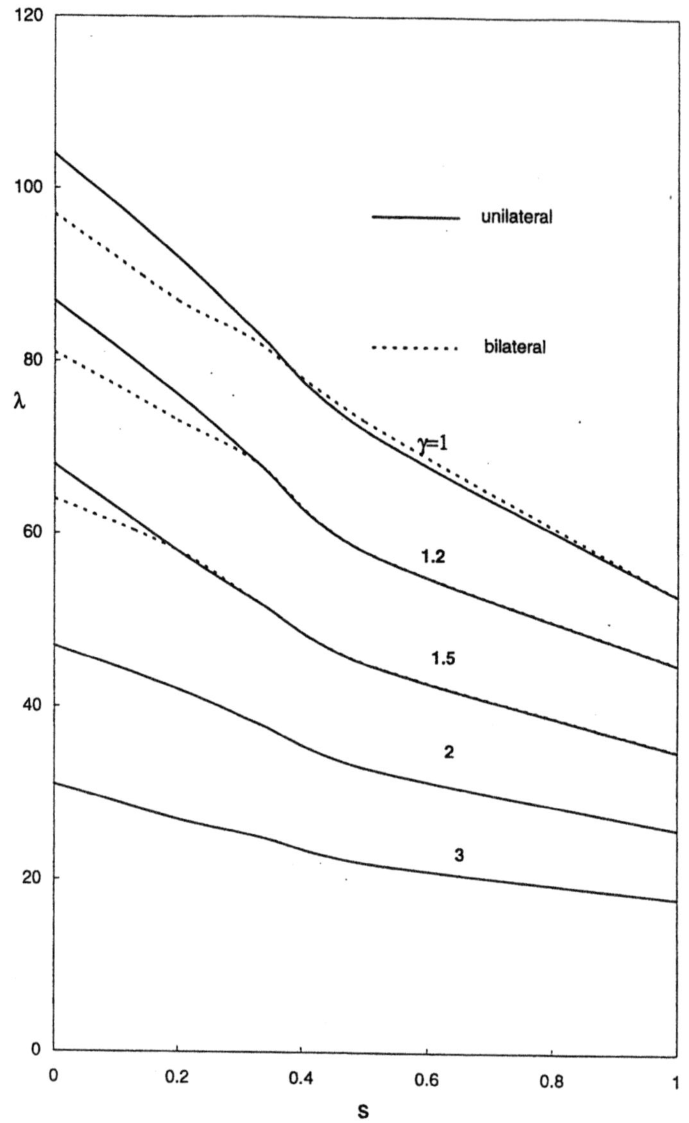

Fig. 8. Interaction of a shear and compression for a C-F-C-F plate.

combination for the lowest slenderness limit is pure bending (Fig. 4) and pure compression (Fig. 8). In both cases, for values of the stress ratio parameter greater than about 0.5 (corresponding to equality of the shear and bending stress or to the shear and compressive stress) the unilateral and bilateral limits are the same, indicating that as the shear decreases and bending or compression dominate, the presence of the unilateral constraint does not affect the slenderness limit.

4.4. Plate with two clamped edges, and one edge simply supported and one edge free

Fig. 9 shows the aspect ratio-dependent slenderness limit when the compressed edges are clamped, and one longitudinal edge is free and the other longitudinal edge

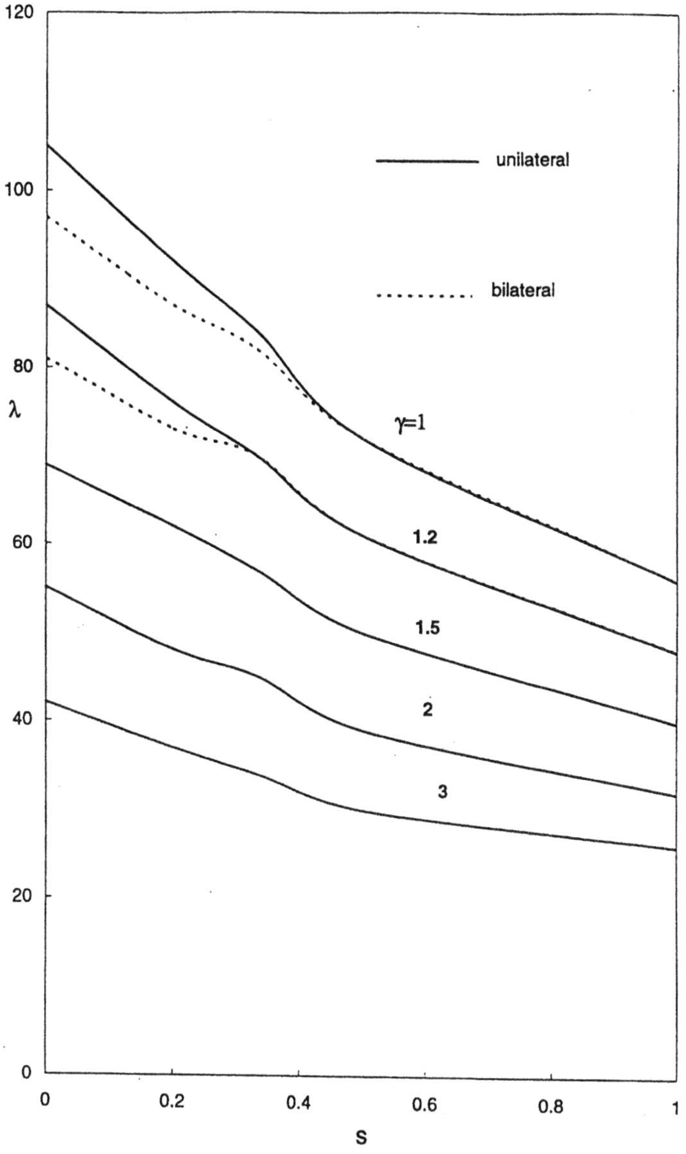

Fig. 9. Interaction of shear and compression for a C-F-C-S plate.

is simply supported. The corresponding case for bending and shear is shown in Fig. 5. The behaviour depicted in Fig. 9 is somewhat similar, with the critical loading case being that for pure compression. Two points are worth noting from this particular study. Firstly, the unilateral and bilateral modes limits are identical (except for low aspect ratios in pure shear), and secondly the buckling behaviour shown in Fig. 9 is similar to that shown in Fig. 8 in which the simply supported longitudinal edge is replaced by a free longitudinal edge.

5. Interaction of bending, compression and shear

In general, all three critical stresses $\sigma_{b,cr}$ in Eq. (11), $\sigma_{c,cr}$ in Eq. (16) and τ_{cr} in Eq. (12) may interact, so that the yield slenderness limit λ_{ey} is a function of $k_{b,sc}$, $k_{c,bs}$ and $k_{s,bc}$ which are the elastic buckling coefficients for bending in the presence of shear and compression, compression in the presence of bending and shear and shear in the presence of bending and compression respectively. This relationship may be determined for both unilateral and bilateral buckling using the numerical method described in this paper. In the plated beam illustration, the plate will generally be subjected to a combination of pure bending, compressive and shear actions.

Again, the Von Mises yield condition may be invoked to determine the interaction of the three buckling actions at first yield. In presenting the results, the shear local buckling coefficient in the presence of bending and compression $k_{s,bc}$ is presented as a fraction of the pure shear unilateral buckling coefficient k_s by $k_{s,bc} = \rho k_s$, and the stress loading parameter expressed as $Q = q/(q+1)$, where $q = k_{b,sc}/k_{c,bs}$. Hence $Q=0$ corresponds to pure bending while $Q=1$ corresponds to pure compression. The first yield condition in Eq. (14) then takes the form

$$26.89\{(k_{b,sc}+k_{c,bs})^2+3(\rho k_s)^2\}^{1/4}=\frac{b}{t}\sqrt{\frac{f_y}{250}}=\lambda_{ey} \tag{17}$$

The values of k_s have been determined by the authors [16], and for the cases considered below are given in Table 1.

The yield slenderness limits for a plate with four edges simply supported are shown in Fig. 10 for an aspect ratio of $\gamma=1$. In all cases, for a given value of the shear buckling coefficient, the most critical loading (for which λ_{ey} is least) is that of pure bending. For $Q<0.75$, increasing the shear decreases the bending and compression local buckling coefficients to the extent that $\sqrt{3}\rho k_s$ in Eq. (17) dominates $(k_{b,cs}+k_{c,bs})$, and the slenderness limit increases.

Fig. 11 plots the relationship between the yield slenderness limit and bending and compression in the presence of shear for a plate with its two ends subjected to longitudinal stress clamped, and the two longitudinal edges free. The plot is aspect ratio-dependent, and is given in Fig. 11 for $\gamma=1.0$. It can be seen that as the shear increases, the influence of the interaction of bending and compressive actions (reflected in Q) is decreased, and the yield slenderness limit is only dependent on the shear. A similar relationship can be seen in Fig. 12, where the same interaction is shown for the free edge that is subjected to the lesser compressive stress being simply supported. As the shear increases, the relationship between λ_{ey} and Q shown

Table 1
Unilateral shear buckling coefficients for $\gamma=1$ [16]

S-S-S-S	C-F-C-F	C-S-C-F
9.355	8.709	8.710

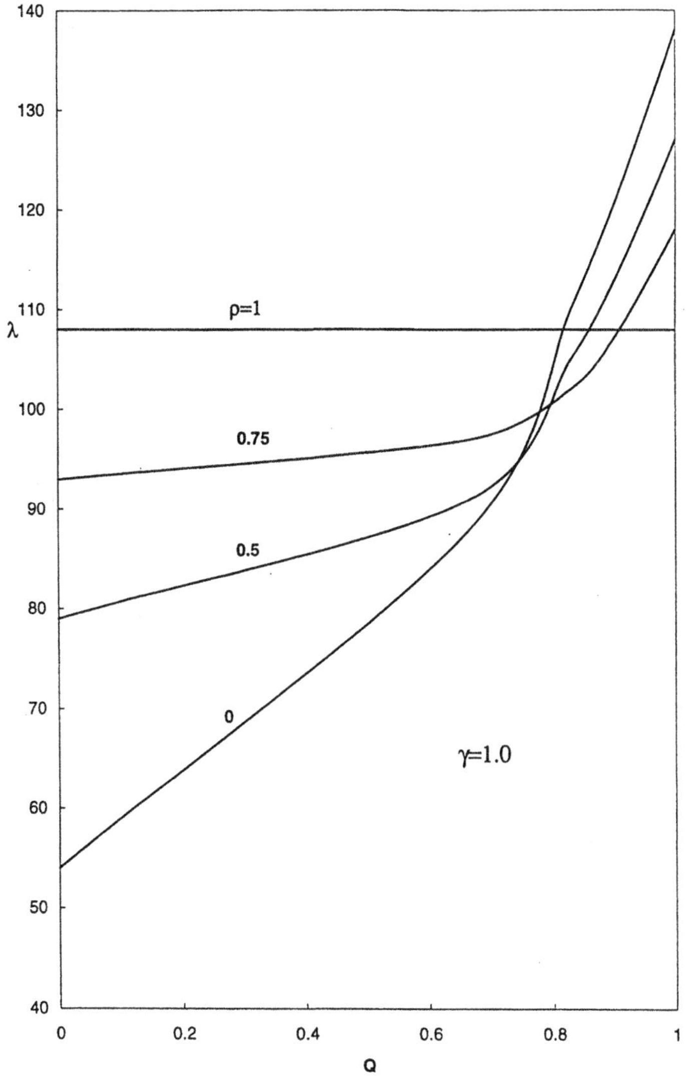

Fig. 10. Interaction of bending, compression and shear for a S-S-S-S plate.

in Fig. 11 becomes nearly identical to that shown in Fig. 12. This is because a free edge in high shear and low levels of longitudinal compression behaves, as would be expected, in a similar fashion to a free edge in high shear and low levels of longitudinal compression.

6. Concluding remarks

A numerical method for studying the behaviour of unilaterally restrained plates, based on the Rayleigh-Ritz procedure that uses a penalty formulation, has been described. Steel plates subjected to this unilateral condition of restraint are finding

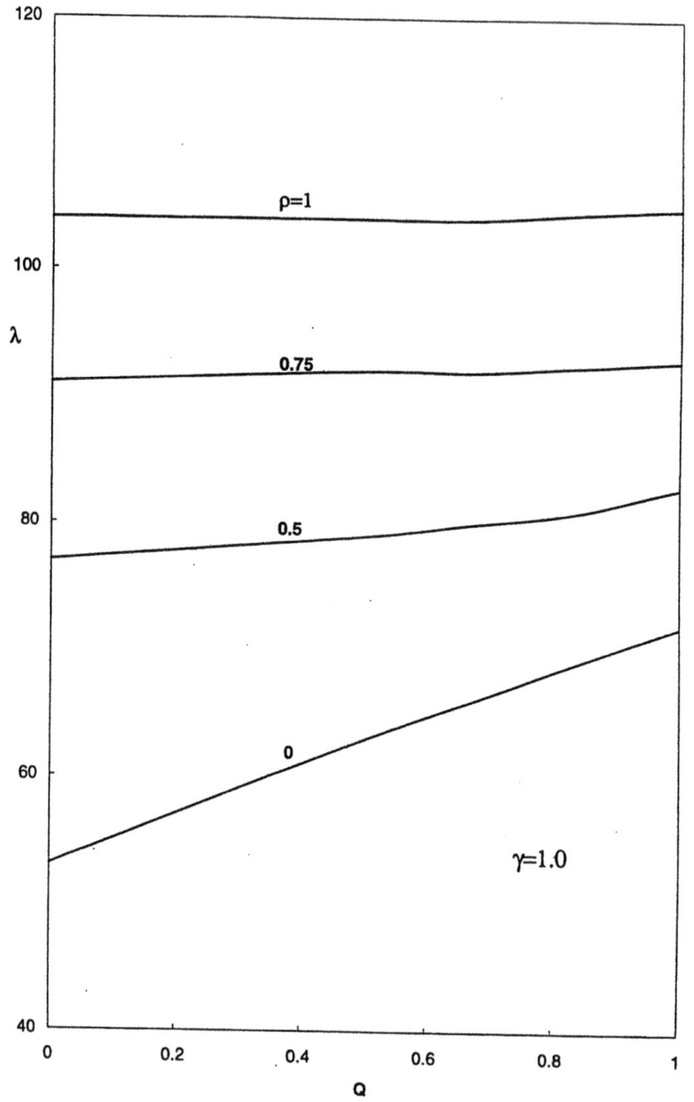

Fig. 11. Interaction of bending, compression and shear for a C-F-C-F plate.

widespread application in the strengthening and/or stiffening of deteriorating reinforced concrete beams.

Plates may buckle in a unilateral local mode in bending, shear and compression. In the design of steel plates, it is often essential to ensure that first yield will precede elastic local buckling. The attainment of this condition defines the semi-compact width to thickness ratio of a plate, that depends on the ratio of the width to thickness and also on the yield stress. The slenderness limit that defines a semi-compact plate depends on the interaction of bending, compression and shear. Based on a first yield condition that is expressed as the well-known Von Mises yield criterion, the paper has presented the yield slenderness limits for the unilateral buckling of mild steel plates that are subjected to a combination of these three loading actions. These results

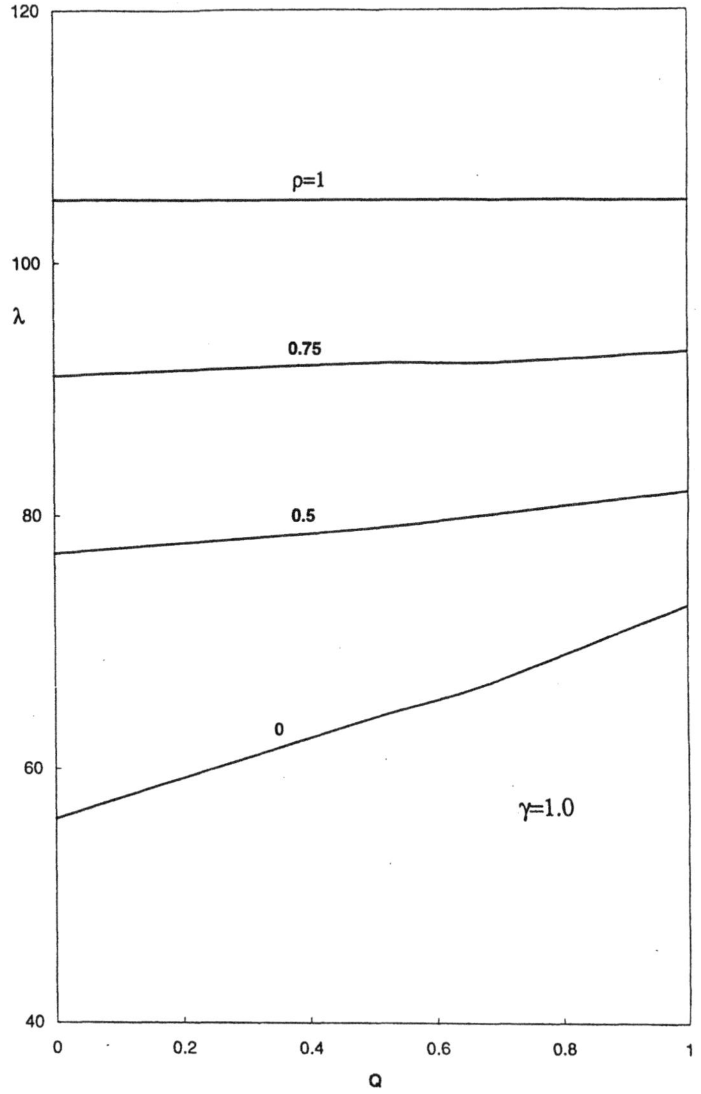

Fig. 12. Interaction of bending, compression and shear for a C-S-C-F plate.

are important for the proportioning of plates used to retrofit concrete beams, as in general they will be subjected to combinations of bending, shear and compression. These results are essential for distributing the bolt shear connectors in bolted-plated beam construction when steel plates are bolted to the sides of existing reinforced concrete beams to increase their strength and stiffness [24,25]. Tests have shown [25] that deep plated beams are prone to buckling, because the relatively few shear connectors required to transfer both the longitudinal and transverse shear may not provide too sufficient density to prevent buckling and, hence, local buckling will often control the numbers of connectors required.

Finally, the yield slenderness limits given in design codes are based on semi-empirical formulations, where the slenderness is reduced below the first yield value to allow for the effects of residual stresses and geometric imperfections. In the absence of any available test data, the slenderness limits presented in the figures in this paper are based on first yield only. More experimental work is thus needed in calibrating the yield slenderness limits presented in this paper.

Acknowledgements

The work described in this paper was supported by a grant made available to the Universities of New South Wales and Adelaide under the Australian Research Council's Large Grants Scheme. The second author was funded by an Australian Postgraduate Award and a Faculty of Engineering Dean's Scholarship.

References

[1] Ohtake K, Oden JT, Kikuchi N. Analysis of certain unilateral problems in Von Karman plate theory by a penalty method — Part 1. A variational principle with penalty. Computer Methods in Applied Mechanics and Engineering 1980;24:187–213.
[2] Ohtake K, Oden JT, Kikuchi N. Analysis of certain unilateral problems in Von Karman plate theory by a penalty method — Part 2. Approximation and numerical analysis. Computer Methods in Applied Mechanics and Engineering 1980;24:317–37.
[3] Oden JT, Kikuchi N. Finite element methods for constrained problems in elasticity. International Journal for Numerical Methods in Engineering 1982;18:701–25.
[4] Bathe KJ, Chaudhary A. A solution method for planar and axisymmetric contact problems. International Journal for Numerical Methods in Engineering 1985;21:65–88.
[5] Simo JC, Laursen TA. An augmented Lagrangian treatment of contact problems involving friction. Computers and Structures 1992;42:97–116.
[6] Stein E, Wriggers P. Stability of rods with unilateral constraints, a finite element solution. Computers and Structures 1984;19:205–11.
[7] Cheung YK, Zienkiewicz OC. Plates and tanks on elastic foundations — an application of the finite element method. International Journal of Solids and Structures 1965;1:451–61.
[8] Cheung YK, Nag DK. Plates and beams on elastic foundations — linear and nonlinear behaviour. Geotechnique 1968;18:250–60.
[9] Ascione L, Grimalid A. Unilateral contact between a plate and an elastic foundation. Meccanica 1984;19:223–33.
[10] Lewandowski R, Switka R. Unilateral plate contact with the elastic-plastic Winkler-type foundation. Computers and Structures 1991;39:641–51.
[11] Khathlan AA. Large deformation analysis of plates on unilateral elastic foundation. Journal of Engineering Mechanics, ASCE 1994;120(8):1820–7.
[12] Kastikadelis JT, Kallivokas LE. Clamped plates on Pasternak-type elastic foundation by the boundary element method. Journal of Applied Mechanics, ASME 1986;53:909–17.
[13] Timoshenko SP, Gere JM. Theory of elastic stability. New York: McGraw-Hill, 1961.
[14] Liew KM, Lam KY, Chow ST. Free vibration analysis of rectangular plates using orthogonal plate functions. Computers and Structures 1990;34:79–85.
[15] Liew KM, Wang CM. pb-2 Rayleigh-Ritz method for general plate analysis. Engineering Structures 1993;15:55–60.
[16] Smith ST, Bradford MA, Oehlers DJ. Elastic buckling of unilaterally constrained rectangular plates in pure shear. Engineering Structures 1999;21(5):443–53.
[17] Smith ST, Bradford MA, Oehlers DJ. Local buckling of side-plated reinforced concrete beams. 1: Theoretical study. Journal of Structural Engineering, ASCE 1999;125(6):622–34.

[18] Smith ST, Bradford MA, Oehlers DJ. Local buckling of side-plated reinforced concrete beams. 2: Experimental Study. Journal of Structural Engineering, ASCE 1999;125(6):635–45.
[19] Trahair NS, Bradford MA. The behaviour and design of steel structures to AS4100. 3rd ed. London: E & FN Spon, 1998.
[20] Azhari M, Bradford MA. Inelastic local buckling of plates with and without residual stresses. Engineering Structures 1993;15:31–9.
[21] Bradford MA, Azhari M. Local buckling of I-sections bent about the minor axis. Journal of Constructional Steel Research 1994;31:73–89.
[22] Swartz SE, O'Neill RJ. Linear elastic buckling of plates subjected to combined loads. Thin-Walled Structures 1995;21:1–15.
[23] Standards Australia. AS4100 Steel Structures. Sydney: SA, 1998.
[24] Oehlers DJ, Nguyen NT, Ahmed M, Bradford MA. Lateral and longitudinal partial interaction in composite bolted side-plated reinforced-concrete beams. Structural Engineering and Mechanics 1997;5(5):553–64.
[25] Oehlers DJ, Ahmed M, Nguyen NT, Bradford MA. Bolting steel plates to the sides of R.C. beams. 2. Transverse interaction and rigid plastic design, submitted for publication.

Papers from

Thin-Walled Structures

A Koiter's perturbation strategy for the imperfection sensitivity analysis of thin-walled structures with residual stresses

A.D. Lanzo *

DiSGG, Università della Basilicata, 85100 Potenza, Italy

Received 1 January 1999; received in revised form 13 January 2000; accepted 31 January 2000

Abstract

This paper suggests a strategy for the imperfection sensitivity analysis of elastic thin-walled structures with notable residual stresses. The analysis is carried out by means of a Koiter's perturbation approach. The concept of imperfection, traditionally associated with geometric and load factors, is extended in this paper to the residual stresses. The strategy is implemented in a FEM code. A comparison of the obtained results allows a discussion on the accuracy and the influence of the different coefficients connected to the asymptotic analysis of the residual stresses. © 2000 Elsevier Science Ltd. All rights reserved.

Keywords: Elastic stability; Imperfection sensitivity analysis; Perturbation method; FEM; Thin-walled structures

1. Introduction

In the analysis of the critical and post-critical behavior of elastic structures, the role of the initial residual stresses of the manufacturing processes is generally considered only through the influence of the initial deflections induced from them (first-order imperfection sensitivity, see [1]). The analysis of metallic thin-walled structures obtained by a special process of assemblage and welding of the panels [2] instead deserves a different discussion. This kind of structure is practically unaffected by initial deflections, whereas the initial residual stresses can have very relevant values,

* Corresponding author. Tel.: +390-971-205055; fax: +390-971-205070.
E-mail address: lanzo@disgg.unibas.it (A.D. Lanzo).

Reprinted from *Thin-Walled Structures* **37 (1)**, 77-95 (2000)

of the same order as the buckling critical stresses, and such as to affect the critical and post-critical behavior of the structures: therefore they have to be considered directly in the analysis (second-order imperfection sensitivity).

Nevertheless to be modelled into an analysis strategy, the residual stresses must discount the aleatory character of their distribution, as the analysis has to be performed for a fairly wide spectrum of possible distributions. Because of the high computational costs, it penalises incremental-iterative strategies of analysis: the residual stresses being an initial datum of the problem, the analysis of the structure must be performed ex-novo for each new possible distribution. The fundamental feature required for an optimal analysis strategy is that the influence of the residual stresses can be valued after the structure analysis in the absence of residual stresses, with negligible additional cost. The treatment of the *imperfections* in the perturbation strategy of analysis based on Koiter's theory of elastic stability is well-suited to this approach [3–6].

This analysis strategy is articulated in two distinct phases. In the first phase the behavior of the structure in absence of imperfections (the *perfect* structure), typically connected to bifurcations of the relative equilibrium paths, is analyzed. In the second phase the influence is considered, especially in terms of load carrying capacity, of all those factors (the imperfections) that contaminate the scheme of perfect structure, generally evolving the bifurcative character of the behavior in a phenomenology of snapping and limit load. It is the first phase of the analysis that requires higher computational costs (they are substantially equivalent to a linear stability analysis and then, in any case, of some lower orders of an individual step-by-step analysis), reducing the last phase to the evaluation of some scalar coefficients and to the resolution of a nonlinear algebraic system with a reduced number of equations. This simplicity in dealing with additional imperfections compensates the aleatory character of their distribution, the analysis being capable of repetition for a fairly wide spectrum of possible imperfections with additional negligible costs. In this scheme, the imperfection sensitivity analysis is therefore performed in an effective way.

Imperfections are traditionally associated with geometric and load factors. The aim of this paper is to extend the concept of imperfection of Koiter's theory to the residual stresses, leading again to the simplicity and efficiency of an imperfection sensitivity analysis of the perturbation strategy.

The paper is organized as follows. The fundamental lines of a perturbation strategy oriented to the FEM analysis of thin-walled structures, presented by the author in Refs [7–14], are initially summarized. Afterwards the presence of residual stresses, with reference to membranal states, is brought within a variational formulation based on the stationary condition of the total potential energy and then framed in the perturbation strategy of imperfection sensitivity analysis. In particular the energy contributions connected to the residual stresses are represented by means of additional terms, while the influence is evaluated with coefficients (of imperfection) to the several perturbation orders of the equilibrium equations of the structure. Finally, the quantitative weights of such coefficients are tested with reference to some sample problems, giving some practical indication of their treatment within a FEM-perturbation analysis algorithm.

2. The perturbation strategy

For structures subjected to conservative loads increasing according to a λ parameter ($p[\lambda]=\lambda\hat{p}$), characterized by a strain energy $\Phi[u]$ (where u denotes the displacements field) and a load potential $p[\lambda]u$ linear in u, the curves of the equilibrium paths (λ,u) could be reconstructed by means of a Koiter's perturbation strategy [3–5] with reference to the variational formulation of stationariety of the total potential energy

$$\Pi[\lambda,u]=\Phi[u]-p[\lambda]u=stat_u \tag{1}$$

expressed in virtual work terms by

$$\Pi'[\lambda,u]\delta u=\Phi'[u]\delta u-p[\lambda]\delta u=0,\ \forall\delta u^1 \tag{2}$$

For thin-walled structures in the presence of interactions among multiple buckling modes, an accurate and reliable perturbation strategy, oriented to a FEM automatic approach, is presented by the author in papers [7–14]. This strategy is articulated in the following points.

1. Linear extrapolation in λ of the fundamental path $u^f[\lambda]=\lambda\hat{u}$ by means of the resolution of the linear problem:

$$\Phi''_o\hat{u}\delta u-\hat{p}\delta u=0,\ \forall\delta u \tag{3}$$

2. Determination of bifurcation points and primary modes of buckling (λ_i,$\dot{v}_i$) along the fundamental path resolving the eigenvalue problem:

$$\Phi''[\lambda_i\hat{u}]\dot{v}_i\delta u=0,\ \forall\delta u,\ -\Phi'''_b\hat{u}\dot{v}_i\dot{v}_j=\begin{cases}1 & i=j\\ 0 & i\neq j\end{cases},\ (i,j=\{1,\ldots,m\}) \tag{4}$$

3. Calculation of the secondary modes of buckling $\ddot{w}_{ij}$ associated with the primary modes $\dot{v}_i$ as solutions of the linear problems:

$$\Phi''_b\ddot{w}_{ij}\delta u+\Phi'''_b\dot{v}_i\dot{v}_j\delta u=0,\ \forall\delta u,\ \Phi'''_b\hat{u}\ddot{w}_{ij}\dot{v}_k=0,\ (i,j,k=\{1,\ldots,m\}) \tag{5}$$

4. Calculation of the coefficients connected to the third and fourth variation of the strain energy

$$\mathscr{A}_{ijk}=\Phi'''_b\dot{v}_i\dot{v}_j\dot{v}_k \tag{6}$$

$$\mathscr{B}_{ijhk}=\Phi''''_b\dot{v}_i\dot{v}_j\dot{v}_h\dot{v}_k-\Phi''_b(\ddot{w}_{ij}\ddot{w}_{hk}+\ddot{w}_{ih}\ddot{w}_{jk}+\ddot{w}_{ik}\ddot{w}_{jj})$$

[1] We use the synthetic notation of the functional analysis as defined in Ref. [15]. In particular a prime denotes Frechét's derivative with respect to the field u.

and of the implicit imperfection coefficients

$$\mu_k[\lambda]=\frac{1}{2}\lambda^2\Phi_o^{\prime\prime\prime}\hat{u}^2\dot{v}_k \tag{7}$$

5. Determination of the relationship among the parameters $(\lambda,\xi_1,\ldots,\xi_m)$ of representation of the equilibrium path by means of resolution of the nonlinear algebric equation system:

$$\mu_k[\lambda]=-\xi_k(\lambda-\lambda_k)+\frac{1}{2}\sum_{i,j=1}^{m}\xi_i\xi_j\mathcal{A}_{ijk}+\frac{1}{6}\sum_{i,j,h=1}^{m}\xi_i\xi_j\xi_h\mathcal{B}_{ijhk}=0,\ (k=\{1,\ldots,m\}) \tag{8}$$

6. Reconstruction of the equilibrium paths of the structure in the manifold

$$u=\lambda\hat{u}+\sum_{i=1}^{m}\xi_i\dot{v}_i+\frac{1}{2}\sum_{i,j=1}^{m}\xi_i\xi_j\dot{w}_{ij} \tag{9}$$

The scheme of perfect structure, i.e. characterized by bifurcative phenomena along its natural equilibrium path, is destroyed by the presence of small additional imperfections. They do not modify the dominant nature of the structural behavior, closely connected to the behavior of the perfect structure, even if they can strongly affect notable aspects, with generally a reduction in the load carrying capacity of the structure (imperfection sensitivity). For imperfections related to the geometry of the structure $\varepsilon\tilde{u}$ or to the applied loads $\varepsilon\tilde{p}$, such an influence can be evaluated after the perturbation analysis of the perfect structure by means of a simple redefinition of the relationship given in Eq. (8) among the parameters $(\lambda,\xi_1,\ldots,\xi_m)$, obtained respectively by adding the coefficients

$$\varepsilon\lambda\Phi_b^{\prime\prime\prime}\tilde{u}\hat{u}\dot{v}_k,\ \varepsilon\tilde{p}[\lambda]\dot{v}_k \tag{10}$$

and reconstructing the equilibrium path in the same bifurcation manifold (Eq. (9)) set for the perfect structure. The work aims to realize an analogous treatment for imperfections connected to initial manifacturing processes residual stresses.

3. Variational formulation with residual stresses

Denoting with ϵ the strain field of the body in a generic configuration, we set the following nonlinear kinematics relation with the displacement field u measured from the initial reference configuration $\mathcal{K}$:

$$\epsilon=l_1(u)+\frac{1}{2}l_2(u) \tag{11}$$

where l_1 and l_2 stand for a linear and a quadratic operator, respectively. For Eq. (11), each virtual variation δu of the displacements field yields a virtual variation of the strain given by

$$\delta\epsilon = l_1(\delta u) + l_{11}(u,\delta u) \tag{12}$$

being the bilinear operator l_{11} set by

$$l_2(u+v) = l_2(u) + 2l_{11}(u,v) + l_2(v) \tag{13}$$

The equilibrium relationship between the external applied loads $p[\lambda]$ and the internal stresses σ is stated in virtual work terms by the following equation (see [15])

$$\sigma\cdot\delta\epsilon = p[\lambda]\delta u,\ \forall\delta u \tag{14}$$

valid for each virtual variation δu.

Let σ^o be a state of residual stresses in the reference configuration $\mathscr{K}$, i.e. balanced in this configuration with null external loads. We set a linear hyperelastic constitutive relation

$$\sigma = \sigma^o + \mathscr{C}\epsilon \tag{15}$$

with the elastic coefficient tensor $\mathscr{C}$ endowed of appropriate symmetry features. Replacing this relation, the equilibrium equation Eq. (14) changes into:

$$\sigma^o\cdot\delta\epsilon + \mathscr{C}\epsilon\cdot\delta\varepsilon = p[\lambda]\delta u,\ \forall\delta u \tag{16}$$

Observing that in the reference configuration the equilibrium of σ^o with null external loads is expressed by the following relation[2]

$$\sigma_o\cdot l_1(\delta u) = 0 \tag{17}$$

Eq. (16) is simplified into

$$\sigma^o\cdot l_{11}(u,\delta u) + \mathscr{C}\epsilon\cdot\delta\epsilon = p[\lambda]\delta u,\ \forall\delta u \tag{18}$$

This relation is expressed in variational form by the stationary condition of the total potential energy functional

$$\Pi[\sigma^o,u] = \Phi[\sigma^o,u] - p[\lambda]u = staz_u \tag{19}$$

taking care to replace the usual definition of the strain energy

$$\Phi[u] = \frac{1}{2}\mathscr{C}\epsilon\cdot\epsilon \tag{20}$$

with the following

[2] The relation comes from Eq. (14) for null external load, taking into account that in the reference configuration, being $u=0$, is $l_{11}(0,\delta u)=0$ and then, for Eq. (12), $\delta\epsilon^o = l_1(\delta u)$.

$$\Phi[\sigma^o,u]=\frac{1}{2}\mathcal{C}\epsilon\cdot\epsilon+\frac{1}{2}\sigma^o\cdot l_2(u) \tag{21}$$

i.e. enriched by the term

$$\frac{1}{2}\sigma^o\cdot l_2(u) \tag{22}$$

in order to take into account the presence of residual stresses σ^o.

For the structures considered in the present work, obtained as assemblages of thin rectangular plates interconnected along their longitudinal sides, we associate the residual stress to membranal distributions ($\sigma^o\equiv N^o_{ij}$) in the plane (x_1,x_2) of the panels and use the following nonlinear plate model (see [8]):

$$\varepsilon_{ij}=\frac{1}{2}\left(u_{i,j}+u_{j,i}+\sum_k^3 u_{k,i}u_{k,j}\right),\ \chi_{ij}=u_{k,ij},\ (i,j=\{1,2\})^3 \tag{23}$$

together with the constitutive relations

$$N_{ij}=N^o_{ij}+\sum_{hk}C_{ijhk}\varepsilon_{ij},\ M_{ij}=\sum_{hk}D_{ijhk}\chi_{hk},\ (i,j,h,k=\{1,2\}) \tag{24}$$

Therefore, on the basis of the definition in Eq. (21), the strain energy is expressed by

$$\Phi[\sigma^o,u]=\frac{1}{2}\int_A\left\{\sum_{ijhk}^2 C_{ijhk}\varepsilon_{ij}\varepsilon_{hk}+\sum_{ijhk}^2 D_{ijhk}\chi_{ij}\chi_{hk}+\sum_{ijk}^2 N^o_{ij}u_{k,i}u_{k,j}\right\}dx_1dx_2 \tag{25}$$

with the contribution connected to the residual stresses σ^o evaluated according to

$$\frac{1}{2}\int_A\left\{\sum_{ijk}^2 N^o_{ij}u_{k,i}u_{k,j}\right\}dx_1dx_2 \tag{26}$$

4. Perturbation treatment of residual stresses

The variational formulation Eq. (19), and thus the definitions Eqs. (21) and (25) of the strain energy, allows an extension of the perturbation approach to thin-walled

---

[3] To applied purposes the kinematical model can be simplified in some of its nonlinear terms according to the following

$$\sum_k^3 u_{k,1}u_{k,1}\approx u^2_{2,1}+u^2_{3,1},\ \sum_k^3 u_{k,2}u_{k,2}\approx u^2_{1,2}+u^2_{3,2},\ \sum_k^3 u_{k,1}u_{k,2}\approx u_{3,1}u_{3,2}$$

structures with residual stresses. This formulation takes their influence into account in an accurate way, but is unsuitable to the purpose of a sensitivity analysis of the structures. In fact, the term in Eq. (26) connected to the residual stresses directly affects the definition of the perfect structure through the relative strain energy: this means that the perturbation analysis, like a step-by-step analysis, must be entirely repeated for new stress distributions, with a notable increase in computational costs. A different formulation, less accurate but more suitable for a sensitivity analysis, is suggested. In this formulation the influence of the energy contributions connected to the residual stresses is considered only in the final phase, after the analysis of the perfect structure, then allowing the treatment as a term of imperfection. To this end it is necessary that the energy quota of the residual stresses appears directly in the variational formulation by means of an additional term that does not alter the energy definition of the perfect structure, that is

$$\Pi[\sigma^o,u]=\Phi[u]-p[\lambda]u+\tilde{\Phi}[\sigma^o,u]=stat_u \tag{27}$$

with

$$\Phi[u]=\frac{1}{2}\int_A \sum_{ijhk}^{2}\{C_{ijhk}\varepsilon_{ij}\varepsilon_{hk}+D_{ijhk}\chi_{ij}\chi_{hk}\}\mathrm{d}x_1\mathrm{d}x_2 \tag{28}$$

$$\tilde{\Phi}[\sigma^o,u]=\frac{1}{2}\int_A \sum_{ijk}^{2}\{N^o_{ij}u_{k,i}u_{k,j}\}\mathrm{d}x_1\mathrm{d}x_2 \tag{29}$$

The solution of the equilibrium problem in Eq. (27) of the imperfect structure is then sought on the same bifurcation manifold defined by the resolution of the structure in the absence of imperfections. With reference to the equivalent form in virtual works terms

$$\Phi'[u]\delta u-p[\lambda]u+\tilde{\Phi}'[\sigma^o,u]\delta u=0,\ \forall\delta u \tag{30}$$

the equilibrium problem presents the following Galerkin formulation

$$\begin{cases}\Phi'[u]\dot{v}_k-p[\lambda]\dot{v}_k+\tilde{\Phi}'[\sigma^o,u]\dot{v}_k=0,\ (k=\{1,\dots,m\}),\\ u=\lambda\hat{u}+\sum_{i=1}^{m}\xi_i\dot{v}_i+\frac{1}{2}\sum_{i,j=1}^{m}\xi_i\xi_j\dot{w}_{ij}\end{cases} \tag{31}$$

where, because of the definition in Eq. (29), the imperfection terms $\tilde{\Phi}'[\sigma^o,u]\dot{v}_k$ are bilinear in (σ^o,u). The series expansion up to the terms containing the fourth variation of the energy provides, in approximate form, the relationship among the representation parameters $(\lambda,\xi_1,\dots,\xi_m)$ of the curves of the equilibrium path. Taking into account the definitions of Eqs. (3)–(9) and in the same simplifying hypothesis used in [8,9] (we refer to these references for a deeper insight into the implementation details of the computation of the coefficient of the energetic variation), these relationships are expressed by

$$\mu_k[\lambda]=\lambda\tilde{\Phi}''\hat{u}\dot{v}_k+\frac{1}{2}\sum_{i=1}^{m}\xi_i\tilde{\Phi}''\dot{v}_i\dot{v}_k+\sum_{i,j=1}^{m}\xi_i\xi_j\tilde{\Phi}''\ddot{w}_{ij}\dot{v}_k-\xi_k(\lambda-\lambda_k)+\frac{1}{2}\sum_{i,j=1}^{m}\xi_i\xi_j\mathcal{A}_{ijk} \tag{32}$$

$$+\frac{1}{6}\sum_{i,j,h=1}^{m}\xi_i\xi_j\xi_h\mathcal{B}_{ijhk}=0,\ (k=\{1,\ldots,m\})$$

Therefore the additional energy terms connected to the residual stresses, like geometric or load coefficients in Eq. (10), contribute only by a redefinition of the relationship among the parameters $(\lambda,\xi_1,\ldots,\xi_m)$: this allows the performance of a simple and efficient sensitivity analysis of the structure to residual stresses.

In particular, in the presence of residual stresses, the relationships among the representation parameters are redefinined by the addition of the three groups of coefficients $(\tilde{\Phi}''\hat{u}\dot{v}_k,\tilde{\Phi}''\dot{v}_i\dot{v}_k,\tilde{\Phi}''\ddot{w}_{ij}\dot{v}_k)$ to the different order of the series expansion. It is worth noting that:

- The term

  $$\lambda\tilde{\Phi}''\hat{u}\dot{v}_k \tag{33}$$

  is closely connected with the implicit imperfection coefficient of Eq. (7) (it can be considered a correction at the λ linear order). In the same way, in fact, the respective weight affects the behavior of the structures as much as the extrapolated fundamental path $u^f[\lambda]=\lambda\hat{u}$ differs from the natural equilibrium path of the structure.
- The term

  $$\sum_{i=1}^{m}\xi_i\tilde{\Phi}''\dot{v}_i\dot{v}_k \tag{34}$$

  is essentially connected to a correction to the critical values of λ at bifurcations of the fundamental path. In the same way, the term

  $$\frac{1}{2}\sum_{i,j=1}^{m}\xi_i\xi_j\tilde{\Phi}''\ddot{w}_{ij}\dot{v}_k \tag{35}$$

  is essentially connected to a correction to the coefficients that define the post-critical slopes along the paths bifurcating from the fundamental one. In fact, set $\mu[\lambda]=\tilde{\Phi}''\hat{u}\dot{v}_k=0$ (structure with bifurcation of the equilibrium path) and referring to the simple bifurcation case (only one buckling mode $m=1$, $\lambda_1=\lambda_b$), the relationship Eq. (32) changes in

$$0=\xi\tilde{\Phi}''\dot{v}_b\dot{v}_b+\frac{1}{2}\xi^2\tilde{\Phi}''\ddot{w}_b\dot{v}_b+\xi(\lambda-\lambda_b)\Phi_b'''\hat{u}\dot{v}_b^2+\frac{1}{2}\xi^2\Phi_b'''\dot{v}_b^3+\frac{1}{6}\xi^3(\Phi_b''''\dot{v}^4-3\Phi_b''\ddot{w}_b^2) \tag{36}$$

and then in

$$\lambda=\left(\lambda_b-\frac{\tilde{\Phi}''\dot{v}_b\dot{v}_b}{\Phi_b'''\hat{u}\dot{v}_b^2}\right)+\xi\left(\dot{\lambda}_b-\frac{\tilde{\Phi}''\ddot{w}_b\dot{v}_b}{\Phi_b'''\hat{u}\dot{v}_b^2}\right)+\frac{1}{2}\xi^2\ddot{\lambda}_b \tag{37}$$

where the critical and post-critical parameters in the absence of residual stresses are expressed by

$$\lambda_b,\quad \dot{\lambda}_b=-\frac{1}{2}\frac{\Phi_b'''\dot{v}_b^3}{\Phi_b'''\hat{u}\dot{v}_b^2},\quad \ddot{\lambda}_b=-\frac{1}{3}\frac{\Phi_b''''\dot{v}^4-3\Phi_b'\ddot{w}_b^2}{\Phi_b'''\hat{u}\dot{v}_b^2} \tag{38}$$

5. Numerical results

With the aim of investigating the influence of the residual stresses on the buckling and post-buckling structural behavior, this section reports the results achieved in the analysis of two examples of thin-walled structures, widely studied in Ref. [9] in the absence of residual stresses. The first example deals with a beam of 'C' cross-section uniformly loaded in compression. The second example presents a beam of 'T' cross-section loaded by a concentrate force at mid-length. The beams are simple supported at the ends and made of isotopic material. Oriented to model distribution of residual stresses of practical interest [2], the analysis takes into account only membranal distributions with flow lines parallel to the beam longitudinal axes ($N^o_{22}=N^o_{12}=0$, $N^o_{11}\neq 0$).

The analysis is performed by a FEM code based on a perturbation approach. High continuity (HC) finite elements are used [16]. We refer to Refs [8,9] for more implementation details. With respect to these references, third and fourth energy variation terms are now computed by means of a Gauss numerical integration with four points for each element.

In the following the results obtained with two different treatments of the residual stresses in the perturbation strategy are compared. In the first (which we named the *complete* treatment) the residual stresses are considered as an initial datum of the problem and their influence is evaluated accurately on the basis of Eq. (19), i.e. by means of a redefinition of the perfect structure (and then of its fundamental path) used in the perturbation analysis. The second treatment (which we named *approximated*), more suitable to implement an effective sensitivity analysis, refers to a perfect structure without residual stresses, taking into account their influence in an approximate way on the basis of Eq. (27) by means of imperfection coefficients from Eqs. (33)–(35). This allows a useful investigation of the weight of these coefficients and of the general accuracy of the approximate treatment of residual stresses.

5.1. Example 1: beam of 'C' cross-section

This example is described in its geometric and material aspects in Fig. 1. The load is applied at its ends in such a way as to cause the uniform compression condition

$$N_{11}=\lambda\hat{N}_{11},\ N_{22}=N_{12}=0$$

with $\hat{N}_{11}=1$ ($\hat{p}=125$). The residual stresses are given by the distribuition of components N^{o}_{11} reported in Fig. 1.

The analysis refers to a discretization of 75×(5+9+5)=1800 HC finite elements. The first 10 buckling modes (see Fig. 2) have been taken into account. Table 1 shows the relative critical values of the load in the absence of initial stresses (column 'no-init. stress').[4] As reported in the same table (column 'with-init. stresses'), the presence of initial stresses in the definition of the fundamental path alters the buckling loads.[5] It should be observed that these changes $\Delta\lambda_{crit.}$ are more notable for distortional modes (the local modes), less notable for modes characterised by a rigid displacement of the beam cross-section (the flexural global mode $\dot{v}_4$). Table 1 also shows the imperfection coefficients $\tilde{\Phi}''\dot{v}_i\dot{v}_k$ that can be considered a good estimate of the differences $\Delta\lambda_{crit.}$ of critical values, proving a statement made in the previous section.

In order to study the influence of residual stresses on the post-buckling behavior,

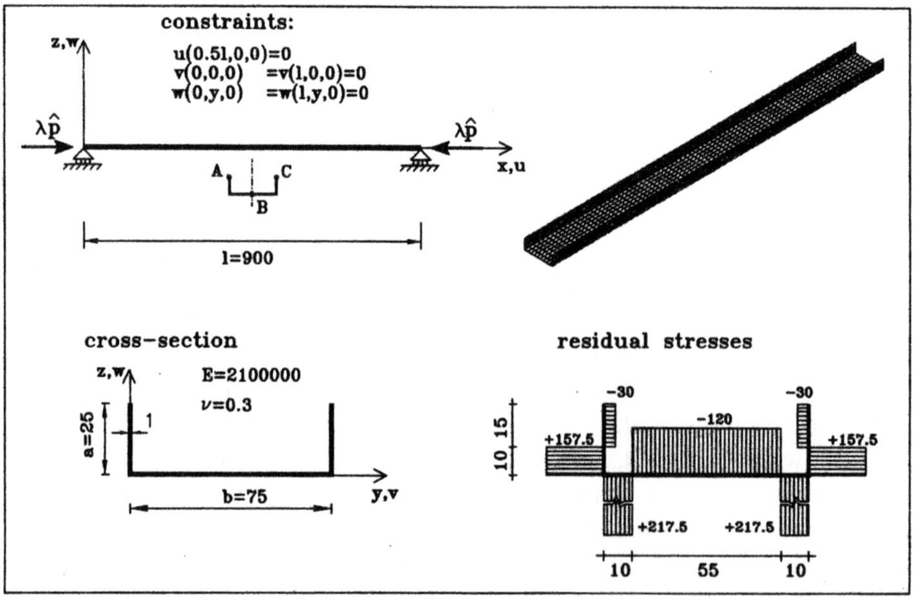

Fig. 1. Beam of 'C' cross-section (Example 1).

---

[4] Slight differences can be observed with respect to the values reported in Ref. [9] because of the different numerical integration technique now used.

[5] For the distribution of initial stresses considered in the analysis, no changes are observed in the shape of the buckling modes.

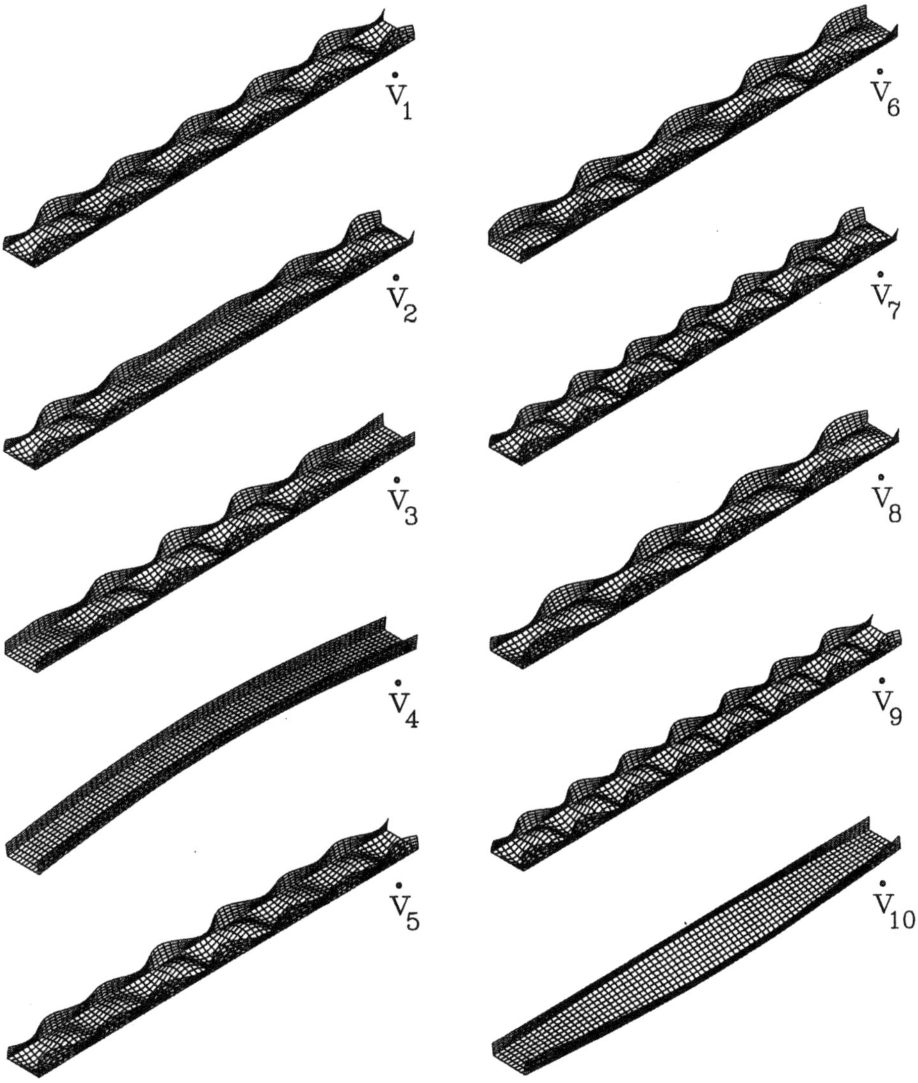

Fig. 2. Buckling modes of Example 1.

the analyses in Figs. 3–6 refer to a structure affected also by an initial small geometric imperfection set by a combination of a local shape (along the mode $\dot{v}_1$) and a flexural shape (along the mode $\dot{v}_4$)

$$\tilde{u}=\tilde{\xi}_1\dot{v}_1+\tilde{\xi}_4\dot{v}_4$$

with $\tilde{\xi}_1\dot{v}_1$ such as to have the displacement $v_C=-0.01$ and with $\tilde{\xi}_4\dot{v}_4$ such as to have $w_B=-0.1$ (the points A, B and C are located in the mid-length cross-section, see Fig. 1). It is worth noting that, for this geometric imperfection, the post-buckling structural behavior is affected by the interaction between the flexural mode $\dot{v}_4$ and the different local modes, while no influence of the torsional mode $\dot{v}_{10}$ is observed.

Table 1
Critical values of Example 1

Mode	$\lambda_{crit.}$		$\tilde{\Phi}''\dot{v}_i\dot{v}_k$	$(\Delta\lambda_{crit.})$
	No-init. stress	With-init. stresses		
$\dot{v}_1$	1441.29	1360.81	−79.76	(−80.48)
$\dot{v}_2$	1446.74	1365.76	−80.04	(−80.98)
$\dot{v}_3$	1454.91	1375.30	−79.25	(−79.61)
$\dot{v}_4$	1462.59	1461.58	−0.62	(−01.01)
$\dot{v}_5$	1473.64	1390.89	−81.84	(−82.75)
$\dot{v}_6$	1484.75	1407.27	−77.08	(−77.48)
$\dot{v}_7$	1509.49	1425.15	−83.54	(−84.34)
$\dot{v}_8$	1548.96	1473.23	−75.29	(−75.73)
$\dot{v}_9$	1555.10	1469.42	−84.83	(−85.68)
$\dot{v}_{10}$	1608.64	1649.31	+40.48	(+40.67)

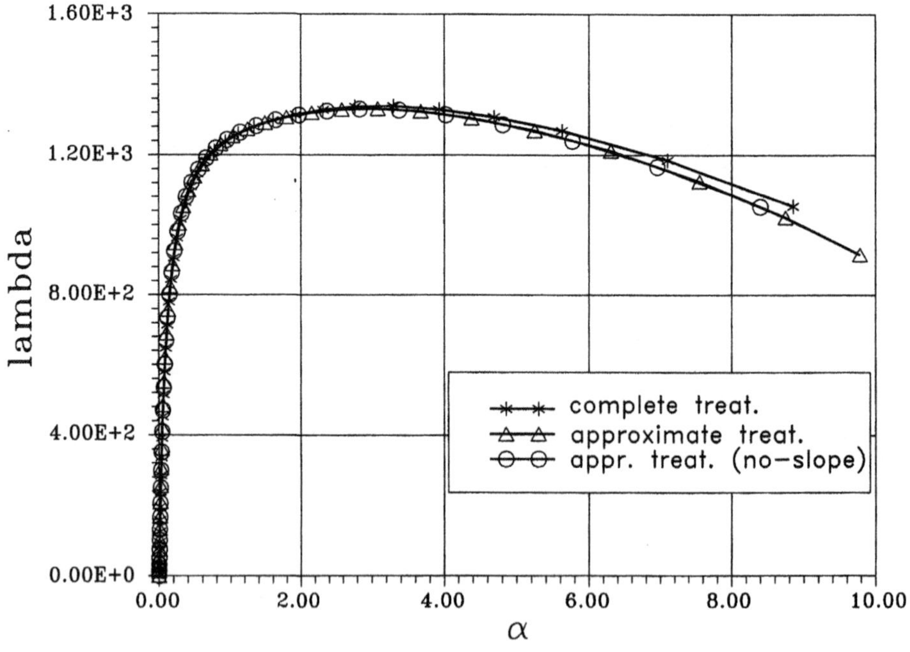

Fig. 3. Load versus α-parameter of Example 1 (complete and approximate residual stress treatment).

The equilibrium paths are represented in Figs. 3–6 by plotting the load parameter versus the displacement w_A and versus an asymptotic parameter α set by

$$\alpha = \sqrt{\sum_i^m \xi_i^2}$$

where ξ_i is the ith buckling mode component of the equilibrium path.

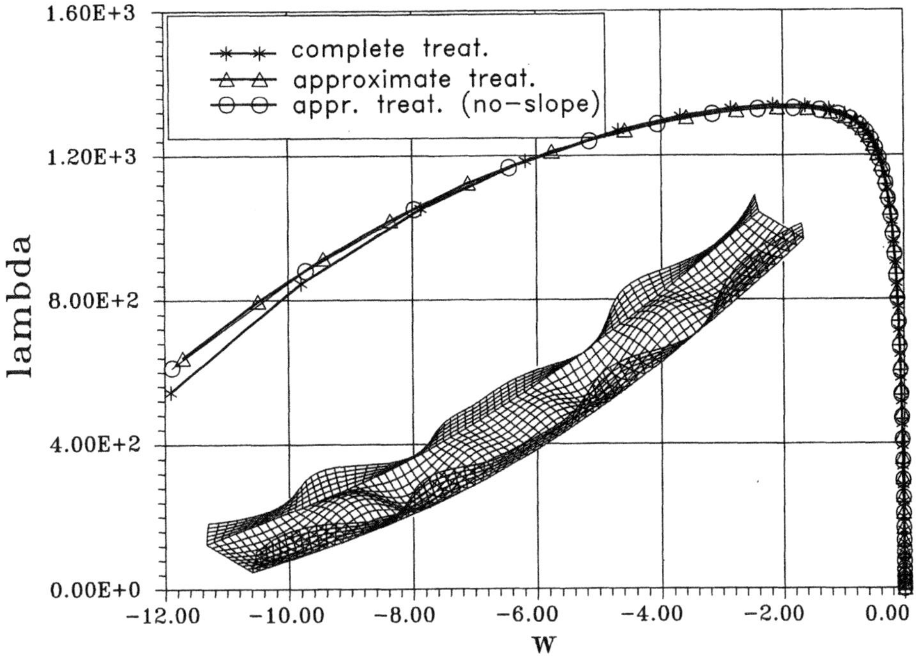

Fig. 4. Load versus displacement of Example 1 (complete and approximate residual stress treatment).

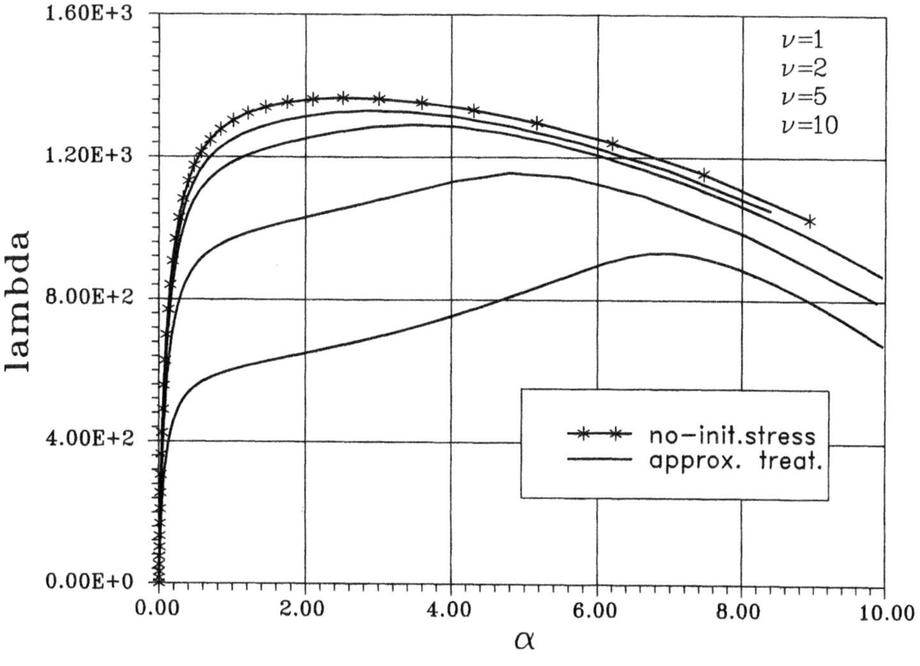

Fig. 5. Load versus α-parameter of Example 1 (residual stress sensitivity analysis).

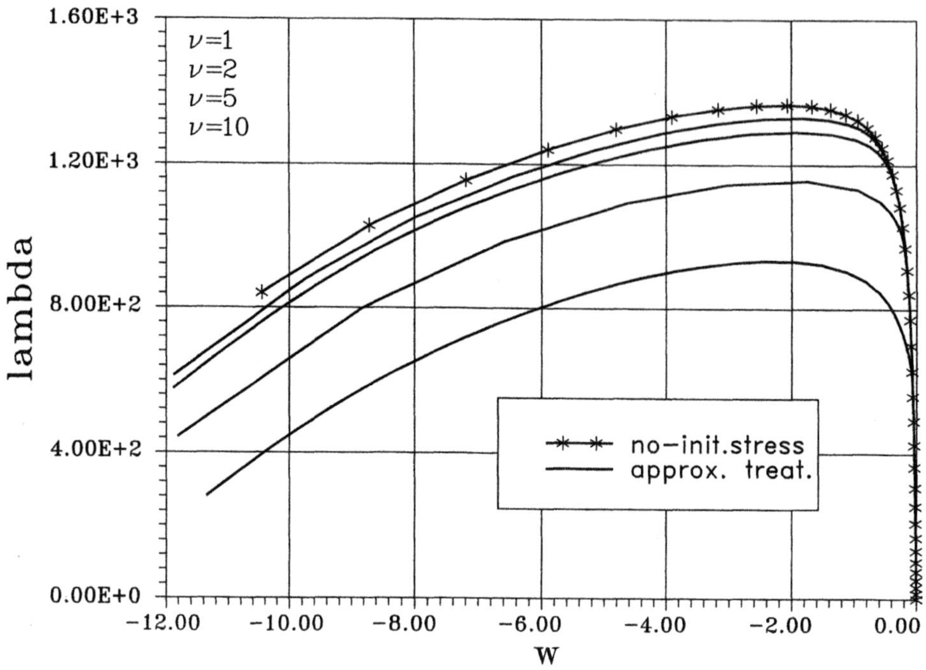

Fig. 6. Load versus displacement of Example 1 (residual stress sensitivity analysis).

In Figs. 3 and 4 the equilibrium paths obtained by means of the complete treatment of the residual stresses (curve *complete treat.*) are compared with the curves obtained by means of the approximate treatment taking into account the influence of all the imperfection coefficients (curve *approximate treat.*). A good accordance of the results can be observed. However, it should be noted that the excellence of this agreement is also connected with the particular residual stress distribution considered in the analysis. Lesser accordance is expected for initial stresses able to alter noticeably not only the critical values, but also the shapes of the buckling modes.

The third equilibrium path plotted in Figs. 3 and 4 (curve *no-slope*) refers to the approximate treatment of the residual stresses, but neglecting the contribution of the coefficients $\tilde{\Phi}''\ddot{w}_{ij}\dot{v}_k$ (the alteration in the post-critical slope). The perfect agreement of this curve with the curve *approximate treat.* proves the neglecting influence of the coefficients $\frac{1}{2}\sum_{i,j=1}^{m}\xi_1\xi_j\tilde{\Phi}''\ddot{w}_{ij}\dot{v}_k$, being the influence of the residual stresses on the structural behavior essentially related to the coefficients $\sum_{i=1}^{m}\xi_i\tilde{\Phi}''\dot{v}_i\dot{v}_k$ (the alteration of the critical values).

The structural sensitivity to residual stresses is highlighted in Figs. 5 and 6 where the equilibrium paths (obtained by means of the approximate treatment), are reported for rising values of the stresses νN^o_{11} (ν=1,2,5,10) and compared with the values of zero initial stresses case (curve *no-init. stress*).

5.2. Example 2: beam of 'T' cross-section

This example refers to a simply supported beam of 'T' cross-section loaded with a concentrate force λ at mid-length. Its geometric and material details, with the residual stresses N^o_{11} considered in the analysis, are described in Fig. 7.

The discretization used is based on 75×(6+6+12)=1650 HC finite elements. The first four buckling modes shown in Fig. 8 are taken into account in the perturbation analysis (nevertheless the behavior observed is connected essentially to the interaction only of the first three modes). The comparison in Table 2 between the critical values of the load parameter proves the observations given in the previous example on the meaning of the terms $\tilde{\Phi}''\dot{v}_i\dot{v}_k$ as a reliable estimate of the differences of values $\Delta\lambda_{crit.}$ obtained with and without residual stresses in the definition of the fundamental path for the perturbation analysis.

The equilibrium path of the structure affected also by a load inperfecton $\tilde{p}$=0.001λ applied on mid-length section point C normally to the plane (x,z) of the problem is reported in Figs. 9–12 plotting the load parameter versus the displacement v_A m and versus the asymptotic parameter α defined above. Also in this example, the comparisons in Figs. 9 and 10 prove the accuracy of approximate treatment of the residual stresses and the neglegible effects of the coefficients $\tilde{\Phi}''\ddot{w}_{ij}\dot{v}_k$. Finally, the sensitivity analysis of the structure for increasing values of initial stress νN^o_{11} (ν=1,2,5,10) is shown in Figs. 11 and 12.

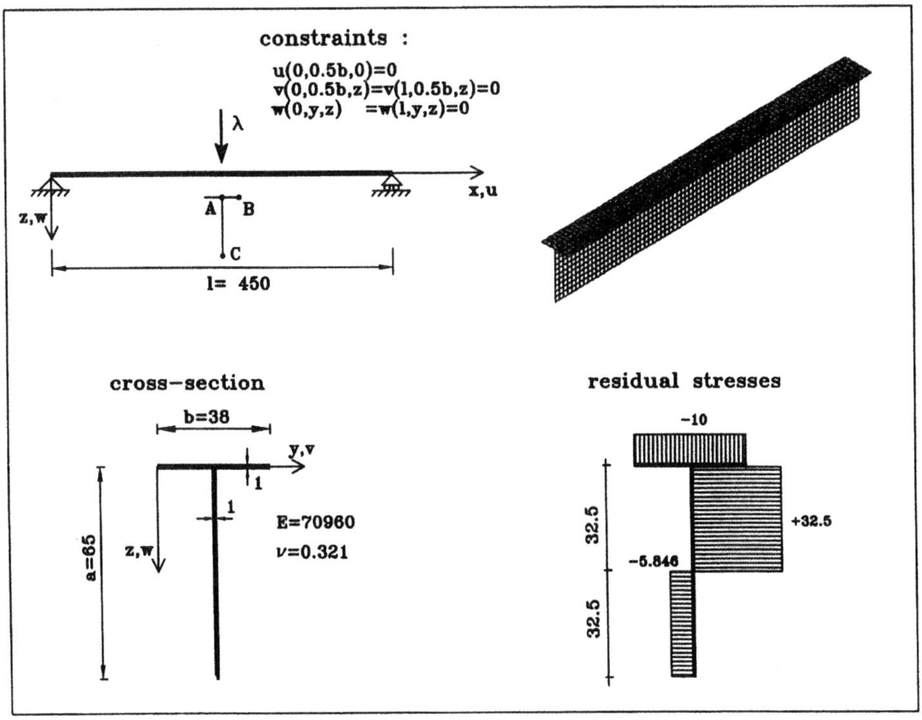

Fig. 7. Beam of 'T' cross-section (Example 2).

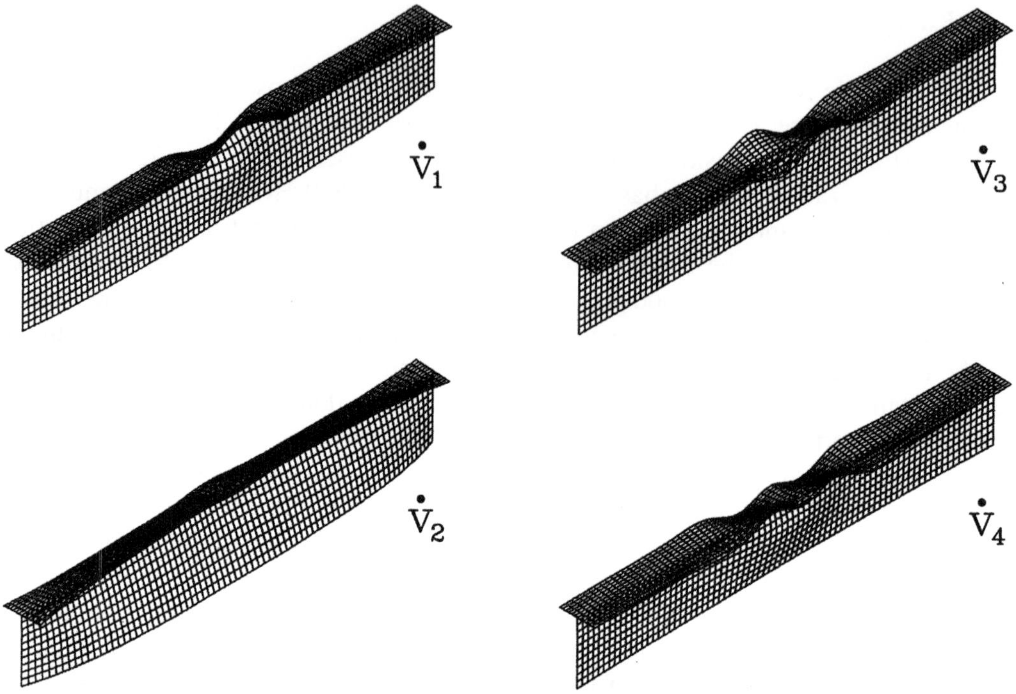

Fig. 8. Buckling modes of Example 2.

Table 2
Critical values of Example 2

Mode	$\lambda_{crit.}$		$\tilde{\Phi}''\dot{v}_i\dot{v}_k$	$(\Delta\lambda_{crit.})$
	No-init. stress	With-init. stresses		
$\dot{v}_1$	2846.87	2765.13	−62.60	(−81.74)
$\dot{v}_2$	3102.79	2930.92	−162.87	(−171.87)
$\dot{v}_3$	3524.87	3326.86	−194.54	(−198.01)
$\dot{v}_4$	3918.35	3750.48	−165.42	(−167.87)

6. Conclusions

This paper suggests an efficient strategy for structural sensitivity analysis for a class of thin-walled structures with remarkable residual stresses [2]. The strategy is implemented in a FEM-perturbation computer code presented by the author in Ref. [9]. The strategy extends Koiter's imperfection analysis to initial stresses. In this scheme, the influence of the residual stresses is taken into account after the analysis of the structure in the absence of initial stresses with a negligible additional computational cost. In particular, for the structures considered in the present paper, the results of the analysis highlight that the residual stresses condition the post-buckling

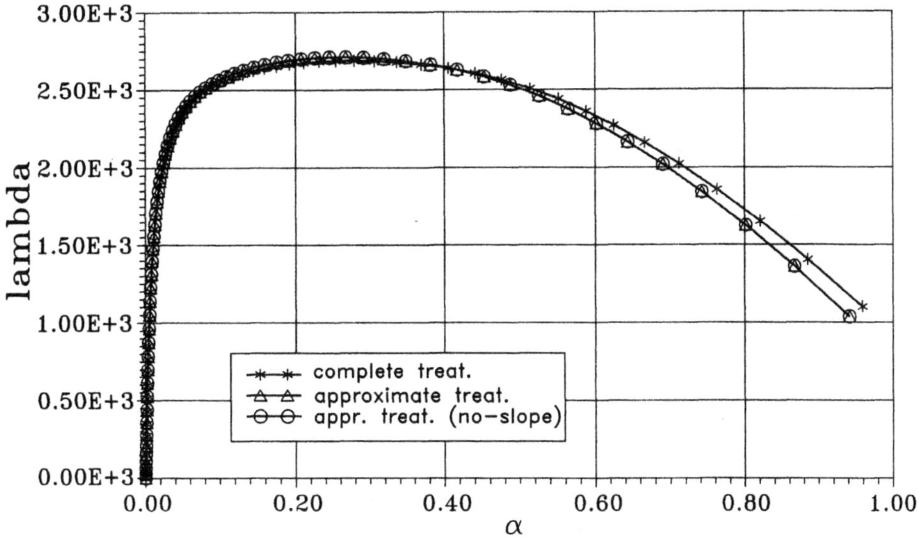

Fig. 9. Load versus α-parameter of Example 2 (complete and approximate residual stress treatment).

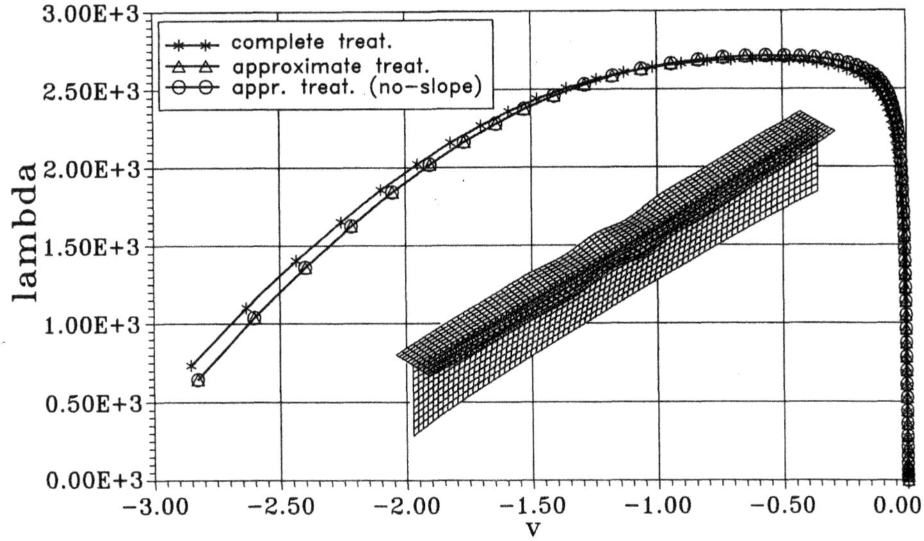

Fig. 10. Load versus displacement of Example 2 (complete and approximate residual stress treatment).

behavior essentially by means of an alteration in the critical values of the load, linear with the initial stress intensity, in accordance with that observed in Ref. [1].

Acknowledgements

This work has been partly supported by a M.U.R.S.T. 40% grant and by a BRITE-EURAM (APRICOS) grant. This support is gratefully acknowledged. A particular

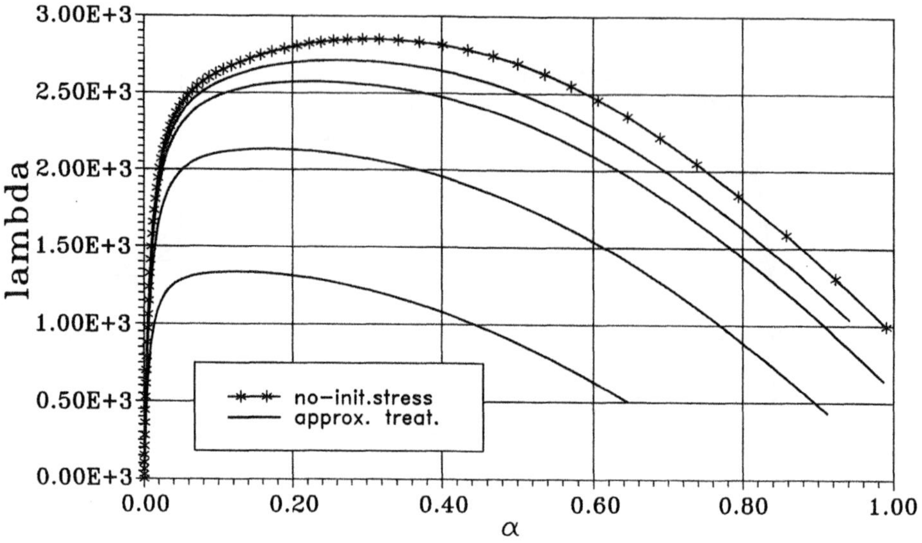

Fig. 11. Load versus α-parameter of Example 2 (residual stress sensitivity analysis).

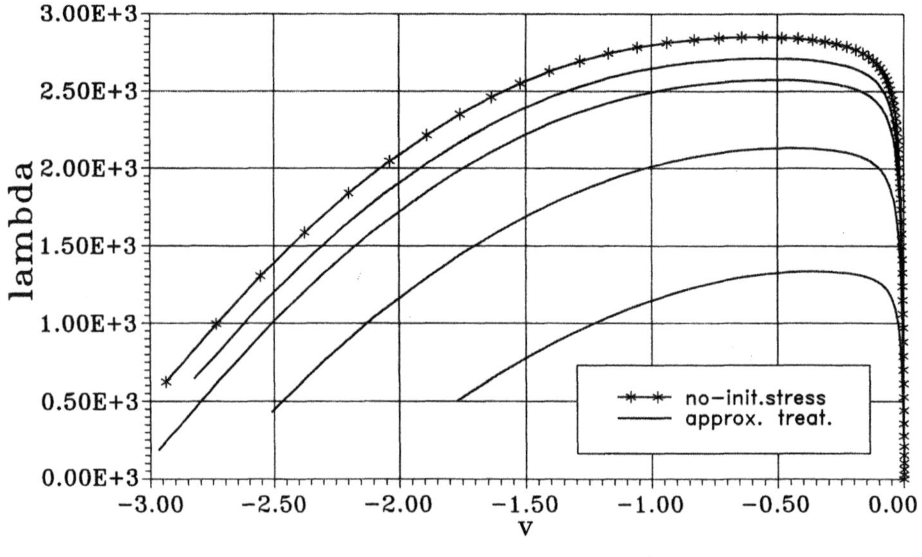

Fig. 12. Load versus displacement of Example 2 (residual stress sensitivity analysis).

acknowledgement is due to Prof. C.R. Calladine from the University of Cambridge (England) for his precious suggestions.

References

[1] Thompson JMT. Imperfection-sensitivity uninfluenced by pre-stress. Int J Mech Sci 1978;20:57–8.
[2] Rasmussen KJR, Hancock GJ. Nonlinear analyses of thin-walled channel section columns. Thin-Walled Struct 1991;13:145–76.

[3] Koiter WT. On the stability of elastic equilibrium, PhD Thesis, Delft, 1945 (English transl. NASA TT-F10, 883, 1967 and AFFDL TR70-25, 1970).
[4] Casciaro R, Lanzo AD, Salerno G. Computational problems in elastic structural stability. In: Carmignani C, Maino G, editors. Proceedings of the ENEA Workshops on Nonlinear Dynamics, 4, Rome, Italy, Nonlinear Problems in Engineering. Singapore: World Scientific Publications, 1991.
[5] Casciaro R, Salerno G, Lanzo AD. Finite element asymptotic analysis of slender elastic structures: a simple approach. Int J Num Meth Engng 1992;35:1397–426.
[6] Salerno G, Casciaro R. Mode jumping and attractive paths in multimode elastic buckling. Int J Num Meth Engng 1997;40:833–61.
[7] Lanzo AD. Analisi perturbativa di sistemi elastici bidimensionali mediante elementi finiti ad alta continuità, (in Italian). In: Tesi di Dottorato di Ricerca. Roma, Italy: Università di Roma 'La Sapienza', 1992.
[8] Lanzo AD, Garcea G, Casciaro R. Asymptotic post-buckling analysis of rectangular plates by HC finite elements. Int J Num Meth Engng 1995;38:2325–45.
[9] Lanzo AD, Garcea G. Koiter's analysis of thin-walled structures by s finite element approach. Int J Num Meth Engng 1996;39:3007–31.
[10] Lanzo AD, Garcea G, Casciaro R. Analisi perturbativa ad elementi finiti HC di strutture a pannelli sottili, (in Italian). In: Proceedings of XI Congresso Nazionale AIMETA, Trento, Italy, 1992.
[11] Lanzo AD, Garcea G, Casciaro R. Analisi perturbativa multimodale ad elementi finiti di strutture in parete sottile, (in Italian). In: Proceedings of III Convegno di Meccanica Computazionale, Trieste, Italy, 1993.
[12] Lanzo AD, Garcea G. Asymptotic post-buckling analysis of thin-walled structures. In: Proceedings of the International Conference on Mechanics of Solids and Materials Engineering, Singapore, 5–7 June, 1995.
[13] Garcea G, Lanzo AD. Validazione mediante analisi step-by-step di un codice FEM per l'analisi multimodale di strutture elastiche in parete sottile, (in Italian). In: Proceedings of XII Congresso AIMETA, Napoli, Italy, ottobre, 1995.
[14] Lanzo AD, Garcea G. Coupled instabilities in thin-walled structures by a FEM implementation of Koiter's approach. In: Proceedings of the Second International Conference on Coupled Instabilities in Metal Structures, Liege, Belgium, 5–7 September, 1996.
[15] Budiansky B. Theory of buckling and post-buckling of elastic structures. Adv Appl Mech 1974;14:1–65.
[16] Aristodemo M. A high-continuity finite element model for two-dimensional elastic problems. Comp Struct 1985;21:987–93.

Structural Engineering Compendium I

Dynamic buckling of thin isotropic plates subjected to in-plane impact

D. Petry *, G. Fahlbusch

Universität der Bundeswehr München, Institut für Luftfahrttechnik und Leichtbau, D-85579 Neubiberg, Germany

Received 21 March 2000; accepted 31 August 2000

Abstract

The dynamic stability behaviour of imperfect simply supported plates subjected to in-plane pulse loading is investigated. For the calculation of dynamic buckling loads a stress failure criterion is applied. The large-deflection plate equations are solved by a Galerkin method by using Navier's double Fourier series. In this paper the dynamic load factor (DLF) is redefined and plots that are useful for the design of plate structures are presented. Parametric studies are performed in which the influences of the pulse duration, shock function, imperfection, geometric dimensions and limit stress of the material are discussed. Comparison between the dynamic buckling loads, which are obtained by the commonly used criterion of Budiansky and Hutchinson [Proceedings of the 11th International Congress of Applied Mechanics (1964) 636], and the dynamic elastic limit loads, which are computed by the stress failure criterion, shows that the latter criterion is more useful for the design of lightweight structures. © 2001 Elsevier Science Ltd. All rights reserved.

Keywords: Dynamic buckling; Plates; In-plane impact; Stress failure; Galerkin method

1. Introduction

The dynamic buckling analysis of in-plane loaded structures is a problem of dynamic response, in which imperfections are necessary to cause out-of-plane motion.

For those configurations that have an instable postbuckling branch (i.e., cylindrical

* Corresponding author. Tel.: +49-89-6004-4139; fax: +49-89-6004-4103.
E-mail address: dirk.petry@unibw-muenchen.de (D. Petry).

Reprinted from *Thin-Walled Structures* **38 (3)**, 267-283 (2000)

shells), critical conditions for defining a dynamic buckling load can be found by using the Equation of Motion Approach or Energy Approaches [2]. But there does not exist any standard criterion for the investigation of structures with stable postbuckling behaviour like plates. Therefore it is necessary to establish critical conditions for finding a dynamic buckling load.

In different publications [3–5] dynamic buckling loads are determined by considering the stability criterion of Budiansky and Hutchinson [1]: a dynamically critical condition is defined if some characteristic value increases rapidly with the loading amplitude. In these works the quotient of the dynamic buckling load and the load of bifurcation is defined as the dynamic load-amplification factor (DLF). The disadvantage of this criterion is caused by the fact that the load-carrying capacity of the structure is not taken into account. Therefore such a criterion is not very useful for proper design of structures with a stable postbuckling branch.

In this paper the dynamic buckling of plates is investigated by using a stress failure criterion and the effects of the shock function, imperfections, geometric dimensions and limit stress of the material are studied.

2. Dynamic stability of plates

As will be shown in the following, the application of the Budiansky–Hutchinson criterion [1] leads to very conservative dynamic buckling loads. Therefore a dynamic buckling criterion that is founded on stress analysis is used: a stress failure occurs if the effective stress σ_E exceeds the limit stress σ_L of the material. A dynamic response caused by an impact is defined to be dynamically stable if

$$\sigma_E \leq \sigma_L \tag{1}$$

is fulfilled at every time everywhere in the structure. Even if local yielding does not result in global failure of the structure, it is practical for ductile materials to use the yield stress σ_Y as limit stress σ_L. The application of this stress failure criterion results in a unique failure load depending on the shock function. The corresponding amplitude of the impact function is called the dynamic failure load, N_F^{dyn}.

A dynamic load factor has been defined in the literature as the quotient of the dynamic buckling (here: failure) load, N_F^{dyn}, and the classical bifurcation load, N_{crit}:

$$\mathrm{DLF}_{crit} = \frac{N_F^{dyn}}{N_{crit}}. \tag{2}$$

Considering the fact that the static failure load of a plate could exceed its static bifurcation load several times [6], it seems better to compare the dynamic failure load N_F^{dyn} with the static one, N_F^{stat}. In redefining the dynamic load factor by

$$\mathrm{DLF} = \frac{N_F^{dyn}}{N_F^{stat}}, \tag{3}$$

this load-amplifying quotient only describes the dynamic behaviour of the structure under impact loading.

3. Analysis

Consider a rectangular plate of length a, width b and constant thickness h subjected to pulse loading $\bar{N}_x(t)$, $\bar{N}_y(t)$ and $\bar{N}_{xy}(t)$ (Fig. 1). The plate has an initial imperfection w_0 in the z-direction. The total transverse displacement is defined by

$$w = w_{el} + w_0, \tag{4}$$

where w_{el} is the flexible displacement. Applying Kirchhoff's hypothesis, the in-plane displacements $\bar{u}$ and $\bar{v}$ are taken as

$$\bar{u} = u(x, y) - z\frac{\partial w}{\partial x} \tag{5}$$

and

$$\bar{v} = v(x, y) - z\frac{\partial w}{\partial y}. \tag{6}$$

The derivatives of the displacements $\bar{u}$ and $\bar{v}$ — i.e., $\partial\bar{u}/\partial x$, $\partial\bar{u}/\partial y$, $\partial\bar{v}/\partial x$ and $\partial\bar{v}/\partial y$ — are small compared with $\partial w/\partial x$ and $\partial w/\partial y$. Accordingly, their squares are negligible against the other terms in the non-linear strain–displacement relations. Considering the imperfection they can be written as [7]:

$$\bar{\varepsilon}_x = \frac{\partial \bar{u}}{\partial x} + \frac{1}{2}\left(\frac{\partial w}{\partial x}\right)^2 - \frac{1}{2}\left(\frac{\partial w_0}{\partial x}\right)^2, \tag{7}$$

$$\bar{\varepsilon}_y = \frac{\partial \bar{v}}{\partial y} + \frac{1}{2}\left(\frac{\partial w}{\partial y}\right)^2 - \frac{1}{2}\left(\frac{\partial w_0}{\partial y}\right)^2 \tag{8}$$

and

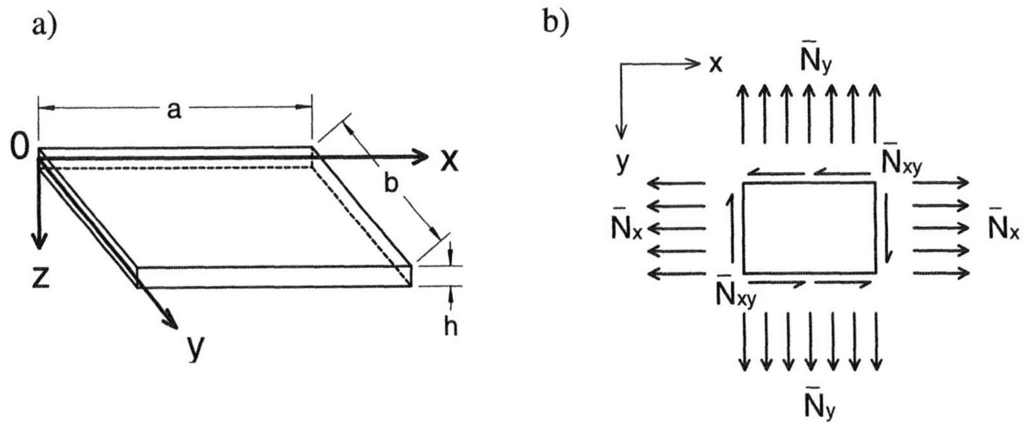

Fig. 1. (a) Plate dimensions and coordinates; (b) definition of the average boundary membrane forces.

$$\bar{\gamma}_{xy}=\frac{\partial\bar{u}}{\partial y}+\frac{\partial\bar{v}}{\partial x}+\frac{\partial w}{\partial x}\frac{\partial w}{\partial y}-\frac{\partial w_0}{\partial x}\frac{\partial w_0}{\partial y}. \tag{9}$$

As in the classic plate theory, the membrane forces N_x, N_y, N_{xy} and bending or twisting moments M_x, M_y, M_{xy} are defined by

$$\{N_x, N_y, N_{xy}\}=\int_{-h/2}^{+h/2}\{\sigma_x, \sigma_y, \tau_{xy}\}\,\mathrm{d}z \tag{10}$$

and

$$\{M_x, M_y, M_{xy}\}=\int_{-h/2}^{+h/2}z\{\sigma_x, \sigma_y, \tau_{xy}\}\,\mathrm{d}z. \tag{11}$$

Substituting the linear elastic Hooke's law for isotropic materials

$$\begin{Bmatrix}\sigma_x\\ \sigma_y\\ \tau_{xy}\end{Bmatrix}=\frac{E}{1-\nu^2}\begin{bmatrix}1 & \nu & 0\\ \nu & 1 & 0\\ 0 & 0 & \frac{1-\nu}{2}\end{bmatrix}\begin{Bmatrix}\bar{\varepsilon}_x\\ \bar{\varepsilon}_y\\ \bar{\gamma}_{xy}\end{Bmatrix} \tag{12}$$

into Eqs. (10) and (11), the membrane forces and the moments in terms of the three displacements can be derived. In Eq. (12), E is Young's modulus and ν is Poisson's ratio. Neglecting the terms of plane and rotary inertia, the applied strain–displacement relations lead to the following equation of motion

$$\rho h\ddot{w}=N_x\frac{\partial^2 w}{\partial x^2}+N_y\frac{\partial^2 w}{\partial y^2}+2N_{xy}\frac{\partial^2 w}{\partial x\partial y}-K(\Delta\Delta w-\Delta\Delta w_0)+q_z, \tag{13}$$

where Δ is the Laplacian operator, q_z is the load per area in the z-direction and the plate stiffness K is defined by

$$K=\frac{Eh^3}{12(1-\nu^2)}. \tag{14}$$

By combining the middle-plane strains $\varepsilon_x=\bar{\varepsilon}_x(z=0)$, $\varepsilon_y=\bar{\varepsilon}_y(z=0)$ and $\gamma_{xy}=\bar{\gamma}_{xy}(z=0)$, the compatibility equation

$$\frac{\partial^2\varepsilon_x}{\partial y^2}+\frac{\partial^2\varepsilon_y}{\partial x^2}-\frac{\partial^2\gamma_{xy}}{\partial x\partial x}=\left(\frac{\partial^2 w}{\partial x\partial y}\right)^2-\frac{\partial^2 w}{\partial x^2}\frac{\partial^2 w}{\partial y^2}-\left\{\left(\frac{\partial^2 w_0}{\partial x\partial y}\right)^2-\frac{\partial^2 w_0}{\partial x^2}\frac{\partial^2 w_0}{\partial y^2}\right\} \tag{15}$$

can be found. Regarding the definitions of membrane forces, by introducing Airy's stress function ϕ which satisfies

$$N_x=\frac{\partial^2\phi}{\partial y^2},\ N_y=\frac{\partial^2\phi}{\partial x^2}\ \text{and}\ N_{xy}=-\frac{\partial^2\phi}{\partial x\partial y},\tag{16}$$

it follows that

$$\Delta\Delta\Phi=Eh\left[\left(\frac{\partial^2 w}{\partial x\partial y}\right)^2-\frac{\partial^2 w}{\partial x^2}\frac{\partial^2 w}{\partial y^2}-\left\{\left(\frac{\partial^2 w_0}{\partial x\partial y}\right)^2-\frac{\partial^2\omega_0}{\partial x^2}\frac{\partial^2 w_0}{\partial y^2}\right\}\right].\tag{17}$$

This non-linear equation of compatibility takes the plane problem into consideration.

4. Solution of the plate equations

The boundary conditions for w and w_0 of a simply supported plate where the edges remain straight after buckling are:

$$\left.\begin{array}{r} w=w_0=0\\ \frac{\partial^2 w}{\partial x^2}=\frac{\partial^2 w_0}{\partial x^2}=0\\ \frac{\partial u}{\partial y}=C_1\end{array}\right\}\text{ at }x=0,a;\quad \left.\begin{array}{r} w=w_0=0\\ \frac{\partial^2 w}{\partial y^2}=\frac{\partial^2 w_0}{\partial y^2}=0\\ \frac{\partial v}{\partial x}=C_2\end{array}\right\}\text{ at }y=0,b.\tag{18}$$

If $\bar{N}_x$, $\bar{N}_y$ and $\bar{N}_{xy}$ are the average membrane forces at the edges, the stress function has to satisfy the following relations:

$$x=0,d:\quad \frac{1}{b}\int_0^b\frac{\partial^2\phi}{\partial y^2}\,\mathrm{d}y=\bar{N}_x;\quad -\frac{1}{b}\int_0^b\frac{\partial^2\phi}{\partial x\partial y}\,\mathrm{d}y=\bar{N}_{xy}\tag{19}$$

and

$$y=0,b:\quad \frac{1}{a}\int_0^a\frac{\partial^2\phi}{\partial x^2}\,\mathrm{d}x=\bar{N}_y;\quad -\frac{1}{a}\int_0^a\frac{\partial^2\phi}{\partial x\partial y}\,\mathrm{d}x=\bar{N}_{xy}.\tag{20}$$

In accordance with the boundary conditions, Navier's double Fourier series with the coefficients ${}^tW_{mn}$ and ${}^0W_{mn}$ are chosen to describe the displacement function $w(x, y)$ and the geometric imperfection $w_0(x, y)$:

$$w=\sum_{m=1}^{k}\sum_{n=1}^{l}{}^tW_{mn}\sin\frac{m\pi x}{a}\sin\frac{n\pi y}{b}\tag{21}$$

and

$$w_0=\sum_{m=1}^{k}\sum_{n=1}^{l}{}^0W_{mn}\sin\frac{m\pi x}{a}\sin\frac{n\pi y}{b}.\tag{22}$$

By inserting these two relations into the differential equation of compatibility [Eq. (17)] the following equation is obtained:

$$\Delta\Delta\Phi=Eh\left\{\sum_{i=1}^{k}\sum_{j=1}^{l}\sum_{m=1}^{k}\sum_{n=1}^{l}\left(\frac{ijmn\pi^4}{a^2b^2}\cos\frac{i\pi x}{a}\cos\frac{j\pi y}{b}\cos\frac{m\pi x}{a}\cos\frac{n\pi y}{b}\right.\right. \tag{23}$$

$$\left.\left.-\frac{i^2n^2\pi^4}{a^2b^2}\sin\frac{i\pi x}{a}\sin\frac{j\pi y}{b}\sin\frac{m\pi x}{a}\sin\frac{n\pi y}{b}\right)[{}^tW_{ij}{}^tW_{mn}-{}^0W_{ij}{}^0W_{mn}]\right\}.$$

Using different trigonometric relations, Airy's stress function can be derived as:

$$\Phi=\frac{1}{2}\bar{N}_xy^2+\frac{1}{2}\bar{N}_yx^2-\bar{N}_{xy}xy+Eh\sum_{i=1}^{k}\sum_{j=1}^{l}\left(A_1^{ij}\cos\frac{2i\pi x}{a}+A_2^{ij}\cos\frac{2j\pi y}{b}\right)[{}^tW_{ij}^2$$

$$-{}^0W_{ij}^2]+Eh\sum_{i=1}^{k}\sum_{j=1}^{l}\sum_{m=1}^{k}\sum_{\substack{n=1\\(i,j)\neq(m,n)}}^{l}\left\{A_3^{ijmn}\cos\frac{(i-m)\pi x}{a}\cos\frac{(j-n)\pi y}{b}\right.$$

$$+A_4^{ijmn}\cos\frac{(i-m)\pi x}{a}\cos\frac{(j+n)\pi y}{b}+A_5^{ijmn}\cos\frac{(i+m)\pi x}{a}\cos\frac{(j-n)\pi y}{b}$$

$$\left.+A_6^{ijmn}\cos\frac{(i+m)\pi x}{a}\cos\frac{(j+n)\pi y}{b}\right\}[{}^tW_{ij}{}^tW_{mn}-{}^0W_{ij}{}^0W_{mn}], \tag{24}$$

with the shortenings:

$$A_1^{ij}=\frac{a^2j^2}{32b^2i^2},\quad A_3^{ijmn}=\frac{(ijmn-i^2n^2)a^2b^2}{4[(i-m)^2b^2+(j-n)^2a^2]^2},\quad A_5^{ijmn}=\frac{(ijmn+i^2n^2)a^2b^2}{4[(i+m)^2b^2+(j-n)^2a^2]^2}$$

$$A_2^{ij}=\frac{b^2i^2}{32a^2j^2},\quad A_4^{ijmn}=\frac{(ijmn+i^2n^2)a^2b^2}{4[(i-m)^2b^2+(j+n)^2a^2]^2},\quad A_6^{ijmn}=\frac{(ijmn-i^2n^2)a^2b^2}{4[(i+m)^2b^2+(j+n)^2a^2]^2}.$$

With the definitions (16), the membrane forces N_x, N_y and N_{xy} are computable by this solution of ϕ and the boundary conditions (19) and (20) are satisfied. By using Eqs. (7, 8) and (12) one can show that the boundary conditions $\partial u/\partial y|_{x=0,\,a}$ and $\partial v/\partial x|_{y=0,\,b}$ [Eq. (18)] are valid too. Together with the double Fourier series [Eqs. (21) and (22)] the membrane forces are inserted into the equation of motion [Eq. (13)]. The relation obtained is multiplied by

$$\sin\frac{r\pi x}{a}\sin\frac{s\pi y}{b},\quad r=1,2,\ldots,k;\ s=1,2,\ldots,l \tag{25}$$

and next integrated over the plate area. A system of $k\times l$ second-order ordinary differential equations is obtained to compute the coefficients ${}^tW_{rs}$ of the double Fourier series:

$${}^t\ddot{W}_{rs}+\frac{K}{\rho h}\left(\frac{r^2\pi^2}{a^2}+\frac{s^2\pi^2}{b^2}\right)^2[{}^tW_{rs}-{}^0W_{rs}]-\frac{E}{\rho}\sum_{i=1}^{k}\sum_{j=1}^{l}\sum_{p=1}^{k}\sum_{q=1}^{l}(B_1+B_2)[{}^tW_{ij}^2$$

$$-{}^{0}W_{ij}^{2}]^{t}W_{pq}-\frac{E}{\rho}\sum_{i=1}^{k}\sum_{j=1}^{l}\sum_{m=1}^{k}\sum_{n=1}^{l}\sum_{p=1}^{k}\sum_{q=1}^{l}{}^{(i,j)\neq(m,n)}\{B_3+B_4+B_5+B_6\}[{}^{t}W_{ij}{}^{t}W_{mn}$$

$$-{}^{0}W_{ij}{}^{0}W_{mn}]^{t}W_{pq}+\frac{1}{\rho h}\frac{r^2\pi^2}{a^2}\bar{N}_x{}^{t}W_{rs}+\frac{1}{\rho h}\frac{s^2\pi^2}{b^2}\bar{N}_y{}^{t}W_{rs}$$

$$-\sum_{p=1}^{k}\sum_{q=1}^{l}\frac{8}{\rho h}\frac{pq\pi^2}{ab}XY(p,q,r,s)\bar{N}_{xy}{}^{t}W_{pq}=0, \tag{26}$$

$r=1, 2, \ldots, k$ respectively $s=1, 2, \ldots, l$.

The abbreviations B_1 to B_6 and XY are defined in Appendix A. The corresponding system of equations for static problems results from Eq. (26) by omitting ${}^{t}\ddot{W}_{rs}$. For dynamic buckling investigations q_z has not been considered.

A computer code that solves the dynamic problem in dependence on the imperfection coefficients ${}^{0}W_{mn}$ and the loading functions $\bar{N}_x(t)$, $\bar{N}_y(t)$ and $\bar{N}_{xy}(t)$ has been developed in FORTRAN. For time integration a fourth-order Runge–Kutta method has been used. The initial conditions are set as:

$${}^{t}\dot{W}_{mn}(t=0)=0 \tag{27}$$

and

$${}^{t}W_{mn}(t=0)={}^{0}W_{mn}. \tag{28}$$

To solve the coefficients of the deflection function in the static case, a quadratic convergent Newton method is applied.

Compared with calculations performed by the finite element code MSC-Nastran for static as well as for dynamic analysis, the applied Galerkin method is very effective in relation to deflection and stress analysis because only a few modes have to be employed for the investigations [8]. The CPU time increases nearly quadratically with the number of terms of the Fourier series.

5. Results and discussion

In this section results will be presented for aluminium alloy plates with various geometric dimensions and imperfections under unidirectional impact loading, $\bar{N}_x(t)$. In order to analyse the behaviour of dynamically loaded plates for different loading durations the amplitude of the loading function $\bar{N}_x(t)$, the exceeding of which leads to stress failure in the plate, is searched. This computation is performed by a bisection method which varies the pulse amplitude. For every bisection step the system of coupled differential equations (26) has to be solved. The effective stress σ_E is calculated by using the von Mises yield criterion [9]:

$$\sigma_E=\sqrt{\sigma_x^2+\sigma_y^2-\sigma_x\sigma_y+3(\tau_{xy}^2+\tau_{yz}^2+\tau_{xz}^2)}. \tag{29}$$

The results will be presented in form of DLF versus pulse duration (T_S) charts. To compute the DLF the static failure load, which is obtained by a static postbuckling calculation, is necessary. Unless otherwise stated, a one-term imperfection ${}^0W_{11}$, which corresponds to the basic buckling mode of the plates investigated, is used for the calculations. Because of the symmetric imperfection shape all antisymmetric terms of the Fourier series become zero. Therefore Navier's double series can be reduced to a summation over odd indices, and so the CPU time decreases considerably. The calculations, the results of which are presented in this paper, are performed by using a (k=5)×(l=3) series that results in convergent solutions.

5.1. Influence of pulse duration

The dependence of the dynamic behaviour of a structure on the pulse duration could be described by a shock spectrum. In such a chart the residual response amplitudes and maximax response amplitudes are plotted over the duration of impact. The residual response amplitude is defined by the free vibration amplitude after loading, while the maximax response amplitude results from the maximum of deflection during the motion of a structure caused by the loading [10]. For a sinusoidial impact function (S) that is described by

$$\bar{N}_x^S(t)=\begin{cases} N_x^{max}\sin\dfrac{\pi t}{T_S}, & 0\le t\le T_S \\ 0, & \text{otherwise} \end{cases}, \tag{30}$$

the shock spectrum of a plate (a/b=1; h/b=0.005; ${}^0W_{11}/h$=0.2; σ_L=100 MPa) subjected to a loading amplitude of $N_x^{max}=3N_{crit}$ is presented in Fig. 2, in which the mid-point transverse deflection has been normalized to the corresponding static deflection, w_{stat}.

It is shown in Fig. 2 that the maximum of response is reached for a shock duration close to the period time of transverse vibration, T_P. The DLF plot of the plate is shown in Fig. 3. Depending on the shock duration, this chart yields the factor by which to multiply the static failure load to obtain the dynamic failure load. Accordingly the DLF plot is useful to get the dynamic failure load from the corresponding static load. If the DLF plot is known, the design of plates relating to in-plane dynamic loading is reduced to a static postbuckling analysis. In design practice such computations are performed by the simple method of effective width [6].

As demonstrated in the plot, high loads are possible without the occurrence of failure for short-time pulses while for higher pulse durations — because of the dynamic overshooting of displacements — the dynamic load has to be reduced in relation to the static failure load.

For a wide range of shock durations loads smaller than the static failure load lead to stress failure in the plate. For high ratios of T_S/T_P the quasi-static limit load is reached asymptotically.

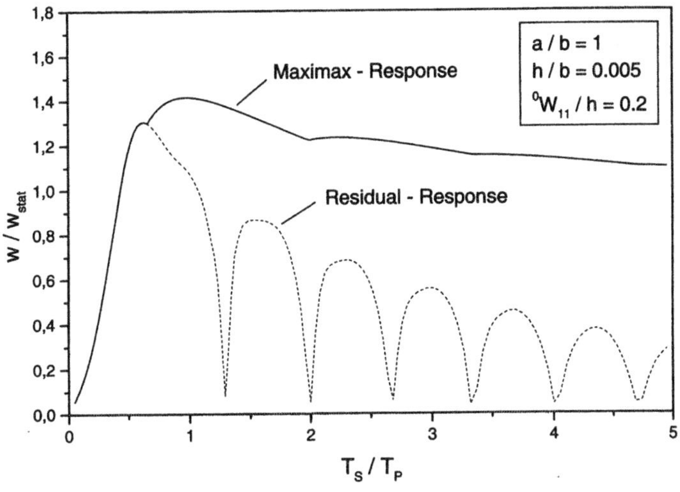

Fig. 2. Shock spectrum of a plate subjected to unidirectional dynamic loading ($N_x^{max}=3N_{crit}$).

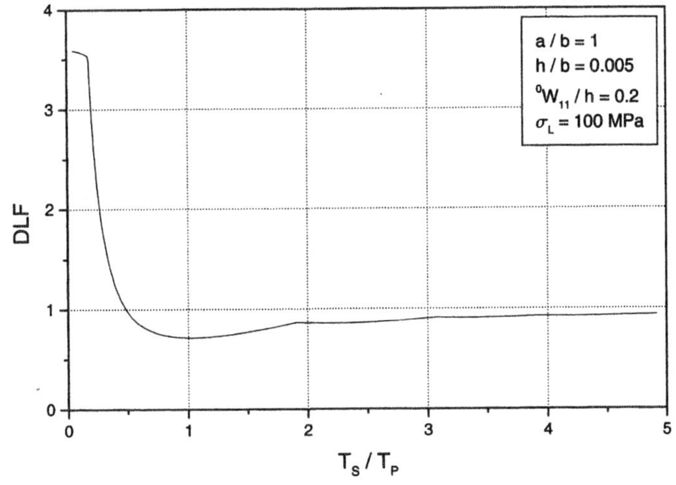

Fig. 3. DLF plot of a plate subjected to unidirectional dynamic loading.

As in the shock spectrum, the effects of transverse inertia affect the DLF chart differently depending on the pulse duration. The DLF plot corresponds with the shock spectrum. Thus high deflections in the shock spectrum result in low DLF values. On account of the non-linear behaviour, local maxima or minima in the DLF plot and the shock spectrum are not reached for exactly the same shock durations.

Fig. 3 also shows that the dynamic load-amplification factor is limited for short pulse durations by the flat plates problem:

$$\mathrm{DLF}_{\max}=\frac{\sigma_L h}{|N_F^{\mathrm{stat}}|}. \tag{31}$$

5.2. *Influence of the shock function*

To examine the influence of different loading functions on the dynamic behaviour of imperfect plates, in addition to the sinusoidial impact rectangular (R) and triangular (T) pulses will be investigated. These two functions are described by:

$$\bar{N}_x^{R}(t)=\begin{cases} N_x^{max}, & 0\le t\le T_S \\ 0, & \text{otherwise} \end{cases} \tag{32}$$

and

$$\bar{N}_x^{T}(t)=\begin{cases} 2N_x^{max}\dfrac{t}{T_S}, & 0\le t\le \dfrac{T_S}{2} \\ 2N_x^{max}\left(1-\dfrac{t}{T_S}\right), & \dfrac{T_S}{2}<t\le T_S \\ 0, & \text{otherwise} \end{cases}. \tag{33}$$

The corresponding DLF plots (Fig. 4) show that the particular force function influences the dynamic behaviour of stability exceptionally.

In the region of residual response, which is characterized by the visible drop in DLF, the plot moves to shorter pulse durations for shock functions with higher pulse area (i.e., impulse) if the loading parameters (T_S, N_x^{max}) are equal. In the same way the DLF reaches lower values. While shock functions having finite rise time — like the sinusoidial or the triangular pulse — approach quasi-static behaviour for higher shock durations, the rectangular pulse having infinite rise time is constant after reaching its minimum. It has been demonstrated that the position of the DLF plots to one another is defined by the impulse ratio of the shock functions which have equal loading parameters [11].

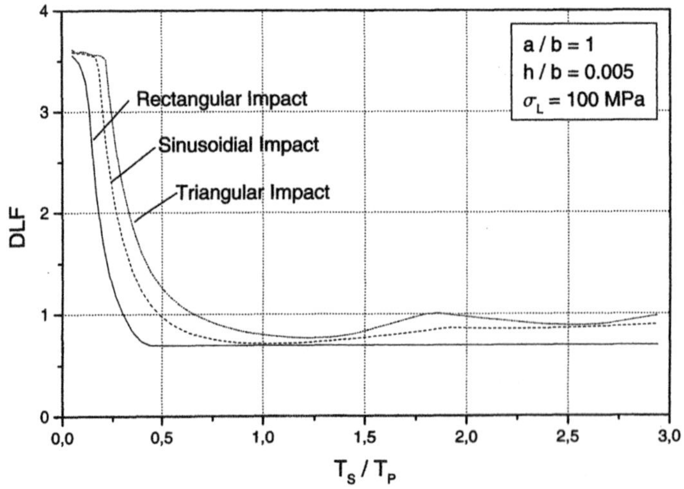

Fig. 4. Influence of the shock function on dynamic buckling.

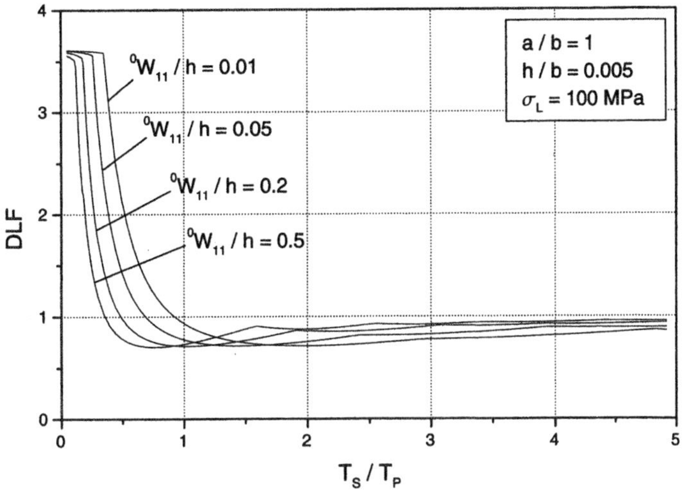

Fig. 5. DLF plots for different imperfection magnitudes.

5.3. Imperfection sensitivity

Many publications have been pointed out the imperfection sensitivity of dynamic stability behaviour [3,12]. The DLF plots of four plates having different geometric imperfections ($^0W_{11}/h$=0.01, 0.05, 0.2, 0.5) are compared in Fig. 5. As shown, higher imperfection leads to smaller dynamic load-amplification factors for short pulse durations while, for higher durations, the DLF of plates having minor imperfection is comparatively small. The smaller the magnitude of imperfection, the more the dynamic deflection lags behind the static one, which is shown in Fig. 6. This is why the location of the DLF minimum is dependent on the size of imperfection.

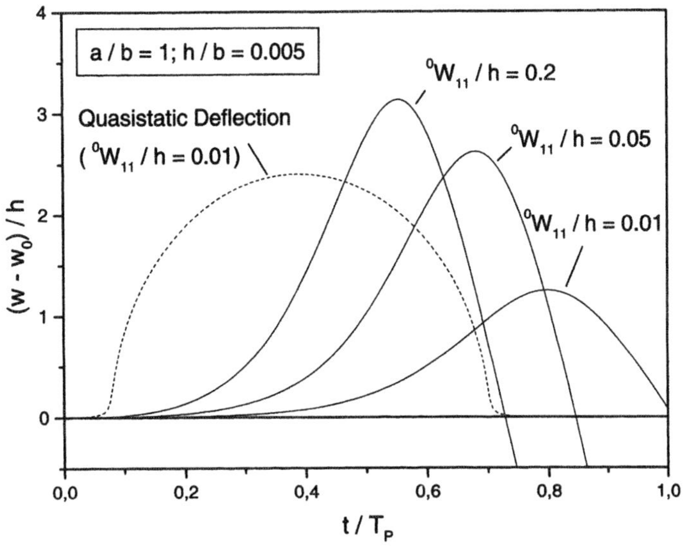

Fig. 6. Response curves for different imperfection magnitudes (T_S/T_P=0.78; N_x^{max}=0.7N_F^{stat}; σ_F=100 MPa).

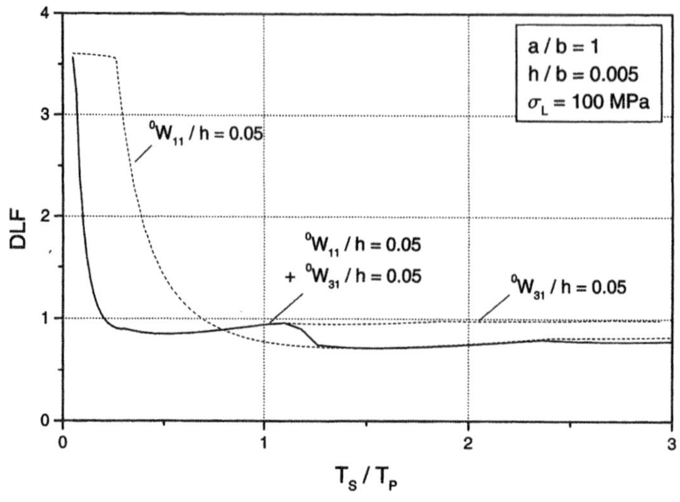

Fig. 7. Influence of geometric imperfection shapes on dynamic buckling.

In previous publications on plates subjected to in-plane pulse loading only one term of the imperfection function (22) has been considered. But if there exist more wave terms of imperfection, the dynamic behaviour of stability is influenced differently for varying pulse durations. Thus higher undulating imperfections influence the dynamic behaviour of short-duration impacts considerably, while for long-duration pulses imperfections of high wavelength dominate the plate's dynamic behaviour. Fig. 7 shows this effect of imperfection superposition.

5.4. *Influence of geometric dimensions*

DLF plots of three plates with different thickness (*h*/*b*=0.0025, 0.00375, 0.005) are demonstrated in Fig. 8. Because of the variable thickness also the $^0W_{11}/h$ ratio differs. All other plate parameters are identical.

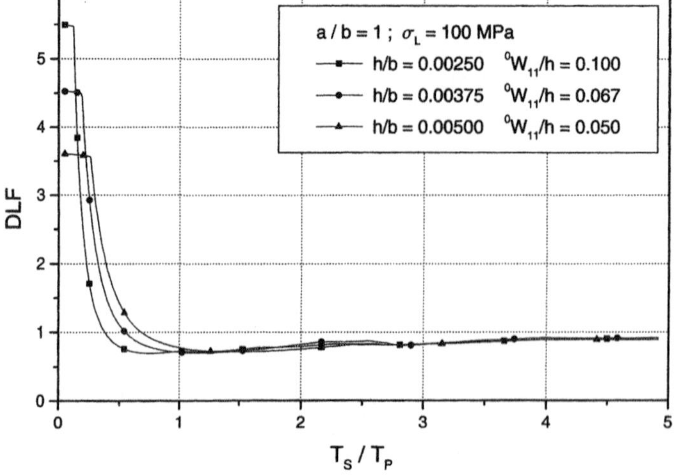

Fig. 8. Influence of the plate thickness on dynamic buckling.

For higher pulse durations the thickness of a plate does not affect its dynamic stability much. But there exist differences in dynamic behaviour for short-time impacts. Investigations proved that the influence of varying ${}^0W_{11}/h$ ratio is not responsible for the change in the DLF plot. The thinner the plates, the more the drop in DLF moves to shorter pulse durations. The thicker the plates, the more the dynamic deflections lag behind the quasi-static deflections (cf. influence of the magnitude of the imperfection).

The static failure load decreases overproportionally while the dynamic failure load falls proportionally by reducing the thickness [Eq. (31)], thus higher maxima of DLF are reached for thinner plates.

Fig. 9 shows three DLF plots of plates with the same static basic buckling mode. As demonstrated, the aspect ratio of the plate does not affect the dynamic stability much.

5.5. Influence of failure stress

Fig. 10 shows DLF charts of plates made of aluminium alloys that have the same mechanical properties but different limit stresses (AlMg 5454: σ_L=80 MPa; AlMg 5086: 100 MPa; AlMg 5083: 130 MPa). All plates show nearly the same dynamic behaviour of stability. There is only a difference for short pulse durations: because of the non-linear behaviour, raising the limit load of the material σ_L increases the static failure load underproportionally. Considering that for short-time impacts the dynamic failure load is proportional to σ_L [Eq. (31)], the DLF increases.

5.6. Influence of the criterion of stability

In this section an example is given to show how the dynamic buckling load is affected by the criterion of stability.

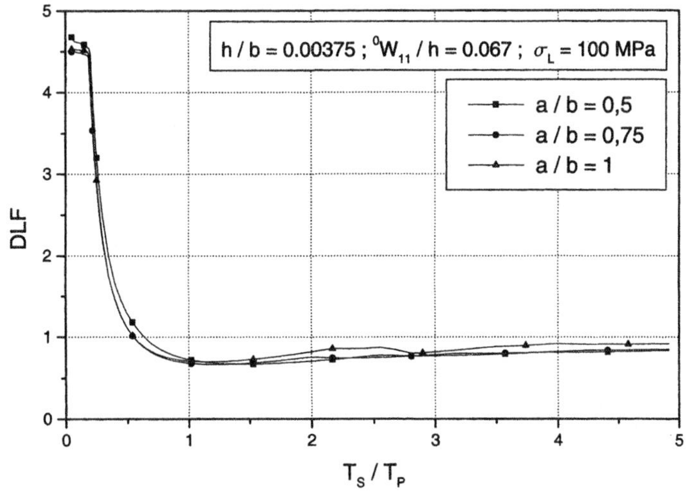

Fig. 9. DLF plots for different aspect ratios *a/b*.

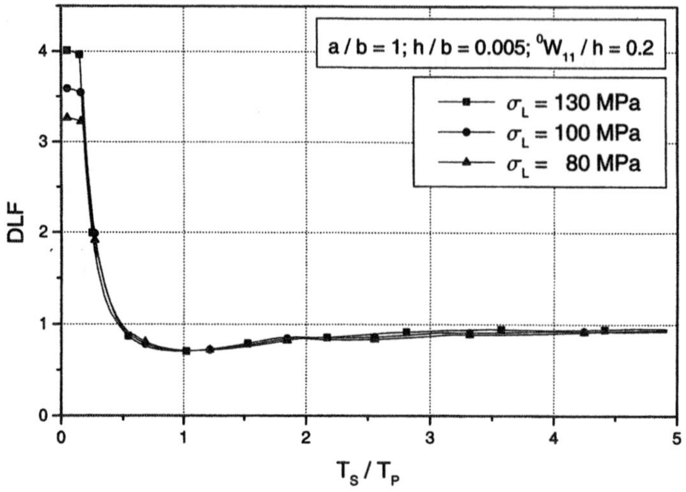

Fig. 10. Influence of the material's limit stress σ_F on dynamic buckling.

For a shock duration T_S/T_P=1, the dynamic buckling load is computed by the Budiansky–Hutchinson [1] criterion and the stress failure criterion (1). To determine the dynamic buckling load by the Budiansky–Hutchinson criterion [1], the maximum deflection in the z-direction caused by the impact is investigated by variation of the loading amplitude. The dynamic buckling load based on the Budiansky–Hutchinson criterion [1] is defined by the maximum of the gradient $\partial w_{max}/\partial N_x^{max}$. In Fig. 11 the corresponding dynamic buckling load and the region of instability are marked. The dynamic buckling load computed by the stress failure criterion applied in this paper is shown as well. Hence it appears in the example that the load-carrying capability is only exploited for nearly 50% by the Budiansky–Hutchinson criterion [1].

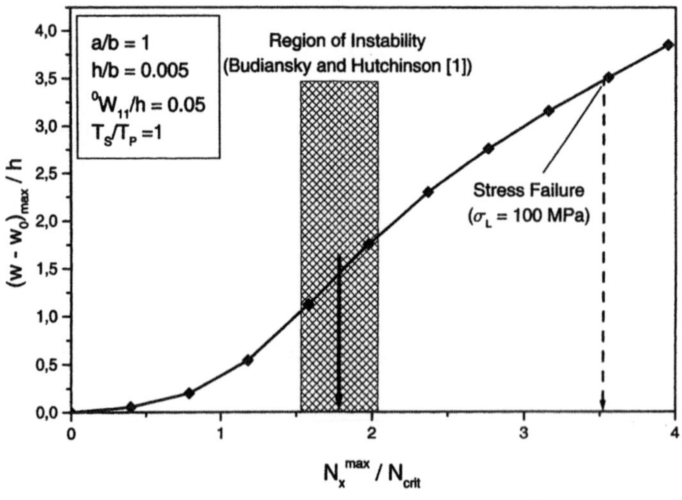

Fig. 11. Comparison of Budiansky–Hutchinson [1] and stress failure criteria.

6. Conclusions

A study of dynamic buckling of thin isotropic plates under unidirectional loading has been presented in this paper. For this purpose a stress failure criterion has been applied which considers the load-carrying capability of the structure. It has been demonstrated that application of the criterion of Budiansky and Hutchinson [1] is not suitable for the design of plates considering dynamic buckling in lightweight structures.

The dynamic load factor (DLF) has been defined such that it describes only dynamic effects in dependence on the shock duration. If the DLF is known, the design of plates relating to dynamic buckling is reduced to a non-linear static post-buckling analysis. For a wide range of shock durations DLFs smaller than unity are caused by the dynamic overshooting of the structure. As demonstrated in this paper, the dynamic behaviour of a plate subjected to in-plane dynamic loading is essentially influenced by the loading function, the duration of impact and the geometric imperfections. Not only the magnitude of the imperfection but also its shape affect dynamic buckling considerably. Regarding the variation of the other parameters (i.e., ultimate stress, thickness and aspect ratio), plates show only a modest sensitivity.

Appendix A

Shortenings $B_1, B_2, \ldots, B_6$ of Eq. (26):

$$B_1=\frac{i^2p^2\pi^4}{2a^4}X_1(p, r)Y_2(j, q, s),$$

$$B_2=\frac{j^2q^2\pi^4}{2b^4}X_2(i, p, r)Y_1(q, s),$$

$$B_3=\frac{(ijmn-i^2n^2)\pi^4}{[(i-m)^2b^2+(j-n)^2a^2]^2}\{[(j-n)^2p^2+(i-m)^2q^2]X_3(i, m, p, r)Y_3(j, n, q, s)-2(i-m)(j-n)pqX_4(i, m, p, r)Y_4(j, n, q, s)\},$$

$$B_4=\frac{(ijmn-i^2n^2)\pi^4}{[(i-m)^2b^2+(j+n)^2a^2]^2}\{[(j+n)^2p^2+(i-m)^2q^2]X_3(i, m, p, r)Y_5(j, n, q, s)-2(i-m)(j+n)pqX_4(i, m, p, r)Y_6(j, n, q, s)\},$$

$$B_5=\frac{(ijmn+i^2n^2)\pi^4}{[(i+m)^2b^2+(j-n)^2a^2]^2}\{[(j-n)^2p^2+(i+m)^2q^2]X_5(i, m, p, r)Y_3(j, n, q, s)-2(i+m)(j-n)pqX_6(i, m, p, r)Y_4(j, n, q, s)\}$$

and

$$B_6=\frac{(ijmn-i^2n^2)\pi^4}{[(i+m)^2b^2+(j+n)^2a^2]^2}\{[(j+n)^2p^2+(i+m)^2q^2]X_5(i, m, p, r)Y_5(j, n, q, s)-2(i$$

$+m)(j+n)pqX_6(i, m, p, r)Y_6(j, n, q, s)\}$,

where $X_1, X_2, \ldots, X_6$ respectively $Y_1, Y_2, \ldots, Y_6$ and XY are defined by integrations of trigonometric functions:

$$XY(p, q, r, s)=\frac{1}{ab}\int_0^a\int_0^b \cos\frac{p\pi x}{a}\cos\frac{q\pi y}{b}\sin\frac{r\pi x}{a}\sin\frac{s\pi y}{b}\,dx\,dy,$$

$$X_1(p, r)=\frac{1}{a}\int_0^a \sin\frac{p\pi x}{a}\sin\frac{r\pi x}{a}\,dx,$$

$$Y_1(q, s)=\frac{1}{b}\int_0^b \sin\frac{q\pi y}{b}\sin\frac{s\pi y}{b}\,dy,$$

$$X_2(i, p, r)=\frac{1}{a}\int_0^a \cos\frac{2i\pi x}{a}\sin\frac{p\pi x}{a}\sin\frac{r\pi x}{a}\,dx,$$

$$Y_2(j, q, s)=\frac{1}{b}\int_0^b \cos\frac{2j\pi y}{b}\sin\frac{q\pi y}{b}\sin\frac{s\pi y}{b}\,dy,$$

$$X_3(i, m, p, r)=\frac{1}{a}\int_0^a \cos\frac{(i-m)\pi x}{a}\sin\frac{p\pi x}{a}\sin\frac{r\pi x}{a}\,dx,$$

$$Y_3(j, n, q, s)=\frac{1}{b}\int_0^b \cos\frac{(j-n)\pi y}{b}\sin\frac{q\pi y}{b}\sin\frac{s\pi y}{b}\,dy,$$

$$X_4(i, m, p, r)=\frac{1}{a}\int_0^a \sin\frac{(i-m)\pi x}{a}\cos\frac{p\pi x}{a}\sin\frac{r\pi x}{a}\,dx,$$

$$Y_4(j, n, q, s)=\frac{1}{b}\int_0^b \sin\frac{(j-n)\pi y}{b}\cos\frac{q\pi y}{b}\sin\frac{s\pi y}{b}\,dy,$$

$$X_5(i, m, p, r)=\frac{1}{a}\int_0^a \cos\frac{(i+m)\pi x}{a}\sin\frac{p\pi x}{a}\sin\frac{r\pi x}{a}\,dx,$$

$$Y_5(j, n, q, s)=\frac{1}{b}\int_0^b \cos\frac{(j+n)\pi y}{b}\sin\frac{q\pi y}{b}\sin\frac{s\pi y}{b}\,dy,$$

$$X_6(i, m, p, r)=\frac{1}{a}\int_0^a \sin\frac{(i+m)\pi x}{a}\sin\frac{p\pi x}{a}\sin\frac{r\pi x}{a}\,dx$$

and

$$Y_6(j, n, q, s)=\frac{1}{b}\int_0^b \sin\frac{(j+n)\pi y}{b}\sin\frac{q\pi y}{b}\sin\frac{s\pi y}{b}\,dy.$$

References

[1] Budiansky B, Hutchinson JW. Dynamic buckling of imperfection-sensitive structures. In: Applied Mechanics Proceedings of the 11th International Congress of Applied Mechanics. Berlin: Springer, 1964. p. 636–51.

[2] Huyan X, Simitses G. Dynamic buckling of imperfect cylindrical shells under axial compression and bending moment. AIAA J 1997;35(8):1404–12.

[3] Ari-Gur J, Singer J, Weller T. Dynamic buckling of plates under longitudinal impact. In: TAE No. 430. Haifa: Technion Israel Institute of Technology, Department of Aeronautical Engineering, 1981.

[4] Ari-Gur J, Simonetta SR. Dynamic pulse buckling of rectangular composite plates. Composites, Part B 1997;28B:301–8.

[5] Ari-Gur J, Elishakoff I. Effects of shear deformation and rotary inertia on the pulse buckling of imperfect plates. In: Recent Advances in Structural Mechanics PVP-vol. 227/NE-vol. 7. ASME, 1991.

[6] Singer J, Arbocz J, Weller T. Buckling experiments: experimental methods in buckling of thin-walled structures. Chichester (UK): John Wiley & Sons, 1998.

[7] Wolmir AS. Biegsame Platten und Schalen. Berlin: VEB Verlag für Bauwesen, 1962.

[8] Petry D. Imperfekte isotrope Platten unter ebener Stoßbelastung (in German). Research Report 99-01. Munich: Universität der Bundeswehr München, Institut für Luftfahrttechnik und Leichtbau, 1999.

[9] Pilkey WD. Formulas for stress, strain, and structural matrices. New York: John Wiley & Sons, 1994.

[10] Harris CM, Crede CE. Shock and vibration handbook. 2nd ed. New York: McGraw-Hill, 1976.

[11] Petry D. Einfluss der Stoßfunktion auf das dynamische Stabilitätsverhalten von Platten unter ebener Impactbelastung (in German). In: Technical Note TN99-01. Munich: Universität der Bundeswehr München, Institut für Luftfahrttechnik und Leichtbau, 1999.

[12] Ekstrom RE. Dynamic buckling of a rectangular orthotropic plate. AIAA J 1973;11(12):1655–9.

Papers from

Engineering Structures

Modeling the effects of residual stresses on defects in welds of steel frame connections

C.G. Matos, R.H. Dodds Jr *

Department of Civil Engineering, University of Illinois, Urbana, IL 61801, USA

Received 17 February 1999; received in revised form 10 June 1999; accepted 10 June 1999

Abstract

This work describes an eigenstrain approach to impose realistic residual stress fields on 3-D finite element models for the lower-flange welds of the type found in large, beam-column connections. The 3-D models incorporate the full geometry of a pull-plate specimen designed to reduce the cost of testing full connections while retaining essential features of the fracture prone, weld region. The 3-D analyses examine pull-plate configurations with a variety of geometric parameters for both through-width and semi-elliptical cracks in the weld root pass. The numerical results include tables of values for the non-dimensional "geometry" factors (Y) needed to compute stress intensity factors. The residual stress fields generated by the proposed eigenstrain functions closely match those computed using 2-D, thermo-mechanical simulation of the welding process. Stress intensity factors, obtained from corresponding J-integral computations, show clearly the significant fracture toughness demand from residual stresses prior to application of the structural loads. An alternative design for a grooved backup bar shows considerable promise to reduce fracture toughness demands by re-positioning the weld root pass below the beam flange and thus outside the primary load path. Analyses here indicate approximately a 65% reduction in K_I-values for tension loading in the flange. © 2000 Elsevier Science Ltd. All rights reserved.

Keywords: Residual stresses; Finite elements; Eigenstrains; Stress-intensity factors; Welding

1. Introduction

When properly designed and constructed, steel frames with welded, moment resistant connections provide a safe and economical structural system to resist strong lateral loads from earthquake events. Such structures often have large, welded beam-to-column connections of the type illustrated in Fig. 1. Geometric discontinuities in the connection (90° angles, welding access holes, backup bars, etc.) create complex load transfer paths through the connection. Inspections following the 1994 Northridge (California) Earthquake [1] found fractures in more than 100 older and newer buildings of this type (including some under construction). Failure of the connections by fracture most often occurred before reaching the levels of inelastic deformation anticipated in the design process. In response to these events, new research initiatives (e.g. the SAC Joint Venture[1]) have been formed to establish causes for the failures, propose repair and new design procedures, and to quantify hazards for similar buildings located in earthquake prone areas.

Most fractures initiated at the lower flange weld in the connections. Full-scale, laboratory tests of these type connections [2–4] following the earthquake exhibited very similar fractures, now generally attributed to a combination of factors including mechanical and metallurgical defects created by manual, on-site welding; use of low-toughness electrodes; heavy plate thicknesses and high stresses; and the various geometric discontinuities. Chi et al. [5] performed extensive 2-D and limited 3-D finite element analyses using models with pre-existing cracks to establish the fracture toughness demands in terms of K_I and CTOD values needed to prevent propagation for a range of connection parameters. El-Tawil et

* Corresponding author. Tel.: +1-217-333-3276; fax: +1-217-333-9464.

E-mail address: r-dodds@uiuc.edu (R.H. Dodds Jr).

[1] The SAC Joint Venture is a partnership of the Structural Engineers Association of California, the Applied Technology Council, and the California State Universities for Earthquake Engineering.

Reprinted from *Engineering Structures* **22 (9)**, 1103-1120 (2000)

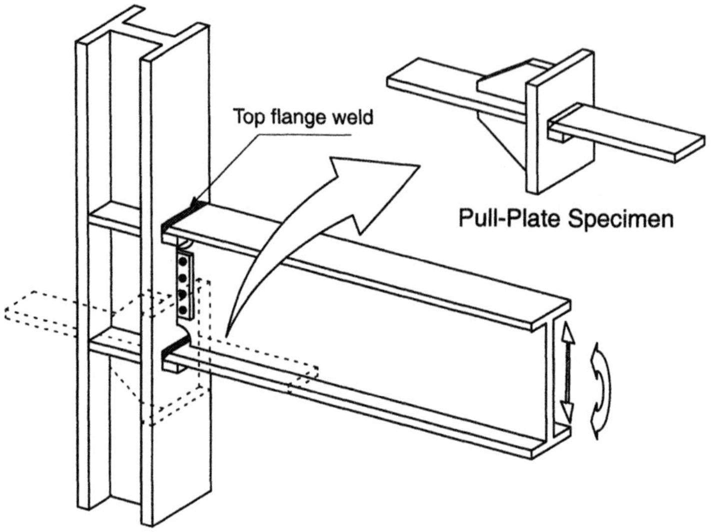

Fig. 1. Schematic of typical beam-column connection and the simplified pull-plate specimen developed by Kauffman and Fisher [4].

al. [6] followed with extensive 3-D nonlinear analyses using coupled shell-solid models (without cracks) to provide a much broader understanding of local stress and deformation fields at the lower flange connection. Neither of these efforts addressed the impact of welding induced residual stresses on the fracture response of the connection. Zhang and Dong [7] used (2-D) thermo-mechanical simulations of the multi-pass welding procedure for the lower flange connection to provide the first quantitative descriptions of the residual stress fields. Their accompanying 2-D fracture analyses illustrate the detrimental effects of residual stresses on fracture demands for cracks located in the weld root pass.

Full-scale laboratory tests of these connections are quite expensive. Kauffman and Fisher [8] developed the so-called "pull-plate" test specimen (see Fig. 1) which isolates the fracture prone lower flange weld from other (geometric) parameters involved in the connection behavior. The pull-plate test enables rapid, more economical evaluation of alternative welding procedures. Static and high-rate tests performed on the pull-plate specimen revealed the same kind of failures found in full-scale connections. Axial loads applied to the pull-plate specimen do not impose a secondary (local) bending at the weld-column flange interface as predicted by models of the full connections due to the web access hole (see Fig. 1). This reduces constraint at crack fronts and must be addressed to transfer toughness values measured in the pull-plate specimens to assess similar cracks in full connections. Additional tests in-progress using this specimen are investigating a broad range of welding procedures, backup bar designs and loading rates [9].

In the present work, we investigate the effects of welding induced residual stresses and applied axial forces on cracks located in the lower-flange weld region using fully 3-D analyses of the pull-plate specimen. This first in a series of papers on this topic describes the modeling procedures to incorporate residual stress fields into the (geometrically) complex weld configuration, and the resulting stress intensity factors for through-width and part-through (surface) cracks in the weld root region. Subsequent papers describe applications of recently improved Weibull stress models [10] to quantify the effects of residual stresses on the probability of cleavage fracture early in the loading history, when plastic deformation remains confined in the weld region. Both static loading and loading rates expected during major earthquakes are studied. Ductile tearing prior to cleavage observed in some of these fractures further complicates those analyses.

The eigenstrain [11–13] approach adopted here provides a convenient, straightforward procedure to incorporate residual stresses in the finite element models of cracked components. Panontin and Hill [14] employed this same approach in 2-D analyses of cracks in welded pipes to explore the effects of residual stresses on cleavage fracture and the initiation of ductile tearing using simple (deterministic) micro-mechanical models for fracture initiation. The welding simulations of Zhang and Dong [7] provide a key contribution in the present work to establish *representative* residual stress fields expected to develop in the lower-flange weld in the beam-column connection.

The plan of the paper is as follows. Section 2 describes the eigenstrain approach to model residual stresses, details of the functions developed specifically for the pull-plate specimen, and their implementation through anisotropic thermal expansion coefficients. Section 3 summarizes the geometric configurations studied and details of the 3-D finite element models. Section 4 describes the residual stress fields generated by the eigenstrain formulation developed here and provides tables of stress intensity factors for both axial force and residual stress loading for three types of cracks in the pull-plate specimen (through crack, semi-elliptical surface crack, and the unfused backup bar gap as the only initial crack). The section concludes with a brief examination of a proposed alternative design for a "grooved" backup bar. Appendix A describes modifications needed in the domain integral formulation for J-integral computations to maintain path independence when materials have anisotropic thermal expansion coefficients.

2. 3-D modeling of residual stresses

2.1. Features of the eigenstrain approach to model residual stresses

The multi-pass welding procedures employed to fabricate large structural joints introduce a field of strains and associated deformations (thermal, transformation and plastic) not associated with (externally) applied mechanical loading [7,11,12]. These strain fields generally do not satisfy the geometric compatibility relations. Consequently, *residual* stresses must be present to eliminate the incompatibility, thereby restoring geometric continuity of the welded component. Incompatible strain fields, denoted here by ϵ^*_{ij}, are often referred to as *eigenstrains* [11] or *inherent* strains [12]. The incompatible property is described mathematically by the six strain compatibility equations in terms of the three independent displacement components (u_i) in Cartesian coordinates,

$$R_{pq}=e_{pki}e_{qlj}\epsilon^*_{ij,kl}\ (i,j,k,l,p,q=1,2,3) \tag{1}$$

where the usual index notation with implied summation is employed with commas denoting differentiation. Here, e_{pki} denotes the third order alternating tensor. When the symmetric tensor R_{pq} vanishes for a given field of eigenstrains, ϵ^*_{ij}, no residual stresses are required to restore geometric compatibility. The total (compatible) strain tensor can be decomposed as

$$\epsilon_{ij}=\epsilon^e_{ij}+\epsilon^*_{ij} \tag{2}$$

where ϵ^e_{ij} denotes the elastic strain tensor necessary to restore compatibility created by the eigenstrain tensor ϵ^*_{ij}, and the strains arising from separately applied mechanical loads. Under such conditions, the final linear-elastic stresses are then given by

$$\sigma_{ij}=D_{ijkl}(\epsilon_{kl}-\epsilon^*_{kl}). \tag{3}$$

In the absence of mechanically applied loads (or restraints), the response of the component to the eigenstrains must generate a residual stress field, σ^*_{ij}, that satisfies equilibrium

$$\sigma^*_{ij}=D_{ijkl}\epsilon^e_{kl} \tag{4}$$

$$\sigma^*_{ij,j}=0,\ \sigma^*_{ij}n_j=0 \tag{5}$$

and compatibility

$$R_{pq}=e_{pki}e_{qlj}(\epsilon^e_{ij}+e^*_{ij})_{,kl}\equiv 0. \tag{6}$$

The form of Eq. (3) suggests modeling of the eigenstrains simply as thermal strains imposed in a finite element analysis. However, the welding procedures and weld/component geometry generally combine to produce a strongly, non-isotropic field ϵ^*_{ij}. To accommodate these conditions, we adopt the procedure suggested by Hill and Nelson [13]: (1) impose a unit temperature increase everywhere on the model, and (2) define a spatial distribution of anisotropic thermal expansion coefficients, a_{ij}, such that $a_{ij}=\epsilon^*_{ij}$ at each material point.

The eigenstrain approach described here proves especially convenient to model the effects of residual stresses on fracture mechanics parameters (K_I,J) as follows. To measure residual stresses, common experimental procedures employ various cutting or hole drilling methods to measure (mechanically) specific components of the strains ϵ^e_{ij}, from Eq. (4). Unfortunately, the cutting process frequently alters the distribution of residual stresses being measured. However, the (ideal) mechanical sectioning of the body does *not* alter the eigenstrains [12]. Hill and Nelson [13] exploit this key feature of the eigenstrains to develop a coupled experimental-computational method to infer the actual residual stress distribution in the body. They solve the inverse problem to find e^*_{ij} corresponding to measured values of ϵ^e_{ij} produced by mechanical sectioning. In the present fracture analyses, this same feature of the eigenstrains simplifies the modeling procedure, i.e., the introduction of a crack with traction free surfaces does not alter the specified eigenstrains for the welded component. The same (final) stress field and thus fracture parameters result by (1) imposing ϵ^*_{ij} and then "cutting" in the crack, or (2) starting with a model that contains the crack and imposing ϵ^*_{ij}; this second approach provides the more straightforward numerical procedure.

2.2. Eigenstrain functions for the beam-column, lower-flange weld

The lower-flange welds in beam-column connections and the same welds in simpler pull plate specimens pose a difficult configuration to model the residual stresses. Difficulties include: (1) the non-symmetric single-vee geometry, (2) the presence of a backup bar left in-place and (3) the influence of an adjacent (large) column flange. Results of 2-D, thermo-mechanical simulations of the welding procedure, performed by Zhang and Dong [7] for similar geometrical configurations, provide qualitative spatial features of the residual stress field, especially the through thickness variation of longitudinal stress in the beam flange. As a starting point to construct the eigenstrain field that generates an approximation to the Zhang and Dong [7] residual stresses, we adopt the symmetric eigenstrain field proposed by Hill and Nelson [13]. This eigenstrain field generates residual stresses applicable for a long, parallel-sided (butt) weld in a flat plate produced by a multi-pass welding process. Panontin and Hill [14] modified these simple eigenstrain functions to model residual stresses in double-vee circumferential welds for large diameter pipes. Their final eigenstrain field generates through-wall, axial and hoop residual stresses symmetric about the mid-thickness with tension on the inner and outer surfaces that matches commonly measured residual stress profiles [15]. With these eigenstrain fields available, this section describes our proposed modifications to accommodate the non-symmetric weld geometry, the backup bar and the thick column flange. The resulting eigenstrain fields suitable for use in 3-D finite element analyses have the form $\epsilon^*_{ij}=[kf(x,y,z)]_{ij}$, where k represents an amplitude parameter to scale the intensity of the corresponding residual stresses.

Fig. 2 describes the 2-D eigenstrain functions developed by Hill and Nelson [13] applicable for very long (uniform) and symmetric groove welds in flat plates. The rectangular parellelopiped, $[-a,a]\times[-b,b]\times[-c,c]$ where $a\gg(b,c)$, defines the region of non-zero eigenstrains with (x_0,y_0,z_0) located at its center. This region extends outside the weld to include heat affected zones (HAZs), base metal and possibly any reinforcing cap (Hill and Nelson [13] used $b=2c$). Normalized coordinates (ξ,η,ζ) introduced at this point simplify subsequent expressions for non-symmetric welds with backup bars in the structural connections,

$$\xi=\frac{x-x_0}{a},\ \eta=\frac{y-y_0}{b},\ \zeta=\frac{z-z_0}{c}. \tag{7}$$

Hill and Nelson [13] define the two symmetric functions f_1 and g_1 by

$$f_1(\eta)=\tfrac{1}{2}(1+\cos\pi\eta) \tag{8}$$

$$g_1(\zeta)=\tfrac{1}{2}(1-\cos\pi\zeta) \tag{9}$$

and the two anti-symmetric functions f_2 and g_2 by

$$f_2(\eta)=\sqrt{3}(1-|\eta|)\sin\pi\eta \tag{10}$$

$$g_2(\zeta)=\tfrac{1}{2}\sin\left(\frac{\pi\zeta}{2}\right). \tag{11}$$

Fig. 2(b) shows these four functions. They provide the basis to construct eigenstrains that exhibit: (1) no dependence on ξ, (2) deformational symmetry about η,ζ, (3) residual stresses that vanish away from the weldline, and (4) symmetric normal strains and anti-symmetric shear strains. Thus, Hill and Nelson [13] suggested the following forms for ϵ^*_{ij} applicable in this simple geometry

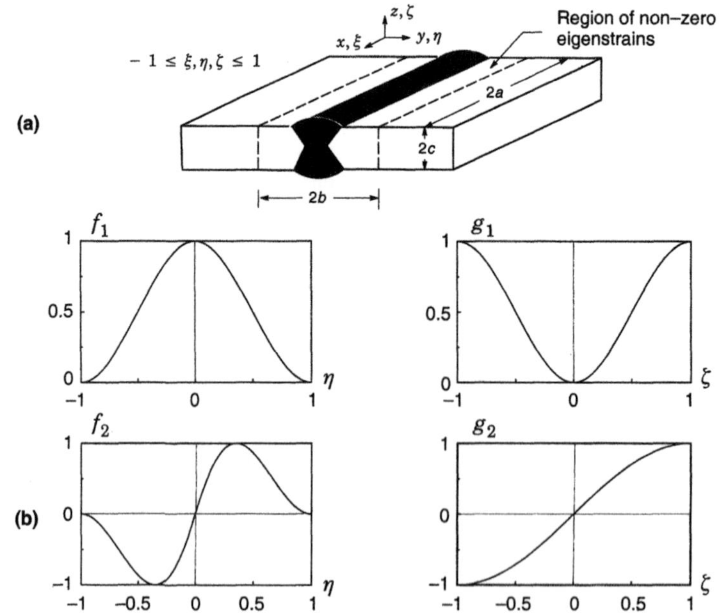

Fig. 2. (a) Welded plate with a symmetric double-vee groove (b) 2-D eigenstrain functions suggested by Hill and Nelson applicable when $a\gg c$ [13].

$$\epsilon_{11}^*=\epsilon_{22}^*=\epsilon_{33}^*=-\frac{\sigma^*}{E}f_1(\eta)g_1(\zeta) \tag{12}$$

$$\epsilon_{23}^*=-\frac{\sigma^*}{E}f_2(\eta)g_2(\zeta) \tag{13}$$

$\epsilon_{12}^*=\epsilon_{13}^*\doteq 0$ (due to the weld length relative to plate thickness) (14)

where σ^* denotes a convenient stress measure to scale the resulting field of residual stresses. Hill and Nelson [13] and Panontin and Hill [14] demonstrate through axisymmetric and 3-D finite element analyses the realistic form of residual stresses induced by these eigenstrain functions.

To introduce the effect on σ_{22}^* of a finite width beam flange, $2a = b_f$, experimental studies [15] suggest the use of a simple parabolic function

$$h_1(\xi)=1-\frac{\xi^2}{c_h} \tag{15}$$

which provides a peak σ_{22}^* residual stress at the flange mid-width, ξ=0. The coefficient, c_h, introduced above controls the eventual σ_{22}^*, for example, at $\xi = \pm 1$. c_h is taken here as 2.0 to reflect the restraining effects of the column flange during welding. The 2.0 value maintains tensile σ_{22}^* stresses at the outside edges of the beam flange, which would be generally compressive in a simple plate with no column flange present. With this modification, ϵ_{22}^* becomes

$$\epsilon_{22}^*=-\frac{\sigma^*}{E}C_2h_1(\xi)f_1(\eta)g_1(\zeta) \tag{16}$$

where the new scaling parameter C_2 enables specification of different maximum values for each normal strain ϵ_{ii}^* (Hill and Nelson [13] used equal values for the three normal components in their exploratory work). The product of functions $h_1(\xi)f_1(\eta)g_1(\zeta)$ has a maximum value of one.

For the lower-flange weld in the structural connection (see Fig. 3 for details), further modifications become necessary to reflect: (1) the distortion of single-vee welds not modelled by the symmetric Hill and Nelson [13] functions [see Fig. 4(b)]; and (2) the region of non-zero eigenstrains takes on a trapezoidal shape in the (η,ζ) plane and extends partially into the backup bar, into the column and into the weld cap [see Figs. 4 (c) and (d)]. The trapezoidal eigenstrain region introduces additional parameters required to define the shape (θ_1,θ_2) as shown in Fig. 4(d) with the top edge of the trapezoid (ζ=1) assigned the length $2b$. Reference points A, B are located at mid-sides of the top and bottom edges. The function f_1 now depends on both (η,ζ) but its *shape* retains the form $f_1(\eta,\zeta)\propto\frac{1}{2}(1+\cos[\pi\lambda(\eta,\zeta)])$, where $\lambda(\eta,\zeta)$ represents a simple shifting and scaling function, i.e., f_1 becomes very small away from the line A-B (vanishing at the edges η=1) with maximum values at the center (see Fig. 4(d)). Functions g_1, f_2 and g_2 remain unchanged from those of Hill and Nelson [13].

At this point we performed preliminary analyses of the pull-plate specimen to compute residual stress fields for various forms of $\lambda(\eta,\zeta)$, and the eigenstrain region shape/size and position relative to the weld. The qualitative form of the Zhang and Dong [7] residual stress fields obtained from weld process simulations helped to guide these computations. After some iterations, the following form for the modified f_1 seems satisfactory

$$f_1(\eta,\zeta)=\begin{cases}\frac{1}{2}\left[1+\cos\left(\pi\frac{\eta-\bar{\eta}(\zeta)}{\eta^*(\zeta)}\right)\right] & \textit{iff}\left|\frac{\eta-\bar{\eta}(\zeta)}{\eta^*(\zeta)}\right|\leq 1\\ 0 & \text{otherwise}\end{cases} \tag{17}$$

where $\bar{\eta}(\zeta)$ applies a (horizontal) shift and $\eta^*(\zeta)$ applies a (horizontal) stretching to reflect the changing width of the trapezoid. $\bar{\eta}(\zeta)$ defines an equation for the line of peak values of the key function f_1. Fig. 4(c) suggests, and the full pull-plate analyses confirm, that both of these two functions can be linear in ζ. These functions have non-zero values in the trapezoidal (dashed) region shown in Fig. 4(d), represented by the above inequality.

From Fig. 4(c) and simple trigonometric arguments, $\bar{\eta}(\zeta)$ and $\eta^*(\zeta)$ then have the forms

$$\bar{\eta}(\zeta)=\frac{1}{2}\zeta\left(\frac{1}{\tan\theta_1}-\frac{1}{\tan\theta_2}\right) \tag{18}$$

$$\eta^*(\zeta)=\frac{1}{2}\left(2-\frac{1}{\tan\theta_1}-\frac{1}{\tan\theta_2}\right)+\frac{1}{2}\zeta\left(\frac{1}{\tan\theta_1}+\frac{1}{\tan\theta_2}\right) \tag{19}$$

where θ_1 and θ_2 denote angles for sides of the trapezoid. For the beam-column connections and the pull-plate specimens, we simply take θ_2=90°.

A final set of calibration analyses yield recommended values for the geometric and amplitude scaling parameters. The criteria adopted to select the best values include: (1) a description for σ_{22}^* over the flange thickness on the (x=0) symmetry plane that closely matches the Zhang and Dong [7] weld simulation (2-D) values [see Fig. 7(b) and (c)]; (2) reproduction of the typically observed profiles for σ_{22}^* along the longitudinal (x) and transverse (y) directions for the weld line [6]; and (3) a stress σ_{33}^* that follows commonly measured profiles [15] for thick plates (t_f>25 mm). Key geometric parameters of the eigenstrain region emerged as: a=b_f/2, b=t_f, and c=0.6t_f. The thickness of the eigenstrain zone, $2c$=1.2t_f, exceeds the flange thickness and includes an affected part of the backup bar and weld cap, see Fig. 4(c). The absolute position of the trapezoidal region has x_0=0, y_0=0.7t_f and z_0=0.5t_f, where these offsets are relative the (x,y,z) system having the origin located at the intersec-

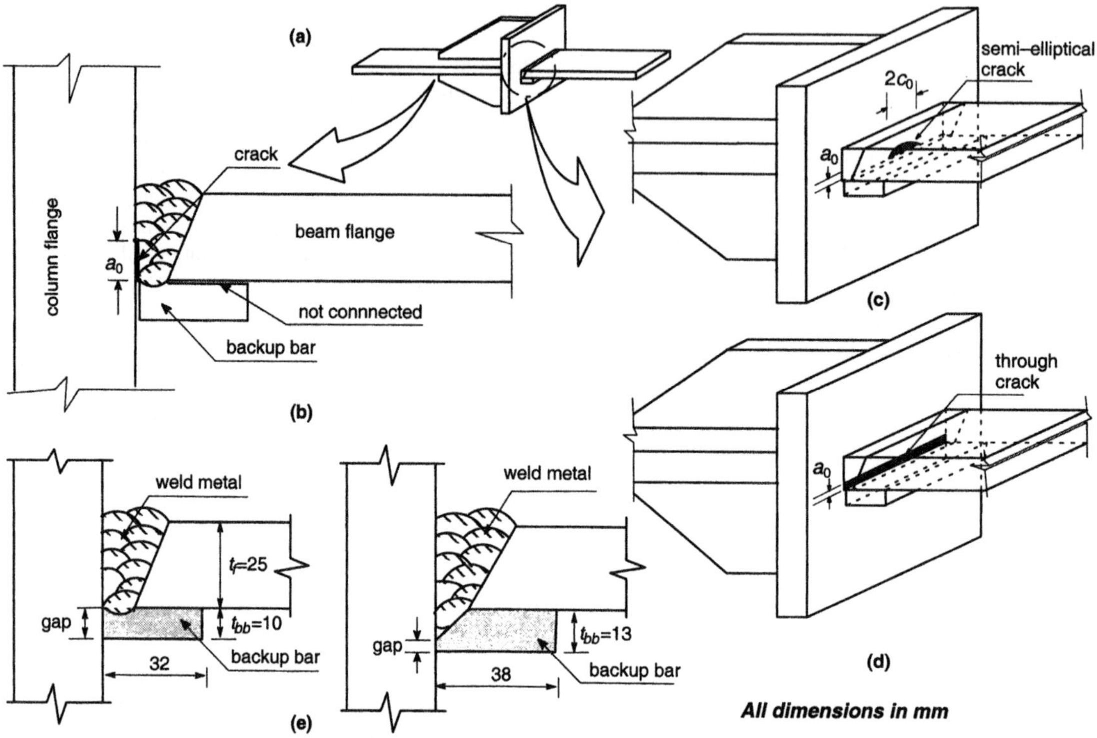

Fig. 3. (a) Schematic of pull-plate specimen; (b) weld geometry showing conventional backup bar (c) a semi-elliptical crack centered horizontally across the flange width that extends beyond the backup bar gap; (d) a constant depth through crack that extends beyond the backup bar gap; (e) pre-Northridge backup bar configuration and a proposed "grooved" configuration.

tion of the lower surface of the beam flange bottom surface and column flange center as indicated in Fig. 4(c). The assigned z_0 location proves crucial to ensure tension residual stress, σ_{22}^*, at the crack-tip location along the backup bar/flange depth (and thus crack opening rather than closing). Finally, the best value for θ_1 is given by $\tan\theta_1=5.5\ (c/b)$ for the pull-plate (weld) configuration.

With these geometric values, the shifting and stretching functions, $\bar{\eta}(\zeta)$ and $\eta^*(\zeta)$, have the final forms

$$\bar{\eta}(\zeta)=0.15\zeta \tag{20}$$

$$\eta^*(\zeta)=0.85+0.15\zeta \tag{21}$$

and the inequality in Eq. (17) becomes

$$-0.85\leq\eta\leq0.85+0.30\zeta.$$

Evaluation of the above equations at $\zeta=\pm1$ shows that: the top width of the trapezoidal region of non-zero eigenstrains is $2b+2t_f$, the bottom width is $1.4\times t_f$, and the eigenstrain zone extends a distance of $0.15\times t_f$ into the column flange. The "global" scaling amplitude, σ^*, appearing in all ϵ_{ij}^* terms, and the coefficients C_1 (for ϵ_{11}^*) and C_2 generally have values assigned to enforce maximum residual stresses equal to the weld metal yield stress (σ_{ys}^{wm}). If $\sigma^*=\sigma_{ys}^{wm}$, then $C_1=0.75$ and $C_2=1.25$ for this weld-flange geometry.

In summary, the eigenstrain functions to generate a representative residual stress field for analysis of the lower-flange, beam-column connection and the simple pull-plate specimen are

$$\epsilon_{11}^*=-\frac{\sigma^*}{E}C_1f_1(\eta,\zeta)g_1(\zeta) \tag{23}$$

$$\epsilon_{22}^*=-\frac{\sigma^*}{E}C_2h_1(\xi)f_1(\eta,\zeta)g_1(\zeta) \tag{24}$$

$$\epsilon_{33}^*=-\frac{\sigma^*}{E}f_1(\eta,\zeta)g_1(\zeta) \tag{25}$$

$$\epsilon_{23}^*=-\frac{\sigma^*}{E}f_2(\eta,\zeta)g_2(\zeta) \tag{26}$$

$$\epsilon_{12}^*=\epsilon_{13}^*=0 \tag{27}$$

with functions h_1, f_1, g_1, f_2, and g_2 defined by Eqs. (15, 17, 9–11). The scaling stress applied to all values is $\sigma^*=\sigma_{ys}^{wm}$. The calibrated coefficients then have values as follows

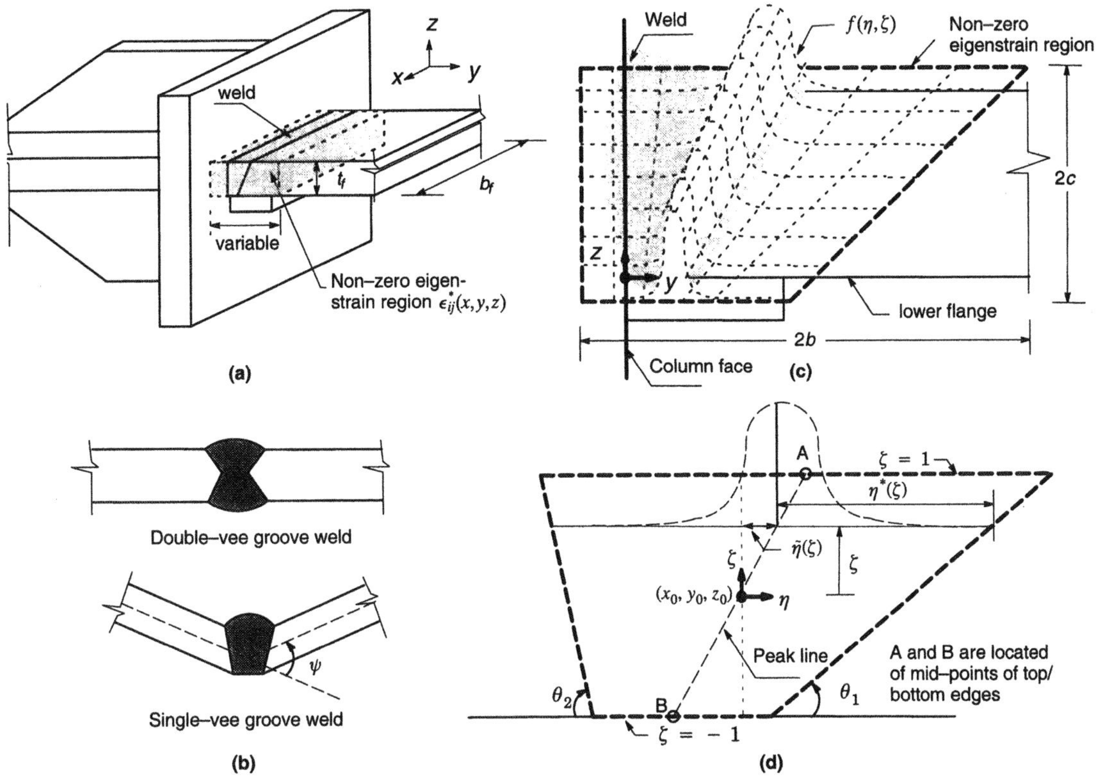

Fig. 4. (a) Location on eigenstrain region; (b) deformation induced by single-vee weld; (c) trapezoidal eigenstrain region and parametric coordinates (note origin of *X-Y-Z* coordinates used to define size/positioning of eigenstrain region and stress components); (d) geometric parameters to define shape/size of eigenstrain region and the shifting/scaling functions.

$$C_1=0.75,\ C_2=1.25 \tag{28}$$

$$a=\frac{b_f}{2},\ b=t_f,\ c=0.6t_f \tag{29}$$

$$\theta_1=\frac{\pi}{2},\ \tan\theta_2=5.5\left(\frac{c}{b}\right) \tag{30}$$

$$x_0=0,\ y_0=0.7t_f,\ z_0=0.5t_f. \tag{31}$$

2.3. *3-D finite element analyses including eigenstrains*

Anisotropic thermal expansion coefficients, a_{ij}, provide a simple mechanism to introduce eigenstrains into the computational model. A spatially uniform temperature change $\Delta\Theta(x,y,z)\equiv$ constant is imposed everywhere in the model such that

$$\epsilon^*_{ij}(x,y,z)=\Delta\Theta a_{ij}(x,y,z) \tag{32}$$

where a unit increase $\Delta\Theta=1°$ leads simply to $a_{ij}(x,y,z)=\epsilon^*_{ij}(x,y,z)$. Based on the (x,y,z) coordinates of the element center in the usual isoparametric coordinates, each element in the model is assigned a single set of values a_{ij}.

Pointwise values for the Mode I stress intensity factor, K_I, along crack fronts are computed from values of the J-integral given by [17,18]

$$J=\lim_{\Gamma\to 0}\int_{\Gamma}\left[Wn_1-\sigma_{ij}\frac{\partial u_i}{\partial x_1}n_j\right]d\Gamma \tag{33}$$

where W denotes the stress-work density per unit volume, Γ is a vanishingly small contour that lies in the principal normal plane at location s on the front, n is the unit vector normal to Γ, and σ_{ij} denotes the usual (Cauchy) stress tensor that is work conjugate to the displacement gradient, $\partial u_i/\partial x_j$. All field quantities here are expressed in a local orthogonal coordinate system (x_1,x_2,x_3) at location s on the crack front. This important result considers only mechanical energy balance for a local translation of the crack front in the x_1 direction (Mode I). Any form of loading (including temperature changes) and arbitrary material behavior (anisotropic thermal response) are permitted when $\Gamma\to 0$. Moran and

Shih [19,20] have proven the *local* path independence of J on the actual shape of Γ in the limit as $\Gamma \to 0^+$.

The geometric difficulties encountered in defining a contour Γ that passes through integration points in a 3-D finite element model led Li et al. [21] to develop a domain integral approach to evaluate Eq. (33); values of J computed over volumetric domains enclosing a point s on the crack front exhibit domain independence. Shih et al. [22] describe extensions of the formulation to include the effects of temperature loading for thermally isotropic materials. Appendix A here describes additional modifications required to maintain domain independence of J-values for the thermally anisotropic materials needed in this work to model residual stresses.

Away from traction-free ends of the crack front, pointwise K_I values follow by using a plane-strain conversion such that

$$K_I^{mech} = \sqrt{\frac{EJ_{mech}}{(1-\nu^2)}} \qquad K_I^{rs} = \sqrt{\frac{EJ_{rs}}{(1-\nu^2)}} \tag{34}$$

where *mech* and *rs* denote the separate effects of applied mechanical loading and residual stresses, respectively. For the linear-elastic conditions analyzed here, the total K_I is then

$$K_I = K_I^{mech} + K_I^{rs}. \tag{35}$$

Finite element analyses of the pull-plate specimen to compute only the effects of residual stresses have one remote end with specified zero displacements. The computed zero values for reactions at those nodes verify the self-equilibrating residual stress field in the weld region (far from the fixed end).

3. Geometric configurations studied and analysis procedures

3.1. Specimen and crack configurations

Inspections of welded connections in buildings and of full-scale joints fabricated under laboratory conditions [16] reveal a range of crack-like defects along the root pass of the weld. The most commonly used flux cored arc welding procedures employ pre-heating but no post-weld treatments [23]. In construction practice prior to the 1994 Northridge earthquake, backup bars were routinely left in-place during service. Lack-of-fusion type defects of variable depth can extend the full width of the weld. Larger defects can be found locally at the discontinuity introduced by the requirement of welding from both sides of the beam web through an access hole. Post-fracture inspections have reported crack-like defects in the welds with depths up to 10 mm (in addition to the backup bar thickness). Even when no weld defects exist, the unfused gap between the backup bar and the column flange creates a crack-like defect as illustrated in Fig. 3(b).

These observations led us to study the following geometric and crack configurations for the pull-plate specimen (shown in Fig. 3):

- specimens having the commonly used, pre-Northridge backup bar design shown in Fig. 3(e), a "grooved" backup bar design that reduces the initial crack depth (shown in the same figure), and a configuration with the backup bar removed;
- a uniform depth, sharp crack that extends the full width of the beam flange with depth equal only to the unfused gap between the backup bar and the column face (9.5 mm for the older, rectangular backup bar design and 3.2 mm for the grooved design);
- a uniform depth, sharp crack that extends the full width of the beam flange with depth equal to the unfused backup bar gap (9.5 or 3.2 mm) plus an additional a_0=5.7 mm to reflect a typical bead size for the root pass;
- a semi-elliptical, sharp crack of depth a_0=5.7 mm (a_0/t_f =0.225) and aspect ratio $2c_0/a_0$=6 located at the center of weld (in addition to the unfused backup bar gap that extends the full width of the beam flange).

The backup bars have a 10 mm fused length with the beam flange. In practice, variations in beam lengths/alignments and column alignment frequently alter this (nominal) dimension.

In addition to these configurations, we also analyzed various constructions of the pull-plate specimen, for example, the beam flange plate only with and without backup bar. Fig. 5 defines six such cases. Case 3 identifies the pull-plate specimen as tested with the backup bar removed; Case 4 identifies the full pull-plate specimen including the backup bar.

3.2. Loading and boundary conditions

Computed values for the stress intensity factors are presented for two loadings: (1) residual stresses acting alone and (2) a remotely applied tension acting alone. For both loadings, Fig. 5(a) illustrates the displacement boundary conditions applied at the remote ends of the pull-plate specimen. These idealized conditions approximate the fixed-grip testing arrangement reported by Kaufmann and Fisher [8].

Results for the residual stress loading thus represent the following sequence: (1) the cracked specimen model is placed into the fixed-grip condition, (2) eigenstrains are imposed to produce residual stresses. This sequence approximates a post-weld straightening procedure that reduces welding induced distortions. However, analyses conducted with one end of the specimen fully restrained and the other fully free show that restraint of the poten-

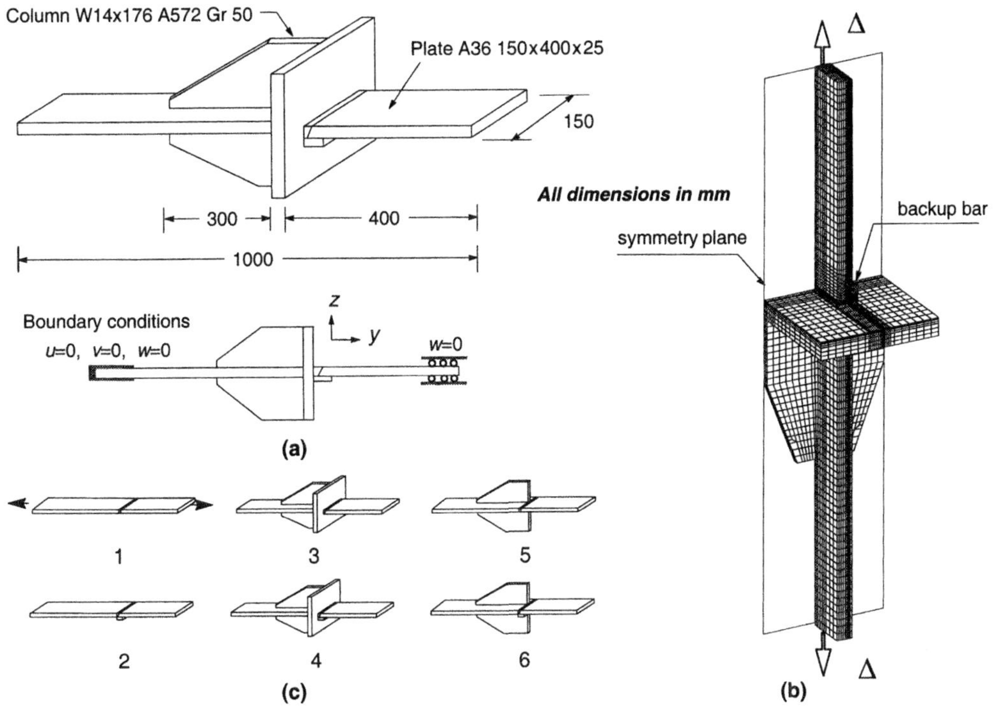

Fig. 5. (a) Pull-plate specimen with dimensions; (b) 3-D finite element model (a one-quarter mesh is actually used in analyses); (c) definitions for the six (6) "cases" analyzed and displacement boundary conditions imposed in the analyses.

tial distortions lowers the K_Γ-values by less than 3%. To support consistent comparisons of the computed K_Γ-values, the applied eigenstrain field defined by Eqs. (23)–(31) remains fixed for all six of the geometric configurations analyzed.

As indicated in Fig. 5(a), uniform axial (v) displacements applied over the flange thickness at the ends impose the remote tension loading. Corresponding stress intensity factors are normalized by the resulting average axial stress that develops remote from the weld region.

3.3. Finite element models

Fig. 6 illustrates key features of the near-tip regions of the finite element models. Fig. 5(b) shows the meshing away from the weld region. The heavy lines indicate the non-fused regions between the backup bars and the beam/column flanges. The 3-D models are constructed using eight-node isoparametric elements with the $\bar{B}$ formulation to suppress volumetric locking. The symmetric geometry enables modeling only one-half of the pull-plate specimen as shown in Fig. 5(b). All crack tip meshes have 16 elements defined in the crack front θ direction ($0 \leq \theta \leq 2\pi$) as shown in Fig. 5(d). A minimum of 15 concentric rings of elements with increasing size extend outward from the tip at each location along the front. This supports J-integral evaluation over a large number of domains at each crack front position to verify the path independence. Meshes typically have 20–30 000 elements depending on the backup bar and crack configuration. For these linear elastic analyses, mechanical constitutive properties for the plates and weld are taken to be isotropic with Poisson's ratio of 0.3.

To check the adequacy of this mesh refinement, K_Γ-values are compared to the Newman and Raju [24] expressions for our Case 1 in Fig. 5(c) (a semi-elliptical crack in a simple flat plate subjected to axial tension). The differences lie below 2% along the crack front. Similarly, analyses for a more refined set of meshes than those shown in Fig. 5 reveal less than a 1% change in displacements, stresses and K_Γ-values.

Analyses are performed using the WARP3D fracture mechanics code [25]. This code supports linear and non-linear fracture analysis of 3-D solids with fast solvers on parallel computers, crack growth analyses, and a very general domain integral capability to compute J-values including thermal, inertial and crack face loadings. Each analysis requires 1–2 minutes on a desktop HP (Unix) workstation.

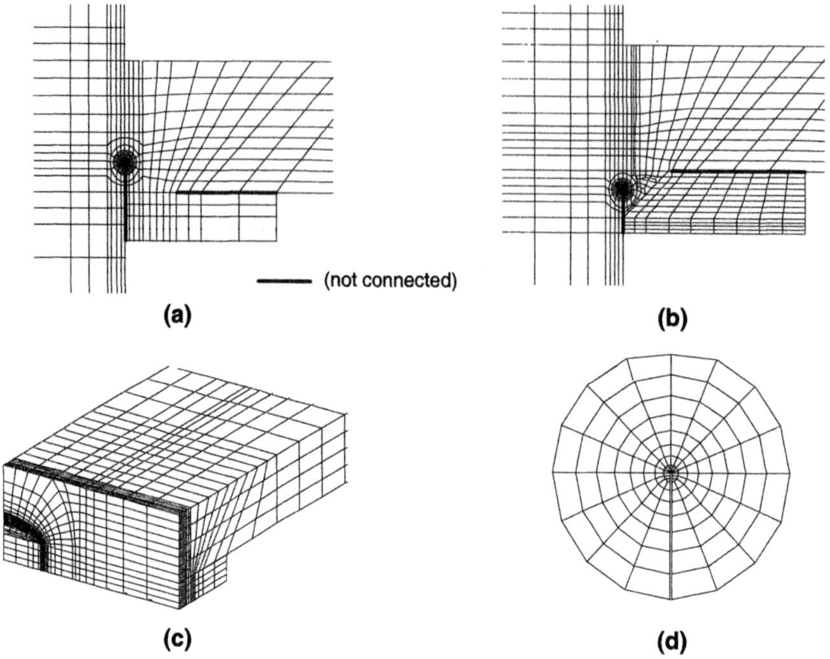

Fig. 6. Features of the near-tip finite element meshes for each pull-plate configuration analyzed. (a) the pre-Northridge backup bar design; (b) the alternative "grooved" backup bar design; (c) the semi-elliptical crack centered on the beam flange; (d) typical mesh refinement at the crack tip.

4. Results and discussion

4.1. Residual stress fields

Figs. 7 and 8 summarize residual stress fields computed for the pull-plate configuration with the pre-Northridge backup bar. These figures show stress components plotted along various reference lines in the weld region. The origin of the coordinate system to locate these lines lies at mid-width of the beam flange where the bottom surface of the plate intersects the column flange. Fig. 7(a) compares the profile of transverse (σ_{yy}) stress over the beam flange thickness at mid-width obtained by Zhang and Dong [7] and the profile obtained by the present eigenstrain approach. The sharp breaks in the Zhang and Dong [7] profile arise from their bead-by-bead simulation of the welding procedure which used 9 beads. The models have no crack beyond the backup bar gap. The profile produced by the eigenstrain approach yields an expected smooth curve that follows both the general shape and amplitude of the Zhang and Dhong profile [7]. Fig. 7(b) shows a fringe plot of the σ_{22} stress for the eigenstrain model over the general weld region on the x=0 plane. All of these plots indicate clearly the strong and localized transverse tensile stress created by the backup bar gap. The computed elastic stress values at the gap termination point (z=0) for both models simply reflect the relative levels of mesh refinement.

Fig. 7 shows the key distributions of computed residual stresses in the uncracked configuration (but with the standard backup bar in place). Each plot has a pull-plate schematic with a heavy line drawn for reference. The longitudinal stresses (σ_{xx}) shown in Fig. 8(a) along the top surface of the beam flange (x=0, z=t_f) reach maximum values equal to the weld metal yield stress [recall from Section 2.2. that σ^*=σ_{ys}^{wm} sets the overall amplitude of the imposed eigenstrain field specified in the analysis]. The peak values develop very near the fusion line. The typical longitudinal stress field for a symmetric weld in a flat plate [15] without a crack is provided for comparison. Quite good agreement is evident for the tensile part of the response; the pull-plate specimen reveals a negligible compressive region most likely due to the influence of the column flange.

The variation of transverse stress along the top edge of the beam flange (y=0, z=t_f) is shown in Fig. 8(b). Here, the influence of the h_1 function appearing in Eq. (24) becomes apparent to define a general parabolic reduction of values from a maximum at the center flange location. Figs. 8(c) and (d) provide the through-thickness variations for the longitudinal and perpendicular stresses at the flange mid-width location (x=0). Stress fields on y=0 exhibit a strong singularity at z=0 created by the crack-

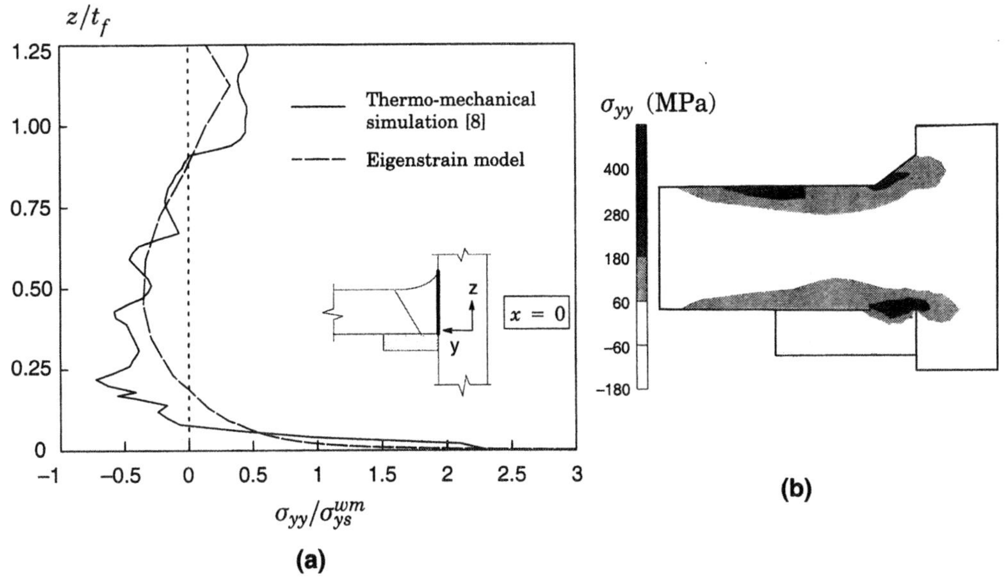

Fig. 7. Distribution of stress transverse to the weld (σ_{yy}) from thermo-mechanical simulation [7] and present eigenstrain model; (a) over depth of flange at flange-column interface as indicated; (b) stress field σ_{yy} in weld region from eigenstrain model.

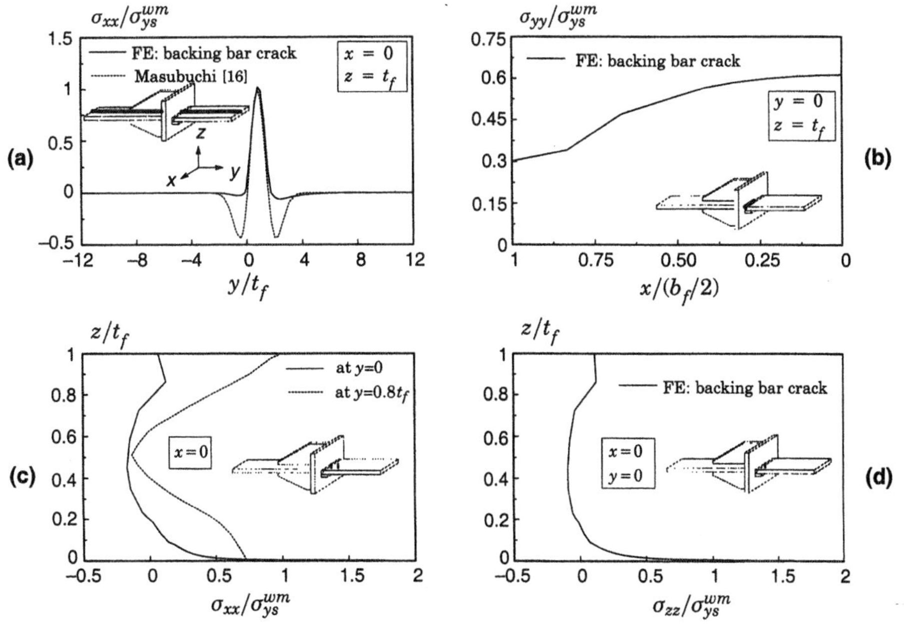

Fig. 8. Residual stresses computed using 3-D eigenstrain model for the pre-Northridge backup bar configuration (heavy lines provide location): (a) stress σ_{xx} along the specimen axis; (b) stress σ_{yy} along flange width; (c) stress σ_{xx} over flange depth; (d) stress σ_{zz} along flange depth.

like discontinuity from the backup bar. In Figs. 8(a) and (c), the restraining effect of the column flange clearly lowers the otherwise strong (longitudinal) tensile stress on the top surface of the flange (compare with the longitudinal stress shown at y=0.8t_f). The perpendicular stress (σ_{zz}) decreases to zero at the flange top surface as required by the traction free boundary conditions.

4.2. Stress intensity factors

The maximum stress intensity factor for all geometric configurations and both loadings (applied axial tension force and residual stresses) occurs at the point of maximum crack depth on x=y=0 (mid-width of the beam flange). To simplify discussions, the various geometric configurations are identified using the numbers assigned in Fig. 5. The average value of the J-integral computed over 15 domains at this location varies less than 1% from the average. Application of Eq. (34) yields the K_I-value using this average J-value. The K_I-values for the different configurations and loadings are normalized in the usual way using a (non-dimensional) geometric factor, Y, such that

$$Y=\frac{K_I}{\bar{\sigma}\sqrt{\pi\bar{a}}} \tag{36}$$

where $\bar{\sigma}$ and $\bar{a}$ denote suitable measures for the nominal applied stress and for the crack length, respectively. Y factors for the axial tension loading are denoted Y_{mech} and those for residual stress loading by Y_{rs}. For the axial tension loading, $\bar{\sigma}$ equals the average stress on the beam flange remote from the crack plane (computed numerically using reaction forces that arise from the enforced axial extension). For the eigenstrain loading to generate the residual stresses, $\bar{\sigma}$ equals the global scaling parameter, σ^*, appearing in Eqs. (23)–(26). The crack length measure used in Eq. (36) takes on the following values:

- when only the gap created by the backup bar provides the effective crack, $\bar{a}$ equals the length of the unfused gap between the backup bar and the column face. For the pre-Northridge (rectangular) backup bar, $\bar{a}$=t_{bb}, while for the grooved backup bar, $\bar{a}$=t_{bb}/4 (see Fig. 3(e));
- when an additional crack exists, $\bar{a}$ denotes the depth of the additional crack only, a_0, and does not include the length of the unfused gap between the backup bar and the column face.
- For the semi-elliptical crack, a_0 denotes the maximum depth;
- in all cases examined here, a_0=5.7 mm, i.e., a_0/t_f=0.225.

The Y values shown here apply only to these specific geometric dimensions and for those other configurations obtained by uniform, geometric scaling of the entire specimen.

Consider first the residual stress loading imposed on the uniform, through-width crack with the pre-Northridge backup bar (Case 4). Zhang and Dong [7] analyzed a similar case in their welding simulation using a 2-D, plane strain model but with a_0=2.5 mm (compared to 5.7 mm here). To gauge the relative significance of stress intensity factors caused by the residual stresses, consider the pull-plate with dimensions shown in Fig. 3(e) and with $\sigma^*=\sigma_{ys}^{wm}$=414 MPa. The corresponding K_I-values are 20 MPa $\sqrt{m}$ for the Zhang and Dong [7] analysis and 23 MPa $\sqrt{m}$ for the present eigenstrain analysis. The reported toughness levels for the commonly used pre-Northridge weld metal (E70T-4) range from 44–66 MPa $\sqrt{m}$ at the service temperature [16]. Consequently, the residual stresses alone consume a major portion of the available fracture toughness prior to application of the structural load. We experimented with a range of the geometric parameters to define the residual stresses fields, i.e., θ_1 and θ_2, to assess the relative sensitivity of the eigenstrain approach. Stress intensity factors varied at most 20%.

To check the K_I-values for the through-width crack configuration, we performed analyses using 2-D models for the pull-plate. The 2-D Y-values for comparison with the 3-D values listed in Table 1 are summarized as follows (3-D values from Table 1 are given in square brackets []): (a) Case 3 mechanical loading 1.56 [1.65], (b) Case 4 mechanical loading 1.56 [1.66], and (c) Case 4 residual stress loading 0.36 [0.40]. In addition, standard handbooks provide Y=1.43 for the through-width crack of Case 1 under mechanical loading compared to Y=1.41 from the present 3-D analyses. These small differences under linear-elastic conditions between 2-D and 3-D analyses are not surprising. Once plasticity develops, significant differences can be expected between 2-D and 3-D model predictions of constraint effects near the mid-width of the flange.

Table 1
Stress intensity geometry factors for the pre-Northridge, pull-plate specimen

Case	Mechanical loading (Y_{mech})		Residual stress loading (Y_{rs})	
	Through crack	Semi-elliptical crack	Through crack	Semi-elliptical crack
1	1.41	0.98	0.39	0.24
2	1.43	0.88	0.40	0.19
3	1.65	1.03	0.40	0.20
4	1.66	0.94	0.40	0.14
5	1.94	1.46	0.41	0.24
6	1.97	1.39	0.41	0.19

Table 1 summarizes the geometric Y factors obtained from 3-D analyses for axial tension and residual stress loadings applied to the pull-plate specimen having the pre-Northridge characteristics. The following features are observed:

- the uniform, through-width crack represents the most severe case for all configurations and for both loadings;
- K_I-values caused by the residual stresses for the semi-elliptical crack are roughly half those for the uniform, through-width crack—the relative difference for the two crack types is somewhat less for the tension loading;
- Cases 3 and 4 for the through-crack highlight the negligible influence of the backup bar for this configuration (the effect would be greater for a smaller a_0/t_f ratio where the crack tip lies in the stress concentration caused by the backup bar gap) — Zhang and Dong [7] observe very similar trends;
- Cases 3 and 4, compared to 5 and 6, illustrate the strong effect of the column flange to reduce stress intensity factors for the axial tension loading by providing a more uniform load path through the connection;
- Cases 3 (no backup bar) and 4 (backup bar present) for the semi-elliptical flaw indicate that the presence of the backup bar reduces the stress intensity factor for both mechanical and residual stress loadings. The continuous connection of the backup bar along the bottom flange (except over the length of the surface crack) does alter the longitudinal stress distribution (σ_{xx}) at the column face. The backup bar creates a stress concentration along the continuous connection thereby slightly decreasing the stress concentration at the maximum depth of the surface crack (from equilibrium requirements over the section). K_I-values thus decrease slightly at the center of the surface crack compared to the no backup bar configuration.

In all of the above cases, the backup bar has a 10 mm fused length to the beam flange. Two additional analyses of the pre-Northridge configuration (Case 4) for mechanical loading of the through crack show a negligible effect of increasing the fused length to 20 and 30 mm. In Table 1, Y_{mech}=1.66 for the 10 mm fused length; the additional analyses yield values of 1.65 for both the 20 and 30 mm fused lengths.

Table 2 provides the geometric Y factors for the pre-Northridge configurations with no crack other than the backup bar gap. Here, only Cases 4 and 6 apply. K_I-values show roughly a 50% reduction relative to values including the additional cracks listed in Table 1.

Table 2
Stress intensity geometry factors for the pre-Northridge, pull-plate specimen with no crack beyond the unfused gap between the backup bar and the column flange ($\bar{a}=t_{bb}$)

Case	Mechanical loading (Y_{mech})	Residual stress loading (Y_{rs})
4	0.69	0.21
6	1.04	0.27

Table 3
Stress intensity geometry factors for pull-plate specimens with the alternative, "grooved" backup bar design

Case	Mechanical loading (Y_{mech})		Residual stress loading (Y_{rs})	
	Through crack	Semi-elliptical crack	Through crack	Semi-elliptical crack
4	0.58	0.33	0.22	0.09
6	0.65	0.57	Not analyzed	Not analyzed

4.3. Alternative, "grooved" backup bar design

Tables 3 and 4 summarize the geometry factors for tension and residual stress loadings applied to the pull-plate specimen having the "grooved" backup bar design, see Fig. 3(e). Here, the values differ from values in Tables 1 and 2 only for Cases 4 and 6 that have the backup bar included. The grooved backup bar has been proposed as an alternative to the more expensive option of removing the backup bars altogether. The slightly larger, grooved bar re-positions the critical root pass, moving it away from the direct tensile force in the beam flange. The finite element models in Figs. 6(a) and (b) also show this re-positioning (the crack size in both backup bars remains 5.7 mm; a representative size of the root pass). When no crack forms in the root pass, the natural crack formed by the unfused gap between the backup bar and the column flange has a smaller size, $t_{bb}/4$ rather than t_{bb}, as a consequence of the groove.

For tension loading of the flange, Table 3 shows a

Table 4
Stress intensity geometry factors for pull-plate specimen having the alternative, "grooved" backup bar design with no crack beyond the unfused gap between the backup bar and the column face ($\bar{a}=t_{bb}/4$)

Case	Mechanical loading (Y_{mech})	Residual stress loading (Y_{rs})
4	0.38	0.14
6	0.56	Not analyzed

reduction of K_I-values to nearly 35% of the values for the standard backup bar. Both the through-width and semi-elliptical cracks exhibit this reduction. As is the case for the standard backup bar design, the through-width crack again creates larger K_I-values. Similarly, reduced K_I-values for the residual stress loading have only 50% of the values in Table 1.

Tables 2 and 4 enable comparisons between the two backup bar configurations when the crack occurs only from the gap created by the backup bar. Note that crack lengths for normalization of K_I-values differ for the two backup bar designs, $\bar{a}=t_{bb}$ vs. $t_{bb}/4$. The geometric Y-values thus do not provide direct comparisons of the changes in K_I-values, i.e., actual K_I-ratios must be computed using the provided Y-values and the actual crack lengths. Using the values in Tables 2 and 4 for the mechanical loading, the grooved backup bar K_I-values are only 30% of values for the standard backup bar design. For residual stress loading, K_I-values for the grooved bar are only 36% of values for the standard backup bar design, due to the lowered position of the crack front in the fixed eigenstrain fields used for all analyses.

The grooved backup bar design clearly shows promise to reduce the driving force for fracture due to tension loading in the beam flange. However, construction costs may become an issue as the "perfect" alignment of the backup bar groove and the lower-flange chamfer (prior to welding) may prove difficult to achieve, and the sensitivity to misalignment requires examination. Further study for the residual stress loading is warranted to determine if a residual stress field, estimated by welding simulations specifically for the grooved bar design, would alter the trends suggested by Table 4.

5. Summary and conclusions

This work develops an eigenstrain approach to impose a representative residual stress field on 3-D finite element models for the lower-flange welds of the type found in large, beam-column connections. The 3-D finite element models employed here incorporate the full geometry of the pull-plate specimen [8] designed to reduce the cost of testing full connections while retaining essential features of the fracture prone, weld region. The 3-D analyses examine pull-plate configurations that include: (1) a cracked root pass across the full width of the beam flange; (2) a localized, semi-elliptical crack in the root pass, centered at the beam web-lower flange intersection where a preponderance of lack of fusion defects has been observed; and (3) no cracks other than the "gap" between the backup bar and the adjacent column flange. The numerical results include tables of values for the non-dimensional "geometry" factors (Y) needed to compute stress intensity factors (K_I-values) for geometrically similar configurations.

The residual stress fields generated by the proposed eigenstrain functions closely match those computed using 2-D, thermo-mechanical simulation of the multi-pass welding process [7]. Stress intensity factors, obtained from corresponding J-integral computations, show clearly the significant fracture toughness demand placed on the material by residual stresses prior to application of the structural loads. The present analyses indicate that residual stresses alone impose K_I-values equal to approximately one-third of the available fracture toughness measured in typical weld metal of pre-Northridge connections.

For both axial force and residual stress loadings, the 3-D analyses demonstrate that the cracked root pass across the full beam flange represents a more severe condition than the localized, semi-elliptical crack in the root pass. This can simplify the future parametric studies of connection geometries since the through-crack configuration leads to far less mesh generation effort.

To eliminate the cost of removing backup bars altogether, an alternative design for a grooved backup bar shows considerable promise. The groove re-positions the weld root pass below the beam flange and thus outside the primary load path. Analyses here indicate approximately a 65% reduction in K_I-values for tension loading in the flange. However, reliable estimates for the effects of residual stresses on K_I-values await a more realistic description of residual stress fields specifically for the grooved backup bar geometry.

The linear-elastic analyses and results presented here represent our initial assessment of the fracture toughness demands in connections having these welds for expected residual stresses and structural loads. The next phase employs a probabilistic micro-mechanics model for cleavage fracture in fully nonlinear, 3-D analyses. Those studies focus on quantifying the coupled effects of residual stresses and axial force loading of the beam flange under a range of plastic deformation levels, loading rates and mismatch of flow properties between the weld-base metal. The eigenstrain approach developed in this work to incorporate residual stresses provides the necessary starting point for those studies.

Acknowledgements

This work was sponsored in part by the NASA-Ames Research Center through grants NAG 2-1031 and 2-1126. The authors wish to acknowledge the helpful and simulating discussions with Prof. Greg Deierlein of Stanford University, Dr. Pingsha Dong of Battelle Memorial Institute, Columbus, Ohio, and Prof. Mike Hill of the University of California-Davis.

Appendix A

A.1. Domain integral form for J in 3-D

In Eq. (33), the stress-work density (W) per unit volume may be defined in terms of the mechanical strains as

$$W=\int_0^t \boldsymbol{\sigma}:(\boldsymbol{\epsilon}-\boldsymbol{\epsilon}^{th})dt \tag{A1}$$

where $\boldsymbol{\sigma}$ denotes the (symmetric) stress tensor, $\boldsymbol{\epsilon}$ is the symmetric (small) strain tensor, $\boldsymbol{\epsilon}^{th}$ denotes the contribution arising from specified thermal strains, and t defines a monotonically increasing loading parameter.

By using the divergence theorem and a vector weight function denoted by $\boldsymbol{q}$, which may be interpreted as a virtual displacement field (or *virtual* crack advance), the contour integral of Eq. (33) is converted into an area integral for two dimensions and into a volume integral for three dimensions (Li et al. [21], Shih et al. [22]). Fig. 9(a) illustrates such a volume (i.e. *domain*) centered at point b between points a and c on the crack front. The resulting expressions can be written in the form:

$$\bar{J}_{a-c}=\int_{s_a}^{s_c}[J(s)q_t(s)]ds=\bar{J}_1+\bar{J}_2 \tag{A2}$$

where each integral is defined by

$$\bar{J}_1=\int_V\left(\sigma_{ij}\frac{\partial u_i}{\partial X_k}\frac{\partial q_k}{\partial X_j}-W\frac{\partial q_k}{\partial X_k}\right)dV \tag{A3}$$

$$\bar{J}_2=-\int_V\left(\frac{\partial W}{\partial X_k}-\sigma_{ij}\frac{\partial^2 u_i}{\partial X_j\partial X_k}\right)q_k dV \tag{A4}$$

Here, q_k denotes a component of the vector weight function in the k coordinate direction, $q_t(s)$ represents the resultant magnitude of the weight function at point s on the crack front, and V represents the volume of the domain surrounding the crack front. Inertial and body forces are assumed to be zero for simplicity. $J(s)$ thus defines the local energy release rate that corresponds to the crack front perturbation at s, $q_t(s)$.

The vector function $\boldsymbol{q}$ has direction parallel to the local crack growth extension. For Mode I, all field quantities can be transformed to the local crack front coordinate system at point s, such that only the q_1 term remains non-zero. In subsequent discussions, the k subscript on q terms is thus dropped with q alone implying the q_1 term.

The q-function must vanish on the surfaces A_1, A_2 and A_3 in Fig. 9(a) for the development of Eqs. (A2)–(A4) from Eq. (33). This requirement makes area integrals (line integrals in two dimensions) defined on these surfaces vanish. Fig. 9(b) shows the variation in amplitude of a valid q-function for the domain. The value of q at each point in the volume, V, is readily interpreted as the displacement of a material point due to the virtual extension of the crack front, $q_t(s)$.

An approximate value of $J(s_b)$ follows by application of the mean-value theorem over the interval $s_a \leq s \leq s_c$. The pointwise value of the J-integral at s_b is given by, see Fig. 9(b):

$$J(s=b)\approx\frac{\bar{J}}{A_q};\ A_q=\int_{s_a}^{s_c}q_t(s)ds \tag{A5}$$

where $\bar{J}$ is the energy released due to the crack-front perturbation, $q_t(s)$. The increase in crack-area corresponding to this perturbation, A_q, is simply the integral of $q_t(s)$ along the crack front from s_a to s_c. In the finite element implementation, points a, b, and c correspond to corner nodes for two elements along the front.

A.2. Thermal loading effects

The $\bar{J}_2$ integral vanishes for an elastic material (linear or nonlinear) in the absence of thermal strains. After exchanging the order of differentiation and using symmetry of σ_{ij}, the second term in Eq. (A4) is rewritten as:

$$-\sigma_{ij}\frac{\partial^2 u_i}{\partial X_j\partial X_1}=-\sigma_{ij}\frac{\partial}{\partial X_1}\left(\frac{\partial u_i}{\partial X_j}\right)=-\sigma_{ij}\frac{\partial \epsilon_{ij}}{\partial X_1}. \tag{A6}$$

The chain rule is now evoked to expand the first term in Eq. (A4), using small-displacement gradients. The derivative of strain energy density with respect to strain is the stress tensor for elastic materials. The result is:

$$\frac{\partial W}{\partial X_1}=\frac{\partial W}{\partial \epsilon_{ij}}\frac{\partial \epsilon_{ij}}{\partial X_1}=\sigma_{ij}\frac{\partial \epsilon_{ij}}{\partial X_1}. \tag{A7}$$

The two terms defining the integrand of $\bar{J}_2$ thus sum to zero for elastic materials in the absence of thermal strains (which leads to the path independence of the original J-contour integral).

Now consider the influence of initial strains from imposed thermal loading. Eq. (A6) remains unchanged; however, Eq. (A7) must be re-written more explicitly as

$$\frac{\partial W}{\partial X_1}=\frac{\partial W}{\partial \epsilon^e_{ij}}\frac{\partial \epsilon^e_{ij}}{\partial X_1}=\sigma_{ij}\frac{\partial \epsilon^e_{ij}}{\partial X_1}=\sigma_{ij}\frac{\partial}{\partial X_1}(\epsilon_{ij}-\epsilon^{th}_{ij}) \tag{A8}$$

where the *total* strain is now given by elastic (including nonlinear) and thermal components such that $\epsilon_{ij}=\epsilon^e_{ij}+e^{th}_{ij}$. Upon combining Eqs. (A6) and (A8), we have

$$\bar{J}_2=\int_V \sigma_{ij}\frac{\partial \epsilon^{th}_{ij}}{\partial X_1}q\, dV. \tag{A9}$$

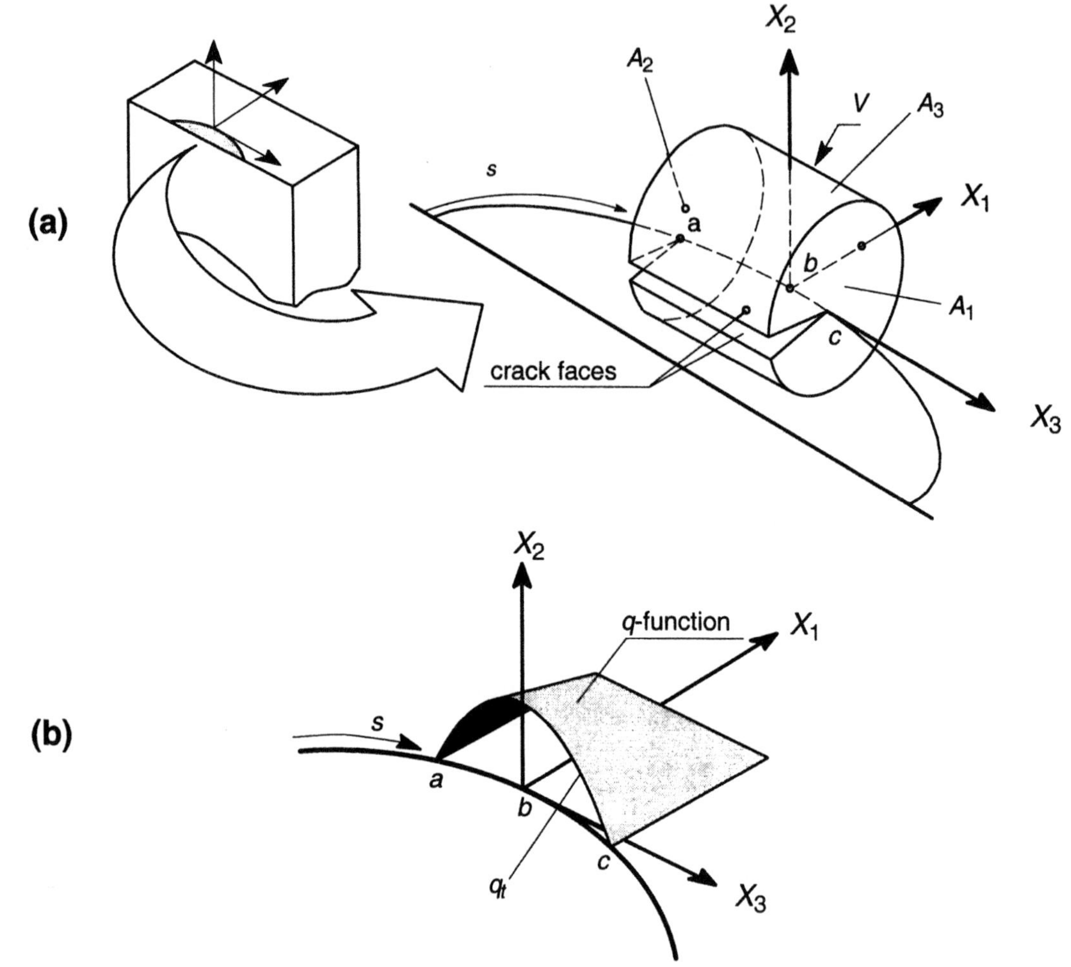

Fig. 9. (a) Finite volume for domain integral computations at an interior location along curved crack front showing local, crack front coordinates axes; (b) variation of weight function, q, over volume at crack front.

Thermal strains can be written as $\epsilon_{ij}^{th}=a_{ij}\Theta$, where Θ denotes the temperature change. The above expression becomes

$$\bar{J}_2=\int_V \sigma_{ij}\left[a_{ij}\frac{\partial\Theta}{\partial X_1}+\frac{\partial a_{ij}}{\partial X_1}\Theta\right]q\,dV. \quad \text{(A10)}$$

which somewhat simplifies implementation in a finite element context since temperatures are generally known at element nodes. Computation of the temperature derivative in the (local) Cartesian coordinates at the crack front location s follows standard finite element procedures.

If the thermal expansion coefficients are homogeneous in the (local) crack direction, X_1, over the domain volume V, then the second term above vanishes. In the eigenstrain approach to model residual stresses, coefficients a_{ij} most often vary strongly near the crack front and thus within the domains of integration to compute $\bar{J}_2$. Consequently, the second term above must be included to maintain path independence of the computed J-values.

Cartesian derivatives of the thermal expansion coefficients with respect to the crack direction (X_1) must be obtained to evaluate Eq. (A10) and this does complicate the computations. Prior to this operation, the symmetric tensors of thermal expansion coefficients, a_{ij}, require transformation into crack front coordinates. The numerical procedure implemented in the WARP3D code [25] is as follows: (1) the analyst specifies a tensor a_{ij} in the structure Cartesian system for each element in the model, (2) nodal average values, $\bar{a}_{ij}$, are computed over the model from the element contributions; (3) $\bar{a}_{ij}$ tensors are transformed into (local) crack front coordinates

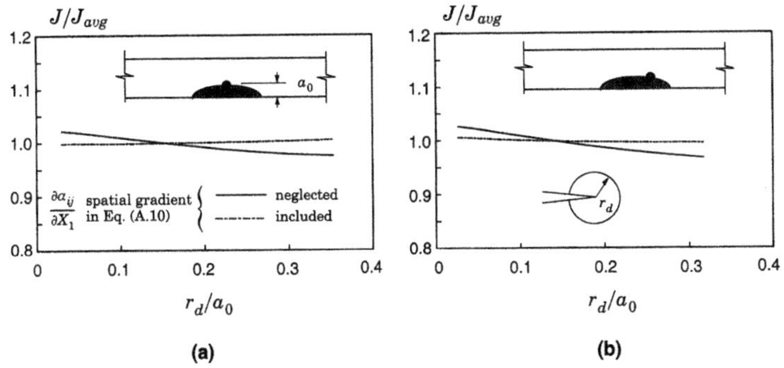

Fig. 10. Effects of including spatial gradient of thermal expansion coefficients on path (domain) dependence of the 3-D J-integral values. Domain radius normalized by maximum crack depth of semi-elliptical crack: (a) location of maximum crack depth; (b) location nearer the free surface.

defined for the domain at point s on the front; (4) using $\bar{a}_{ij}$ at the nodes, the Cartesian derivatives w.r.t. X_1 at element Gauss point locations are computed using standard isoparametric procedures; and (5) Gauss quadrature over each element then provides the contribution to $\bar{J}_2$. This procedure defines approximate gradients of a_{ij} over elements in the domain. Alternatively, analysts could specify $\bar{a}_{ij}$ values directly at structure nodes but this does not fit well into the design of most codes where material properties are associated with elements not nodes.

A.3. Illustrative example

Fig. 10 illustrates the effects of gradients of a_{ij} on the computed J-integral values for residual stress loading via the eigenstrain approach. The model for illustration has a semi-elliptical crack located in a flat plate with loading by the imposed eigenstrains described in Eqs. (23)–(27). This corresponds to Case 1 in Fig. 5. The finite element mesh is that shown in Fig. 6(c) with the backup bar removed.

J-values are plotted for each of 15 domains at two locations on the crack front. We typically discard values for the first 2–3 domains at the crack tip. The radius of each semi-circular domain, denoted r_d, is normalized by the maximum crack depth. Values with the a_{ij} gradients included show essentially no path dependence at each front location. J-values without the gradients decrease monotonically with distance from the crack front with a 5–7% change over these distances normal to the front. In other cases with more severe gradients, a much larger degree of domain dependency can be anticipated.

References

[1] Interim Guidelines: Evaluation, Repair, Modification and design of Welded Steel Moment Frame Structures, FEMA-267, August 1995.

[2] Popov EP, Yang T, Chang S. Design of Steel MRF connections before and after 1994 Northridge Earthquake. International Conference on Advanced in Steel Structures, December, 1996.

[3] Engelhardt MD, Sabol TA, Aboutaha RS. Testing of steel ductile frame joints. In: Proceedings, 13th Structures Congress, Part 2 (of 2). New York (NY, USA), ASCE, 1995;1411–4.

[4] Stojadinovic B, Goel SC, Lee K-H. Welded steel moment connections — old and new. In: Proceedings, ASCE Structures Congress, New Orleans, April 1999.

[5] Chi W, Deierlein GG, Ingraffea AR. Finite element fracture mechanics investigation of welded beam-column connections. Report No. SAC/BD-97/05, December 1997.

[6] El-Tawil S, Mikesell TD, Vidarsson E, Kunnath SK. Inelastic behavior and design of steel panel zones. J of Struct Engng, ASCE, February 1999.

[7] Zhang J, Dong P, Residual stress in welded moment frames and implications on structural performance. J. of Struct Engng, ASCE, 1999 [in press].

[8] Kauffmann EJ, Fisher JW. A study of the effects of material and welding factors on moment frame weld joint performance using a small scale tension specimen. Technical Report SAC 95-08, 1995.

[9] Ricles J. Private communication, Lehigh University, 1999.

[10] Gao X, Ruggieri C, Dodds RH. Calibration of Weibull stress models using fracture toughness data. Int J Fracture 1998;92:175–200.

[11] Mura T. Micromechanics of defects in solids. The Netherlands: Kluwer Academy Publishers, 1991.

[12] Ueda Y, Fukuda K. New measuring method of three dimensional residual stress in long welded joints using inherent strains as parameters-L_z method. Journal of Engineering Materials and Technology 1987;111:1–8.

[13] Hill MR, Nelson DV. The inherent strain method for residual stress determination and its application to a long welded joint. Structural Integrity of Pressure Vessels, Piping, and Components ASME 1995;318:343–52.

[14] Panontin TL, Hill MR. The effect of residual stress on brittle and ductile fracture initiation predicted by micromechanical models. International Journal of Fracture 1996;82:317–33.

[15] Masubuchi K. Analysis of welded structures. New York: Pergamon Press, 1980.

[16] Kauffmann EJ, Fisher JW. Fracture analysis of failed moment frame weld joints produced in full-scale laboratory tests and buildings damaged in the Northridge earthquake. Technical Report SAC 95-08, Part 2, 1996;1-1 to 1-21.

[17] Eshelby JD. Energy relations and the energy momentum tensor in continuum mechanics. In: Kanninen MF et al, editors. Inelastic behavior of solids. New York: McGraw-Hill, 1970.

[18] Cherepanov GP. Crack propagation in a continuous medium. Journal of Applied Mathematics and Mechanics 1967;31:467–88.
[19] Moran B, Shih CF. A general treatment of crack tip contour integrals. International Journal of Fracture 1987;35:295–310.
[20] Moran B, Shih CF. Crack tip and associated domain integrals from momentum and energy balance. Engineering Fracture Mechanics 1987;27:615–42.
[21] Li FZ, Shih CF, Needleman A. A comparison of methods for calculating energy release rates. Engineering Fracture Mechanics 1985;21:405–21.
[22] Shih CF, Moran B, Nakamura T. Energy release rate along a three-dimensional crack front in a thermally stressed body. International Journal of Fracture 1986;30:79–102.
[23] Blodgett OW, Funderburk RS, Miller DK. Fabricator's and erector's guide to welded steel construction. The James F. Lincoln Arc Welding Foundation, JFLF-845, 1997.
[24] Raju IS, Newman JC. Stress-intensity factors for a wide range of semi-elliptical surface cracks in finite-thickness plates. Engineering Fracture Mechanics 1979;11:817–29.
[25] Koppenhoefer K, Gullerud A, Ruggieri C, Dodds R, Healy B. WARP3D: dynamic nonlinear analysis of solids using a preconditioned conjugate gradient software architecture. Structural Research Series (SRS) 596, UILU-ENG-94-2017, University of Illinois at Urbana-Champaign, 1998.

Active aerodynamic bidirectional control of structures I: modeling and experiments

Himanshu Gupta [a], T.T. Soong [b,*], G.F. Dargush [b]

[a] *Spars International Inc., Houston, TX 77079, USA*
[b] *Department of Civil and Environmental Engineering, State University of New York at Buffalo, Buffalo, NY 14260, USA*

Received 16 May 1997; received in revised form 23 September 1998; accepted 24 September 1998

Abstract

A new aerodynamic method for controlling bidirectional wind-induced motion of bluff bodies is proposed. A generic bluff body model is suggested to develop a mathematical description of the control system dynamics. The motion in the along-wind direction is assumed to be generated by incident turbulence, while vortex-induced resonant vibrations are assumed for the across-wind direction. The control system equations are derived and the closed-loop algorithms are presented. Experiments conducted to verify the basic assumptions and to study the basic fluid dynamical features of the new device are also reported. © 1999 Elsevier Science Ltd. All rights reserved.

Keywords: Structural control; Vortex shedding; Wind tunnel experiments; Nonlinear control; Bluff body aerodynamics; Parametric vibrations

Nomenclature

ρ	= air density (1.2593 kg/m^3)
ω	= natural frequency
ξ	= mechanical damping ratio (% of critical)
λ	= trade-off parameter
θ_0	= steady controller angle (45°)
$\Delta\theta$	= angular displacement of controller
AFS	= Aerodynamic Flap System
BSF	= Binary Switching Function
ESF	= Exponential Switching Function
a_v	= tip translation amplitude of flaps
d	= width of the control flap
f	= frequency
m	= mass per unit length
u	= turbulence
v	= feedback of wake due to flap control
x	= along-wind coordinate
y	= across-wind coordinate
A	= projected area of bluff body
A_p	= projected area of flaps
B	= control influence parameter
C_D	= drag coefficient
C_k	= surface drag coefficient (0.04)
D	= dimension of the bluff body
L_c	= length of controller
L_{ux}	= scale of turbulence (1200 m)
R	= reduced frequency
St	= Strouhal number for square section (0.15)
S	= switching function value
S_{uu}	= turbulence spectra
U	= mean wind speed
Y_1, Y_2	= parameters of across wind motion

* Corresponding author. Tel.: + 1-716-645-2114; Fax: + 1-716-645-3733

1. Introduction

Wind-induced vibrations of bluff bodies have been a subject of interest for the past several decades. Vibrations of flexible structures, due to fluid forces, are of importance in civil, mechanical and aerospace engineering applications. A bluff body suspended in a fluid flow can be excited into motion due to turbulence in the approach flow or due to fluid-elastic instabilities. The resulting vibrations may occur in the direction of flow or in the direction perpendicular to the flow. The wind-induced vibrations are undesirable in most of the situ-

Reprinted from *Engineering Structures* **22 (4)**, 379-388 (2000)

ations and methods of suppressing them are frequently desired. This paper presents an active aerodynamic control method to mitigate bidirectional wind-induced vibrations of bluff bodies. The paper discusses mathematical modeling and experimental results for a generic bluff body excited into motion due to incident turbulence of the approach flow in the along-wind direction and due to vortex resonance in the across-wind direction.

1.1. Aerodynamic control of bluff bodies

In the past two decades tremendous progress has been made in structural control research. A number of control devices, algorithms and design methodologies have been developed over these years. Since the control concepts developed in the past address basic dynamics of a structure, they are equally applicable to either earthquake or wind-excited motion of structures. A comprehensive overview of structural control theory has been presented in Soong [17]. Housner et al. [8] and Kobori [10] provide a review of more recent developments in this area. In general, structural control as applied to wind-excited motion of structures can be divided into two distinct approaches: structural, where mechanical properties of the structure are altered for control purposes; and aerodynamic, where aerodynamic modifications are made to the structural geometry.

The aerodynamic control approach for mitigation of wind-induced motion provides a distinct advantage over the traditional structural approach. Aerodynamic control devices are usually energy efficient since the energy in the flow is used to produce the desired control forces. Several passive aerodynamic devices have been developed in the past. Zdravkovich [18], for example, provides an excellent review of vortex suppression devices. These aerodynamic devices (or spoilers) destroy or reduce the coherence of shed vortices and thereby alleviate vortex induced oscillations. Ogawa et al. [14] used a device formed by circular guide vanes set asymmetrically at the rectangular section corners. Shiraishi et al. [15] performed experiments on two-dimensional rectangular cross-sections modified by cutting the corners. Kubo et al. [11] have proposed the use of rotors on separation points of square sections to control the separated shear flow. Soong and Skinner [16] performed experiments on an active device, called an aerodynamic appendage, for controlling along-wind motion of tall buildings. More recently, Mukai et al. [13] used a modified appendage mechanism to accommodate the changes in the angle of attack of the wind flow with respect to the building.

2. Proposed active control device

The proposed control device, hereafter called the Aerodynamic Flap System (or AFS for short), consists of two flaps or appendages placed symmetrically at the leading separation edges of a bluff body or any other structure (Fig. 1). The flaps rotate about the leading edges of the structure like the flapping of a bird's wing. AFS is an active system driven by a feedback control algorithm which guides its operation based on information obtained from the vibration sensors. The area of the flaps and angular amplitude of rotation are the principal design parameters of AFS.

AFS is based on the idea of controlling the free shear layers, and thus vortices emerging from the separation edges of the structure, by applying time varying geometrical perturbations at these edges. Several passive control devices are based on a similar concept of altering the vortex patterns through changes made on the separation edge geometry. AFS, however, has the capacity of controlling the bidirectional (along-wind and across-wind) motion of structures. It produces an extra drag force through the exposed area (of the flap), a feature that can be exploited for controlling the along-wind motion. Experiments performed by Soong and Skinner [16] fundamentally support the action of AFS in the along-wind direction. AFS also acts as a vortex generator, with the capacity of producing vortices at a desired frequency and phase. This feature can be exploited to control the across-wind motion. Experiments performed by Mizota and Okazima [14] provide insight on physical mechanisms for the across-wind action of AFS.

3. Mathematical modeling

Consider the bluff body shown in Fig. 1 (although a square cross-section is shown, any other cross-sectional shape is permissible.) From a mechanical standpoint, this bluff body is a mass attached to a spring and a damper in two directions, x (along-wind) and y (across-wind).

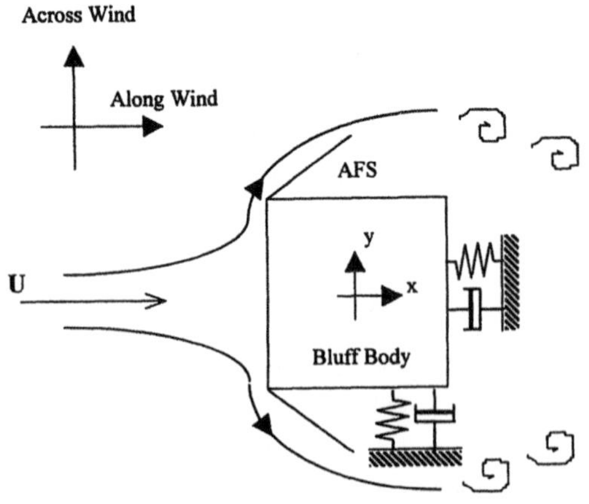

Fig. 1. Generic bluff body.

For simplicity, the stiffness and damping are assumed to be the same in both directions. Thus the body can vibrate in either of the two or both directions simultaneously if a dynamic force is applied to it. It is assumed that the applied aerodynamic force acts at the elastic center of the body, thus eliminating any torsional motion. From an aerodynamical point of view, the bluff body has a two-dimensional cross-section. Therefore, the flow around the body is idealized as two-dimensional (i.e., velocity component of the fluid particles in the direction normal to the plane of the bluff body in Fig. 1 is zero). It is also assumed that there is no mechanical or aerodynamic coupling in the two directions of motion and hence the equations can be developed separately in the two directions. These assumptions not only simplify the physics involved in mathematical modeling but also serve the purpose of developing generic equations. The equations developed for the dynamic motion of this two-dimensional bluff body can be extended to incorporate three-dimensional flow effects and any three dimensional structural geometry.

3.1. Along-wind dynamics

The motion in the along-wind direction primarily results from fluctuations in the upstream flow and, in terms of the drag coefficient C_D, the net force $F(t)$ acting in the along-wind direction can be written as

$$F(t) = \frac{1}{2}\rho A C_D (U + u(t))^2, \qquad (1)$$

where ρ is the air density, A is the projected area, while U and u represent the mean and fluctuating components of the wind speed, respectively. This can be simplified to

$$F(t) = \frac{1}{2}\rho A C_D U^2 + \rho A C_D U u(t), \qquad (2)$$

based on the assumption that the nonlinear velocity term in Eq. (1) contributes very little to the response. The wind speed fluctuation, $u(t)$, is usually assumed to be a stationary Gaussian process. Several spectral descriptions of $u(t)$ are available in empirical form, e.g., the Davenport spectrum, defined as Davenport [3]

$$S_{uu}(\hat{\omega}) = \frac{2C_k L_{ux}^2|\hat{\omega}|}{\pi^2\left[1 + \left(\dfrac{L_{ux}\hat{\omega}}{\pi U_{10}}\right)^2\right]^{4/3}}, \qquad (3)$$

where C_k is the surface drag coefficient, L_{ux} is the scale of turbulence, and U_{10} is the wind speed at the 10 m height. The time-dependent part of the along-wind force excites the flexible bluff body into motion.

The major contribution of AFS to the along-wind dynamics comes from the exposed surface area of the flaps. Thus for the along-wind direction, AFS acts as a force producing device. The force generated by AFS in the along-wind direction can be expressed using the drag coefficient and the surface area of the flaps exposed to the wind. This procedure is similar to that developed in Soong and Skinner [16]. The expression for force generated in the along-wind direction due to AFS can be written as

$$F_c(t) = 2\left[\frac{1}{2}\rho A_p(t) C_D (U + u(t))^2 S\right]. \qquad (4)$$

The factor two appears because of two symmetrically attached flaps. The exposed surface area of each flap $A_p(t)$ is a function of time since the flaps rotate about the separation edges. For small flap oscillations, it can be expressed as

$$A_p(t) = L(d \sin\theta_0 + v_f(t)) \qquad (5)$$

where L is length of the flap, d is the width of the flap, θ_0 is the mean flap angle (around which the flap oscillates), and $v_f(t)$ is the tip displacement of the flap ($= \Delta\theta(t)d$). In Eq. (4), S is the control switching function which guides the flap operation. Thus, the dynamic equation of motion in the along-wind direction can be expressed simply as

$$\ddot{x} + 2\xi\omega\dot{x} + \omega^2 x = \frac{1}{m}\left[\frac{1}{2}\rho A C_D U^2 + \rho A C_D U u(t) + 2\frac{1}{2}\rho A_p(t) S C_D (U + u(t))^2\right] \qquad (6)$$

In adopting Eq. (6), it is assumed that the drag coefficient for the bluff body is not significantly influenced by the presence of the AFS.

3.2. Across-wind dynamics

The major contribution to the across-wind motion of bluff bodies comes from three factors; incident turbulence in the across-wind direction, fluctuations in the wake (vortex shedding, etc.), and aeroelastic amplification due to building motion. The atmospheric flows have very little incident turbulence in the across-wind direction and consequently it can be neglected. The next two factors form the main part of across-wind excitations. The vortex shedding process is greatly amplified by the structural motion near the resonance conditions and produces potentially hazardous vibrations. Therefore, resonant vortex shedding vibrations in the across-wind directions are of primary interest in this study. Numerous studies on vortex shedding vibrations, both theoretical and experimental, reported in past several decades have been reviewed extensively in Bearman [1]. A more recent semi-empirical model proposed by Goswami et al. [6], developed from synthesis of a coupled model proposed by Billah [2], will be used to describe across-wind vibrations in this study.

The motion of the flaps changes the vortex shedding pattern and subsequently the wake. The wake, an oscillator in itself as suggested by Billah [2], responds only at the natural vortex shedding frequency of the bluff body in the absence of the flaps. It was observed by Mizota and Okazima [12] that oscillation of the leading edge flaps shows two frequency components in the wake; one of these was the natural vortex shedding frequency and the second was the frequency of the flap. The preceding observation suggests that the wake oscillator can be modeled as a forced system in the presence of the flaps. Hence, the flap action (or actuator motion) can be modeled as a forcing term on the right-hand side of the wake equation of Billah [2]. It is reasonable to assume that flaps have no direct influence on the structural vibrations, as the flaps modify only the aerodynamics of the flow (i.e., wake and vortex shedding features). In light of these arguments, Billah's equations can be rewritten as

$$\ddot{y} + 2\xi\omega\dot{y} + \omega^2 y + 2\alpha y q + 4\beta y^3 q + (\gamma y + \epsilon y^3)\dot{q}^2 = 0 \quad (7)$$

$$\ddot{q} + f(q,\dot{q}) + (2\omega_s)^2 q + \alpha y^2 + \beta y^4 - 2(\gamma y + \epsilon y^3)\dot{y}\dot{q} = bv(t) \quad (8)$$

Here y is the across-wind displacement, q is formation length (a wake variable), ω is the natural frequency of the structure, ξ is the damping ratio, ω_s is the vortex shedding frequency, α, β, γ and ϵ are model parameters, $f(q, \dot{q})$ is a nonlinear damping pair, b is the control influence coefficient and $v(t)$ is the actuator feedback. It is desirable to reduce these two equations into a single equation. Following a procedure similar to Goswami et al. [7], Eqs. (7) and (8) can be reduced to

$$\ddot{y} + 2\xi\omega\dot{y} + \omega^2 y = \frac{\rho U^2 D}{2m}\left[Y_1(R)\frac{\dot{y}}{U} + Y_2(R)\frac{y^2\dot{y}}{D^2 U} + B(R)v(t)\frac{y}{D^2}S\right], \quad (9)$$

where $Y_1(R)$ is the aerodynamic damping coefficient, $Y_2(R)$ is the nonlinear aerodynamic response coefficient and $B(R)$ is the non-dimensional control influence coefficient. These coefficients depend on $R(=\omega D/(2\pi U))$, the reduced frequency. The term involving $v(t)$ is the influence of AFS on across-wind dynamics of the bluff body. It is noted that the control term in Eq. (9) is assumed to be that of parametric excitation. In Eq. (9), S is the switching function. The switching function S introduces bidirectional coupling in the two vibratory directions through the control term.

The bidirectional coupling through the control term introduces a two-way interaction in the dynamics of the bluff body. The switching function in the along-wind direction directly influences the operational time of AFS in the across-wind direction. The oscillation of the flaps for the across-wind direction changes surface area of the appendage exposed to the wind, thus changes the forces AFS can generate in the along-wind direction. Thus, the bidirectional coupling introduces interesting features to the AFS. On the one hand, it opens the possibility of controlling two independent vibratory directions by using a single controller and on the other, it necessitates a trade-off between the degrees of control that can be achieved in the two directions.

4. Control algorithms for AFS

4.1. Across-wind direction

In this study, a control approach suggested by Fujino et al. [5] will be adopted. Fujino et al. [5] derived a controller that uses dissipative effects of parametric systems. It is well known that parametric resonance occurs most readily when frequency of the parametric excitation is twice that of the natural frequency of the bluff body or more generally, the system is out of phase with the motion. Based on this observation, Fujino et al. [5] suggested that if the response is $y(t) = y_0\cos(\omega t)$ then $v(t)$ should be defined as

$$v(t) = a_v\cos(2\omega t + \pi/2), \quad (10)$$

for optimal energy dissipation. When amplitude and frequency of y varies slowly with time, Eq. (10) can be better implemented in a feedback form as

$$v(t) = \frac{2a_v}{\omega}\frac{y\dot{y}}{\left[y^2 + \frac{\dot{y}^2}{\omega^2}\right]} \quad (11)$$

where a_v is the control gain.

The control algorithm, $v(t)$ in Eq. (11), is feedback of the wake (and the vortices in it) to the structural motion. The wake of the bluff body being controlled by the flaps provides a feedback at twice the natural frequency of motion and is out of phase with the structural motion. The uncontrolled body (i.e., no flaps) sheds vortices from its separation edges at twice the frequency but shedding is in phase with the motion producing resonant vibrations. Vortex shedding from flaps occurs at the optimal frequency required for parametric vibrations (i.e., twice) but the vortices are shed at an opposite phase, thus damping out the motion.

The algorithm in Eq. (11) is a closed-loop feedback algorithm and requires structural displacement and velocity information. However, this information is used only to adjust the phase and frequency of the controller motion with respect to the structural motion. The controller amplitude a_v is the stroke of the flap and is equal to the length of the flap multiplied by the angular amplitude. The flaps are initially positioned at an angle of 45° and the rotations take place around this angle. The motion of the individual flaps is at the frequency of ω, the natural frequency of motion. The two flaps attached to the two separation edges shed vortices alternately and thus create two vortices in one oscillation of the body. This feature is similar to vortex shedding from an uncontrolled body, where both the separation edges shed one vortex each creating two vortices in the wake during one complete oscillation of the body. The feedback form of the individual flap motion can be determined as

$$v_f(t) = \frac{a_v y}{\sqrt{y^2 + \dfrac{\dot{y}^2}{\omega^2}}}. \qquad (12)$$

4.2. Along-wind direction

The force generated by AFS in the along-wind direction can be utilized for control purposes through the switching function S defined in Eqs. (6) and (9). The switching function adopted in this study for along-wind direction control is a simple on-off type function, hereafter called Binary Switching Function (BSF). This function was introduced by Klein et al. [9] and later verified experimentally by Soong and Skinner [16]. The form of the function is

$$S = 1, \dot{x} \leq 0; \text{ and } S = 0, \dot{x} > 0. \qquad (13)$$

The idea here is to switch on the flaps when the structure sways in the direction opposite to that of the wind, thus creating an extra push, and switch it off when the structure sways in the direction of wind. It can be seen that this function is in a feedback form and requires the sign of the building velocity in the along-wind direction. The switching function in Eq. (13) can also be written as

$$S = \frac{1}{2}[1 - sgn(\dot{x})], \qquad (14)$$

where *sgn* is the signum function.

4.3. Bidirectional control

In Sections 4.1 and 4.2, control was considered with only one direction of motion at a time. The power of AFS in producing bidirectional control can be exploited if the switching function can be modified to create an optimal trade-off. For this purpose, an Exponential Switching Function (ESF) will be introduced. The form of the function is

$$S = \exp[-\lambda\dot{x}(1 + sgn(\dot{x}))]. \qquad (15)$$

where λ is a parameter of ESF. Moreover, ESF collapses to BSF when $\lambda \rightarrow \infty$ and becomes one when $\lambda = 0$. The above behavior of ESF makes it useful for trade-off between the along-wind control and across-wind control. When the value of λ is zero, all the control effort is dedicated to the across-wind direction and, when the value of λ approaches ∞, the control effort is dedicated to the along-wind direction.

5. Experimental verification

Experimental verification of the control system is an important part of AFS development. Very few experiments of similar nature have been reported. The experiments performed during this research involved aerodynamic and aeroelastic testing of a square cross-section model in a wind tunnel. The focus of the experiments was to demonstrate the effectiveness of the present control technique in response reduction and to study the fundamental features of the flow as modified by the AFS. The experiments were restricted to verification of AFS in the across-wind direction. The action of AFS in the along-wind direction is fundamentally supported by experiments performed by Soong and Skinner [16].

5.1. Experimental setup

A low-speed open-circuit wind tunnel was used for the experiments. The wind tunnel operates in an open loop with a maximum flow rate of 25 000 ft^3/min. The wind speed at the test section is fairly uniform and has a low initial turbulence intensity (about 2%). The test section of the wind tunnel is about 7.5 m long and has a cross section of 1 m by 1 m. The model can be placed at 5 m downstream of the air entrance. Hot wire anemometer was used for measurement of flow downstream of the model. A pitot tube was used for measuring the inlet flow speed of the wind tunnel.

Section model of a tall building, used for the experiments, was a square cylinder made of aluminum. The cylinder had a cross-section of 152.5 mm and length of 762 mm. The control flaps were made of 3.2 mm thick aluminum plate and were attached to the leading edges of the cylinder with piano hinges. The flaps were 762 mm long (covering the entire length of the model) and 51 mm wide. The left end of the model was equipped with motor and flap actuation system. The ends of the model were covered with 203 mm square aluminum plates to serve as end plates. The purpose of providing end plates was to preserve the two-dimensionality of the flow over the model. The model ends had 13 mm diameter threaded rods projecting outwards. Two different test configurations were developed: aerodynamic (or rigid) and aeroelastic (or flexible) support system. The drag coefficient, C_D, of the section was determined as 2.6 and the Strouhal number *St*, which provides a nondimensional measure of the vortex shedding frequency, was determined as 0.15.

The aerodynamic support system (rigid setup) was used to attach the model and the force balance to the wind tunnel rigidly. The support system consisted of two legs, anchored to the floor of the tunnel section, on either side of the model. A force balance was rigidly mounted on the legs and the model was suspended through the force balance. The height of the force balance on the support legs was adjusted such that center of the model was exactly at the center of the wind tunnel cross-section. This support system was anchored 5 m downstream of the wind tunnel inlet. The support system was sufficiently stiff so that the model would not experience any vibration during the wind flow. A picture of the aerodynamic test setup is shown in Fig. 2.

An aeroelastic support system was developed to support the model using an arrangement of springs, so as to permit only a vertical degree of freedom. As shown in Fig. 3, two pairs of springs separated by 1 m were used on either side of the model (total of eight springs). One end of every spring was attached to a horizontal rod, the center of which was supporting the model. The other end of the spring was connected to vertical rods attached to the wind tunnel roof or floor. All the eight springs were prestretched by pulling the free end of the vertical rods. The model was then attached to the spring arrangement. The centerline of the model was adjusted to coincide with the centerline of the wind tunnel cross-section. The springs had stiffness of 1431 N/m each with a free length of 102 mm. The natural frequency of the system was 3.69 Hz, the damping ratio was 0.4% of critical and system mass was 13 kg.

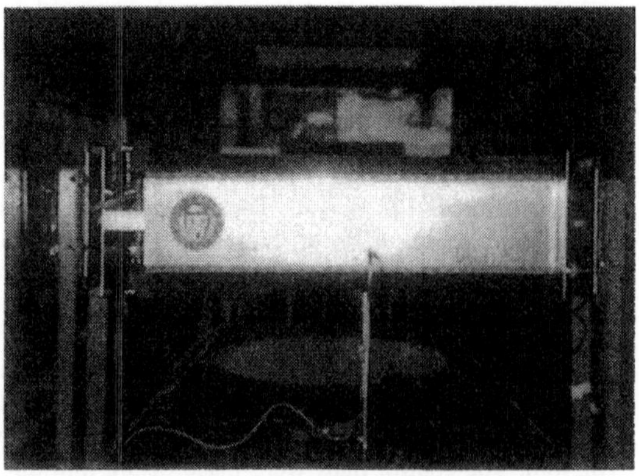

Fig. 2. Aerodynamic support system.

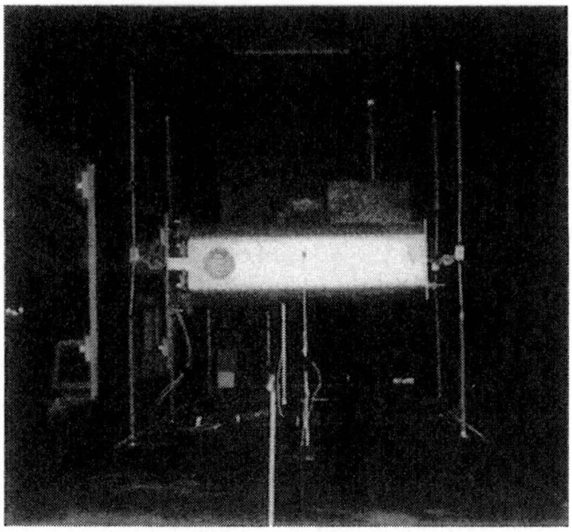

Fig. 3. Aeroelastic support system.

The flap actuation system was developed to perform the task of generating the desired motion of the control flaps. The main elements of the flap actuation system were: the motor control subsystem, the link actuation subsystem, and control and data acquisition software. The motor control subsystem consisted of a DC servo motor, a tachometer, a motor controller/amplifier and a DC power supply. The feedback device used in this setup was a tachometer to measure the actual RPM of the motor. The shaft of the tachometer was coupled to the rear end shaft projection from the motor.

The link actuation system transformed the rotary motion of the motor to a sinusoidal motion at the control flaps. As shown in Fig. 4, this was accomplished by three gears and links, which form a four-bar linkage mechanism. The three identical gears used had 48 teeth with a pitch diameter of 51 mm. The central gear (primary) was attached to the motor shaft and the lower and upper (secondary) gears were meshed to the central gear and attached to a dowel pin fitted in a ball bearing and supported on the model. The secondary gear had a small slot bridge which passes through the center of the gear. One end of the link was attached rigidly to this slot. The links are made of two pieces of small 6 mm threaded rods which have right-hand and left-hand threads, respectively. These rods were connected together through a turnbuckle, which allows length adjustment of the links. The other end of this link is fixed on a slot

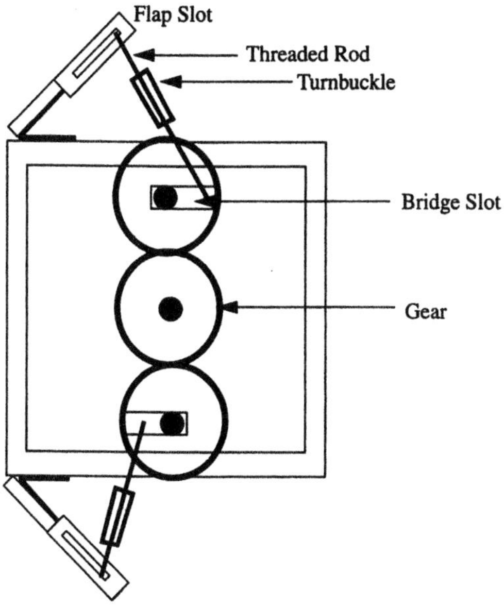

Fig. 4. Schematic of link actuation system.

attached to the control flaps. The amplitude of the flap rotation can be adjusted by changing position of the link ends on the bridge slot, or flap slot, or both. The mean position of flap rotation can be adjusted by changing the length of the links using the turnbuckle.

The control system diagram linking all the various elements described above is shown in Fig. 5. The computer sends a command voltage corresponding to the desired motor RPM to the controller. The controller was also supplied with the tachometer feedback signal. The task of the controller was to minimize the error between the command signal (desired RPM) and the tachometer output (actual RPM of the motor). The controller sends a voltage and current signal to the motor so as to speed it up or down in order to reduce this error. The control process becomes especially important when wind acts on the flaps and produces fluctuating loads on the flaps. These loads are converted to a fluctuating torque load at

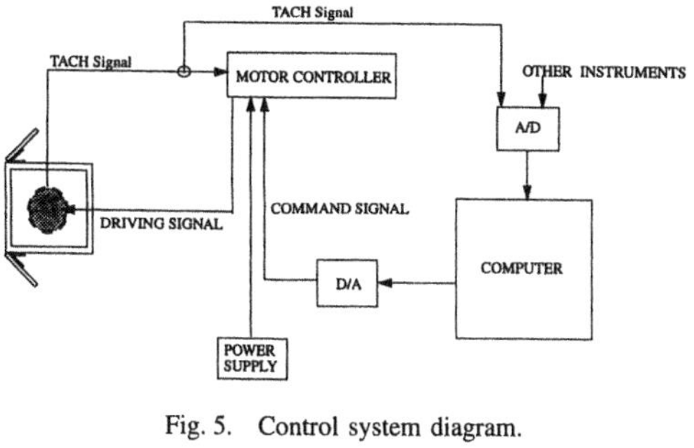

Fig. 5. Control system diagram.

the motor, which makes the motor shift continuously from its steady rotation speed.

5.2. Experimental results

5.2.1. Unsteady forces due to flaps

Lift forces generated on the section model due to control flaps were studied in the aerodynamic test setup. Fig. 6 shows the behavior of lift coefficient versus the wind speed for a given flap frequency. The data is plotted using an unsteady lift coefficient, defined as,

$$C_L = \frac{F_L}{\frac{1}{2}\rho U^2 DL} \tag{16}$$

where F_L is the peak of the spectrum at the lift force frequency. It is observed that the force coefficient peaks at the resonant wind velocity dictated by the Strouhal relationship. This feature suggests that significant resonant forces can be generated on the model by operating the flaps at the resonant condition. The increase in the flap amplitude is seen to increase the magnitude of the peak, suggesting that a higher amplitude of flap oscillation would exert higher forces. A similar behavior was observed for the wake velocity downstream of the model. These resonance features demonstrate that changes in the wake velocity and lift forces have similar characteristics to the ones with the whole model in heaving oscillations. The flaps were thus observed to generate forces and vortices at the controller frequency. These forces can be used effectively to damp out the model oscillations (control of vibrations) if the phase of these forces is adjusted accordingly with respect to the model oscillations.

5.2.2. Control of oscillations

The square-section model was observed to oscillate at $U = 5.1$ m/s in a steady state in accordance with the

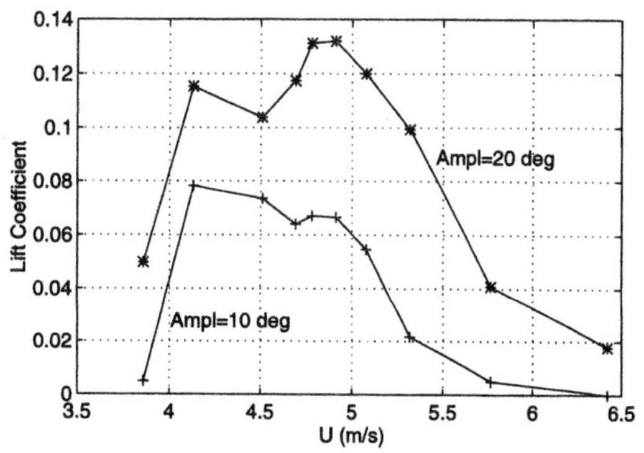

Fig. 6. Lift coefficient (flap freq = 6 Hz, Ampl is the flap oscillation amplitude).

Strouhal relationship. The wind speed when reduced or increased from this value resulted in damped oscillations, confirming existence of lock-in vortex shedding oscillations. The effect of control flaps in reducing the steady state vortex shedding oscillations was studied for different flap frequencies and amplitudes. The switching 'on' of control flaps when the model was operating in the steady state vortex-induced vibrations was seen to reduce the amplitude and dampen the vibrations eventually. Thus a net positive damping was added to the system by the controller, changing the system from limit cycle vibrations to an asymptotically stable system. Fig. 7 shows the effect of an increase in the amplitude of control flaps on the oscillations of the model. The effective damping in the system increases due to increases in control flap amplitude. Lift forces generated on the rigid model show similar characteristics. A higher amplitude of flap oscillations thus produces a stronger wake to oppose the oscillations of the model.

Vortex shedding steady-state oscillations occur when the shedding frequency in the wake is in resonance with the natural frequency of the system, creating a wake feedback to the model at its natural frequency. It was observed that the wake is forced at the same frequency as the control flap frequency, thus changing the feedback frequency of the wake. Hence, when the model is vibrating in resonance and the control is switched on at any frequency, the wake feedback frequency changes slowly from the resonant frequency to the flap frequency and this results in a shift from the resonant condition and an eventual diminishing of the oscillations. Fig. 8 shows this behavior. It is observed that effective damping in this case is very low. The maximum damping is created when the flap shedding frequency is twice of the model frequency.

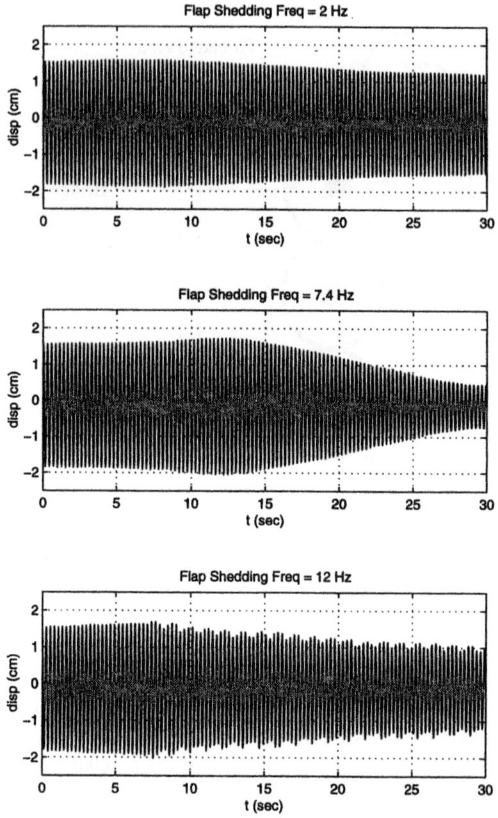

Fig. 8. Effect of flap shedding frequency on controlled response.

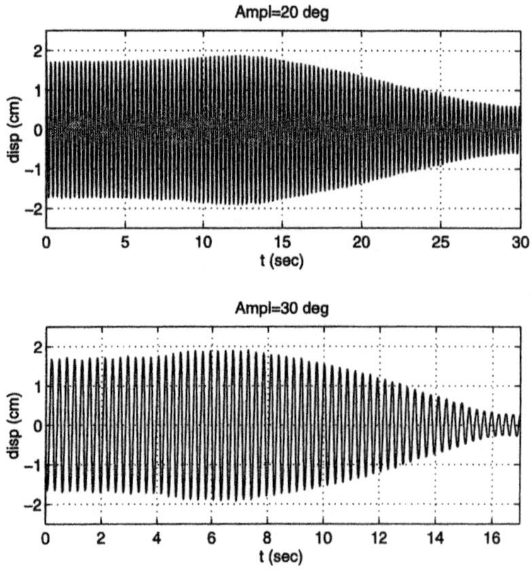

Fig. 7. Effect of amplitude on controlled response.

Operation of the flaps at resonance was seen to create maximum forces on the rigid model. Hence, flap operation should be adjusted out of phase with model oscillation to create the maximum damping in the system. It was observed that the effect of the flaps, when operating under the resonance condition, is to change the phase of vortex shedding with respect to model oscillations. The phase of vortex shedding (wake velocity signal from hot wire) with respect to the model oscillations in the uncontrolled case was found to be 2°. The corresponding phase angle in the controlled case was found to be 155°, 106°, 178°, and 115° for flap amplitudes of 10°, 15°, 20°, and 30°, respectively. Thus, the uncontrolled model has vortex shedding in phase with model oscillations whereas switching of control flaps forces vortex shedding to be out of phase with the model oscillations.

5.3. Parameter identification

The uncontrolled part ($v(t) = 0$) in Eq. (9) is a standard Vander Pol oscillator equation, i.e.,

$$\ddot{y} + 2\xi\omega\dot{y} + \omega^2 y = \frac{\rho U^2 D}{2m} \left[Y_1(R)\frac{\dot{y}}{U} + Y_2(R)\frac{y^2\dot{y}}{D^2 U} \right]. \tag{17}$$

Amplitude ratio method (Ehsan et al. [4]) can be used to identify the parameters of this equation. The equation for Y_2 is given by

$$Y_2 = \frac{4m\omega D}{n\pi\rho U a_y^2} \ln\left[\frac{A_0^2 R_n^2 - a_y^2}{R_n^2(A_0^2 - a_y^2)}\right], \tag{18}$$

where R_n is the ratio of amplitudes n cycles apart, A_0 is the initial amplitude, and a_y is the steady state vibration amplitude. The parameter Y_1 is given by

$$Y_1 = \frac{m}{2\rho D^2}\left[-\frac{\rho Y_2 a_y^2}{2m} + \frac{8\xi\omega D}{U} \right]. \tag{19}$$

Using Eqs. (18) and (19) for $a_y = 17.37$ mm, $A_0 = 16.0$ mm and $A(n = 22) = 17.29$ mm obtained from experimental time series yields

$$Y_1 = 26 \text{ and } Y_2 = -5382 \text{ for } R = 1.02. \tag{20}$$

It will be assumed, for identification of parameter B, that the values of Y_1 and Y_2 do not change when control is operational. This condition implies that net energy dissipation of the system is zero before the control starts (i.e., during steady-state vibrations) and all the energy dissipation thereafter is due to the control term. Now, consider per cycle energy production of different terms in Eq. (9). One has

$$2\xi\omega\dot{y} \text{ term} \rightarrow E_1 = \int_0^{2\pi/\omega} -2\xi\omega\dot{y}\dot{y}\,dt = -2\xi\omega^2\pi a_y^2$$

$$Y_1(R) \text{ term} \rightarrow E_2 = \frac{\rho U^2 D Y_1}{2mU}\int_0^{2\pi/\omega} \dot{y}\dot{y}\,dt = \frac{\rho U^2 D Y_1}{2mU}\omega\pi a_y^2 \tag{21}$$

$$Y_2(R) \text{ term} \rightarrow E_3 = \frac{\rho U^2 D Y_2}{2mD^2 U}\int_0^{2\pi/\omega} y^2\dot{y}\dot{y}\,dt = \frac{\rho U^2 D Y_2}{2mD^2 U}\omega\pi a_y^4/4$$

$$B(R) \text{ term} \rightarrow E_4 = \frac{\rho U^2 DB}{2mD^2}\int_0^{2\pi/\omega} v\dot{y}\dot{y}\,dt = \frac{\rho U^2 DB}{2mD^2} a_v \pi a_y^2$$

where a_v is the controller amplitude. The net energy $E_{net} = E_4$ because $E_1 + E_2 + E_3 = 0$, which can be verified using the values of Y_1 and Y_2. The net energy (E_{net}) of the system can be equated to energy produced by an equivalent linear damping term

$$E_d = -2\xi_a\omega^2\pi a_y^2, \tag{22}$$

where ξ_a is the additional equivalent linear damping coefficient. The simplification of the equation $E_{net} = E_4$ for B yields

$$B = -\frac{4\xi_a\omega^2 m D_{eff}}{\rho U^2 a_v}, \tag{23}$$

where D_{eff} ($= 0.224$ m) is the effective depth of the controlled cross-section (i.e., original D plus the projection of the flaps). The average values of ξ_a were calculated from the experimental time series for different controller amplitudes and parameter B was estimated using Eq. (23). The values of ξ_a were 0.3%, 0.6%, 0.9%, and 1.38% for $\Delta\theta = 10°$, 15°, 20°, and 30°, respectively. It was found that variation of B values for different amplitudes is small and hence a mean value of $B = -250$ is proposed for all the controller amplitudes.

6. Conclusions

A new active aerodynamic control device has been proposed in this paper. The device is simple to implement and has the capability of controlling the along-wind as well as the across-wind vibrations. In the along-wind direction, the device produces control forces based on the exposed surface area of the flaps. A switching function was presented to effectively use the force produced by the device for controlling the along-wind motion. Across-wind vibrations were modeled as vortex shedding resonant vibrations. The effect of the control device is to produce vortices at the desired phase and frequency. The influence of the device on the across-wind vibrations was derived as a parametric stiffness term. An energy optimal control algorithm was proposed for the across-wind direction. A bidirectional control algorithm was also derived to control the two vibratory directions simultaneously. Wind tunnel experiments were performed on a rigid model as well as a flexible model. The action of the proposed device in the across-wind direction was verified during these experiments. The fundamental features of the device were examined during the experiments. The experimental results verified the assumptions made in the mathematical modeling and demonstrated the effectiveness of the proposed control device. The control device proposed in this paper can be used for any bluff body in situations involving flow-induced vibrations.

Acknowledgements

This work was supported in part by the National Science Foundation under Grant No. CMS 9700387. The authors wish to acknowledge the help of Mark Pitman and Richard Cizdziel during the course of the experiments. The help of Scott Woodward in operating the wind tunnel is also acknowledged.

References

[1] Bearman PW. Vortex shedding from oscillating bluff bodies. Ann Rev Fluid Mech 1984;16:195–222.
[2] Billah KYR. A study of vortex-induced vibration. Ph.D. Dissertation, Princeton University, Princeton (New Jersey), 1989.
[3] Davenport AG. The application of statistical concepts to wind loading on structures. Proc Inst of Civ Engrg 1961;19:449–72.
[4] Ehsan F, Scanlan RH. Vortex induced vibrations of flexible bridges. Journal of Engr Mechanics 1990;116:1392–411.
[5] Fujino Y, Warnitchai P, Pacheo BM. Active stiffness control of cable vibration. Journal of Appl Mech 1993;60:948–53.
[6] Goswami I, Scanlan RH, Jones NP. Vortex induced vibrations of circular cylinders I: experimental data. Journal of Eng Mech 1993;119:2270–87.
[7] Goswami I, Scanlan RH, Jones NP. Vortex induced vibrations of circular cylinders II: new model. Journal of Eng Mech 1993;119:2289–302.
[8] Housner GW, Soong TT, Masri SF. Second generation of active structural control in civil engineering. Keynote Paper, Proceedings of First World Conference on Structural Control, Pasadena (CA), 1994.
[9] Klein RE, Cusano C, Slukel JV. Investigation of a method to stabilize wind induced oscillations in large structures. Paper No. 72-WA/AUT-11, ASME Annual Meeting (NY), 1972.
[10] Kobori T. Future directions on research and development of seismic response controlled structure. Keynote Paper, Proceedings of First World Conference on Structural Control, Pasadena (CA), 1994.
[11] Kubo Y, Modi VJ, Yasuda H, Kato K. On the suppression of aerodynamic instabilities through the moving surface boundary layer control. Proceedings Eighth International Conference on Wind Engineering, Part 1, London (Canada), 1992.
[12] Mizota T, Okazima A. Unsteady aerodynamic forces and wakes of rectangular prisms with oscillating flaps at leading edges. Proceedings Eighth International Conference On Wind Engineering, London (Canada), 1992:727–38.
[13] Mukai Y, Tachibana E, Inoue Y. Experimental study of active fin system for wind induced structural vibrations. Proceedings of First World Conference on Structural Control, Pasadena (CA), 1994.
[14] Ogawa K, Sakai Y, Sakai F. Aerodynamic device for suppressing wind induced vibrations of rectangular sections. Proceedings Seventh International Wind Engineering Conference, Aachen (Germany), 1987:2.
[15] Shiraishi N, Matsumoto M, Shirato H, Ishizaki H. On the aerodynamic stability effects for bluff rectangular cylinders with their corner-cut. Proceedings Seventh International Wind Engineering Conference, Aachen (Germany), 1987:2.
[16] Soong TT, Skinner GT. Experimental study of active structural control. J. of Eng. Mech. 1981;107:1057–68.
[17] Soong TT. Active structural control: theory and practice. London: Longman Scientific and Technical, and New York: Wiley, 1990.
[18] Zdravkovich MM. Review and classification of various aerodynamic and hydrodynamic means for suppression of vortex shedding. Journal of Wind Engineering and Industrial Aerodynamics 1981;7:145–89.

Performance of reinforced concrete frames using force and displacement based seismic assessment methods

A.M. Chandler [a], P.A. Mendis [b,*]

[a] *Department of Civil Engineering, The University of Hong Kong, Pokfulam Road, Hong Kong*
[b] *Department of Civil and Environmental Engineering, The University of Melbourne, Parkville, Victoria 3052, Australia*

Received 23 January 1998; received in revised form 6 October 1998; accepted 6 October 1998

Abstract

This paper reviews the traditional force-based (FB) seismic design method and the newly proposed displacement-based (DB) seismic assessment approach. A case study is presented for reinforced concrete (RC) moment-resisting frames designed and detailed according to European and Australian earthquake code provisions, having low, medium and high ductility capacity. The aim is to assess the performance characteristics of these frames, using the well known El Centro NS earthquake ground motion as the seismic input. Overall ductility demands have been computed for the force-based analyses conducted on the typical design frames. In the second part of the paper, the performance of the case study frames has been re-evaluated in the light of displacement-based principles. A recently proposed method for displacement-based seismic assessment of existing RC frame structures has been implemented for this purpose, from which it has been concluded that the displacement-based approach predicts very similar overall displacement demands for such frames. These results, whilst limited to the consideration of a small number of seismic frame structures and a single, typical strong earthquake ground motion, nevertheless give confidence that the displacement-based approach can rapidly and easily facilitate a seismic assessment of an existing RC structure, without the necessity to undertake detailed inelastic dynamic analyses. © 1999 Elsevier Science Ltd. All rights reserved.

Keywords: RC frames; Seismic assessment; Yielding; Ductility demands; Displacements

1. Introduction

Reinforced concrete (RC) moment resisting frames are a common lateral force resisting structural system in low to medium rise buildings in seismically active parts of Europe, Australia, Western U.S.A. and many other parts of the world. In Europe and Australia, with the introduction of the new unified seismic code EC8 [1] and the new earthquake standard AS 1170.4 [2], most of the medium to low rise buildings will be subjected to seismic design.

The major difference in the design approach for seismic forces (as opposed to wind forces) as a lateral force is that the designer is allowed to utilise the ductile capacity of the structure and design for reduced lateral forces. In this way, the elastic design strengths can be substantially reduced on the provision of adequate ductility capacity of the structure, to sustain an appreciable amount of plastic deformation under a maximum credible earthquake condition. Both EC8 and AS1170.4 specify three levels of lateral forces and corresponding ductility ratios for the design of RC moment resisting frames. The ductility demands experienced at critical locations in such frames when they form a plastic mechanism under sustained inelastic loading (as generated by severe earthquakes), have traditionally been regarded as a key measure of potential seismic structural damage. Recently however there has been a shift of attention away from such traditional methods of seismic design, based on the view that a strong correlation may be observed between the global and local (storey) displacements or deformations of a structure, and the damage recorded in earthquakes [3–7]. The development of the so-called displacement-based (DB) design methods has been stimulated largely by the view that both damage and ultimate failure of an earthquake resistant structure are more fundamentally dependent on the exceedance of displacement or ductility capacity than the exceedance of

* Corresponding author. Tel: + 61-3-9344-6789; fax: + 61-3-9344-4616.

Reprinted from *Engineering Structures* **22** **(4)**, 352-363 (2000)

strength capacity. The latter is embodied in all existing seismic design codes which traditionally focus the designer's attention on achieving required strength, rather than the required combination of elastic stiffness and inelastic energy-dissipating capacity, which is the focus of displacement-based (DB) approaches.

Essentially, for standard types of building structures such as RC moment resisting frames, both the traditional force-based (FB) design approach [8] and the newly proposed DB design methods have similar overall objectives which are to give an acceptable performance of the structure by limiting structural damage and preventing overall collapse under the designated ultimate limit state earthquake, conventionally based on the 1 in 500 year event. But because the two methods tend to approach the problem from opposite ends of the design process, questions arise as to the comparability of the approaches and their relative effectiveness in achieving these aims. In the force-based method, the structural displacements and element beam/column ductility demands are end products of the procedure which are not directly controllable by the designer. In the displacement-based method, by contrast, the displacements and ductility demands become fundamental design parameters and the procedure aims to ensure that the design targets or capacities set for these parameters will not be exceeded under the design-level earthquake ground motion.

This paper briefly reviews the requirements of earthquake standards, EC8 and AS1170.4 for design and detailing of RC frame structures under the traditional force-based approach. A case study is then presented for frames designed with low, medium and high ductility capacity, to assess the performance levels and the yielding mechanisms of these frames, using the well known El Centro NS earthquake ground motion as the seismic input. The spectral characteristics of this earthquake are known to match closely the design spectral shapes adopted in the European [1], Australian [2] and U.S.A. [9] earthquake codes for earthquakes on firm soil sites. From both static inelastic push-over analysis and dynamic time history analyses, it has been concluded that the seismic frames with low and intermediate ductility both form column hinging collapse mechanisms when subjected to the El Centro earthquake. The special moment resisting frame (SMRF) detailed for high ductility capacity generally has been found to develop plastic hinges in the beams rather than the columns, which is consistent with the ACI 318 [10] strong column-weak beam detailing philosophy used in the design of this SMRF. Overall ductility demands have also been computed for the FB analyses conducted on the typical design frames.

In the second part of the paper, the performance of the designed frames has been re-evaluated in the light of displacement-based principles. The method recently proposed by researchers such as Priestley [7] for displacement-based assessment of RC frame structures has been implemented for this purpose, from which it has been concluded that the displacement-based approach predicts very similar overall displacement demands for such frames.

2. Structural response modification

The response modification factor, R_f, as implied by the seismic standards is the ratio of the total elastic strength demand under the design earthquake, C_{eu}, and the first yield load level, C_s (Fig. 1).

$$R_f = \frac{C_{eu}}{C_s} = \frac{C_{eu}}{C_y} \times \frac{C_y}{C_s} \tag{1}$$

As seen from Eq. (1), the response modification factor depends on two parameters:

(a) The available ductility, which is expressed by ductility reduction factor R_μ, defined as:

$$R_\mu = \frac{C_{eu}}{C_y} \tag{2}$$

Where, $R_\mu = \mu$ for equal displacement $R_\mu = \sqrt{2\mu - 1}$ for equal energy and,

(b) The reserve strength that exists between the actual structural yield and the first member yield (assuming elastic-plastic idealisation), which is expressed by the over-strength factor, Ω, defined as:

$$\Omega = \frac{C_y}{C_s} \tag{3}$$

Eq. (1) can be rewritten in terms of the over-strength factor and the ductility reduction factor as given by Eq. (4).

$$R_f = \frac{C_{eu}}{C_y} \times \frac{C_y}{C_s} = R_\mu \Omega \tag{4}$$

It is evident that in redundant structures, the over-strength of the structural system can be equally as important as the ductility, in the selection of appropriate structural response modification factors.

In addition to these factors, extra over-strength is provided to the structure indirectly by the strength reduction factor, ϕ, given in AS 3600 [11]. In effect at the design stage, these strength reduction factors (eg. $\phi = 0.8$ for bending) reduce the dependable elastic strength of the overall structure from C_s to C_e (Fig. 1).

$$C_e = \frac{C_s}{1.25} \tag{5}$$

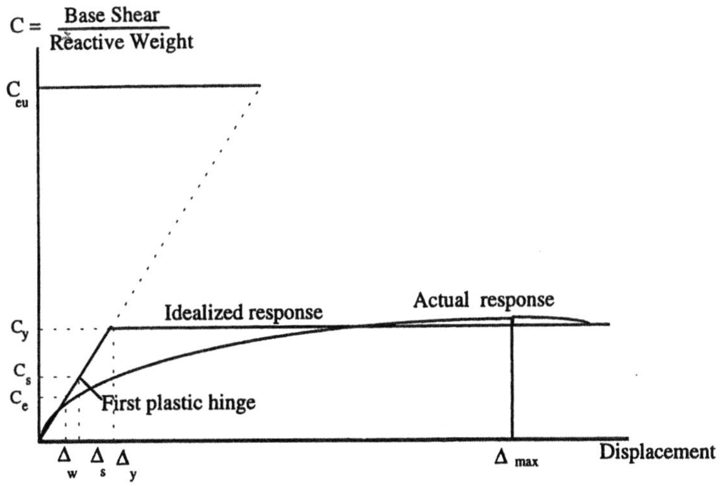

Fig. 1. Full range structural response.

Both EC8 and the Australian earthquake standard AS1170.4 specify three levels of design elastic strengths, C_s, with appropriate ductility requirements for RC moment resisting frames. Table 1 gives a summary of the Australian Standard requirements for seismic RC frames, where C is the seismic design base shear coefficient, S the soil parameter (= 1.0 for rock sites and 1.2 for stiff soil sites) and I, the structure importance factor (= 1.25 for structures with an important post-earthquake function).

In the European seismic code EC8, with regard to the required hysteretic dissipation capacity three ductility classes DC"L" (low ductility), DC"M" (medium ductility) and DC"H" (high ductility) are distinguished for concrete structures:

DC"L". Ductility Class "L" corresponds to structures designed and dimensioned according to EC2 [12], supplemented by rules enhancing available ductility.

DC"M". Ductility Class "M" corresponds to structures designed, dimensioned and detailed according to specific earthquake resistant provisions, enabling the structure to enter well within the inelastic range under repeated reversed loading, without suffering brittle failures.

DC"H". Ductility Class "H" corresponds to structures for which the design, dimensioning and detailing provisions are such as to ensure, in response to the seismic excitation, the development of chosen stable mechanisms associated with large dissipation of hysteretic energy.

In order to provide the appropriate amount of ductility in the three ductility classes, specific provisions for all structural elements must be satisfied in each class [13]. In correspondence with the different available ductility in the three ductility classes, different values of the behaviour factor, q, are used for each class. The behaviour factor is the equivalent of the ductility reduction factor R_μ in Eq. (4).

In low seismicity zones such as in northern Europe, concrete buildings may be designed under the seismic load combination following only the rules of EC2 [12] and neglecting the specific provisions given in EC8 [1] provided a specific assessment of the behaviour factor q is made based on the principles of EC8.

Table 1
AS 1170.4 requirements for reinforced concrete frames

Type	Seismicity	C_s	R_f	Detailing
Ordinary Moment Resisting Frames (OMRF)	All areas $h < 50$ m for $a.S >$ 0.10	C_s = CSI/4 25% of C_{eu}	4	AS 3600 (low ductile)
Intermediate Moment Resisting Frames (IMRF)	All areas	C_s = CSI/6 17% of C_{eu}	6	AS 3600, Appendix A (limited ductility)
Special Moment Resisting Frames (SMRF)	All areas	C_s = CSI/8 12.5% of C_{eu}	8	ACI 318–95 Ch.21 (fully ductile)

a = design ground acceleration (g).

The behaviour factor q, introduced fundamentally to account for energy dissipation capacity is derived for each design direction as follows:

$$q = q_0 k_D k_R k_W \geq 1.5 \quad (6)$$

where, q_0 = basic value of the behaviour factor, dependent on the structural type, k_D = factor reflecting the ductility class, k_R = factor reflecting the structural regularity in elevation, and k_W = factor reflecting the prevailing failure mode in structural systems with walls.

For regular RC moment-resisting frames, the factors k_R and k_W are both equal to unity and the basic value of the behaviour factor, q_0 is 5.0. The factor k_D reflecting the ductility class is taken as follows:

$$k_D = 1.00 \text{ for DC"H"}, \quad (7)$$

$$0.75 \text{ for DC"M" and } 0.50 \text{ for DC"L"}$$

and hence for the equivalent three categories of regular RC frame structures given in Table 1, the EC8 values for q are as follows:

$$\text{OMRF } q = 2.5$$

$$\text{IMRF } q = 3.75 \quad (8)$$

$$\text{SMRF } q = 5.0$$

Hence, comparing the EC8 q-values with the Australian code R_f values in Table 1, together with Eq. (4), it may be deduced that RC frame systems with over-strength factor, $\Omega = 1.6$ will have identical lateral design strengths (base shear) when designed according to either EC8 or AS1170.4, assuming equal soil and importance factors. In single degree of freedom systems, the response modification factor and the displacement ductility ratio are the same since the over-strength factor, Ω, equals unity [14]. However these parameters are difficult to quantify in redundant structures. Consequently the current practice used for predicting the non linear response of complex redundant structures subjected to dynamic loading is based on the simple scaling of the elastic response of an equivalent single degree of freedom system by the structural response modification factor. This response modification factor, R_f, is independent of ground motion and structural dynamic characteristics. Further discussion of these points and estimates of the over-strength factor, Ω, for the RC frame systems examined in the case studies of this paper are given elsewhere [15].

It is important that the level of ductility assumed in the design can be delivered by the structural system. The overall displacement ductility μ is defined as $\Delta_{max}/\Delta_{yield}$ where Δ_{max} and Δ_{yield} are the maximum and yield displacements at the top of the structure. This overall displacement ductility is generally considerably less than the local ductility demand of individual members defined by the rotational ductility, $\theta_{max}/\theta_{yield}$ or the curvature ductility, ϕ_{max}/ϕ_{yield} (θ_{yield} and ϕ_{yield} are rotation and curvature at the yield level and θ_{max} and ϕ_{max} are the maximum rotation and curvature). The detailing requirements and the drift limitations specified in Australia (AS3600 [11] and AS1170.4 [2]) and in Europe (EC8 Part 1.3 [1]) attempt to ensure that the local ductility demands of members can be achieved so that the overall displacement ductility is not compromised.

A discussion of design and detailing requirements for the three classes of RC frames studied herein, has been given by DeSilva et al. [16]. Overall, very similar requirements pertain to the design of seismic RC frames in Europe, as detailed in CEB [13].

In the case of special moment resisting frames (SMRF), the recommended earthquake resistant design approach is to ensure that a rational yielding mechanism develops when the structure is subjected to severe inelastic deformations. The concept of a strong column-weak beam is recommended by codes such as ACI 318 [10] in regions of high seismicity. Generally, the curvature ductility demands are greater if the columns commence yielding prior to the beams, particularly if plastic hinges form in the columns of one storey creating a 'soft' storey. The above design and detailing requirements have been specified in ACI 318–95 to encourage plastic hinges to develop in beams. However it should be noted that these requirements do not guarantee the prevention of plastic hinges forming in columns.

3. Case study 1: force-based (FB) seismic assessment

A case study has been carried out to assess the performance levels of different types of moment resisting frames recommended in AS 1170.4 and EC8. A typical 6 storey building shown in Fig. 2 was chosen for the case study and designed for three different ductility levels (R_f = 4, 6 and 8) in accordance with the recommendations of AS3600 and AS 1170.4. As noted above, these three ductility levels correspond very closely to those designated in EC8 as DC "L", "M" and "H", respectively, see Eqs. (7) and (8). An additional frame structure was designed for gravity and wind loads, ignoring seismic loads. This structure is labeled as an existing structure as most of the existing structures in Australia are designed for gravity and wind forces only. A brief description of the design procedure, member reinforcement details, response and yielding mechanisms are presented in this paper. A complete description is given by DeSilva [17].

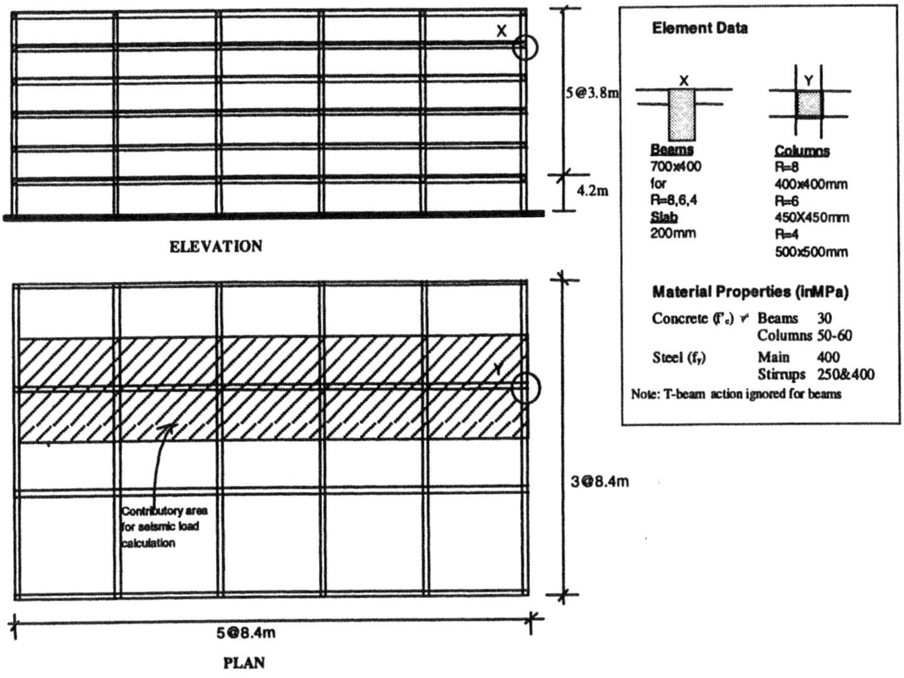

Fig. 2. Structure configuration and member properties.

3.1. Design procedure

The preliminary estimation of member properties was based on the equivalent static method specified in AS 1170.4 [2]. Main design parameters are summarised in Table 2. The seismicity is taken typical of that for a low to moderately active seismic region, with a design peak ground acceleration (500 year return period) of a = 0.11*g*, as further discussed below.

Table 2
Design parameters, based on the Australian earthquake code AS 1170.4

Lateral load resisting system	100% Moment resisting frames
Dead Load due to self weight	5.5 kPa
Dead Load due to fixtures	1.3 kPa
Live Load (Floors)	3 kPa
Live Load (Roof)	1 kPa
Load combinations	1.25DL + 1.5LL (All Spans) 1.25DL + 1.5LL and 1.0DL (Alt. Spans) 0.8DL + 1.5LL 1.1DL + 0.4LL 1.25DL + 0.4LL + W_u 0.8DL + W_u 1.0DL + 0.4LL + F_{eq} 0.8 (1.0DL + 0.4LL) + F_{eq}
Seismicity, Importance factor, Site factor and Response modification factors	a = 0.11, S = 1.0, R_f = 4 (OMRF), R_f = 6 (IMRF), R_f = 8 (SMRF) I = 1.25 (Structure with a post disaster function)

3.2. Estimation of seismic forces

The fundamental or the natural period of a structure is required to estimate the seismic loads. The accurate estimation of the natural period depends on the mass and stiffness distribution of the structure which in turn cannot be determined without the knowledge of the design loads. Consequently AS 1170.4 provides an empirical formula for the calculation of the fundamental natural period based on the height of the structure.

$$T = \frac{h}{46} \tag{9}$$

Where h is the height of the building in metres.

For the frame structure considered in this study, Eq. (9) gives a period estimated at 0.5 seconds.

In EC8 [1], for buildings with heights up to 80 m, the value of the fundamental period T may be approximated from the following formula.

$$T = C_t h^{\frac{3}{4}} \tag{10}$$

Where C_t = 0.075 for moment resisting concrete

frames as studied here. This empirical method gives a period $T = 0.8$ sec, considerably larger than that estimated from Eq. (9).

Significant changes to the calculated design earthquake loads can result from inaccuracies in the estimation of the natural period of the structure. However the lateral load estimation, design, and detailing of these frames are based on the equivalent static method of AS 1170.4, with the intention of representing general design practice, not only in Australia but also in other regions of low to moderate seismicity.

The equivalent static base shear was estimated using the AS1170.4 code formula for three ductility levels:

$$V = C_s G_g \tag{11}$$

Where C_s is given in Table 1 and G_g is the gravity force, and defined in AS 1170.4 as

$$G_g = G + \psi_c Q \tag{12}$$

where, G is the dead load, Q is the live load, and ψ_c is the factor for strength and stability limit state ($\psi_c = 0.4$).

Lateral seismic load at each storey level was calculated using the inverse triangular distribution recommended in AS 1170.4. A static analysis was performed for all the load combinations given in Table 2. Comparison of overturning moments due to wind and earthquake forces are given in Fig. 3. The members were designed according to AS 3600 and detailed to the AS 1170.4 requirements described in the previous section. Designed cross sections and reinforcement details are presented in Table 3.

A push-over analysis was used in this study [15] to confirm the over-strengths of the frames. The over-strengths (Ω) calculated by Eq. (3) for OMRF, IMRF and SMRF were 1.3, 1.4 and 1.8 respectively. These values confirm that RC frame systems will approximately have identical lateral design strengths (base shear) when designed according to either EC8 or AS1170.4, assuming equal soil and importance factors (see Section 2).

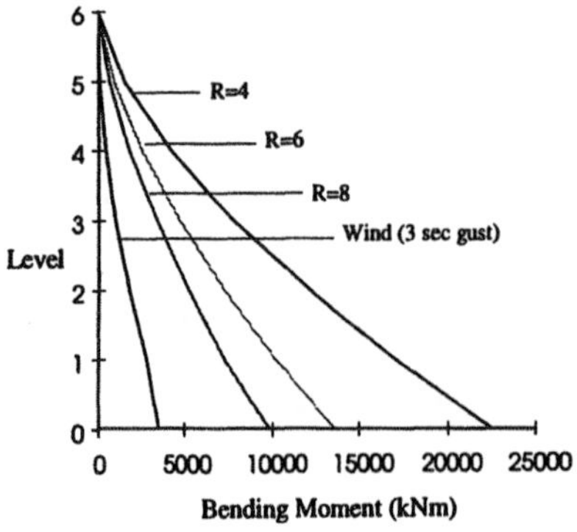

Fig. 3. Overturning moments due to lateral forces.

3.3. Ground motions

The El Centro earthquake (PGA = 0.33g) was selected to represent a major inter-plate earthquake (Fig. 4). The time history analyses were performed using the nonlinear dynamic structural analysis program DRAIN2D [18]. The beams were modelled by the Modified Takeda Model provided in the program. The column behaviour was modelled by the axial force-moment interaction relationship of the column cross-section. A ground motion was also developed for the time history analyses in the form of a synthetic accelerogram generated using the earthquake simulation program SIMQKE [19]. This synthetic ground motion was generated with the AS1170.4 design spectrum as the target spectrum. The details of this latter analysis are given elsewhere [15].

3.4. Results from force-based method

The following response parameters are presented here to illustrate the dynamic behavioural characteristics of the structures designed for different ductility levels. The maximum storey displacements of frames with different ductility levels are compared in Fig. 5. The roof displacement time history results of three moment resisting frames under the first 10 seconds of the El Centro excitation are compared in Fig. 6. The roof displacements of the three frame types do not differ substantially and are in all three cases less than 0.6% of the building height ($0.006h$ = 13 cm). It has been suggested by researchers such as De Stefano et al. [20] that the onset of severe structural damage occurs approximately at an overall (roof) displacement of $0.01h$. Hence the damage levels in the considered frames are expected to be moderate, when subjected to the El Centro earthquake ground motion.

The yielding mechanisms of three moment resisting frames under El Centro earthquake are shown in Fig. 7. The yielding mechanism of the "existing" frame (designed only for wind forces) is also compared in Fig. 7. All the structures other than the special moment resisting frame, which was designed using the capacity design concept, exhibited a column side sway mechanism, as expected. Finally a comparison of the critical displacement ductility ratios are given in Table 4 under the El Centro earthquake. The yield displacement, Δ_y, in Table 4 is the displacement at the top of the structure, when the first plastic hinge is formed. The displacement ductility demand of the special moment resisting frame was about three times more than that of the ordinary moment resisting frame. These additional ductility demands have resulted from the 50% reduction (Table 1) in design elastic lateral force of the special moment

Table 3
Reinforcement details of members

Beams	GL–L2	L2–L4	L4–L6		GL–L2	L2–L4	L4–L6
Existing	T–3Y28 + 3Y24 B–2Y24 R10 150 c/c	T–3Y28 + 3Y20 B–2Y24 R10 150 c/c	T–5Y24 B–2Y24 R10 150 c/c	IMRF (R_f = 6)	T–4Y28 + 4Y24 B–4Y24 R10 150 c/c	T–8Y24 B–4Y24 R10 150 c/c	T–4Y28 B–2Y28 R10 150 c/c
OMRF (R_f = 4)	T–4Y28 + 4Y32 B–2Y28 R10 150 c/c	T–4Y24 + 4Y28 B–2Y28 R10 150 c/c	T–3Y28 + 3Y24 B–2Y24 R10 150 c/c	SMRF (R_f = 8)	T–3Y28 + 3Y20 B–4Y24 R10 150 c/c	T–3Y28 + 3Y20 B–4Y24 R10 150 c/c	T–5Y24 B–3Y24 R10 150 c/c

	External columns				Internal columns			
		GL–L2	L2–L4	L4–L6		GL–L2	L2–L4	L4–L6
Existing		400 × 400 mm 8Y32 R10 150 c/c	400 × 400 mm 8Y32 R10150 c/c	400 × 400 mm 8Y32 R10 150 c/c		400 × 400 mm 12Y32 R10 150 c/c	400 × 400 mm 12Y32 R10 150 c/c	400 × 400 mm 12Y32 R10 150 c/c
OMRF (R_f = 4)		500 × 500 mm 6Y28 + 8Y24 R10 150 c/c	500 × 500 mm 8Y32 R10 150 c/c	500 × 500 mm 8Y32 R10 150 c/c		500 × 500 mm 8Y32 + 4Y28 R10 150 c/c	500 × 500 mm 12Y20 R10 150 c/c	500 × 500 mm 12Y20 R10 150 c/c
IMRF (R_f = 6)		450 × 450 mm 8Y28 R10 100 c/c	450 × 450 mm 8Y28 R10 100 c/c	450 × 450 mm 8Y28 R10 100 c/c		450 × 450 mm 12Y28 R10 100 c/c	450 × 450 mm 12Y16 R10 100 c/c	450 × 450 mm 12Y16 R10 100 c/c
SMRF (R_f = 8)		400 × 400 mm 8Y32 2Y12 60 c/c	400 × 400 mm 8Y32 2Y12 60 c/c	400 × 400 mm 8Y32 2Y12 60 c/c		400 × 400 mm 12Y32 2Y12 60 c/c	400 × 400 mm 12Y32 2Y12 60 c/c	400 × 400 mm 12Y32 2Y12 60 c/c

Note: Beam dimensions are given in Fig. 2.
Main Rft: "Y"—Deformed Bars, Yield Strength = 400 MPa.
Lateral Rft: "R"—Plain Bars, Yield Strength = 250 MPa.

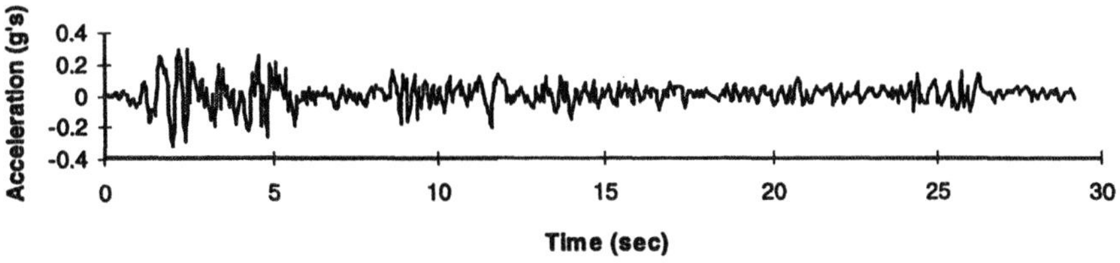

Fig. 4. El Centro earthquake, PGA = 0.33*g*.

Table 4
Ductility demands under El Centro earthquake

Frame type	Available displacement ductility ($R = \mu$)	Displacement ductility demand (Δ_{max}/Δ_y)	Mechanism
OMRF	4	1.13	column
IMRF	6	1.62	beam and column
SMRF	8	3.21	beam and column

resisting frame compared with the ordinary moment resisting frame.

4. Case study 2: displacement-based (DB) seismic assessment

The moment-resisting frames used in case study 1 have also been assessed using a displacement based (DB) method. The objective is to obtain a comparison with the results of the earlier force-based assessment (Section 3) and hence to provide some preliminary

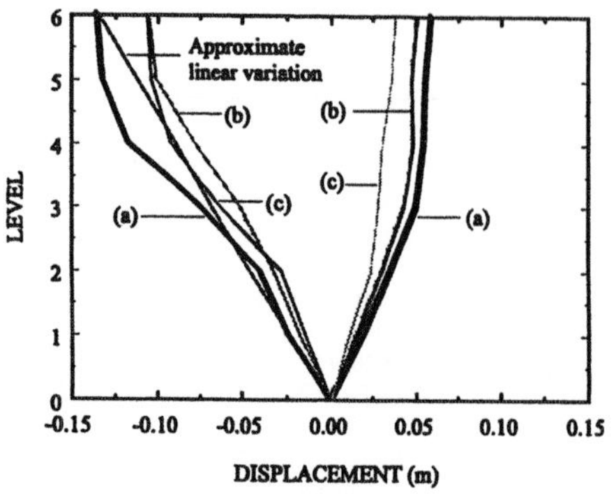

Fig. 5. Floor displacements of frames under El Centro earthquake (a) SMRF ($R_f = 8$) (b) IMRF ($R_f = 6$) (c) OMRF ($R_f = 4$).

evaluation of the newly proposed displacement-based approach, the methodology of which has been discussed in the literature [3–7], but without direct comparison with force-based results obtained by rigorous dynamic analysis. The displacement-based method was applied to the case of the structure subjected to the El Centro earthquake ground motion, which for the 3 frames (OMRF, IMRF, SMRF) produces varying degrees of inelastic behaviour and contrasting yielding mechanisms (Section 3.4).

4.1. Outline of displacement-based (DB) methodology

4.1.1. Effective SDOF period

Fig. 8 defines the stiffnesses for both force-based and displacement-based assessment, defined in terms of an idealised elastic perfectly plastic system (EPP). The

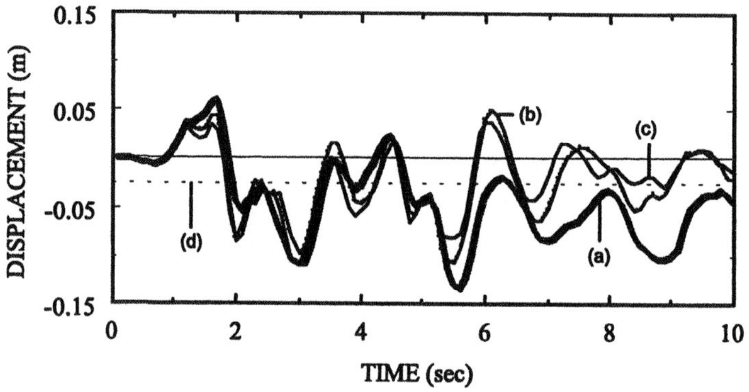

Fig. 6. Roof displacement time history of frames under El Centro earthquake (a) SMRF ($R_f = 8$) (b) IMRF ($R_f = 6$) (c) OMRF ($R_f = 4$) (d) Wind (3 sec gust).

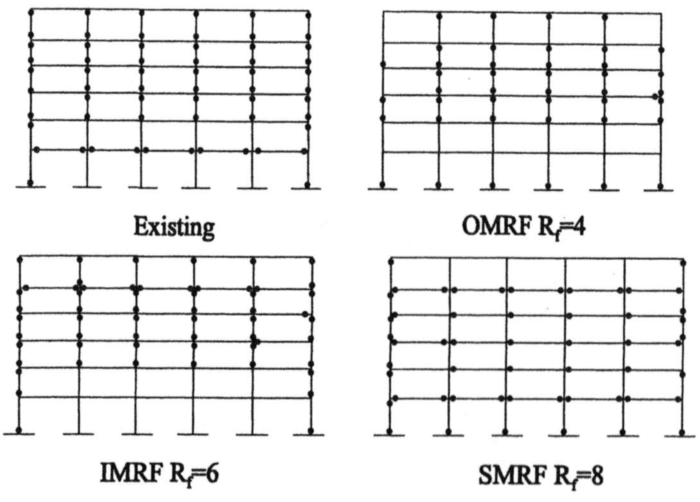

Fig. 7. Dynamic yielding mechanisms of moment resisting frames under El Centro earthquake (PGA = 0.33*g*).

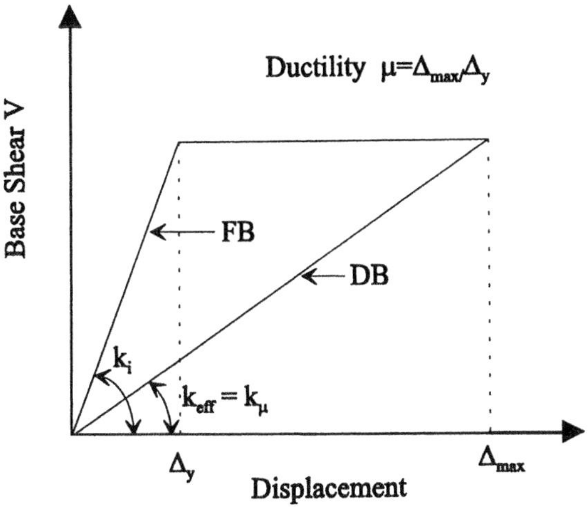

Fig. 8. Stiffness for force and displacement based assessment.

initial or elastic stiffness k_i may be accurately computed for the FB method, and subsequently determines the fundamental structural period, T_i. For a yielding system, the effective stiffness k_{eff} (Fig. 8) is related to both the shape of the loading curve (EPP) and the ductility demand μ, interpreted here as the ratio Δ_{max}/Δ_y (Δ_{max} = maximum roof displacement and Δ_y = roof displacement at first plastic hinge formation). For the displacement-based method, with the same base shear strength and a displacement ductility factor μ,—the effective stiffness at maximum response is

$$k_{eff} = k_m = \frac{l}{\mu} k_i \tag{13}$$

The effective period at peak response is thus

$$T_{eff} = T_e = (T_i)(\mu)^{1/2} \tag{14}$$

Although not attempted in this study, the push-over test results can be used to refine the effective stiffnesses.

4.1.2. Effective (equivalent) damping

In the displacement-based assessment method, a relationship is required to define the viscous damping equivalent to hysteresis and ductility. For an EPP system as in Fig. 8, the following relationship holds, under the hypothesis of a resonant response with all cycles at maximum amplitude (with ξ viscous damping ratio and μ ductility demand):

$$\xi_u = \frac{2\left(1 - \frac{1}{\mu}\right)}{\pi} \tag{15}$$

More complex relationships can be developed for other hysteresis rules, commonly needed for reinforced concrete members, as shown in Fig. 9 [after Calvi and Pavese [3]].

The relationship of Eq. (15) is shown in Fig. 9 as the upper bound (UB) curve. As seen from Fig. 9, there is a considerable amount of scatter in the results obtained by different researchers. This topic requires further research. For this study, the simple relationships (Fig. 10) presented by Priestley [7] are used. As shown in Fig. 10, the energy dissipated in beam plastic hinges is typically about 60% larger than in column plastic hinges, and this should be recognised in the estimation of equivalent viscous damping. The damping associated with a column yielding mechanism is indicated in Fig. 10 to be about 40% of ξ_u, the latter as given by Eq. (15).

4.2. Smeared damping ratio, ξ_e

For a structural frame undergoing a plastic mechanism under seismic loadings (Fig. 7), the application of the displacement-based assessment method requires an effective or smeared damping ratio ξ_e to be assigned to the structure as a whole. The empirical procedure adopted in this paper for this purpose has been to assign a damping value to each storey level, based on the mechanism formed (if any) and whether it is formed by beam or column hinging, or a combination of these effects. The yielding mechanisms shown in Fig. 7, obtained by dynamic analysis using the El Centro earthquake input, have been combined with the associated displacement ductility demands in Table 4 to assign an equivalent damping ratio to each storey level, according to the methodology outlined in Section 4.1.2 above. The key components of this methodology are the relationships defined in Eq. (15) and Fig. 10. Based on these storey damping ratios, an average value has been obtained for the equivalent damping ξ_e associated with the structure as a whole. The equivalent damping includes an allowance of 5% ($\xi = 0.05$) to allow for viscous effects in the elastic range of response as also assumed in the force-based dynamic analysis described in Section 3 above. A summary of the damping ratios computed by this methodology, for the case study frames, is given in Table 5.

The damping values given in the last column of Table 5 are in close agreement with similar calculations on effective damping ratios for frames undergoing beam and column yielding mechanisms, described by Calvi and Pavese [3].

4.3. Calculation of spectral displacements

Using the methodology of Section 4.1, along with Eq. (14), the effective periods T_e for the case study frames have been calculated, as shown in Table 6. The effective

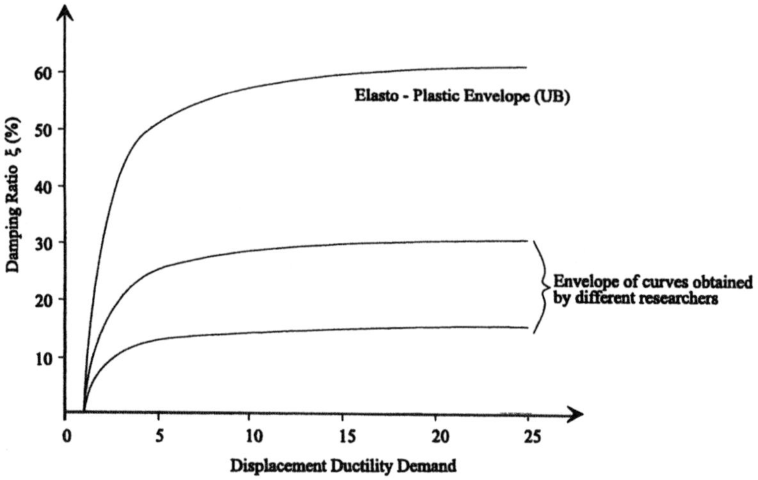

Fig. 9. Typical relations between equivalent viscous damping and ductility, depending on the actual structural response (after Calvi and Pavese [3]).

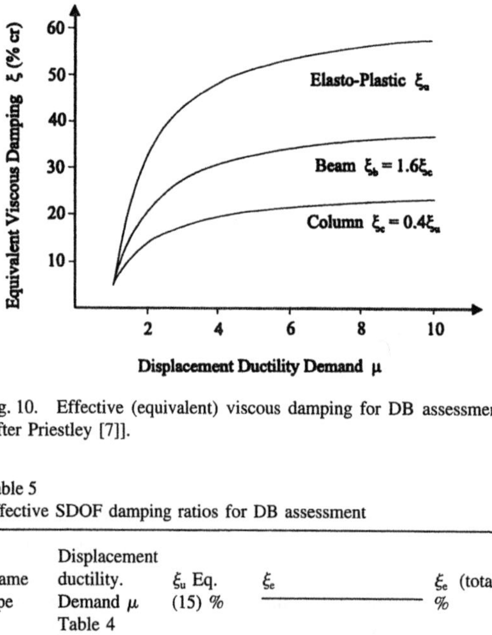

Fig. 10. Effective (equivalent) viscous damping for DB assessment [after Priestley [7]].

Table 5
Effective SDOF damping ratios for DB assessment

Frame type	Displacement ductility. Demand μ Table 4	ξ_u Eq. (15) %	ξ_e Viscous (elastic)	ξ_e Hysteretic (plastic)	ξ_e (total) %
OMRF	1.13	7.3	5	2	7
IMRF	1.62	24.4	5	8	13
SMRF	3.21	43.8	5	20	25

spectral displacements given in the last column of Table 6 have been determined from the 5% damped elastic displacement response spectrum (DRS) of the El Centro NS earthquake ground motion, illustrated in Fig. 11. The periods T_e from Table 6 have been used for this purpose.

Table 6
Effective period and 5% damped spectral displacements

Frame Type	T_i (sec) Calculated	Disp. ductility demand μ (as for Table 4)	T_e (sec)	$(S_D)_e$(mm)**
OMRF	1.3	1.13	1.38	90
IMRF	1.5	1.62	1.91	112
SMRF	1.7	3.21	3.05	265

**$(S_D)_e$ calculated from 5% damped El Centro spectrum at 2/3 of height (ie. Level 4).

The next stage of the displacement-based assessment of the case study frames has been to adjust the displacements $(S_D)_e$ from Table 6, to allow for the effective damping ratios ξ_e (total) computed in Table 5. For this purpose, displacement spectra for damping values different from 5% have been obtained by multiplying the 5% displacement spectrum (El Centro earthquake) in Fig. 11 by the factor:

$$\mu = \left(\frac{7}{2+\xi}\right)^{\frac{1}{2}} \tag{16}$$

Eq. (16) has been used in developing displacement spectra for the European seismic code EC8 [1].

In Table 7, the spectral displacements $(S_D)_e$ from Table 6 have been corrected according to Eq. (16), using the effective damping values ξ_e computed above (Table 5).

Under the development of plastic hinge yielding mechanisms as in Fig. 7, it has been suggested [7] that the displacement profile under severe seismic ground

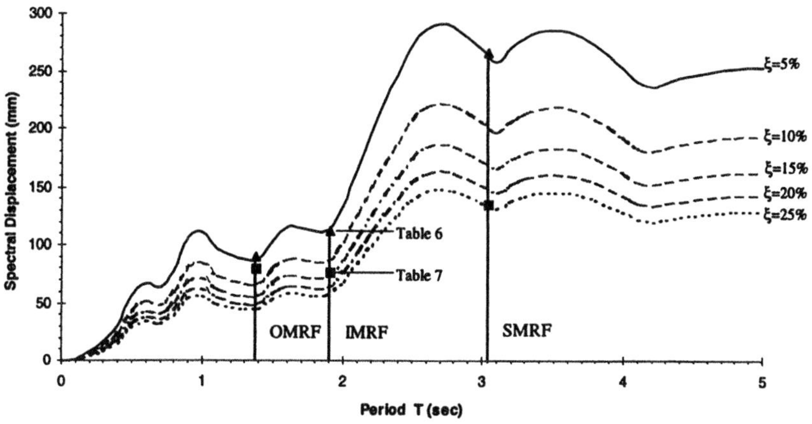

Fig. 11. Displacement spectra for the El Centro NS earthquake.

Table 7
Damping correction to determine predicted (effective) displacements

Frame type	5% damped $(S_D)_e$ [Table 6] (a) mm	$\xi_e(\%)$ [Table 5]	Damping correction factor η [Eq. (15)] (b)	Predicted displacements at Level 4 (a) × (b) mm
OMRF	90	7	0.88	79
IMRF	112	13	0.68	76
SMRF	265	25	0.51	135

motions may be approximated as a linear variation. The results from the force-based dynamic analyses shown in Fig. 5 indicate that this approximation is an adequate one, for the case studies carried out. The assumption of a linear variation of maximum floor displacements with height implies that the effective inertial mass of the frame may be assumed to be at two-thirds height, namely at level 4. Hence a direct comparison may be made between the displacements in Table 7 as predicted by the displacement-based equivalent SDOF methodology, and the calculated displacements shown in Fig. 5 at Level 4 of the three case study frames. This comparison has been made in Table 8.

It is seen from Table 8 that the two methods have been found to give displacement predictions that match to within 20%. The conventional force-based method gives a slightly larger than predicted displacements for the OMRF, whilst the two methods give almost identical results for the IMRF. For the seismic capacity designed SMRF, the displacement-based method conservatively estimates the structural displacement, the predicted value exceeding that from force-based analysis by 11%. In this study Eq. (16) was used to convert spectral displacements from 5% level to effective damping values of the different structures. Further refinements can be made by generating elastic displacement response spectra for other damping ratios. This work is presently underway in a joint project being undertaken at the University of Melbourne and the University of Hong Kong.

Table 8
Comparison of displacement-based and force-based Methods for Case Study Frames under El Centro excitation

Frame type	DB (predicted) mm	FB (calculated) mm	FB/DB
OMRF	79	95	1.20
IMRF	76	78	1.03
SMRF	135	120	0.89

Note: the quoted displacements are at level 4 of the structures considered.

5. Discussion and concluding remarks

1. The earthquake force combination governed the design of all frames except the upper storeys of the special moment resisting frame ($R_f = 8$) in which the design was governed by the gravity dead and live loads.
2. The structure response factors R_f specified in AS1170.4 are empirically based and reflect both the ductility and over-strength (Ω) of the structural framing system. The EC8 values (q) are comparable when Ω is included.

3. The ‘existing’ reinforced concrete frame designed for wind forces only, developed plastic hinges in the columns under the 0.33g El Centro earthquake. Such column sway mechanisms are discouraged in regions of higher seismicity due to very high curvature ductility demands, associated with this type of yielding.
4. Both OMRF and IMRF developed plastic hinges in the columns under the El Centro earthquake. The SMRF generally developed plastic hinges in the beams rather than the columns. This was consistent with the ACI 318–95 strong column-weak beam detailing philosophy used in the design of this SMRF.
5. The displacement ductility and rotation ductility demands of the SMRF during the 0.33g El Centro earthquake were some 3 times that of the OMRF, demonstrating the trade off between the ductility demand and the level of structural strength provided to resist lateral loading.
6. Complex analyses, such as non-linear time history analysis of multi-degree-of-freedom systems are not practical enough for everyday design use and are not appropriate for specification in design standards. A recently proposed method for displacement-based (DB) seismic assessment of existing RC frame structures has been implemented for this purpose, from which it has been concluded that the displacement-based approach predicts accurately the overall displacement demands for such frames. These results, whilst limited to the consideration of a small number of seismic frame structures and a single, typical strong earthquake ground motion, nevertheless give confidence that the displacement-based approach can rapidly and easily facilitate a seismic assessment of an existing RC structure, without the necessity to undertake detailed inelastic dynamic analyses.
7. This conclusion gives further stimulus to the investigation of a more detailed programme of research into the displacement-based approach, leading eventually to its incorporation in codified form in the seismic design procedures of national and international building standards. A detailed investigation should include refinements such as the use of more accurate methods to find the effective stiffnesses, generation of displacement spectra for different damping ratios, consideration of stiffness and strength degradation of concrete elements, a sensitivity analysis including more earthquakes and a variety of structural models such as irregular frames and frame-wall seismic systems. However the displacement-based method used in this paper is a very versatile method that can be easily implemented in a design office.

Acknowledgements

The dynamic analysis reported in this paper was performed by Saman De Silva as part of his Ph.D. project at the University of Melbourne. His assistance is gratefully acknowledged by the authors. The contribution of Chris Panagopoulos in helping to prepare the drawings and Mr. John Wilson and Dr. Nelson Lam for stimulating discussion are also greatfully acknowledged.

References

[1] CEN. Eurocode 8: Earthquake resistant design of structures—Part 1: General Rules, 1995.
[2] AS 1170.4. The design of earthquake resistant buildings, Standards Association of Australia, 1993.
[3] Calvi GM, Pavese A. Displacement based design of building structures. Proceedings of the European Seismic Design Practice, Rotterdam: Balkema, 1995:127–32.
[4] Fardis MN. Current trends in earthquake resistant design of reinforced concrete structures. Editor. In: Elnashai A, European Seismic Design Practice, 1995:375–82.
[5] Mayes RL. Interstorey drift design and damage control issues. The Struct Desn of Tall Bldgs 1995;4:15–25.
[6] Moehle JP. Displacement-based design of RC structures subjected to earthquakes. Earthquake Spectra 1992;8(2):403–28.
[7] Priestley MJN. Displacement-Based seismic assessment of existing reinforced concrete buildings. Proceedings of the Pacific Conference on Earthquake Engineering, Australia, 1995:225–44.
[8] Park R. A static force-based procedure for the seismic assessment of existing reinforced concrete moment resisting frames. Proceedings NZ National Conf on Earthquake Engineering, New Plymouth, 1996.
[9] UBC. Earthquake regulations. Uniform Building Code. Whittier, California, 1994.
[10] ACI 318. Building code Requirements for Reinforced Concrete, 1995.
[11] AS 3600. Concrete Structures. Standards Association of Australia, 1994.
[12] CEN. Eurocode 2: Design of concrete structures—Part 1: General Rules, 1992.
[13] CEB. RC frames under earthquake loading. Thomas Telford Publishing, 1996.
[14] Paulay T, Priestley MJN. Seismic Design of Reinforced Concrete and Masonry Buildings. New York: J. Wiley and Sons, 1992.
[15] Chandler A, Mendis P. A comparison of force and displacement based seismic assessment methods for reinforced concrete frames. Structural Engineering Research Report 2/97, Dept. of Civil and Environmental Engineering, University of Melbourne, 1997.
[16] De Silva S, Mendis PA, Wilson J. Design and detailing of R/C structures in accordance with the new Australian Earthquake Standard. Australian Civil Eng Trans 1994;36(3):245–56.
[17] De Silva S. Response of strength degrading limited ductile R/C frames under intraplate earthquakes. PhD thesis, University of Melbourne, 1997.
[18] Kannan AE, Powell GH. Drain-2D, A general computer program for dynamic analysis of inelastic plane structures. Report No. EERC 73–22, University of California, Berkeley, California, 1975.
[19] SIMQKE. A program for artificial motion generation, users manual and documentation. Department of Civil engineering, Massachusetts Institute of Technology, 1976.
[20] De Stefano M, Faella G, Realfonzo R. Seismic response of 3D RC frames: Effect on plan irregularity. Editor Elnashai A. In: European Seismic Design Practice, 1995:219–26.

Wrinkling on stretched circular membrane under in-plane torsion: Bifurcation analyses and experiments

Tomoshi Miyamura *

Institute of Environmental Studies, Graduate School of Frontier Sciences, The University of Tokyo, Tokyo, Japan

Received 21 June 1999; received in revised form 3 November 1999; accepted 10 November 1999

Abstract

Wrinkling on a stretched circular membrane under in-plane torsion is a typical and classical problem in the study of wrinkling on membrane structures. This paper presents the experiment and bifurcation analysis concerning this problem. Both isotropic polyester film and orthotropic PVC coated textile are considered. The nonlinear finite element method is used in the analysis. In the experiment, a method to measure stresses in the wrinkled membranes is proposed, and stresses are measured. The stress distributions and the shape of wrinkles observed in the experiments agreed well with those obtained by the analyses. The results of experiments and analyses also show that the tension field is formed in the wrinkled membranes. © 2000 Elsevier Science Ltd. All rights reserved.

Keywords: Wrinkling; Membrane; Bifurcation; Experiment

1. Introduction

An important characteristic of membranes is their thin, light quality. This is why membranes are often used in large span spatial structures. However, because of their small bending rigidity, they cannot support compressive stresses. When membranes deform in a compressive direction, buckling occurs and wrinkles are formed. Therefore, studying the wrinkling on stretched membranes is an important subject from the viewpoint of designing membrane structures.

Wrinkling on stretched circular membranes is a typical problem in the studies of wrinkling, i.e. a benchmark problem. Many analytical or experimental studies have been conducted, including several by the present author [1–4].

In analytical studies there are two different approaches: the bifurcation analysis and the tension field analysis. First, studies based on the bifurcation theory are considered. Because wrinkles are deformations made by the bifurcation buckling, it makes sense to analyze the wrinkling by employing the bifurcation buckling theory. The shape of wrinkles can be obtained precisely by this approach. The number of wrinkles can be determined by considering the bending rigidity of membranes. The bifurcation analysis can also be performed when the bending rigidity is omitted and only the geometrical stiffness is considered.

Kármán derived the governing equation of the isotropic circular plates that takes into account the effect of initial stresses. Dean [5] approximately solved Kármán's equation under the condition of in-plane torsion by using trigonometric functions to present the solution. In the 1980s many researchers solved Kármán's equation by using the finite difference method [6–8]. In those studies, the elasto-plasticity and polar-orthotropic elasticity are considered. Suzuki et al. [9] analyzed the circular stretched membrane without bending rigidity but with geometrical stiffness by using the Rayleigh Ritz method. They also analyzed the membrane with bending rigidity but without initial tension, and compared the results. In [5–9], only the bifurcation buckling modes were obtained. Cheng et al. [10,11] analyzed the post-wrinkling configurations by using the finite difference method. They also considered the large in-plane rotation based on the Reissner's equation [12]. The result was compared with that of Kármán's equation.

* Correspondence address: c/o Department of Quantum Engineering and Systems Science, School of Engineering, The University of Tokyo, 7-3-1 Hongo, Bunkyo-ku, Tokyo 113-8656, Japan. Tel.: +81-3-5841-8976; fax: +81-3-5800-2429.

E-mail address: miyamura@garlic.q.t.u-tokyo.ac.jp (T. Miyamura).

Reprinted from *Engineering Structures* **22 (11)**, 1407-1425 (2000)

The tension field theory is another method to analyze the wrinkling. In this theory, the assumption that membranes do not support compressive stresses is introduced. There are many studies concerning tension field theory, and in these the problem of the circular membrane is often analyzed. Kondo [13], Reissner [14], and Iai [15] separately derived the fundamental equations of the tension field theory. They analyzed the circular membrane without initial tension under in-plane torsion. When initial tension is not introduced, a whole domain becomes a tension field immediately. Stein [16] proposed a theory that can analyze partly wrinkled membranes. As a typical example for his theory, a stretched infinite membrane under in-plane torsion was analyzed. Mikulas [17] analyzed the stretched circular membrane by using Stein's theory, and also carried out experiments and compared the results of the analyses with those of the experiments. There have been many other theoretical studies based on the tension field theory [18–22]. The theory is also used with the finite element method [23,24]. Rodemann et al. [25,26] presented a general theory that can consider geometrical nonlinearity, anisotropic material, and material nonlinearity. The theory was discretized by using the finite element method; the orthotropic elastic circular membrane under in-plane torsion was analyzed as an illustrative example [26].

There have been some experimental studies on wrinkling. As described above, Mikulas [17] carried out precise experiments on the circular membrane, and compared the results with the theoretical analysis based on Stein's theory. Photos of the wrinkles are found in [5,16]. Experiments on wrinkling in other problems can be found in [18,20,27–30].

The above survey shows that few studies exist that consider the following points: (1) wrinkling on orthotropic membranes, (2) experiments involving measurement of stresses in wrinkled membranes, and (3) bifurcation analysis based on the finite element method of wrinkles on stretched membranes.

In this paper, both isotropic and orthotropic membranes are considered, and experiments and bifurcation analyses concerning them are conducted. In the experiment described, a method to measure stresses in thin membranes is proposed. The results of measured stresses in the wrinkled membrane are then compared with the stresses obtained by the analyses. The shapes of wrinkles are also compared.

2. Stretched circular membrane under in-plane torsion

Fig. 1 shows a circular membrane investigated in the present study. In the center of the membrane, a rigid hub is attached. Displacements in the radial direction are prescribed at the outer boundary for introducing the initial tension. Then, the center hub is rotated to introduce in-plane torsion. Wrinkles are partly formed around the center hub, and the wrinkled region gradually spreads as the rotation increases.

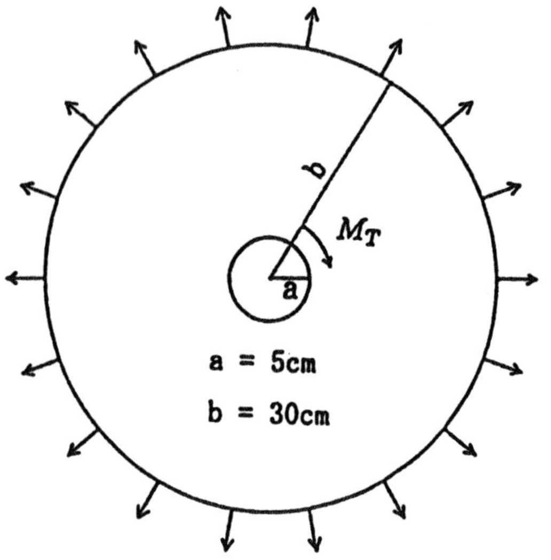

Fig. 1. Stretched circular membrane under in-plane torsion (1 cm=10 mm).

3. Test apparatus and method to measure stress

3.1. Overview

As described in Section 1, Mikulas carried out precise experiments of the circular membrane. The material of the membrane used in his experiment was isotropic elastic film. In the present experiment, orthotropic membranes are also tested; the membranes used in the spatial architectures are usually orthotropic. A method to measure stresses in thin membrane is proposed in Section 3.4.

3.2. Membranes used in the experiment

Table 1 illustrates membranes used in the present experiment. The polyester film is isotropic and almost elastic. The PVC (polyvinyl chloride) coated polyester textile is orthotropic. The stress–strain relationship of this membrane is not linear; that is, initial rigidity is small, but it becomes large after the textile is stretched enough [31].

Young's modulus and Poisson's ratio of the polyester film are measured by uniaxial tensile test. In the test, strain is calculated from the displacement measured by the potential meter attached to the film. Stress is calculated from the applied load. The PVC coated textile does not behave in an elastic manner, especially in the first

Table 1
Membranes used in tests for stress measurement[a]

Material	t (thickness)	E_1t (Et)	$\nu_{12}(\nu)$	E_2t	ν_{21}	Gt
Polyester film	0.018	1022	0.267	–	–	404
PVC coated textile	0.050	391	0.0	194	0.36	11

[a] t: thickness (cm); E_it (Et): Young's modulus×thickness (kgf/cm); Gt: shear coefficient×thickness (kgf/cm); ν_{ij} (ν): Poisson's ratio; i,j=1: warp, 2: Fill (1 kgf=9.8 N, 1 cm=10 mm).

loading. However, if the creep strain is eliminated, it behaves like an elastic membrane and Young's moduli and Poisson's ratios of orthotropic elasticity can be measured. The material properties measured by Segawa et al. [31] are shown in Table 1 These properties depend on the ratio of initial stresses in warp and fill directions. The ratio for the properties in Table 1 is 1:1. Shear elastic modulus is measured by the test described in [32].

3.3. Test apparatus

The test apparatus is illustrated in Plates 1and 2 and Fig. 2 and 3. Eight jigs with chucks are attached to the outer boundary of the test specimens, and the jigs are pulled in radial directions by counterweights to introduce the initial tension (Plate 2). The hub is attached to the center of the test specimen; it is connected to the center axis supported by ball bearings. The in-plane torsion is introduced by using a bar connected to the center axis. The bar is pulled by counterweights whose total weight is measured by a load cell. The in-plane torsion is calculated from this weight and the length of the bar. The test apparatus including the jigs and the hub are made of steel.

3.4. A method to measure stress in thin wrinkled membrane

3.4.1. Concept of the method

Membranes used in spatial structures are generally not elastic. However, a part of thin membranes where a strain gage is glued behaves in an almost elastic manner; that is, stresses can be measured by the glued strain gage. Hence, when elastic properties of the part of a thin membrane with a strain gage are measured by tensile tests; then the stresses can be calculated from the strains measured by the strain gage. The part of the membrane where the strain gage is attached can be considered a 'composite material' made of membrane, strain gage, and adhesive; it is a kind of load cell embedded in the membrane.

Plate 1. Test apparatus.

The elastic properties of the 'composite material' can be evaluated by tensile tests using slender specimens as illustrated in Fig. 3 and Plate 3. In orthotropic membranes, the composite material is assumed to be orthotropic elastic. Therefore, the tests of the following three specimens are carried out to measure five elastic moduli under a plane stress condition: two specimens whose axes coincide with warp and fill directions, and a specimen whose axis and warp or fill intersect at 45°.

The stress field around the part where the gage is glued may be slightly disturbed, and this disturbance may induce some errors. However, as Iai [15] and Mansfield [18] described, "the tensile stresses in the wrinkled membrane would not be affected if the membrane were to be cut along the tension rays (lines)." This fact shows that a wrinkled membrane can be represented by an assemblage of the slender specimen as in Fig. 3. Since the stress concentration is not likely to occur in the uniaxial stress field, the present method can precisely measure the stresses, especially in the wrinkled membranes.

3.4.2. Specimens for the tensile test

The shape of a test specimen is illustrated in Fig. 3. Two circular strain gages are glued on both surfaces of the specimen at the same position. Strains in two directions are measured on the both surfaces, and the values are averaged in each direction. The width of the specimen is 11 mm; that coincides with the diameter of the strain gage, so that the whole load is applied to the portion with the strain gages.

3.4.3. Results of the tensile tests and evaluation of the elastic modulus

The tensile tests for the polyester film and the PVC coated textile are carried out twice. The results of the two tests for each membrane are almost identical. Fig. 4 illustrates the stress–strain relationships of the poly-

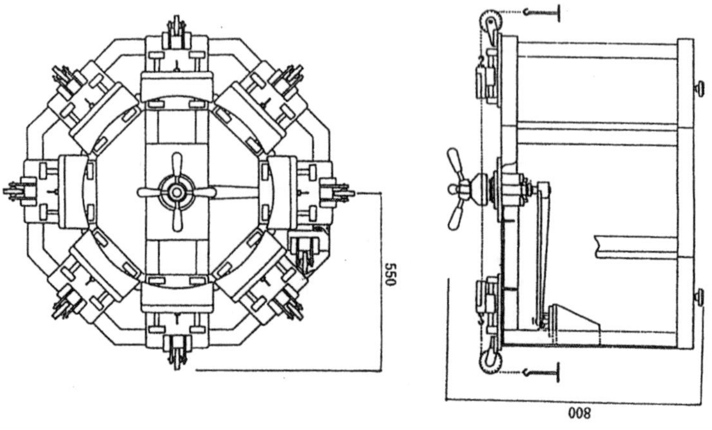

Fig. 2. Test apparatus. (a) Plan; (b) elevation.

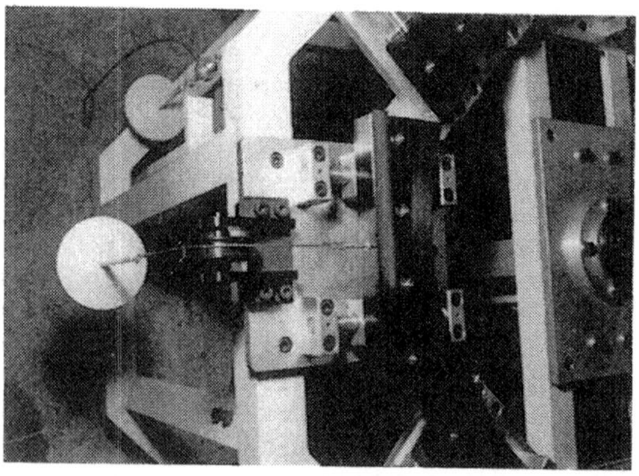

Plate 2. Jigs at outer boundary.

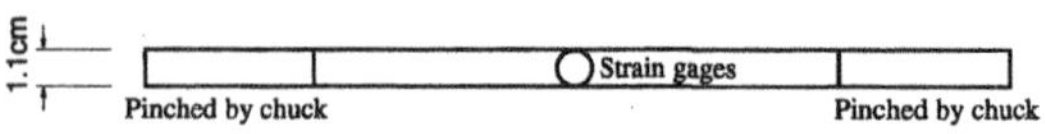

Fig. 3. Test specimen to measure the rigidity of the 'composite material' that consists of membrane, strain gages and adhesive (1 cm=1 mm).

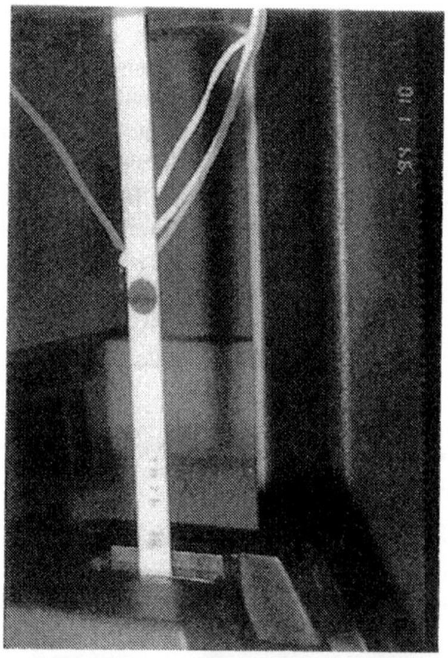

Plate 3. Test specimen with strain gages.

ester film in two directions, and Figs. 5–7 illustrate those of the three specimens of PVC coated textile. The stresses are actually stress resultants obtained from the applied load divided by the initial width of the specimens. Strains are measured by the strain gages as described in the previous section. Although the PVC coated textile is not elastic, the 'composite material' at the strain gages shows elastic behavior. The evaluated elastic moduli are shown in Table 2. Overlines denote that these properties are used only in the present method. The elastic properties of the PVC coated textile do not satisfy the condition $\bar{\nu}_{12}/\bar{E}_1=\bar{\nu}_{21}/\bar{E}_2$. However, these values are used in the following sections.

3.4.4. Evaluation of the proposed method

The proposed method is applied to a simple model as shown in Fig. 8, which is a rectangular specimen. The strain gages are glued in the center of the specimen, and a tensile load is applied. The axial stress is directly obtained from the applied load divided by the initial width. In Fig. 9 solid lines illustrate the relationship between this 'exact' stresses and the measured strains. The dotted lines are the results calculated by using $\bar{E}_1 t$ and $\bar{\nu}_{12}$ for the PVC coated textile in Table 2.

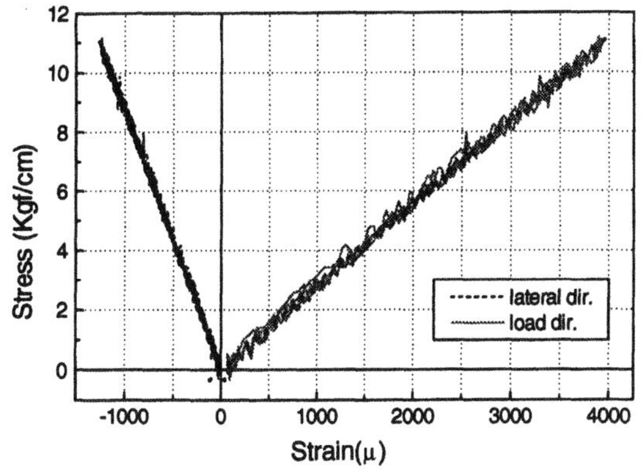

Fig. 4. Stress–strain relationships for the stress measurement (polyester film) (1 kgf=9.8 N, 1 cm=10 mm).

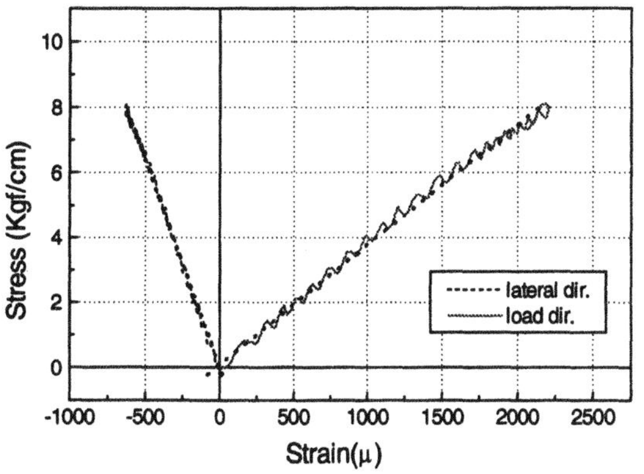

Fig. 5. Stress–strain relationships for the stress measurement (PVC coated textile, warp direction) (1 kgf=9.8 N, 1 cm=10 mm).

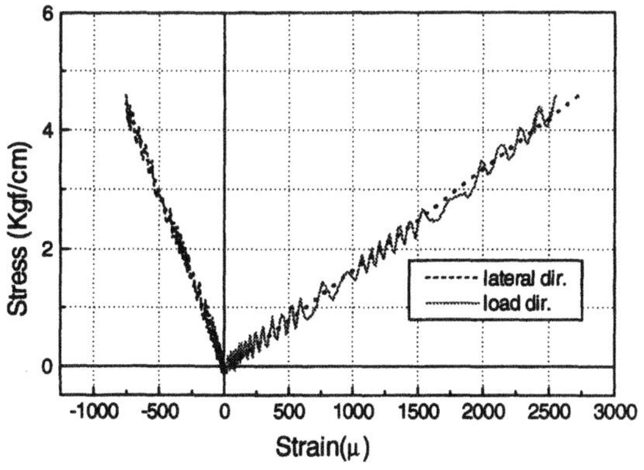

Fig. 6. Stress–strain relationships for the stress measurement (PVC coated textile, fill direction) (1 kgf=9.8 N, 1 cm =10 mm).

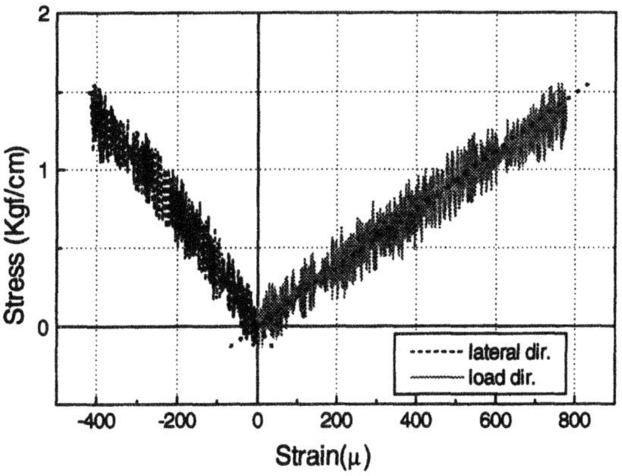

Fig. 7. Stress–strain relationships for the stress measurement (PVC coated textile, 45° direction) (1 kgf=9.8 N, 1 cm=10 mm).

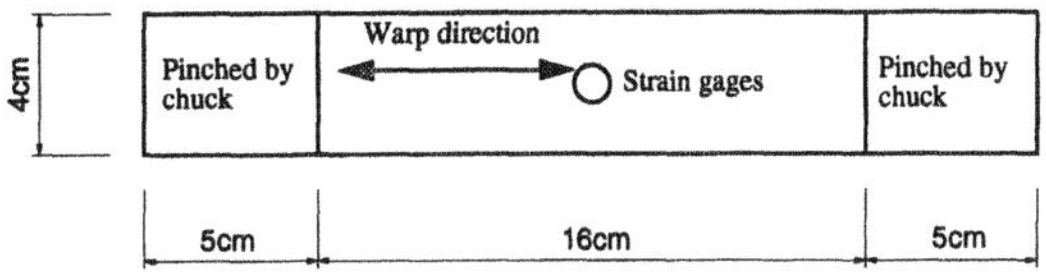

Fig. 8. Simple specimen for evaluation of the present method (1 cm=10 mm).

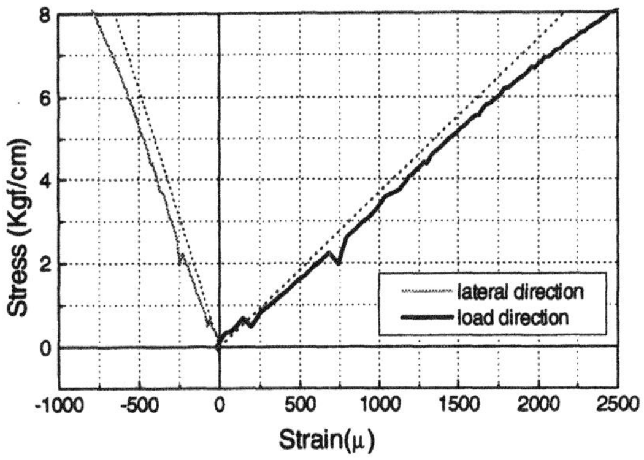

Fig. 9. Comparison between measured 'exact' stress–measured strain relationships and relationships obtained from the properties in Table 2 (1 kgf=9.8 N, 1 cm=10 mm).

The axial stress measured by the proposed method is obtained from this dotted line. This stress, the one measured by the proposed method, is slightly higher than the stress obtained from a solid line, that is, the stress calculated from the applied load. However, this difference is small, and the proposed method appropriately evaluated the stresses. Note that, in Fig. 9, the reduction of the width due to Poisson's ratio is not taken into account to obtain the solid lines; that means the difference is actually smaller.

Table 2
Elastic properties for the present method for stress measurement[a]

Material	E_1t (Et)	$\nu_{12}(\nu)$	E_2t	ν_{21}	Gt
Polyester film	2668	0.315	–	–	1015
PVC coated textile	3690	0.298	1908	0.337	588

[a] t: thickness (cm); $\bar{E}_it$ (Et): Young's modulus×thickness (kgf/cm); $\bar{G}t$: shear coefficient×thickness (kgf/cm), $\bar{\nu}_{ij}(\bar{\nu})$: Poisson's ratio; i, j=1: warp, 2: fill) (1 kgf=9.8 N, 1 cm=10 mm).

4. Measurement of stresses in the wrinkled circular membranes

4.1. Specimens

Strain gages that measure strains in three directions are glued to both surfaces of the different test specimens as illustrated in Figs. 10–12. In Fig. 12, positions of gages A and C are on the tangents of the center hub whose directions coincide with the directions of warp and fill. Results of the numerical analysis that will be described in Section 5 show that the tensile stresses on these tangents are larger than those in the other region. Directions of the gages are shown in Fig. 13. Table 3 illustrates the materials and the numbers of the figures showing the positions of the gages.

4.2. Measurement of stresses

The initial tension is introduced by counterweights that pull the jigs of the outer boundary. First, each jig is pulled by a 10 kgf (1 kgf=9.8 N) counterweight, then 4 kgf counterweights are added to each jig one by one. The total weight for each jig is 18 kgf. The torsion is also introduced by counterweights, that is, 1 kgf counter weights are successively added one by one.

Figs. 14 and 15 illustrate relationships between the in-plane torsion and the values measured by the strain

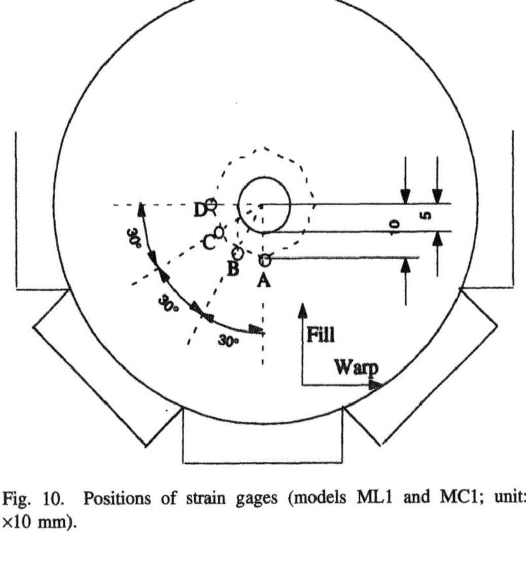

Fig. 10. Positions of strain gages (models ML1 and MC1; unit: ×10 mm).

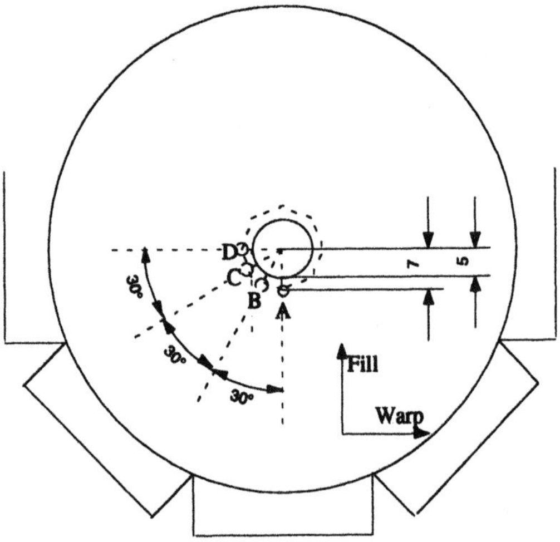

Fig. 11. Positions of strain gages (model ML2; unit: ×10 mm).

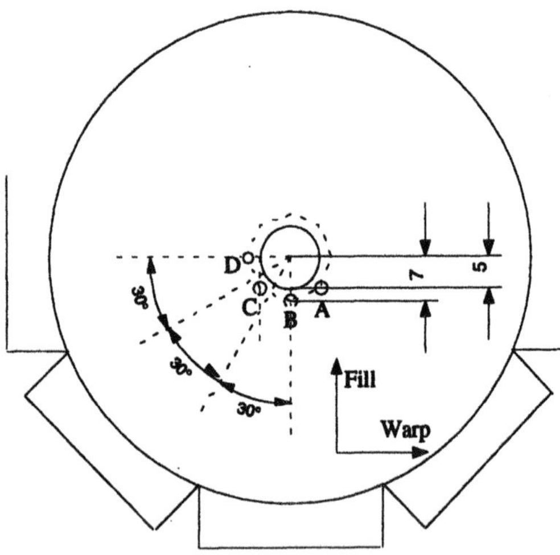

Fig. 12. Positions of strain gages (model MC2; unit: ×10 mm).

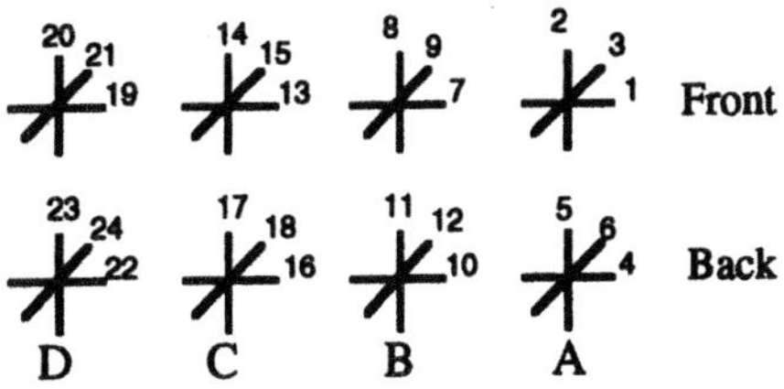

Fig. 13. Directions of the strain gages.

Table 3
Test specimens

Specimen	Membrane	Positions of strain gages
ML1	Polyester film	Fig. 10
ML2	Polyester film	Fig. 11
MC1	PVC coated textile	Fig. 10
MC2	PVC coated textile	Fig. 12

gages. The results of the specimens ML2 and MC2 are shown. The strains are the average of the strains measured by the two gages glued on both surfaces of the membranes. Gradients of these curves change when wrinkling occurs. In polyester film (ML2) the change of the gradients occurs at the same load level, since the homogeneous wrinkles are formed in a circumferential direction. In the orthotropic PVC coated textile (MC2), however, the change for each curve occurs at a different load level because wrinkles are not homogeneous in a circumferential direction. Creep deformations also develop in this material. As described in Section 3, the measured strains are almost proportional to the stresses, and the fluctuation of the stresses due to the creep is small because the stresses are in equilibrium with constant loads introduced by the counter weights. Hence, in Fig. 15, differences between the curves for the loading and those for the unloading are small.

The shear strains can be calculated from the strains in three directions. The stresses are calculated using the values in Table 2.

4.3. Principal stresses

The principal stresses are depicted in Section 5 (Figs. 23–28) together with the results of the bifurcation analyses. The results of the specimens ML1 and ML2, and those of the specimens MC1 and MC2 are depicted in the same figure. The magnitudes of the in-plane torsion in which wrinkles first formed are: 160 kgf cm in the polyester film and 80 kgf cm in the PVC coated textile. These levels do not agree with the points in Figs. 14 and 15 where gradients of the strain–torque relationships change. This is because the wrinkles are first formed around the center hub, and the wrinkled region gradually

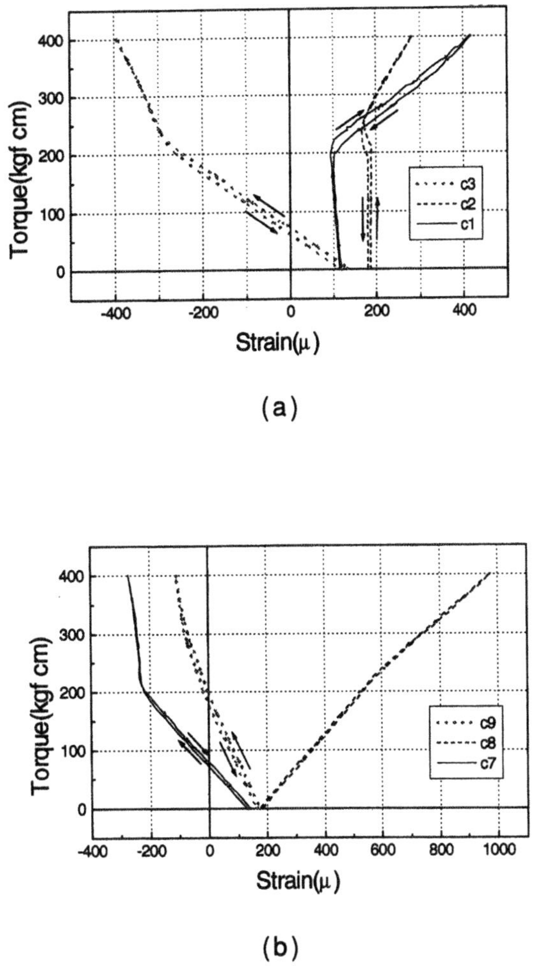

Fig. 14. Relationships between in-plane torsion (torque) and the measured strains (model ML2=polyester film) (1 kgf=9.8 N, 1 cm=10 mm); c[n] is gage number, [n] for the front surface in Fig. 13.

spreads, as the magnitude of the in-plane torsion becomes larger.

Distributions of the principal stresses when the initial tension is introduced are disturbed because they are introduced by pulling the eight jigs from eight directions. Frictions in the slide parts of the jigs are not negligible. The measured initial stresses are compared with the theoretical solutions under the prescribed radial displacements at the outer boundary; it is estimated that about 80% of the loads are transmitted to the membranes. This fact should be considered in the following analyses that simulate the experiments.

Figs. 23–28 show that no compressive stresses exist even when the in-plane torsion is applied, which means the tension field is formed in the membranes.

In orthotropic PVC coated textiles, the directions of

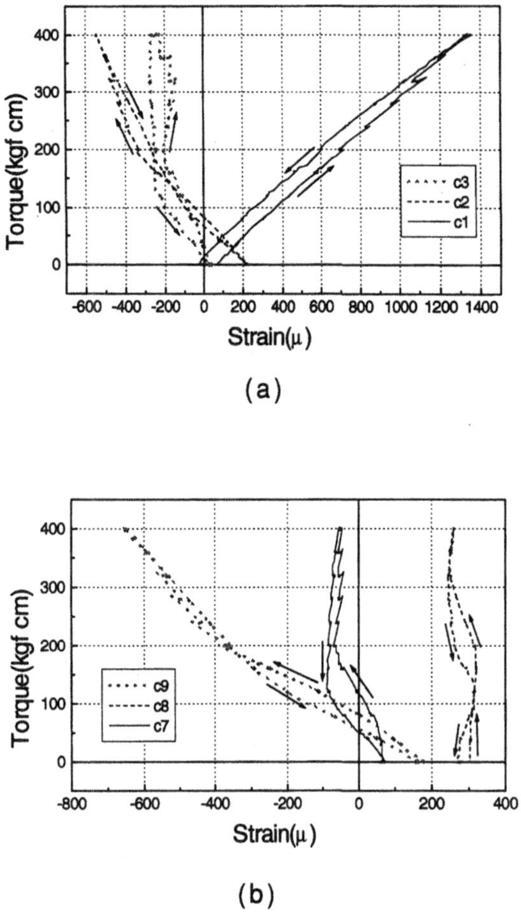

Fig. 15. Relationships between in-plane torsion (torque) and the measured strains (model MC2= PVC coated textile) (1 kgf=9.8 N, 1 cm=10 mm); c[n] is gage number, [n] for the front surface in Fig. 13.

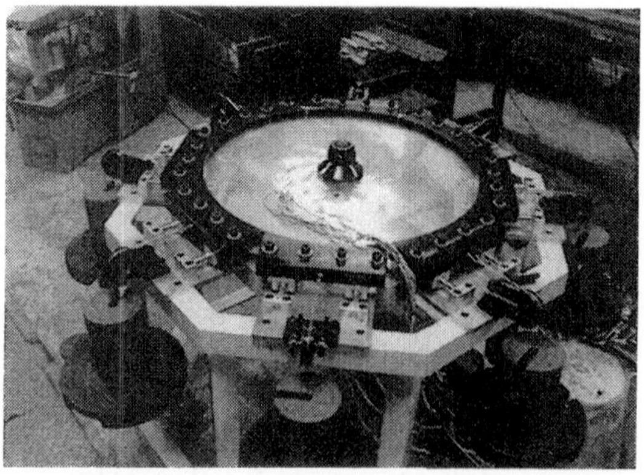

Plate 4. Test specimen (ML1: polyester film); in-plane torsion=400 kgf cm (1 kgf=9.8 N, 1 cm=10 mm).

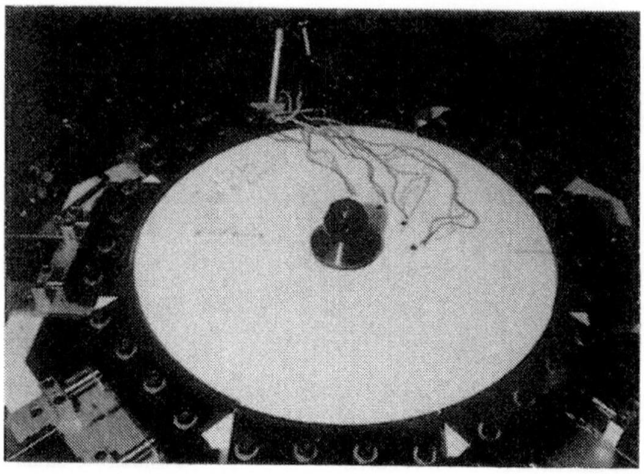

Plate 5. Test specimen (MC1: the PVC coated textile); in-plane torsion=400 kgf cm (1 kgf=9.8 N, 1 cm=10 mm).

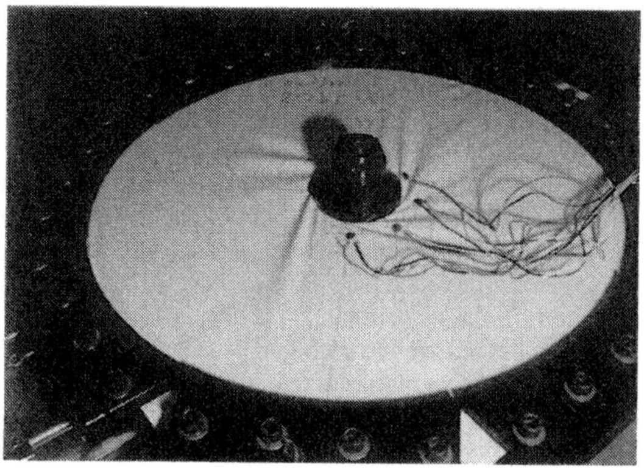

Plate 6. Wrinkles around the strain gages (the PVC coated textile).

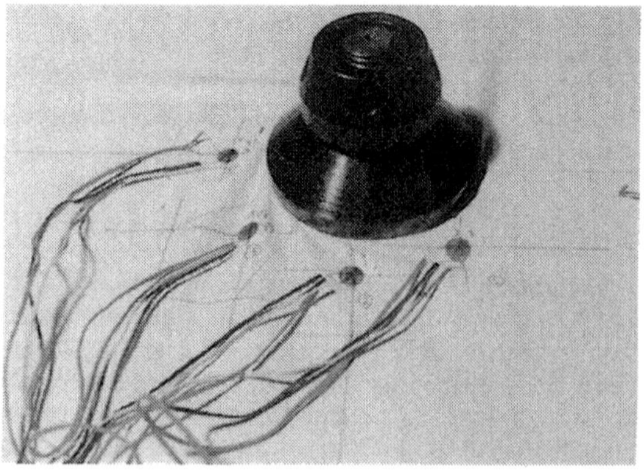

Plate 7. Close-up of the specimen in Plate 6.

Plate 8. Rubber (isotropic; thickness=1.0 mm).

Plate 11. Thin coated film (orthotropic; thickness=0.15 mm; initial tension=10 kgf counter weight/outer boundary jig) (1 kgf=9.8 N).

Plate 9. Teflon film (isotropic; thickness=0.10 mm).

Plate 10. Membrane for balloons (orthotropic; thickness=0.10 mm; fibers are embedded).

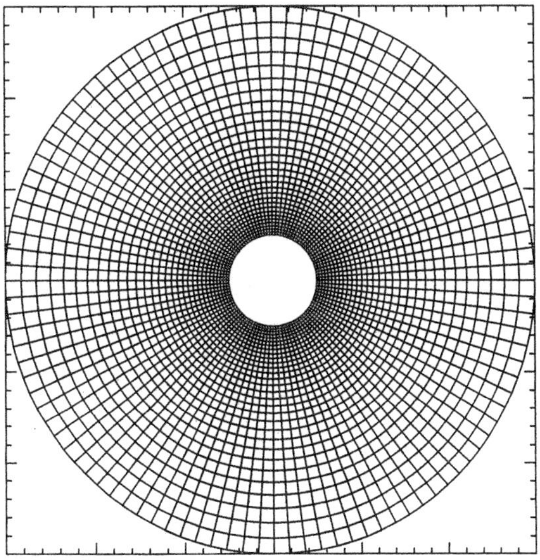

Fig. 16. Mesh subdivision (3120 elements, total degrees of freedom is 9048).

Table 4
Analysis conditions (1 kgf=9.8 N, 1 cm=10 mm)

Prescribed radial displacement at outer boundary (unit: cm)	0.0108
Factor α in Eq. (10)	1.0, 0.05, 0.0
Young's modulus×thickness, $Et(=E_1t=E_2t)$ (unit: kgf/cm)	1022
Poisson's ration, $\nu(=\nu_{12}=\nu_{21})$	0.267

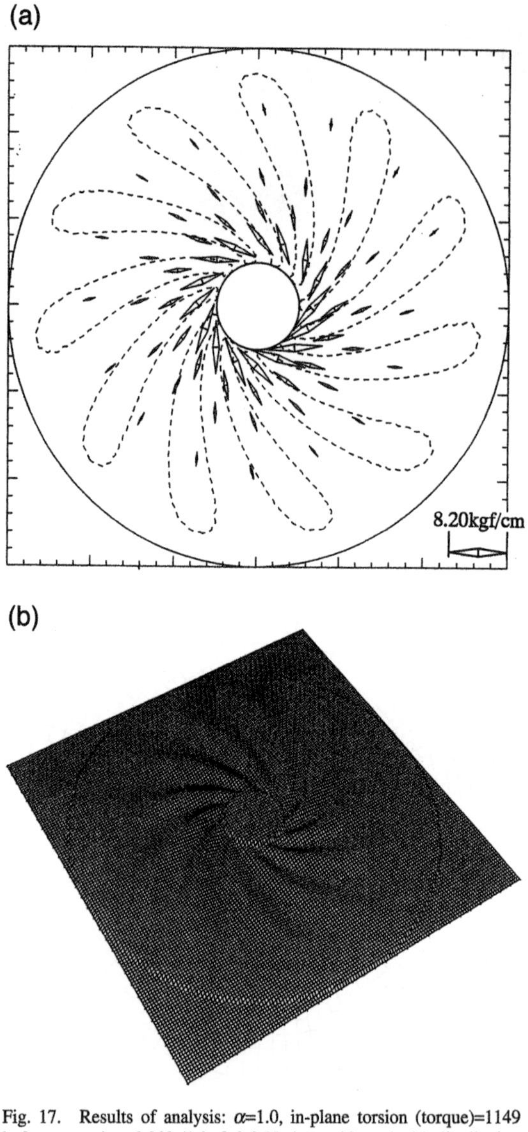

Fig. 17. Results of analysis: α=1.0, in-plane torsion (torque)=1149 kgf cm; rotation=0.92° (1 kgf=9.8 N, 1 cm=10 mm); (a) distribution of principal stresses and contour lines for out-of-plane displacement=0.067 mm; (b) bird's eye view (magnified by three times).

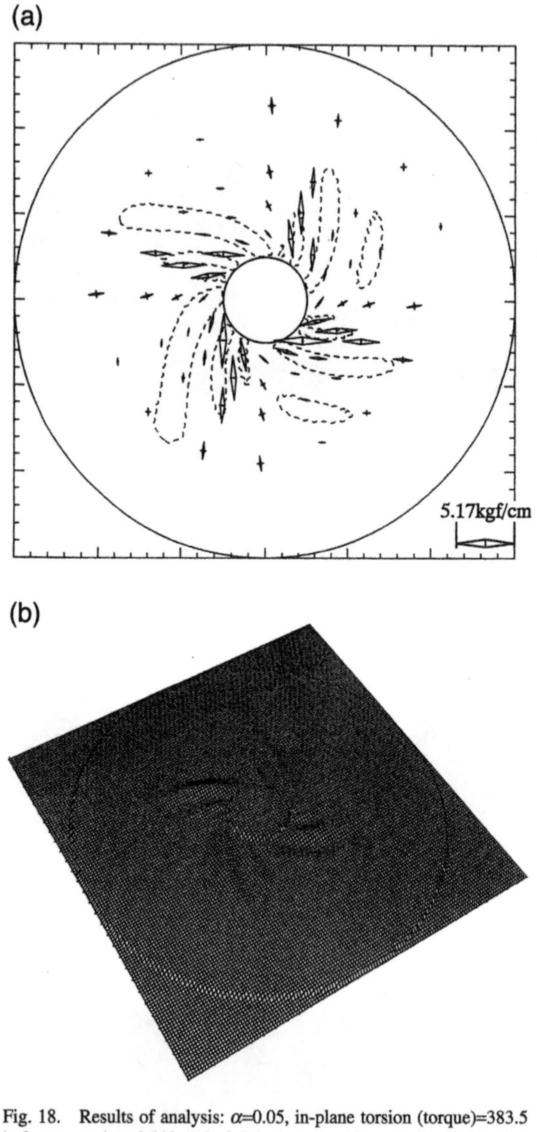

Fig. 18. Results of analysis: α=0.05, in-plane torsion (torque)=383.5 kgf cm; rotation=0.92° (1 kgf=9.8 N, 1 cm=10 mm); (a) distribution of principal stresses and contour lines for out-of-plane displacement=0.067 mm; (b) bird's eye view (magnified by three times).

the principal stresses and those of the wrinkles almost coincide with the directions of the textiles, that is, warp and fill.

Plates 4 and 5 show the specimens. Plates 6 and 7 show wrinkles around the strain gages glued to the PVC coated textile. The disturbances due to the gages are observed, but these are small and not essential.

4.4. Wrinkles in various membranes

Tests for other membranes are also carried out. Plates 8–11 show wrinkles on rubber (isotropic), Teflon film (isotropic), the membrane for balloons (orthotropic), and a thin coated textile (orthotropic), respectively. In isotropic membranes, the wrinkles are homogeneous in circumferential direction. In contrast, in orthotropic membranes the wrinkles are formed along the directions of textiles as observed in the PVC coated textile. Because

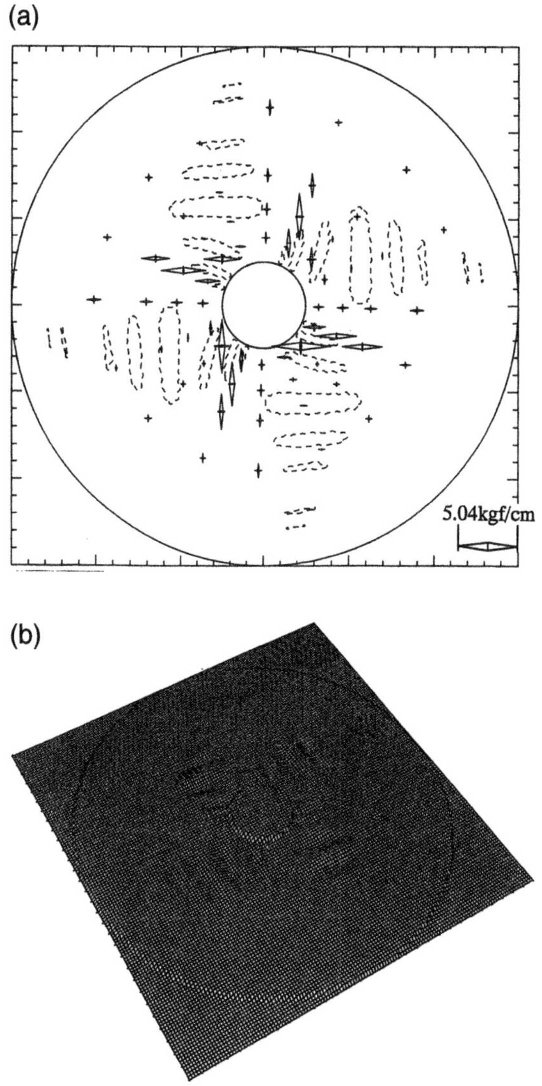

Fig. 19. Results of analysis: α=0.0, in-plane torsion (torque)=217.4 kgf cm; rotation=0.92° (1 kgf=9.8 N, 1 cm=10 mm); (a) distribution of principal stresses and contour lines for out-of-plane displacement=0.067 mm; (b) bird's eye view (magnified by three times).

the orthotropic property of the membrane for balloons is not so strong, directions of wrinkles do not completely coincide with those of fibers embedded in the membrane. The number of wrinkles depends on the thickness and material of each membrane.

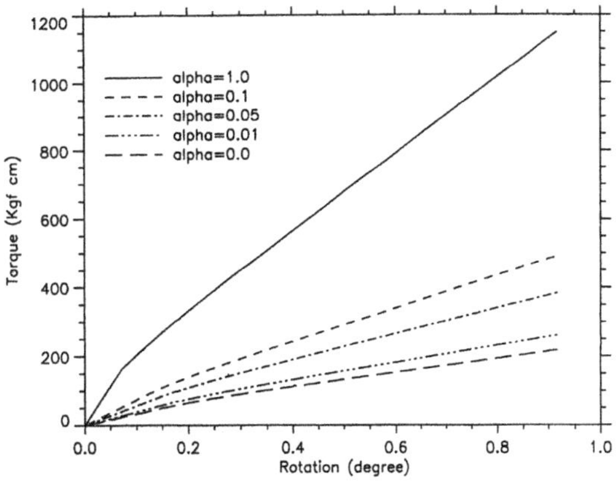

Fig. 20. Relationships between in-plane torsion (torque) and rotation (1 kgf=9.8 N, 1 cm=10 mm).

5. Bifurcation analysis of wrinkling

5.1. Overview

In this section, the bifurcation analyses of wrinkling on isotropic or orthotropic elastic membranes are presented. A few studies concerning wrinkling on orthotropic membranes were found [19,25,26]. Roddeman [26] analyzed the wrinkling of the stretched circular membrane with orthotropic elastic material. He used the finite element method based on the theory proposed in [25], which is a kind of tension field theory.

In our study, a four-node membrane element that considers precisely the geometrical nonlinearity is used. From the theoretical point of view, an infinitesimal number of wrinkles have to be formed due to the bifurcation buckling when the bending rigidity is ignored. However, in the finite element analysis, the number of wrinkles depends on the fineness of mesh subdivision. Therefore, a finite number of wrinkles are obtained by the bifurcation analysis.

5.2. Four-node isoparametric membrane element

The formulation of the four-node isoparametric membrane element is briefly illustrated in this section. In the formulation, the convective coordinate system is used (see, for example, [33–35]). The summation convention is used in the following formulation. The superscripts represent contravariant components and the subscripts represent the covariant components. The total Lagrangian method is adopted to describe deformations.

The position vector of the initial configuration $\mathbf{x}$ and the displacement vector $\mathbf{u}$ are represented by using the bilinear shape functions in the following way:

$$\mathbf{x}(r^1,r^2)=\frac{1}{4}\{(1-r^1-r^2+r^1r^2)\mathbf{x}_1+(1+r^1-r^2-r^1r^2)\mathbf{x}_2 \quad (1)$$

Table 5
Material properties used in the analyses (1 kgf=9.8 N, 1 cm=10 mm)

Membrane (Unit)	E_1t (Et) (kgf/cm)	$\nu_{12}(\nu)$ –	E_2t (kgf/cm)	ν_{21} –	Gt (kgf/cm)
Polyester film	1022	0.267	–	–	404
PVC coated textile (measured)	391	0.0	194	0.36	11
PVC coated textile (used in analyses)	390	0.49	190	0.36	11

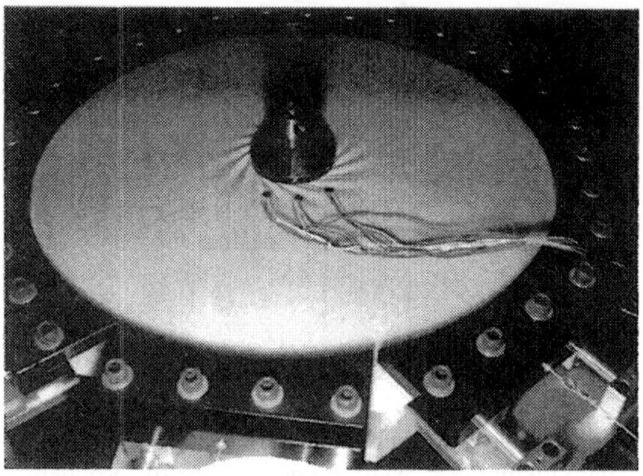

Plate 12. Wrinkles on isotropic Teflon film.

$$+(1+r^1+r^2+r^1r^2)\mathbf{x}_3+(1-r^1+r^2-r^1r^2)\mathbf{x}_4\}$$

$$\mathbf{u}(r^1,r^2)=\frac{1}{4}\{(1-r^1-r^2+r^1r^2)\mathbf{u}_1+(1+r^1-r^2-r^1r^2)\mathbf{u}_2 \quad (2)$$

$$+(1+r^1+r^2+r^1r^2)\mathbf{u}_3+(1-r^1+r^2-r^1r^2)\mathbf{u}_4\}$$

where $\mathbf{x}_k$ $(k=1–4)$ represent the initial nodal position vectors, $\mathbf{u}_k$ $(k=1–4)$ represent the nodal displacement vectors and r^1 and r^2 are components of the two dimensional natural coordinate system. The following Green–Lagrange strain tensor referring to the initial configuration is employed to represent the kinematic relationships:

$$ {}_0^t\mathbf{E}=\frac{1}{2}({}^t\mathbf{g}_i\cdot{}^t\mathbf{g}_j-\mathbf{G}_i\cdot\mathbf{G}_j)\mathbf{G}^i\otimes\mathbf{G}^j \quad (i \text{ and } j=1,2,3) \qquad (3)$$

where t denotes current time, $\mathbf{G}_i$ and ${}^t\mathbf{g}_i$ are the basis vectors concerning with the initial configuration and the deformed configuration at time t, respectively, and $\otimes$ denotes the tensor product. Note that the dyadic notation is adopted (see for example [35]). The second Piola–Kirchhoff stress tensor referring to the initial configuration is used, and the relationship of the stress tensor and the Green–Lagrange strain tensor (the constitutive equations) is described as follows:

$$ {}_0^tS^{mn}=C^{mnpq}\,{}_0^tE_{pq} \quad (m,n,p \text{ and } q=1,2) \qquad (4)$$

where C^{mnpq} is a component of the forth order elastic tensor and ${}_0^tS^{mn}$ is a component of the second Piola–Kirchhoff stress tensor. Materials are assumed elastic. The elastic tensor for the orthotropic membrane is conveniently represented in the Cartesian coordinate system when directions of the basis vectors coincide with the directions of the textile (warp and fill). Components of the elastic tensor in the Cartesian coordinate system are described by matrix form as follows:

$$\begin{bmatrix} \bar{C}_{1111} & \bar{C}_{1122} & \bar{C}_{1112} \\ \bar{C}_{2211} & \bar{C}_{2222} & \bar{C}_{2212} \\ \bar{C}_{1211} & \bar{C}_{1222} & \bar{C}_{1212} \end{bmatrix} \qquad (5)$$

$$=\begin{bmatrix} E_1/(1-\nu_{12}\nu_{21}) & \nu_{21}E_1/(1-\nu_{12}\nu_{21}) & 0 \\ \nu_{12}E_2/(1-\nu_{12}\nu_{21}) & E_1/(1-\nu_{12}\nu_{21}) & 0 \\ 0 & 0 & G \end{bmatrix}$$

where E_1 and E_2 are Young's moduli and ν_{12} and ν_{21} are Poisson's ratios. The plane stress condition is assumed. The relationship between C^{mnpq} in Eq. (4) and $\bar{C}_{ijkl}$ in Eq. (5) is represented as follows:

$$C^{mnpq}=\bar{C}_{ijkl}(\mathbf{e}_i\cdot\mathbf{G}^m)(\mathbf{e}_j\cdot\mathbf{G}^n)(\mathbf{e}_k\cdot\mathbf{G}^p)(\mathbf{e}_l\cdot\mathbf{G}^q) \qquad (6)$$

where $\mathbf{e}_i$ $(i=1, 2)$ are the basis vectors of the 'textile' Cartesian coordinate system whose directions coincide with the directions of warp and fill. In Eq. (6), summation of dummy indices is calculated for 1 and 2.

5.3. Introduction of initial tension

Initial tension is introduced by prescribing the radial displacements at the outer boundary of the circular membrane. However, since membranes are unstable when the initial tension is not introduced, the tangent stiffness matrix is singular in the initial configuration. A technique to avoid this singularity in introducing the initial tension is described in this section.

The tangent stiffness matrix can be decomposed into

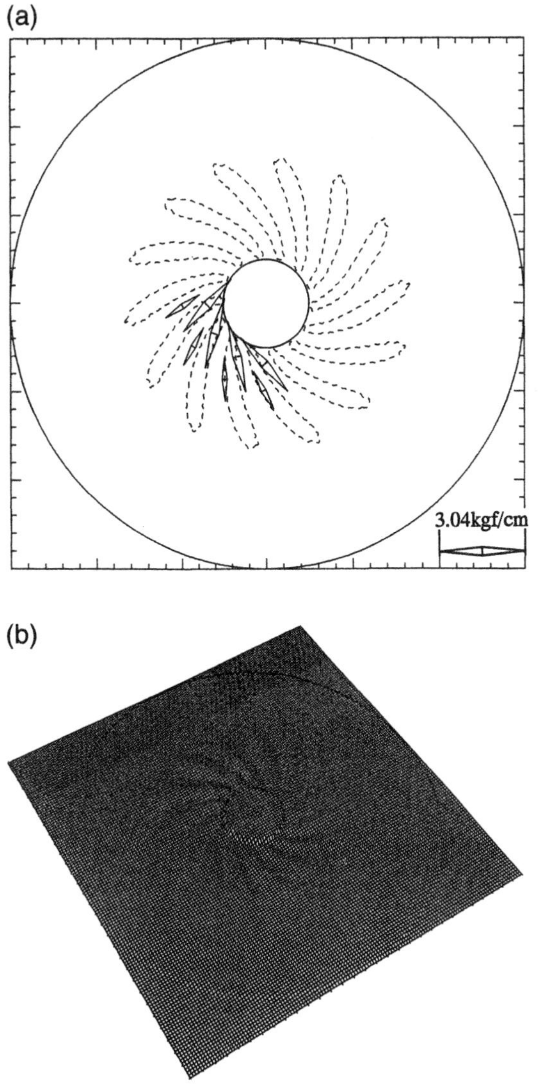

Fig. 21. Analysis result for polyester film (in-plane torsion=401 kgf cm; out-of-plane deformation is magnified by three; corresponding to Plate 12) (1 kgf=9.8 N, 1 cm=10 mm).

the elastic stiffness $\mathbf{K}_E$ and the geometrical stiffness $\mathbf{K}_G$. The incremental equilibrium equations for a discrete system are described as follows:

$$(\mathbf{K}_E+\mathbf{K}_G)\mathbf{U}=\lambda\mathbf{P} \qquad (7)$$

where $\mathbf{U}$ is an unknown incremental displacement vector, $\mathbf{P}$ is a load vector, and λ is a load parameter. When membrane elements are employed, $\mathbf{K}_E$ is singular and the inverse cannot be calculated. This singularity shows that membrane structures are unstable when the initial tension is not introduced. Therefore, $\mathbf{U}$ cannot be calculated in the initial stage where $\mathbf{K}_G=\mathbf{O}$. After initial tension is introduced, some components of $\mathbf{K}_G$ become nonzero, and $\mathbf{K}_G$ stabilizes the system. If the Moor–Penrose generalized inverse of $\mathbf{K}_E$, i.e. $(\mathbf{K}_E)^+$, is employed, a solution vector can be calculated even in the first incremental step as follows (see [37]):

$$\mathbf{u}=\lambda(\mathbf{K}_E)^+\mathbf{p} \qquad (8)$$

In large scale problems, however, calculating the generalized inverse is very expensive or impossible. In this study the stiffness matrix is stabilized by adding a penalty term $\mathbf{I}_s$ as follows:

$$\mathbf{u}=\lambda(\mathbf{K}_E+\mathbf{I}_s)^{-1}\mathbf{p} \qquad (9)$$

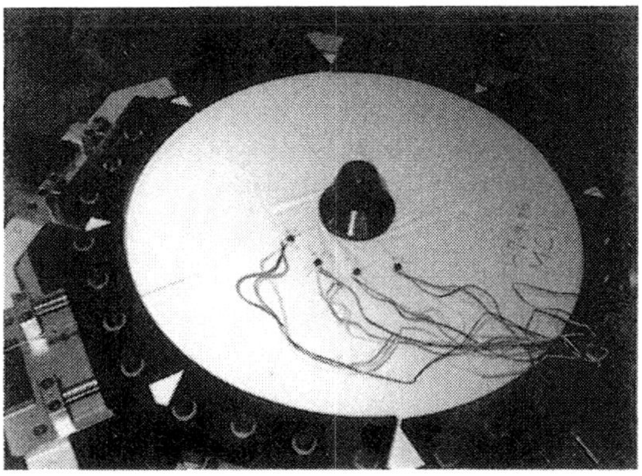

Plate 13. Wrinkles on the PVC coated textile.

In the case of flat membranes, a diagonal matrix can be used as $\mathbf{I}_s$ in which the diagonal components corresponding to the out-of-plane displacements have a finite value (i.e. the penalty constant), and the other components are set to zero. In the case of curved membrane structures, it is difficult to find appropriate $\mathbf{I}_s$. In those structures, however, analyses can usually be done without problems because the analysis models are stabilized by the discretization errors in the finite element method.

5.4. Wrinkling analysis based on the bifurcation analysis

The nonlinear equilibrium paths (both the main path and the bifurcation paths) are traced by using the incremental method combined with the Newton–Raphson method. The bifurcation point, i.e. the point where wrinkling occurs, is searched by using the linearized buckling technique [34] in each incremental step. The minimum eigenvalue and corresponding eigenvector of the tangent stiffness matrix are calculated by the inverse iteration. The bifurcation point is calculated by the bisection method. When the bifurcation point is obtained, the

(a)

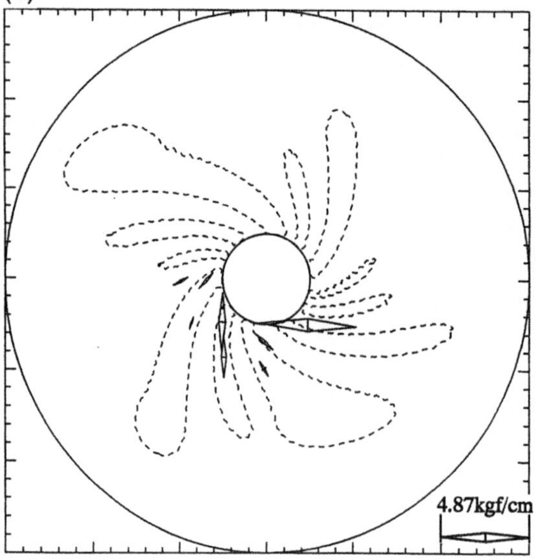

(b)

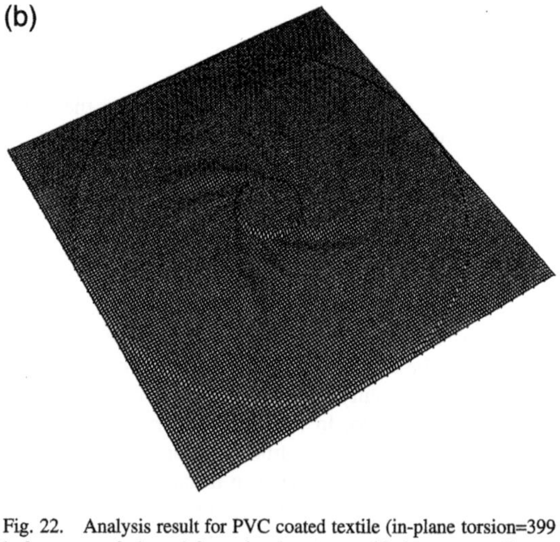

Fig. 22. Analysis result for PVC coated textile (in-plane torsion=399 kgf cm; out-of-plane deformation is not magnified; corresponding to Plate 13) (1 kgf=9.8 N, 1 cm=10 mm).

path switching is conducted by using the eigenvector corresponding to zero eigenvalue.

6. Effect of shear modulus on wrinkling

An important parameter in orthotropic elastic membranes is the shear modulus. Here, the bifurcation analyses of wrinkling for various kinds of shear modulus are performed. The shear modulus G is given by introducing a parameter α as follows:

$$G=\alpha\frac{E}{2(1+\nu)} \tag{10}$$

The two Young's moduli and the two Poisson's ratios in Eq. (5) are supposed equal to E and ν, respectively. The following results show that α determines the characteristic of the wrinkles on orthotropic membranes.

Fig. 16 shows the mesh subdivision. The number of elements is 3120 and the total degrees of freedom is 9048. Table 4 shows sizes and material properties. Figs. 17–19 show results: distributions of principal stresses, contours, and perspectives of the wrinkles. As α becomes smaller, differences between directions of textiles and those of wrinkles become smaller. A similar characteristic is observed for directions of principal stresses. A trajectory of the points on which the directions of principal stresses are constant is called the tension ray [18,19] or the line of tension [25]. By using the tension field theory, it can be proved that the tension rays are straight. This property is observed in the present results. However, although tangential direction of a wrinkle coincides with the direction of the tension ray (i.e. direction of the principal stress), wrinkles are not formed along a tension ray, i.e. they are not straight but curved.

In Fig. 18, the symmetry of the shapes about the axes whose directions coincide with the directions of warp and fill is a little disturbed. This may be because the mesh subdivision of the analysis model is not appropriately fine.

Fig. 20 illustrates the relationships between the in-plane torsion and the rotation of the center hub. At bifurcation points (i.e. at the point wrinkling occurs), gradients of the curves discontinuously change. As α becomes smaller, the in-plane shear rigidity also becomes smaller.

7. Comparison between experiments and bifurcation analyses

7.1. Overview

In this section the bifurcation analyses of wrinkling corresponding to the experiments in Section 4 are presented. Both the polyester film and the PVC coated textile are analyzed. Material properties used in the analyses are illustrated in Table 5. Although the PVC coated textile is actually not elastic, it is regarded as near-elastic. Because the measured elastic properties do not satisfy the relationship $\nu_{12}/E_1=\nu_{21}/E_2$, these values are modified as shown in Table 5.

In the analyses, the initial tension is introduced by prescribing the uniform radial displacements at the outer boundary. This condition is different from the actual boundary condition of the experiments in which the specimens are fixed to the eight jigs and pulled from eight directions as described in Section 3.3.

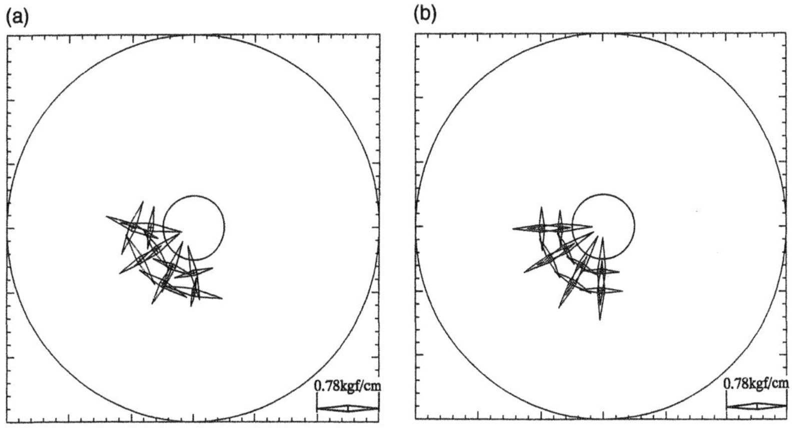

Fig. 23. Distribution of principal stresses (polyester film; isotropic; after initial tension is introduced) (1 kgf=9.8 N, 1 cm=10 mm); (a) experiment (in-plane torsion=0.0 kgf cm; maximum stress resultant =0.782 kgf/cm); (b) bifurcation analysis (in-plane torsion=0.0 kgf cm; maximum stress resultant=0.810 kgf/cm).

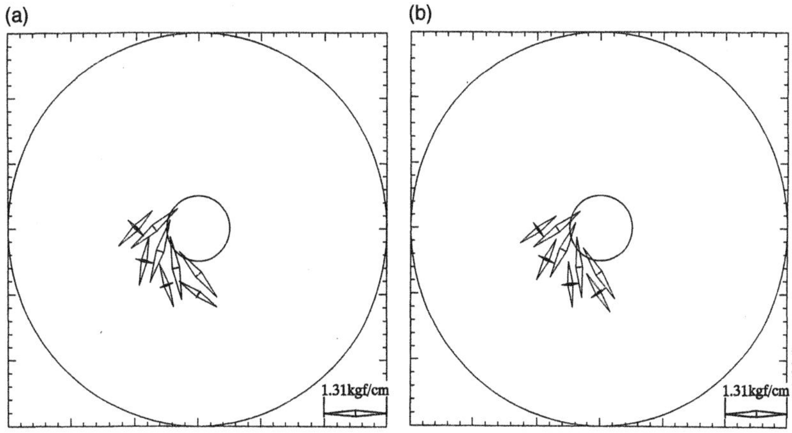

Fig. 24. Distribution of principal stresses (polyester film; isotropic; before wrinkling occurs) (1 kgf=9.8 N, 1 cm=10 mm); (a) experiment (in-plane torsion=160 kgf cm, maximum stress resultant=1.31 kgf/cm); (b) bifurcation analysis (in-plane torsion=162 kgf cm, maximum stress resultant=1.20 kgf/cm).

7.2. Comparison of wrinkles

The numbers of wrinkles obtained in the analyses depend on the fineness of mesh subdivisions because the membrane elements have no bending rigidity. Therefore, comparisons of the number of wrinkles have no meaning. Only characteristics of wrinkles can be compared. Note that the effect of the bending rigidity was investigated for isotropic elastic membranes by the present author using finite strip elements [36].

Plate 12 shows wrinkles formed on isotropic Teflon film. It is shown instead of a photo of polyester film because the polyester film is transparent and consequently presents difficulty in observing the wrinkles in photos (see Plate 4). Characteristics of the shape of wrinkles on those two membranes are almost identical. Fig. 21 depicts the shape of the wrinkles and distribution of the principal stresses obtained by the bifurcation analysis for the polyester film. Plate 13 shows the wrinkles on the PVC coated textile. Fig. 22 shows the corresponding results of the analysis. Characteristics of the shapes of wrinkles obtained by the analyses for the two membranes agree with those observed in the experiments, although the numbers of wrinkles are different.

7.3. Comparison of the principal stresses

Figs. 23–28 illustrate the distributions of the principal stresses, where (a) shows the result of the experiment and (b) shows that of the analysis. The stresses are actu-

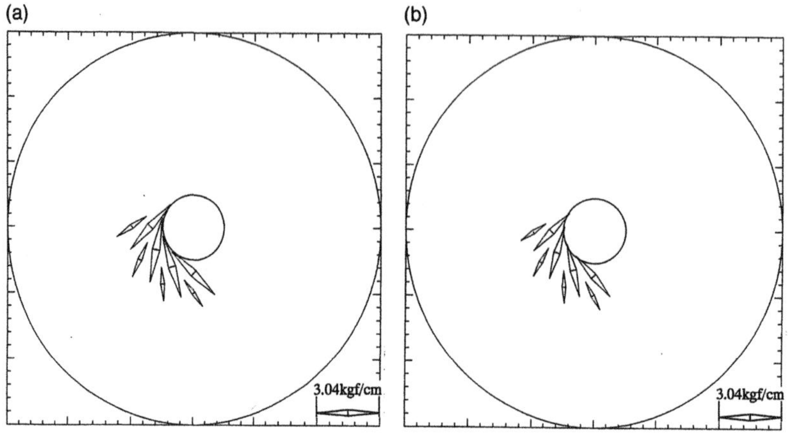

Fig. 25. Distribution of principal stresses (polyester film, isotropic, after wrinkling occurs) (1 kgf=9.8 N, 1 cm=10 mm); (a) experiment (in-plane torsion=400 kgf cm, maximum stress resultant=3.04 kgf/cm); (b) bifurcation analysis (in-plane torsion=401 kgf cm, maximum stress resultant=2.51 kgf/cm).

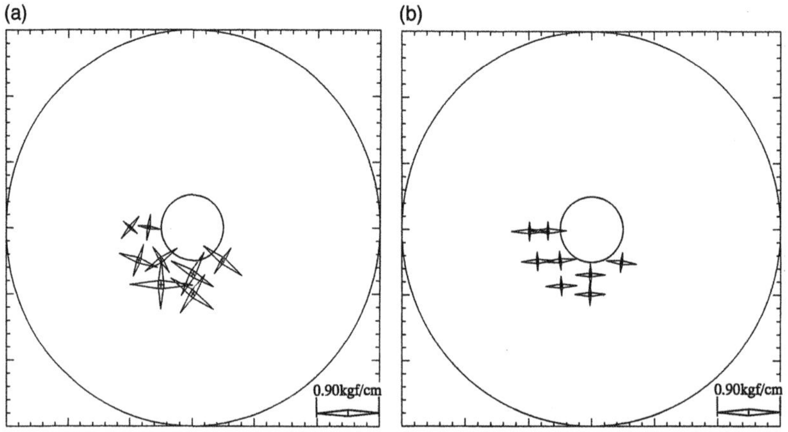

Fig. 26. Distribution of principal stresses (PVC coated textile; orthotropic; after initial tension is introduced) (1 kgf=9.8 N, 1 cm=10 mm); (a) experiment (in-plane torsion=0.0 kgf cm; maximum stress resultant=0.901 kgf/cm); (b) bifurcation analysis (in-plane torsion=0.0 kgf cm; maximum stress resultant=0.519 kgf/cm).

ally the stress resultants, that is, the stresses are multiplied by the thickness of specimens. Figs. 23–25 are the results for the isotropic polyester film, and Figs. 26–28 are those for the PVC coated textile. The results for three different levels of the in-plane torsion are shown for each these two membranes. Some considerations regarding the measurement of principal stresses have already been described in Section 4.3.

In the results of the analyses, principal stresses and their directions of the second Piola–Kirchhoff stress tensors are depicted on the initial configuration of the membranes. Note that the second Piola–Kirchhoff stress tensor does not change under the rigid body rotation (see, for example, [34]). In the experiments, stresses are measured by the strain gages that rotate together with the membranes, which means that they correspond to the components of the Cauchy stress tensor in the corotational system. Principal stresses and their directions are computed from these components, and they are depicted on the initial configuration of the membranes. This means that the effect of the rigid body rotation on the Cauchy stress tensor is eliminated. If stretch or shear deformation is small, the above principal stresses and their directions obtained in the experiments and analyses can be compared with each other.

The results of the experiments and the analyses when the in-plane torsion is zero do not agree well because the conditions for introducing the initial tension are dif-

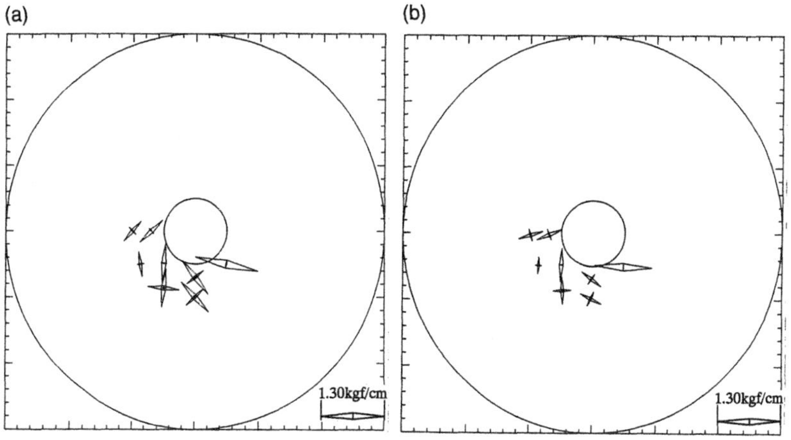

Fig. 27. Distribution of principal stresses (PVC coated textile; orthotropic; before wrinkling occurs) (1 kgf=9.8 N, 1 cm=10 mm); (a) Experiment (in-plane torsion=80.0 kgf cm; maximum stress resultant=1.30 kgf/cm); (b) bifurcation analysis (in-plane torsion=81.6 kgf cm; maximum stress resultant=1.17 kgf/cm).

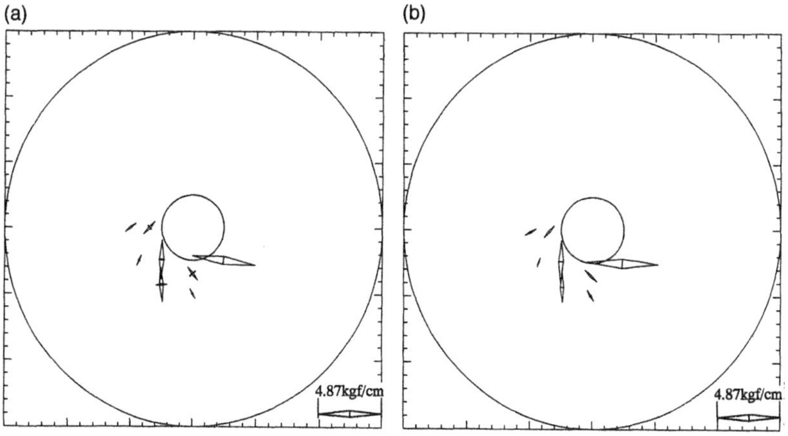

Fig. 28. Distribution of principal stresses (PVC coated textile, orthotropic, after wrinkling occurs) (1 kgf=9.8 N, 1 cm=10 mm); (a) experiment (in-plane torsion=400 kgf cm; maximum stress resultant=4.87 kgf/cm); (b) bifurcation analysis (in-plane torsion=399 kgf cm; maximum stress resultant=5.49 kgf/cm).

ferent. When the in-plane torsion becomes larger, the results of the experiments and those of the analyses agree well because the effect of the initial tension becomes relatively small. Although the actually inelastic PVC coated textile is analyzed as an elastic membrane, the bifurcation analyses can quantitatively simulate the direction and amplitude of the principal stresses. The material nonlinearity and creep do not significantly affect the distributions of stresses. This is because stresses are in equilibrium with the in-plane torsion. By comparing the results in this section and those in Section 6, the parameter α in Eq. (10) is considered a most important material property.

In the PVC coated textile, the stresses along warps are larger than those along fills. Compressive stresses are not observed in all results. Both the experiments and the bifurcation analyses show that the tension field is really formed in the wrinkled membranes.

8. Concluding remarks

The experiments and bifurcation analyses of wrinkling on circular stretched membranes under in-plane torsion are presented here. This example is a very typical example in the study of wrinkling.

In the experiment, first, a simple method to measure stresses in thin wrinkled membranes is proposed. Then, stresses in the wrinkled membranes are measured by using the present method. The tested membranes are the orthotropic PVC coated textile and the isotropic polyester film.

Some techniques of the bifurcation analysis of wrinkling are described. The principal stresses measured in the experiments are compared with the results of the bifurcation analyses. These results agree well, that is, directions and amplitude of the principal stresses agree well. Shapes of the wrinkles are also compared and the characteristics of the shape agree well. However, detailed shapes of the wrinkles cannot be compared because the bending rigidity is ignored in the analysis and the fineness of the mesh subdivision is not adequate.

From the experiments and the analyses, the following conclusions are obtained:

1. No compressive stresses are observed in both the experiments and the analyses, that is, the tension field is really formed.
2. In orthotropic membrane, the directions of principal stresses almost coincide with those of textiles especially when shear rigidity is small.
3. Material nonlinearities and creep do not significantly affect the distributions of stresses. Hence, elastic analysis is still an effective tool to understand the wrinkling.
4. The shape of wrinkles is determined by the bending rigidity. Therefore, the analysis using the shell element with fine mesh subdivision is necessary. This is an important subject for future study.

Acknowledgements

This paper is part of the author's doctoral dissertation, written under the supervision of Professor Yasuhiko Hangai, in the Institute of Industrial Science, the University of Tokyo. Professor Hangai passed away on August 9, 1998. The author was supported by JSPS (Japan Society for the Promotion of Science) Research Fellowships for Young Scientists. This support is gratefully acknowledged.

References

[1] Miyamura T, Hangai Y. Wrinkling on isotropic and orthotropic elastic circular membranes under in-plane torsion. IASS—Symposium 1995, Milano, Italy, 1995;2:851–8.

[2] Miyamura T, Hangai Y. Wrinkling on stretched circular membranes under inplane torsion—experiments with polyester films and coated fabrics (in Japanese). J Struct Constr Eng AIJ 1997;494:83–90.

[3] Miyamura T, Hangai Y. Wrinkling on stretched circular membranes under inplane torsion—comparison between bifurcation analyses and experiments (in Japanese). J Struct Constr Eng AIJ 1997;494:91–8.

[4] Miyamura T. Bifurcation analysis and experiment for wrinkling on stretched membrane (in Japanese). Doctoral dissertation, University of Tokyo, 1995.

[5] Dean WR. The elastic stability of an annular plate. Proc Roy Soc Lond A 1924;106:268–84.

[6] Hamada M, Harima T. In-plane torsional buckling of an annular plate. Bull JSME 1986;29(250):1089–95.

[7] Durban D, Stavsky Y. Elastic buckling of polar-orthotropic annular plates in shear. Int J Solids Struct 1982;18(1):51–8.

[8] Ore E, Durban D. Elastoplastic buckling of annular plates in pure shear. J Appl Mech 1989;56(3):644–51.

[9] Suzuki T, Ogawa T, Motoyui S, Sueoka T. Investigation on wrinkling problem of membrane structure. Proc. of IASS–MSU Symposium on Domes from Antiquity to the Present, Istanbul, 1988:695–702.

[10] Cheng C, Lui X. Buckling and post-buckling of annular plates in shearing, part I: buckling. Comput Meth Appl Mech Eng 1991;92(2):157–72.

[11] Cheng C, Lui X. Buckling and post-buckling of annular plates in shearing, part II: post-buckling. Comput Meth Appl Mech Eng 1991;92(2):173–91.

[12] Cheng C, Shang X. The effect of large rotation on post-buckling of annular plates. Acta Mech Solida Sinca 1992;5(3):277–84.

[13] Kondo K. J Soc Aero Sci Nipp 1938;5(41):285–99.

[14] Reissner E. On tension field theory. Proceedings of the 5th International Congress for Applied Mechanics Harvard University and MIT. 1938:88–92.

[15] Iai T. J Soc Aero Sci Nippon 1943;10(96):158–80.

[16] Stein M, Hedgepeth JH. Analysis of partly wrinkled membrane. NASA TN D-813, 1961.

[17] Mikulas MM Jr. Behavior of a flat stretched membrane wrinkled by the rotation of an attached hub. NASA TN D-2456, 1964.

[18] Mansfield EH. Load transfer via a wrinkled membrane. Proc Roy Soc Lond A 1970;316:269–89.

[19] Mansfield EH. Analysis of wrinkled membranes with anisotropic and nonlinear elastic properties. Proc Roy Soc Lond A 1977;353:475–98.

[20] Moriya K, Uemura M. An analysis of the tension field after wrinkling in flat membrane structures. Proceedings IASS Pacific Symposium Part 2 on Tension Structures and Space Frames, Tokyo and Kyoto, 1971:189–98.

[21] Wu CH, Canfield TR. Wrinkling in finite plane-stress theory. Q Appl Math 1981;39(2):179–99.

[22] Li X, Steigmann DJ. Finite plane twist of an annular membrane. Q J Mech Appl Math 1993;46(4):601–25.

[23] Miller RK, Hedgepeth JM et al. Finite element analysis of partly wrinkled membranes. Comput Struct 1985;20(1–3):631–9.

[24] Nishimura T, Tosaka N, Honma T. Finite element techniques in tension field problems (in Japanese). J Struct Constr Eng AIJ 1985;351:76–83.

[25] Roddeman DG, Drukker J, Oomens CWJ, Janssen JD. The wrinkling of thin membranes: part I—theory, part II—numerical analysis. Trans ASME 1987;54:884–92.

[26] Roddeman DG. Finite element analysis of wrinkling membranes. Commun Appl Num Meth 1991;7(4):299–307.

[27] Roddeman DG, van Hout MC, Oomens CWJ, Janssen JD, Drukker J. The wrinkling of thin membranes—comparison between experiments and theory. ASME AMD (Appl Mech Div) 1987;84:267–70.

[28] Kikuchi S. Summary Technical Papers Annual Meeting AIJ (in Japanese). 1973:621–2.

[29] Kikuchi S. Summary Technical Papers Annual Meeting AIJ (in Japanese). 1974:797–8.

[30] Kawaguchi M. The shallowest possible pneumatic forms, IASS bulletin. 1977;63.

[31] Segawa S, Mitsui Y, Sasagawa A. An experimental study on the evaluation of membrane material constants in which are distinguished creep phenomenon from the stress–strain curve (in Japanese). Kouzou Kougaku Ronbunsyu AIJ 1995;41B:259–69.

[32] Standard of Membrane Structures Association of Japan. Testing method for in-plane shear properties of membrane materials (in Japanese). MSAJ/M-01-1993, Membrane Structures Association of Japan, 1993.

[33] Noguchi H, Hisada T. Integrated FEM formulation for total/updated-Lagrangian method in geometrically nonlinear problems. JSME Int J Jap Soc Mech Eng Ser I 1995;38(1):23–9.

[34] Bathe KJ. Finite element procedures. Englewood Cliffs: Prentice Hall, 1996.

[35] Hisada T. Foundations of tensor analysis for nonlinear finite element method. Maruzen, 1992.

[36] Miyamura T, Hangai Y. Finite element analysis of wrinkles on membrane structures of revolution (in Japanese). J Struct Constr Eng AIJ 1996;481:63–70.

[37] Hangai Y. Numerical analysis in the vicinity of critical points by the generalized inverse. Bull Int Assoc Shell Spat Struct 1988;18(3):23–6.

Papers from

Computers and Structures

A comprehensive study of a multiplicative elastoplasticity model coupled to damage including parameter identification

Rolf Mahnken*

Institut für Baumechanik und Numerische Mechanik, University of Hannover, Appelstrasse 9a, 30167, Hannover, Germany

Received 25 May 1997; accepted 19 November 1998

Abstract

In this contribution various aspects for a plasticity model coupled to damage are considered. The formulation of the model is performed in the intermediate configuration which occurs as a consequence of the multiplicative decomposition of the deformation gradient. We will resort to thermodynamic consistency, continuous tangent operator, algorithmic tangent operator and sensitivity analysis for parameter identification. Furthermore, for the discretized constitutive problem a robust iteration scheme with a two-level algorithm is proposed. In the numerical example material parameters are determined by least-squares minimization based on experimental data obtained with an optical method. © 1999 Elsevier Science Ltd. All rights reserved.

Keywords: Multiplicative plasticity; Damage; Parameter identification; Two-level local iteration; Linearization; Sensitivity analysis

1. Introduction

The extensive loading of metallic structures leads to degradation of mechanical properties up to complete failure, and this progressive physical process is commonly referred to as damage. Various manifestations of damage have been described in the literature, such as creep damage, low cycle fatigue, high cycle fatigue and brittle damage [1]. The present paper is concerned with isotropic ductile damage, which is induced by large plastic deformations.

Metallographic studies [2,3] demonstrate that ductile damage is basically characterized by three mechanisms of void growth: (i) nucleation of voids due to fracture of particle–matrix interface, failure of the particle or micro-cracking of the matrix surrounding the inclusion; (ii) growth of voids, thus leading to an enlargement of existing holes; and (iii) coalescence or micro-cracks linking neighbouring voids, thus leading to vanishing load carrying capacity of the material, as the void volume fraction approaches unity.

It appears that mainly two different conceptions can be found in the literature in order to model ductile damage effects: *micro-mechanical damage models* and *phenomenological damage models*. A model of the first kind is formulated by Gurson [4], where he derived a yield potential for porous plastic materials from simple cell models. Modifications have been proposed to improve the predictions at low void volume fractions [5] and to provide a better representation of final void coalescence [6]. In this way micro-mechanical models are based on physical soundness, and various applications have modelled void growth and ductile rupture, see e.g. [5–8]. However the identification and determination of the associated micromechanical material parameters is still a rather new approach with no generally accepted recommendations. It can be done by combining metallurgical examinations with cell modelling and macroscopic testing results [8,9].

* Tel.: +49-511-762-2220; fax: +49-511-762-5496.

Reprinted from *Computers and Structures* **74 (2)**, 179-200 (2000)

Phenomenological damage models are based on the concept of Kachanov [10], who was the first to introduce for the isotropic case a one-dimensional variable, which might be interpreted as the effective surface density of microdefects per unit volume ([1], p. 12). The damage variable is combined with the *effective stress concept* in order to relate the effective stresses, acting on the undamaged material, to the nominal stresses, acting on the damaged area. This allows the application of the *strain equivalence principle* in order to formulate constitutive equations in terms of the effective stresses without modifications of known constitutive models (see Lemaitre [1], p. 13ff). Because of the irreversible nature of the damage process thermodynamic concepts have been proposed for the geometric linear case, in which the damage variable is regarded as an internal variable (see e.g. [11–13]). Extensions to the framework of finite deformations are presented e.g. by De Souza et al. [14] or Steinmann et al. [15].

The model of this work is a modification of a previous representation for finite elasto-plasticity described by Miehe [16], whereby in the above-mentioned three approaches are incorporated: (i) a scalar damage variable; (ii) the *effective stress concept*; and (iii) the *strain equivalence principle*. The formulation is obtained relative to the intermediate configuration which occurs as a consequence of the multiplicative decomposition for the deformation gradient into an elastic and a plastic part. The elastic response is formulated in terms of elastic logarithmic strains, where the total elastic part of the free energy is coupled to damage. The choice of a proper dissipation potential leads to rate equations for the evolution of plastic gliding, isotropic hardening and damage. The above-mentioned choice for the coupling of elasticity and damage is regarded as an extension to the formulation of Steinmann et al. [15], where only the isochoric part of the free energy is coupled to damage. As a consequence, the evolution equation for damage is able to simulate the experimental observation that void growth is highly influenced by the triaxiality ratio which is an important feature in the process of the rupture of materials [3,17].

Furthermore thermodynamic consistency of the constitutive relations can be easily verified, and we derive the continuous tangent operator relative to the intermediate configuration, which relates the rate of Mandel stresses to its work conjugate velocity gradient.

The use of an exponential type integration scheme for the flow rule, as proposed by Weber and Anand [18] and Eterovic and Bathe [19] results in a nonlinear problem in the principal directions of the elastic left Cauchy Green trial tensor. This problem can be reduced to a two-dimensional system of equations and is solved with a two-level scheme as proposed by Johansson et al. [20]. The strategy allows the combination of (i) one-dimensional iteration schemes (bisection or pegasus method), which show superior global convergence properties, with (ii) the Newton scheme, which shows only a superior local convergence property.

The additional task of finding the material parameters for the model—which in the mathematical terminology is an inverse problem [21–23]—is based on experimental testing. In classical approaches the specimen is loaded at the heads (force or displacement controlled), and then experimental data are obtained at certain points of the sample assuming *uniform* distribution of all stress or strain quantities within the sample. However, very often the uniformness is difficult to preserve during the experiment (see e.g. Lemaitre [1] p. 22). Therefore, in this contribution spatially distributed data are used, which are obtained with an optical method [24,25]. Then it is the object to minimize the distance of these data to spatially distributed data obtained from finite element computations in a least-squares sense by varying the material parameters. This approach, including the corresponding sensitivity analysis is described e.g. by Mahnken and Stein [26] for the geometric linear theory, and in Mahnken and Stein [27] it has been extended to the geometric nonlinear case based on an algorithm described by Simo [28].

An outline of this work is as follows: in Section 2 the constitutive equations of the damage model by use of tensor quantities relative to the intermediate configuration are summarized. Then thermodynamic consistency of the model is shown and a continuous tangent operator, also relative to the intermediate configuration, is derived. In Section 3 a robust two-level algorithm is proposed, in order to solve the discretized constitutive problem. The algorithmic tangent operator, needed for the application of a Newton-scheme in the equilibrium iteration, and the sensitivity load term, needed for the application of a gradient-based optimization scheme (e.g. Gauss–Newton method, quasi-Newton method) for parameter identification, are derived in Section 4. For illustrative purposes, in Section 5 material parameters are determined by least-squares minimization based on experimental data obtained with an optical method.

1.1. Notations

Square brackets $[\,\cdot\,]$ are used throughout the paper to denote 'function of' in order to distinguish from mathematical groupings with parentheses $(\,\cdot\,)$.

2. Formulation of the model equations

2.1. *Notation and kinematics of multiplicative elastoplasticity*

Let $\mathscr{B}_0 \subset \mathbb{R}^3$ be the reference configuration of a continuum body $\mathbf{B}$ with smooth boundary $\partial\mathscr{B}_0$, $\mathscr{I}=[t_0, T] \in \mathbb{R}_+$ the time interval of interest and $\mathscr{K}$ a (vector) space of admissible material parameters. We consider the configuration field $\boldsymbol{\varphi}[\mathbf{X},t,\boldsymbol{\kappa}]$ in terms of the three independent variables $\mathbf{X} \subset \mathscr{B}_0$, $t \subset \mathscr{I}$ and $\boldsymbol{\kappa} \subset \mathscr{K}$. Thus $\boldsymbol{\varphi}[\cdot,t,\boldsymbol{\kappa}]$: $\mathscr{B}_0 \to \mathbb{R}^3$ defines a nonlinear deformation map at time $t \subset \mathscr{I}$ for given material parameters $\boldsymbol{\kappa} \subset \mathscr{K}$ mapping particles $\mathbf{X} \subset \mathscr{B}_0$ to their actual position $\mathbf{x} \subset \mathscr{B}$ in the deformed configuration. The associated nonlinear deformation gradient $\mathbf{F} = \nabla_X \boldsymbol{\varphi}[\mathbf{X},t,\boldsymbol{\kappa}]$ with $J = \det\mathbf{F}$ defines a mapping of increments $\mathrm{d}\mathbf{X} \subset T\mathscr{B}_0$ of a locally defined tangent space $T\mathscr{B}_0$ associated with the undeformed configuration to increments $\mathrm{d}\mathbf{x} \subset T\mathscr{B}$ of a locally defined tangent space $T\mathscr{B}$ associated with the deformed configuration. Here and if not stated otherwise also in the subsequent presentation, explicit indication of the arguments t and $\boldsymbol{\kappa}$ is omitted. Furthermore we endow the tangent spaces $T\mathscr{B}_0$ and $T\mathscr{B}$ with co-variant metric tensors $\mathbf{G}^\flat$ and $\mathbf{g}^\flat$, respectively.

The time and parameter derivatives of the configuration field are denoted by $\dot{\boldsymbol{\varphi}} = \partial_t\boldsymbol{\varphi}$ and $\partial_{\boldsymbol{\kappa}}\boldsymbol{\varphi}$, respectively. For the subsequent representation, we will also introduce kinematic expressions by replacing the time derivative $\partial_t(\,\cdot\,)$ or the parameter derivative $\partial_{\boldsymbol{\kappa}}(\cdot)$ by its variational counterparts $\partial_\delta(\,\cdot\,)$ and $\partial_\Delta(\,\cdot\,)$, respectively, where now $\partial_\delta(\,\cdot\,)$ and $\partial_\Delta(\,\cdot\,)$ are short-hand notations for the Gâteaux derivative at the point $\mathbf{x} = \boldsymbol{\varphi}(\mathbf{X},t,\boldsymbol{\kappa})$ in the direction of the virtual variation $\delta\mathbf{u}$ and increment for linearization $\Delta\mathbf{u}$, respectively. With the above notation different kinematic variables and associated derivatives can be defined [27]. Some examples for velocity gradients, rate of deformation tensors and Lie derivative operators are as follows:

$$\mathbf{l} = \nabla_\mathbf{x}\partial_t\boldsymbol{\varphi} \quad \mathbf{d} = \mathrm{sym}(\mathbf{g}^\flat\cdot\mathbf{l})$$
$$\mathscr{L}_t^\#(\cdot) = \mathbf{F}\cdot\partial_t(\mathbf{F}^{-1}\cdot(\cdot)^\#\cdot\mathbf{F}^{-t})\cdot\mathbf{F}^t$$
$$\mathscr{L}_t^\flat(\cdot) = \mathbf{F}^{-t}\cdot\partial_t(\mathbf{F}^t\cdot(\cdot)^\flat\cdot\mathbf{F})\cdot\mathbf{F}^{-1}$$

$$\mathbf{l}_{\boldsymbol{\kappa}} = \nabla_\mathbf{x}\partial_{\boldsymbol{\kappa}}\boldsymbol{\varphi} \quad \mathbf{d}_{\boldsymbol{\kappa}} = \mathrm{sym}(\mathbf{g}^\flat\cdot\mathbf{l}_{\boldsymbol{\kappa}})$$
$$\mathscr{L}_{\boldsymbol{\kappa}}^\#(\cdot) = \mathbf{F}\cdot\partial_{\boldsymbol{\kappa}}(\mathbf{F}^{-1}\cdot(\cdot)^\#\cdot\mathbf{F}^{-t})\cdot\mathbf{F}^t$$
$$\mathscr{L}_{\boldsymbol{\kappa}}^\flat(\cdot) = \mathbf{F}^{-t}\cdot\partial_{\boldsymbol{\kappa}}(\mathbf{F}^t\cdot(\cdot)^\flat\cdot\mathbf{F})\cdot\mathbf{F}^{-1}$$

$$\mathbf{l}_\delta = \nabla_\mathbf{x}\partial_\delta\boldsymbol{\varphi} \quad \mathbf{d}_\delta = \mathrm{sym}(\mathbf{g}^\flat\cdot\mathbf{l}_\delta)$$
$$\mathscr{L}_\delta^\#(\cdot) = \mathbf{F}\cdot\partial_\delta(\mathbf{F}^{-1}\cdot(\cdot)^\#\cdot\mathbf{F}^{-t})\cdot\mathbf{F}^t$$
$$\mathscr{L}_\delta^\flat(\cdot) = \mathbf{F}^{-t}\cdot\partial_\delta(\mathbf{F}^t\cdot(\cdot)^\flat\cdot\mathbf{F})\cdot\mathbf{F}^{-1}$$

$$\mathbf{l}_\Delta = \nabla_\mathbf{x}\partial_\Delta\boldsymbol{\varphi} \quad \mathbf{d}_\Delta = \mathrm{sym}(\mathbf{g}^\flat\cdot\mathbf{l}_\Delta)$$
$$\mathscr{L}_\Delta^\#(\cdot) = \mathbf{F}\cdot\partial_\Delta(\mathbf{F}^{-1}\cdot(\cdot)^\#\cdot\mathbf{F}^{-t})\cdot\mathbf{F}^t \tag{1}$$
$$\mathscr{L}_\Delta^\flat(\cdot) = \mathbf{F}^{-t}\cdot\partial_\Delta(\mathbf{F}^t\cdot(\cdot)^\flat\cdot\mathbf{F})\cdot\mathbf{F}^{-1}$$

where $(\,\cdot\,)^\#$ and $(\,\cdot\,)^\flat$ denote a contra-variant and co-variant tensor object, respectively. The underlying concept of multiplicative elastoplasticity assumes the decomposition

$$\mathbf{F} = \mathbf{F}_e\cdot\mathbf{F}_p \tag{2}$$

where $\mathbf{F}_e$ and $\mathbf{F}_p$ represent the elastic and plastic part of $\mathbf{F}$, respectively, and which implies a plastic intermediate configuration $\mathscr{B}_p$ as macro stress free. Based on the assumption (2) an elastic pull-back of the velocity gradient $\mathbf{l}=\dot{\mathbf{F}}\cdot\mathbf{F}^{-1}=\dot{\mathbf{F}}_e\cdot\mathbf{F}_e^{-1}+\mathbf{F}_e\cdot\dot{\mathbf{F}}_p\cdot\mathbf{F}_p^{-1}\cdot\mathbf{F}_e^{-1}$ yields for the velocity gradient relative to the intermediate configuration $\bar{\mathbf{L}}=\mathbf{F}_e^{-1}\cdot\mathbf{l}\cdot\mathbf{F}_e$ with additive decomposition

$$\bar{\mathbf{L}} = \bar{\mathbf{L}}_e + \bar{\mathbf{L}}_p \tag{3a}$$

where

$$\bar{\mathbf{L}}_e := \mathbf{F}_e^{-1}\cdot\dot{\mathbf{F}}_e \tag{3b}$$

and

$$\bar{\mathbf{L}}_p := \dot{\mathbf{F}}_p\cdot\mathbf{F}_p^{-1} \tag{3c}$$

As noted by Miehe [16], this decoupled representation for the evolution of elastic and plastic deformation is possible within a geometric setting relative to the intermediate configuration in terms of mixed-variant (contra-covariant) tensor fields $\bar{\mathbf{L}}_e$ and $\bar{\mathbf{L}}_p$.

In what follows we will also consider the elastic right Cauchy–Green tensor defined on the intermediate configuration

$$\mathbf{C}_e = \mathbf{F}_e^t\cdot\mathbf{g}^\flat\cdot\mathbf{F}_e \tag{4}$$

and the multiplicative split

$$\mathbf{C}_e = J_e^{2/3}\hat{\mathbf{C}}_e, \quad \text{where} \quad J_e = (\det\mathbf{C}_e)^{1/2} \tag{5}$$

Consequently $\hat{\mathbf{C}}_e$ and J_e represent the isochoric and volumetric part of the elastic deformation, respectively.

2.2. Free energy and thermodynamic forces

The starting point for a thermodynamic consistent model suitable for ductile materials is the following Helmholtz free energy:

$$\Psi = (1-\alpha)\Psi^{el}[\mathbf{C}_e] + \Psi^{p}[q] \tag{6a}$$

where

$$\Psi^{el} = \Psi^{vol}[J_e] + \Psi^{iso}[\hat{\mathbf{C}}_e] \tag{6b}$$

$$\Psi^{vol} = \tfrac{1}{2}K(\ln J_e)^2 \tag{6c}$$

$$\Psi^{iso} = \frac{G}{4}(\mathrm{tr}(\ln \hat{\mathbf{C}}_e)^2) \tag{6d}$$

$$\Psi^{p} = c\left(q + \frac{1}{b}\exp(-bq)\right) \tag{6e}$$

Here the decoupled form for the elastic part is based on the volumetric-isochoric split, Eq. (5), for the elastic right Cauchy–Green tensor with corresponding bulk modulus K and shear modulus G, respectively. The damage variable α represents local degradation of the elastic properties, and q is a strain-like internal variable describing the state of the material at the micro level induced by dislocations with associated material parameters c and b. The above formulation is the analog of the Helmholtz free energy within a geometrically linear 'state kinetic coupling theory' formulated by Lemaitre ([1], p. 42). In particular, state coupling of damage with elastic strain as shown experimentally is assumed, but there is no state coupling either between plasticity and elasticity or between damage and plasticity. The formulation (6e) gives the classical expression for isotropic hardening with saturation for large plastic strain.

Thermodynamnic formulations for isothermal processes are based on the principle of positive dissipation

$$\mathscr{D} = \mathscr{P} - \dot{\Psi} \geq 0 \tag{7}$$

Using tensor quantities relative to the actual configuration the stress-power is given as a dual pairing $\mathscr{P} = \boldsymbol{\tau}{:}\mathbf{d}$ with the (contravariant) Kirchhoff stress tensor $\boldsymbol{\tau}$ and the (covariant) spatial rate of deformation tensor $\mathbf{d} = \mathrm{sym}(\mathbf{g}^{\flat}\cdot\mathbf{l})$, whereas with tensor quantities relative to the reference configuration we have $\mathscr{P} = \mathbf{S}{:}\mathbf{D}$ in terms of the (contravariant) second Piola–Kirchhoff stress tensor $\mathbf{S} = \mathbf{F}^{-1}\cdot\boldsymbol{\tau}\cdot\mathbf{F}^{-t}$ and a (covariant) material rate of deformation tensor $\mathbf{D} = \mathbf{F}^{t}\cdot\mathbf{d}\cdot\mathbf{F} = \mathrm{sym}(\dot{\mathbf{F}}^{t}\cdot\mathbf{F})$. Following Miehe [16], alternatively the stress power can be written relative to the intermediate configuration as $\mathscr{P} = \bar{\mathbf{T}}{:}\bar{\mathbf{L}}$, where $\bar{\mathbf{T}} = \mathbf{F}_e^{t}\cdot\mathbf{g}^{\flat}\cdot\boldsymbol{\tau}\cdot\mathbf{F}_e^{-t}$ is the mixed variant (co-contravariant) Mandel stress tensor, and $\bar{\mathbf{L}}$ is the total velocity gradient introduced in the previous section. Consequently, using the additive decomposition, Eqs. (3a)–(3c), and the identity $\partial\Psi/\partial\mathbf{C}_e{:}\dot{\mathbf{C}}_e = 2(\mathbf{C}_e\cdot\partial\Psi/\partial\mathbf{C}_e){:}\bar{\mathbf{L}}_e$ the dissipation (7) results into

$$\mathscr{D} = \left[\bar{\mathbf{T}} - 2\mathbf{C}_e\cdot\frac{\partial\Psi}{\partial\mathbf{C}_e}\right]{:}\bar{\mathbf{L}} + 2\mathbf{C}_e\cdot\frac{\partial\Psi}{\partial\mathbf{C}_e}{:}\bar{\mathbf{L}}_p - \frac{\partial\Psi}{\partial q}\dot{q} - \frac{\partial\Psi}{\partial\alpha}\dot{\alpha} \geq 0 \tag{8}$$

Employing the standard argument of thermodynamics, that the above relation holds for all processes $\bar{\mathbf{L}}$, implies $\bar{\mathbf{T}} = 2\mathbf{C}_e\cdot\partial\Psi/\partial\mathbf{C}_e$ and thus for the logarithmic model, Eqs. (6a)–(6c) it follows:

$$\bar{\mathbf{T}} = (1-\alpha)\hat{\bar{\mathbf{T}}}, \quad \text{with} \quad \hat{\bar{\mathbf{T}}} = \hat{\bar{\mathbf{T}}}^{vol} + \hat{\bar{\mathbf{T}}}^{iso} \quad \text{and} \tag{9a}$$

$$\hat{\bar{\mathbf{T}}}^{vol} = p\mathbf{1}, \quad p = K\ln J_e \tag{9b}$$

$$\hat{\bar{\mathbf{T}}}^{iso} = G\ln\hat{C}_e = G\,\mathrm{dev}\ln\mathbf{C}_e = \hat{\bar{\mathbf{T}}}^{dev} \tag{9c}$$

The relations (9a)–(9c) express the *effective stress concept* (see Lemaitre [1], p. 42), thus relating the 'nominal' Mandel stress tensor $\bar{\mathbf{T}}$ to the effective Mandel stress tensor $\hat{\bar{\mathbf{T}}}$ acting on the remaining undamaged material.

For subsequent purposes we define the hardening variable Q and the damage energy release rate A as the thermodynamic forces conjugate to the internal variables q and α as

$$Q = \frac{\partial\Psi}{\partial q} = c(1 - \exp(-bq))$$

$$A = \frac{\partial\Psi}{\partial\alpha} = -\frac{p^2}{2K} - \frac{1}{4G}\|\hat{\bar{\mathbf{T}}}^{dev}\|^2 \tag{10}$$

Here the relations $\Psi^{vol} = (1/2)K(\ln J_e)^2 = p^2/(2K)$ and $\Psi^{iso} = (1/4)G\mathrm{tr}(\ln\hat{\mathbf{C}}_e)^2 = \|\hat{\bar{\mathbf{T}}}^{dev}\|^2/(4G)$ have been exploited.

2.3. Dissipation potential and evolution of internal variables

In order to describe plasticity of ductile materials a separate potential of dissipation $\Phi*$ is introduced depending on the effective Mandel stress $\hat{\bar{\mathbf{T}}}$, the hardening variable Q and the damage energy release rate A as a sum of two functions

$$\Phi^{*} = \Phi_p^{*}[\hat{\bar{\mathbf{T}}}, Q] + \Phi_d^{*}[A] \tag{11}$$

Note, that the formulation for Φ_p^* is in accordance with the *principle of strain equivalence*, whereby the yield criterion is written in the same way as for the nondamaged material except that the stress tensor $\bar{\mathbf{T}}$ is replaced by the effective stress tensor $\hat{\bar{\mathbf{T}}}$. For definiteness we choose for the plastic potential Φ_p^* and the damage potential Φ_d^*

$$\Phi_p^* = \|\hat{\bar{\mathbf{T}}}^{\mathrm{dev}}\| - \sqrt{\frac{2}{3}}(Y_0 + Q)$$

$$\Phi_d^* = \frac{(-A)^2}{2S(1-\alpha)^m}\mathcal{T}[q] \tag{12}$$

where the threshold function $\mathcal{T}[q]$ has been introduced, in order to activate damage only if a certain limit ϵ_1 has been obtained for q (see Lemaitre [1], p. 96). A possible choice for $\mathcal{T}$ is the Hermitian polynomial

$$\mathcal{T}[q;\epsilon_1,\epsilon_2] = 0 \quad \text{if} \quad q \leq \epsilon_1$$

$$\mathcal{T}[q;\epsilon_1,\epsilon_2] = \frac{(q-\epsilon_1)^2}{(\epsilon_2-\epsilon_1)^2}\left(3 - 2\frac{q-\epsilon_1}{\epsilon_2-\epsilon_1}\right) \quad \text{if} \quad \epsilon_1 < q < \epsilon_2$$

$$\mathcal{T}[q;\epsilon_1,\epsilon_2] = 1 \quad \text{if} \quad \epsilon_2 \leq q \tag{13}$$

which due to its smoothness is of advantage for the numerical implementation, see Johansson et al. [20]. It follows that ϵ_1,ϵ_2 can be regarded as material parameters.

The evolution for the set of internal variables $[\bar{\mathbf{L}}_p,\dot{q},\dot{\alpha}]$ is now proposed based on the principle of generalized normalities as

$$\bar{\mathbf{L}}_p = \dot{\lambda}\frac{\partial\Phi^*}{\partial\bar{\mathbf{T}}} = \frac{\dot{\lambda}}{1-\alpha}\bar{\mathbf{M}}, \quad \text{where} \qquad \bar{\mathbf{M}} = \frac{\partial\Phi_p^*}{\partial\hat{\bar{\mathbf{T}}}} = \frac{(\hat{\bar{\mathbf{T}}}^{\mathrm{dev}})^{\mathrm{t}}}{\|\hat{\bar{\mathbf{T}}}^{\mathrm{dev}}\|} \tag{14a}$$

$$\dot{q} = -\dot{\lambda}\frac{\partial\Phi^*}{\partial Q} = -\dot{\lambda}M_q, \quad \text{where} \qquad M_q = \frac{\partial\Phi_p^*}{\partial Q} = -\sqrt{\frac{2}{3}} \tag{14b}$$

$$\dot{\alpha} = -\dot{\lambda}\frac{\partial\Phi^*}{\partial A} = -\dot{\lambda}M_\alpha, \quad \text{where} \qquad M_\alpha = \frac{\partial\Phi_d^*}{\partial A} = \frac{A}{S(1-\alpha)^m}\mathcal{T}[q] \tag{14c}$$

The scalar $\dot{\lambda}$ is the plastic multiplier obtained from the loading and unloading conditions

$$\dot{\lambda} \geq 0$$

$$\Phi_Y \leq 0$$

$$\dot{\lambda}\Phi_Y = 0 \tag{15}$$

where Φ_Y is the yield function. In the simplest case we set

$$\Phi_Y = \Phi_p^* = \|\hat{\bar{\mathbf{T}}}^{\mathrm{dev}}\| - \sqrt{\frac{2}{3}}(Y_0 + Q) \tag{16}$$

Remark 1. Note, that due to the choices, Eq. (12), the above evolution equations (14a)–(14c) exhibit an associated flow rule for the evolution of plastic flow and hardening evolution but a nonassociative flow rule for the damage evolution.

Remark 2. The evolution rule α specified by Eq. (14c) predicts the rate of α to be proportional to the conjugate thermodynamic force A. This implies the identical growth of α for both positive and negative hydrostatic stress. However for certain materials (e.g. brittle materials) and certain loading conditions the defect may close in compression and reopen in tension. Following Lemaitre [1], this 'microcrack closure reopening' (MCR) effect can be incorporated into the above constitutive equations by a straightforward modification. An example within the geometric linear setting is proposed by Johansson et al. [20].

Remark 3. Upon defining the invariants

$$\sigma_v := \sqrt{\frac{3}{2}}\|\bar{\mathbf{T}}^{\mathrm{dev}}\|$$

$$\sigma_h := \frac{1}{3}\mathrm{tr}(\bar{\mathbf{T}})$$

$$\dot{e}_v := \sqrt{\frac{2}{3}}\|\bar{\mathbf{L}}_p\| \tag{17}$$

for the Mandel stress tensor and the plastic part of the velocity gradient relative to the intermediate configuration, the evolution equation for the damage variable (14c) can be rewritten as

$$\dot{\alpha} = -\dot{e}_v\frac{\bar{A}}{S(1-\alpha)^{(m-1)}}\mathcal{T}[q]$$

where

$$\bar{A} = -\frac{\sigma_v^2}{(1-\alpha)^2}\left(\frac{1}{2K}\frac{\sigma_h^2}{\sigma_v^2} + \frac{1}{6G}\right) \tag{18}$$

The presentation (18) reveals that evolution of damage is proportional to the accumulated plastic strain rate $\dot{e}_v$, which is in accordance with the analytical investigations for ductile growth of voids by Rice and Tracey [17]. Furthermore, as noted by Lemaitre ([1], p. 44 and p. 97), the triaxiality ratio σ_h/σ_v, plays a very important role in the rupture of materials (see also Rice and Tracey [17]). The presentation (18) shows an increase of damage evolution with increasing σ_h/σ_v, which is in agreement with the experimental observation, that the measured ductility at fracture decreases as the triaxiality ratio increases [2,3]. In this respect the formulation (14c) is regarded as an important modification to the formulation presented by Steinmann et al. [15], and, as it will be seen later, has some consequences in the numerical implementation. Lastly we note that, due to the factors $(1-\alpha)$ in the denominators of the presentation (18), an increase of damage evolution with increasing α is obtained for $m > 0$.

2.4. Thermodynamic consistency

According to the second law of thermodynamics the dissipation inequality (7) must be satisfied for the evolution equations (14a)–(14c). Here only the loading case $\dot{\lambda} > 0$, $\Phi_Y = 0$ shall be considered, since the unloading case with $\dot{\lambda} = 0$ is trivial. Combining the relations (14a)–(14c), (10), (8) entails writing

$$\mathscr{D} = \bar{\mathbf{T}}:\bar{\mathbf{L}}_p - Q\dot{q} - A\dot{\alpha}$$

$$= \dot{\lambda}\left(\frac{1-\alpha}{1-\alpha}\hat{\bar{\mathbf{T}}}^{dev} : \frac{\hat{\bar{\mathbf{T}}}^{dev}}{\|\hat{\bar{\mathbf{T}}}^{dev}\|} - Q\sqrt{\frac{2}{3}} + \frac{A^2}{S(1-\alpha)^m}\mathscr{T}(q)\right)$$

$$= \dot{\lambda}\left(\sqrt{\frac{2}{3}}Y_0 + \frac{A^2}{S(1-\alpha)^m}\mathscr{T}(q)\right) > 0$$

The last relation is obtained from the yield function (16) for the case of loading with $\Phi_Y = 0$.

2.5. Summary of material parameters

We are now in a position to summarize all material constants of the model, characterizing the inelastic behavior, which, apart from the elastic constants K and G, have to be calibrated based on experimental data

$$\boldsymbol{\kappa} = [Y_0, c, b, S, m, \epsilon_1, \epsilon_2]^t \tag{19}$$

In order to be physically meaningful the material parameters are restricted to lower and upper bounds a_i, b_i, respectively. These constraints then define the feasible domain $\mathscr{K}$, such that

$$\boldsymbol{\kappa} \in \mathscr{K} \subset \mathbb{R}^{n_p}$$

$$\mathscr{K} := \{\boldsymbol{\kappa}: a_i \le \kappa_i \le b_i, i = 1, \dots, n_p\} \tag{20}$$

where $n_p = \dim(\boldsymbol{\kappa}) = 7$ denotes the number of material parameters.

2.6. Continuous tangent operator

Continuous consistent tangent operators relate time derivatives of some stress tensor to its work conjugate rate of deformation tensor. In the framework of multiplicative plasticity these relations, which can be regarded as counterparts of so-called *Prandtl–Reuss tensors* of the geometric linear theory, have been firstly derived by Miehe [16] for associative flow rules without hardening and furthermore by Steinmann [29], where additionally damage has been taken into account. In what follows we will derive an operator $\mathscr{A}_0^{pl}$, which relates the time derivative of Mandel stresses $\dot{\bar{\mathbf{T}}}$ to the velocity gradient $\bar{\mathbf{L}}$ relative to the intermediate configuration as

$$\dot{\bar{\mathbf{T}}} = \mathscr{A}_0^{pl}:\bar{\mathbf{L}}^t \tag{21}$$

Starting with $\hat{\bar{\mathbf{T}}} = 2\mathbf{C}_e \cdot \partial\Psi^{el}/\partial\mathbf{C}_e$ and using the relation $\dot{\mathbf{C}}_e = \text{sym}(\mathbf{C}_e \cdot \bar{\mathbf{L}}_e)$ the total time derivative of the effective Mandel stress is

$$\dot{\hat{\bar{\mathbf{T}}}} = \mathscr{A}_0^{el}:\bar{\mathbf{L}}_e^t$$

$$\text{where } \mathscr{A}_0^{el} = \mathscr{A}_2^{el} + \mathbf{I}\otimes\hat{\bar{\mathbf{T}}}^t + \mathbf{C}_e\underline{\otimes}\hat{\bar{\mathbf{T}}}^t \cdot \mathbf{C}_e^{-1}$$

$$\mathscr{A}_2^{el} = \mathbf{C}_e \cdot \bar{\mathbb{C}} \cdot \mathbf{C}_e$$

$$\bar{\mathbb{C}} = 4\frac{\partial^2\Psi^{el}}{\partial\mathbf{C}_e\partial\mathbf{C}_e} \tag{22}$$

which defines a fourth-order hyperelastic tensor $\mathscr{A}_0^{el}$. (Here $\otimes$ and $\underline{\otimes}$ define non-standard tensor products such that $(\mathbf{a}\otimes\mathbf{b}):\mathbf{c} = \mathbf{a}\cdot\mathbf{c}\cdot\mathbf{b}^t$, $(\mathbf{a}\underline{\otimes}\mathbf{b}):\mathbf{c} = \mathbf{a}\cdot\mathbf{c}^t\cdot\mathbf{b}^t$, see e.g. Steinmann [29].) Likewise, the time derivatives of the stress like internal variables are

$$\dot{Q} = -\dot{\lambda}H_q M_q, \quad \text{where } H_q = \frac{\partial^2\Psi}{\partial q\partial q}$$

$$\dot{A} = -\dot{\lambda} H_\alpha M_\alpha, \quad \text{where } H_\alpha = \frac{\partial^2\Psi}{\partial\alpha\partial\alpha} \tag{23}$$

Upon introducing the definitions

$$\bar{\mathbf{N}} = \frac{\partial\Phi_Y}{\partial\hat{\bar{\mathbf{T}}}} \tag{24a}$$

$$N_q = \frac{\partial\Phi_Y}{\partial Q} \tag{24b}$$

$$N_\alpha = \frac{\partial\Phi_Y}{\partial A} \tag{24c}$$

and combining the relations (22) and (23) and $\bar{\mathbf{L}}_e = \bar{\mathbf{L}} - \dot{\lambda}/(1-\alpha)\bar{\mathbf{M}}$ the consistency condition $\dot{\Phi}_Y = \bar{\mathbf{N}}:\dot{\hat{\bar{\mathbf{T}}}} + N_q\dot{Q} + N_\alpha\dot{A} = 0$ yields the plastic multiplicator as

$$\dot{\lambda} = \frac{1}{h}\bar{\mathbf{N}}:\mathcal{A}_0^{\mathrm{el}}:\bar{\mathbf{L}}^{\mathrm{t}}, \quad \text{where}$$
$$h = \frac{1}{1-\alpha}\bar{\mathbf{N}}:\mathcal{A}_0^{\mathrm{el}}:\bar{\mathbf{M}} + N_q H_q M_q + N_\alpha H_\alpha M_\alpha \tag{25}$$

From Eq. (9a) it follows

$$\dot{\bar{\mathbf{T}}} = (1-\alpha)\dot{\hat{\bar{\mathbf{T}}}} - \hat{\bar{\mathbf{T}}}\dot{\alpha} \tag{26}$$

which combined with Eqs. (14c), (22) and (25) yields the hyperelastic–plastic tangent operator in the intermediate configuration as

$$\mathcal{A}_0^{\mathrm{pl}} = (1-\alpha)\mathcal{A}_0^{\mathrm{el}} - \frac{1}{h}(\mathcal{A}_0^{\mathrm{el}}:\bar{\mathbf{M}}^{\mathrm{t}} + \hat{\bar{\mathbf{T}}}M_\alpha)\otimes\bar{\mathbf{N}}:\mathcal{A}_0^{\mathrm{el}} \tag{27}$$

and which in the general case exhibits a nonsymmetric structure.

For the logarithmic model, Eqs. (6a)–(6e), use of the relations (see e.g. Steinmann [29])

$$\dot{\overline{\ln J_e}} = \mathbf{1}:\bar{\mathbf{L}}_e^{\mathrm{t}}$$

$$\dot{\overline{\mathrm{tr}\ln\mathbf{C}_e}} = 2\mathbf{1}:\bar{\mathbf{L}}_e^{\mathrm{t}} \tag{28}$$

expands the elastic operator $\mathcal{A}_0^{\mathrm{el}}$ from Eqs. (9a)–(9c) as

$$\mathcal{A}_0^{\mathrm{el}} = \mathcal{A}_0^{\mathrm{el,vol}} + \mathcal{A}_0^{\mathrm{el,dev}}, \quad \text{where } \mathcal{A}_0^{\mathrm{el,vol}} = K\mathbf{1}\otimes\mathbf{1}$$

$$\mathcal{A}_0^{\mathrm{el,dev}} = 2G\left[\frac{\partial\ \ln\ \mathbf{C}_e}{\partial\mathbf{C}_e}\cdot\mathbf{C}_e - \frac{1}{3}\mathbf{1}\otimes\mathbf{1}\right] \tag{29}$$

Furthermore, for isotropy we have $\bar{\mathbf{N}}:\mathcal{A}_0^{\mathrm{el}}:\bar{\mathbf{L}}_e^{\mathrm{t}} = G\bar{\mathbf{N}}:\dot{\overline{\ln\ \mathbf{C}_e}} = 2G\bar{\mathbf{N}}:\bar{\mathbf{L}}_e^{\mathrm{t}}$ and thus $\bar{\mathbf{N}}:\mathcal{A}_0^{\mathrm{el}} = 2G\bar{\mathbf{N}}$ and $\bar{\mathbf{N}}:\mathcal{A}_0^{\mathrm{el}}:\bar{\mathbf{N}} = 2G$. From this, Eq. (27) reveals the hyperelastic–plastic tangent operator for the logarithmic model as

$$\mathcal{A}_0^{\mathrm{pl}} = (1-\alpha)\mathcal{A}_0^{\mathrm{el}} - \frac{2G}{h}(2G + \|\hat{\bar{\mathbf{T}}}\|M_\alpha)\bar{\mathbf{N}}\otimes\bar{\mathbf{N}} \tag{30}$$

Note, that due to the associative flow rule, the above operator shows a symmetric structure.

3. Numerical implementation

3.1. Integration scheme

In this section the numerical integration of the evolution equations (14a)–(14c) is described over a finite time step $\Delta t = {}^{n+1}t - {}^{n}t$ for given initial data ${}^{n}q$, ${}^{n}\alpha$, ${}^{n}\mathbf{C}_{\mathrm{p}}^{-1}$ and deformation gradient ${}^{n+1}\mathbf{F}$, where ${}^{n}\mathbf{C}_{\mathrm{p}} = {}^{n}\mathbf{F}_{\mathrm{p}}^{\mathrm{t}}\cdot{}^{n}\mathbf{F}_{\mathrm{p}}$ is a plastic right Cauchy–Green strain tensor. The starting point for numerical integration of the flow rule, Eq. (14a) is the representation

$$\bar{\mathbf{L}}_{\mathrm{p}} = \dot{\mathbf{F}}_{\mathrm{p}}\cdot\mathbf{F}_{\mathrm{p}}^{-1} = \frac{\dot{\lambda}}{1-\alpha}\bar{\mathbf{M}}$$

$$\bar{\mathbf{M}} = \bar{\mathbf{N}} \leadsto \dot{\mathbf{F}}_{\mathrm{p}} = \frac{\dot{\lambda}}{1-\alpha}\bar{\mathbf{N}}\cdot\mathbf{F}_{\mathrm{p}} \tag{31}$$

where the relations (3c), (14a), (24a) and (16) have been combined. Then, using an exponential type of integration rule for $\mathbf{F}_{\mathrm{p}}$ [18,19] and a backward Euler scheme for q and α we have

$${}^{n+1}\mathbf{F}_{\mathrm{p}} = \exp\left(\frac{\Delta\lambda}{1-\alpha}\bar{\mathbf{N}}\right)\cdot{}^{n}\mathbf{F}_{\mathrm{p}} \tag{32a}$$

$${}^{n+1}q = {}^{n}q + \Delta\lambda\sqrt{\tfrac{2}{3}} \tag{32b}$$

$${}^{n+1}\alpha = {}^{n}\alpha - \Delta\lambda\frac{{}^{n+1}A}{S(1-\alpha)^m}\mathcal{T}[{}^{n+1}q] \tag{32c}$$

From now on, if confusion is out of danger we will neglect the index $n+1$ referring to the actual time step. Then, we introduce the following 'trial' quantities

$$\mathbf{F}^{\mathrm{tr}} = \mathbf{F}\cdot{}^{n}\mathbf{F}_{\mathrm{p}}^{-1} \leadsto \mathbf{C}^{\mathrm{tr}} = \mathbf{F}^{\mathrm{tr}^{\mathrm{t}}}\cdot\mathbf{F}^{\mathrm{tr}} \leadsto \mathbf{C}_e$$
$$= \mathbf{F}_e^{\mathrm{t}}\cdot\mathbf{F}_e = \exp\left(\frac{-\Delta\lambda}{1-\alpha}\bar{\mathbf{N}}\right)\cdot\mathbf{C}^{\mathrm{tr}}$$
$$\cdot\exp\left(\frac{-\Delta\lambda}{1-\alpha}\bar{\mathbf{N}}\right) \tag{33}$$

Using the fact, that $\mathbf{C}_e$, $\mathbf{C}^{\mathrm{tr}}$ and $\bar{\mathbf{N}}$ commute (i.e. have identical principal axes) and decomposing the logarithmic part of $\mathbf{C}_e$ into its deviatoric and its volumetric part yields

$$\ln \hat{\mathbf{C}}_e = \ln \hat{\mathbf{C}}^{tr} - \frac{2\Delta\lambda}{1-\alpha}\bar{\mathbf{N}}$$

where $\hat{\mathbf{C}}^{tr} = (J^{tr})^{-2/3}\mathbf{C}^{tr}$

$$J^{tr} = \det(\mathbf{F}^{tr})$$

$$\ln J_e = \ln J^{tr} \tag{34}$$

Then, by use of Eqs. (9a)–(9c), the Mandel stresses are decomposed as

$$\mathbf{T} = (1-\alpha)\hat{\bar{\mathbf{T}}}, \quad \text{with } \hat{\bar{\mathbf{T}}} = \hat{\bar{\mathbf{T}}}^{vol} + \hat{\bar{\mathbf{T}}}^{dev} \text{ and}$$

$$\hat{\bar{\mathbf{T}}}^{vol} = p\mathbf{1}, \quad p = K \ln J^{tr}$$

$$\hat{\bar{\mathbf{T}}}^{dev} = \hat{\bar{\mathbf{T}}}^{dev,tr} - \frac{2G\Delta\lambda}{1-\alpha}\bar{\mathbf{N}}, \quad \text{where } \hat{\bar{\mathbf{T}}}^{dev,tr} = G \ln \hat{\mathbf{C}}^{tr} \tag{35}$$

A more advantageous formulation for the finite element implementation is obtained relative to the actual configuration with the Kirchhoff stresses $\boldsymbol{\tau} = \mathbf{g}^{\#} \cdot \mathbf{F}_e^{-t} \cdot \bar{\mathbf{T}} \cdot \mathbf{F}_e^{t}$. Thus, by use of the left Cauchy–Green trial tensor

$$\mathbf{b}^{tr} = \mathbf{F}^{tr} \cdot (\mathbf{F}^{tr})^t = \mathbf{F} \cdot^{n} \mathbf{C}_p^{-1} \cdot \mathbf{F}^t \tag{36}$$

the Kirchhoff stresses are obtained from the relations

$$\boldsymbol{\tau} = (1-\alpha)\hat{\boldsymbol{\tau}}, \text{ with } \hat{\boldsymbol{\tau}} = \hat{\boldsymbol{\tau}}^{vol} + \hat{\boldsymbol{\tau}}^{dev} \text{ and}$$

$$\hat{\boldsymbol{\tau}}^{vol} = p\mathbf{g}^{\#}, p = K \ln J^{tr}$$

$$\hat{\boldsymbol{\tau}}^{dev} = \hat{\boldsymbol{\tau}}^{dev,tr} - \frac{2G\Delta\lambda}{1-\alpha}\mathbf{n}, \text{ where } \hat{\boldsymbol{\tau}}^{dev,tr} = G \operatorname{dev} \ln \mathbf{b}^{tr}$$

$$\mathbf{n} = \frac{\hat{\boldsymbol{\tau}}^{dev}}{\|\hat{\boldsymbol{\tau}}^{dev}\|} \tag{37}$$

These sets of equations can be regarded as the counterpart of the relation (35) relative to the actual configuration.

3.2. Spectral decomposition

Upon using a spectral decomposition of the left elastic Cauchy–Green trial tensor and using the fact that due to isotropy $\mathbf{b}^{tr}$ and $\boldsymbol{\tau}$ commute, we have

$$\mathbf{b}^{tr} = \sum_{A=1}^{3}(\lambda_A^{tr})^2\mathbf{m}_A \rightsquigarrow \boldsymbol{\tau} = \sum_{A=1}^{3}\beta_A\mathbf{m}_A \tag{38}$$

Here λ_A^{tr}, $\mathbf{m}_A$, $A = 1,2,3$ are the eigenvectors and eigenbasis of $\mathbf{b}^{tr}$, respectively, and β_A, $A = 1,2,3$ are the principal values of the effective Kirchhoff stresses which by use of the vector/matrix notations

$$\underline{\varepsilon}^{tr} := \begin{bmatrix} \ln \lambda_1^{tr} \\ \ln \lambda_2^{tr} \\ \ln \lambda_3^{tr} \end{bmatrix}$$

$$\underline{\beta} := \begin{bmatrix} \beta_1 \\ \beta_2 \\ \beta_3 \end{bmatrix}$$

$$\underline{1} := \begin{bmatrix} 1 \\ 1 \\ 1 \end{bmatrix}$$

$$\underline{I}_3 := \begin{bmatrix} 1 & & \\ & 1 & \\ & & 1 \end{bmatrix}$$

$$\underline{I}_3^{dev} := \underline{I}_3 - \frac{1}{3}\underline{1} \otimes \underline{1} \tag{39}$$

are obtained from the relations

$$\underline{\beta} = (1-\alpha)\hat{\underline{\beta}}, \text{ with } \hat{\underline{\beta}} = \hat{\underline{\beta}}^{vol} + \hat{\underline{\beta}}^{dev} \text{ and}$$

$$\hat{\underline{\beta}}^{vol} = p\underline{1}, p = K\underline{1} \cdot \underline{\varepsilon}^{tr}$$

$$\hat{\underline{\beta}}^{dev} = \hat{\underline{\beta}}^{dev,tr} - \frac{2G\Delta\lambda}{1-\alpha}\underline{\nu}, \text{ where } \hat{\underline{\beta}}^{dev,tr} = 2G\underline{I}_3^{dev} \cdot \underline{\varepsilon}^{tr}$$

$$\underline{\nu} = \frac{\hat{\underline{\beta}}^{dev}}{\|\hat{\underline{\beta}}^{dev}\|} \tag{40}$$

These sets of equations can be regarded as the counterpart of the relations (35) and (37) in the principal directions. Note, that the above structure for the principal Kirchhoff stresses is identical to the geometric linear theory, see e.g. Simo [28].

3.3. Local iteration: two level algorithm

The two unknowns $\underline{x} = [\Delta\lambda, \alpha]^t$ appearing in Eq. (40) are obtained from a nonlinear system of equations $\underline{r} = [r_1, r_2]^t = \underline{0}$, where

$$r_1 = \|\hat{\underline{\beta}}^{dev,tr}\| - \frac{2G\Delta\lambda}{1-\alpha} - \sqrt{\frac{2}{3}}(Y_0 + Q[q[\Delta\lambda]]) \tag{41a}$$

$$r_2 = -\alpha + {}^n\alpha + \frac{\Delta\lambda}{S(1-\alpha)^m} \frac{1}{2}\left(\frac{(Q+Y_0)^2}{3G} + \frac{p^2}{K}\right)\mathcal{T}[q[\Delta\lambda]] \quad (41b)$$

Here Eq. (41a) expresses the yield constraint and Eq. (41b) is obtained from the update scheme (32c) for the damage variable. A further constraint—which can lead to severe numerical difficulties—arises from the fact, that the variables contained in the vector $\underline{x}$ are restricted to lower and upper bounds a_i, b_i, $i = 1,2$. In particular we have $0 \leqslant \Delta\lambda < \infty$ and ${}^n\alpha \leqslant \alpha < 1$. We are therefore confronted with the following problem:

$$\text{Find } \underline{x} \in \mathbb{R}^2, \text{ such that } \underline{r}[\underline{x}] = \underline{0}, \quad a_i \leq x_i \leq b_i \quad i = 1,2 \quad (42)$$

If the starting point $\underline{x}^{(k=0)}$ is close enough to the solution point (assuming that it exists) the problem (42) can be solved iteratively with a Newton method

$$\underline{x}^{(k+1)} = \underline{x}^{(k)} - [\underline{J}^{(k)}]^{-1}\underline{r}^{(k)} \quad k = 0,1,2,\ldots \quad (43)$$

where the Jacobian

$$\underline{J} = \frac{\partial \underline{r}}{\partial \underline{x}} = \begin{bmatrix} \frac{\partial r_1}{\partial x_1} & \frac{\partial r_1}{\partial x_2} \\ \frac{\partial r_2}{\partial x_1} & \frac{\partial r_2}{\partial x_2} \end{bmatrix} \quad (44)$$

is required at each iteration point.

However, the superior local convergence property of the Newton method does not necessarily imply a good global convergence property in the case of improper starting values, see Johansson et al. [20] for a numerical example. Therefore, in order to gain more control and robustness of the iteration process a two-level strategy is adopted, whereby the two-dimensional problem (42) is solved by a sequence of one-dimensional problems. This allows the advantages of various one-dimensional solution schemes to be combined, e.g. the superior global convergence properties of a bisection method or a pegasus method, respectively (see e.g. Engelin–Müllges and Reuter [30]), with the superior local convergence property of a Newton method.

The iteration scheme is explained as follows [20]: firstly, for given (fixed) x_2 we use the one-dimensional iteration $x_1^{(k+1)} = x_1^{(k)} + \Delta x_1^{(k)}$, $k = 0,1,2,..$ until the condition $r_1[x_1^{(k)}, x_2] = 0$ is satisfied. However, the solution

$$x_1[x_2] = \arg\{r_1[x_1[x_2], x_2] = 0\} \quad (45)$$

in general does not satisfy the second condition $r_2[x_1[x_2], x_2] = 0$ so that an additional (outer) iteration $x_2^{(k+1)} = x_2^{(k)} + \Delta x_2^{(k)}$, $k = 0,1,2,.$. becomes necessary. The basic idea consists now in determining a solution $x_1[x_2]$ according to Eq. (45) whenever x_2 is modified. In this manner the two-dimensional problem, Eq. (42), is converted into a sequence of one-dimensional problems.

In Table 1 the algorithm is summarized. For determination of the increment Δx_i, $i = 1,2$ several choices are possible:

- Bisection method: here the increment is obtained simply by

$$\Delta x_i^{(k)} = \tfrac{1}{2}(x_i^{\min} - x_i^{(k)}), \quad x_i^{\max} = x_i^{(k)}, \quad \text{if} \quad r_i(x_i^{(k)}) \cdot r_i(x_i^{\max}) > 0$$

$$\Delta x_i^{(k)} = \tfrac{1}{2}(x_i^{\max} - x_i^{(k)}), \quad x_i^{\min} = x_i^{(k)}, \quad \text{else} \quad (46)$$

- Newton method: in this case the increment is calculated according to

$$\Delta x_i^{(k)} = -\left[\frac{\mathrm{d}r_i^{(k)}}{\mathrm{d}x_i}\right]^{-1} r_i^{(k)} \quad (47)$$

thus requiring the derivative of the residual r_i. For $i = 2$ we have

$$\frac{\mathrm{d}r_2}{\mathrm{d}x_2} = \frac{\partial r_2}{\partial x_2} + \frac{\partial r_2}{\partial x_1} \cdot \frac{\mathrm{d}x_1}{\mathrm{d}x_2}$$

Table 1
Two-level algorithm for nonlinear two-dimensional problem

FIND-ZERO(i):
Object: determine x_i such that $r_1[x_1[x_2], x_2] = 0$ if $i = 1$ or $r_2[x_1[x_2], x_2] = 0$, if $i = 2$
Algorithm:

0. Initialize:	$k = 0$; $x_i^{(k=0)}$
1. Change level:	If ($i = 2$) call FIND-ZERO($i - 1$)
2. Residual:	r_i
3. Check tolerance:	If $\lvert r_i \rvert <$ tol, RETURN
4. Determine increment:	$\Delta x_i^{(k)}$
5. Update:	$x_i^{(k+1)} = x_i^{(k)} + \Delta x_i^{(k)}$, $k = k + 1$, GOTO 1

For evaluation of dx_1/dx_2 we exploit Step 1 of the algorithm in Table 1 which insures the condition

$$r_1[x_1[x_2],x_2] = 0 \rightsquigarrow \frac{dr_1}{dx_2} = \frac{\partial r_1}{\partial x_2} + \frac{\partial r_1}{\partial x_1}\frac{dx_1}{dx_2} = 0$$

The result for the derivative needed for the increment Eq. (47) is thus summarized as

$$i = 1: \quad \frac{dr_1}{dx_1} = \frac{\partial r_1}{\partial x_1} \tag{48a}$$

$$i = 2: \quad \frac{dr_2}{dx_2} = \frac{\partial r_2}{\partial x_2} - \frac{\partial r_2}{\partial x_1} \cdot \left[\frac{\partial r_1}{\partial x_1}\right]^{-1} \cdot \frac{\partial r_1}{\partial x_2} \tag{48b}$$

Observe, that the result (48b) is simply obtained by static condensation of the Jacobian, Eq. (44).

Alternative common iteration methods for one-dimensional problems are e.g. Regula Falsi, the pegasus method, and we refer to Engelin-Müllges [30] for the specific iteration schemes.

4. Equilibrium problem and associated derivatives

4.1. Weak formulation

Denoting $\boldsymbol{\varphi}$ as the configuration field at time ${}^{n+1}t$ for given parameters $\boldsymbol{\kappa} \in \mathscr{K}$ and using the notation $\langle \cdot , \cdot \rangle$ for the L_2 dual pairing on $\mathscr{B}_0$ of functions, vectors or tensor fields, the equilibrium problem as the classical weak form of momentum at time ${}^{n+1}t$ with spatial quantities reads

$$\text{Find } \boldsymbol{\varphi}: g[\boldsymbol{\varphi}] = \langle \boldsymbol{\tau}:\mathbf{d}_\delta\rangle - \bar{g} = 0 \; \forall \delta\mathbf{u} \text{ for given } \boldsymbol{\kappa} \in \mathscr{K} \tag{49}$$

Here the spatial rate of deformation tensor $\mathbf{d}_\delta$ induced by the virtual displacement $\delta\mathbf{u}$ is defined in the third part of Eq. (1), and $\bar{g} := \langle \bar{\mathbf{B}} \cdot \delta\mathbf{u}\rangle + \langle \bar{\mathbf{T}} \cdot \delta\mathbf{u}\rangle_{\partial_\sigma \mathscr{B}}$ designates the external part of the weak form for the case of dead loading with dual pairing $\langle \cdot , \cdot \rangle_{\partial_\sigma \mathscr{B}}$ on the boundary $\partial_\sigma \mathscr{B}$ and volume forces $\bar{\mathbf{B}}$ and surface forces $\bar{\mathbf{T}}$.

4.2. General concept: directional derivative and sensitivity operator

Before calculating derivatives of the weak form, Eq. (49), it is useful to consider the dependencies of some quantities w.r.t. the configuration $\boldsymbol{\varphi}$ and the material parameters $\boldsymbol{\kappa}$, e.g. from the definition of the deformation gradient $\mathbf{F} = \nabla_X \boldsymbol{\varphi}[\mathbf{X}, {}^{n+1}t, \boldsymbol{\kappa}]$ we can write $\mathbf{F} = \mathbf{F}[\boldsymbol{\varphi}[\boldsymbol{\kappa}]]$. However, the plastic part of the deformation gradient at time ${}^{n}t$ is *not* dependent on the actual configuration, and therefore we have $\mathbf{F}_p = \mathbf{F}_p[\boldsymbol{\kappa}]$. Both quantities define the left Cauchy–Green trial tensor, and thus

$$\mathbf{b}^{tr} = \mathbf{b}^{tr}[\boldsymbol{\varphi}[\boldsymbol{\kappa}],\boldsymbol{\kappa}] \rightsquigarrow \boldsymbol{\tau} = \boldsymbol{\tau}[\boldsymbol{\varphi}[\boldsymbol{\kappa}],\boldsymbol{\kappa}] \tag{50}$$

where the last relation for the Kirchhoff stresses is due to Eq. (37).

The above relations motivate the definitions of the following operators for any (scalar, vector or tensor valued) function $\mathbf{w}[\boldsymbol{\varphi}[\boldsymbol{\kappa}],\boldsymbol{\kappa}]$, which is dependent on the material parameters both implicitly via the configuration $\boldsymbol{\varphi}$ and explicitly: firstly, we introduce the standard *directional derivative (Gateaux) operator*

$$\partial_\Delta \mathbf{w} = \frac{d\mathbf{w}}{d\boldsymbol{\varphi}} \cdot \Delta\mathbf{u} = \frac{d}{d\epsilon}\{\mathbf{w}[\boldsymbol{\varphi}[\boldsymbol{\kappa}] + \epsilon\Delta\mathbf{u},\boldsymbol{\kappa}]\}_{\epsilon=0} \tag{51}$$

necessary for linearization of $\mathbf{w}$. Secondly, upon using the notation $\mathbf{v}_\kappa = d\boldsymbol{\varphi}/d\boldsymbol{\kappa}$ we define a *sensitivity operator*

$$\partial_\kappa \mathbf{w} = \partial_\kappa^\varphi \mathbf{w} + \partial_\kappa^p \mathbf{w} \tag{52a}$$

where

$$\partial_\kappa^\varphi \mathbf{w} = \frac{d\mathbf{w}}{d\boldsymbol{\varphi}} \cdot \mathbf{v}_\kappa = \frac{d}{d\epsilon}\{\mathbf{w}[\boldsymbol{\varphi}[\boldsymbol{\kappa}] + \epsilon\mathbf{v}_\kappa,\boldsymbol{\kappa}]\}_{\epsilon=0} \tag{52b}$$

$$\partial_\kappa^p \mathbf{w} = \frac{d\mathbf{w}(\cdot,\boldsymbol{\kappa})}{d\boldsymbol{\kappa}} = \left(\lim_{\Delta\kappa_i \to 0} \frac{\mathbf{w}[\boldsymbol{\varphi}[\boldsymbol{\kappa}],\kappa_i + \Delta\kappa_i] - \mathbf{w}[\boldsymbol{\varphi}[\boldsymbol{\kappa}],\kappa_i]}{\Delta\kappa_i}\right)_{i=1}^{n_p} \tag{52c}$$

which defines the total derivative $\partial_\kappa \mathbf{w} = d\mathbf{w}/d\boldsymbol{\kappa}$. Note, that the term $\partial_\kappa^\varphi \mathbf{w}$ has the same structure as $\partial_\Delta \mathbf{w}$ and can thus be obtained from the results for linearization by simply exchanging $\Delta\mathbf{u}$ with $\mathbf{v}_\kappa$. The second term $\partial_\kappa^p \mathbf{w}$ basically excludes the implicit dependence of $\boldsymbol{\kappa}$ via the configuration at the actual time (or load) step ${}^{n+1}t$ and will subsequently be called the *partial parameter derivative*. Applying the above concept to the Kirchhoff stresses we obtain

$$\partial_\Delta \boldsymbol{\tau} = \mathscr{L}_\Delta^\# \boldsymbol{\tau} + 2\,\mathrm{sym}(\boldsymbol{\tau} \cdot \mathbf{l}_\Delta^t)$$

$$\text{where } \mathscr{L}_\Delta^\# \boldsymbol{\tau} = 2\frac{\partial \boldsymbol{\tau}}{\partial \mathbf{g}^\flat} : \frac{1}{2}\mathscr{L}_\Delta^\flat \mathbf{g}^\flat = \mathbb{C}:\mathbf{d}_\Delta \tag{53a}$$

$$\partial_\kappa \boldsymbol{\tau} = \mathscr{L}_\kappa^\# \boldsymbol{\tau} + 2\,\mathrm{sym}(\boldsymbol{\tau} \cdot \mathbf{l}_\kappa^t) + \partial_\kappa^p \boldsymbol{\tau}$$

$$\text{where } \mathscr{L}_\kappa^\# \boldsymbol{\tau} = 2\frac{\partial \boldsymbol{\tau}}{\partial \mathbf{g}^\flat} : \frac{1}{2}\mathscr{L}_\kappa^\flat \mathbf{g}^\flat = \mathbb{C}:\mathbf{d}_\kappa \tag{53b}$$

Here the spatial rate of deformation tensors $\mathbf{d}_\Delta$ and $\mathbf{d}_\kappa$ induced by the linearization displacement $\Delta\mathbf{u}$ and configuration sensitivity $\mathbf{v}_\kappa$, respectively, are defined in Eq.

(1). Furthermore, $\mathbb{c}:=2\partial\boldsymbol{\tau}/\partial\mathbf{g}^\flat$ is the fourth-order algorithmic spatial operator, and as alluded to above, the partial parameter derivative $\partial^{\mathrm{p}}_{\boldsymbol{\kappa}}\boldsymbol{\tau}$ excludes the implicit dependence of $\boldsymbol{\kappa}$ via the configuration $\boldsymbol{\varphi}$. Finally, upon using the relations

$$\partial_\Delta \mathbf{d}_\delta = -\mathrm{sym}(\mathbf{g}^\flat \cdot \mathbf{l}_\delta \cdot \mathbf{l}_\Delta)$$

$$\partial_{\boldsymbol{\kappa}} \mathbf{d}_\delta = -\mathrm{sym}(\mathbf{g}^\flat \cdot \mathbf{l}_\delta \cdot \mathbf{l}_{\boldsymbol{\kappa}}) \tag{54}$$

we obtain the associated derivatives of the weak form as

$$\partial_\Delta g[\boldsymbol{\varphi}] = \langle(\mathbb{c}{:}\mathbf{d}_\Delta){:}\mathbf{d}_\delta + \mathbf{l}_\Delta \cdot \boldsymbol{\tau}{:}\mathbf{d}_\delta\rangle$$

$$\partial_{\boldsymbol{\kappa}} g[\boldsymbol{\varphi}] = \langle(\mathbb{c}{:}\mathbf{d}_{\boldsymbol{\kappa}}){:}\mathbf{d}_\delta + \mathbf{l}_{\boldsymbol{\kappa}} \cdot \boldsymbol{\tau}{:}\mathbf{d}_\delta + \partial^{\mathrm{p}}_{\boldsymbol{\kappa}}\boldsymbol{\tau}{:}\mathbf{d}_\delta\rangle \tag{55}$$

The first term is used for computation of an increment $\Delta\mathbf{u}$ within a Newton iteration step for iterative solution of the weak form, and the second term is used for computation of the configuration sensitivity $\mathbf{v}_{\boldsymbol{\kappa}}$ within an iteration step for minimization of some least-squares functional.

The next task consists in determination of the algorithmic spatial tangent operator $\mathbb{c}$ and the partial parameter derivative of Kirchhoff stresses $\partial^{\mathrm{p}}_{\boldsymbol{\kappa}}\boldsymbol{\tau}$, consistent with the integration scheme of Section 3.

4.3. Spatial algorithmic tangent operator

For the Kirchhoff stress tensor with structure of the right side of Eq. (38) the spatial algorithmic tangent operator $\mathbb{c}$ of Eq. (53) is obtained as (see e.g. Simo [28])

$$\mathbb{c} = \sum_{A=1}^{3}\sum_{B=1}^{3}\frac{\mathrm{d}\beta_A}{\mathrm{d}\varepsilon_B^{\mathrm{tr}}}\mathbf{m}_A\otimes\mathbf{m}_B + \sum_{A=1}^{3}2\beta_A\frac{\mathrm{d}\mathbf{m}_A}{\mathrm{d}\mathbf{g}^\flat} \tag{56}$$

where the expression for $\mathrm{d}\mathbf{m}_A/\mathrm{d}\mathbf{g}^\flat$ is presented in Simo [28]. The starting point for determination of the quantities $\mathrm{d}\beta_A/\mathrm{d}\varepsilon_B^{\mathrm{tr}}$ is the representation (40). Using vector notation we have

$$\begin{aligned}\frac{\mathrm{d}\underline{\beta}}{\mathrm{d}\underline{\varepsilon}^{\mathrm{tr}}} &= (1-\alpha)\frac{\mathrm{d}}{\mathrm{d}\underline{\varepsilon}^{\mathrm{tr}}}(\underline{\hat{\beta}}^{\mathrm{vol}} + \underline{\hat{\beta}}^{\mathrm{dev,tr}}) \\ &\quad - (\underline{\hat{\beta}}^{\mathrm{vol}} + \underline{\hat{\beta}}^{\mathrm{dev,tr}})\otimes\frac{\mathrm{d}\alpha}{\mathrm{d}\underline{\varepsilon}^{\mathrm{tr}}} \\ &\quad - 2G\left(\underline{\nu}\otimes\frac{\mathrm{d}\Delta\lambda}{\mathrm{d}\underline{\varepsilon}^{\mathrm{tr}}} + \Delta\lambda\frac{\mathrm{d}\underline{\nu}}{\mathrm{d}\underline{\varepsilon}^{\mathrm{tr}}}\right)\end{aligned} \tag{57}$$

For determination of $\mathrm{d}\Delta\lambda/\mathrm{d}\underline{\varepsilon}^{\mathrm{tr}}$ and $\mathrm{d}\alpha/\mathrm{d}\underline{\varepsilon}^{\mathrm{tr}}$ we consider the local problem (41) as an implicit function and conclude

$$\begin{aligned}&\underline{r}[\underline{\varepsilon}^{\mathrm{tr}},\underline{x}[\underline{\varepsilon}^{\mathrm{tr}}]] = \underline{0} \rightsquigarrow \frac{\mathrm{d}\underline{r}}{\mathrm{d}\underline{\varepsilon}^{\mathrm{tr}}} = \frac{\partial\underline{r}}{\partial\underline{\varepsilon}^{\mathrm{tr}}} + \frac{\partial\underline{r}}{\partial\underline{x}}\frac{\mathrm{d}\underline{x}}{\mathrm{d}\underline{\varepsilon}^{\mathrm{tr}}} = \underline{0} \\ &\rightsquigarrow \frac{\mathrm{d}\underline{x}}{\mathrm{d}\underline{\varepsilon}^{\mathrm{tr}}} = -\underline{J}^{-1}\frac{\partial\underline{r}}{\partial\underline{\varepsilon}^{\mathrm{tr}}}\end{aligned} \tag{58}$$

where $\underline{J}$ is the Jacobian, Eq. (44). Upon defining

$$\begin{bmatrix}\delta_{11} & \delta_{12}\\ \delta_{21} & \delta_{22}\end{bmatrix} = \frac{1}{\det \underline{J}}\begin{bmatrix}-2GJ_{22} & J_{12}h\\ 2GJ_{21} & J_{11}h\end{bmatrix}$$

$$\text{where } h = \frac{\Delta\lambda}{(1-\alpha)^m}\frac{\mathscr{T}p}{S} \tag{59}$$

the following results are obtained after some algebra

$$\frac{\mathrm{d}\Delta\lambda}{\mathrm{d}\underline{\varepsilon}^{\mathrm{tr}}} = \delta_{11}\underline{\nu} + \delta_{12}\underline{1}$$

$$\frac{\mathrm{d}\alpha}{\mathrm{d}\underline{\varepsilon}^{\mathrm{tr}}} = \delta_{21}\underline{\nu} + \delta_{22}\underline{1} \tag{60}$$

Furthermore, employing the results

$$\frac{\mathrm{d}\underline{\nu}}{\mathrm{d}\underline{\varepsilon}^{\mathrm{tr}}} = \frac{2G}{\|\underline{\hat{\beta}}^{\mathrm{dev,tr}}\|}(\underline{I}_3^{\mathrm{dev}} - \underline{\nu}\otimes\underline{\nu})$$

$$\frac{\mathrm{d}}{\mathrm{d}\underline{\varepsilon}^{\mathrm{tr}}}(\underline{\hat{\beta}}^{\mathrm{vol}} + \underline{\hat{\beta}}^{\mathrm{dev,tr}}) = K\underline{1}\otimes\underline{1} + 2G\underline{I}_3^{\mathrm{dev}} \tag{61}$$

expands the expression (57) as

$$\frac{\mathrm{d}\underline{\beta}}{\mathrm{d}\underline{\varepsilon}^{\mathrm{tr}}} = \alpha_1\underline{1}\otimes\underline{1} + \alpha_2\underline{I}_3^{\mathrm{dev}} + \alpha_3\underline{\nu}\otimes\underline{\nu} + \alpha_4\underline{\nu}\otimes\underline{1} + \alpha_5\underline{1}\otimes\underline{\nu}$$

$$\text{where } \alpha_1 = (1-d)K - p\delta_{22}$$

$$\alpha_2 = 2G\left(1 - \alpha - \frac{2G\Delta\lambda}{\|\underline{\hat{\beta}}^{\mathrm{dev,tr}}\|}\right)$$

$$\alpha_3 = \frac{(2G)^2\Delta\lambda}{\|\underline{\hat{\beta}}^{\mathrm{dev,tr}}\|} - 2G\delta_{11} - \|\underline{\hat{\beta}}^{\mathrm{dev,tr}}\|\delta_{21}$$

$$\alpha_4 = -2G\delta_{12} - \|\underline{\hat{\beta}}^{\mathrm{dev,tr}}\|\delta_{22}$$

$$\alpha_5 = -p\delta_{21} \tag{62}$$

Thus, due to the nonsymmetry of the above expression the algorithmic tangent operator (56) exhibits also a nonsymmetric structure contrary to the hyperelastic–plastic tangent operator (30). This is a consequence of the complete coupling between elasticity and damage

Table 2
Large strain problems formulated in principal directions: partial parameter sensitivity for the Kitchhoff stresses and the parameter sensitivity of right Cauchy–Green plastic strain (material independent part)

(a) Partial parameter sensitivity for Kitchhoff stresses (pre-processing) input: ${}^n\mathbf{C}_p^{-1}$	
$\partial^p_{\boldsymbol{\kappa}}\mathbf{b}^{tr} =$	$\mathscr{L}^{\#}_{\boldsymbol{\kappa}}\mathbf{b}^{tr} = \mathbf{F}\cdot\partial^n_{\boldsymbol{\kappa}}\mathbf{C}_p^{-1}\cdot\mathbf{F}^t$
$\partial^p_{\boldsymbol{\kappa}}\varepsilon^{tr}_A =$	$\frac{1}{2}\frac{1}{(\lambda^{tr}_A)^2}\mathbf{m}_A:\partial^p_{\boldsymbol{\kappa}}\mathbf{b}^{tr}$, $A = 1,2,3$
for $\partial^p_{\boldsymbol{\kappa}}\beta_A$:	see material-dependent part in Table 3
$\partial^p_{\boldsymbol{\kappa}}\mathbf{m}_A =$	$\partial_{\mathbf{b}^{tr}}\mathbf{m}_A:\partial^p_{\boldsymbol{\kappa}}\mathbf{b}^{tr}$, $A = 1,2,3$ (for $\partial_{\mathbf{b}^{tr}}\mathbf{m}_A$ see Ref. [31])
$\partial^p_{\boldsymbol{\kappa}}\boldsymbol{\tau} =$	$\sum_{A=1}^3\partial^p_{\boldsymbol{\kappa}}\beta_A\mathbf{m}_A + \sum_{A=1}^3\beta_A\partial^p_{\boldsymbol{\kappa}}\mathbf{m}_A$
(b) Parameter sensitivity for the internal strain-like variables (post-processing) input: ${}^n\mathbf{C}_p^{-1}$, $\mathbf{v}_{\boldsymbol{\kappa}}$	
$\mathbf{l}_{\boldsymbol{\kappa}} =$	$\partial_{\boldsymbol{\kappa}}\mathbf{F}\cdot\mathbf{F}^{-1}$, $\partial_{\boldsymbol{\kappa}}\mathbf{F} = \frac{\partial\mathbf{v}_{\boldsymbol{\kappa}}}{\partial\mathbf{X}}$
$\partial_{\boldsymbol{\kappa}}\mathbf{b}^{tr} =$	$\mathscr{L}^{\#}_{\boldsymbol{\kappa}}\mathbf{b}^{tr} + 2\text{sym}(\mathbf{l}_{\boldsymbol{\kappa}}\cdot\mathbf{b}^{tr})$, where $\mathscr{L}^{\#}_{\boldsymbol{\kappa}}\mathbf{b}^{tr} = \mathbf{F}\cdot\partial^n_{\boldsymbol{\kappa}}\mathbf{C}_p^{-1}\cdot\mathbf{F}^t$
$\partial_{\boldsymbol{\kappa}}\varepsilon^{tr}_A =$	$\frac{1}{2}\frac{1}{(\lambda^{tr}_A)^2}\mathbf{m}_A:\partial_{\boldsymbol{\kappa}}\mathbf{b}^{tr}$
for $\partial_{\boldsymbol{\kappa}}\varepsilon^{el}_A$:	see material-dependent part in Table 3
$\partial_{\boldsymbol{\kappa}}\lambda^{el^2}_A =$	$2\lambda^{el^2}_A\partial_{\boldsymbol{\kappa}}\varepsilon^{el}_A$, $A = 1,2,3$
$\partial_{\boldsymbol{\kappa}}\mathbf{m}_A =$	$\partial_{\mathbf{b}^{tr}}\mathbf{m}_A:\partial_{\boldsymbol{\kappa}}\mathbf{b}^{tr}$, $A = 1,2,3$ (for $\partial_{\mathbf{b}^{tr}}\mathbf{m}_A$ see Ref. [31])
$\partial_{\boldsymbol{\kappa}}\mathbf{b}^{el} =$	$\sum_{A=1}^3(\partial_{\boldsymbol{\kappa}}\lambda^{el^2}_A\mathbf{m}_A + \lambda^{el^2}_A\partial_{\boldsymbol{\kappa}}\mathbf{m}_A)$
$\partial_{\boldsymbol{\kappa}}\mathbf{C}_p^{-1} =$	$-2\text{sym}(\mathbf{F}^{-1}\cdot\partial_{\boldsymbol{\kappa}}\mathbf{F}\cdot{}^n\mathbf{C}_p^{-1}) + \mathbf{F}^{-1}\cdot\partial_{\boldsymbol{\kappa}}\mathbf{b}^{el}\cdot\mathbf{F}^{-t}$

as introduced in Eq. (6), and is different to the presentation by Steinmann et al. [15], where only the elastic isochoric part of the free energy is coupled to damage.

4.4. *Partial parameter derivative of Kirchhoff stresses*

4.4.1. *General remarks*

Similar as to the material operator $\mathbb{c}$, Eq. (56), the partial parameter derivative of Kirchhoff stresses appearing in Eq. (53) consists of two parts (see e.g. [27]):

$$\frac{\partial^p\boldsymbol{\tau}}{\partial\boldsymbol{\kappa}} = \sum_{A=1}^{3}\mathbf{m}_A\otimes\frac{\partial^p\beta_A}{\partial\boldsymbol{\kappa}} + \sum_{A=1}^{3}\beta_A\frac{\partial^p\mathbf{m}_A}{\partial\boldsymbol{\kappa}} \tag{63}$$

The expressions necessary for determination of $\partial^p\boldsymbol{\tau}/\partial\boldsymbol{\kappa}$ are summarized in Tables 2 and 3, respectively. In Table 2 those terms are recalled from Mahnken and Stein in [27], which are valid for any model formulated in principal directions. Table 3 contains the expressions resulting from the specific model formulation in Section 2.

From the results of a *pre-processing part* in Tables 2 and 3, respectively, it can be seen that calculation of $\partial^p\boldsymbol{\tau}/\partial\boldsymbol{\kappa}$ also involves expressions for the parameter derivative of state variables $d^nq/d\boldsymbol{\kappa}$, $d^n\alpha/d\boldsymbol{\kappa}$, $d^n\mathbf{C}_p^{-1}/d\boldsymbol{\kappa}$ at the *previous time step*. Therefore, after having solved the linear problem in the second equation of Eq. (55) for $\mathbf{v}_{\boldsymbol{\kappa}}$, it becomes necessary to determine $d^{n+1}q/d\boldsymbol{\kappa}$, $d^{n+1}\alpha/d\boldsymbol{\kappa}$, $d^{n+1}\mathbf{C}_p^{-1}/d\boldsymbol{\kappa}$ at the actual time step in a *post-processing part*, in order to make them available for the next time step.

As already mentioned, all expressions of Table 2 are derived and explained in Mahnken and Stein [27]. In what follows, we will briefly comment on the expressions of Table 3.

4.4.2. *Pre-processing*

The starting point for determination of $\partial^p\beta_A/\partial\boldsymbol{\kappa}$ in Eq. (63) is the representation (40). Using vector notation analogously to Eq. (57) we have

$$\begin{aligned}\frac{\partial^p\underline{\beta}}{\partial\boldsymbol{\kappa}} &= (1-\alpha)\frac{\partial^p}{\partial\boldsymbol{\kappa}}(\hat{\underline{\beta}}^{vol} + \hat{\underline{\beta}}^{dev,tr}) \\ &\quad - (\hat{\underline{\beta}}^{vol} + \hat{\underline{\beta}}^{dev,tr})\otimes\frac{\partial^p\alpha}{\partial\boldsymbol{\kappa}} \\ &\quad - 2G\left(\underline{\nu}\otimes\frac{\partial^p\Delta\lambda}{\partial\boldsymbol{\kappa}} + \Delta\lambda\frac{\partial^p\underline{\nu}}{\partial\boldsymbol{\kappa}}\right)\end{aligned} \tag{64}$$

The determination of $\partial^p\Delta\lambda/\partial\boldsymbol{\kappa}$ and $\partial^p\alpha/\partial\boldsymbol{\kappa}$ is contained in the following result: we consider the local problem (41) as an implicit function and derive

$$\begin{aligned}&\underline{r}[\boldsymbol{\kappa},\underline{x},{}^nq,{}^n\alpha,\underline{\varepsilon}^{tr}] = \underline{0} \rightsquigarrow \frac{d\underline{r}}{d\boldsymbol{\kappa}} = \frac{\overline{\partial}\underline{r}}{\partial\boldsymbol{\kappa}} + \frac{\partial\underline{r}}{\partial\underline{x}}\frac{d\underline{x}}{d\boldsymbol{\kappa}} = \underline{0} \\ &\rightsquigarrow \frac{d\underline{x}}{d\boldsymbol{\kappa}} = -\underline{J}^{-1}\frac{\overline{\partial}\underline{r}}{\partial\boldsymbol{\kappa}}\end{aligned} \tag{65}$$

for the total derivative. Here $\underline{J}$ is the Jacobian (44) and the notation

$$\frac{\overline{\partial} \underline{r}}{\partial \boldsymbol{\kappa}} = \frac{\partial \underline{r}}{\partial \boldsymbol{\kappa}} + \frac{\partial \underline{r}}{\partial^n q} \frac{\mathrm{d}^n q}{\mathrm{d}\boldsymbol{\kappa}} + \frac{\partial \underline{r}}{\partial^n \alpha} \frac{\mathrm{d}^n \alpha}{\mathrm{d}\boldsymbol{\kappa}} + \frac{\partial \underline{r}}{\partial \underline{\varepsilon}^{\mathrm{tr}}} \frac{\mathrm{d}\underline{\varepsilon}^{\mathrm{tr}}}{\mathrm{d}\boldsymbol{\kappa}} \tag{66}$$

has been used. The result for $\partial^{\mathrm{p}}\underline{x}/\partial\boldsymbol{\kappa}$ is then obtained from Eq. (65) by exchanging $\mathrm{d}\underline{\varepsilon}^{\mathrm{tr}}/\mathrm{d}\boldsymbol{\kappa}$ with $\partial^{\mathrm{p}}\underline{\varepsilon}^{\mathrm{tr}}/\partial\boldsymbol{\kappa}$ and is given in the pre-processing part of Table 3. Using this result, the final expression for $\partial^{\mathrm{p}}\beta_A/\partial\boldsymbol{\kappa}$ is obtained from Eq. (64) and is also summarized in the pre-processing part of Table 3.

4.4.3. Post-processing

For determination of $\mathrm{d}^{n+1}q/\mathrm{d}\boldsymbol{\kappa}$, $\mathrm{d}^{n+1}\alpha/\mathrm{d}\boldsymbol{\kappa}$, $\mathrm{d}^{n+1}\mathbf{C}_{\mathrm{p}}^{-1}/\mathrm{d}\boldsymbol{\kappa}$ in Table 3 firstly $\mathrm{d}\Delta\lambda/\mathrm{d}\boldsymbol{\kappa}$ and $\mathrm{d}\alpha/\mathrm{d}\boldsymbol{\kappa}$ are calculated from Eq. (65). The detailed expressions are given in the post-processing part of Table 3. Then, from Eq. (32b) we have

$$\frac{\mathrm{d}q}{\mathrm{d}\boldsymbol{\kappa}} = \frac{\mathrm{d}^n q}{\mathrm{d}\boldsymbol{\kappa}} + \sqrt{\frac{2}{3}} \frac{\mathrm{d}\Delta\lambda}{\mathrm{d}\boldsymbol{\kappa}} \tag{67}$$

For determination of $\mathrm{d}^{n+1}\mathbf{C}_{\mathrm{p}}^{-1}/\mathrm{d}\boldsymbol{\kappa}$ in Table 2 we need $\mathrm{d}\underline{\varepsilon}^{\mathrm{el}}/\mathrm{d}\boldsymbol{\kappa}$, where

$$\underline{\varepsilon}^{\mathrm{el}} = \underline{\varepsilon}^{\mathrm{tr}} - \frac{\Delta\lambda}{1-\alpha}\underline{\nu} \tag{68}$$

(see Mahnken and Stein [27], Proposition 5.8). Differentiation of Eq. (68) w.r.t. $\boldsymbol{\kappa}$ yields after some algebra the result presented in the post-processing part of Table 3.

5. Plane sheet with two notches

A plane sheet with two notches is considered with geometry as shown in Fig. 1. The material of the specimen is a mild steel, Baustahl St52 due to the German regulations for construction steel. This example was investigated experimentally in the context of the german research network Sonderforschungsbereich 319 (SFB 319): 'Stoffgesetze für das inelastische Verhalten metallischer Werkstoffe—Entwicklung und Anwendung' University of Braunschweig, Germany. In particular spatially distributed data were obtained with an optical method, a *grating method* (see Andresen and Hübner [24] and Bergmann et al. [25]). To this end a grating is positioned on the surface as shown in Fig. 2. This is photographed with a digital camera at consecutive observation states during the displacement controlled experiment with load sizes according to Table 4. Finally the data are analyzed digitally, thus leading to highly resolved spatially distributed data for displacement fields.

The numerical simulation of this plane stress example is performed with a plane stress element presented by Steinmann et al. [32]. This formulation allows the incorporation of general 3D constitutive

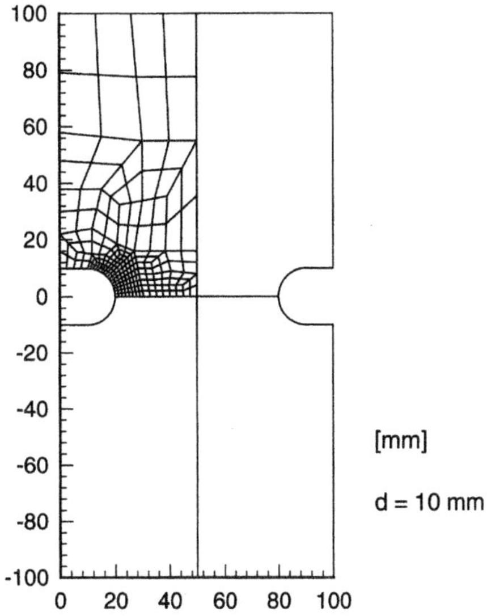

Fig. 1. Plane sheet with two notches: geometry and discretization.

models without any plane stress constraints. Instead, the constraint $\tau_{33}=0$ is satisfied in a weak sense. In our simulation a quadratic nine node element has been used. The spatial discretization of the sample is shown in Fig. 1.

The material is assumed to be elastoplastic J_2-flow theory, combined with isotropic damage and nonlinear isotropic hardening according to Section 2. Integration is performed with the algorithm described in Section 3 in $N = 205$ load steps.

The input data for the identification process consists of displacements in x- and/or y-direction at different, arbitrarily chosen points near the notch region, so that the total number of data at each observation state is $n_{x\mathrm{dat}}=51$. The total number of observation states is $n_{\mathrm{tdat}}=7$ according to Table 4.

The object is to identify the six parameters Y_0, b, c, S, m, ϵ_1 of Eq. (19) which characterize the inelastic behavior of the material. Two elastic parameters were pre-determined as $E = 20600$ kN/cm^2 and $\nu=0.3$ for Young's modulus and Poisson's ratio, respectively, and the parameter ϵ_1 was set to 0.01. In this respect, the following objective function of least-squares type is examined:

$$f(\boldsymbol{\kappa}) = \frac{1}{2}\sum_{i=1}^{n_{\mathrm{tdat}}=7}\sum_{j=1}^{n_{x\mathrm{dat}}=51}(u_{ij}-\bar{u}_{ij})^2 + \frac{1}{2}\sum_{i=1}^{n_{\mathrm{tdat}}=7} w_i(F_i-\bar{F}_i)^2 \to \min_{\boldsymbol{\kappa}} \tag{69}$$

where here u_{ij} denotes the displacement in either x- or

Table 3
Large strain multiplicative von Mises elastoplasticity formulated in principal directions: algorithmic constants for the partial parameter sensitivity of the stress variables (pre-processing) and the parameter sensitivity of internal strain-like variables (post-processing)

Input for pre-processing: $\partial^n_{\boldsymbol{\kappa}} q$, $\partial^n_{\boldsymbol{\kappa}} \alpha$, replace $\partial_{\boldsymbol{\kappa}} \underline{\varepsilon}^{\mathrm{tr}}$ by $\partial^p_{\boldsymbol{\kappa}} \underline{\varepsilon}^{\mathrm{tr}}$ from Table 2, part (a)
Input for post processing: $\partial^n_{\boldsymbol{\kappa}} q$, $\partial^n_{\boldsymbol{\kappa}} \alpha$, $\partial_{\boldsymbol{\kappa}} \underline{\varepsilon}^{\mathrm{tr}}$ from Table 2, part (b)
Effective principal stresses:

$$\partial_{\boldsymbol{\kappa}} p = K \operatorname{tr}(\partial_{\boldsymbol{\kappa}} \underline{\varepsilon}^{\mathrm{tr}}) \rightsquigarrow \partial_{\boldsymbol{\kappa}} \hat{\underline{\beta}}^{\mathrm{vol}} = \partial_{\boldsymbol{\kappa}} p \underline{1}, \quad \partial_{\boldsymbol{\kappa}} \hat{\underline{\beta}}^{\mathrm{dev,tr}} = 2G \operatorname{dev}(\partial_{\boldsymbol{\kappa}} \underline{\varepsilon}^{\mathrm{tr}})$$

Normal vector:

$$\partial_{\boldsymbol{\kappa}} \underline{\nu} = \frac{1}{\|\hat{\underline{\beta}}^{\mathrm{dev,tr}}\|} [\underline{I}_3^{\mathrm{dev}} - \underline{\nu} \otimes \underline{\nu}] \partial_{\boldsymbol{\kappa}} \hat{\underline{\beta}}^{\mathrm{dev,tr}}$$

Residual r_1:

$$\partial_{\boldsymbol{\kappa}} Q = \begin{bmatrix} 0 \\ cq \exp(-bq) \\ 1 - \exp(-bq) \\ 0 \\ 0 \\ 0 \\ 0 \end{bmatrix}, \partial_{\boldsymbol{\kappa}} Y_0 = \begin{bmatrix} 1 \\ 0 \\ 0 \\ 0 \\ 0 \\ 0 \\ 0 \end{bmatrix}, \frac{\partial Q}{\partial^n q} = cb \exp(-bq)$$

$$\bar{\partial}_{\boldsymbol{\kappa}} r_1 = \underline{\nu} : \partial_{\boldsymbol{\kappa}} \hat{\underline{\beta}}^{\mathrm{dev,tr}} - \sqrt{\frac{2}{3}} \left(\partial_{\boldsymbol{\kappa}} Y_0 + \partial_{\boldsymbol{\kappa}} Q + \frac{\partial Q}{\partial^n q} \partial^n_{\boldsymbol{\kappa}} q \right)$$

Residual r_2:

$$\partial_{\boldsymbol{\kappa}} S = \begin{bmatrix} 0 \\ 0 \\ 0 \\ 1 \\ 0 \\ 0 \\ 0 \end{bmatrix}, \partial_{\boldsymbol{\kappa}} m = \begin{bmatrix} 0 \\ 0 \\ 0 \\ 0 \\ 1 \\ 0 \\ 0 \end{bmatrix}, \partial_{\boldsymbol{\kappa}} \mathcal{T} = \begin{bmatrix} 0 \\ 0 \\ 0 \\ 0 \\ 0 \\ \dfrac{-6(q - \varepsilon_1)}{(\epsilon_2 - \epsilon_1)^2} + \dfrac{6(q - \epsilon_1)}{(\epsilon_2 - \epsilon_1)^3} \\ \dfrac{-6(q - \epsilon_1)^2}{(\epsilon_2 - \epsilon_1)^3} + \dfrac{6(q - \epsilon_1)^3}{(\epsilon_2 - \epsilon_1)^4} \end{bmatrix} \mathit{if}\, \epsilon_1 < q < \epsilon_2\, \mathit{else}\, \partial_{\boldsymbol{\kappa}} \mathcal{T} = \underline{0}$$

$$\frac{\partial \mathcal{T}}{\partial^n q} = \frac{6(q - \epsilon_1)(\epsilon_2 - q)}{(\epsilon_2 - \epsilon_1)^3}, \quad \bar{\partial}_{\boldsymbol{\kappa}} \mathcal{T} = \frac{\partial \mathcal{T}}{\partial^n q} \partial^n_{\boldsymbol{\kappa}} q + \partial_{\boldsymbol{\kappa}} \mathcal{T}$$

$$\bar{\partial}_{\boldsymbol{\kappa}} r_2 = \partial^n_{\boldsymbol{\kappa}}\alpha + \frac{\Delta\lambda}{(1-\alpha)^m}\frac{1}{2}\left(\frac{(Q+Y_0)^2}{3G} + \frac{p^2}{K}\right)\left(\frac{1}{S}\bar{\partial}_{\boldsymbol{\kappa}}\mathscr{T} - \frac{\mathscr{T}}{S^2}\partial_{\boldsymbol{\kappa}} S\right) + \frac{\Delta\lambda}{(1-\alpha)^m}\frac{\mathscr{T}}{S}\frac{1}{2}\left(\frac{Q+Y_0}{3G}\partial_{\boldsymbol{\kappa}}(Q+Y_0) + \frac{p}{K}\partial_{\boldsymbol{\kappa}} p\right) - \frac{\Delta\lambda}{(1-\alpha)^m}\frac{\mathscr{T}}{S}\frac{1}{2}\left(\frac{(Q+Y_0)^2}{3G} + \frac{p^2}{K}\right)\log(1-\alpha)\partial_{\boldsymbol{\kappa}} m$$

$$\partial_{\boldsymbol{\kappa}}\Delta\lambda = \frac{-1}{\det J}(\bar{\partial}_{\boldsymbol{\kappa}} r_1 J_{22} - \bar{\partial}_{\boldsymbol{\kappa}} r_2 J_{12}), \quad \partial_{\boldsymbol{\kappa}}\alpha = \frac{-1}{\det J}(\bar{\partial}_{\boldsymbol{\kappa}} r_2 J_{11} - \bar{\partial}_{\boldsymbol{\kappa}} r_1 J_{21})$$

If pre-processing then

$$h_1 = 1 - \alpha - \frac{2G\Delta\lambda}{\|\hat{\underline{\beta}}^{\text{dev,tr}}\|}$$

$$H^1_{\boldsymbol{\kappa}} = \frac{2G\Delta\lambda}{\|\hat{\underline{\beta}}^{\text{dev,tr}}\|}\mathbf{n}:\partial_{\boldsymbol{\kappa}}\hat{\underline{\beta}}^{\text{dev,tr}} - \|\hat{\underline{\beta}}^{\text{dev,tr}}\|\partial_{\boldsymbol{\kappa}}\alpha - 2G\partial_{\boldsymbol{\kappa}}\Delta\lambda$$

$$H^2_{\boldsymbol{\kappa}} = (1-\alpha)\partial_{\boldsymbol{\kappa}} p - p\partial_{\boldsymbol{\kappa}}\alpha$$

$$\partial_{\boldsymbol{\kappa}}\hat{\underline{\beta}} = h_1\partial_{\boldsymbol{\kappa}}\hat{\underline{\beta}}^{\text{dev,tr}} + H^1_{\boldsymbol{\kappa}}\underline{\nu} + H^2_{\boldsymbol{\kappa}}\underline{1}$$

else

$$\partial_{\boldsymbol{\kappa}} q = \partial^n_{\boldsymbol{\kappa}} q + \sqrt{\frac{2}{3}}\partial_{\boldsymbol{\kappa}}\Delta\lambda$$

$$H^3_{\boldsymbol{\kappa}} = \frac{\Delta\lambda}{\|\hat{\underline{\beta}}^{\text{dev,tr}}\|}\mathbf{n}:\partial_{\boldsymbol{\kappa}}\hat{\underline{\beta}}^{\text{dev,tr}} - \frac{\partial_{\boldsymbol{\kappa}}\Delta\lambda}{1-\alpha} - \frac{\Delta\lambda\partial_{\boldsymbol{\kappa}}\alpha}{(1-\alpha)^2}$$

$$\partial_{\boldsymbol{\kappa}}\underline{\varepsilon}^{\text{el}} = \operatorname{dev}\partial_{\boldsymbol{\kappa}}\underline{\varepsilon}^{\text{tr}} + H^3_{\boldsymbol{\kappa}}\underline{\nu} - \frac{\Delta\lambda}{(1-\alpha)\|\hat{\underline{\beta}}^{\text{dev,tr}}\|}\partial_{\boldsymbol{\kappa}}\hat{\underline{\beta}}^{\text{dev,tr}}$$

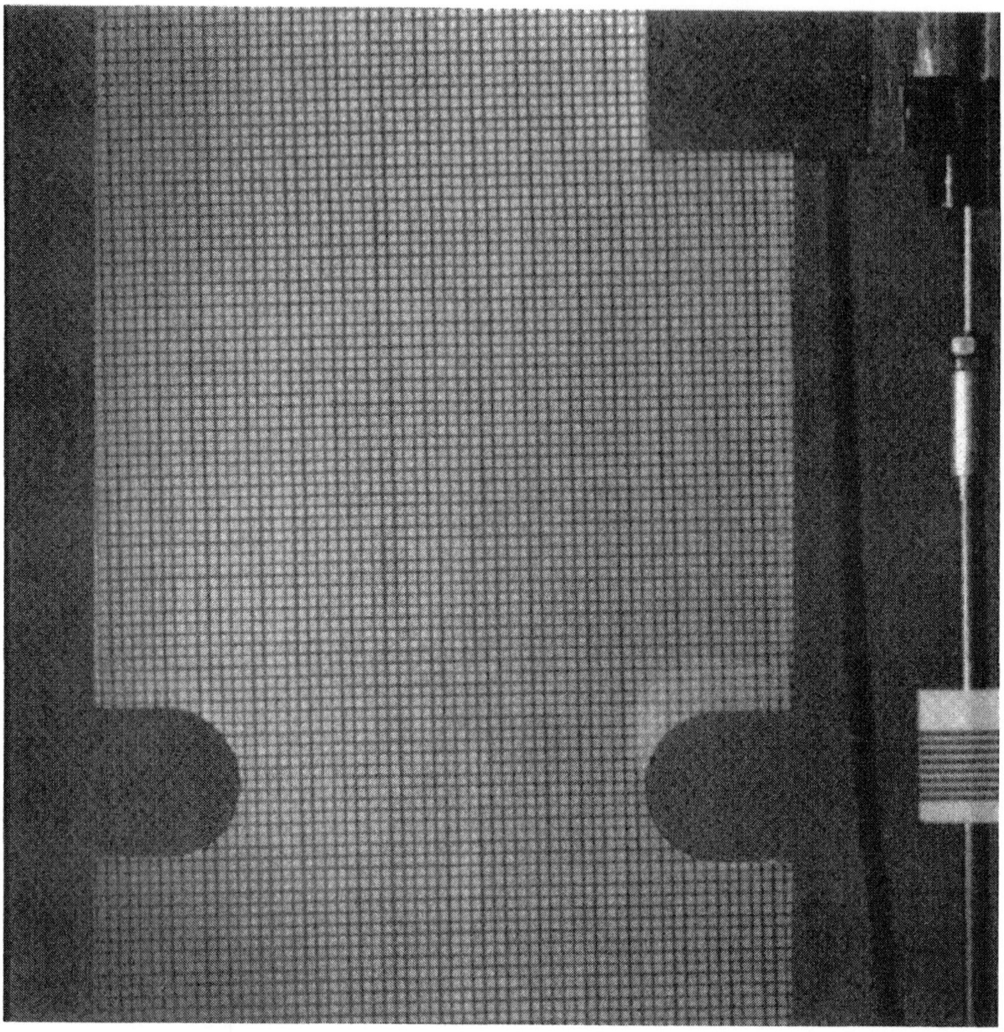

Fig. 2. Plane sheet with two notches: photography of the sample with grating on the surface.

y-direction and F_j the total tension load acting on the specimen. w is a weighting factor and is set to 2×10^{-8}. For minimization of the least-squares functional (69) an algorithm due to Bertsekas [33] is used, where the iteration matrix is obtained from the update rule for the BFGS formula. Further details of the algorithm are explained in e.g. Mahnken and Stein [26,27].

Table 4
Plane sheet with two notches: sizes of displacement at the top of the sample at different states of optical observation

NLST	1	6	10	15	18	21	28
Load size (mm)	0.0	0.44	1.0	2.0	3.0	4.0	5.3

The starting point and the results of the optimization process for the material parameters are given in Table 5. The result has been obtained after 18 iteration steps. For comparison also the results of [27] are recalled, which are obtained from identification of a different experiment with the same material, however from a different production charge. This difference in the specimen causes some type of scattering and is regarded as the main reason for the differences in the results. A larger amount of data would become necessary in order to apply a stochastic approach to this type of model uncertainty [34,35].

In Fig. 3 the total load versus the upper displacement is depicted for both for simulation and experiment, which reveals a very good agreement. In Figs. 4–6 we compare the contours for the displacements u

Table 5
Plane sheet with two notches: starting and obtained values for the material parameters of a mild steel Baustahl St52

	Starting	Solution	Ref. [27]
Y_0 (N/mm^2)	300.0	351.17	360.26
b	10.0	7.223	3.95
c (N/mm^2)	800.0	355.8	416.51
S	10.0	8.068	–
m	0.4	0.499	–
ϵ_1	0.4	0.405	–

and v, respectively, at the observation states $NSLT = 6$ and $NSLT = 28$ of Table 4 in the grating region. Again, a very substantial agreement between experimental and simulated data is observed.

6. Summary

In this work a model for plasticity coupled to damage formulated in the plastic intermediate configuration has been proposed. The formulation is based on complete coupling between the elastic part of the free energy function and the damage variable. As a consequence an evolution equation for the damage variable is obtained which is proportional to the accumulated plastic strain rate and which is in accordance with the analytical investigations for ductile growth of voids by Rice and Tracey [17]. Furthermore it preserves an increase of damage evolution with increasing triaxiality ratio, which is in agreement with the experimental observation that the measured ductility at fracture decreases as the triaxiality ratio increases [2,3]. In this respect the formulation is regarded as an important modification to the presentation by Steinmann et al. [15], where only the elastic isochoric part of the free energy function is coupled to damage.

Furthermore, issues of thermodynamic consistency, continuous tangent operator, and algorithmic tangent operator have been discussed. Due to the complete coupling of elasticity and damage in the free energy function, the algorithmic tangent operator becomes non-symmetric. Furthermore, for the discretized constitutive problem a robust iteration scheme with a two-level algorithm is proposed. A gradient based optimization algorithm is used for minimizing the associative least-squares functional for parameter identification, and to this end the associated sensitivity analysis con-

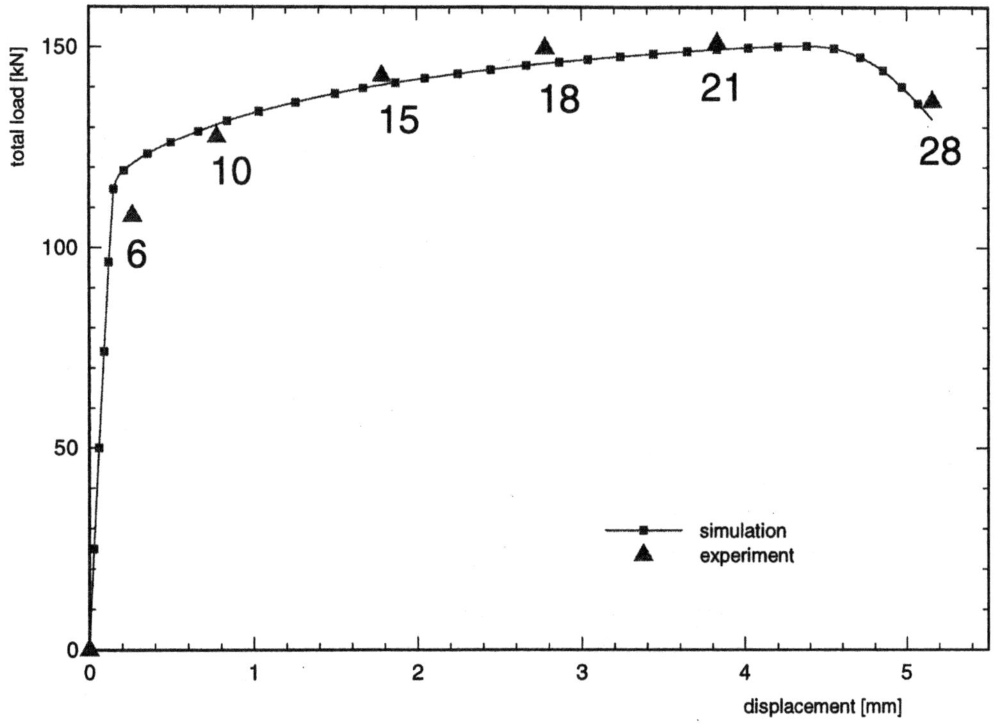

Fig. 3. Plane sheet with two notches: load versus displacement for simulation and experiment.

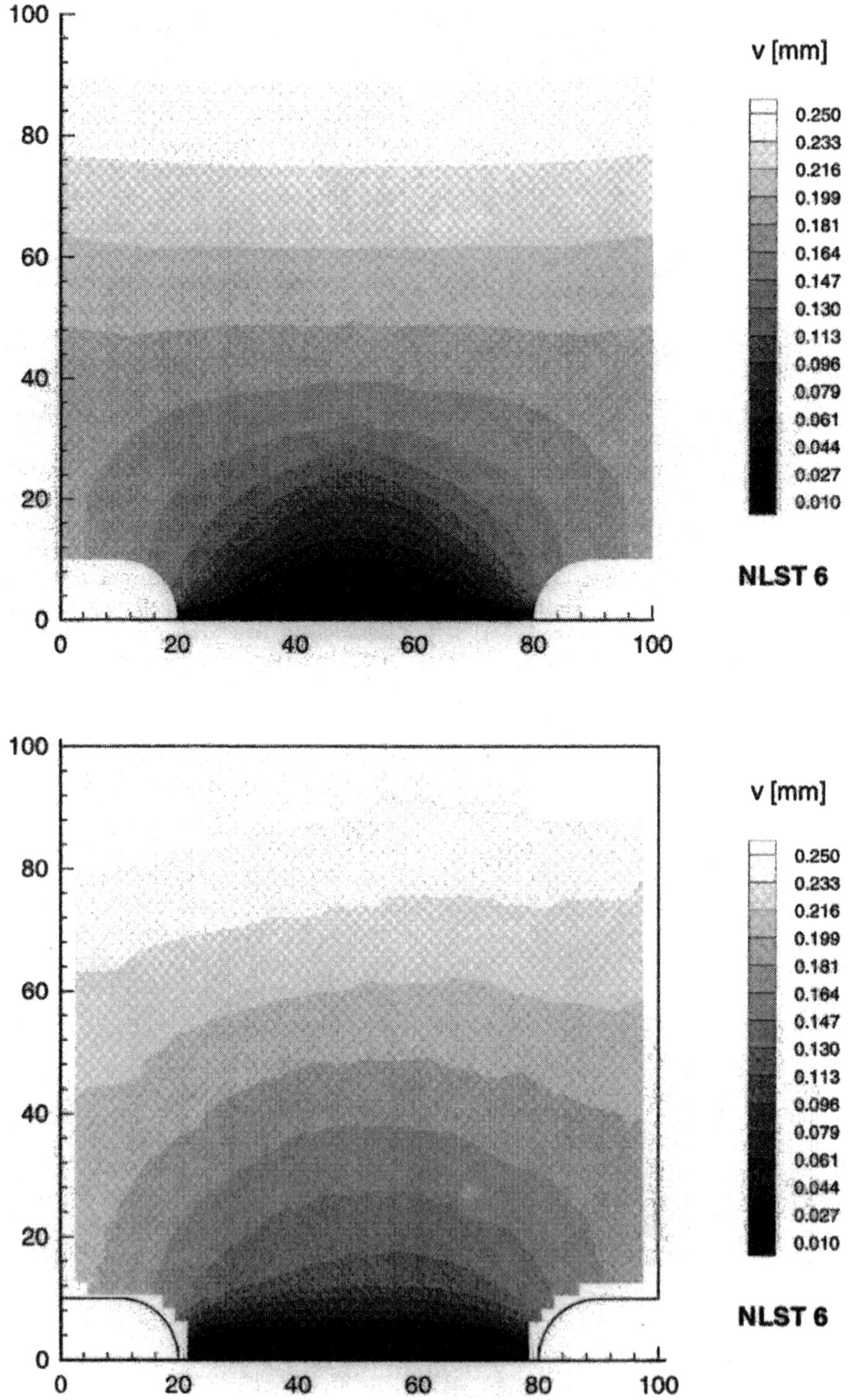

Fig. 4. Plane sheet with two notches: contours of displacements v for simulation (top) and experiment (bottom) at observation state 6 in the grating region.

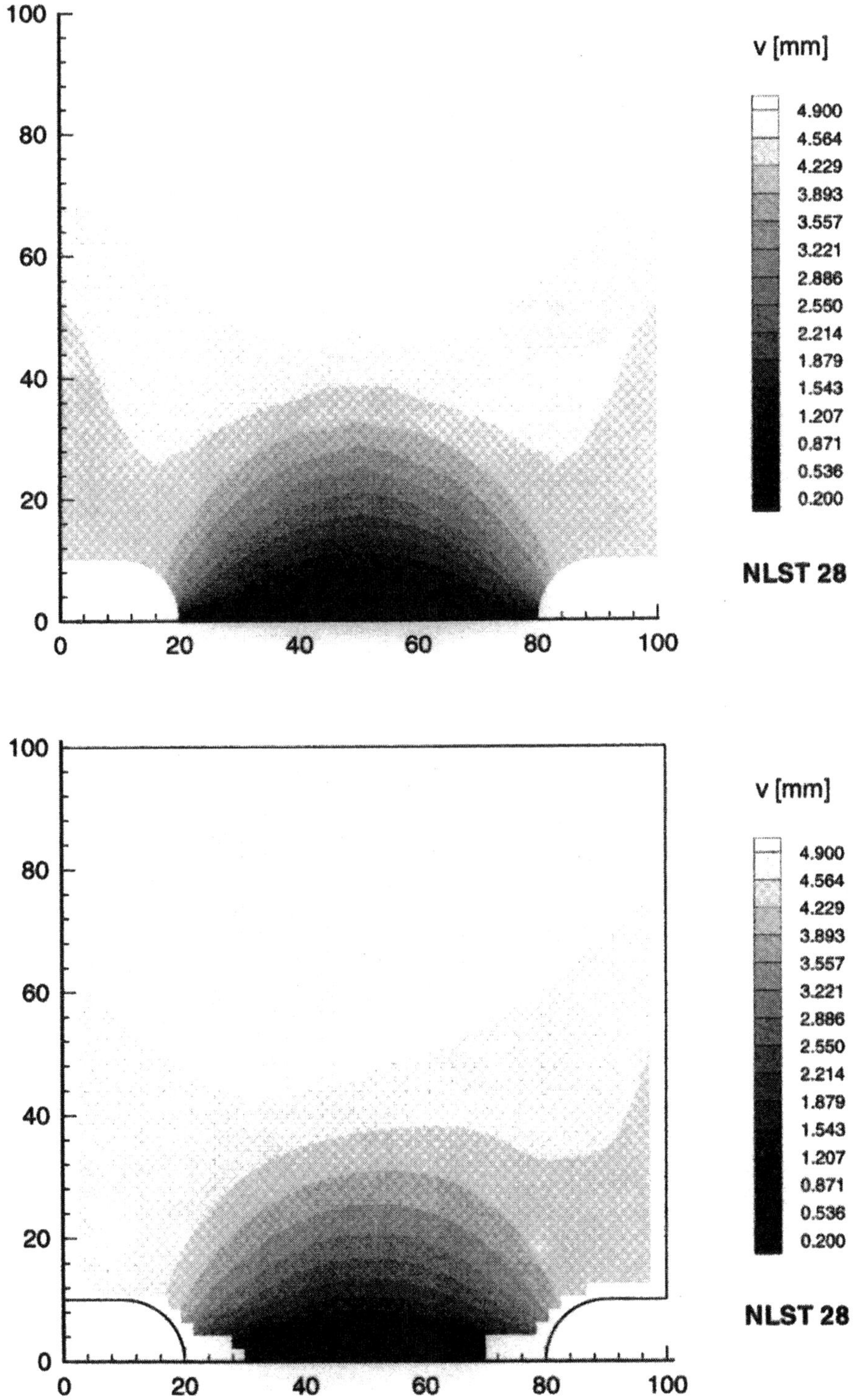

Fig. 5. Plane sheet with two notches: contours of displacements v for simulation (top) and experiment (bottom) at observation state 28 in the grating region.

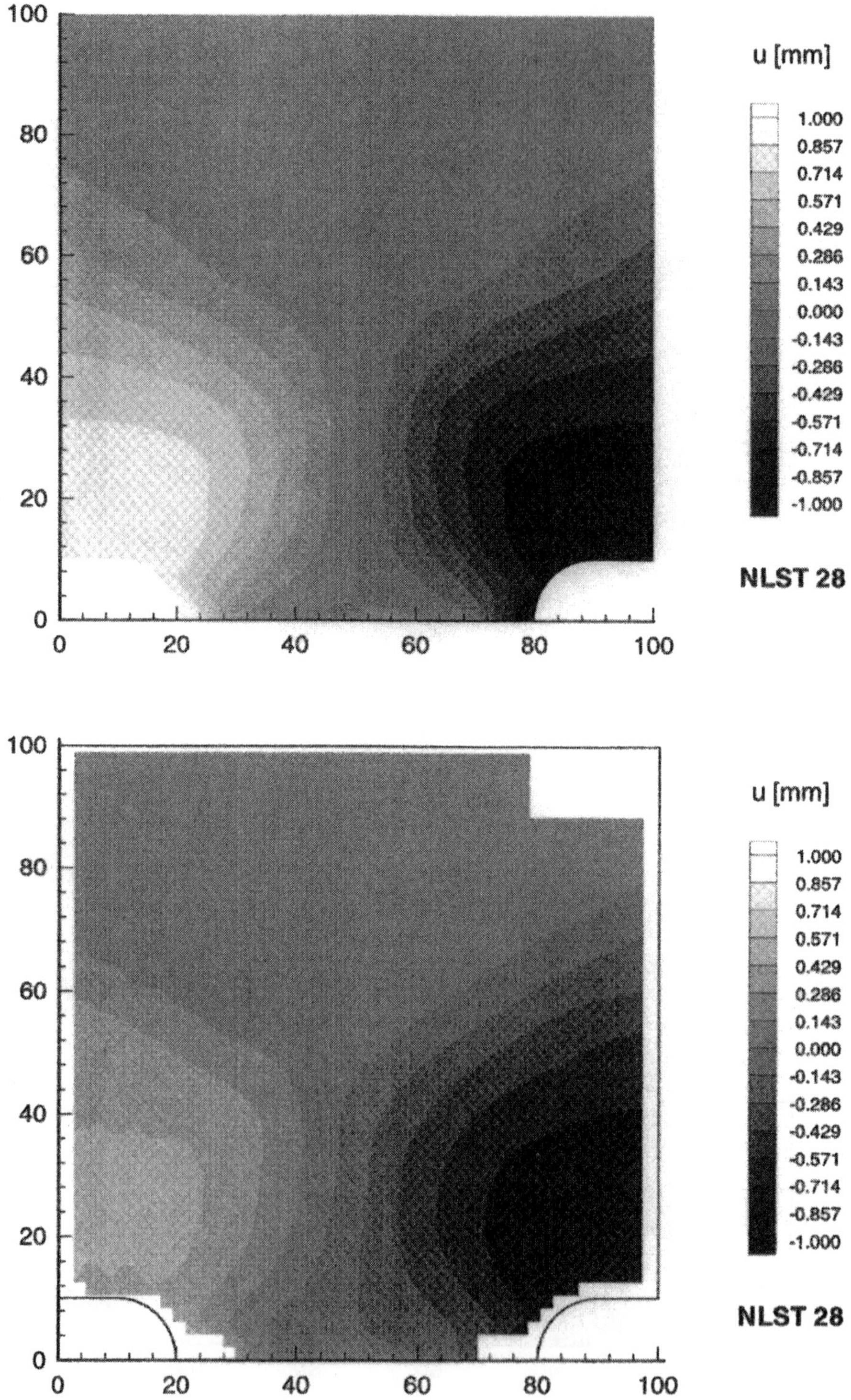

Fig. 6. Plane sheet with two notches: contours of displacements u for simulation (top) and experiment (bottom) at observation state 28 in the grating region.

sistent with the integration scheme has been described. In the example experimental data obtained with a grating method were used, in order to determine the material parameters of the model, thus leading to very good agreement between numerical simulation and experimental observations.

Acknowledgements

This work was performed in the context of the German research network Sonderforschungsbereich 319 (SFB 319): 'Stoffgesetze für das inelastische Verhalten metallischer Werkstoffe—Entwicklung und Anwendung'. In this respect the cooperative collaboration with the following institutes and associated persons from the University of Braunschweig is gratefully acknowledged: Professor Dr-Ing. U. Peil and Dipl.-Ing. S. Dannemeier from the Institut für Stahlbau, who performed the experiment; Professor Dr-Ing. R. Ritter and Dipl.-Ing. H. Friebe from the Institut für experimentelle Mechanik, who did the digital photographs. The financial support for this research was provided by the Deutsche Forschungsgemeinschaft and is gratefully acknowledged.

References

[1] Lemaitre J. A course on damage mechanics, 2nd ed. Berlin: Springer, 1996.

[2] Thomason PF. Ductile fracture of metals. Oxford: Pergamon Press, 1990.

[3] Hancock JW, Mackenzi AC. On the mechanisms of ductile failure in the high-strength steels subjected to multi-axial stress-states. J Mech Phys Solids 1976;24:147–69.

[4] Gurson AL. Continuum theory of ductile rupture by void nucleation and growth—I. Yield criteria and flow rules for porous ductile media. Engng Mater Technol 1977;99:2–15.

[5] Tvergaard V. Influence of voids on shear band instabilities under plane strain conditions. Int J Fract 1981;17:389–407.

[6] Tvergaard V, Needleman A. Analysis of the cup–cone fracture in a round tensile bar. Acta Metall 1984;32:157–69.

[7] Bennani B, Picart P, Oudin J. Some basic finite element analysis of microvoid nucleation, growth and coalescencs. Engng Comput 1993;10:409–21.

[8] Steglich D, Brocks W. Micromechanical modelling of the behaviour of ductile materials including particles. Comput Mater Sci 1997;9:7–17.

[9] Mahnken R. Aspects on the finite element implementation of the Gurson model including parameter identification (submitted).

[10] Kachanov LM. Time to the rupture process under creep conditions. TVZ Akad Nauk SSR Otd Tech Nauk 1958;8:26–31.

[11] Chaboche JL. Continuum damage mechanics: part I—general concepts. J Appl Mech 1988;55:59–72.

[12] Simo JC, Ju JW. Strain- and stress-based continuum damage models—I. Formulation. Int J Solids Struct 1987;23:821–40.

[13] Lemaitre J. Coupled elasto-plasticity and damage constitutive equations. Comput Meth Appl Mech Engng 1985;51:31–49.

[14] De Souza EA, Peric D, Owen DR. A model for elasto-plastic damage at finite strains: algorithmic issues and applications. Engng Comput 1994;11:257–81.

[15] Steinmann P, Miehe Ch, Stein E. Comparison of different finite deformation inelastic damage models within multiplicative plasticity for ductile materials. Int J Comput Mech 1993;13:458–74.

[16] Miehe Ch. On the representation of Prandtl–Reuss tensors within the framework of multiplicative elastoplasticity. Int J Plast 1994;10:609–21.

[17] Rice, Tracey. On the ductile enlargement of voids in triaxial stress fields. J Mech Phys Solids 1969;17:201–17.

[18] Weber G, Anand L. Finite deformation constitutive equations and a time integration procedure for isotropic, hyperelastic–viscoplastic solids. Comput Meth Appl Mech Engng 1990;79:173–202.

[19] Eterovic AL, Bathe KJ. A hyperelastic based large strain elasto-plastic constitutive formulation with combined isotropic-kinematic hardening using the logarithmic stress and strain measures. Int J Numer Meth Engng 1990;30:1099–114.

[20] Johansson M, Mahnken R, Runesson K. Efficient integration for generalized viscoplasticity coupled to damage, Int J Numer Meth Engng (in press).

[21] Banks HT, Kunisch K. Estimation techniques for distributed parameter systems. Boston: Birkhäuser, 1989.

[22] Bui HD. Inverse problems in the mechanics of materials, an introduction. Boca Raton, FL: CRC Press, 1994.

[23] Bui HD, Tanaka M, editors. Inverse problems in engineering mechanics. Rotterdam: Balkema, 1994.

[24] Andresen K, Hübner B. Calculation of strain from an object grating on a Reseau film by a correlation method. Exp Mech 1992;32:96–101.

[25] Bergmann D, Galanulis K, Ritter R. Optical field methods for 3D-deformation measurement in fracture mechanics. In: Carpinteri A, editor. Size–Scale Effects in the Failure Mechanics of Materials and Structure. London: E & FN SPON, 1996. p. 524–536.

[26] Mahnken R, Stein E. A unified approach for parameter identification of inelastic material models in the frame of the finite element method. Comput Meth Appl Mech Engng 1996;136:225–58.

[27] Mahnken R, Stein E. Parameter identification for finite deformation elastoplasticity in principal directions. Comput Meth Appl Mech Engng 1997;147:17–39.

[28] Simo JC. Algorithms for static and dynamic multiplicative plasticity that preserve the classical return mapping schemes of the infinitesimal theory. Comput Meth Appl Mech Engng 1992;99:61–112.

[29] Steinmann P. 1997 Modellierung und Numerik duktiler Werkstoffe. Habilitation, Forschungs-und Seminarberichte aus dem Bereich der Mechanik der Universität Hannover, F97/1.

[30] Engelin-Müllges G, Reuter F. Formelsammlung zur numerischen Mathematik mit Standard-FORTRAN 77 Programmen. Mannheim: BI-Wissenschaftsverlag, 1988.

[31] Miehe C. Computation of isotropic tensor functions. Commun Appl Numer Meth 1993;9:889–96.

[32] Steinmann P, Betsch P, Stein E. FE plane stress analysis incorporating arbitrary 3D large strain constitutive models. Engng Comput 1997;14:175–201.

[33] Bertsekas DP. Projected Newton methods for optimization problems with simple constraints. SIAM J Con Opt 1982;20:221–46.

[34] Pugachew VS. Probability theory and mathematical statistics for engineers. Oxford: Pergamon Press, 1984.

[35] Stein E, Ohnimus S, Mahnken R. Adaptive modeling and computation of elastic and inelastic structures. In: Natke HG, Ben-Haim Y, editors. Uncertainty: Models and Measures, Proceedings of the International Workshop, Lamprecht, Germany, 22–24 July 1996. Chichester: Wiley, 1997.

An evaluation of the MITC shell elements

Klaus-Jürgen Bathe[a,*], Alexander Iosilevich[a], Dominique Chapelle[b]

[a]*Massachusetts Institute of Technology, Department of Mechanical Engineering, Cambridge, MA 02139, USA*
[b]*INRIA-Rocquencourt, BP 105, 78153 Le Chesnay Cedex, France*

Accepted 10 August 1999

Abstract

Based on fundamental considerations for the finite element analysis of shells, we evaluate in the present paper the performance of the MITC general shell elements. We give the results obtained in the analysis of judiciously selected test problems and conclude that the elements are effective for general engineering applications. © 2000 Elsevier Science Ltd. All rights reserved.

Keywords: Shell elements; Mixed interpolation; MITC elements; Benchmark problems

1. Introduction

Shell structures are encountered in many engineering designs, and the accurate stress analysis of these structures is frequently required. A major difficulty in such analyses is that different shell structures can behave very differently depending on the shell geometry and boundary conditions used [1]. Finite elements for the analysis of shells must therefore be able to effectively capture different shell behaviors with varied and complex stress conditions. Because of these challenges in shell analyses, numerous finite elements have been proposed as improved analysis procedures. However, only very few elements can be recommended for general use in engineering practice.

In an attempt to provide a basis for a deeper study of the currently available shell finite elements, and the development of improved discretization procedures, we presented some fundamental considerations for the finite element analysis of shell structures in Ref. [1]. We summarized the different characteristic behaviors of shell structures, discussed the difficulties encountered in finite element analysis of such structures, and finally presented an evaluation strategy of finite element procedures. Our objective in this paper is to apply this evaluation strategy to the MITC shell elements [2–4], and discuss the results obtained.

In the following sections of the paper, we first briefly review the difficulties encountered in the analysis of shell structures and in the development of general shell finite elements. We then evaluate the MITC shell elements in the light of the general shell analysis requirements by solving the test problems proposed in Ref. [1]. The results of these analyses show that the MITC elements provide effective discretization schemes for the general analysis of shell structures.

2. The basic shell analysis problem

When studying the shell analysis problem, it is expedient to consider first the underlying mathematical model and then the finite element solution of that model [1,2].

* Corresponding author. Tel.: +1-617-253-6645; fax: +1-617-253-2275.

Reprinted from *Computers and Structures* **75 (1)**, 1-30 (2000)

2.1. On the mathematical model

The general finite element analysis approach for shell structures is to use shell elements that are formulated based on general three-dimensional continuum theory and kinematic and stress assumptions [2]. The shell assumptions are those of Naghdi shell theory, that the normal stress through the thickness of the shell vanishes and that straight fibers originally normal to the midsurface remain straight during the deformations of the shell [5]. The "underlying shell mathematical model" of the general continuum-mechanics based finite element discretizations is derived in Ref. [6], where the differences to the Naghdi shell model are also enumerated.

In practice, the difficulties in shell analysis are most pronounced when the shell is thin. For this reason, in theoretical discussions, the case of the shell thickness t being small is considered (including the limit case $t \rightarrow 0$), and in the numerical evaluation of shell elements, thin shell analysis is considered. Since the underlying shell mathematical model of the general shell finite element analysis approach is equivalent to the Naghdi shell theory when t is small (and indeed the same limit problems are obtained when $t \rightarrow 0$, see Ref. [6]), we can use the Naghdi shell theory to identify the analysis difficulties and develop an appropriate evaluation strategy for shell elements.

Using the Naghdi shell theory, the general shell analysis problem is:

Find $\boldsymbol{U}_t \in \mathscr{U}$ such that

$$t^3 A(\boldsymbol{U}_t, \boldsymbol{V}) + tD(\boldsymbol{U}_t, \boldsymbol{V}) = G(\boldsymbol{V}) \quad \forall \boldsymbol{V} \in \mathscr{U} \tag{1}$$

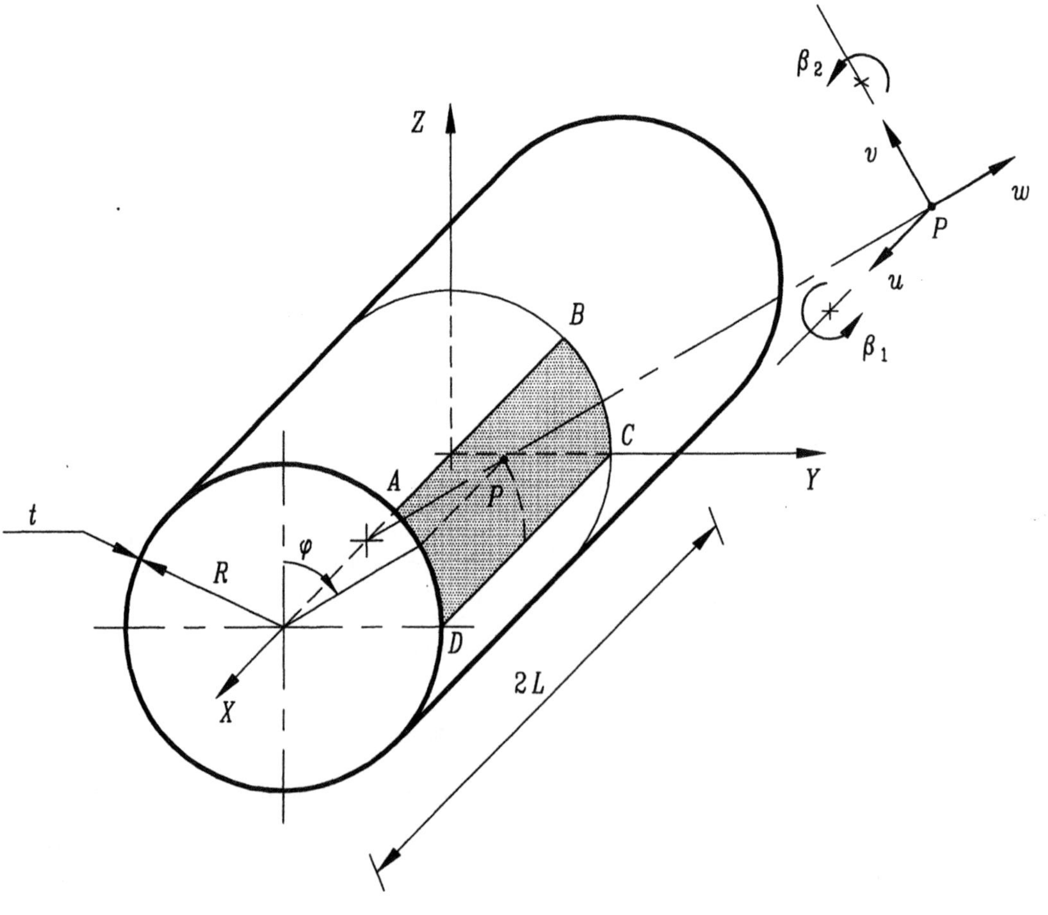

Fig. 1. Cylindrical shell.

where

$$\mathcal{U} = \left\{ V = (v, \eta), v \in \left[H^1(\Omega)\right]^3, \eta \in \left[H^1(\Omega)\right]^2 \right\} \cap \mathcal{BC} \tag{2}$$

and v is the displacement vector of the shell midsurface, η lists rotations of the sections (originally normal to the midsurface), and $\mathcal{BC}$ symbolically denotes the essential boundary conditions. The bilinear forms $A(\cdot, \cdot)$ and $D(\cdot, \cdot)$ capture, respectively, the bending and membrane/shear strain energies.

As we have discussed in Ref. [1], in the asymptotic analysis (i.e., when $t \rightarrow 0$), the following subspace, $\mathcal{U}_0$ takes on a crucial role:

$$\mathcal{U}_0 \stackrel{\text{def}}{=} \{ V \in \mathcal{U} \mid D(V, V) = 0 \}. \tag{3}$$

This subspace contains all those displacement patterns for which the membrane and shear strains are zero, hence, it is the subspace of pure bending displacements (also referred to as the subspace of inextensional displacements).

An essential specificity of shells is that $\mathcal{U}_0$ may be trivial, i.e., $\mathcal{U}_0 = \{\mathbf{0}\}$. Such a situation is designated to be the case of "inhibited pure bending". The asymptotic behavior of a shell structure is highly dependent on whether or not pure bending is inhibited. Hence, to evaluate shell finite elements, test problems for which pure bending is inhibited ($\mathcal{U}_0 = \mathbf{0}$) and is not inhibited ($\mathcal{U}_0 \neq \{\mathbf{0}\}$) should be considered. In each case, the behavior of the finite element discretization should be measured as the thickness of the shell is decreased.

Considering the case of non-inhibited pure bending, we recognize that for the solution of Eq. (1) to remain both bounded and non-vanishing, we must assume the right-hand side to be of the form

$$G(V) = t^3 F_b(V), \tag{4}$$

so that for each value of t the problem to be solved is:

Find $U_t \in \mathcal{U}$ such that

$$A(U_t, V) + \frac{1}{t^2} D(U_t, V) = F_b(V) \quad \forall V \in \mathcal{U}. \tag{5}$$

When t is very small, the membrane/shear term appears in this problem as a penalty term and for $t \rightarrow 0$ the solution of the following problem is approached:

Find $U_0^b \in \mathcal{U}_0$ such that

$$A\left(U_0^b, V\right) = F_b(V) \quad \forall V \in \mathcal{U}_0. \tag{6}$$

Furthermore, we have [1]

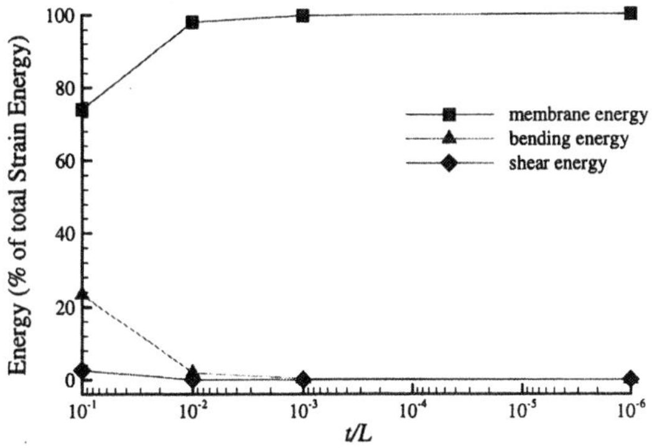

Fig. 2. Distribution of strain energy as a function of t/L for the clamped cylindrical shell.

$$\lim_{t \rightarrow 0} \frac{1}{t^2} D(U_t, U_t) = 0. \tag{7}$$

The above considerations mean in practice, and in the evaluation of shell elements, that considering a pure bending non-inhibited problem, as the thickness approaches zero, the problem remains well-posed and the shear and membrane strains become negligible. Shell structures that carry loads primarily in bending are also referred to as *bending-dominated shells*.

The situation is quite different when a pure bending inhibited shell structure is analyzed. In this case the proper scaling of the loading is

$$G(V) = t F_m(V), \tag{8}$$

because the stiffness of the shell is proportional to the thickness as t becomes small. The problem sequence to solve is:

Find $\mathbf{U}_t \in \mathcal{U}$ such that

$$t^2 A(U_t, V) + D(U_t, V) = F_m(V) \quad \forall V \in \mathcal{U}, \tag{9}$$

and the corresponding limit problem is:

Find $U_0^m \in \mathcal{W}$ such that

$$D\left(U_0^m, V\right) = F_m(V) \quad \forall V \in \mathcal{W}, \tag{10}$$

where $\mathcal{W}$ is a space larger than $\mathcal{U}$ because strictly we no longer need continuity in the transverse displacements and section rotations. However, we require that

$$|F_m(V)|^2 \leq c D(V, V) \quad \forall V \in \mathcal{W}, \tag{11}$$

where c is a constant. This relation ensures that the applied loading can be resisted by membrane stresses

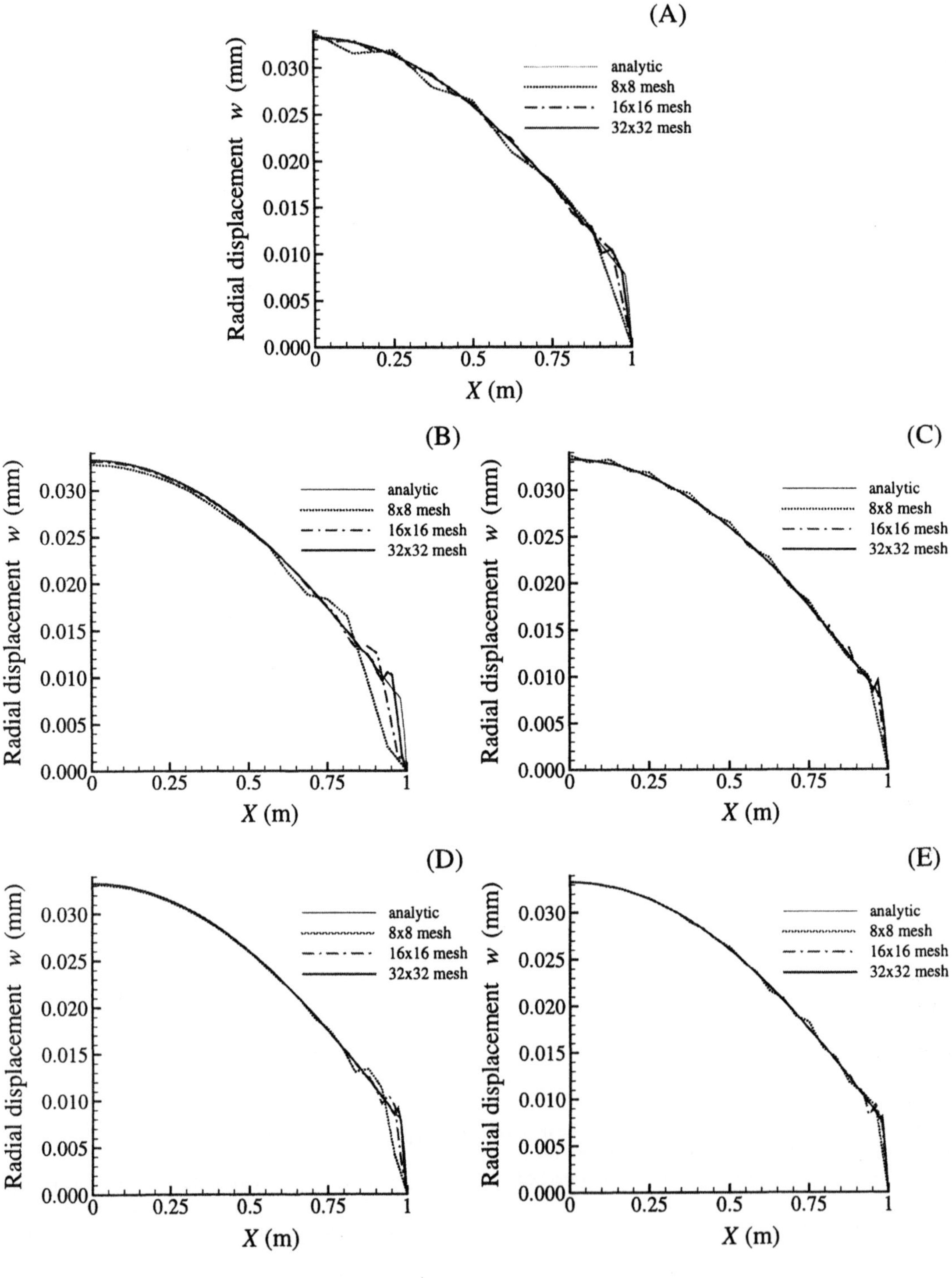

Fig. 3. Axial profile of radial displacement w calculated along line CD for the clamped cylinder problem using $L = R = 1$ m; $t/L = 1/10{,}000$; $E = 2 \times 10^5$ MPa; $\nu = 1/3$; $p_0 = 1$ MPa. (A) MITC4 element; (B) QUAD9; (C) MITC9; (D) QUAD16; (E) MITC16.

only. Shell structures that carry loads primarily in membrane action are also referred to as *membrane-dominated shells.*

2.2. The finite element formulation

For the finite element solution, a natural way to proceed is to use displacement/rotation interpolations. Using conforming interpolations, we select $\mathcal{U}^h \subset \mathcal{U}$, where $\mathcal{U}^h$ signifies the finite element space (h denoting the characteristic element size), and the finite element problem corresponding to Eq. (1) is:

Find $\boldsymbol{U}^h \in \mathcal{U}^h$ such that

$$t^3 A(\boldsymbol{U}^h, \boldsymbol{V}) + tD(\boldsymbol{U}^h, \boldsymbol{V}) = G(\boldsymbol{V}) \quad \forall \boldsymbol{V} \in \mathcal{U}^h. \tag{12}$$

To measure the effectiveness of the solution scheme, we distinguish whether an inhibited or non-inhibited problem is solved. When inhibited problems are considered, the solution using the above approach is effective as proven in [1]. However, when a non-inhibited problem is solved, the numerical phenomenon of "locking" occurs which severely reduces the rate of convergence. Namely, in this case the discrete variational problem considered is

Find $\boldsymbol{U}_t^h \in \mathcal{U}^h$ such that

$$A\left(\boldsymbol{U}_t^h, \boldsymbol{V}\right) + \frac{1}{t^2} D\left(\boldsymbol{U}_t^h, \boldsymbol{V}\right) = F_b(\boldsymbol{V}) \quad \forall \boldsymbol{V} \in \mathcal{U}^h, \tag{13}$$

and the convergence is highly influenced by how rich $\mathcal{U}^h$ is in $\mathcal{U}_0$. In the worst case $\mathcal{U}^h \cap \mathcal{U}_0 = \{\boldsymbol{0}\}$ leading to total loss of convergence

$$\lim_{t \to 0} \boldsymbol{U}_t^h = \boldsymbol{0} \neq \lim_{t \to 0} \boldsymbol{U}_t = \boldsymbol{U}_0^b. \tag{14}$$

In practice, extremely fine meshes are needed for an accurate solution when t is small, and the displacement-based finite element procedure becomes unpractical. The remedy is to use an appropriate *mixed* formulation. The aim with this approach is to interpolate displacements and strains (or stresses) in such a manner as to have no locking of the discretization in the non-inhibited case for any value of thickness t, and to have, as well, a uniformly good behavior for the membrane-dominated case. Ideally, the rate of convergence would be optimal, independent of whether a bending-dominated or membrane-dominated problem is considered and independent of the shell thickness.

The key to reaching this optimal behavior is to use an *appropriate* mixed method with the appropriate "well-balanced" interpolations. The general mixed method that provides the basis of the MITC shell elements is

Find $\boldsymbol{U}^h = (\boldsymbol{u}^h, \boldsymbol{\beta}^h) \in \mathcal{U}^h = (\mathcal{V}^h, \mathcal{B}^h)$

and

$\boldsymbol{E}^h = \left\{\varepsilon_{ij}^h\right\} \in \mathcal{E}^h$ such that

$$\begin{cases} t^3 \tilde{A}(\boldsymbol{E}^h, \boldsymbol{\varepsilon}(\boldsymbol{V})) + t\tilde{D}(\boldsymbol{E}^h, \boldsymbol{\varepsilon}(\boldsymbol{V})) = G(\boldsymbol{V}) \\ \tilde{A}(\boldsymbol{E}^h - \boldsymbol{\varepsilon}(\boldsymbol{U}^h), \boldsymbol{\Psi}) = 0 \\ \tilde{D}(\boldsymbol{E}^h - \boldsymbol{\varepsilon}(\boldsymbol{U}^h), \boldsymbol{\Psi}) = 0 \quad \forall \boldsymbol{V} \in \mathcal{U}^h, \boldsymbol{\Psi} = \{\psi_{ij}\} \in \mathcal{E}^h, \end{cases} \tag{15}$$

where

$$\mathcal{E}^h = \left\{\boldsymbol{\Psi} = \{\psi_{ij}\}, \psi_{ij} \in \mathcal{E}_{ij}^h\right\}, \tag{16}$$

and

$$\tilde{A}(\boldsymbol{\varepsilon}(\boldsymbol{U}), \boldsymbol{\varepsilon}(\boldsymbol{V})) = A(\boldsymbol{U}, \boldsymbol{V})$$

$$\tilde{D}(\boldsymbol{\varepsilon}(\boldsymbol{U}), \boldsymbol{\varepsilon}(\boldsymbol{V})) = D(\boldsymbol{U}, \boldsymbol{V}). \tag{17}$$

Here $\mathcal{E}^h$ is the subspace of assumed strains and includes assumed bending, membrane and shear strains. The displacement and strain interpolations of the elements are summarized in Appendix A. The formulation can be derived from the Hellinger–Reissner variational principle (and therefore, also from the Hu–Washizu variational principle), see Appendix B. However, a stability and convergence analysis is necessary to assess whether the actual discretization is effective. This analysis is difficult to perform analytically for a general setting, therefore we resort to numerical assessments. For the non-inhibited case, we would like that the inf-sup condition be satisfied for the selected interpolations. The results of our inf-sup condition study are presented in Ref. [7] and show that a numerical inf-sup test is satisfied. Our objective in the following section is to give the results obtained in convergence studies.

3. Numerical convergence studies

The test problems we use for the convergence studies are described in Ref. [1], where the reasons for selecting these problems are also given. Using these problems (summarized below), we proceed in each case as follows.

For each of the test problems, we run a sequence of meshes and depict the convergence of the relative error in strain energy of the approximation E_r versus a mesh density indicator N, where

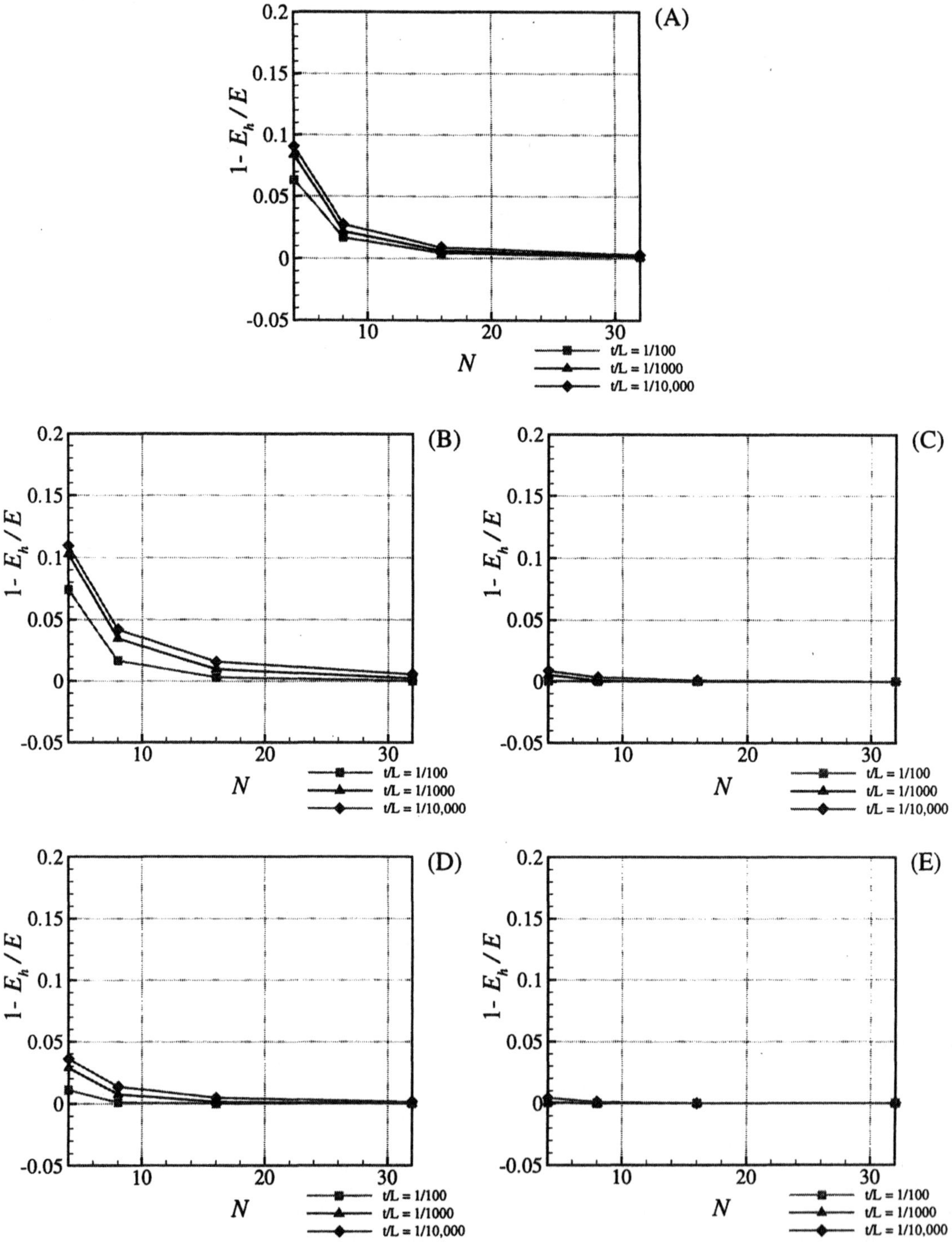

Fig. 4. Convergence in strain energy for the clamped cylinder problem. Uniform meshes. (A) MITC4 element; (B) QUAD9; (C) MITC9; (D) QUAD16; (E) MITC16; (F) TRI6; (G) MITC6.

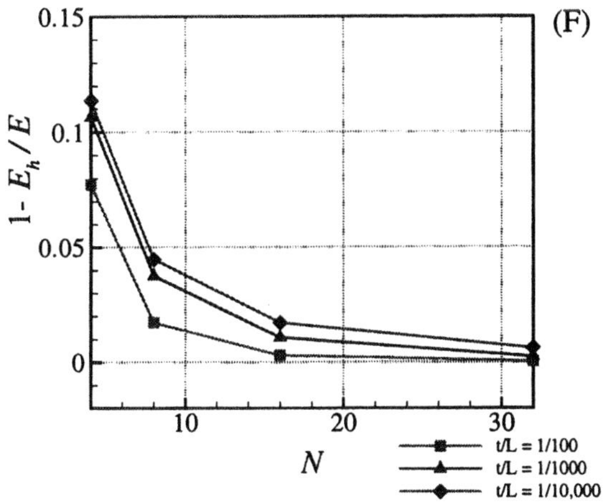

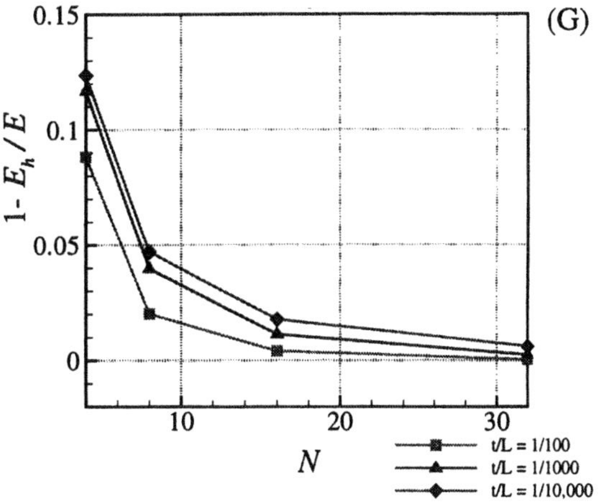

Fig. 4 (*continued*)

$$E_r = \frac{E - E_h}{E}, \tag{18}$$

with E_h being the strain energy of the finite element approximation and E being the strain energy calculated using either the exact solution of the mathematical model, or, when no analytical solution is available, other finite element schemes or very fine meshes of high-order shell elements.

In addition, we plot the E_r values versus the element size h in the logarithmic scale to estimate the convergence constant c and order of convergence $\hat{k}$ of the convergence equation:

$$|E_r| = ch^{2\hat{k}}. \tag{19}$$

The numerical simulations are repeated for a number of thickness parameter values, and changes in the convergence properties of the element are reflected in changes of the convergence constant c and order of convergence $\hat{k}$.

Ideally, we would like to have an element with $\hat{k}$ independent of the nature of the problem (i.e., bending- or membrane-dominated) *and* the thickness t. In addition, the convergence constant c should exhibit no dependence on t for each problem under consideration. Moreover, the order of convergence $\hat{k}$ should be *optimal*. In general, the optimal order of convergence k is prescribed by the order of the discretization scheme under consideration (and the physics of the problem, smoothness of the solution, presence of stress concentrations due to loading, boundary layers, etc.) [2].

3.1. Membrane-dominated tests

3.1.1. Clamped cylindrical shell

We consider a cylindrical shell of uniform thickness t, length $2L$ and radius R loaded by an axially-constant pressure distribution $p(\varphi)$ acting on the outer surface of the shell,

$$p(\varphi) = P_0 \cos(2\varphi). \tag{20}$$

At a point P on the cylinder's midsurface, we select the axial displacement u, tangential displacement v, radial displacement w, rotation of the normal about the X-axis β_1, and rotation of the normal about the tangent vector β_2 as the displacement variables (see Fig. 1).

By symmetry, we can limit calculations to the shaded region ABCD with the following symmetry and boundary conditions imposed:

$u = w = 0$ along AB;

$u = \beta_2 = 0$ along BC; $v = \beta_1 = 0$ along CD;

$$u = v = w = \beta_1 = \beta_2 = 0 \text{ along AD}. \tag{21}$$

The analytical solution for the problem using the Naghdi shell theory has been published for any given value of t (see [8][1]). The problem is membrane-dominated with the combined bending and shear energy being less than 2% even for a moderately thick case of

[1] The values for displacements and deformation energy in Ref. [8] should be multiplied by $(1-\nu^2)$ and $(1-\nu^2)^2$, respectively.

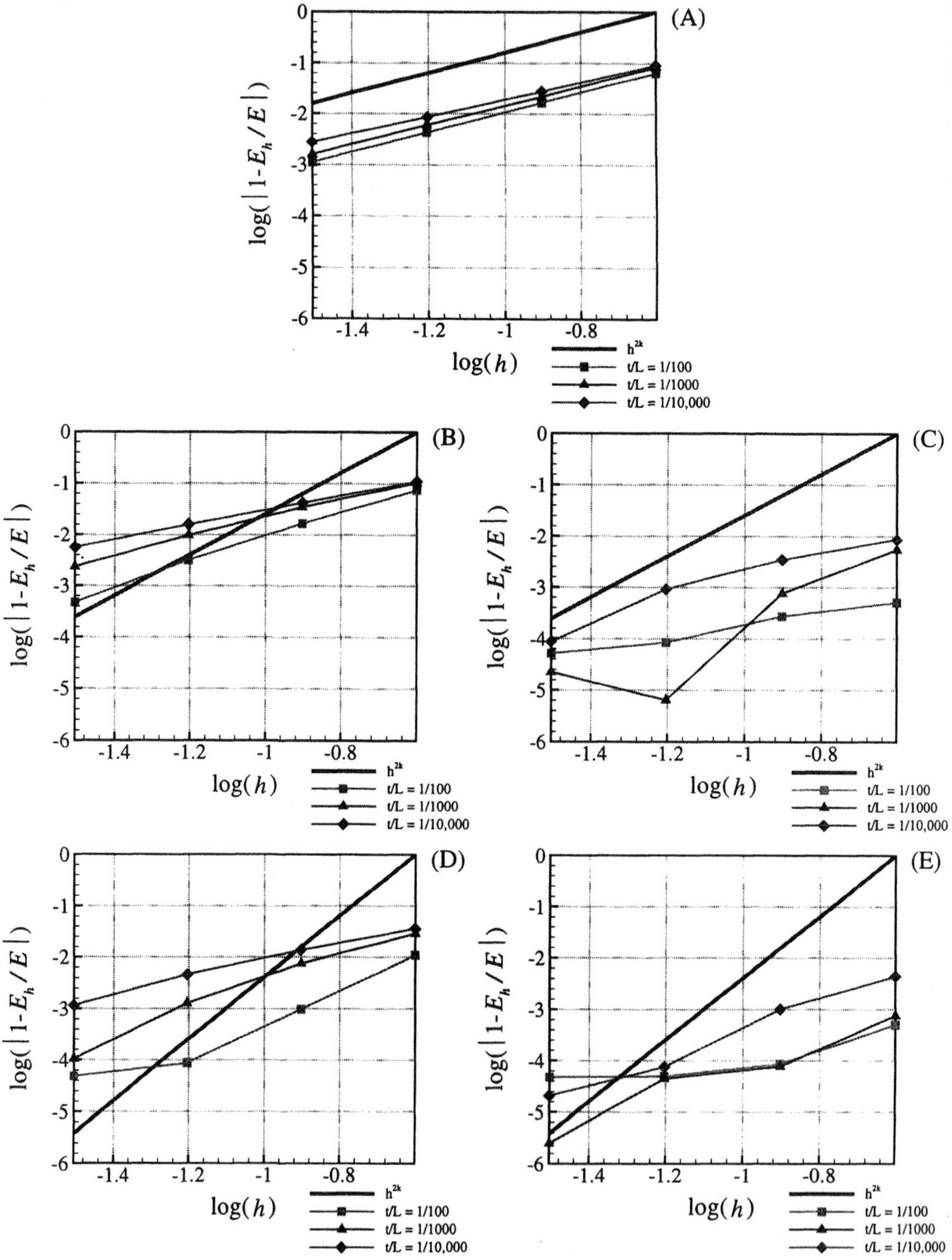

Fig. 5. Convergence in strain energy for the clamped cylinder problem. Uniform meshes. (A) MITC4 element; (B) QUAD9; (C) MITC9; (D) QUAD16; (E) MITC16; (F) TRI6; (G) MITC6.

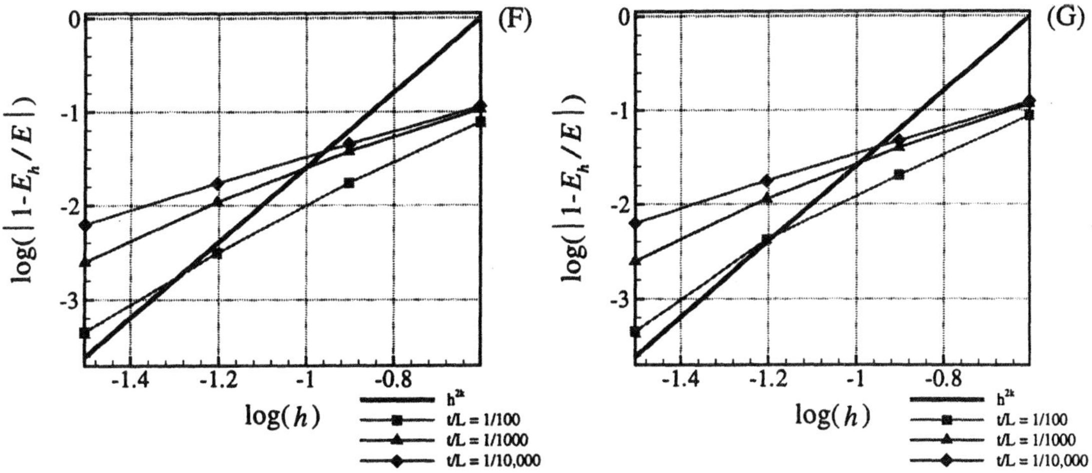

Fig. 5 (*continued*)

$t/L = 1/100$ (see Fig. 2) and hence we scale the applied loading with thickness t as

$$P_0 = p_0 t, \tag{22}$$

where p_0 is a constant independent of t.

First we consider *uniform* $N \times N$ meshes with element sides aligned with the principal directions of curvature, where N is the number of subdivisions per side of the discretized domain (in our tests $N = 4$, 8, 16 and 32). This sequence of meshes is repeated for each tested element for values of dimensionless thickness parameter t/L ranging from 1/100 to 1/10,000.

Fig. 3 gives a comparison of the finite element results with the analytical solution for the radial displacement profile calculated using the elements of the MITC family and the 9- and 16-node displacement-based general shell elements (QUAD9 and QUAD16) for the case of $t/L = 1/10{,}000$.

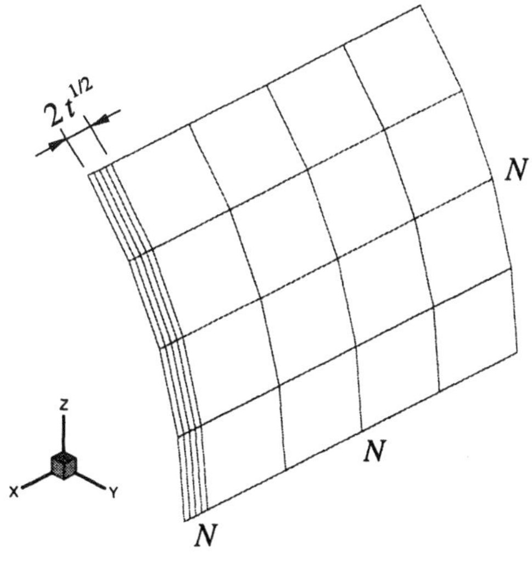

Fig. 6. A mesh grading scheme for the cylindrical shell problem, mesh density $N = 4$.

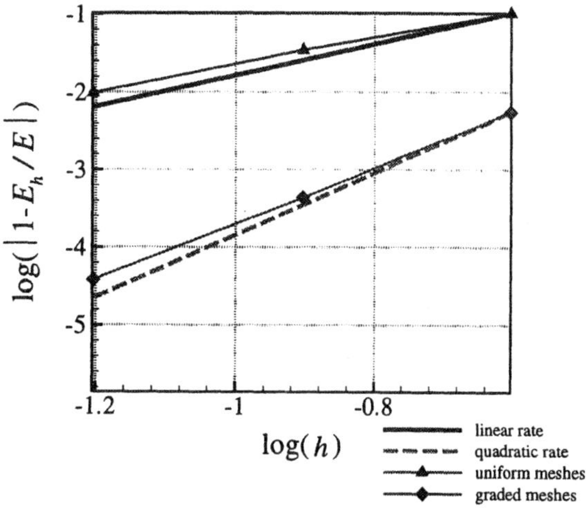

Fig. 7. Convergence in strain energy for the clamped cylinder problem using the nine-node displacement-based element QUAD9 in uniform and graded meshes for the case $t/L = 1/1000$.

Table 1
Benchmark values for the strain energy E and vertical displacement w at point $x = 0$, $y = 0$ for the hemispherical cap under axisymmetric pressure. We use $R = 1$ m; $E = 2 \times 10^5$ MPa; $\nu = 0.3$; $p_0 = 1$ MPa

t/R	1024 1D axisymmetric three-node shell elements	
	Strain energy E (N·mm)	$w(0, 0)$ (mm)
1/100	6.79244×10^{-5}	-7.73688×10^{-3}
1/1000	7.62474×10^{-6}	-8.08931×10^{-3}
1/10,000	7.88885×10^{-7}	-8.19934×10^{-3}

All the elements perform quite well. However, some numerical instabilities especially pronounced for the displacement-based elements can be observed at the clamped end. These instabilities are due to the presence of boundary layers and can be eliminated using graded meshes (as derived in Ref. [8], boundary layers play a dominant role in a $\sim 2\sqrt{t}$ region at the fixed end).

Figs. 4 and 5 give the convergence in relative error E_r for the problem under consideration for a sequence of t/L values. The six-node displacement-based element is referred to as the TRI6 element.

The convergence curves of the MITC4 element stay virtually unaffected by changes in thickness t, and the element's order of convergence $\hat{k}$ is very close to its theoretical value of $k = 1$.

The convergence curves for the other MITC elements have noticeably lower convergence constants c than their displacement-based counterparts and reach the accuracy of 1.0% with the coarsest mesh. Note that for the case of $t/L = 1/100$ the elements do not converge further after reaching the accuracy level of $\sim 0.01\%$, which is explained by differences

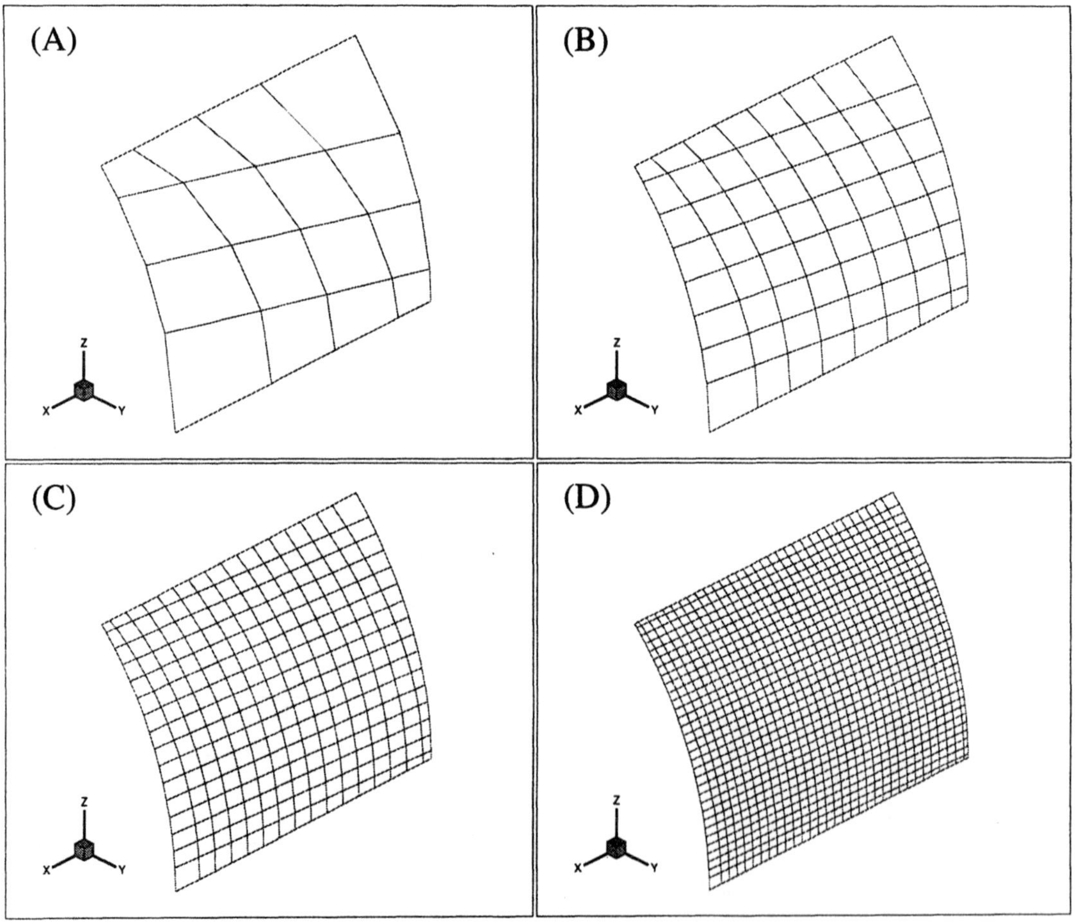

Fig. 8. A sequence of distorted meshes used for cylindrical shell. (A) The 4 × 4 mesh; (B) 8 × 8; (C) 16 × 16; (D) 32 × 32.

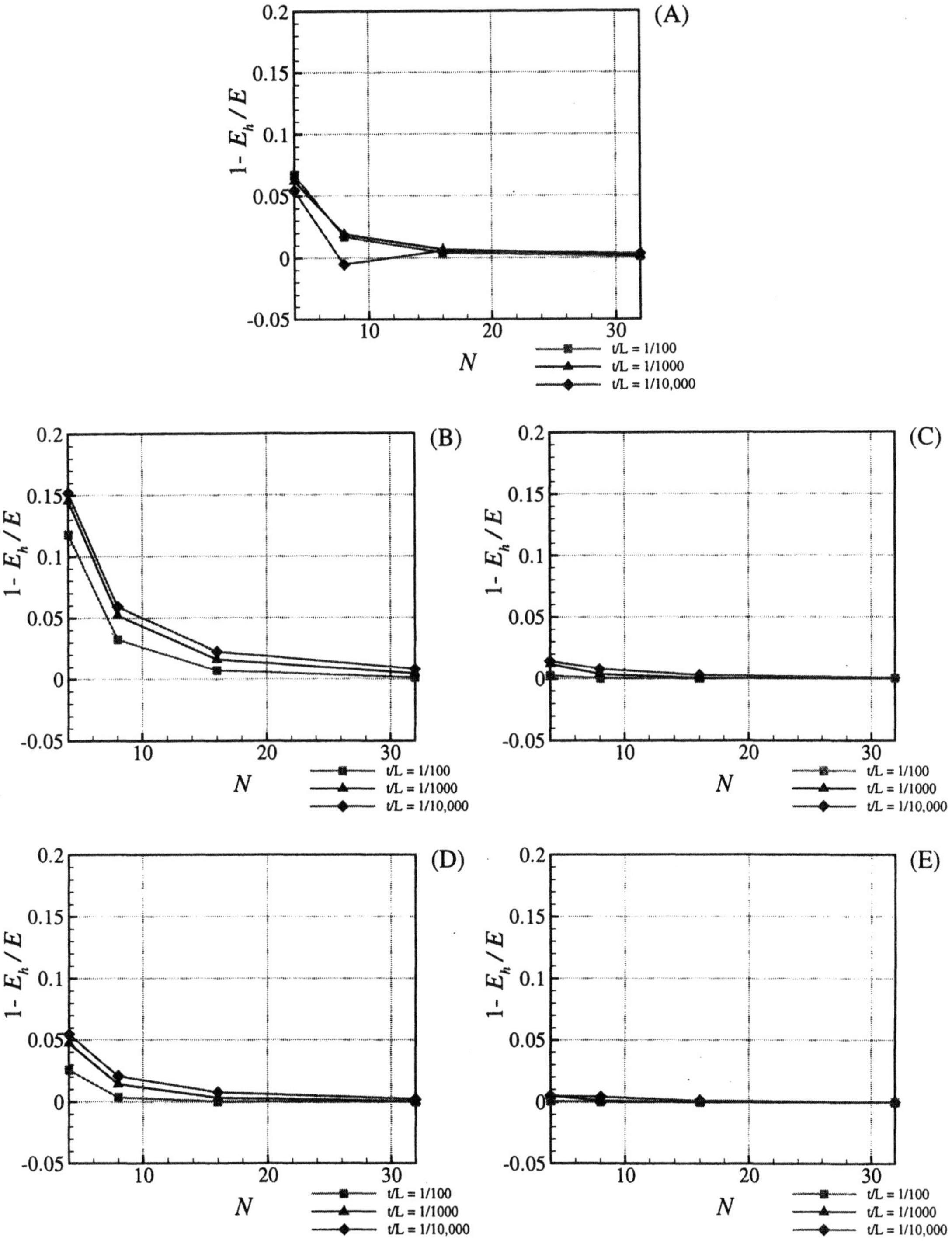

Fig. 9. Convergence in strain energy for the clamped cylinder problem. Distorted meshes. (A) MITC4 element; (B) QUAD9; (C) MITC9; (D) QUAD16; (E) MITC16.

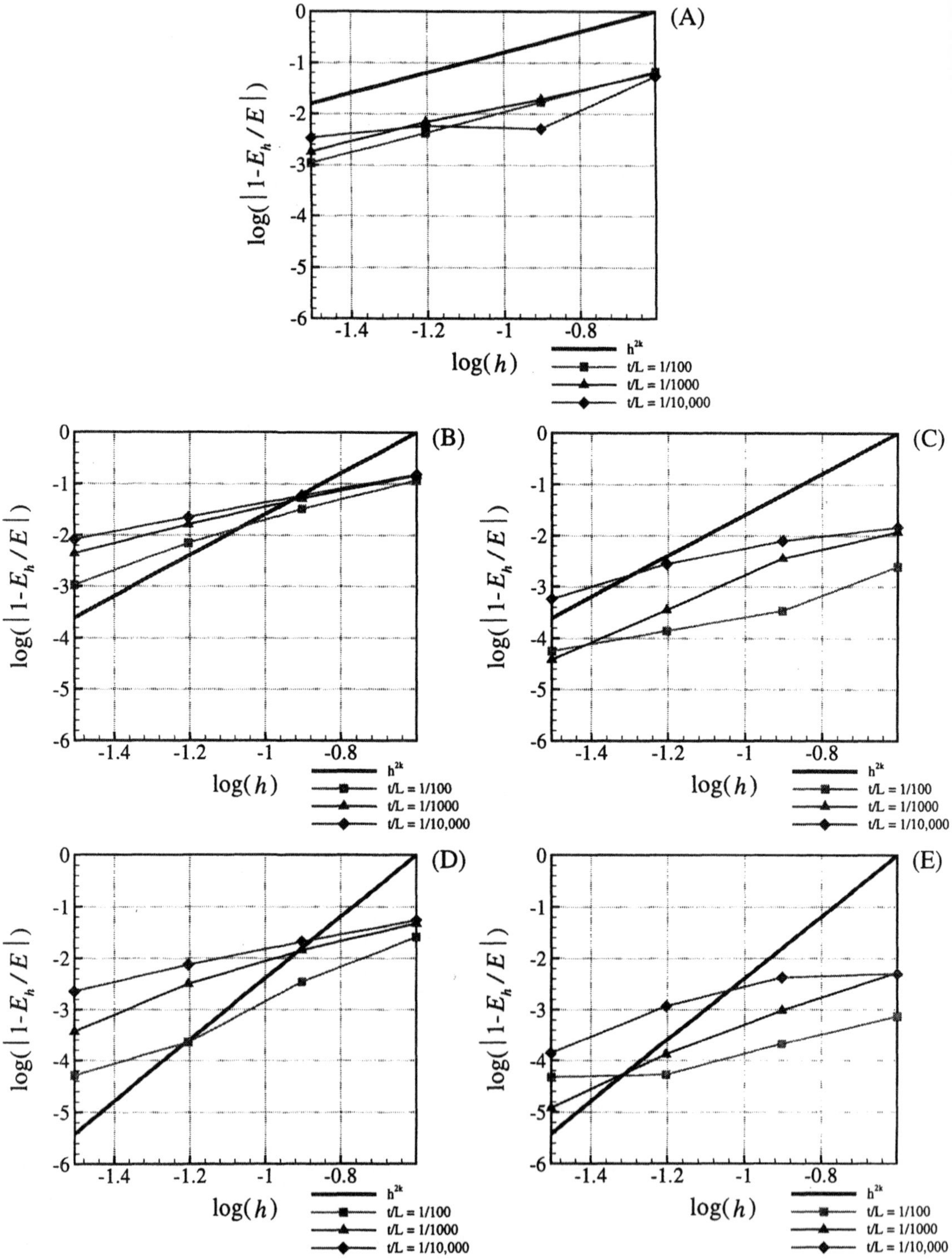

Fig. 10. Convergence in strain energy for the clamped cylinder problem. Distorted meshes. (A) MITC4 element; (B) QUAD9; (C) MITC9; (D) QUAD16; (E) MITC16.

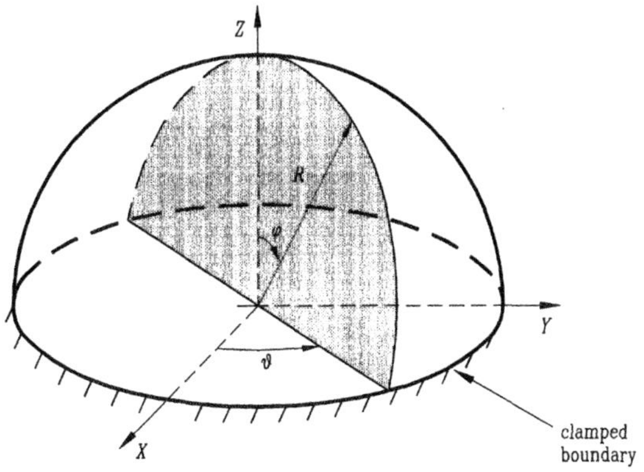

Fig. 11. Clamped hemispherical cap.

between the reference (Naghdi theory) solution and the solution to our underlying shell model [6].

As pointed out in [8], the fact that the observed convergence rates are lower than the respective "optimal" rates (equal to the order of the polynomial interpolations in the elements) may be due to the presence of stress concentration in the boundary layer region and can be improved using *mesh grading*. To identify the effect of mesh grading, we consider the following mesh grading scheme: the discretized domain is separated into two regions — the boundary layer region of width $2\sqrt{t}$ in the X-direction, and the smooth solution region of width equal to $1-2\sqrt{t}$. Both regions are then meshed with the equal mesh density N as shown in Fig. 6.

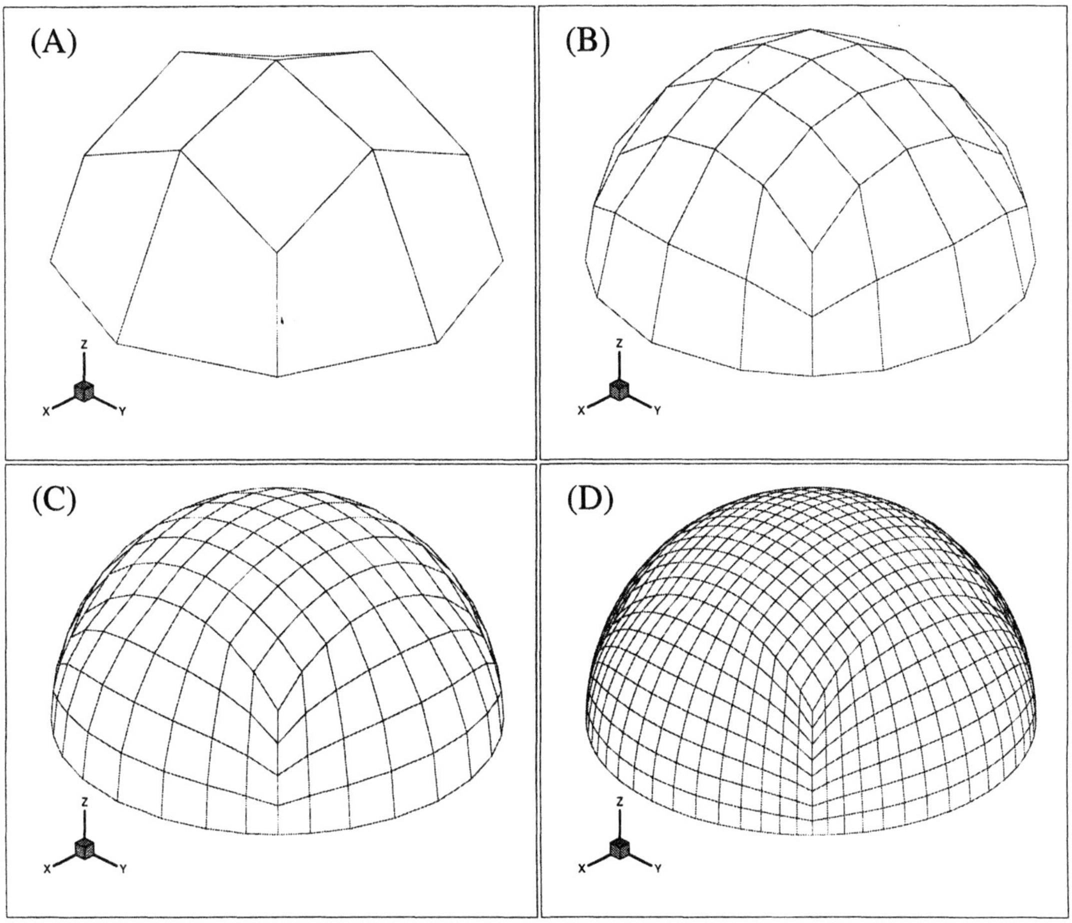

Fig. 12. A sequence of $N_\varphi \times N_\vartheta$ meshes of quadrilateral elements for the hemispherical cap problem. (A) The 2×8 mesh; (B) 4×16; (C) 8×32; (D) 16×64.

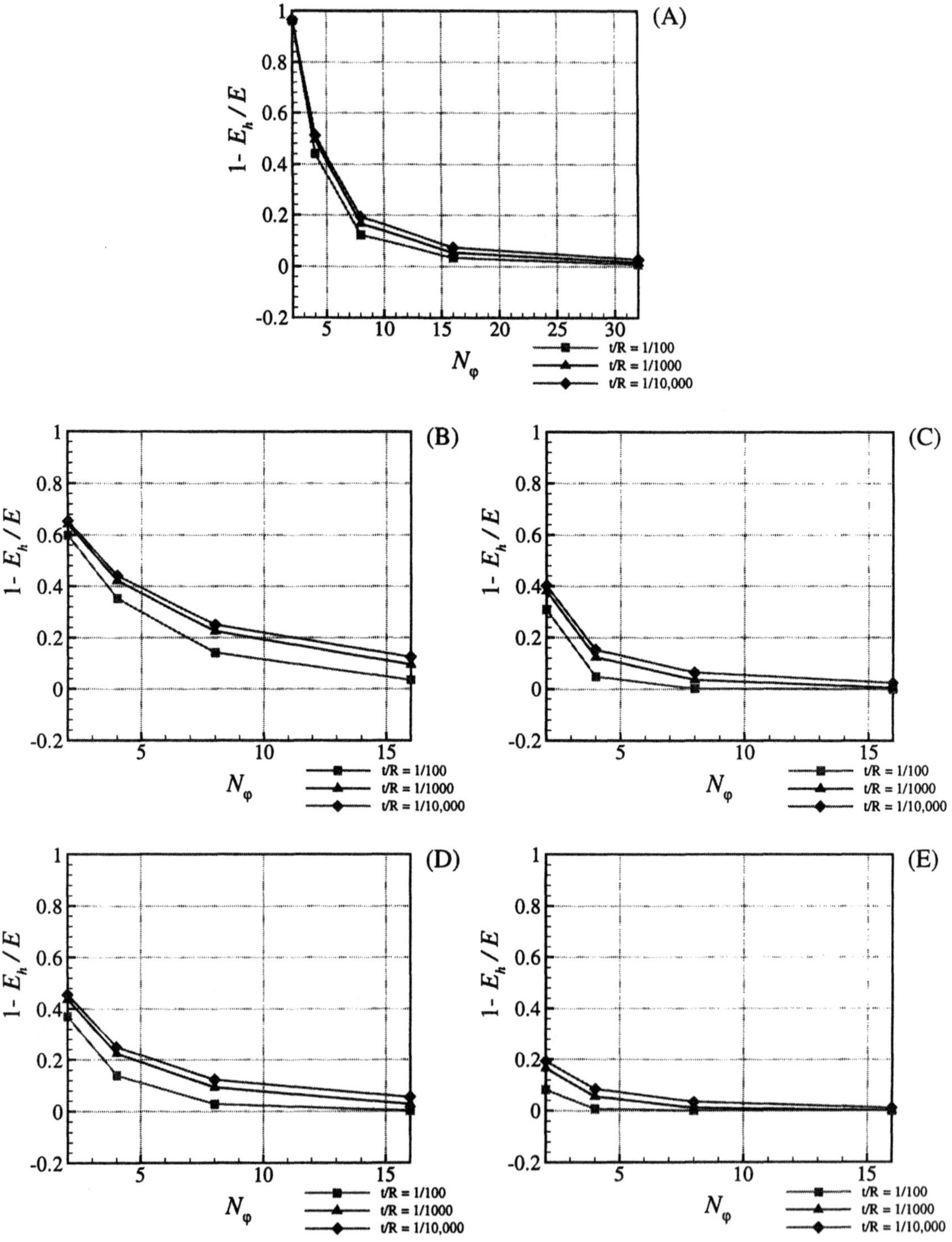

Fig. 13. Convergence in strain energy for the clamped hemispherical cap problem. (A) MITC4 element; (B) QUAD9; (C) MITC9; (D) QUAD16; (E) MITC16; (F) TRI6; (G) MITC6.

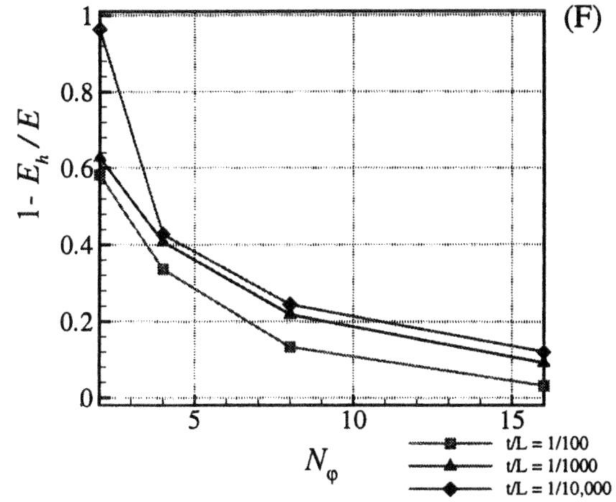

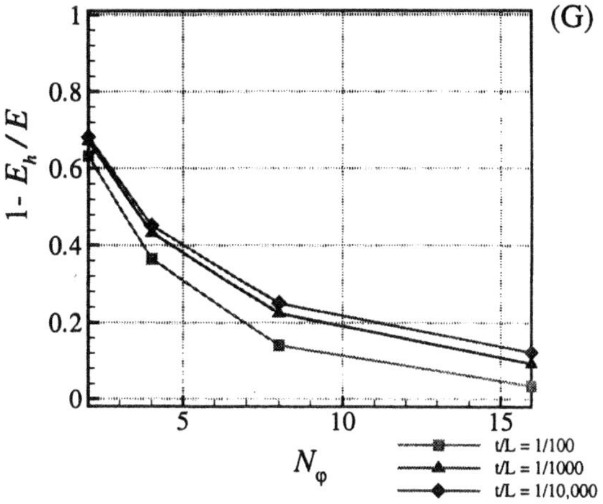

Fig. 13 (*continued*)

Fig. 7 illustrates the change in the convergence rate for the nine-node displacement-based element QUAD9 when a sequence of graded meshes is considered for the case of $t/L = 1/1000$. The order of convergence drastically increases and virtually reaches the optimal rate k, and a pronounceable reduction in the convergence constant c further underlines the importance of mesh grading.

We also test the elements using *distorted* meshes in which element sides are not aligned with the principal directions of curvature. Such sequence is shown in Fig. 8.

Figs. 9 and 10 display convergence results calculated using the distorted meshes of Fig. 8. Overall, no dramatic changes in convergence can be observed, however, the characteristics have worsened for all elements. This effect can be partly attributed to poorer geometry approximation using distorted meshes.

3.1.2. Clamped hemispherical cap

We consider a hemisphere of radius R, fully clamped at $\varphi = \pi/2$ (see Fig. 11), and uniform thickness t loaded with an axisymmetric pressure distribution

$$p(\varphi) = P_0 \cos(2\varphi). \tag{23}$$

The problem is membrane-dominated, hence we scale the applied load p with t.

Since no analytical solution is available for this problem, we use a 1D axisymmetric shell finite element model to obtain benchmark values for the strain energy, see Table 1.

We use a sequence of $N_\varphi \times N_\vartheta$ meshes to study the convergence of the elements for t/R ratios from 1/100 to 1/10,000, where N_φ is the number of divisions in the φ-direction, and N_ϑ is the number of divisions in the ϑ-direction at $\varphi = \pi/2$ (see Fig. 12).

Figs. 13 and 14 report the results of the finite element simulations.

Overall, all the elements have a relatively high value of the convergence constant c and low convergence rates because of severe mesh distortions (the element sides are far from being aligned with the lines of principal curvature) and poor geometry approximation, especially for coarser meshes.

Like for the clamped cylinder problem, the MITC4 element exhibits virtually no dependence on t. The better performance of the other MITC elements (compared to the displacement-based elements) is reflected in lower values of convergence constants.

3.2. Bending-dominated tests

3.2.1. Free cylindrical shell

The geometry and loading are the same as in the

Table 2
Reference values for the strain energy E and vertical displacement w at point $X = L/2$, $Y = 0$ for the hyperbolic paraboloid, calculated using a 48 × 24 mesh of MITC16 elements. We use $L = 1$ m; $E = 2 \times 10^5$ MPa; $\nu = 0.3$; density $\rho = 8000$ kg/m^3

t/L	Strain energy E (N·m)	Displacement $w(L/2, 0)$ (m)
1/100	1.6790×10^{-3}	-9.3355×10^{-5}
1/1000	1.1013×10^{-2}	-6.3941×10^{-3}
1/10,000	8.9867×10^{-2}	-5.2988×10^{-1}

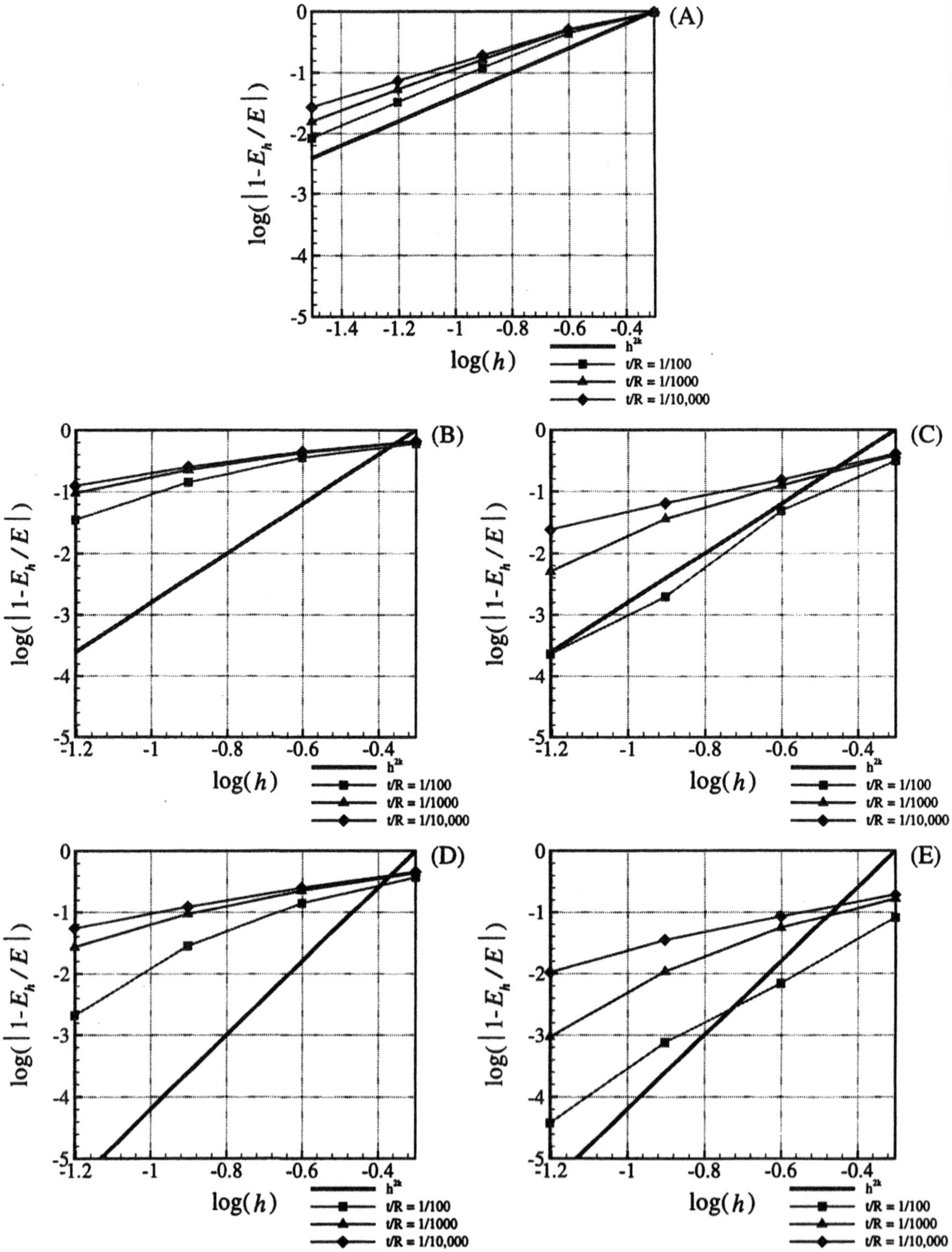

Fig. 14. Convergence in strain energy for the clamped hemispherical cap problem. (A) MITC4 element; (B) QUAD9; (C) MITC9; (D) QUAD16; (E) MITC16; (F) TRI6; (G) MITC6.

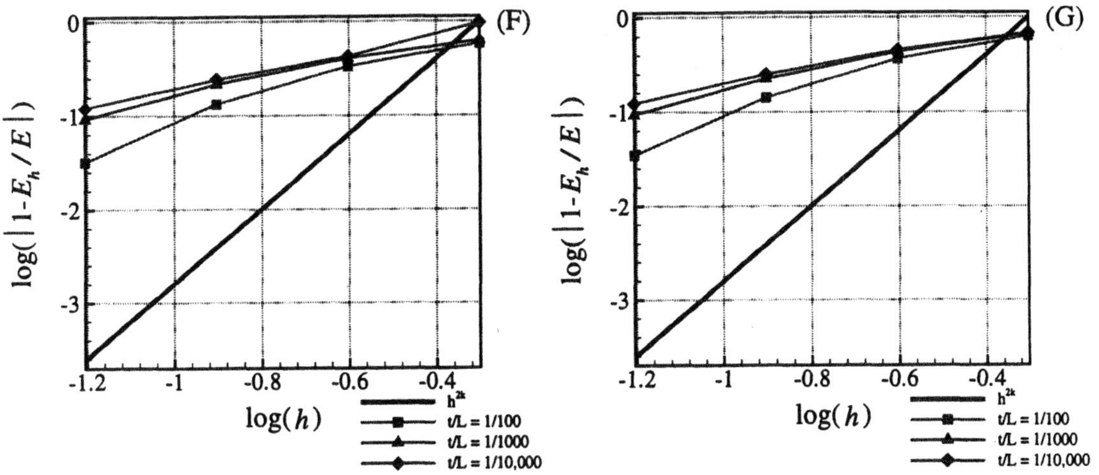

Fig. 14 (*continued*)

clamped cylinder problem considered above. The same boundary and symmetry conditions are applicable but we release all the fixities along the arc AD.

The bending energy prevails even for moderately large t/L ratios — Fig. 15 shows that relative contributions of the shear and membrane terms are essentially negligible.

In the convergence study we use the Naghdi shell theory solution given in Ref. [8], where a general methodology of obtaining the exact solution for any given (positive) value of t is demonstrated. Because bending is dominant in this problem, the load is rescaled as

$$P_0 = p_0 t^3. \qquad (24)$$

Fig. 16 presents a comparison of the analytical solution for the axial profile of the radial displacement w with the finite element simulation results for the $t/L = 1/10{,}000$ case. The nine-node displacement-based element locks and results in zero displacements for coarser meshes. The QUAD16 element starts off quite far from the analytical solution but reaches decent accuracy for finer meshes. The MITC4 element displays robust convergence, and the higher-order MITC elements produce accurate results even for relatively coarse meshes.

The convergence in strain energy using uniform meshes is displayed in Figs. 17 and 18[2]. As in the clamped cylinder case, the MITC4 element convergence curves are "perfect" — they stay parallel to the optimal convergence line, and neither the convergence constant c nor the order $\hat{k}$ are affected by variations in thickness.

The locking of the nine-node displacement-based element can be directly observed — its convergence curves flatten out as t is decreased, and the relative error stays above the 70% level even for very fine meshes when $t/L = 1/10{,}000$.

The behavior of the QUAD16 element is slightly affected by changes in thickness — the convergence constant of the element gets larger, indicating potential difficulties as t is decreased.

The higher-order MITC elements converge to the accuracy of 0.1% in strain energy with the coarsest

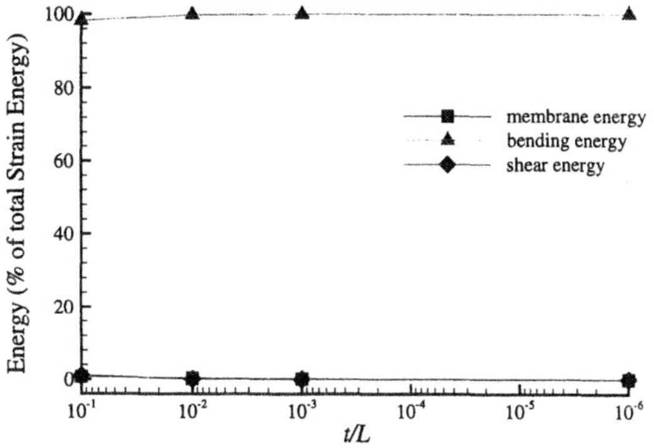

Fig. 15. Distribution of strain energy as a function of t/L for the free cylindrical shell.

[2] Symbols "*" on convergence graphs hereafter indicate that the numerical value of the associated point on the graph might be affected by computational (round-off) errors due to ill-conditioning of the stiffness matrix of the system.

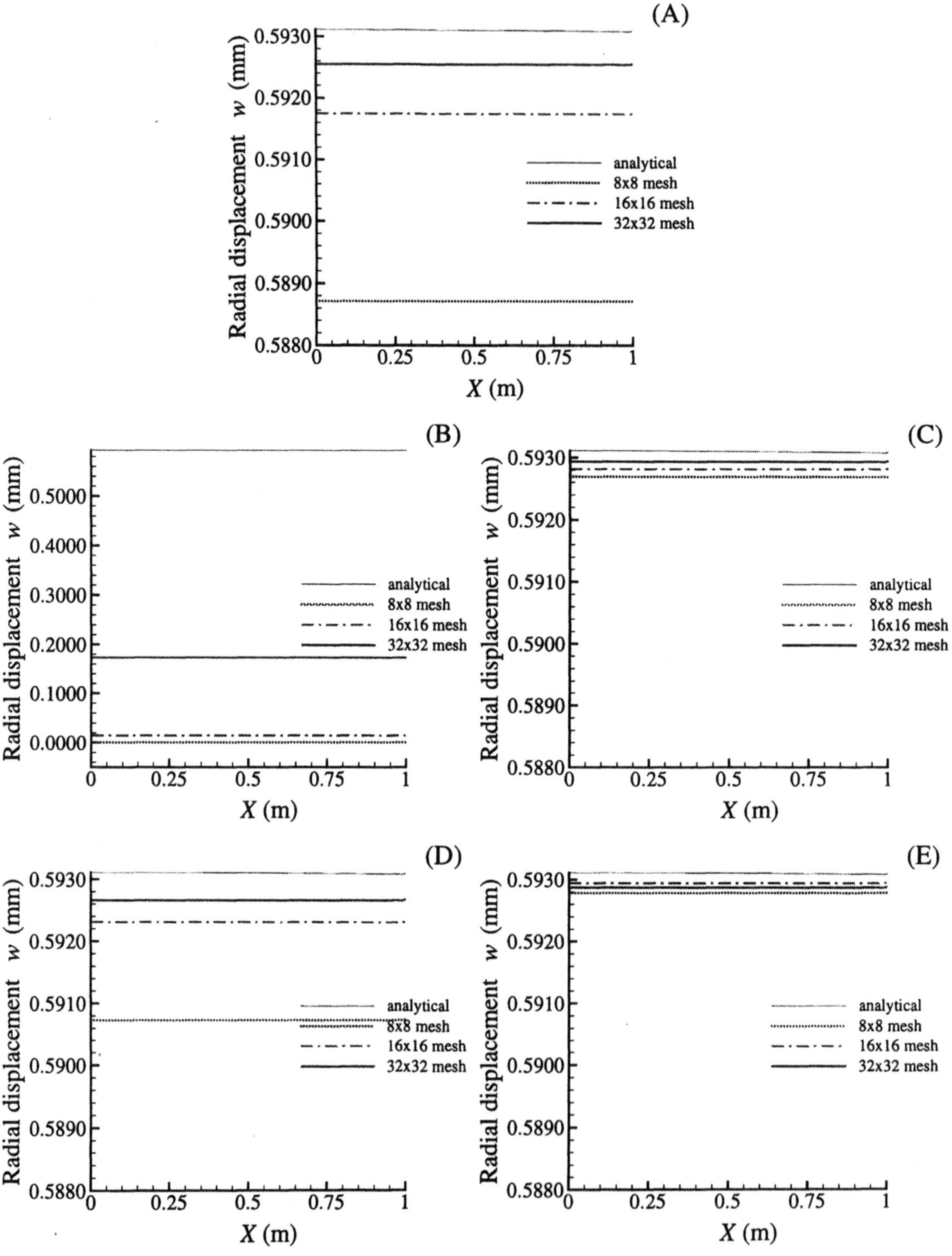

Fig. 16. Axial profile of radial displacement w calculated along line CD for the free cylinder problem using $L = R = 1$ m; $t/L = 1/10{,}000$; $E = 2 \times 10^5$ MPa; $\nu = 1/3$; $p_0 = 1$ MPa. (A) MITC4 element; (B) QUAD9; (C) MITC9; (D) QUAD16; (E) MITC16.

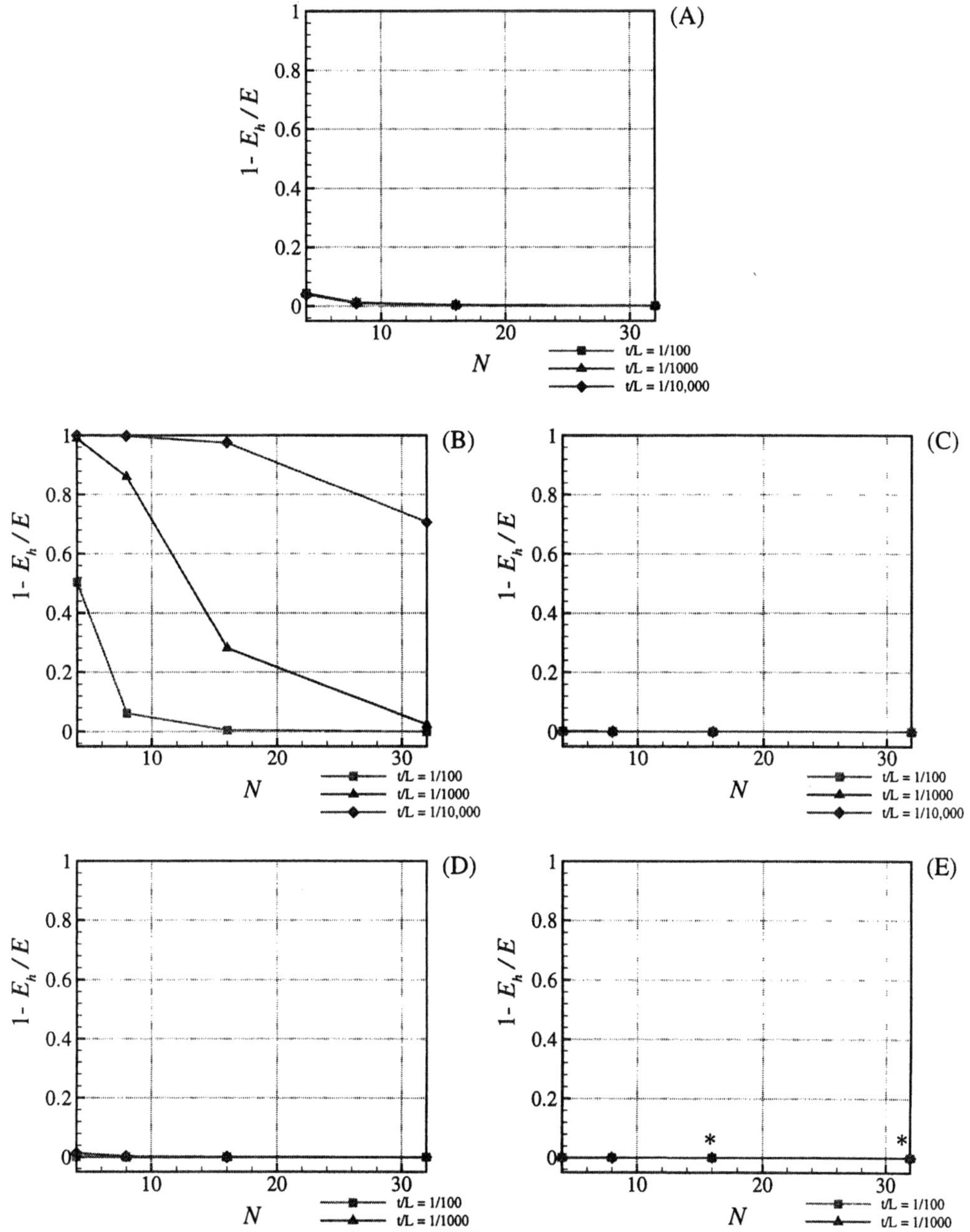

Fig. 17. Convergence in strain energy for the free cylinder problem. Uniform meshes. (A) MITC4 element; (B) QUAD9; (C) MITC9; (D) QUAD16; (E) MITC16; (F) TRI6; (G) MITC6.

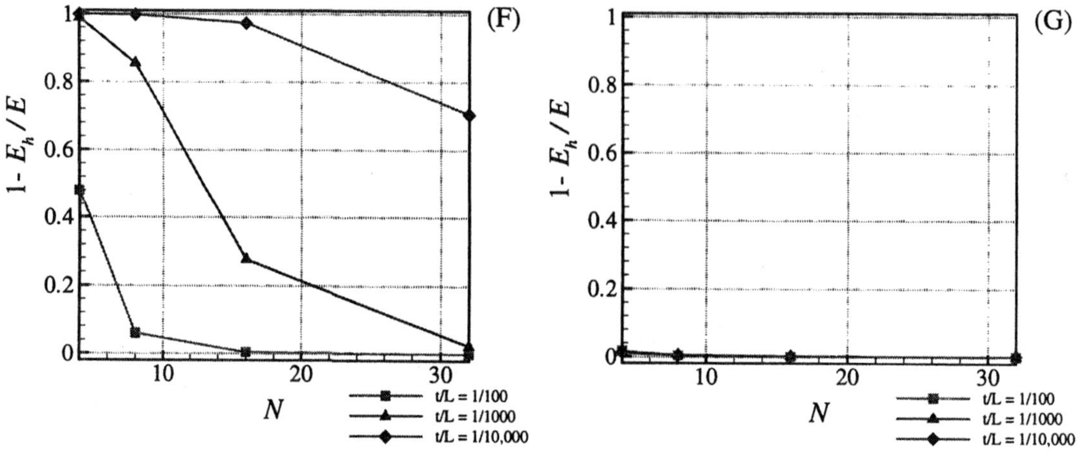

Fig. 17 (*continued*)

meshes for all the thickness values. More accurate results could not be obtained for the case $t/L = 1/10{,}000$ because of round-off errors as a result of ill-conditioning. This solution difficulty is more likely to arise in bending-dominated problems. In these analyses, pure bending and membrane modes are associated with large differences in stiffness values. Note that this difficulty is less likely to arise when using displacement-based elements because these elements do not accurately represent the pure bending modes.

As in the clamped cylinder analysis, for the $t/L = 1/100$ case the convergence curves flatten out at the level of 0.01% because of the discrepancies between the Naghdi theory solution and our three-dimensional shell model.

Figs. 19 and 20 report the performance of the elements when used with the distorted meshes of Fig. 8. Overall, all the discretizations produce worse results compared to uniform meshes. This effect is especially pronounced for coarse meshes with highest distortions. However, as the meshes are refined, the elements reach the accuracy levels obtained with uniform meshes.

3.2.2. Partly clamped hyperbolic paraboloid

The problem was suggested in [1] as a good test for locking behavior. The surface is defined as

$$Z = X^2 - Y^2; (X, Y) \in [(-L/2; L/2)]^2, \tag{25}$$

clamped along the side $X = -L/2$ and loaded by self-weight.

By symmetry, only one half of the surface needs to be considered in the analysis (shaded region ABCD in Fig. 21), with clamped boundary conditions along AD and symmetry conditions along AB.

For the finite element analysis we use sequences of $N \times N/2$ meshes, where N is the number of subdivisions along the X-axis. A typical 16×8 mesh of four-node elements is shown in Fig. 21. These meshes are aligned with the sides of the discretized domain, and not with the asymptotic directions [1].

The analytical solution for this problem is not available, and we have to study the convergence to reference strain energy values obtained with a very fine mesh of shell finite elements. Table 2 reports such results for the strain energy and vertical displacement at point $X = L/2$, $Y = 0$, calculated using a 48×24 mesh of MITC16 elements. Note that since the self-weight loading is proportional to thickness, the strain energy scales as $1/t$, while displacements scale as $1/t^2$.

Figs. 22 and 23 report the convergence in strain energy to the values of Table 2. As in the previous case, the nine-node displacement-based element displays severe locking — for $t/L = 1/10{,}000$ the error in strain energy stays above 90% even for the finest mesh of 32×16 elements, and its convergence constant c noticeably shifts with variations in thickness.

A milder form of locking can be also observed for the QUAD16 element — its convergence constant increases by an order of magnitude for every decrease in t.

The MITC4 element produces good results on the absolute scale, however, the graphs in the logarithmic scale (Fig. 23(A)) show slow convergence. These diffi-

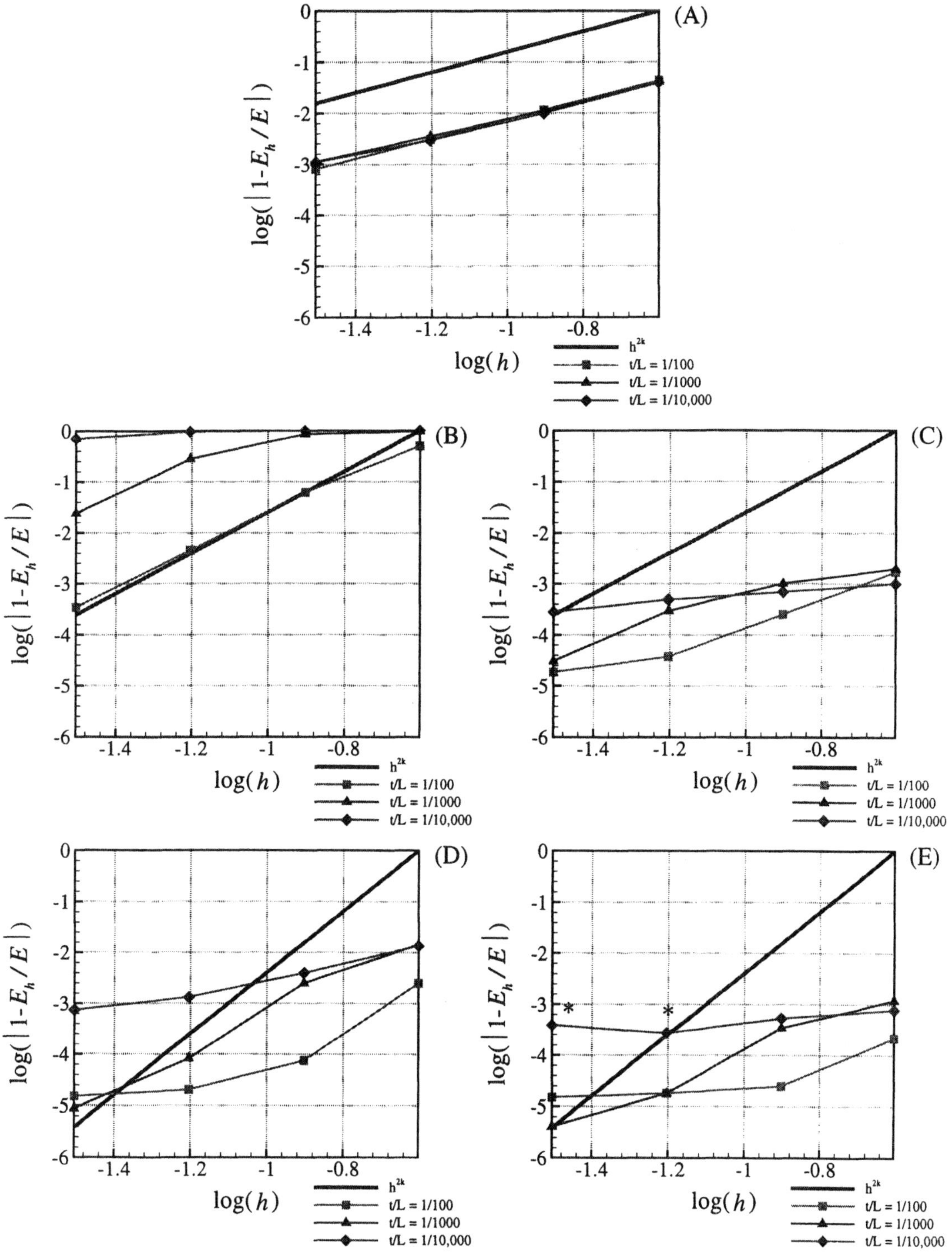

Fig. 18. Convergence in strain energy for the free cylinder problem. Uniform meshes. (A) MITC4 element; (B) QUAD9; (C) MITC9; (D) QUAD16; (E) MITC16; (F) TRI6; (G) MITC6.

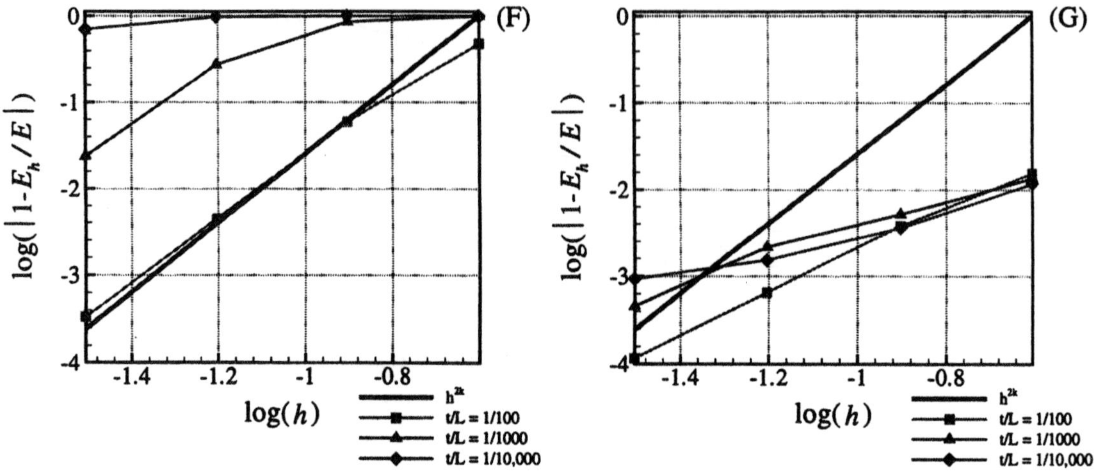

Fig. 18 (*continued*)

culties can be explained by poor geometric approximation properties achieved with low-order elements for complex surface geometries.

The higher-order MITC elements show good convergence with little dependence on changes in the thickness of the shell structure.

4. Concluding remarks

In this paper we have presented a study of the convergence behavior of the MITC shell elements. The problems used in the convergence study were discussed in detail in Ref. [1].

The evaluation of the MITC shell elements shows that the elements are effective in membrane- and bending-dominated shell problems and can thus be employed in general shell analysis situations.

For the *membrane-dominated problems*, the MITC elements produce results no worse and often significantly better than the displacement-based elements, and show robust convergence properties with low dependence on the thickness parameter t.

For the *bending-dominated problems*, displacement-based elements, of course, exhibit severe locking, while the MITC elements show good performance and robustness.

The optimal convergence rate has not necessarily been observed, but this is probably largely due to boundary layers (as shown in the case of the membrane-dominated cylinder problem).

Regarding the test problems, we have made the following general observations:

- The *clamped cylinder* problem has proved to be an effective test for membrane-dominated behavior. However, the presence of boundary layers at the fixed edges might induce (localized) instabilities in numerical solutions and makes interpretation of numerical results somewhat difficult.

  To avoid this complication, graded meshes refined in the boundary layer regions can be employed.
- The *clamped hemispherical cap* problem can be used to test the performance of quadrilateral elements but the elements will be quite distorted if the complete shell is discretized. Of course, only a section of a few degrees could be used. The advantage of the problem is that a doubly-curved shell is considered and in practice, of course, distorted elements are commonly used.
- The *free cylinder* problem should be considered as the *first* and *basic* test for locking in bending-dominated situations. Indeed, the space of pure bending displacements $\mathcal{U}_0$ has a simple functional (polynomial) form, and thus only elements with relatively poor approximation properties would exhibit locking.
- From our experience, the *partly clamped hyperbolic paraboloid* problem is an excellent test for locking in bending-dominated situations. The geometry of the shell is more general than a surface of zero Gaussian curvature and the problem is more realistic than an elliptic surface with free boundary. In addition, the symmetry in the problem allows to use relatively fine meshes, and thus, obtain sound convergence curves.

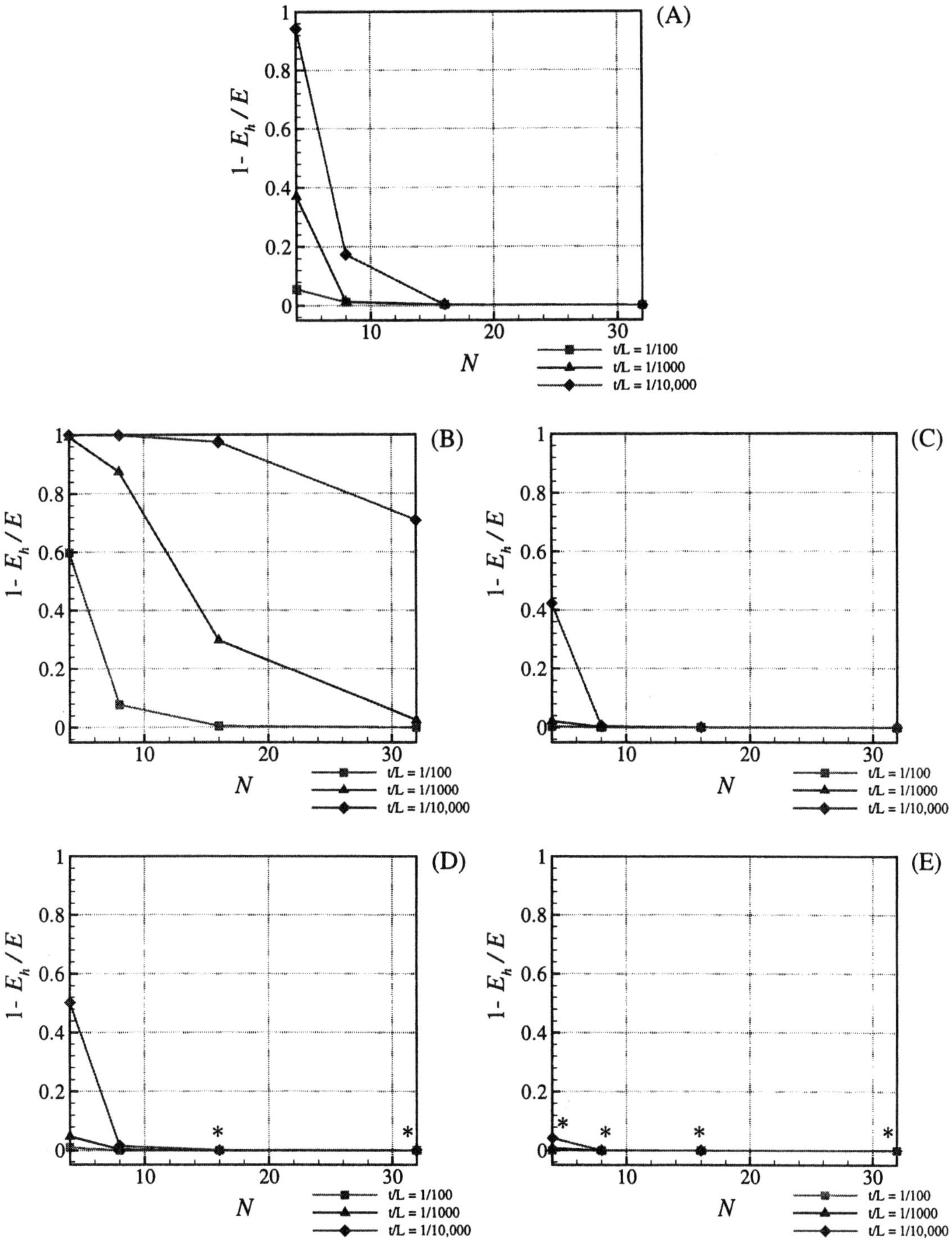

Fig. 19. Convergence in strain energy for the free cylinder problem. Distorted meshes. (A) MITC4 element; (B) QUAD9; (C) MITC9; (D) QUAD16; (E) MITC16.

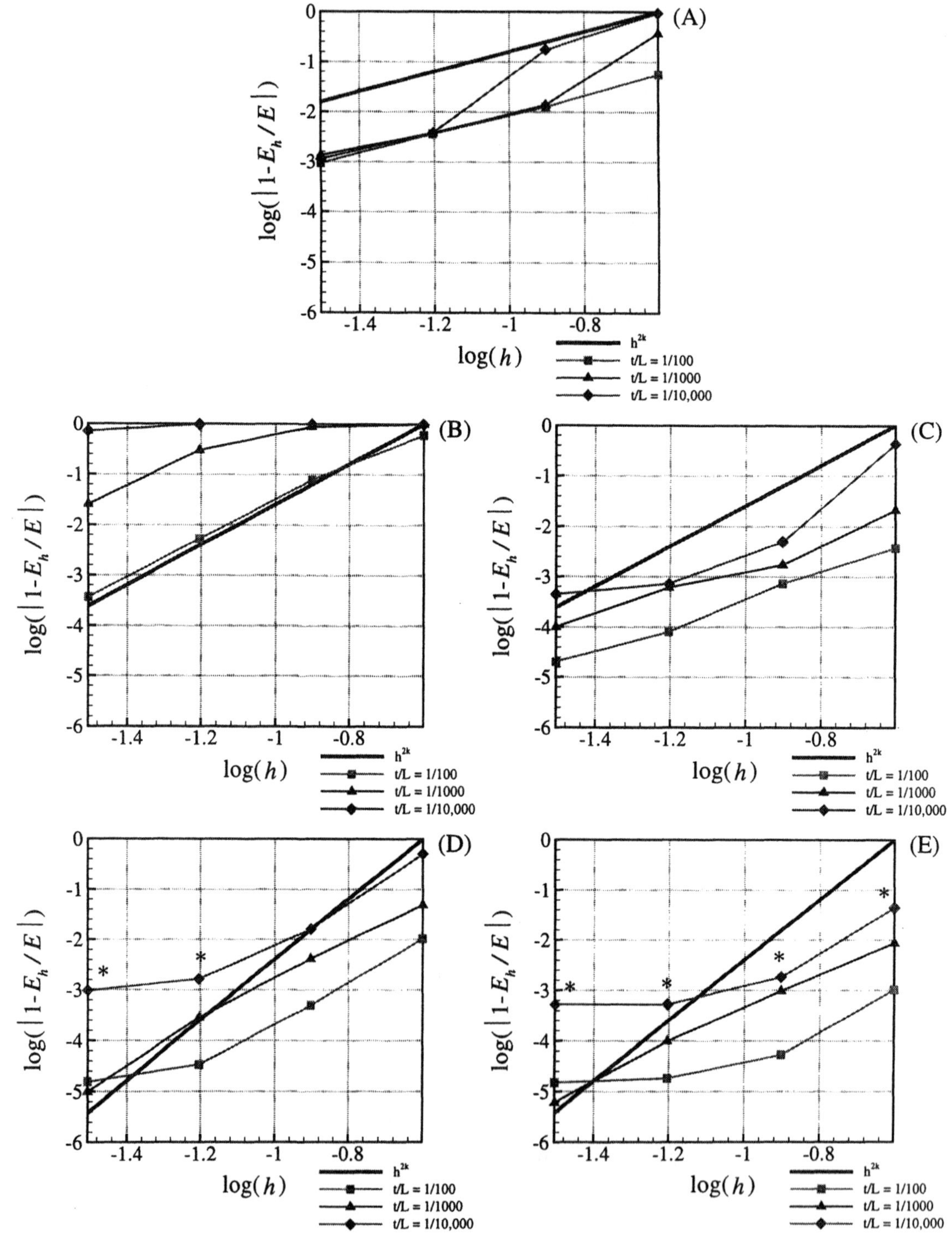

Fig. 20. Convergence in strain energy for the free cylinder problem. Distorted meshes. (A) MITC4 element; (B) QUAD9; (C) MITC9; (D) QUAD16; (E) MITC16.

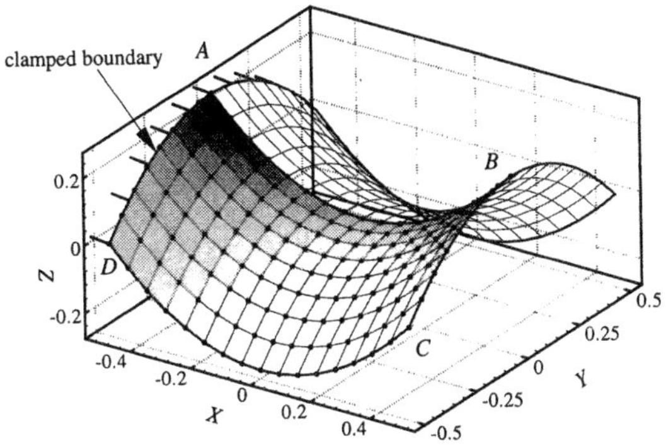

Fig. 21. Partly clamped hyperbolic paraboloid. A 16 × 8 mesh of four-node elements is shown.

Based on the results given in Ref. [1] and this paper, we conclude that the test problems used herein are very valuable to evaluate the performance of general shell elements.

Appendix A. Summary of MITC shell element interpolations

Let $\hat{K}$ denote the area of a reference element. Using the notation already defined, the spaces used for the elements are then as follows.

A.1. The MITC4 element

The following functional spaces are used for the four-node element:

$$\mathscr{V}^h = \left\{ \boldsymbol{v} \in [H^1(\Omega)]^3, \boldsymbol{v}|_{\hat{K}} \in \left[Q_1(\hat{K})\right]^3 \right\},$$

$$\mathscr{B}^h = \left\{ \boldsymbol{\eta} \in [H^1(\Omega)]^2, \boldsymbol{\eta}|_{\hat{K}} \in \left[Q_1(\hat{K})\right]^2 \right\},$$

$$\mathscr{E}^h_{13} = \left\{ \psi \in L^2(\Omega), \psi|_{\hat{K}} \in \mathrm{span}\left\{1, \hat{\xi}_2\right\} \right\},$$

$$\mathscr{E}^h_{23} = \left\{ \psi \in L^2(\Omega), \psi|_{\hat{K}} \in \mathrm{span}\left\{1, \hat{\xi}_1\right\} \right\}.$$

The tying procedure for the transverse shear strains ε_{13} and ε_{23} is shown in Fig. A1(A).

A.2. The MITC9 element

The following functional spaces are used to construct the nine-node element:

$$\mathscr{V}^h = \left\{ \boldsymbol{v} \in [H^1(\Omega)]^3, \boldsymbol{v}|_{\hat{K}} \in \left[Q_2(\hat{K})\right]^3 \right\},$$

$$\mathscr{B}^h = \left\{ \boldsymbol{\eta} \in [H^1(\Omega)]^2, \boldsymbol{\eta}|_{\hat{K}} \in \left[Q_2(\hat{K})\right]^2 \right\},$$

$$\mathscr{E}^h_{11} = \mathscr{E}^h_{13} = \left\{ \psi \in L^2(\Omega), \psi|_{\hat{K}} \in \mathrm{span}\left\{1, \hat{\xi}_1, \hat{\xi}_2, \hat{\xi}_1\hat{\xi}_2, \hat{\xi}_2^2, \hat{\xi}_1\hat{\xi}_2^2\right\} \right\},$$

$$\mathscr{E}^h_{22} = \mathscr{E}^h_{23} = \left\{ \psi \in L^2(\Omega), \psi|_{\hat{K}} \in \mathrm{span}\left\{1, \hat{\xi}_1, \hat{\xi}_2, \hat{\xi}_1\hat{\xi}_2, \hat{\xi}_1^2, \hat{\xi}_1^2\hat{\xi}_2\right\} \right\}.$$

$$\mathscr{E}^h_{12} = \left\{ \psi \in L^2(\Omega), \psi|_{\hat{K}} \in Q_1(\hat{K}) \right\}.$$

The tying procedure for the element is given in Fig. A1(B).

A.3. The MITC16 element

The 16-node cubic element is constructed using:

$$\mathscr{V}^h = \left\{ \boldsymbol{v} \in [H^1(\Omega)]^3, \boldsymbol{v}|_{\hat{K}} \in \left[Q_3(\hat{K})\right]^3 \right\},$$

$$\mathscr{B}^h = \left\{ \boldsymbol{\eta} \in [H^1(\Omega)]^2, \boldsymbol{\eta}|_{\hat{K}} \in \left[Q_3(\hat{K})\right]^2 \right\},$$

$$\mathscr{E}^h_{11} = \mathscr{E}^h_{13} = \left\{ \psi \in L^2(\Omega), \psi|_{\hat{K}} \in \mathrm{span}\left\{1, \hat{\xi}_1, \hat{\xi}_2, \hat{\xi}_1^2, \hat{\xi}_1\hat{\xi}_2, \hat{\xi}_2^2, \hat{\xi}_1^2\hat{\xi}_2, \hat{\xi}_1\hat{\xi}_2^2, \hat{\xi}_2^3, \hat{\xi}_1^2\hat{\xi}_2^2, \hat{\xi}_1\hat{\xi}_2^3, \hat{\xi}_1^2\hat{\xi}_2^3\right\} \right\},$$

$$\mathscr{E}^h_{22} = \mathscr{E}^h_{23} = \left\{ \psi \in L^2(\Omega), \psi|_{\hat{K}} \in \mathrm{span}\left\{1, \hat{\xi}_1, \hat{\xi}_2, \hat{\xi}_1^2, \hat{\xi}_1\hat{\xi}_2, \hat{\xi}_2^2, \hat{\xi}_1^3, \hat{\xi}_1^2\hat{\xi}_2, \hat{\xi}_1\hat{\xi}_2^2, \hat{\xi}_1^3\hat{\xi}_2, \hat{\xi}_1^2\hat{\xi}_2^2, \hat{\xi}_1^3\hat{\xi}_2^2\right\} \right\}.$$

$$\mathscr{E}^h_{12} = \left\{ \psi \in L^2(\Omega), \psi|_{\hat{K}} \in Q_2(\hat{K}) \right\}.$$

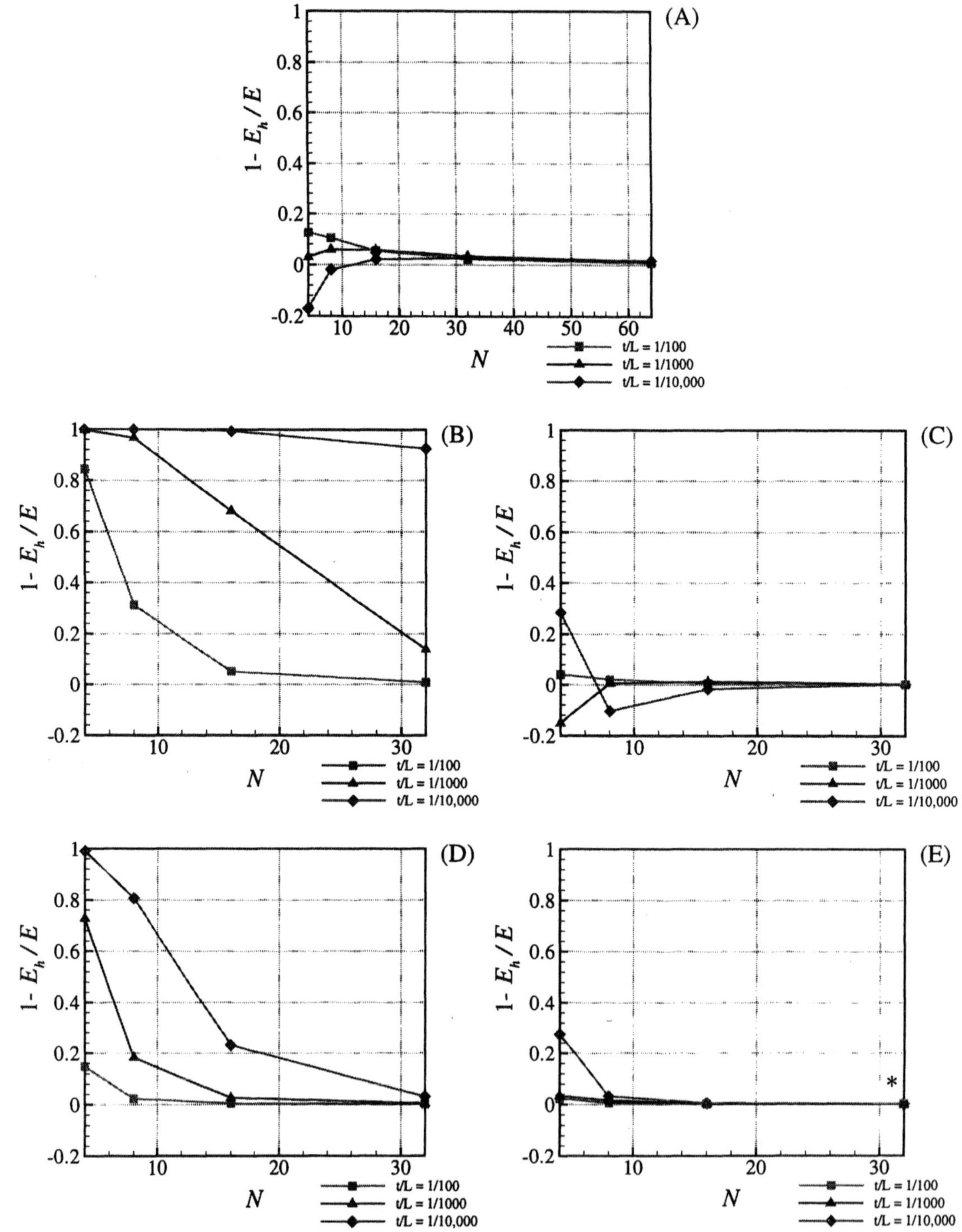

Fig. 22. Convergence in strain energy for the hyperbolic paraboloid problem. (A) MITC4 element; (B) QUAD9; (C) MITC9; (D) QUAD16; (E) MITC16; (F) TRI6; (G) MITC6.

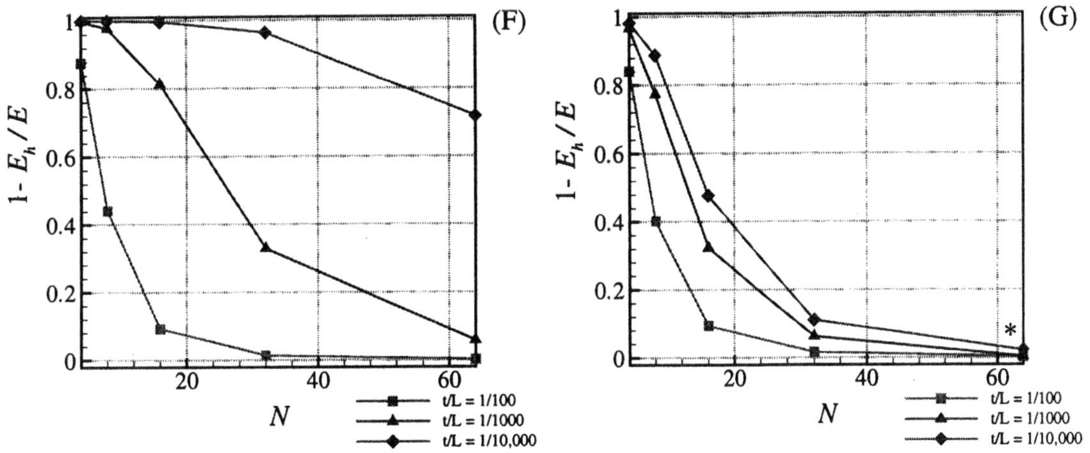

Fig. 22 (*continued*)

The tying procedure for the strain tensor components is presented in Fig. A1(C).

A.4. The MITC6 element

The six-node triangular element is constructed using

$$\mathscr{V}^h = \left\{ \boldsymbol{v} \in [H^1(\Omega)]^3, \boldsymbol{v}|_{\hat{K}} \in \left[P_2(\hat{K})\right]^3 \right\},$$

$$\mathscr{B}^h = \left\{ \boldsymbol{\eta} \in [H^1(\Omega)]^2, \boldsymbol{\eta}|_{\hat{K}} \in \left[P_2(\hat{K})\right]^2 \right\},$$

$$\mathscr{E}^h_{11} = \mathscr{E}^h_{12} = \mathscr{E}^h_{22} = \left\{ \psi \in L^2(\Omega), \psi|_{\hat{K}} \in P_1(\Omega) \right\},$$

$$\mathscr{E}^h_{13} = \left\{ \psi \in L^2(\Omega), \psi|_{\hat{K}} \in \operatorname{span}\left\{1, \hat{\xi}_1, \hat{\xi}_1\hat{\xi}_2, \hat{\xi}_2, \hat{\xi}_2^2\right\} \right\},$$

$$\mathscr{E}^h_{23} = \left\{ \psi \in L^2(\Omega), \psi|_{\hat{K}} \in \operatorname{span}\left\{1, \hat{\xi}_1, \hat{\xi}_1\hat{\xi}_2, \hat{\xi}_2, \hat{\xi}_1^2\right\} \right\}.$$

The choice of interpolation spaces for the transverse shear components proved to be optimal for the mixed-interpolated seven-node plate bending element MITC7 [2]. The tying procedure for the strain tensor components is depicted in Fig. A2.

Appendix B. Mixed formulation for the MITC general shell elements

In this Appendix we demonstrate how the abstract variational problem of type (15) can be derived from the mixed interpolation employed in construction of the general MITC shell elements using the Hellinger–Reissner variational principle.

Using the engineering notation [2], we rewrite the strain tensor in vector form:

$$\boldsymbol{\varepsilon} = [\varepsilon_{11}, \varepsilon_{22}, \varepsilon_{33}, 2\varepsilon_{12}, 2\varepsilon_{23}, 2\varepsilon_{13}]^{\mathrm{T}}, \tag{B1}$$

and assume

$$\boldsymbol{\varepsilon} = \boldsymbol{\varepsilon}^{\mathrm{DB}} + \boldsymbol{\varepsilon}^{\mathrm{AS}}. \tag{B2}$$

The vector $\boldsymbol{\varepsilon}^{\mathrm{DB}}$ contains the components of the strain vector $\boldsymbol{\varepsilon}$ that are *directly* calculated from displacements, while vector $\boldsymbol{\varepsilon}^{\mathrm{AS}}$ contains the components which are obtained using some assumed strain interpolation. Let the complete vector of strains obtained from the displacements be $\bar{\boldsymbol{\varepsilon}}$, that is:

$$\bar{\boldsymbol{\varepsilon}} = \left[\varepsilon_{11}(\boldsymbol{U}), \varepsilon_{22}(\boldsymbol{U}), \varepsilon_{33}(\boldsymbol{U}), 2\varepsilon_{12}(\boldsymbol{U}), 2\varepsilon_{23}(\boldsymbol{U}), 2\varepsilon_{13}(\boldsymbol{U})\right]^{\mathrm{T}} \tag{B3}$$

then

$$\bar{\boldsymbol{\varepsilon}} = \boldsymbol{\varepsilon}^{\mathrm{DB}} + \boldsymbol{\varepsilon}^{\mathrm{TI}}, \tag{B4}$$

where $\boldsymbol{\varepsilon}^{\mathrm{TI}}$ contains the displacement-based strain components to be tied to the assumed strain components $\boldsymbol{\varepsilon}^{\mathrm{AS}}$.

The Hellinger–Reissner variational indicator can be then written as [2] (neglecting the loading terms):

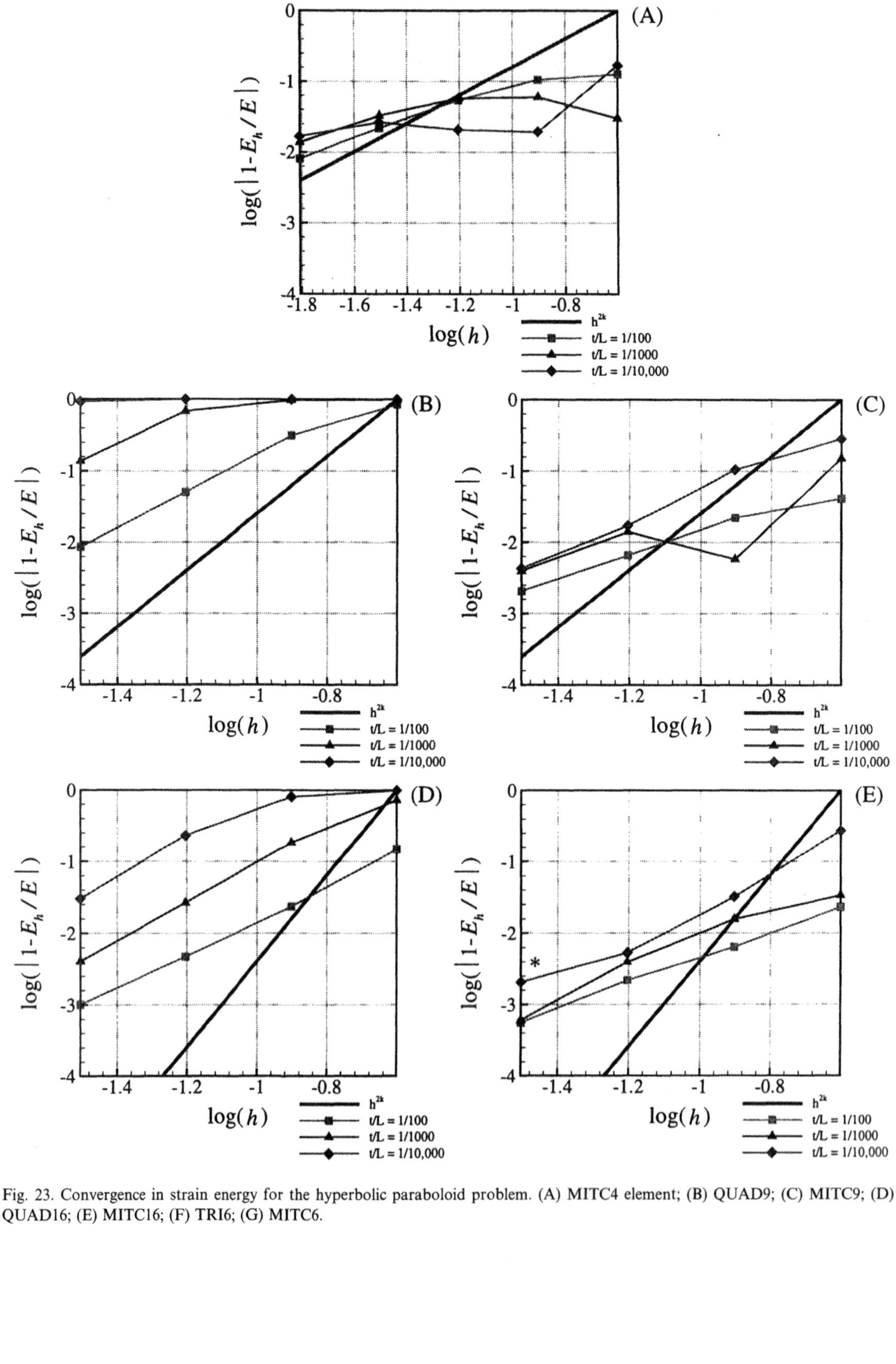

Fig. 23. Convergence in strain energy for the hyperbolic paraboloid problem. (A) MITC4 element; (B) QUAD9; (C) MITC9; (D) QUAD16; (E) MITC16; (F) TRI6; (G) MITC6.

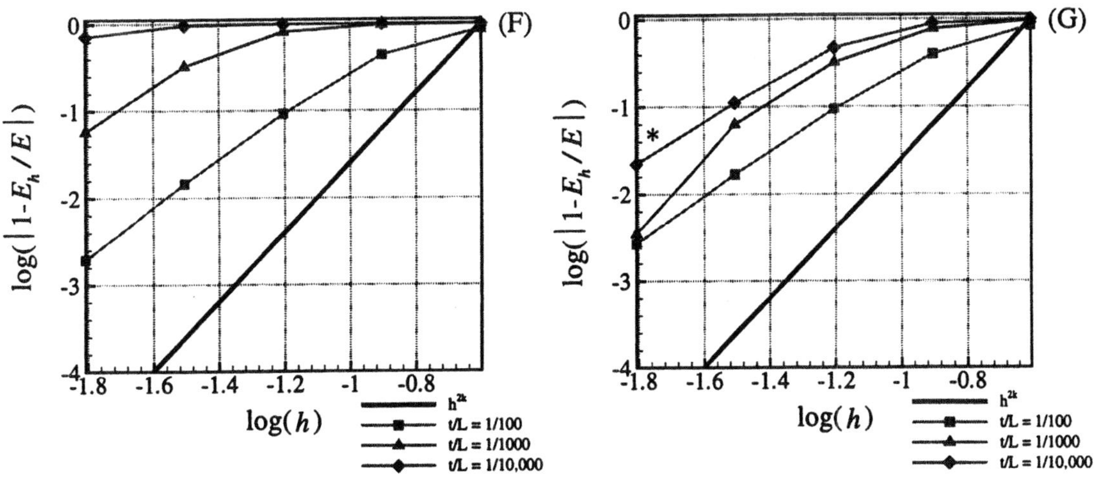

Fig. 23 (*continued*)

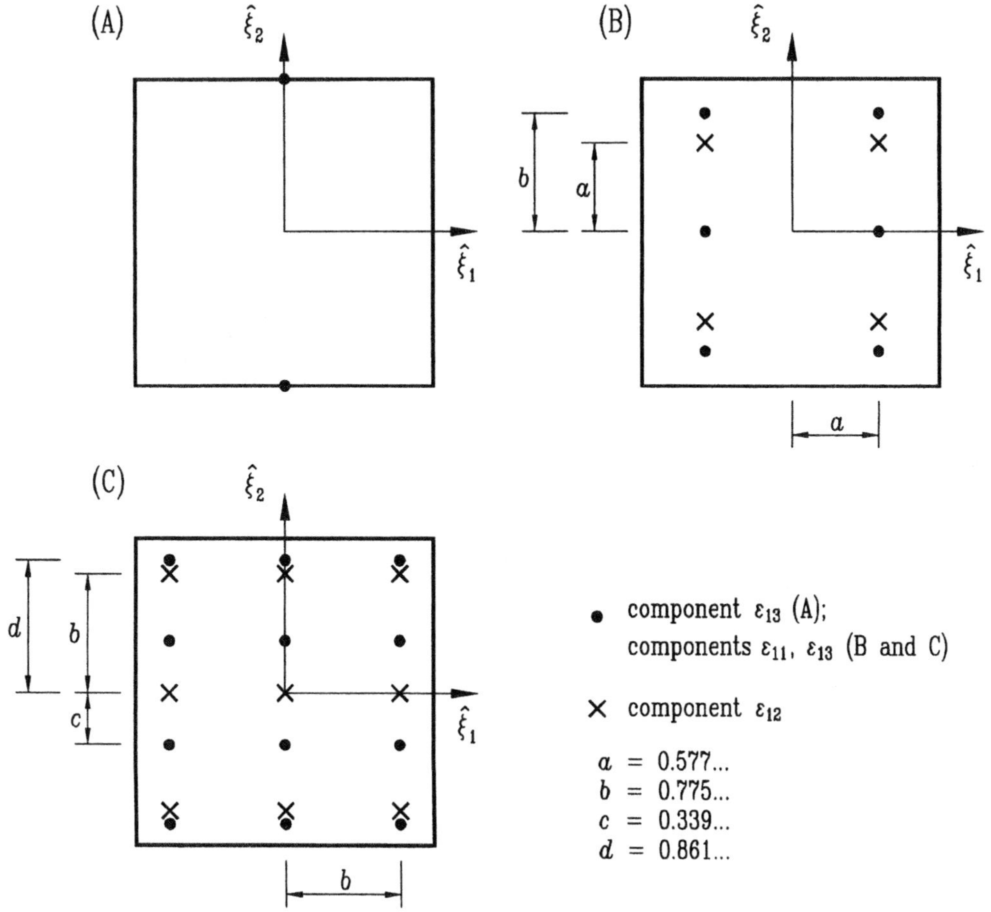

Fig. A1. Tying points for the MITC shell elements. (A) MITC4 element; (B) MITC9 element; (C) MITC16 element.

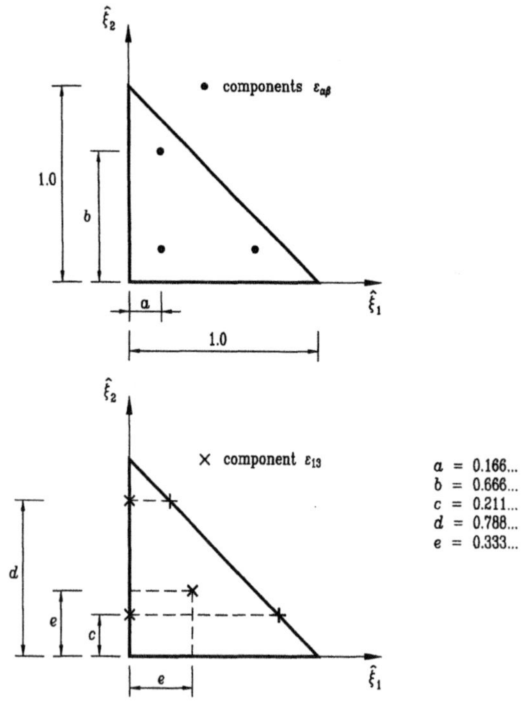

Fig. A2. Tying points for the six-node mixed-interpolated shell element MITC6.

$$\Pi_{HR}(\boldsymbol{U}, \boldsymbol{\varepsilon}) = -\frac{1}{2}\int_V \boldsymbol{\varepsilon}^T \boldsymbol{C}\boldsymbol{\varepsilon}\,dV + \int_V \boldsymbol{\varepsilon}^T \boldsymbol{C}\bar{\boldsymbol{\varepsilon}}\,dV$$

$$= \frac{1}{2}\int_V (\boldsymbol{\varepsilon}^{DB})^T \boldsymbol{C}\boldsymbol{\varepsilon}^{DB}\,dV$$

$$-\frac{1}{2}\int_V (\boldsymbol{\varepsilon}^{AS})^T \boldsymbol{C}\boldsymbol{\varepsilon}^{AS}\,dV$$

$$+\int_V (\boldsymbol{\varepsilon}^{DB})^T \boldsymbol{C}\boldsymbol{\varepsilon}^{TI}\,dV$$

$$+\int_V (\boldsymbol{\varepsilon}^{AS})^T \boldsymbol{C}\boldsymbol{\varepsilon}^{TI}\,dV, \tag{B5}$$

where $\boldsymbol{C}$ is the matrix of constitutive relations.

Invoking the stationarity conditions for Π_{HR} with respect to $\boldsymbol{U}$ and $\boldsymbol{\varepsilon}^{AS}$,

$$\delta\Pi_{HR}(\boldsymbol{U}, \boldsymbol{\varepsilon}^{AS}) = 0 \tag{B6}$$

we obtain:

$$\int_V (\delta\boldsymbol{\varepsilon}^{DB})^T \boldsymbol{C}\bar{\boldsymbol{\varepsilon}}\,dV + \int_V (\delta\boldsymbol{\varepsilon}^{TI})^T \boldsymbol{C}\boldsymbol{\varepsilon}\,dV = 0,$$

$$\int_V (\delta\boldsymbol{\varepsilon}^{AS})^T \boldsymbol{C}(\boldsymbol{\varepsilon}^{TI} - \boldsymbol{\varepsilon}^{AS})\,dV = 0. \tag{B7}$$

Note that when all the components of the strain tensor are tied, i.e., $\boldsymbol{\varepsilon}^{DB} = \delta\boldsymbol{\varepsilon}^{DB} = 0$, the formulation becomes:

$$\int_V (\delta\boldsymbol{\varepsilon}^{TI})^T \boldsymbol{C}\boldsymbol{\varepsilon}^{AS}\,dV = 0,$$

$$\int_V (\delta\boldsymbol{\varepsilon}^{AS})^T \boldsymbol{C}(\boldsymbol{\varepsilon}^{TI} - \boldsymbol{\varepsilon}^{AS})\,dV = 0. \tag{B8}$$

This mixed formulation pertaining to the underlying mathematical model (discussed in Ref. [6]) is the formulation in Eq. (15) for the Naghdi shell model.

The second line of Eq. (B7) defines the "tying" procedure by which $\boldsymbol{\varepsilon}^{AS}$ is obtained from $\boldsymbol{\varepsilon}^{TI}$. For the MITC elements we directly use $\boldsymbol{\varepsilon}^{AS} = \mathcal{I}(\boldsymbol{\varepsilon}^{TI})$, where $\mathcal{I}(\boldsymbol{\varepsilon}^{TI})$ denotes the assumed interpolation of $\boldsymbol{\varepsilon}^{TI}$ in the strain space. Hence, the integral form of Eq. (B8) is equivalent to the MITC formulation if

$$\int_V (\delta\boldsymbol{\varepsilon}^{AS})^T \boldsymbol{C}\boldsymbol{\varepsilon}^{TI}\,dV = \int_V (\delta\boldsymbol{\varepsilon}^{AS})^T \boldsymbol{C}\mathcal{I}(\boldsymbol{\varepsilon}^{TI})\,dV. \tag{B9}$$

This relation holds true if the integral on the left-hand side can be exactly evaluated by using the values of $\delta\boldsymbol{\varepsilon}^{AS}$ and $\boldsymbol{\varepsilon}^{TI}$ at the tying points only. Considering plane elements, this clearly holds for rectangular MITC9 and MITC16 elements for which the tying points correspond to Gauss integration points that can be used to integrate the quantity $(\delta\boldsymbol{\varepsilon}^{AS})^T\boldsymbol{C}\boldsymbol{\varepsilon}^{TI}$ exactly. Eq. (B9) can be shown to hold for the MITC4 and MITC6 elements as well by computing both sides analytically (using symbolic calculus software, for example).

References

[1] Chapelle D, Bathe KJ. Fundamental considerations for the finite element analysis of shell structures. Computers and Structures 1998;66:19–36.

[2] Bathe KJ. Finite element procedures. Englewood Cliffs, NJ: Prentice-Hall, 1996.

[3] Dvorkin EN, Bathe KJ. A continuum mechanics based four-node shell element for general nonlinear analysis. Eng Comput 1984;1:77–88.

[4] Bucalem ML, Bathe KJ. Higher-order MITC general shell elements. Int J Num Meth Eng 1993;36:3729–54.

[5] Naghdi PM. Foundations of elastic shell theory. In: Progress in solid mechanics, vol. 4. Amsterdam: North-Holland, 1963. p. 1–90.

[6] Chapelle D, Bathe KJ. On general shell finite elements and mathematical shell models. In: Topping BHV, editor. Advances in finite element procedures and techniques. Edinburgh, Scotland: Civil-Comp Press, 1998. p. 25–30.

[7] Bathe KJ, Iosilevich A, Chapelle D. An inf-sup test for shell finite elements. Computers and Structures (in press).

[8] Pitkäranta J, Leino Y, Ovaskainen O, Piila J. Shell deformation states and the finite element method: a benchmark study of cylindrical shells. Comp Meth Appl Mech Eng 1995;128:81–121.

Unified topology design of static and vibrating structures using multiobjective optimization

Seungjae Min*, Shinji Nishiwaki, Noboru Kikuchi

Department of Mechanical Engineering and Applied Mechanics, The University of Michigan, Ann Arbor, MI 48105, USA

Received 10 August 1998; accepted 19 January 1999

Abstract

Since structural design is usually required to perform in more than one environment, the ability to consider multiple objectives has to be included within the framework of topology optimization. A unified topology design methodology is proposed to generate structures satisfying both static and vibration performance measures using the multiobjective optimization approach. The weighted sum of conflicting objectives resulting from the norm method is used to generate the optimal compromise solutions, and the decision function is set to select the preferred solution. The objective function is defined by the mean compliance and mean eigenvalue to design a flexible structure which meets both the static and vibration requirements. The optimality conditions of the bicriteria problem are derived based on the modified optimality criteria method. To substantiate this approach, illustrated examples are presented both for verification and application. © 2000 Elsevier Science Ltd. All rights reserved.

Keywords: Topology design; Homogenization design method; Multiobjective optimization

1. Introduction

In practical optimization problems, there exist several, usually conflicting, design criteria that must be considered in the optimal design of structures. For example, reducing the weight and increasing the stiffness of a structure are both typical design goals in a structural design, but these objectives conflict with each other. In order to accommodate many conflicting design goals, the sequential application of each single objective optimization can be considered. This method, however, cannot produce an optimized solution because only a single objective is considered in each optimization process. As a consequence, so-called multiobjective (multicriteria, vector, Pareto) optimization, which considers multiple objectives simultaneously, has recently been regarded as a methodology for solving optimization problems with several objective functions.

A number of techniques and applications of multiobjective optimization have been developed over the past few years. A comprehensive overview of the field of multiobjective optimization in mechanics was introduced by Stadler [1], and multiobjective design optimization was applied to engineering problems, including structural design problems, by Eschenauer et al. [2]. In a general survey, Koski [3] described the state-of-the-art of multiobjective optimization in structural design, listing more than 80 publications.

Multiobjective optimization techniques generally give a set of optimal compromise solutions, a so-called Pareto optimal set. The definition of Pareto optimality

* Corresponding author. Present address: School of Mechanical Engineering, Hanyang University, Korea.

E-mail address: seungjae@email.hanyang.ac.kr (S. Min).

Reprinted from *Computers and Structures* **75 (1)**, 93-116 (2000)

states that the vector is chosen as optimal if no criterion can be improved without worsening at least one other criterion. Since the solutions of the multiobjective problem are usually not determined uniquely, decision making schemes are proposed to find the best optimal solution from the set of compromise solutions. The choice of the best solution can be supported with the help of preference functions. Thus, multiobjective optimization methodology can be divided into two categories: non-preference methods and preference methods. In the non-preference methods, such as the weighting method, constraint method, goal programming, etc., no preference is needed and a set of Pareto optimal solutions is generated. On the other hand, in the preference methods, such as global criterion method, etc., a certain type of preference for the objectives is required and a single optimum solution is produced. Most of the solution schemes for both non-preference and preference methods transform the multiple objectives into a single objective by some sort of scalarization of the vector optimization problem.

There are several works related to non-preference methods. Three commonly used approaches (weighting, noninferior set estimation, and constraint methods) were compared by considering their computational efficiencies and their ability to produce an approximation of the Pareto optimal set [4] and a knowledge-based system was used to select multiobjective optimization algorithms [5]. Koski and Silvennoinen [6] suggested the norm method to obtain the nonconvex optimal set which cannot be achieved by the general weighting method, and proposed the partial weighting method for problems where the number of criteria is large. The weighting method was applied to laminated composite structures where the objective function combined a weight and a strain energy change by Watkins and Morris [7] and the relative importance of each part was reflected by weighting coefficients. Bendse et al. [8] proposed bound formulation so that some continuous and discrete parameterized Min–Max structural design problems were converted into a Min formulation by introducing an additional parameter, and it was shown by Olhoff [9] that bound formulation was a switch from a prescribed resource to a cost minimization formulation by setting the behavioral constraint limit value.

On the topic of preference techniques, several methods are discussed. In a cantilever beam design problem subject to a stochastic base excitation, it was observed by Rao [10] that the game theory approach was superior assuming the proper balance of the various objective functions. A graphic interpretation of the game theory approach was presented by Rao [11] for generating the best compromise solution. The global criterion method was applied to design lightweight, low-cost composite structures of improved dynamic performance including damping [12]. As for recent developments, Grandhi et al. [13] proposed a multiobjective compound scaling (MCS) algorithm for large-scale structures based on the generalized compound scaling method in which the objects are converted to constraints by introducing pseudotargets. The algorithm generates a partial Pareto optimal set in which the optima lie on the constraint surface or at the intersection of constraints, and a reliability- based decision criterion is used for selecting the best compromise design. A generic algorithmic approach was tried to solve the integrated control and structure problems with mixed (continuous and discrete) variables for the zero–one optimization problem [14]. Qualitative optimization based on qualitative and fuzzy formulation was introduced by Arakawa and Yamakawa [15] and an optimum combination of discrete design variables was found by utilizing qualitative sensitivity and an inference algorithm.

Although a variety of multiobjective optimization techniques have been applied in optimal structure design since the late 1970s, the design variables of the multiobjective structural optimization problem are limited to only size variables such as the cross-section area of the truss bar, the thickness of the plate, and other physical dimensions. Such a size optimization problem has the disadvantage that topology known a priori cannot be changed. A new topology optimization approach based on a homogenization method was introduced to structural shape and topology design by Bendsøe and Kikuchi [16]. This new approach can generate arbitrary topologies, a significant advantage over the traditional structural optimization methods. This idea has been successfully applied to single objective optimization. The optimal layout of a linear elastic structure was determined for the static problem in which the objective is to minimize the mean compliance. This method was also extended to the dynamic problem, especially the eigenvalue optimization problem. A solution strategy was proposed by Díaz and Kikuchi [17] to identify the shape and topology of a structure that maximizes a natural frequency. Later the mean eigenvalue defined by multiple eigenvalues was suggested to be maximized by Ma et al. [18]. A kind of multiobjective optimization based on the homogenization design method was proposed by Díaz and Bendse [19] to deal with a multiple loading problem. In this formulation, the objective function was defined by using the weighted average of mean compliances in each loading case, which are the same static measure.

Structures are usually required to be the stiffest from the static viewpoint and the strongest from the dynamic view point. It is, however, observed that the structure generated to meet the static measure of the minimized mean compliance does not comply with the dynamic requirements. Conversely, the structure based on the dynamic measure of the maximized mean eigenvalue does not satisfy the static requirements. This implies that the mean compliance for the static problem and the mean eigenvalue for the dynamic problem are conflicting objectives in the design of a structure.

Therefore, a unified structural design method based on the topology optimization scheme using a multiobjective optimization approach is proposed in order to overcome design limitations resulting from the conflicting objectives. Multiobjective modeling based on the norm method and the decision function is presented to meet both the static and dynamic requirements simultaneously, and the optimality conditions based on the convex approximation are derived. This methodology is applied to obtain the optimal shape and topology design of a structure using the homogenization design method. To verify the design method, two- and three-dimensional plate and three-dimensional (3D) structural design examples are illustrated, and finally, a practical vehicle frame design is presented. It is expected that this approach will provide the structural designer with a tool to create a structure and modify an existing structure satisfying not only static measure but also dynamic measure.

2. Homogenization design method

The main idea of structural topology design is that a solid structural domain is modeled as composite materials with possibly perforated microstructures. As a consequence, the homogenization method is utilized to analyze the composite structure. Thus, the homogenization design method refers to the topology design methodology associated with the homogenization method. By means of the homogenization design method, the structural topology optimization problem is envisioned as the optimal material distribution problem.

The homogenization design method entails finding the optimal material distribution within a prescribed admissible structural domain while the criteria and constraints are satisfied. As shown in Fig. 1, the design domain Ω, with respect to the given boundary conditions and loads, is represented as a porous medium containing infinitely many microscale holes. Microstructures are classified as the void that contains no material (hole size $=$ 1), the solid medium which contains isotropic material (hole size $=$ 0), and the generalized porous medium which contains orthotropic material ($0 <$ hole size < 1). The distribution of void, solid, and porous microstructures indicates the shape and topology of a structure. If the total amount of material of a structure is constant, the material can be moved from one part of a structure to another part during the optimization process. Hence, finding the optimal material distribution within a specified design domain is equivalent to obtaining the optimal structural design.

For structural topology design, the problem is formulated using the homogenized properties, including elasticity matrix, initial stress, body force, and so on, so that it is necessary to solve the unit cell problem to determine the homogenized properties for a given microstructure that can define an appropriate material

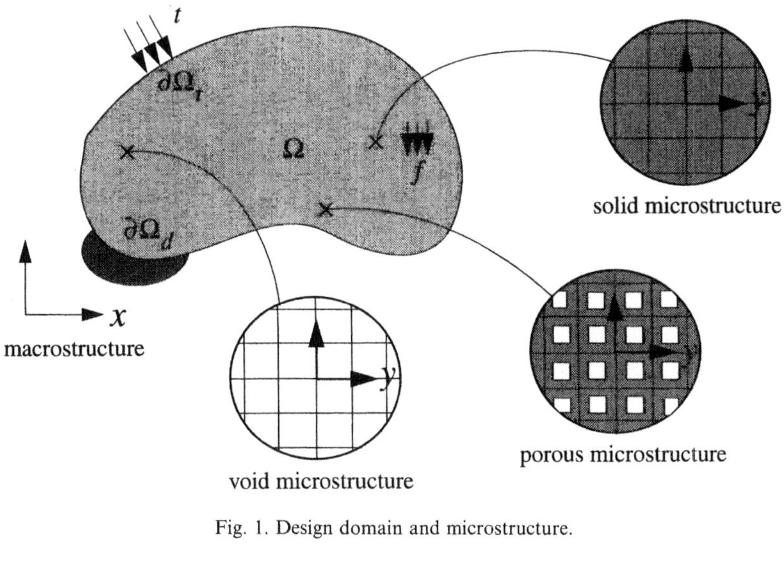

Fig. 1. Design domain and microstructure.

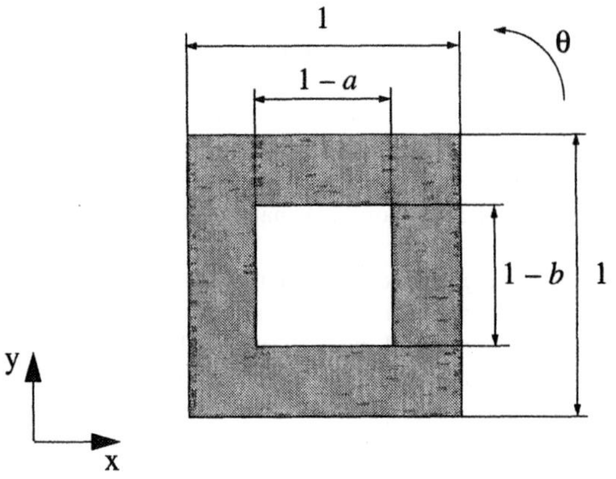

Fig. 2. Unit cell of 2D microstructure.

distribution. Introducing a rectangular hole (a body hole in the 3D case) in the unit cell of microstructures as design variables D, the homogenized elasticity matrix is computed by finite element calculations. The homogenized elasticity matrix must be obtained for a continuously varying size of a rectangular hole in the cell so that the hole size can be a continuous design variable for the optimal material distribution. However, it is impossible to determine the homogenized elastic constant for all the possibilities of the material density which corresponds to the size of a rectangular hole in the cell. Thus, the homogenized elastic matrix for certain sizes of hole is computed, and other values are then interpolated by an appropriate basis function in order to obtain a continuous variation. Thus, the homogenized elasticity matrix $\boldsymbol{E}^{\mathrm{H}}(\boldsymbol{D})$ is determined as a function of design variables—the hole sizes in the unit cell of microstructures.

It is also important to note that if the cell structure is rotated to a fixed reference frame, the transformed elastic tensor is very different from the one for the unrotated case. For example, it is observed that a cell with a large hole will have almost no shearing stiffness, while the same cell rotated 45° has a quite significant shearing stiffness. After defining the rotational matrix $\boldsymbol{R}(\Theta)$ based on the rotation angle Θ, the final elasticity matrix for analysis is computed by

$$\boldsymbol{E}^{\mathrm{G}} = \boldsymbol{R}(\Theta)^{\mathrm{T}}\boldsymbol{E}^{\mathrm{H}}(\boldsymbol{D})\boldsymbol{R}(\Theta) \qquad (1)$$

3. Classification of microstructures

Based on the definition of the problem, three different types of microstructures are introduced:

- 2D microstructure
- 3D plate/shell microstructure
- 3D solid microstructure

3.1. Unit cell of 2D microstructure

As shown in Fig. 2 a unit cell of a 2D microstructure includes a rectangular hole of width $1 - a$ and height $1 - b$. The orientation of the material axes represents the rotation of the microstructure. The design variables are the size of the hole $D = \{a,b\}$ and the orientation $\Theta = \{\theta\}$ of the microstructure. The mass density of a microstructure is defined as $\rho =$

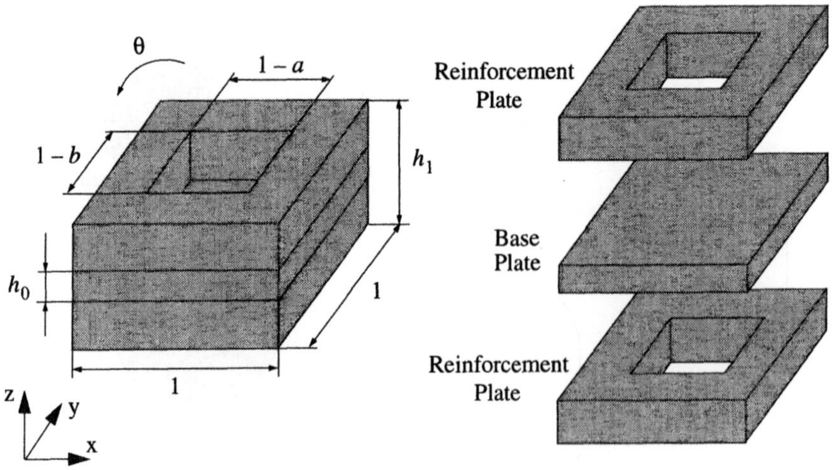

Fig. 3. Unit cell of 3D plate/shell microstructure.

$\rho_0(a+b-ab)$ where ρ_0 is the mass density of the isotropic material.

3.2. Unit cell of 3D plate/shell microstructure

Fig. 3 illustrates a unit cell of 3D plate and shell microstructures. The unit cell consists of an isotropic base plate of thickness and two orthotropic reinforcement plates of thickness $h_1 - h_0$ with rectangular holes. The size and rotation angle of the holes in the two reinforcement plates should be the same to avoid the coupling effect of bending and membrane deformation. The plate/shell model is developed using classical lamination theory which does not consider transverse shear deformations. Also, the structure of this element is approximated to be flat by means of projection onto the x–y plane, in order to avoid the curvature effect. Thus the displacement vector is defined as $U = \{u,v,w,\theta_x,\theta_y\}$ where u, v, and w are the displacements of the neutral plane in the x, y, and z directions, respectively, and θ_x and θ_y are the rotations about the x and y axes using the right-hand convention. Thicknesses h_0 and h_1 are specified and the design variables are the same as in the 2D problem under the plane stress assumption. If ρ_0 is the mass density of the isotropic material, the mass density of a microstructure is defined as $\rho = \rho_0[h_0 + (a+b-ab)(h_1 - h_0)]$.

3.3. Unit cell of 3D solid microstructure

A unit cell of a 3D solid microstructure contains a body hole of width $1-a$, depth $1-b$ and height $1-c$ as shown in Fig. 4. Euler angles are chosen to represent the 3D rotation. The size of the body hole $D = \{a,b,c\}$ and the orientation $\Theta = \{\theta,\psi,\varphi\}$ of the microstructure are used to define design variables. The mass density of a microstructure is defined as $\rho = \rho_0[1-(1-a)(1-b)(1-c)]$ where ρ_0 is the mass density of the isotropic material.

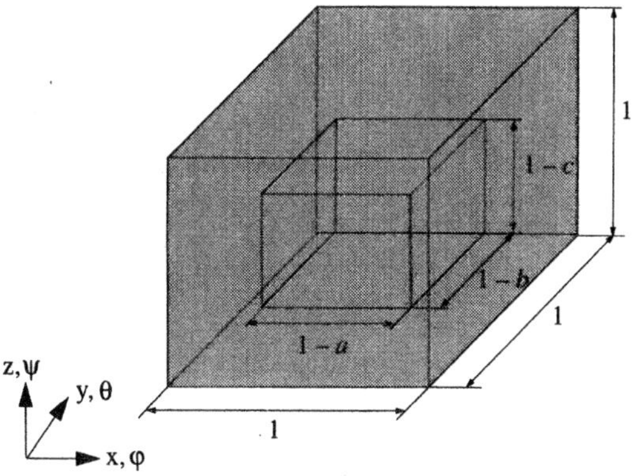

Fig. 4. Unit cell of 3D solid microstructure.

4. Linear static and vibration problem

The topology optimization of the static problem and the vibration problem is reformulated in the uniform context of the principle of virtual displacement to describe the multiobjective optimization problem. A variation of a functional leads to the sensitivity analysis of the objective in each problem which will be substituted into the optimization algorithm.

Consider a linearly elastic structure occupying a domain Ω, that is subjected to the applied body forces $\boldsymbol{f}$ and boundary tractions $\boldsymbol{t}$ as shown in Fig. 1. Let $\boldsymbol{u}$ be the displacement field that defines the equilibrium of a structure through the principle of virtual work. The weak form of the static problem is written as

$$a(\boldsymbol{u},\boldsymbol{v}) = l(\boldsymbol{v}) \quad \forall \boldsymbol{v} \in K \tag{2}$$

where K is the space of kinematically admissible solutions. The energy bilinear form $a(\boldsymbol{u},\boldsymbol{v})$ represents the internal virtual work by an arbitrary virtual displacement $\boldsymbol{v}$ at equilibrium defined by $\boldsymbol{u}$:

$$a(\boldsymbol{u},\boldsymbol{v}) = \int_\Omega \varepsilon \boldsymbol{u}^{\mathrm{T}} \cdot \boldsymbol{E}^{\mathrm{G}} \varepsilon(\boldsymbol{v})\, \mathrm{d}\Omega \tag{3}$$

and the load linear form is as follows:

$$l(\boldsymbol{v}) = \int_\Omega \boldsymbol{f} \cdot \boldsymbol{v}\, \mathrm{d}\Omega + \int_{\partial\Omega} \boldsymbol{t} \cdot \boldsymbol{v}\, \mathrm{d}\partial\Omega \tag{4}$$

In the static problem, the structural topology optimization problem is formulated by minimizing the mean compliance defined by $l(\boldsymbol{u})$. Introducing the total potential energy

$$F_{\mathrm{s}}(\boldsymbol{v}) = \frac{1}{2} a(\boldsymbol{v},\boldsymbol{v}) - l(\boldsymbol{v}) \quad \forall \boldsymbol{v} \in K \tag{5}$$

the following relation is derived at equilibrium

$$F_{\mathrm{s}}(\boldsymbol{u}) = \frac{1}{2} a(\boldsymbol{u},\boldsymbol{u}) - l(\boldsymbol{u}) = -\frac{1}{2} l(\boldsymbol{u}) \tag{6}$$

This implies that minimizing compliance is equivalent to maximizing the total potential energy. Let D be the design variable and take a variation of F_{s} with respect to $\boldsymbol{u}$ and D at equilibrium

$$\delta F_{\mathrm{s}}(\boldsymbol{u},D) = a\left(\boldsymbol{u},\delta\boldsymbol{u} + \frac{\partial \boldsymbol{u}}{\partial D}\delta D\right) - l\left(\delta\boldsymbol{u} + \frac{\partial \boldsymbol{u}}{\partial D}\delta D\right) + \frac{1}{2}\int_\Omega \varepsilon(\boldsymbol{u})^{\mathrm{T}} \frac{\partial \boldsymbol{E}^{\mathrm{G}}}{\partial D} \varepsilon(\boldsymbol{u})\, \mathrm{d}\Omega\, \delta D \tag{7}$$

The first two terms can be cancelled by setting $v = \delta u + \frac{\partial u}{\partial D}\delta D$ and the sensitivity of F_s is defined by

$$\frac{\partial F_s}{\partial D} = \frac{1}{2}\int_\Omega \varepsilon(u)^T \frac{\partial E^G}{\partial D}\varepsilon(u)\, d\Omega \tag{8}$$

Using Eq. 6, the sensitivity of $l(u)$ with respect to D is obtained by

$$\frac{\partial l(u)}{\partial D} = \int_\Omega \varepsilon(u)^T \frac{\partial E^G}{\partial D}\varepsilon(u)\, d\Omega \tag{9}$$

If u is considered as the eigenvector, the weak form of the vibration problem—generalized eigenvalue problem—can be written as

$$a(u,v) = \lambda\, b(u,v) \quad \forall v \in K \tag{10}$$

where $b(u,v)$ represents the inertia effect,

$$b(u,v) = \int_\Omega \rho u v\, d\Omega \tag{11}$$

and the eigenvector $u \in K$ is normalized using

$$b(u,u) = 1 \tag{12}$$

In the vibration problem, the structural topology optimization problem is formulated by maximizing the mean eigenvalue defined by λ. Introducing the following function at equilibrium:

$$F_d(u) = \frac{1}{2}a(u,u) - \lambda\frac{1}{2}b(u,u) \tag{13}$$

the variation of F_d with respect to u and D is derived as

$$\begin{aligned}\delta F_d(u,D) = &\int_\Omega \varepsilon(u)^T E^G \varepsilon\left(\delta u + \frac{\partial u}{\partial D}\delta D\right) d\Omega \\ &- \lambda\int_\Omega \rho u^T\left(\delta u + \frac{\partial u}{\partial D}\delta D\right) d\Omega \\ &- \frac{1}{2}\left(\int_\Omega \varepsilon(u)^T \frac{\partial E^G}{\partial D}\varepsilon(v)\, d\Omega \right.\\ &\left. - \frac{\partial \lambda}{\partial D}\int_\Omega \rho u^T u\, d\Omega - \lambda\int_\Omega \frac{\partial \rho}{\partial D} u^T u\, d\Omega\right)\delta D\end{aligned} \tag{14}$$

The first two terms can be cancelled by setting $v = \delta u + \frac{\partial u}{\partial D}\delta D$ and the sensitivity of λ is defined by

$$\begin{aligned}\frac{\partial \lambda}{\partial D} &= \frac{\int_\Omega \varepsilon(u)^T \frac{\partial E^G}{\partial D}\varepsilon(v)\, d\Omega - \lambda\int_\Omega \frac{\partial \rho}{\partial D}u^T u\, d\Omega}{\int_\Omega \rho u^T u\, d\Omega} \\ &= \int_\Omega \varepsilon(u)^T \frac{\partial E^G}{\partial D}\varepsilon(v)\, d\Omega - \lambda\int_\Omega \frac{\partial \rho}{\partial D}u^T u\, d\Omega\end{aligned} \tag{15}$$

using Eq. 12.

5. Optimization problem

Multiobjective optimization is concerned with a decision making process for satisfying many conflicting objectives, i.e., performance measures. The method usually generates a wide range of solutions, called Pareto optimal solutions, and the decision maker chooses the preferred solution, called the best compromise solution, after the evaluation of solution sets. Thus, the multiobjective optimization scheme includes two basic problems: (1) generation of the optimal solutions and

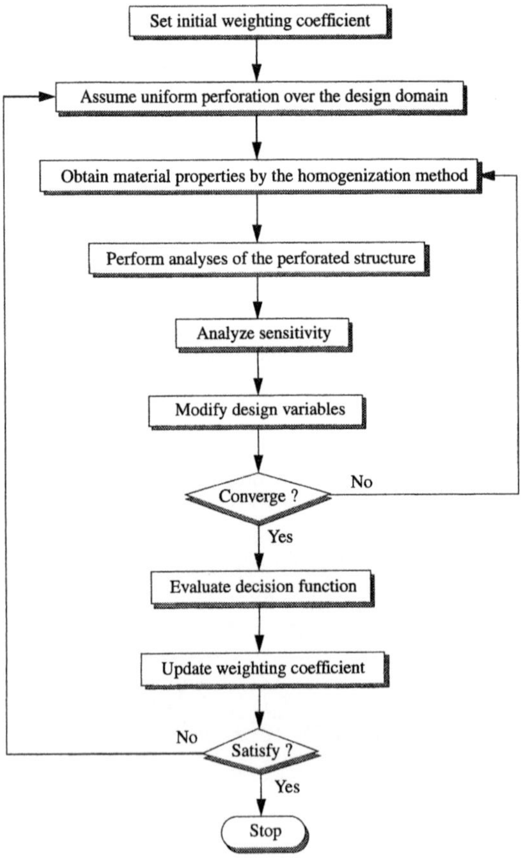

Fig. 5. Multiobjective optimization procedure.

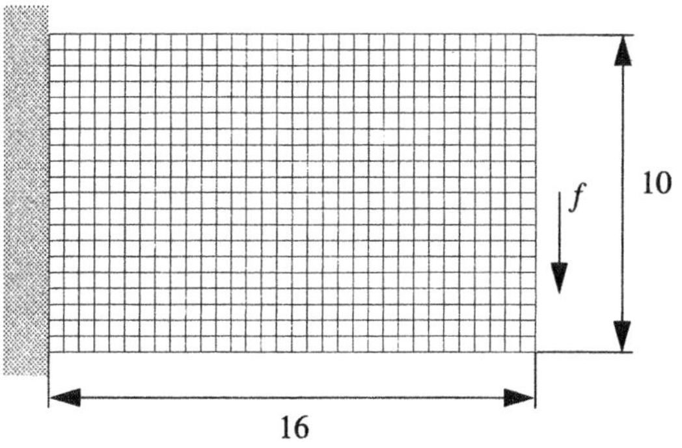

Fig. 6. Discretized design domain of 2D problem.

(2) selection of a preferred solution. The Pareto optimal set is created by the defined objective function and the best compromise solution is determined based on the decision criterion.

5.1. Objective function

The multiple objectives need to be transformed to a single objective by either suitable weighting or by retaining one selected criterion to be optimized, and

treating the remaining objectives as constraints. The choice of constraint limits may be a difficult task because almost all design quantities have a criterion nature rather than a constraint nature. The norm method [6] is frequently used in structural optimization. In this method the distance function is defined as

$$d_p = \left[\sum_i w_i (f_i - z_i)^p \right]^{1/p} \tag{16}$$

where f_i is each objective function, z_i is the chosen reference value, and w_i as well as p are parameters. In most structural design applications z_i and p are fixed, and w_i is the only parameter. Here, the traditional weighting method, obtained from the norm methods by choosing $z_i = 0$ and $p = 1$, is considered. The use of weights to represent relative preferences between the criteria is direct and intuitive, and all of the solutions that are found in the method are Pareto optimal. Such a scalarization has an advantage, in that the formulation is simple and easy to use, and priorities between the competing criteria become explicit. The disadvantage is that this method cannot find Pareto optimal points that lie upon a non-convex boundary of the attainable set. The multiobjective optimization problem

$\Omega_s = 0.2 \times \Omega$　　　　$\Omega_s = 0.4 \times \Omega$　　　　$\Omega_s = 0.6 \times \Omega$

(a) static problem

$\Omega_s = 0.2 \times \Omega$　　　　$\Omega_s = 0.4 \times \Omega$　　　　$\Omega_s = 0.6 \times \Omega$

(b) vibration problem

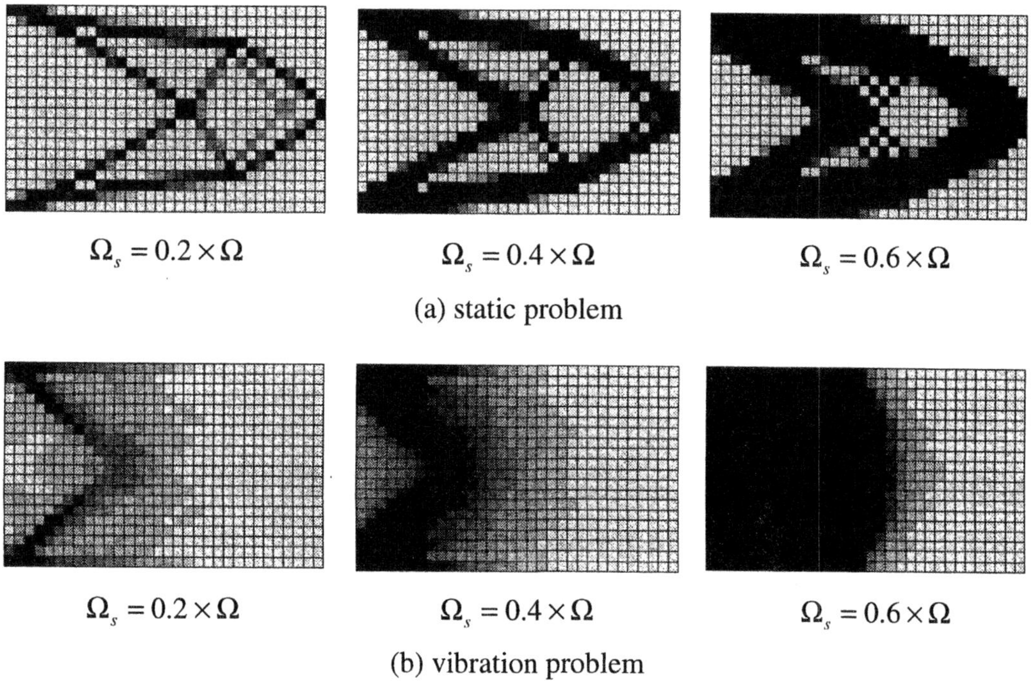

Fig. 7. Optimal 2D structure for single objective optimization.

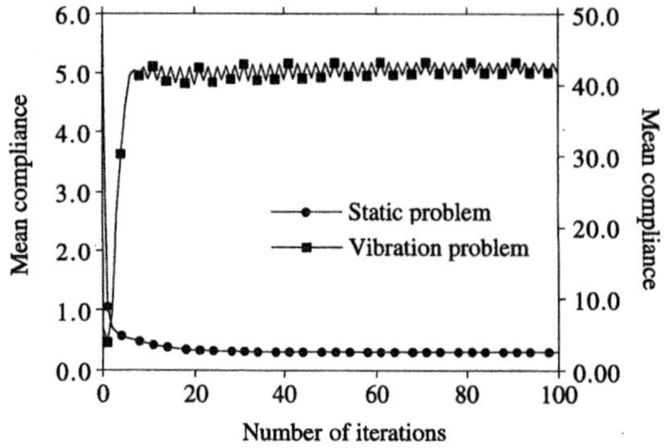

(a) mean compliance

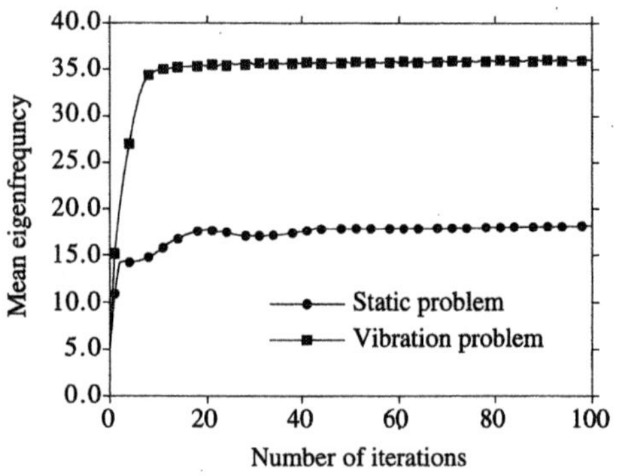

(b) mean eigenfrequency

Fig. 8. Comparison of convergence history ($\Omega_s = 0.4 \times \Omega$).

is transformed into a scalar optimization problem by the formulation

$$f(D,\Theta) = \sum_i w_i f_i(D,\Theta) q_i \tag{17}$$

where $f_i(D,\Theta)$ is the original objective function, D and Θ are the size and orientation design variables, respectively, w_i is the weighting coefficient representing the relative importance of the objectives ($w_i \geq 0$ and $\sum w_i = 1$), and q_i is the constant multiplier. The original single objective functions are minimizing the mean compliance Φ in the static problem and maximizing the mean eigenvalue Λ in the vibration problem, respectively. These are the global performance measures and are defined as

$$\Phi = \int_\Omega \boldsymbol{f} \cdot \boldsymbol{u} \, \mathrm{d}\Omega + \int_{\partial\Omega_t} \boldsymbol{t} \cdot \boldsymbol{u} \, \mathrm{d}\partial\Omega \tag{18}$$

and

$$\Lambda = \lambda_0 + \sum_{i=1}^{m} s_i \left(\sum_{i=1}^{m} \frac{s_i}{\lambda_i - \lambda_{0_i}} \right)^{-1} \tag{19}$$

where λ_i is the chosen eigenvalue, s_i is a given factor, and λ_0 and λ_{0_i} are specified shift parameters. Here, m is the number of eigenvalues associated with the mean eigenvalue. The weighting coefficient w_i does not reflect the relative importance of objective functions proportionally, but these are the factors that control the location in design space. When the multiple loading conditions are considered in the static problem, the present formulation can be extended as

$$\Phi = \underset{i=1,\dots,n}{\mathrm{Max}} \left(\int_\Omega \boldsymbol{f}^i \cdot \boldsymbol{u}^i \, \mathrm{d}\Omega + \int_{\partial\Omega_i} \boldsymbol{t}^i \cdot \boldsymbol{u}^i \, \mathrm{d}\partial\Omega \right) \tag{20}$$

where n represents the number of loading cases. This implies the minimization of the maximum value of the mean compliances of all the loading cases [19].

The objective functions must be normalized because different units of measure of the objectives may hinder interpretation and understanding of the solutions. Thus, the multiplier q_i is included to make each objec-

Table 1
Optimal compromise solutions of the 2D problem

ID number	Weighting coefficient (w)	Mean compliance	Mean eigenfrequency	Decision function
0	0.0	41.7090	35.92700	18574.0
1	0.1	0.54303	26.55000	0.68812
2	0.2	0.37972	25.17800	0.15194
3	0.3	0.35466	24.56800	0.12798
4	0.4	0.34858	23.77400	0.13614
5	0.5	0.33938	23.64400	0.13059
6	0.6	0.32583	22.71200	0.14055
7	0.7	0.32279	21.82700	0.15793
8	0.8	0.30774	21.21600	0.16783
9	0.9	0.30506	19.42600	0.210907
10	1.0	0.30381	18.20200	0.24341

tive function have the dimensionless value of the same order. The mean compliance and the inverse of the mean eigenvalue are combined in a multiobjective function that is to be minimized. The total objective function thus has the form

$$f(D,\Theta) = w(\Phi/\Phi_0) + (1 - w)(\Lambda_0/\Lambda) \quad (21)$$

where Φ_0 and Λ_0 represent the mean compliance and the mean eigenvalue of the uniformly perforated initial structure, respectively. It is noted that the individual objective functions become normalized and have numerical values of the same order so that neither will dominate the solution.

5.2. Constraints

The structural analysis equations can be the equality constraints of the optimization problem. Here, the

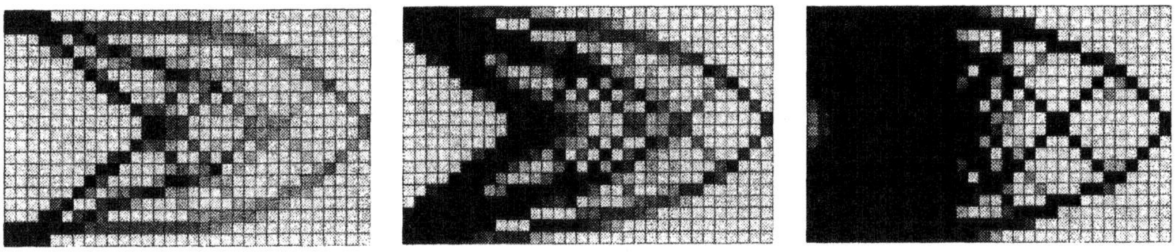

(a) w=0.1

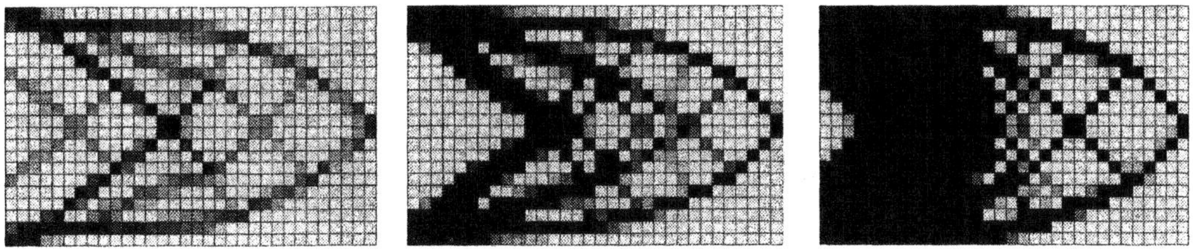

(b) w=0.2

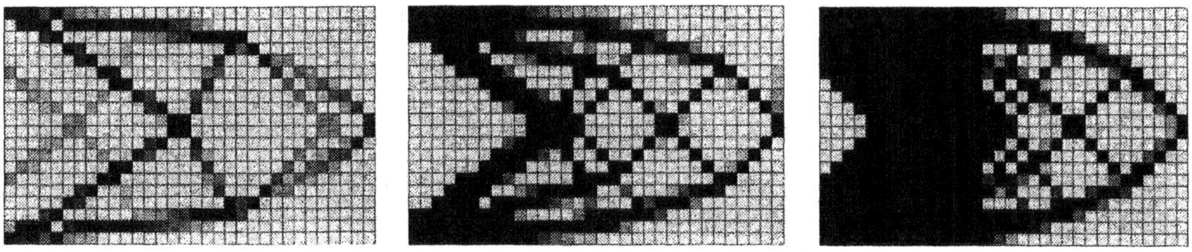

(c) w=0.3

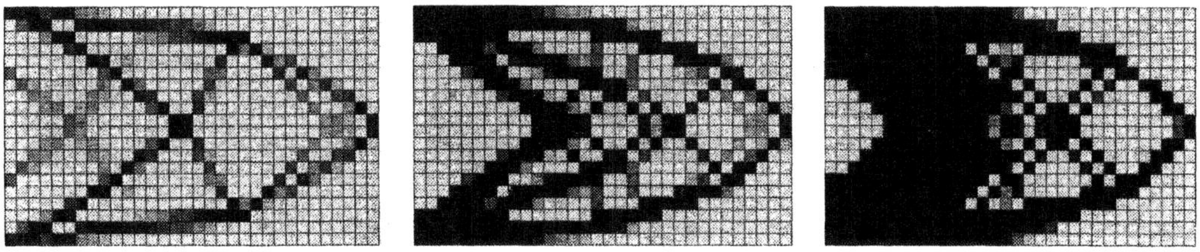

(d) w=0.4

Fig. 9. Optimal solutions of 2D problem.

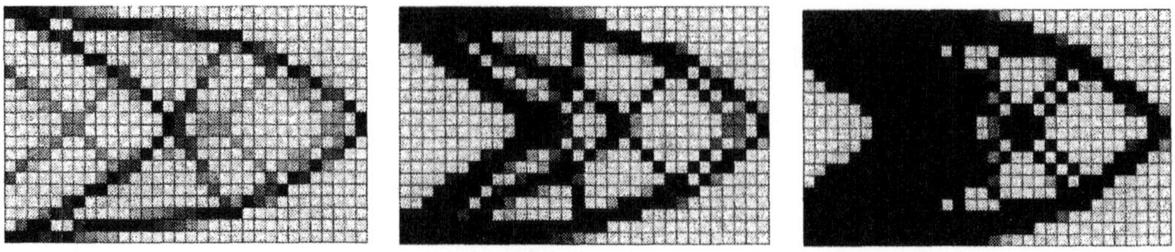

(e) w=0.5

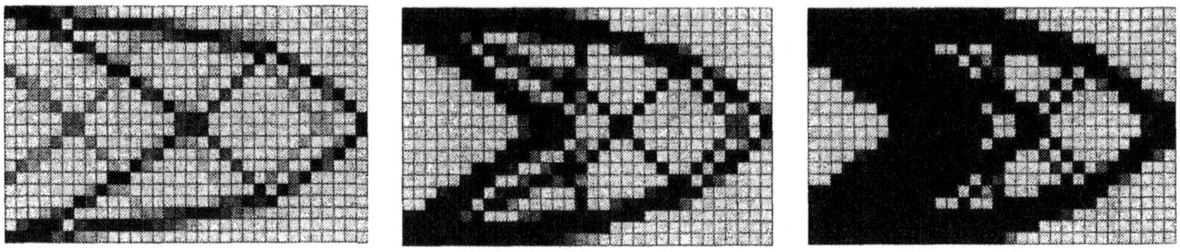

(f) w=0.6

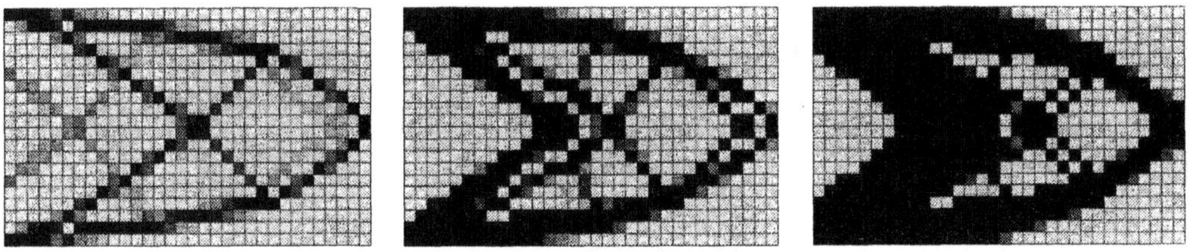

(g) w=0.7

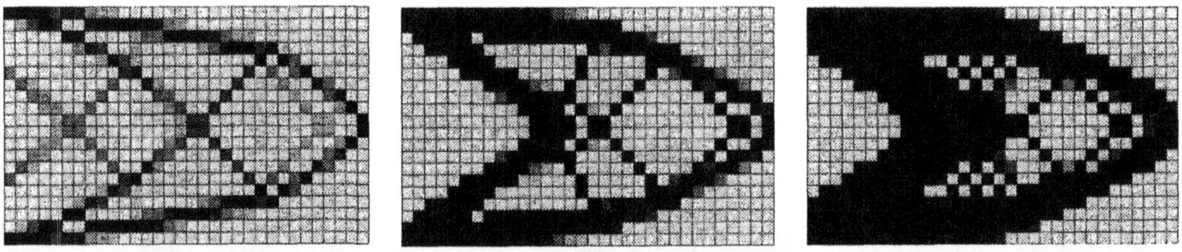

(h) w=0.8

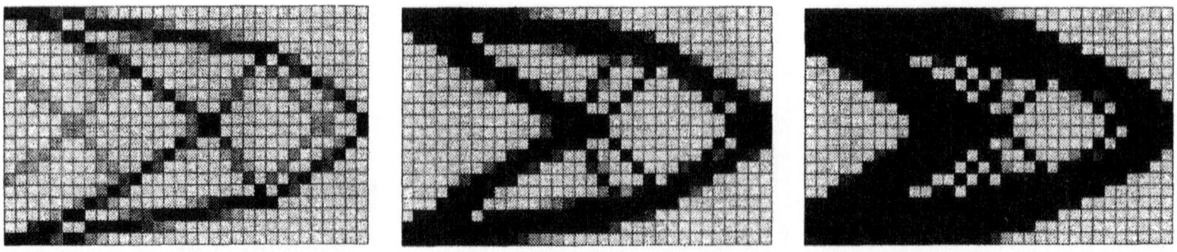

(i) w=0.9

Fig. 9 (*continued*)

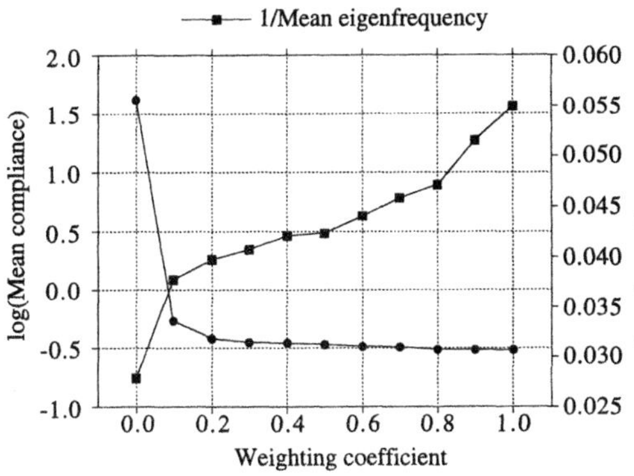

Fig. 10. Conflicting objectives of the 2D problem.

finite element discretization is applied to solve the equilibrium equations. The four-node quadrilateral isoparametric elements are used for the two dimensional and plate/shell problems and eight-node hexagonal elements are used for the three dimensional problem. The homogenized linear elasticity tensor $\boldsymbol{E}^{\mathrm{H}}_{ijkl}$ and the mass density ρ^{H} become functions of size design variables $d_i\,(i = 1,\ldots,N_d)$ and orientation design variables $\theta_i\,(i = 1,\ldots,N_\theta)$, where N_d represents the number of size design variables ($N_d = 2N$ for the 2D and plate/shell problems, $N_d = 3N$ for the 3D problem) and N_θ the number of orientation design variables ($N_\theta = N$ for the 2D and plate/shell problems, $N_\theta = 3N$ for the 3D problem) with finite elements in the discretized design domain. For the static problem the equilibrium equation can be transformed by the finite element method as

$$\boldsymbol{KU} = \boldsymbol{F} \tag{22}$$

where $\boldsymbol{U}$ is the deformation vector, $\boldsymbol{F}$ the load vector, and $\boldsymbol{K}$ the global stiffness matrix. For the free vibration problem the equilibrium equation can be written as

$$(\boldsymbol{K} - \lambda_n \boldsymbol{M})\phi_n = \boldsymbol{0} \tag{23}$$

where λ_n is the nth eigenvalue of the structure, ϕ_n the corresponding eigenvector, and $\boldsymbol{M}$ the global mass matrix. $\boldsymbol{K}$ and $\boldsymbol{M}$ are assembled from the element stiffness and mass matrices, respectively obtained by,

$$K_i = \int_{\Omega_i} B_i^{\mathrm{T}} E_i^{\mathrm{G}} B_i \,\mathrm{d}\Omega, \qquad \boldsymbol{M}_i = \int_{\Omega_i} \rho_i^{\mathrm{H}} N_i^{\mathrm{T}} N_i \mathrm{d}\Omega \tag{24}$$

where B_i and N_i stand for the strain shape function and the shape function, respectively, and E_i^{G} stands for the rotated homogenized stiffness for the ith element.

The inequality constraints considered in the homogenization design method are the total amount of material of the structure and the bounds of design

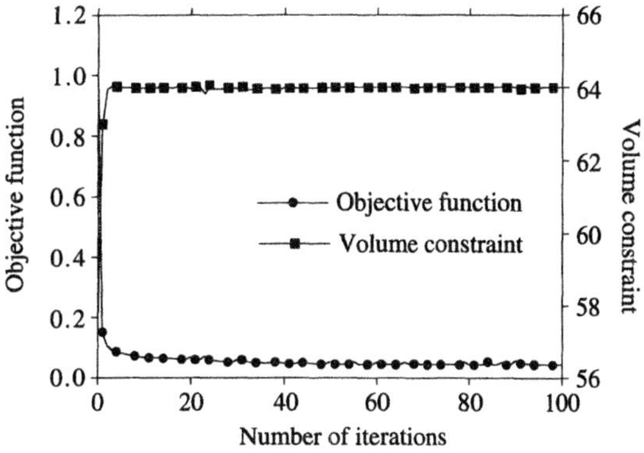

Fig. 12. Convergence history in case of $w = 0.3$ ($\Omega_s = 0.4 \times \Omega$).

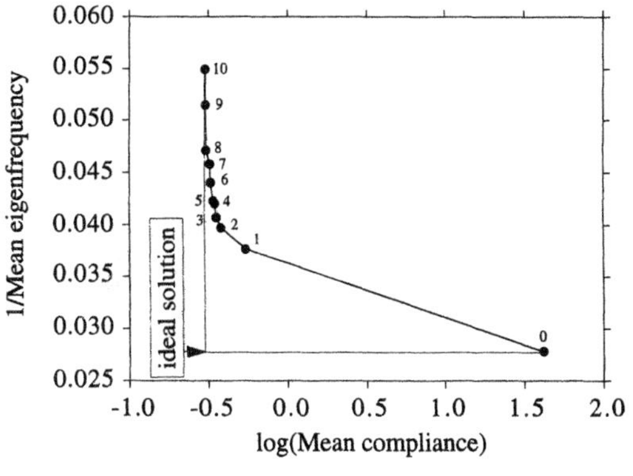

Fig. 11. Pareto optimal solutions set in criteria space.

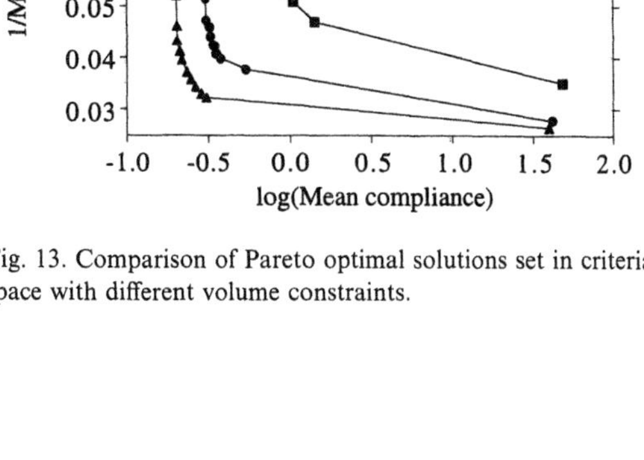

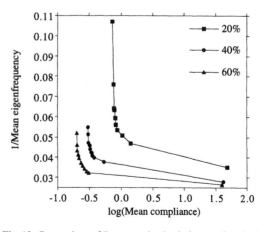

Fig. 13. Comparison of Pareto optimal solutions set in criteria space with different volume constraints.

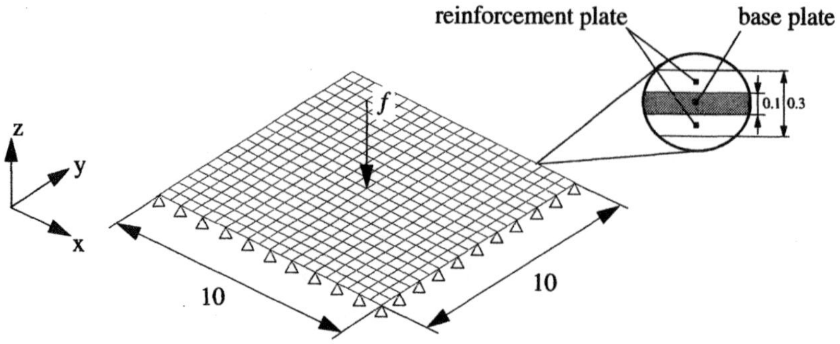

Fig. 14. Discretized design domain of the plate problem.

variables. These can be written as

$$\int_{\Omega} \rho \, d\Omega = \sum_{i=1}^{N} \rho_i \Omega_i \leq \Omega_s \tag{25}$$

$$0 < d_i^{l} \leq d_i \leq d_i^{u} \leq 1 \quad \text{for } i = 1, \ldots, N_d \tag{26}$$

and

$$-\frac{\pi}{2} \leq \theta_i \leq \frac{\pi}{2} \quad \text{for } i = 1, \ldots, N_\theta \tag{27}$$

where ρ_i and Ω_i are the mass density of a microstructure and the volume of ith element, respectively, and Ω_s is the specified amount of material available. Here d_i^{l} and d_i^{u} stand for the lower and upper bounds of the size design variables d_i, respectively.

5.3. Decision function

In practical design problems, it may be difficult for the decision maker to determine appropriate weighting coefficient w_i a priori. In many applications, the weightings do not correspond to familiar quantities. An a priori selection might have been quite different had the designer known what the trade-offs would mean in terms of the final design. Thus, it is preferable to search for the right values of the weighting coefficients interactively during the design process. The best optimal solution must be chosen from the Pareto optimal solutions set. This choice requires that an additional criterion in the form of preference functions be formulated. The weighting coefficient w_i corresponding to the best compromise solution are found by minimizing the sum of relative deviations of the individual objective functions from the feasible ideal solutions. Thus, the decision function model is constructed as

$$DF = \sum_i \left(\frac{f_i^* - f_i}{f_i^*} \right)^p \tag{28}$$

where p is generally taken as 2 and f_i^* is the feasible ideal solutions of the ith objective.

5.4. Optimization method

The multiobjective optimization problem considering

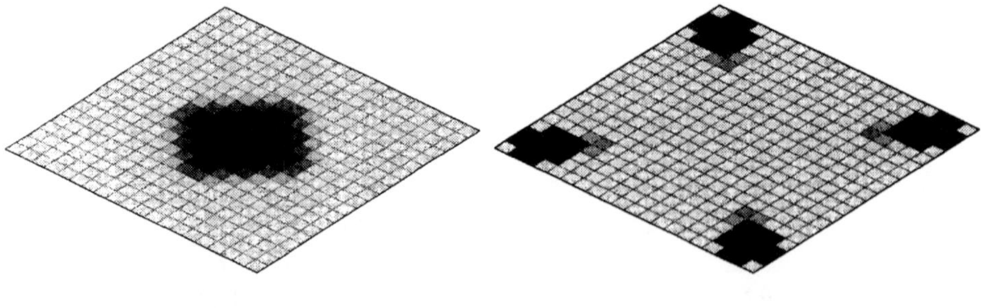

Fig. 15. Optimal reinforcement layouts for single objective optimization.

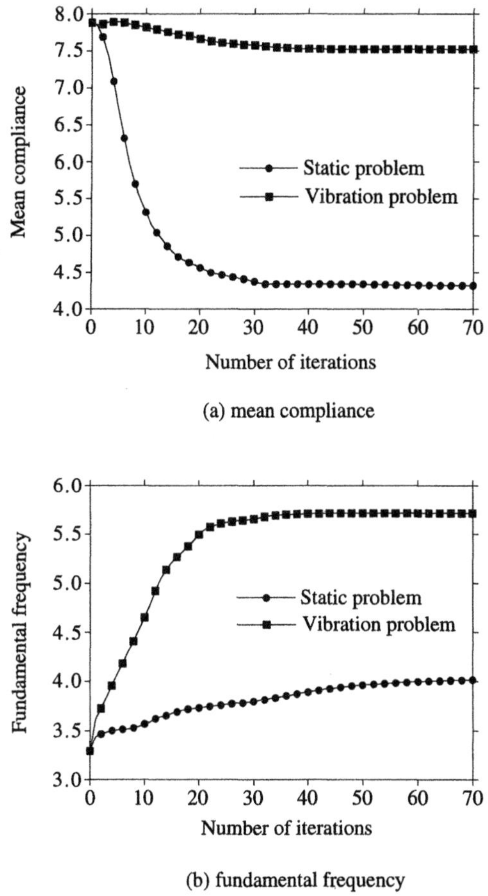

Fig. 16. Comparison of convergence history (plate problem).

both mean compliance and mean eigenvalue may be stated in mathematical form as follows: Determine D and Θ for a given amount of material Ω_s such that

$$\underset{D,\Theta}{\text{minimize}}\ f(\Phi,\Lambda)$$

$$\text{subject to } g = \sum_{i=1}^{N} \rho_i \Omega_i - \Omega_s \leq 0$$
$$0 < d_i^1 \leq d_i \leq d_i^u \leq 1 \tag{29}$$

and subject to equilibrium equations with the boundary conditions for both the static and the vibration problem. In order to derive the optimality condition, the Lagrangian function associated with the constrained optimization problem can be defined in the discretized form as

$$L = f + \Gamma\left(\sum_{i=1}^{N}\rho_i\Omega_i - \Omega_s\right) + \sum_{i=1}^{N_d}\Big[\alpha_i^1(d_i^1 - d_i) + \alpha_i^u(d_i - d_i^u)\Big] \tag{30}$$

where λ, α_i^1 and α_i^u stand for Lagrangian multipliers. Based on the KKT conditions, the necessary conditions of optimality are stationarities of L with respect to design variable d_i and θ_i

$$\frac{\partial L}{\partial d_i} = \frac{\partial f}{\partial d_i} + \Gamma\Omega_i\frac{\partial \rho_i}{\partial d_i} - \alpha_i^1 + \alpha_i^1 + \alpha_i^u = 0 \quad \text{for} \quad i = 1,\ldots,N_d \tag{31}$$

and

$$\frac{\partial L}{\partial \theta_i} = \frac{\partial f}{\partial \theta_i} = 0 \quad \text{for } i = 1,\ldots,N_\theta \tag{32}$$

and the complimentary conditions, which are

$$\Gamma\left(\sum_{i=1}^{N}\rho_i\Omega_i - \Omega_s\right) = 0, \tag{33}$$

Table 2
Compliance value of the optimal layouts in the bridge design problem

ID number	Weighting coefficient (w)	Mean compliance	Fundamental eigenfrequency	Decision function
0	0.0	7.55740	5.66370	0.51426
1	0.1	7.35510	5.63510	0.45048
2	0.2	5.91390	5.47010	0.11930
3	0.3	5.57900	5.28850	0.07600
4	0.4	4.82640	4.95820	0.02485
5	0.5	4.74270	4.85000	0.02666
6	0.6	4.62570	4.75730	0.02821
7	0.7	4.53300	4.49440	0.04352
8	0.8	4.44670	4.34390	0.05441
9	0.9	4.41580	4.31500	0.05672
10	1.0	4.40120	4.29620	0.05830

$$\alpha_i^{l}(d_i^{l} - d_i) = 0 \quad \text{for } i = 1,\ldots,N_d \tag{34}$$

$$\alpha_i^{u}(d_i - d_i^{u}) = 0 \quad \text{for } i = 1,\ldots,N_d \tag{35}$$

with $\lambda \geq 0$, $\alpha_i^{l} \geq 0$ and $\alpha_i^{u} \geq 0$.

It is noted that the material constraint has to be active to make the problem feasible and multipliers α_i^{l} and α_i^{u} become zero if the size design variable is between the lower and upper bounds. Since the sensitivity of such a complex multiobjective function cannot guarantee the positiveness, the modified optimality criteria method [18] is utilized for optimization algorithm. Thus, the updating scheme of size design variables $d_i\ (i = 1,\ldots,N_d)$ for the kth iteration takes the form

$$d_i^{k+1} = d_i^{l} \quad \text{if } i \in I_1^k = \left\{ i | \left(\tilde{e}_i^k\right)^{\eta} d_i^k \leq d_i^{l} \right\}$$

$$d_i^{k+1} = \left(\tilde{e}_i^k\right) d_i^k \quad \text{if } i \in I^k = \left\{ i | d_i^{l} < \left(\tilde{e}_i^k\right)^{\eta} d_i^k < d_i^{u} \right\}$$

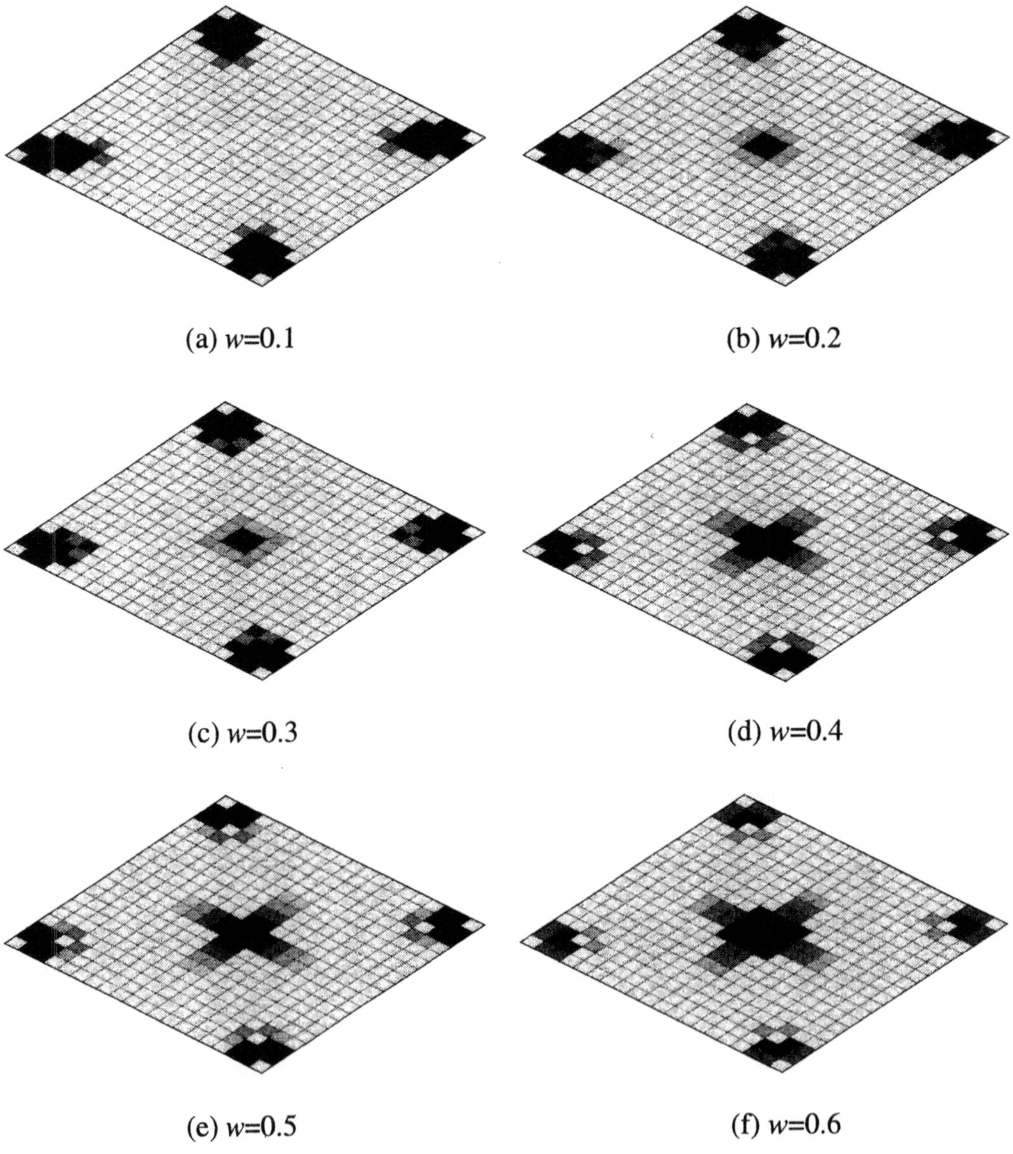

Fig. 17. Optimal reinforcement layouts of the plate problem.

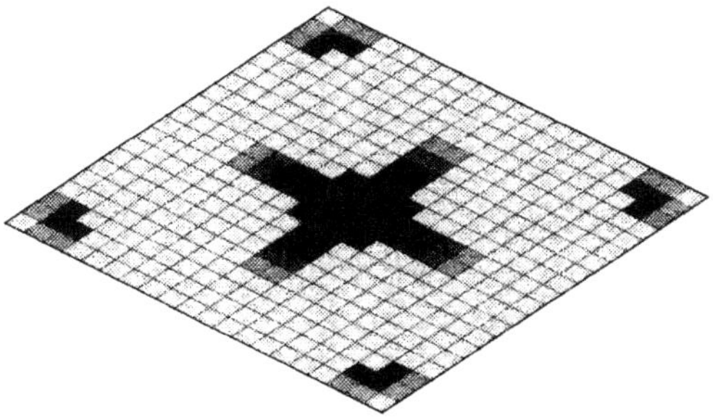

(g) w=0.7

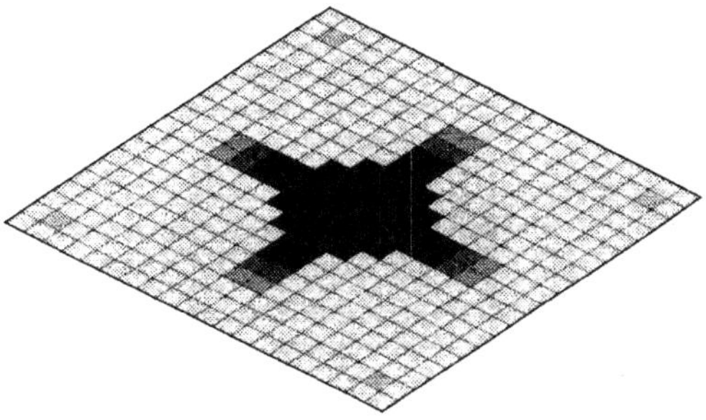

(h) w=0.8

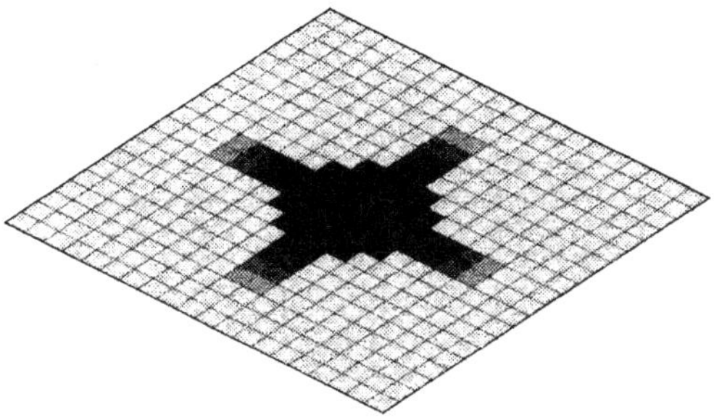

(i) w=0.9

Fig. 17 (*continued*)

$$d_i^{k+1} = d_i^{\mathrm{u}} \quad \text{if } i \in I_u^k = \left\{ i | d_i^{\mathrm{u}} \le \left(\tilde{e}_i^k\right)^\eta d_i^k \right\} \tag{36}$$

where η is the given parameter. The scale factor $\tilde{e}_i^k$, which implies the modified effectiveness of the ith design variable in the kth iteration, is written as

$$\tilde{e}_i^k = \frac{1}{\tilde{\Gamma}^k} \left(\mu^k - \left[\frac{\partial f}{\partial d_i} / \frac{\partial g}{\partial d_i} \right]_{d_i = d_i^k, \theta_i = \theta_i^k} \right) \tag{37}$$

where the shift parameter μ^k in the kth iteration is determined by

$$\mu^k \ge \max_{1 \le i \le N_d} \left\{ \left[\frac{\partial f}{\partial d_i} / \frac{\partial g}{\partial d_i} \right]_{d_i = d_i^k, \theta_i = \theta_i^k} \right\} \tag{38}$$

and the modified Lagrange multiplier $\tilde{\Gamma}^k$ is defined by $\tilde{\Gamma}^k = \Gamma^k + \lambda^k$.

To obtain the updated size design variables, it is necessary to obtain sensitivity information of the objective function and the constraint. According to the sensitivity analysis of a continuum structure described in the problem formulation, the sensitivity of the objective function with respect to size design variable d_i based on the finite element discretization can be derived as

$$\frac{\partial f}{\partial d_i} = -\frac{w}{\Phi_0} \left(\varepsilon_i^{\mathrm{T}} \frac{\partial K_i}{\partial d_i} \varepsilon_i \right) + \frac{(1-w)\Lambda_0 (\Lambda - \lambda_0)^2}{\Lambda^2 \sum_{j=1}^{m} s_j} \sum_{j=1}^{m} \frac{s_j}{(\lambda_j - \lambda_{0_j})^2} \frac{\partial \lambda_j}{\partial d_i} \tag{39}$$

where

$$\frac{\partial \lambda_j}{\partial d_i} = \phi_j^{\mathrm{T}} \left(\frac{\partial K_i}{\partial d_i} - \lambda_j \frac{\partial M_i}{\partial d_i} \right) \phi_j \tag{40}$$

for $i = 1, \ldots, N_d$.

The stationarity of L with respect to the orientation design variable θ_i in Eq. 32 can be interpreted as aligning the angle to the principal stress direction if the material has low shear stiffness. The displacement u^k in the stiffness problem and the eigenvector ϕ^k in the eigenvalue problem are used to calculate stress σ^k in the kth iteration. Thus, the updating scheme for design variables θ_i^k is obtained by computing the principal stress and the angle of principal stress direction.

5.5. Optimization procedure

The overall and iterative procedure of multiobjective optimization is shown in Fig. 5. Firstly, the initial weighting coefficient is set to be an arbitrary value between 0 and 1. Secondly, the optimal structure corresponding to the weighting coefficient is generated through the following steps. In the beginning, a uniformly perforated structure is considered, and the material properties of the structure are obtained by the homogenization method. The static and the dynamic analyses are performed by using these homogenized material properties. The design variables—size and rotation—are modified based on the sensitivity analysis. This process of modifying design variables continues until the solution converges. Thirdly, the decision function is evaluated by the mean compliance and the mean eigenvalue of the final structure, and the weighting coefficient is updated. The entire procedure of the structural design is repeated with the updated weighting coefficient. Finally, the weighting coefficient is obtained for minimizing the decision function.

6. Numerical examples

2D plane, 3D plate, and 3D solid problems are solved to demonstrate the feasibility of the optimal topology design method using the multiobjective optimization approach, and a vehicle frame reinforcement design is considered as a practical example. In the two-dimensional and plate examples, plane stress problems are considered. The properties of the isotropic material are Youngs modulus $E = 100$ N/mm^2, Poissons ratio $\nu = 0.3$, and density $\rho_0 = 7.85 \times 10^{-6}$ kg/mm^3 throughout the problems.

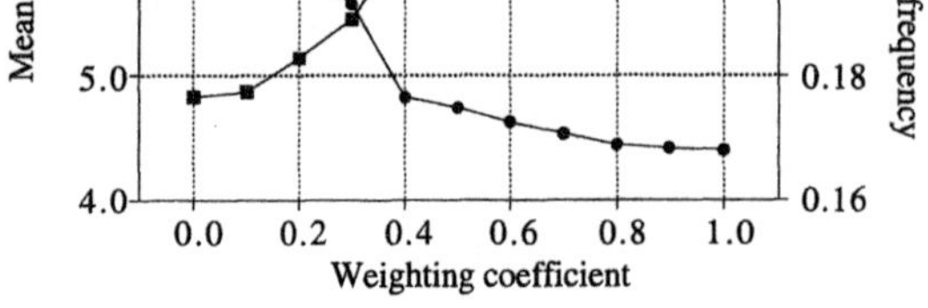

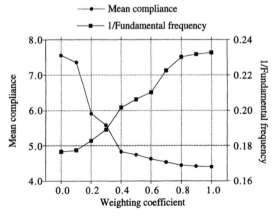

Fig. 18. Conflicting objectives of the plate problem.

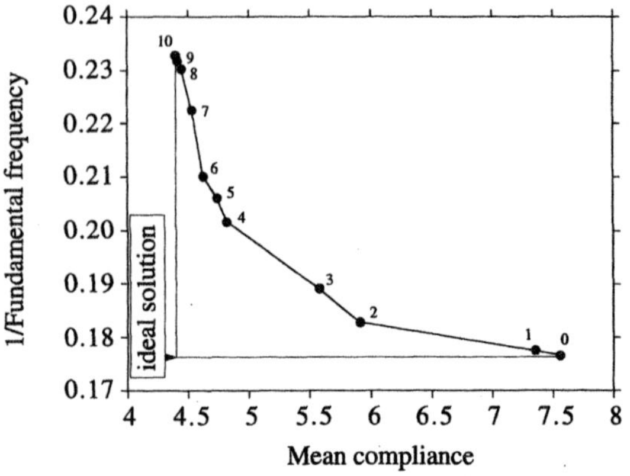

Fig. 19. Pareto optimal solutions set in criteria space (plate problem).

6.1. Example 1: 2D plane problem

The design domain is specified as a rectangle with a horizontal length of 16 by a vertical height of 10 and discretized to 1440 (48 × 30) four-node finite elements with 1519 nodes as shown in Fig. 6. As for boundary conditions, the left side is fixed and the vertical force is applied in the middle of the right side of the design domain. The total amount of material available is 40% of the design domain. Initially, the design variables are imposed to $a_i = b_i = 0.2$ which implies 36% of the total material and $\theta_i = 0.0$ for $i = 1,2,\ldots,N$. The design goal is to obtain the optimal topology of a structure which maximizes the stiffness and the mean eigenvalue, defined by the lowest three eigenvalues of the structure, simultaneously. Fig. 7 shows the optimal structures of the stiffness optimization for the static problem and the mean eigenvalue optimization for the vibration problem when different volume constraints are applied. The stiffest structure against the applied force is generated from the static problem, and the

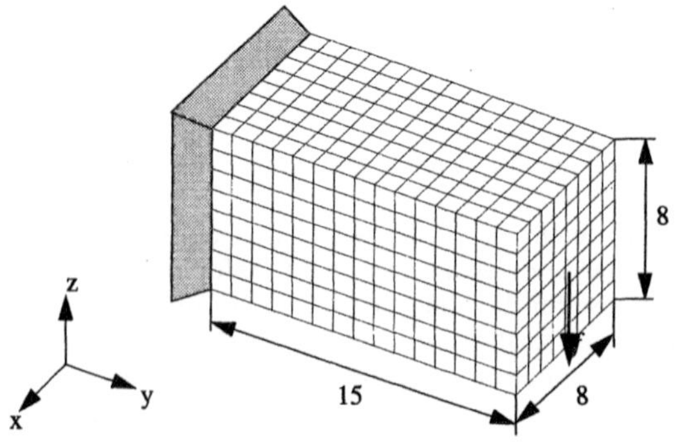

Fig. 20. Discretized design domain of 3D problem.

strongest structure resistant to the free vibration is generated from the vibration problem. Fig. 8 illustrates the convergence of the conflicting objectives in each case when the volume constraint is 40% of the design domain.

Applying 40% volume constraint, the results obtained by multiobjective optimization with decision function are summarized in Table 1. The best compromise design is found in the case of $w = 0.3$ based on the decision function. The Pareto optimal structures are shown in Fig. 9 with different volume constraints, 20, 40 and 60%, respectively. It is noted that the shape and topology of the structure can be controlled by the weighting coefficient which reflects the importance of the objectives. Fig. 10 shows two conflicting criteria in multiobjective optimization problem. In Fig. 11 the Pareto optimal solutions set in criteria space is presented by connecting the members of the optimal set. Fig. 12 illustrates the convergence history of the objective function and the constraint in the case of the best compromise solution ($w = 0.3$). This shows that the non-dimensional value of the objective function converges to the minimum during the optimization process while the constraint is satisfied. The comparison of Pareto optimal solutions set with different volume constraints in Fig. 13 shows that the minimal curve moves to the lower left in criteria space as the amount of material is increased. This implies that the distance between the ideal solution and the Pareto optimal solutions becomes shorter, while the more material is distributed in the design domain.

6.2. 3D plate problem

The reinforcement design of a square plate simply supported at the edges, shown in Fig. 14, is considered for the 3D plate example. A finite element model of the design domain is generated by using 400 (20 × 20) four-node quadrilateral finite elements, and a point transverse force is loaded at the center of the plate. The elements are characterized by the base plate thickness $h_0 = 0.1$, and the total thickness is restricted to

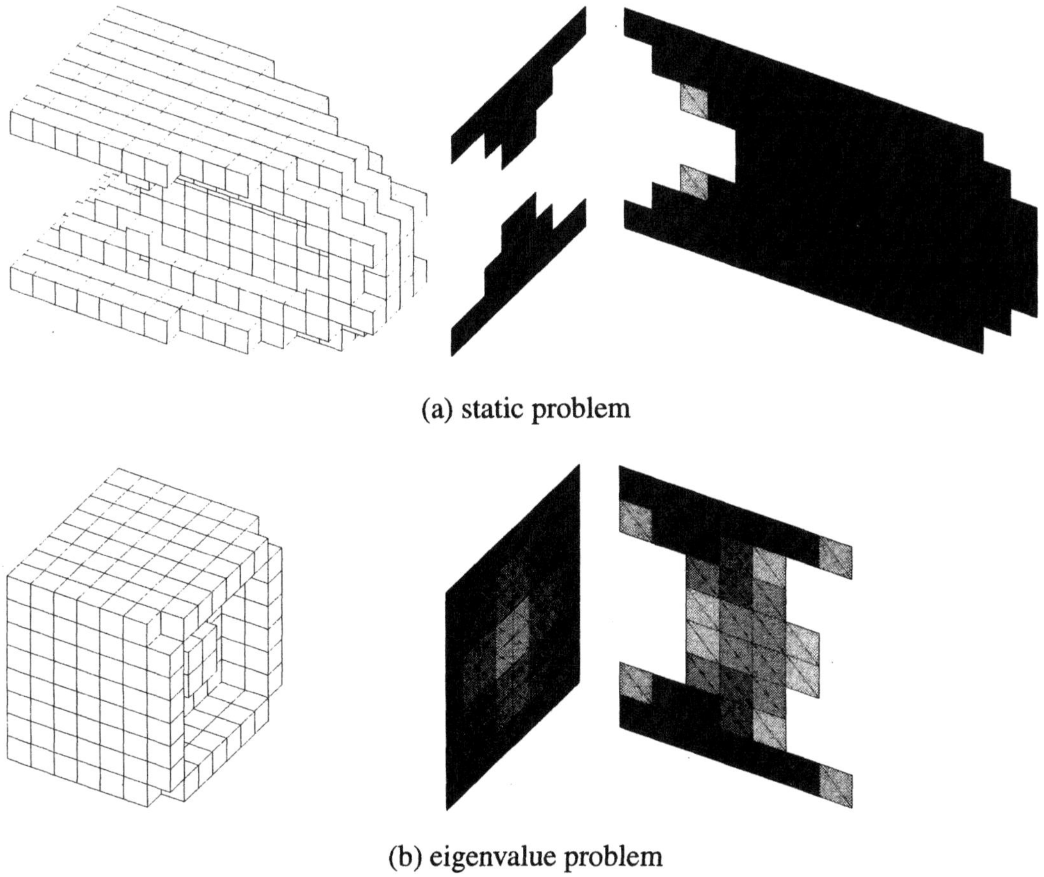

Fig. 21. Optimal 3D structures for single objective optimization.

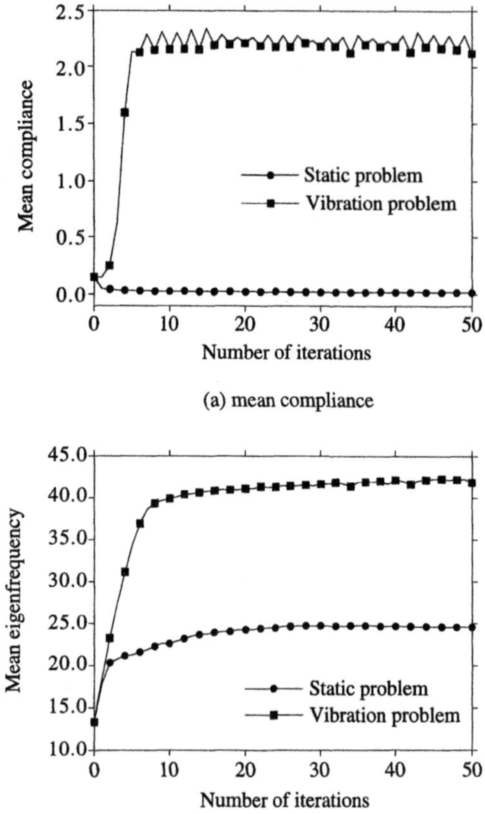

(a) mean compliance

(b) mean eigenfrequency

Fig. 22. Comparison of convergence history (3D problem).

$h_1 = 0.3$. In this example, initial design variables are set to 0.2 for each size and 0.0 for each rotation angle, and the volume constraint is specified as 40% of total material. The goal is to increase both the stiffness and the fundamental frequency of a structure by adding a prescribed amount of material as reinforcement. Structures obtained by the optimization of the individual objective function are shown in Fig. 15. The material is distributed to the boundary in order to resist the vibration for the case considering only the dynamic response, and the stiffest structure considering only the static response is generated by moving the material to the center. The corresponding convergence history of the individual objective function is illustrated in Fig. 16. However, in the static problem the mean compliance is minimized as the number of iterations increases, but the mean eigenvalue is not maximized. Similarly, the mean compliance in the vibration problem is increased, whereas the maximization of the mean eigenvalue is achieved.

The Pareto optimal set generated by multiobjective optimization is summarized in Table 2 and the best compromise solution based on the decision function is found in the case of $w = 0.4$. The optimal structures corresponding to each weighting coefficient are shown in Fig. 17. These layouts, which satisfy both objectives, provide the designer with an idea of the location for reinforcement. Fig. 18 illustrates that two objectives, stiffness and fundamental frequency, have conflicting characteristics since the stiffness of a structure becomes higher as the fundamental frequency becomes lower. The points representing optimal compromise solutions in criteria space are connected as shown in Fig. 19, and it turns out that the Pareto optimal solutions set in the objective space is the convex domain.

6.3. 3D solid problem

As shown in Fig. 20, multiobjective optimization is applied to a simple 3D design of a cantilever beam with one end fixed and a vertical loading at the center of the other end. The design domain is discretized into 960 ($15 \times 8 \times 8$) eight-node solid elements and 40% of the material is given as a constraint. Initially, every

Table 3
Optimal compromise solutions of the 3D problem

ID number	Weighting coefficient (w)	Mean compliance	Mean eigenfrequency	Decision function
0	0.0	7.52120	5.71490	0.56835
1	0.1	7.35480	5.63360	0.51155
2	0.2	6.26070	5.47140	0.21337
3	0.3	5.09400	5.26490	0.04150
4	0.4	4.95840	5.02030	0.03919
5	0.5	4.56360	4.60600	0.04177
6	0.6	4.35790	4.51290	0.04450
7	0.7	4.34970	4.11450	0.07863
8	0.8	4.32010	4.09380	0.08052
9	0.9	4.31060	4.05290	0.08460
10	1.0	4.28830	3.75480	0.11764

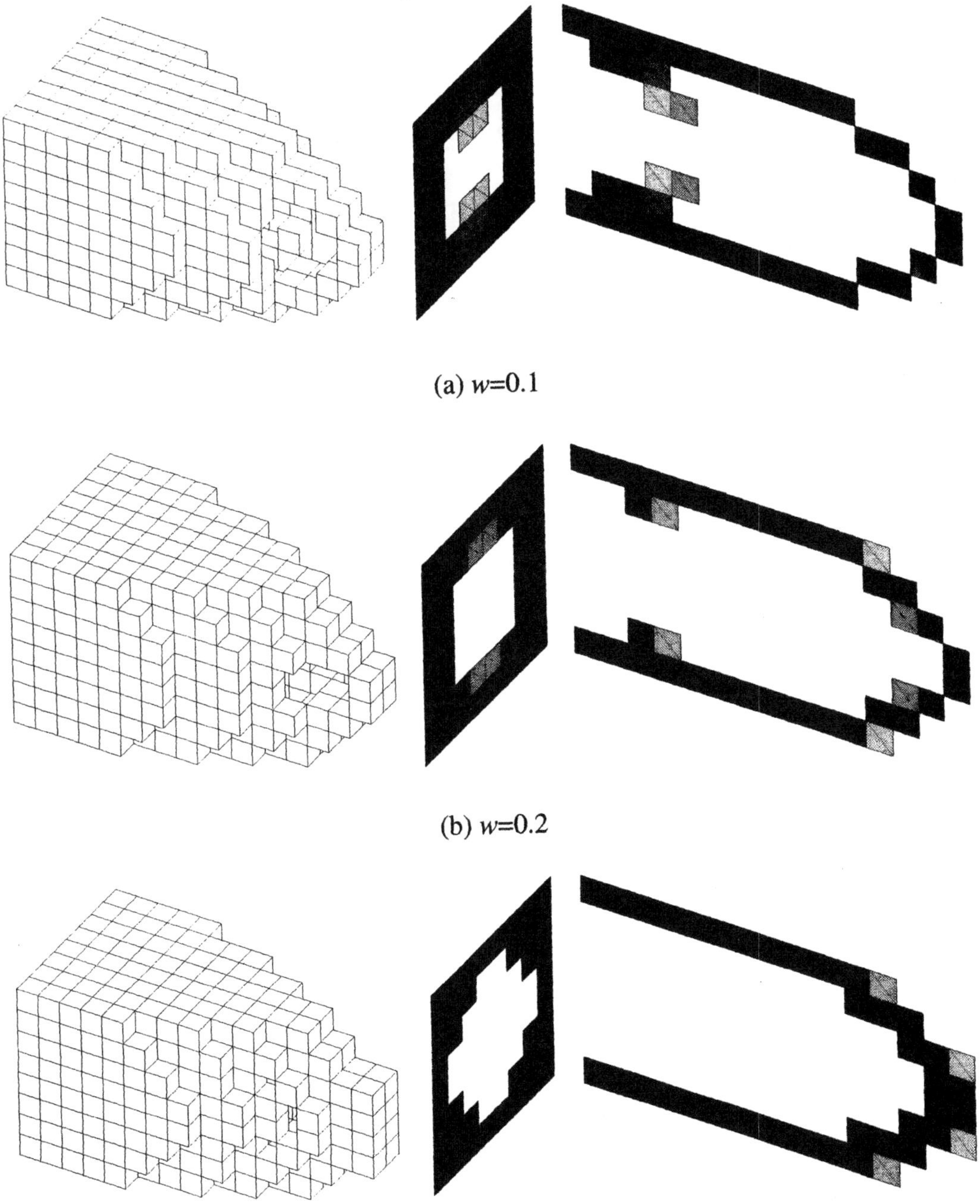

(a) *w*=0.1

(b) *w*=0.2

(c) *w*=0.3

Fig. 23. Optimal solutions of 3D problem.

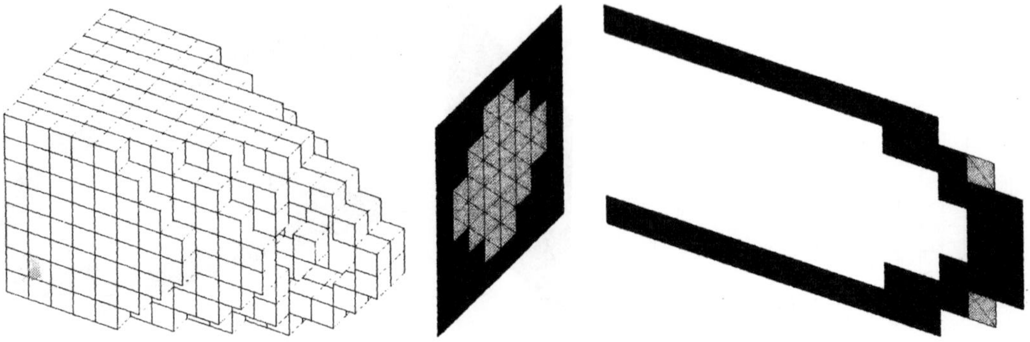

(d) w=0.4

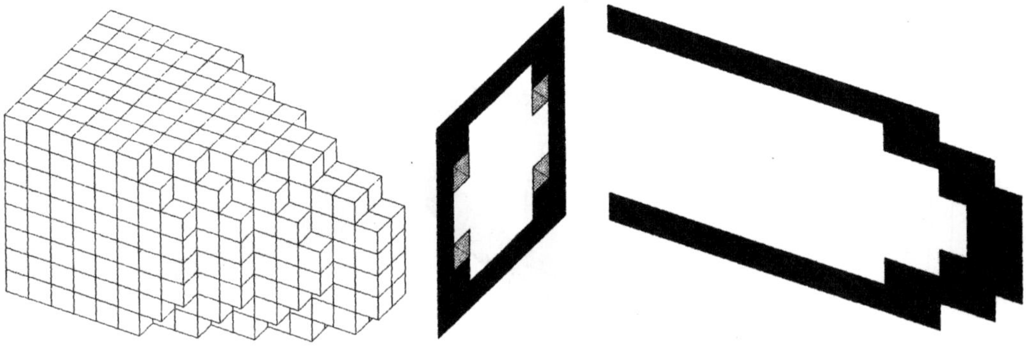

(e) w=0.5

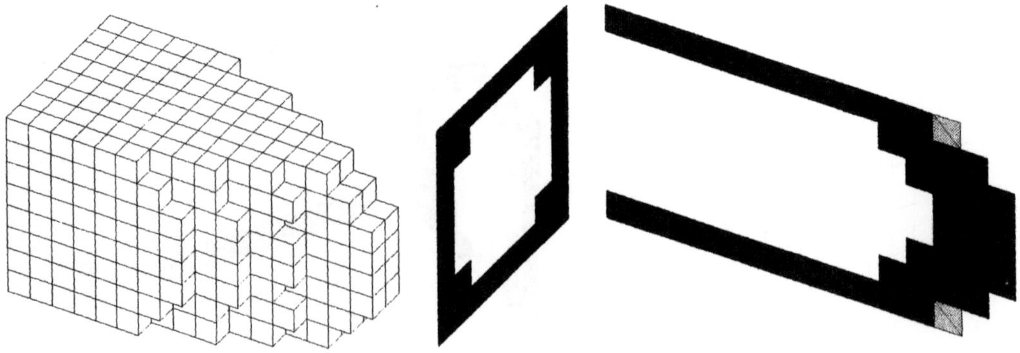

(f) w=0.6

Fig. 23 (*continued*)

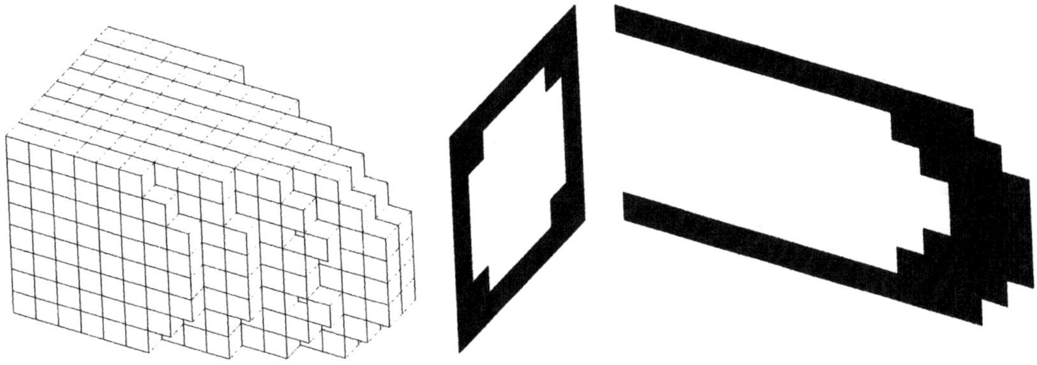

(g) *w*=0.7

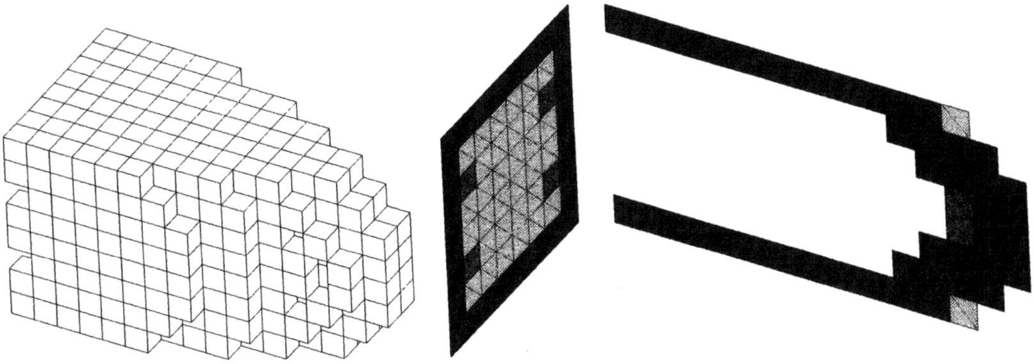

(h) *w*=0.8

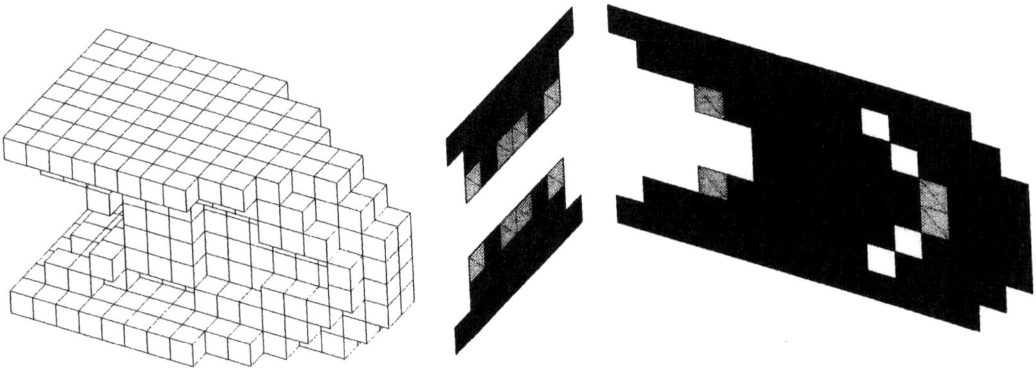

(i) *w*=0.9

Fig. 23 (*continued*)

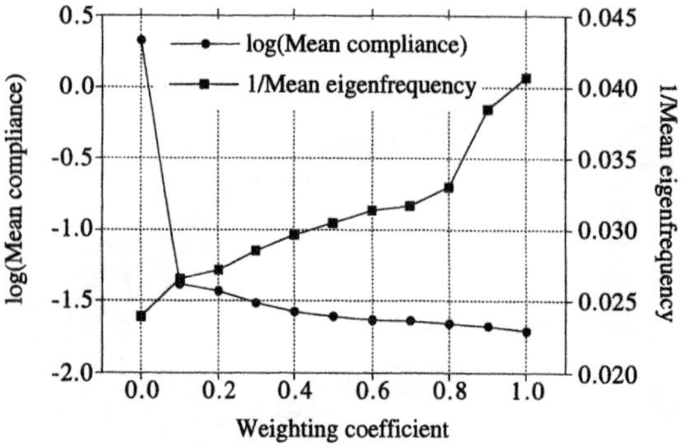

Fig. 24. Conflicting objectives of the 3D problem.

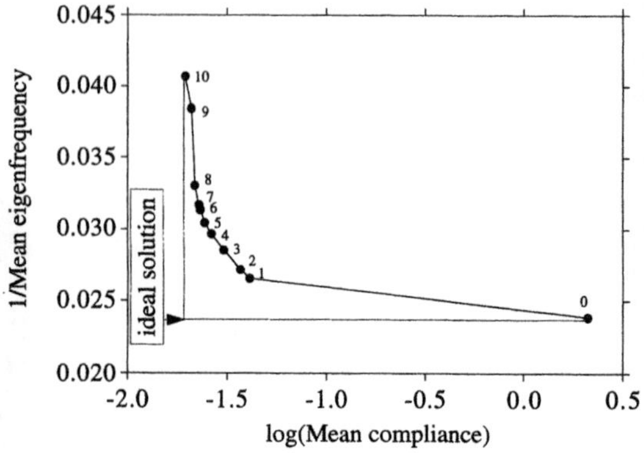

Fig. 25. Pareto optimal solutions set in criteria space (3D problem).

body hole in the solid element has a size of $0.2 \times 0.2 \times 0.2$ and three rotation angles are set to zero. The objective is to minimize the mean compliance and to maximize the mean eigenvalue, defined by the lowest three eigenvalues. Fig. 21 illustrates optimal structures generated by single objective optimization. Each result includes three figures which are the structural configuration, the cross section at the plane of $y = 5$, and the cross section at the plane of $x = 4$, respectively. In the mean compliance minimization problem, a sandwich-type structure, which is good at supporting bending moments, is obtained and in the mean eigenvalue maximization problem, material is distributed to the fixed end to reduce the vibration effect. However, both structures do not meet the static and vibration measures simultaneously as shown in Fig. 22.

The results obtained by multiobjective optimization with the help of the decision function are shown in Table 3 and the optimal structure, with weighting coefficient $w = 0.4$, is chosen as the best compromise solution. Fig. 23 shows the variety of optimal structures corresponding to the weighting coefficients and the possibility that the designer can make a decision from these Pareto optimal solutions. The characteristic of multiobjective optimization can be found in Fig. 24 which illustrates the conflicting objectives. The Pareto optimal set in criteria space is presented in Fig. 25.

6.4. *Vehicle frame reinforcement problem*

The homogenization design method, using a multiobjective optimization, is applied to a practical vehicle frame reinforcement design. The finite element model of a vehicle frame is shown in Fig. 26. The frame model is composed of 307 nodes and 307 shell elements, and the boundary conditions, including restraints and loading conditions, are specified to obtain torsional and bending rigidity under extreme load and off-road conditions. The thicknesses of base plate and reinforce plate are assumed 0.1 and 1.0, re-

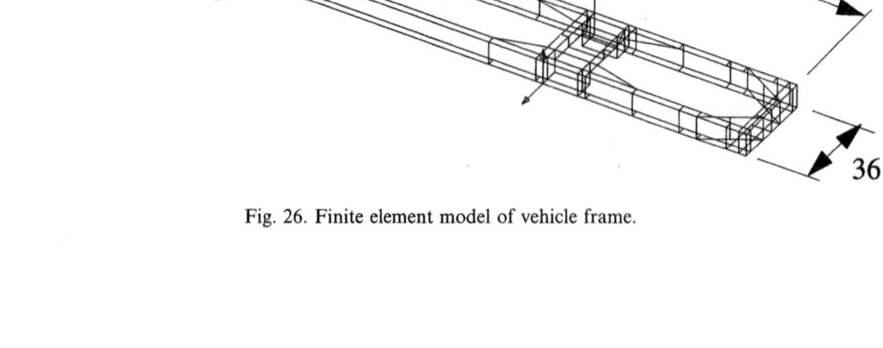

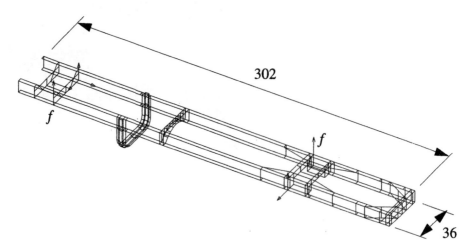

Fig. 26. Finite element model of vehicle frame.

Table 4
Optimal compromise solutions of the vehicle frame problem

Weighting coefficient (w)	Mean compliance	Mean eigenfrequency	Decision function
0.00	7356.79395	17.10608	0.12521
0.25	6688.54688	16.29399	0.05555
0.50	6262.48096	15.51549	0.03189
0.62	5574.10010	14.64098	0.02110
0.80	5533.08447	14.03200	0.03296
1.00	5433.99902	11.28669	0.11573

spectively, and 60% of the total material is specified as a constraint. Unlike previous examples in which the best compromise solution is selected among the optimal solutions set generated by the sequence of weighting coefficients, the weighting coefficient corresponding to the best compromise solution is found by the line search method. The bisection technique is used to locate the minimum value of the decision function by reducing the interval of weighting coefficients. Table 4 shows results obtained from the search process associated with the multiobjective optimization method and shows that the reinforced structure of $w = 0.62$ has the lowest decision function value. The reinforced structures corresponding to each weighting coefficient are

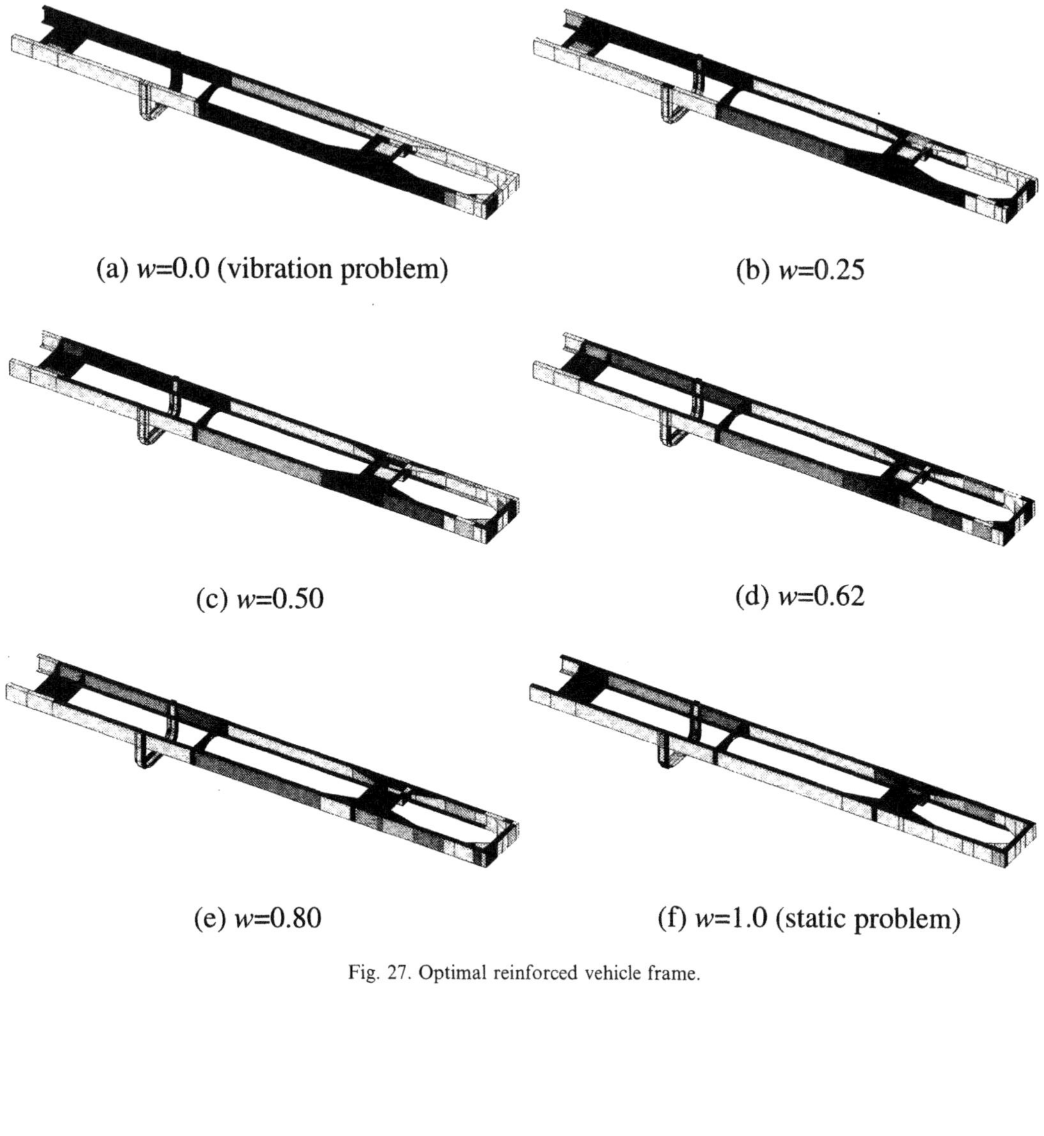

Fig. 27. Optimal reinforced vehicle frame.

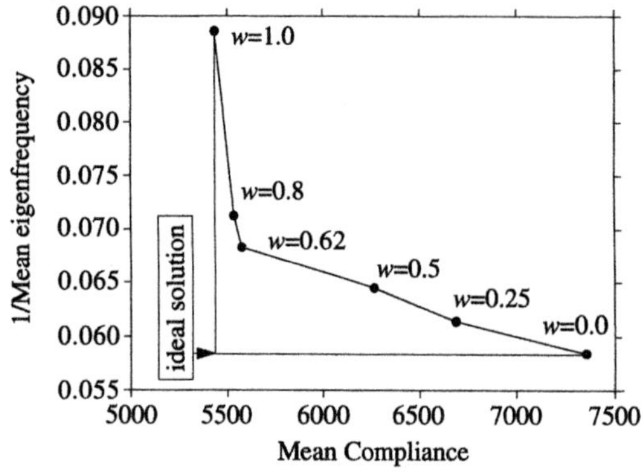

Fig. 28. Pareto optimal solutions set in criteria space (vehicle frame).

shown in Fig. 27. Fig. 28 presents the objective domain in the criteria space with the Pareto optimal solutions set. It is noted that the best compromise solution has the minimum distance from the ideal solution based on the definition of the decision function.

7. Conclusion

Structural topology optimization based on the homogenization design method using a multiobjective optimization approach has been proposed. The objective function has been formulated as the weighted sum of the global objectives to generate alternate optimal solutions, and the decision function is utilized to select the best compromise solution. Two objectives, mean compliance in the static problem and mean eigenvalue in the vibration problem, are considered for dealing with both the static and vibratory responses. The optimization algorithm based on the modified optimality criteria method is utilized to reach the optimal solution to the multiobjective optimization problem. It is shown that the use of the multiobjective optimization technique and the modified optimality criteria algorithm can improve the performance of a structure, not only the mean compliance but also the mean eigenvalue in a weighted average sense. The results illustrate that the multiobjective optimization approach can be the unified topology design methodology for structural design and can provide the designer with a flexible tool.

References

[1] Stadler W. Multicriteria optimization in mechanics. Applied Mechanics Review 1984;37:277–86.

[2] Eschenauer H, Koski J, Osyczka A, editors. Multicriteria design optimization: Procedures and applications. Berlin: Springer, 1990.

[3] Koski J. Multicriteria optimization in structural design: state of the art. In: Proceedings of the 19th Design Automation Conferences. ASME, 1993. p. 621–9.

[4] Balachandran M, Gero JS. A comparison of three methods for generating the pareto optimal set. Engineering Optimization 1984;7:319–36.

[5] Balachandran M, Gero JS. Use of knowledge in selection and control of optimization algorithms. Engineering Optimization 1987;12:163–73.

[6] Koski J, Silvennoinen R. Norm methods and partial weighting in multicriterion optimization of structures. International Journal for Numerical Methods in Engineering 1987;24:1101–21.

[7] Watkins RI, Morris AJ. A multicriteria objective function optimization scheme for laminated composites for use in multilevel structural optimization schemes. Computer Methods in Applied Mechanics and Engineering 1987;60:233–51.

[8] Bendse MP, Olhoff N, Taylor JE. A variational formulation for multicriteria structural optimization. Journal of Structural Mechanics 1983;11:523–44.

[9] Olhoff N. Multicriterion structural optimization via bound formulation and mathematical programming. Structural Optimization 1989;1:11–7.

[10] Rao SS. Multiobjective optimization in structural design with uncertain parameters and stochastic processes. AIAA Journal 1984;22:1670–8.

[11] Rao SS. Game theory approach for multiobjective structural optimization. Computers and Structures 1987;25:119–27.

[12] Saravanos DA, Chamis CC. Multiobjective shape and material optimization of composite structures including damping. AIAA Journal 1992;30:805–13.

[13] Grandhi RV, Bharatram G, Venkayya VB. Multiobjective optimization of large-scale structures. AIAA Journal 1993;31:1329–37.

[14] Rao SS. Genetic algorithm approach for multiobjective optimization of structures. In: Proceedings of the ASME Winter Annual Meeting. ASME, 1993. p. 29–38.

[15] Arakawa M, Yamakawa H. A study on multi-criteria structural optimum design using qualitative and fuzzy reasoning. Finite Elements in Analysis and Design 1993;14:127–42.

[16] Bendsøe MP, Kikuchi N. Generating optimal topologies in structural design using a homogenization method. Comput Methods Appl Mech Eng 1988;71:197–224.

[17] Díaz A, Kikuchi N. Solution to shape and topology eigenvalue optimization problems using a homogenization method. International Journal for Numerical Methods in Engineering 1992;35:1487–502.

[18] Ma Z-D, Kikuchi N, Cheng H-C. Topological design for vibrating structure. Computer Methods in Applied Mechanics and Engineering 1995;121:259–80.

[19] Díaz AR, Bendsøe MP. Shape optimization of structures for multiple loading conditions using homogenization method. Structural Optimization 1992;4:17–22.

Structural Engineering Compendium I

Mechanics of fibre-reinforced cementitious composites

B.L. Karihaloo[a,*], J. Wang[b]

[a]*Cardiff School of Engineering, University of Wales Cardiff, Queen's Buildings, P.O. Box 686, Cardiff CF2 3TB, UK*
[b]*Department of Mechanics, Peking University, Beijing 100871, People's Republic of China*

Abstract

This paper will describe a procedure for modelling the complete macroscopic response (including strain hardening and tension softening) of short fibre reinforced cementitious composites and show how their microstructural parameters influence this response. The strain hardening is essentially due to elastic bridging forces which are proportional to crack opening displacements. On the other hand, an elasto-plastic bridging law governs the initial tension softening response of the composite. To facilitate the investigation of the localisation process as an instability phenomenon perturbations in microcrack sizes are considered. The perturbations may be the result of varying aggregate sizes. The instability caused by the perturbations in the bridging force exerted by the fibres is also studied. © 2000 Elsevier Science Ltd. All rights reserved.

Keywords: Micromechanics; Cementitious composites; Tension softening; Strain hardening; Microcracks; Short fibres; Bridging by fibres; Fracture mechanics

1. Introduction

The nonlinear behaviour of quasi-brittle materials such as concrete and other cementitious composites is most often caused by the growth of pre-existing microcracks and by the nucleation and growth of new microcracks, which eventually lead to crack localisation and failure. One simplified way to model a microcracked solid from a fracture mechanics point of view is to assume that the cracks are arranged in regular patterns. Most studies so far have been devoted to modelling of doubly periodic rectangular arrays of cracks. Sahasakmontri et al. [1], using the pseudo-traction technique and double infinite summations, revealed an anomaly in the use of double infinite series [2,3]. They showed that if a superposition procedure was implemented first for any array of R rows by C columns of cracks, and the number of cracks was then increased while keeping the ratio C/R constant, the mode I and mode II stress intensity factors (SIFs) depended on the chosen ratio C/R. This anomaly was recently resolved by Karihaloo et al. [4] by using a proper superposition procedure and pseudo-traction technique.

In the present paper, we shall present the mathematical details for solving the problem of doubly periodic arrays of cracks when the cracks are free of traction and when they are subjected to a closure pressure. The solution is also based on pseudo-traction formalism, but the superposition procedure makes use of the exact solution for an array (a row) of cracks, thus avoiding divergent double infinite summations [2]. The results

* Corresponding author. Tel.: +44-1222-874597; fax: +44-1222-874597.
E-mail address: KARIHALOOB@cardiff.ac.uk (B.L. Karihaloo).

Reprinted from *Computers and Structures* **76 (1-3)**, 19-34 (2000)

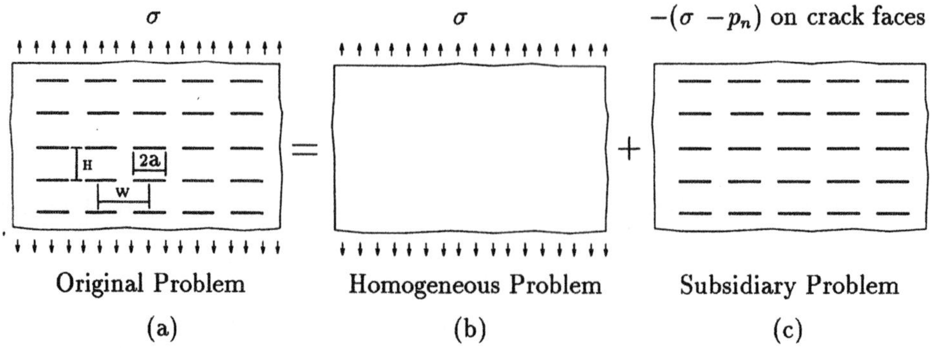

Fig. 1. A doubly periodic array of cracks and the decomposition of the problem.

for unbridged (traction-free) crack arrays are shown to be identical to those obtained by Isida et al. [5] who used the boundary collocation method.

The mathematical solutions are then used to study the influence of microstructure upon the complete macroscopic tensile response, including strain hardening and tension softening, of a conventional fibre-reinforced cementitious composite and a high performance DSP-based fibre composite.

The strain hardening is attributed to the increase in the number of microcracks under increasing tension which are elastically bridged by fibres. The initial tension softening is due to the gradual pull-out of the fibres from the matrix.

The tension softening response will be modelled by a collinear array (row) of cracks subjected to bilinear bridging forces. This model is akin to that used by Horii et al. [6] and Ortiz [7] for the tension softening of unreinforced quasi-brittle materials. It differs from that by Li et al. [8] in that the localised damage in the present model is regarded as being discontinuous, whereas Li et al. [8] assumed that the localised damage has resulted in a through crack. The model developed here is, therefore, suitable for describing the initial post-peak tension softening response of a fibre-reinforced quasi-brittle matrix.

However, as all the cracks are assumed to have the same length and to be subjected to identical bridging tractions so that the problem is inherently symmetric, it is not possible to investigate the process of the localisation of deformation, using the doubly-symmetric model.

Once localisation sets in, the strain level away from the localised zone usually decreases first and then remains more or less at a limited level. Thus, the multiple cracks far away from the localised zone are expected to have much less influence on any major crack than the ones adjacent to it. Therefore, without loss of generality, the present paper will study the stability of a damage zone formed by five dissimilar periodic rows of bridged interacting microcracks, using a superposition procedure and the pseudo-traction technique. The lengths of cracks in the central row and those in the upper and lower rows are assumed to be different to facilitate the investigation of the localisation process itself as an instability phenomenon caused by perturbations in microcrack size. These perturbations may be the result of varying aggregate sizes. The instability caused by the perturbation in the bridging force exerted by the fibres is also studied. In a real cementitious composite these perturbations are unavoidable because of the difficulties in mixing, so that the distribution of short fibres is unlikely to be uniform throughout the composite.

2. Formulation of the problem

An infinite, isotropic elastic solid containing a doubly periodic array of bridged cracks is shown in Fig. 1(a). It is subjected to a uniform remote stress σ normal to the cracks. As the cracks are bridged, their faces are subjected to a normal bridging traction p_n. This traction is not uniformly distributed along the surfaces of a crack but, by symmetry, has the same distribution along all cracks.

In terms of the standard superposition procedure, the original problem shown in Fig. 1(a) can be decomposed into the homogeneous problem of Fig. 1(b) and the subsidiary problem shown in Fig. 1(c). It is only needed to solve the subsidiary problem. Its solution can be obtained by pseudo-traction technique and by proper superposition. To apply the method of the pseudo-tractions [9], the problem 1(c) is further decomposed into a infinite number of sub-problems, designated $1, \ldots, j, \ldots, \infty$ in Fig. 2. Each sub-problem contains only one row of cracks, each of which is subjected to a normal pseudo-traction σ^p. Because of the inherent symmetry of the original problem, the pseudo-tractions are the same for all cracks.

Sahasakmontri et al. [1] decomposed the subsidiary problem of Fig. 1(c) differently. In their decomposition, each sub-problem contained only one crack, so that superposition required summation of two infinite series. The decomposition in the present paper is such that superposition requires the summation of a single infinite series. As we shall see below, the elimination of double infinite series overcomes the problems of divergence and results in highly accurate solutions.

The distribution of unknown pseudo-tractions σ^p will be determined in such a way that the traction-free condition on the crack faces of the problem 2(a) is satisfied when the stress perturbations caused by all the sub-problems are superposed [10]. We shall use the basic solution for an infinite body containing a row of cracks each of which is subjected to two pairs of concentrated normal surface loads of opposite direction as shown in Fig. 3. The exact solution for this problem can be readily deduced from the results in the handbook by Tada et al. [11]. In the sequel, the loads shown in Fig. 3 will be referred to as a set of crack surface tractions at position x.

Because of symmetry, we need to consider only one crack in a sub-problem, e.g. the one labelled with the set of crack surface tractions $\sigma^p(x)$ and co-ordinate system in Fig. 3. The stress σ_{yy} at point x of the crack i is given by the superposition of the contributions from all the sub-problems

$$-\sigma^p(x) + \sum_{j=-\infty,\, j\neq i}^{+\infty} \int_0^{+a} K_\sigma(x, x^j)\sigma^p(x^j)\,\mathrm{d}x^j, \quad x, x^j \in [0, +a) \tag{1}$$

where kernel $K_\sigma(x, x^j)$ represents the stress at x induced by a set of unit crack surface tractions at x^j in sub-problem j [11].

For the crack configuration of Fig. 3, the summation in Eq. (1) is obviously carried out to the geometrical condition

$$y = jH, \quad j = -\infty, \ldots, -1, +1, \ldots, +\infty \tag{2}$$

Because of the symmetry of the problem with respect to the x axis, Eq. (1) can be written as

$$-\sigma^p(x) + 2\sum_{j=1}^{+\infty} \int_0^{+a} K_\sigma(x, x^j)\sigma^p(x^j)\,\mathrm{d}x^j, \quad x, x^j \in [0, +a) \tag{3}$$

By superposition of the sub-problems, the traction-free condition on each crack in the subsidiary problem of Fig. 2(a) can be written as

$$-\sigma^p(x) + 2\sum_{j=1}^{+\infty} \int_0^{+a} K_\sigma(x, x^j)\sigma^p(x^j)\,\mathrm{d}x^j = -(\sigma - p_n), \quad x, x^j \in [0, +a) \tag{4}$$

Eq. (4) is a simplified form of the general formula for two-dimensional interacting cracks derived by Hu et al. [10]. It can be rewritten as

$$\sigma^p(x) - 2\sum_{j=1}^{+\infty} \int_0^{+a} K_\sigma(x, x^j)\sigma^p(x^j)\,\mathrm{d}x^j + p_n = \sigma, \quad x, x^j \in [0, +a) \tag{5}$$

In common with many previous works, the bridging traction p_n is expressed as a function of the crack opening displacement

$$p_n = f_n(v(x)) \tag{6}$$

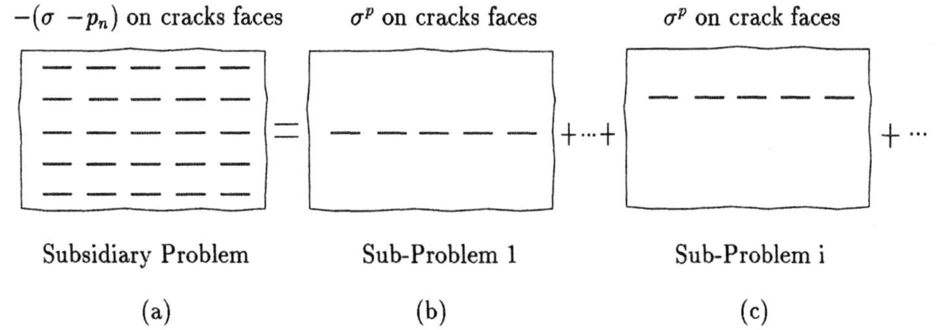

Fig. 2. Decomposition of the subsidiary problem of Fig. 1(c) into an infinite number of sub-problems. σ^p is the pseudo-traction on the faces of each crack in the row in each sub-problem.

where $v(x)$ is solely induced by the pseudo-tractions $\sigma^P(x^i)$ on the cracks in sub-problem i [10]. It can be easily obtained from

$$v(x) = \int_0^{+a} K_v(x, x^i)\sigma^P(x^i)\,dx^i, \quad x, x^i \in [0, +a) \tag{7}$$

where kernel $K_v(x, x^i)$ represents the crack opening displacement at x induced by a set of concentrated unit loads at x^i on the crack surfaces in sub-problem i, see Fig. 4. This kernel can be derived in a closed form from the results in the handbook by Tada et al. [11].

It is seen that by substituting Eq. (5) into Eq. (4), and using the relation between the pseudo-traction and the crack opening displacement (7), we need only to solve an integral equation for the pseudo-traction $\sigma^P(x)$. The kernel $K_\sigma(x, x^j)$ is not singular provided that $H > 0$. The kernel $K_v(x, x^i)$ has an integrable logarithmic singularity at $x = x^i$ which can easily be handled by regularisation.

We shall use Gauss–Legendre quadrature to solve the integral equation (4) and write (with $t = x/a$)

$$\sigma^P(t_l) - 2\sum_{j=1}^{+\infty} a \sum_{k=1}^{N}\left[K_\sigma\left(t_l, t_k^j\right)\sigma^P\left(t_k^j\right)\right]\omega_k + p_n(t_l) = \sigma \tag{8}$$

$$t_l, t_k^j \in (t_l, \ldots, t_N);\ l = 1, \ldots, N$$

The equation for the crack opening displacement (7) can be similarly discretised

$$v(t_l) = a \sum_{k=1,\,k\neq l}^{N} K_v\left(t_l, t_k^i\right)\left[\sigma^P\left(t_k^i\right) - \sigma^P(t_l)\right]\omega_k + \sigma^P(t_l)a\left[\sum_{k=1}^{N} K_v^{ns}\left(t_l, t_k^i\right)\omega_k + \int_0^1 K_v^s\left(t_l, t^i\right)dt^i\right] \tag{9}$$

$$t_l, t_k^i \in (t_l, \ldots, t_N);\ l = 1, \ldots, N$$

In Eqs. (8) and (9), $t_l, \ldots, t_N$ are the collocation points and, at the same time, the integration points of the Gauss–Legendre quadrature. ω_k $(k = 1, \ldots, N)$ are the corresponding weights.

As $j \to +\infty$, $K_\sigma(x, x^j)$ decays exponentially so the infinite summation in Eq. (8) is uniformly convergent for $x \in [0, W/2]$. This feature makes it possible to approximate the infinite summation using the sum of finite terms while the truncation error can be made as small as desired.

3. Unbridged cracks

When the cracks are not bridged ($p_n = 0$), the integral equation (4) is particularly easy to solve with the numerical quadrature (8) and (9), whereafter the mode

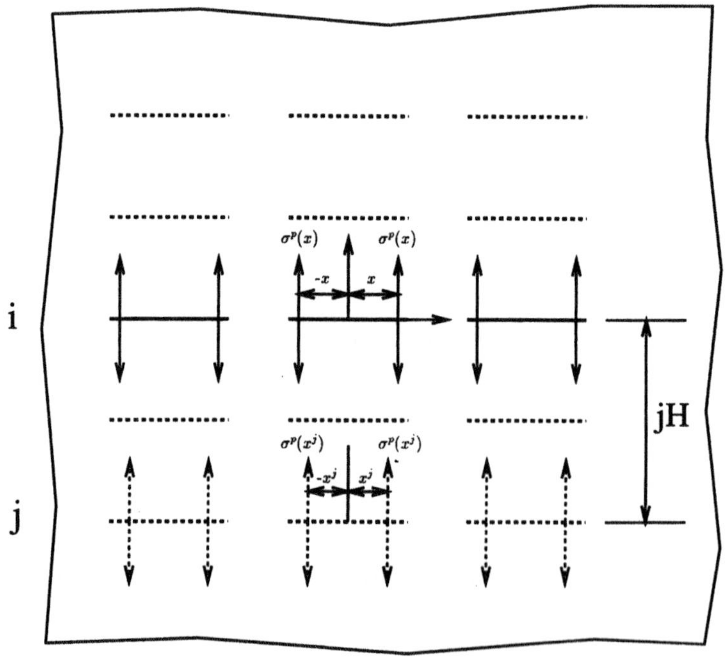

Fig. 3. Superposition of sub-problems, showing pseudo-tractions on faces of cracks in rows i and j. Note that because of symmetry, x and x^j only vary from 0 to a (half-crack length).

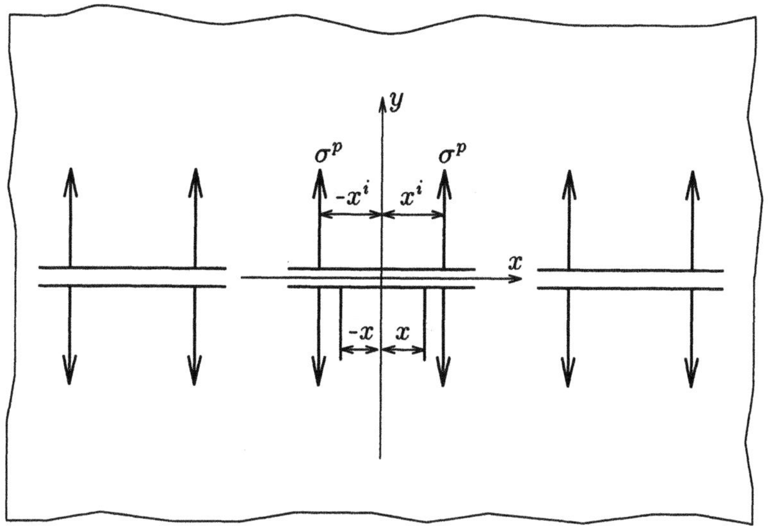

Fig. 4. Sub-problem i, showing a set of pseudo-tractions on faces of a crack in a row. Note that because of symmetry, x^i only varies from 0 to a (half-crack length).

I SIF at each crack tip in the array and the overall effective Young's modulus E_y along the direction of loading can be determined. The non-dimensional mode I SIF is shown in Fig. 5, as a function of the density of cracks defined by $a^2/(WH)$ and the arrangement (or shape) of the rectangular array characterised by H/W. When $H/W \leq 0.8$, the SIF decreases with increasing density of cracks suggesting that the mutual influence of closely spaced cracks reduces the crack driving force. On the other hand, when H/W is large, the mode I SIF increases rapidly with the density of cracks. This is due to the fact that in order to maintain a high density when H is large, the neighbouring crack tips in a row must approach each other.

The overall tangent modulus of the body along the direction of loading is of prime interest. However, it is obvious that the cracked body is no longer a macroscopically isotropic medium but an orthotropic one. Its instantaneous shear modulus in the xy plane can be calculated by subjecting the body to an inplane (xy) shear stress, but we shall not attempt this calculation here. Horii and Nemat-Nasser [9] have given the incre-

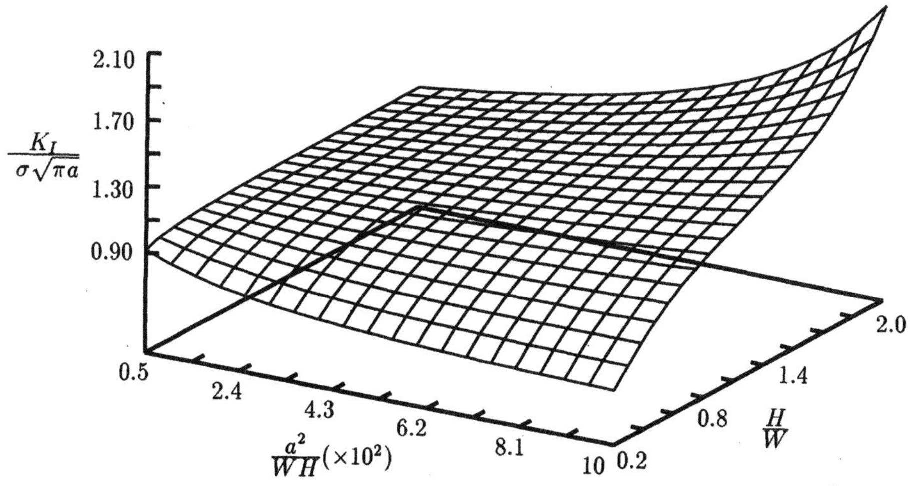

Fig. 5. Variation of normalised mode I stress intensity factor with the density and shape of crack arrays.

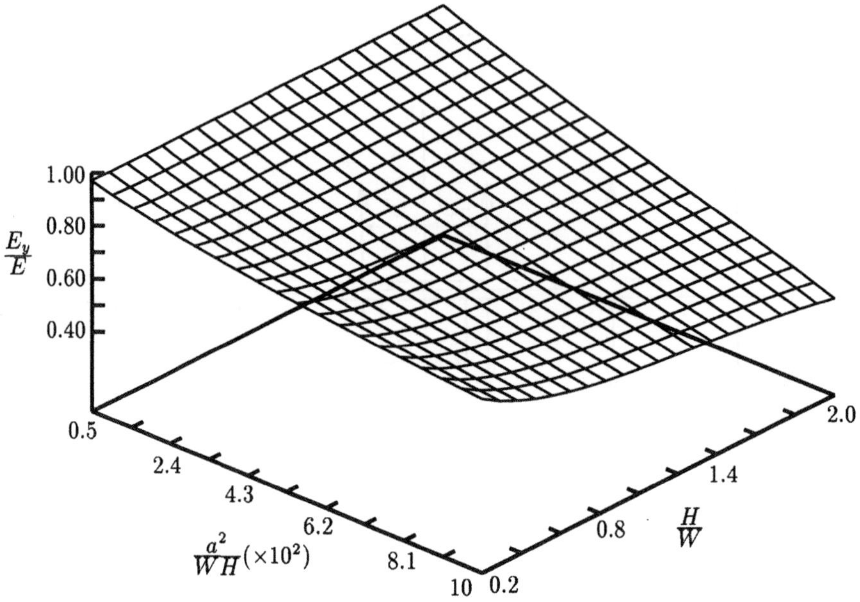

Fig. 6. Variation of normalised tangent modulus with the density and shape of the array of cracks.

mental constitutive relations for a cracked body

$$\Delta\bar{\varepsilon}_{ij} = D_{ijkl}\Delta\bar{\sigma}_{kl} + \frac{1}{V}\int_S \frac{1}{2}\left([v_i]n_j + [v_j]n_i\right) \mathrm{d}S \tag{10}$$

where $[v_j]$ denotes the displacement jump across the crack with unit normal vector n_i, D_{ijkl} is the elastic compliance tensor of the uncracked body, and the integration is carried over the crack surface S contained in a representative volume V of the solid. Application of Eq. (10) to the stress and strain in the loading direction gives the tangent modulus E_y in that direction

$$\frac{E_y}{E} = \frac{1}{1 + \dfrac{8a^2}{WH}\displaystyle\int_0^1 v^*(t)\,\mathrm{d}t} \tag{11}$$

with

$$v^*(t) = \frac{v(t)E}{2a\sigma} \quad t \in [0, 1] \tag{12}$$

$v(t)$ is half crack opening displacement at the position $t = x/a$, and E is plane stress Young's modulus of the uncracked body. Under plane strain, E is replaced with $E/(1 - v^2)$.

Fig. 6 shows the variation of E_y/E with the density and shape of the array of cracks. In contrast to the SIF, the instantaneous elastic modulus in the direction of loading does not seem to be very sensitive to the shape of the array (i.e. H/W). It depends primarily on the crack density (i.e. $a^2/(WH)$). The tangent modulus E_y in the direction of loading for the crack geometry under consideration may be regarded as a lower bound on the value for a solid containing randomly oriented cracks.

4. Bridged cracks

4.1. Elastic bridging

In this section, we shall study the effect of the bridging stiffness on the SIF at the tips of cracks and on the crack opening displacement, assuming that the bridging force is linearly proportional to the latter

$$p_n(t_l) = kv(t_l) \tag{13}$$

Here, k is the bridging stiffness and $t_l\,(l = 1, \ldots, N)$ denote positions on the crack surface in the notation of Eqs. (8) and (9).

Fig. 7 shows the non-dimensional crack opening displacement (for a quarter of the crack) for different values of the non-dimensional bridging stiffness $c = k(2(1 - v^2)/E)a$. The geometry of the array of cracks is described by $W/a = 2.5$ and $H/a = 2.0$. The results indicate that the crack opening displacement is very sensitive to small bridging stiffness. The same is also true for the mode I SIF at the crack tips (Fig. 8) and the tangent modulus (Fig. 9). It appears that the influ-

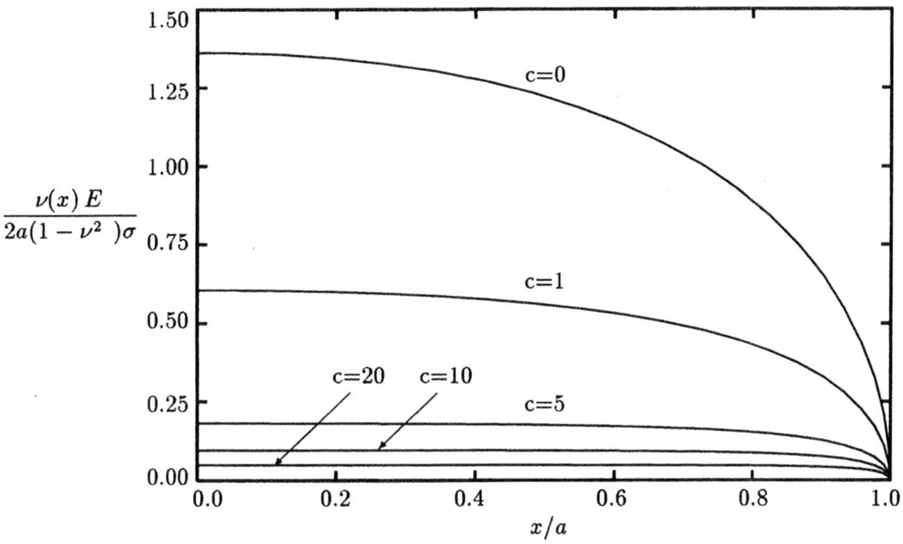

Fig. 7. Influence of bridging stiffness on normalised crack opening displacement.

ence of bridging stiffness peaks at about $c \approx 10$. A similar variation of the mode I SIF for three parallel, but offset bridged cracks is also seen [10].

4.2. Bilinear bridging

It is assumed that when the crack opening displacement reaches a critical value v_{cr}, some of the fibres bridging the crack will debond from the matrix, resulting in a sudden drop in the bridging force (BC in Fig. 10), whereafter these fibres will exert a reduced closure force by frictional pull-out (CD in Fig. 10). If the fibres bridging the crack break then $\alpha = 0$. A bridging law with $\alpha = 0$ has been earlier considered in Ref. [12]. Several analytical models for fibre pull-out in cementitious composites have been developed in Ref. [13–15]. These are based on single fibre pull-out tests giving a relationship between pull-out force P (or inter-

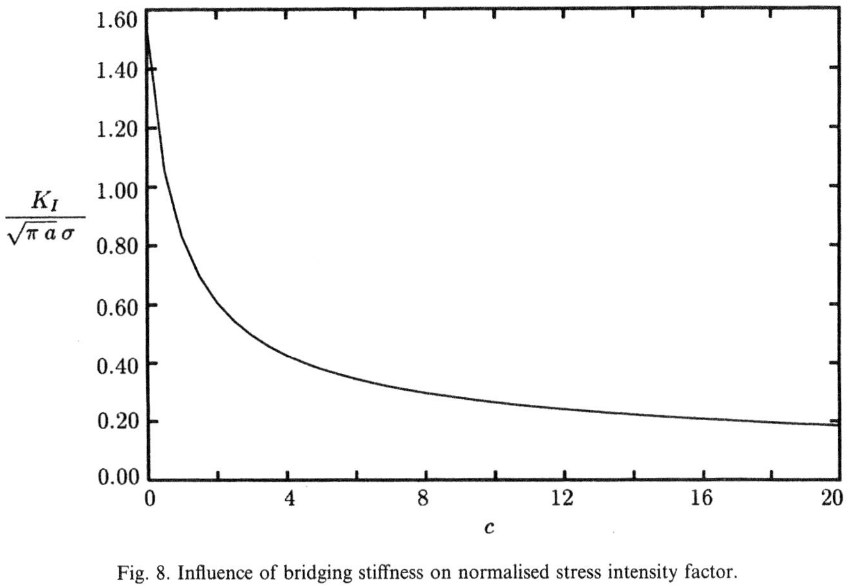

Fig. 8. Influence of bridging stiffness on normalised stress intensity factor.

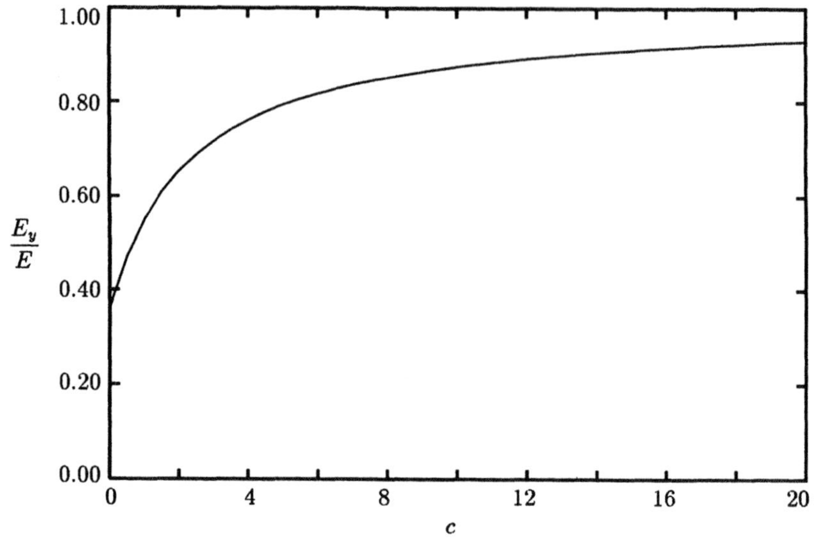

Fig. 9. Influence of bridging stiffness on normalised tangent modulus.

facial shear stress $\tau = P/(\pi d l_e)$ where d is the fibre diameter and l_e its embedded length) and fibre slip v (Fig. 11(a)). From this single-fibre (τ–v) relationship, the average closure force on crack faces bridged by many fibres is obtained through an approximation to the fibre distribution function (Fig. 11(b)). The slope of the elastic part of this averaged pull-out diagram gives directly the parameter k in Eq. (13). When the crack opening exceeds the critical value v_{cr}, the bridging stress $p = kv(t_l)$ drops to the constant frictional pull-out value p_g.

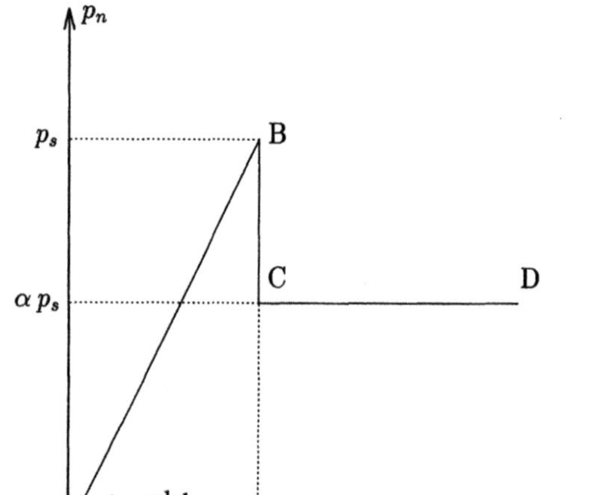

Fig. 10. A linearised bridging law describing the relation between the bridging force and the crack opening displacement.

5. Application to short fibre reinforced concrete

In the following, we shall use the above model to predict the pre-peak (strain hardening) and initial post-peak (tension softening) behaviour of short fibre reinforced conventional and DSP-based cementitious composites. The material parameters are given in Table 1. DSP stands for Densified Systems containing homogeneously arranged ultrafine Particles.

5.1. Strain hardening due to multiple cracking

The decrease in the elastic modulus in the direction of loading induced by multiple elastically bridged cracks can be used to predict the tensile strain hardening behaviour of random short fibre reinforced composites. There is ample experimental evidence from tests on fibre-reinforced brittle [16] and quasi-brittle matrices [17,18] that suggests that the tensile strain hardening in these composites is due to the formation of microcracks whose density increases with increasing tensile/flexural loading until it reaches a saturation level.

To account for the increase in the density (number) of microcracks during the stage of strain hardening, we need a proper evolution law for the multiple cracks. Since the distribution of the size and spacing of multiple cracks in fibre-reinforced composites is of random

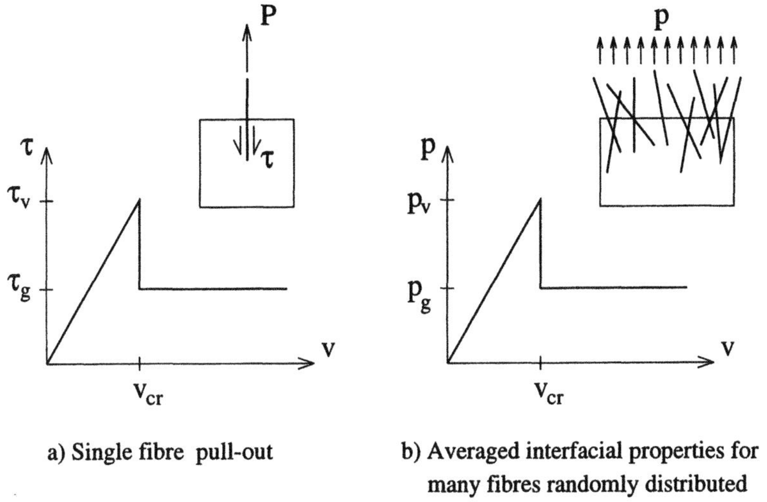

Fig. 11. Conversion from pull-out properties of a single fibre to interfacial properties averaged over the cross-section.

nature, investigations have been conducted to predict the stochastic tendency of the development of the multiple cracks. Zok and Spearing [19] considered a non-linear relation between the spacing of cracks and applied stress. However, experimental observations on glass/ceramic CAS/SiC composite and theoretical predictions of Spearing and Zok [20] based on stochastic simulations showed that the average spacing of cracks is reasonably linear in the applied stress during the course of multiple cracking until a state of saturation is reached

$$\ell = \ell_s \frac{[\sigma_s/\sigma_{mc} - 1]}{[\sigma/\sigma_{mc} - 1]} \tag{14}$$

where σ_{mc} is the stress level at the initiation of the cracks, ℓ is the crack spacing at stress level σ and ℓ_s is the saturation spacing at stress level σ_s.

Following the same line of reasoning as in the works just cited, we propose the following two-dimensional linear evolution law for multiple cracks analogous to the one-dimensional law (14).

$$\frac{\sigma}{\sigma_{mc}} = 1 + \frac{[(WH)/a^2]_s[\sigma_s/\sigma_{mc} - 1]}{(WH)/a^2} \tag{15}$$

where σ_{mc} is the stress level at the initiation of the cracks, and the subscript s denotes the value of the corresponding parameter at the saturation of multiple cracks. The initiation value σ_{mc} may be equated to the tensile strength of the matrix, whereas the saturation value σ_s will represent the ultimate tensile capacity of the composite.

The variation of tangent modulus with the applied stress during the evolution of the multiple cracks can be determined from Eqs. (11) and (15). Once the variation of tangent modulus with the applied stress during the evolution of the multiple cracks is known, the stress–strain relation past the initiation stress level σ_{mc} can be obtained incrementally via

$$\dot{\varepsilon} = \frac{\dot{\sigma}}{E_y(\sigma)} \tag{16}$$

or

Table 1
Material parameters

Parameter	DSP-based FRC	Conventional FRC
E, Young's modulus (GPa)	40	40
σ_m, matrix strength (MPa)	10	3
ν, Poisson's ratio	0.2	0.2
τ_v, interfacial shear strength (MPa)	9	2
τ_g, frictional shear strength (MPa)	5	1
l, fibre length, (mm)	20	20
d, fibre diameter (mm)	0.4	0.4

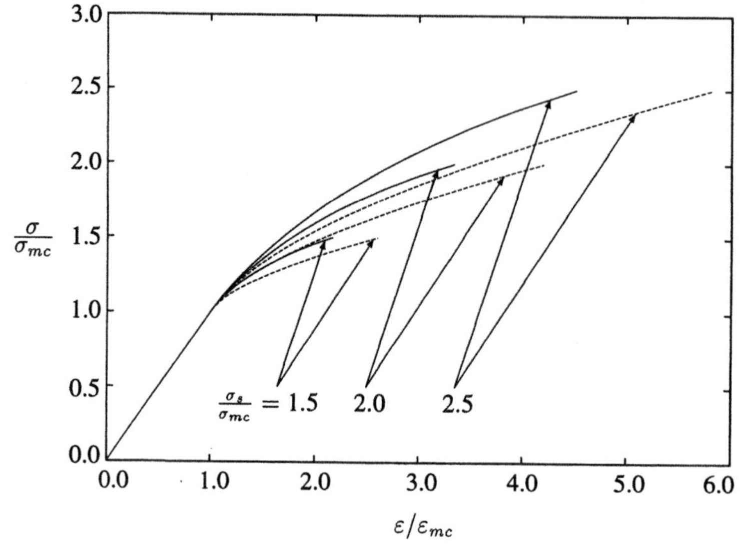

Fig. 12. Normalised stress–strain curves during multiple cracking (strain hardening). Solid line corresponds to Eq. (15), and dashed line to Eq. (14).

$$\frac{\dot{\varepsilon}}{\varepsilon_{mc}} = \frac{\dfrac{\dot{\sigma}}{\sigma_{mc}}}{\dfrac{E_y}{E}} \tag{17}$$

where we have used the definition $\varepsilon_{mc} = \sigma_{mc}/E$. When $\sigma \leq \sigma_{mc}$, $E_y \equiv E$, so that the stress–strain relation is linear. When $\sigma > \sigma_{mc}$, E_y/E changes with σ/σ_{mc}, so that we have to calculate the increment of the strain using the current value of E_y/E at each step. The accuracy of the incremental formulation is the better, the smaller the step length. The normalised stress–strain curves predicted by Eq. (17) via Eqs. (11) and (15) are shown in Fig. 12.

From Fig. 12, it is seen that both evolution laws (14) and (15) produce similar variations of the stress–strain curves. As the final crack densities are assumed to be same for the two evolution processes, the final values of E_y/E are also equal for the same value of σ_s/σ_{mc}. As expected, the two-dimensional evolution law predicts a stiffer response than does the one-dimensional law. The response of a composite with randomly oriented cracks can be expected to be even more stiff.

5.2. Modelling of tension softening

The microcrack model of Horii et al. [6] and Ortiz [7] for modelling the tension softening process of unreinforced cementitious materials will be extended to short fibre reinforced composites. In this model (Fig. 13) the damage at peak load is assumed to localise along the eventual failure plane. The damage is in the form of fragmented cracks interspersed by unbroken material. At peak load each crack is bridged by fibres, some of which are on the verge of pulling out from the matrix. Let the average maximum opening displacement at this instant be $\bar{v}_{cr}$. The initial part of tension softening is a result of the progressive growth of the cracks into the unbroken material, for which two conditions must be simultaneously met: the SIF at the tips of each crack K_I must attain the critical value K_{Ic} of the composite and the fibres must pull out when the crack opening reaches $\bar{v}_{cr}$. Mathematically, these two dynamic conditions for crack growth are

$$K_I(\sigma, a; l) = K_{Ic} \tag{18}$$

$$v(\sigma, a; l) = \bar{v}_{cr} \tag{19}$$

where $v(\sigma, a; l)$ is the crack opening at $x = l$ and σ denotes the reduced tensile carrying capacity. Note that at the onset of tension softening $l = 0$, so that Eq. (12) can be used to calculate the average critical crack opening $\bar{v}_{cr}$. K_{Ic} is related to the fibre content, the critical SIF of the matrix and the interfacial bond strength.

Here, we introduce an equivalent critical SIF K_{Ic} based upon the ultimate tensile strength of short fibre-reinforced concrete. The ultimate tensile strength of a composite containing a moderate fraction of fibre can be calculated using the simple law of mixtures [21]

$$\sigma_c = \alpha\sigma_m(1 - V_f) + \beta\tau V_f\frac{l}{d} \tag{20}$$

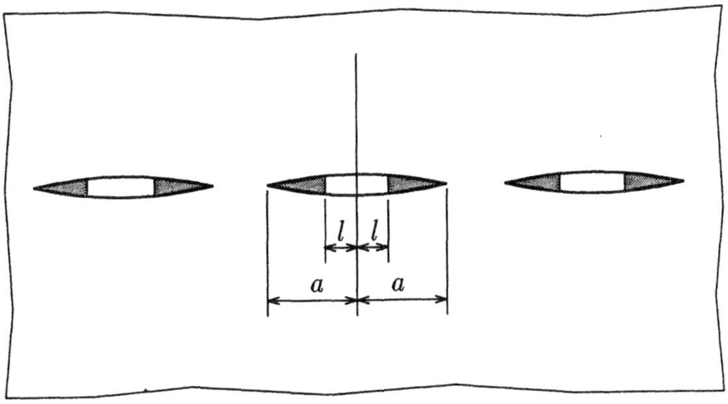

Fig. 13. A microcrack model for tension softening process. The fibres in the central region of length $2l$ of each crack are assumed to pull out, so that the bridging traction in this region is reduced from p_s to αp_s, as shown in Fig. 10. The bridging traction in the shaded regions where the fibres are still fully bonded to the matrix is proportional to crack opening, kv.

where V_f is the volume fraction of fibre, τ is the average bond strength of the fibre-matrix interface, and l/d is the aspect ratio of the fibre. α and β are empirical constants which are to be determined from tests. In order to retrieve $\sigma_c \equiv \sigma_m$ when $V_f = 0$, we take $\alpha = 1$. Aveston et al. [22] have proposed a value $\beta = 0.5$ for randomly distributed short fibres. Note that σ_c equals σ_s in the present model.

In the initial part of tension softening regime the growth of fragmented bridged cracks takes place at a constant SIF equal to the effective fracture toughness of composite K_{Ic}. Its value can, therefore, be calculated using the stress and crack length at the onset of tension softening. The stress equals σ_c ($\equiv \sigma_s$) given by Eq. (20) and the half-crack length is a_0, so that

$$K_{Ic} = \Gamma_0 \sqrt{\pi a_0} \left[\alpha \sigma_m (1 - V_f) + \beta \tau V_f \frac{l}{d} \right] \tag{21}$$

The geometrical factor Γ_0 for the row of elastically

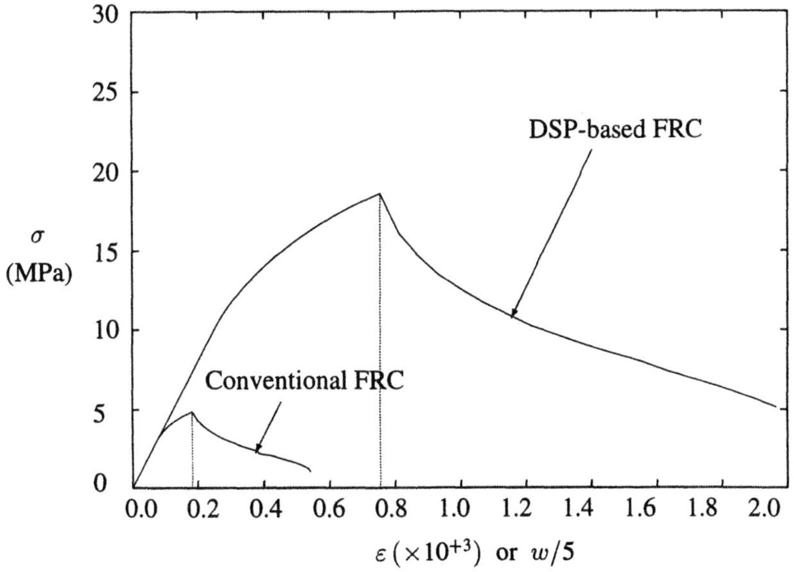

Fig. 14. Complete pre- and post-peak tensile response of the materials of Table 1 and $V_f = 0.04$. Note that the pre-peak response is given in terms of stresses and strains, whereas the post-peak response is via a stress-average crack opening measure.

bridged cracks at the onset of tension softening is calculated using the procedure described in Section 4 with $H \rightarrow +\infty$.

From Eqs. (18), (19) and (21) it is now possible to determine the reduced tensile carrying capacity σ and the length of crack l over which the fibres exert closure force due to frictional pull-out only. To complete the description of tension softening process it remains to determine the average inelastic crack opening along the localised damage band, given by

$$\omega^* \equiv \langle v_i \rangle = \frac{1}{W}\int_{-a}^{a} [v_t]\,dx - \frac{1}{W}\int_{-a_0}^{a_0} [v_e]\,dx \tag{22}$$

where $[v_t]$ is the total opening of the crack faces ($[v_t] = v_t^+ - v_t^-$, with "+" and "−" referring to the upper and lower crack faces), and $[v_e]$ is the elastic (recoverable) opening of the crack faces. Note that over the unbroken ligaments $|x| \leq (W/2) - a$, $[v_t] = 0$.

Fig. 14 shows the complete pre- and post-peak curves for the material parameters of Table 1. The values of W and a_0 in Eq. (22) are assumed to be 50 and 4.83 mm, respectively. β in Eq. (20) is taken as 0.5. For convenience of graphical presentation of strain hardening and tension softening on the same diagram, it is assumed that $\sigma_c = \sigma_s$ and $\sigma_{mc} = \sigma_m$, although in the latter instance it might be more appropriate to set $\sigma_{mc} = \sigma_m/(1 - V_f)$. Moreover, the pre-peak behaviour is presented as a relation between the applied remote stress σ and strain ε, whereas the post-peak behaviour is presented as a relation between the diminishing applied stress σ and the average inelastic crack opening $\omega = \omega^* E/(\pi a_0 \sigma_m)$. Again for ease of comparison of the two response curves σ_m has been arbitrarily chosen equal to 4 MPa. In reality, E of DSP-based FRC is nearly twice that of the conventional FRC (60 GPa compared with 30 GPa), while its σ_m is more than three times that of the latter (see Table 1). The ratio E/σ_m of DSP-based FRC is, therefore, around 6, whereas that of conventional FRC is around 10. This difference should be borne in mind in evaluating the response curves in Fig. 14.

Fig. 14 is in very good agreement with the typical load-elongation responses of these two materials. The stress–strain curve of DSP-based matrix is in excellent agreement with that measured in four-point bending by Tjiptobroto and Hansen [18]. The present model predicts that the improvement in the toughness of the material is primarily due to the enhancement of the ultimate strength (or the equivalent critical SIF K_{Ic}) of the composite. The high strength of the DSP-based matrix itself and the high interfacial bonding strength between the fibre and this matrix makes significant contribution to the improvement in the overall properties of the composite.

6. Localisation study

To study the localisation phenomenon, Fig. 15 shows an infinite elastic solid containing a damage zone simulated by five rows of periodic cracks. The crack configuration shown in Fig. 15 will be referred to as rectangular configuration, following the terminology in the work of Karihaloo [3], who solved a more complicated problem involving plastic relaxation zones around the crack tips in all three modes for rectangular and diamond-shaped doubly periodic arrays of cracks. For brevity, the central row will be referred to as row A. The two identical nearest neighbouring rows to the central row will be referred to as rows B and the two identical neighbouring rows once removed from the central row as rows C. It is assumed that the solid is subjected to a remote unidirectional tension along the y-axis, denoted σ^0. The crack faces are subjected to bridging tractions (closing tractions).

The above problem is solved using the superposition procedure and the pseudo-traction technique described above. It is noted that in the present problem the pseudo-tractions on the crack faces of row A and those on the crack faces of the other rows are different. Let the pseudo-tractions on the crack faces of rows A, B and C be $\sigma_a^P(x)\,(x \in [0, a))$, $\sigma_a^P\,(x \in [0, b))$ and $\sigma_c^P(x)\,(x \in [0, c))$, respectively. For each crack in row A, the traction consistency condition is

$$\sigma_a^P(x) - 2\int_0^b K_{ba}(x, s)\sigma_b^P(s)\,ds - 2\int_0^c K_{ca}(x, s)\sigma_c^P(s)\,ds \tag{23}$$
$$+ p_{na}(x) = \sigma^0, \quad x \in [0, a)$$

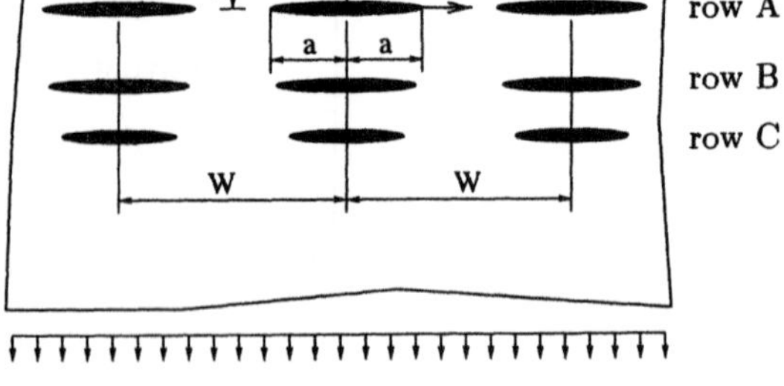

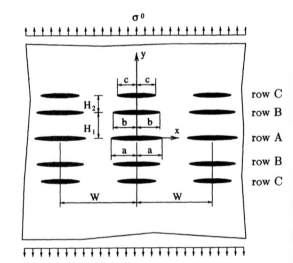

Fig. 15. Five rows of bridged cracks in a solid subjected to unidirectional tension.

where $K_{ba}(x, s)$ represents the perturbation of the stress field at the location x of a crack in row A caused by the pseudo-tractions $\sigma_b^p(s)$ on the faces of each of the cracks in rows B, and $K_{ca}(x, s)$ represents the perturbation caused by the pseudo-tractions $\sigma_c^p(s)$ on the faces of each of the cracks in row C. $K_{ba}(x, s)$ and $K_{ca}(x, s)$ are easily obtained from the formula in the handbook by Tada et al. [11]. $p_{na}(x)$ represents the bridging traction imposed on the faces of the cracks. For linear bridging, it can be written as

$$p_{na}(x) = k_a v_a(x) \tag{24}$$

where k_a is the bridging stiffness and $v_a(x)$ denotes the crack opening displacement.

For each of the cracks in rows B, the traction consistency condition is

$$\begin{aligned}\sigma_b^p(x) &- 2\int_0^a K_{ab}(x, s)\sigma_a^p(s)\,ds - \int_0^b K_{bb}(x, s)\sigma_b^p(s)\,ds \\ &- \int_0^c \left[K_{cb}^1(x, s) + K_{cb}^2(x, s)\right]\sigma_c^p(s)\,ds + p_{nb}(x) \\ &= \sigma^0, \quad x \in [0, b)\end{aligned} \tag{25}$$

The counterpart of Eq. (24) is

$$p_{nb}(x) = k_b v_b(x) \tag{26}$$

For each of the cracks in rows C, the traction consistency condition is

$$\begin{aligned}\sigma_c^p(x) &- \int_0^a K_{ac}(x, s)\sigma_a^p(s)\,ds - \int_0^b \left[K_{bc}^1(x, s)\right. \\ &\left. + K_{bc}^2(x, s)\right]\sigma_b^p(s)\,ds - \int_0^c K_{cc}^1(x, s)\sigma_c^p(s)\,ds \\ &+ p_{nc}(x) = \sigma^0, \quad x \in [0, c)\end{aligned} \tag{27}$$

and the counterpart of Eq. (24) is

$$p_{nc}(x) = k_c v_c(x) \tag{28}$$

In the sequel, we shall use three non-dimensional bridging stiffness parameters which determine the magnitude of the bridging tractions on the faces of the cracks for a given load σ^0

$$\begin{aligned}\kappa_a &= k_a \frac{1 - \nu}{\pi G} a; \qquad \kappa_b = k_b \frac{1 - \nu}{\pi G} b; \\ \kappa_c &= k_c \frac{1 - \nu}{\pi G} a\end{aligned} \tag{29}$$

where G and ν are the shear modulus and Poisson ratio of the solid, respectively.

Noting that the cracks are opened solely by the respective pseudo-tractions on their faces, the opening displacements can be written as

$$v_a(x) = \int_0^a K_{va}(x, s)\sigma_a^p(s)\,ds \quad x \in [0, a) \tag{30}$$

$$v_b(x) = \int_0^b K_{vb}(x, s)\sigma_b^p(s)\,ds \quad x \in [0, b) \tag{31}$$

$$v_c(x) = \int_0^c K_{vc}(x, s)\sigma_c^p(s)\,ds \quad x \in [0, c) \tag{32}$$

where the kernels $K_{av}(x, s)$, $K_{bv}(x, s)$ and $K_{cv}(x, s)$ are defined in the previous work by the authors [4]. They can be obtained in closed forms from the results in the handbook by Tada et al. [11].

Substituting Eqs. (24), (26) and (28) into the traction consistency conditions (23), (25) and (27), respectively, and using Eqs. (30)–(32), we get three coupled integral equations in the unknown pseudo-tractions $\sigma_a^p(x)\,(x \in [0, a))$, $\sigma_b^p(x)\,(x \in [0, b))$ and $\sigma_c^p(x)\,(x \in [0, c))$. As these integral equations are not singular they can be solved accurately using the Gauss–Legendre quadrature, following the procedure outlined in Section 2.

After solving the pseudo-tractions σ_a^p, σ_b^p and σ_c^p, the crack opening displacements can be calculated. The normalised SIFs at the tips of the cracks can be calculated as follows:

$$\begin{aligned}F_a &= \frac{K_I^a}{\sqrt{\pi a}\sigma^0} \\ &= \sqrt{\frac{a}{\pi W}}\sqrt{\tan\frac{\pi a}{W}}\frac{\pi}{M}\sum_{k=1}^{M} G_a\left(\cos\frac{2k-1}{2M}\pi\right)\end{aligned} \tag{33}$$

$$\begin{aligned}F_b &= \frac{K_I^b}{\sqrt{\pi b}\sigma^0} \\ &= \sqrt{\frac{b}{\pi W}}\sqrt{\tan\frac{\pi b}{W}}\frac{\pi}{M}\sum_{k=1}^{M} G_b\left(\cos\frac{2k-1}{2M}\pi\right)\end{aligned} \tag{34}$$

$$\begin{aligned}F_c &= \frac{K_I^c}{\sqrt{\pi c}\sigma^0} \\ &= \sqrt{\frac{c}{\pi W}}\sqrt{\tan\frac{\pi c}{W}}\frac{\pi}{M}\sum_{k=1}^{M} G_c\left(\cos\frac{2k-1}{2M}\pi\right)\end{aligned} \tag{35}$$

where

$$G_a(t) = \frac{\cos\frac{\pi a t}{W}\sqrt{1-t^2}}{\sqrt{\left(\sin\frac{\pi a}{W}\right)^2 - \left(\sin\frac{\pi a t}{W}\right)^2}}\frac{\sigma_a^p(t)}{\sigma^0}, \tag{36}$$

$t \in [0, 1)$

$$G_b(t) = \frac{\cos\frac{\pi b t}{W}\sqrt{1-t^2}}{\sqrt{\left(\sin\frac{\pi b}{W}\right)^2 - \left(\sin\frac{\pi b t}{W}\right)^2}}\frac{\sigma_b^p(t)}{\sigma^0}, \tag{37}$$

$t \in [0, 1)$

$$G_c(t) = \frac{\cos\frac{\pi c t}{W}\sqrt{1-t^2}}{\sqrt{\left(\sin\frac{\pi c}{W}\right)^2 - \left(\sin\frac{\pi c t}{W}\right)^2}}\frac{\sigma_c^p(t)}{\sigma^0}, \tag{38}$$

$t \in [0, 1)$

In the following sub-sections, we will investigate the effect of perturbations in the lengths of the cracks and in the bridging stiffnesses on the propagation of the cracks, in order to shed some light on the localisation of damage in short fibre-reinforced cementitious composites.

6.1. Perturbation in crack lengths

In this section, we shall examine the variations of the normalised mode I SIFs and crack opening displacement of bridged cracks. The geometrical parameters are: $W/b = W/c = 5.0$, $H_1/b = H_2/c = 1.0$, whereas the non-dimensional bridging stiffnesses are $\kappa_b = \kappa_c = 0.50$ and $\kappa_a = \kappa_{b(c)}(a/b)$.

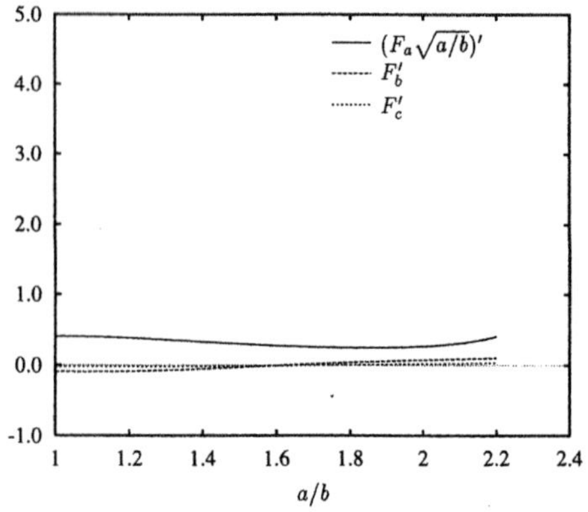

Fig. 17. Variations of the gradients of the stress intensity factors with respect to a/b for bridged cracks.

Figs. 16–18 show that bridging tractions significantly alter SIFs and maximum crack opening displacements in several ways. First, their magnitude, as expected, is considerably reduced in comparison with unbridged cracks. Secondly, they exhibit a more gradual variation with increasing a/b than is the case without bridging. This suggests a less unstable propagation of the cracks. Thirdly, Fig. 18 shows that the maximum opening displacements of the cracks in all rows are almost identical and practically unaffected by changes in a/b. None of the crack rows experiences closure. Therefore, the bridging tractions have induced a uniform stress field in the composites in the later stages of damage. This may explain the difference between the tensile failure of an unreinforced and fibre- reinforced cementitious matrix. It was observed by Balaguru and Shah [23]

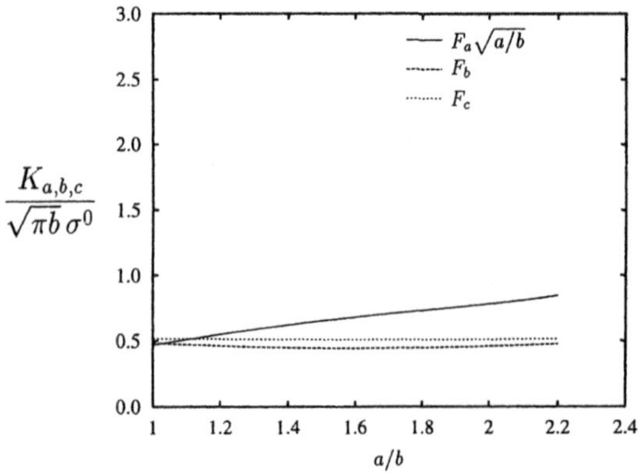

Fig. 16. Variations of the normalised mode I stress intensity factors for bridged cracks.

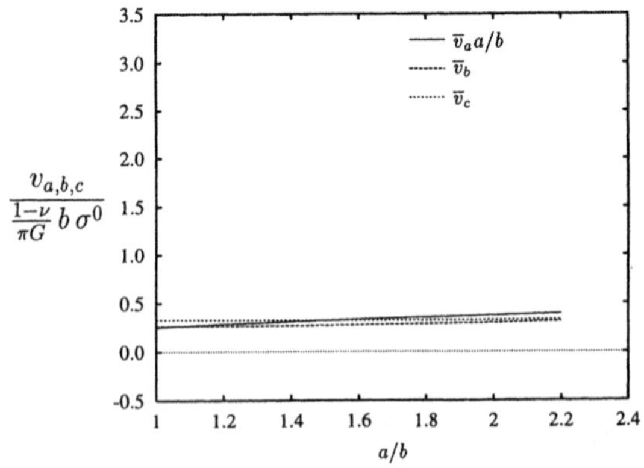

Fig. 18. Variations of the normalised maximum crack opening displacements for bridged cracks.

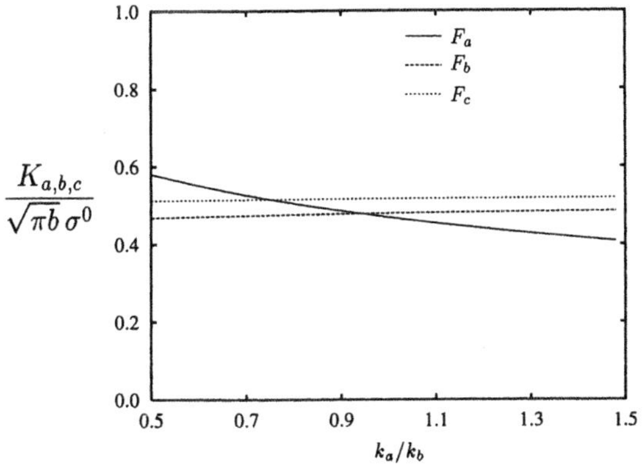

Fig. 19. Variations of the normalised mode I stress intensity factors with the ratio of the bridging stiffnesses k_a/k_b.

that the damage zone in a fibre-reinforced cementitious composite was broader than in a plain cement matrix.

6.2. Perturbation in bridging stiffnesses

In this section, we shall investigate the effect of perturbations in the bridging stiffnesses on the propagation of the crack rows. For simplicity, we assume that the cracks in all rows are identical in length. We shall vary only the bridging stiffness k_a for the central row. In a real cementitious composite, this perturbation is unavoidable because of the difficulties in mixing, so that the distribution of short fibres is unlikely to be uniform throughout the composite.

Fig. 19 shows that the variation of k_a/k_b has only a weak influence on F_b and F_c, but a strong influence on F_a. F_a increases monotonically with a decrease in k_a/k_b. When $k_a/k_b < 0.96$, F_a becomes larger than F_b. Thus, in a fibre-reinforced composite, the damage can localise in a zone where the volume fraction of fibre is just 4% less than the average in the matrix, as the bridging stiffness is a linear function of the volume fraction of fibre [4]. The maximum crack opening displacements in Fig. 20 exhibit similar variations with k_a/k_b to those of the mode I SIFs in Fig. 19.

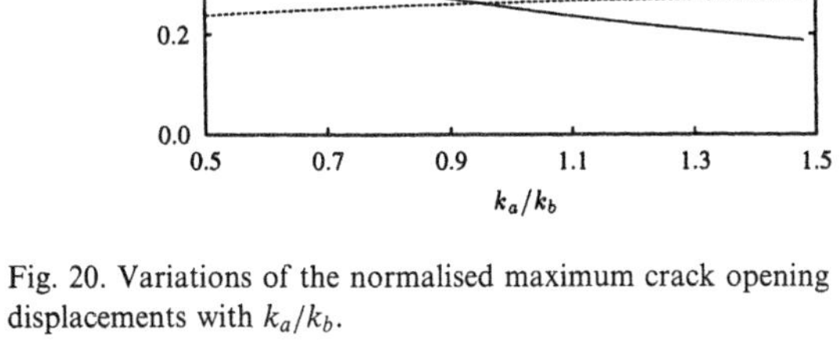

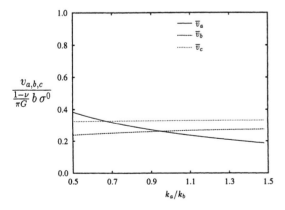

Fig. 20. Variations of the normalised maximum crack opening displacements with k_a/k_b.

7. Conclusions

In short fibre-reinforced cementitious composites, multiple bridged cracks are known to be the source of both strain-hardening and tension-softening. It is shown in this paper that the presence of bridging fibres greatly reduces the SIFs at the tips of the multiple cracks. Therefore, the bridging fibres enhance the toughness of the composite in comparison with monolithic materials.

It has been observed in experiments that once the localisation sets in, the strain level away from the localised zone usually decreases first and then remains more or less at a constant level. Thus, the multiple cracks far away from the damage zone are expected to have much less influence than the ones close to it. Therefore, in the present paper, the stability of a damage zone formed by five dissimilar periodic rows of bridged cracks was investigated. It was confirmed that when the distance between the neighbouring rows of cracks is about two times the crack length, the mutual influence of the rows is indeed insignificant for the considered crack configuration.

The influence of perturbation in the lengths of the cracks and in the bridging stiffnesses was studied. It was found that once the cracks in the central row begin to propagate, they will propagate unstably; whereas the cracks in the neighbouring rows will close, as the applied load drops with the propagation of the cracks in the central rows. This is in agreement with experimental observations. A slightly reduced bridging traction along the cracks in a row will cause the damage to be localised along the plane of this row. Thus, a uniform distribution of fibres in the matrix will prevent premature damage localisation. It was also found that the bridging tractions tend to even out the SIFs in the damage zone, in contrast to the SIFs in an unreinforced matrix in which the SIF at crack tips in the eventual failure plane is far greater than in the adjacent planes. Thus, the failure plane of the plain cement matrix will be narrow and sharp, whereas that of the fibre-reinforced cementitious composite will be diffuse and rough. This is in full agreement with the experiment.

References

[1] Sahasakmontri K, Horii H, Hasegawa A, Nishino F. Mechanical properties of solids containing a doubly periodic rectangular array of cracks. Structural Eng/Earthquake Eng (Japan Society of Civil Engineers) 1987;4:125s–35s.

[2] Delameter WR, Herrmann G, Barnett DM. Weakening of an elastic solid by a rectangular array of cracks. J Appl Mech 1975;42:74–80.

[3] Karihaloo BL. Fracture characteristics of solids containing doubly-periodic arrays of cracks. Proc R Soc Lond 1978;A360:373–87.

[4] Karihaloo BL, Wang J, Grzybowski M. Doubly periodic arrays of bridged cracks and short fibre reinforced cementitious composites. J Mech Phys Solids 1996;44:1565–86.

[5] Isida M, Ushijima N, Kishine N. Rectangular plates, strips and wide plates containing internal cracks under various boundary conditions (in Japanese). Trans Japan Soc Mech Engrs, Series A 1981;47:27–35.

[6] Horii H, Hasegawa A, Nishino F. Process zone model and influencing factors in fracture of concrete. In: Shah SP, Swartz SE, editors. Fracture of concrete and rocks. New York: Springer–Verlag, 1989. p. 205–19.

[7] Ortiz M. Microcrack coalescence and macroscopic crack growth initiation in brittle solids. Int J Solids Structures 1988;24:231–50.

[8] Li VC, Wang Y, Backer S. A micromechanical model of tension-softening and bridging toughening of short random fibre reinforced brittle matrix composites. J Mech Phys Solids 1991;39:607–25.

[9] Horii H, Nemat-Nasser S. Overall moduli of solids with microcracks: load-induced anistropy. J Mech Phys Solids 1983;31:155–71.

[10] Hu KX, Chandra A, Huang Y. On interacting bridged-crack systems. Int J Solids Structures 1994;31:599–611.

[11] Tada H, Paris PC, Irwin GR. The stress analysis of cracks handbook. St. Louis, Missouri: Paris Productions, 1985.

[12] Bao G, Suo Z. Remarks on crack bridging concepts. Appl Mech Reviews 1992;45:355–66.

[13] Gopalaratman VS, Shah SP. Tensile failure of steel fiber-reinforced mortar. ASCE J Eng Mech 1987;113:635–52.

[14] Stang H, Shah SP. Failure of fibre-reinforced composites by pull-out fracture. J Mater Sci 1986;21:953–7.

[15] Lim TY, Paramasivan P, Lee SL. Analytical model for tensile behaviour of steel-fiber concrete. ACI Mater J 1987;84:286–98.

[16] Kim RY, Pagano NJ. Crack initiation in unidirectional brittle-matrix composites. J Am Ceram Soc 1991;74:1082–90.

[17] Mobasher B, Stang H, Shah SP. Microcracking in fibre reinforced concrete. Cement and Concrete Res 1990;20:665–76.

[18] Tjiptobroto P, Hansen W. Tensile strain hardening and multiple cracking in high-performance cement-based composites containing discontinuous fibres. ACI Mater J 1993;90:16–25.

[19] Zok FW, Spearing SM. Matrix crack spacing in brittle matrix composites. Acta Metall Mater 1992;40:2033–43.

[20] Spearing SM, Zok FW. Stochastic aspects of matrix cracking in brittle matrix composites. J Engng Mater Tech 1993;115:314–8.

[21] Karihaloo BL. Fracture mechanics and structural concrete. UK: Addison Wesley Longman, 1995.

[22] Aveston J, Cooper GA, Kelly A. Single and multiple fracture. In: The Properties of Fibre Composites — Conference Proceedings of National Physical Laboratory. Guildford: IPC Science and Technology Press, 1971. p. 15–26.

[23] Balaguru PN, Shah SP. Fibre-reinforced cement composites. In: Basic concepts and mechanical properties: tension. New York: McGraw-Hill, 1992 [Chapter 3].

Failure analysis of R/C columns using a triaxial concrete model

Hong D. Kang, Kaspar Willam *, Benson Shing, Enrico Spacone

CEAE Department, University of Colorado at Boulder, Campus Box 428, Boulder, CO 80309-0428, USA

Received 13 November 1998; accepted 28 December 1999

Abstract

Inelastic failure analysis of concrete structures has been one of the central issues in concrete mechanics. Especially, the effect of confinement has been of great importance to capture the transition from brittle to ductile fracture of concrete under triaxial loading scenarios. Moreover, it has been a challenge to implement numerically material descriptions, which are susceptible to loss of stability and localization. In this article, a novel triaxial concrete model is presented, which captures the full spectrum of triaxial stress and strain histories in reinforced concrete structures. Thereby, inelastic dilatation is controlled by a non-associated flow rule to attain realistic predictions of inelastic volume change at various confinement levels. Different features of distributed and localized failure of the concrete model are examined under confined compression, uniaxial tension, pure shear, and simple shear. The performance at the structural level is illustrated with the example of a reinforced concrete column subjected to combined axial and transverse loading. © 2000 Elsevier Science Ltd. All rights reserved.

Keywords: Triaxial concrete model; Elasto-plastic hardening/softening; Localization properties in tension, compression and shear; R/C column subject to axial loading and shearing

1. Introduction

In recent years, a great number of concrete models have been proposed with the objective to improve numerical failure predictions at the structural level. However, they rarely address the transition from brittle to ductile fracture under a variety of triaxial load scenarios.

In this study, a novel triaxial failure envelope of concrete is presented. Isotropic expansion and contraction describe hardening and softening behavior of concrete, the latter of which is limited to tension and low confined compression below the transition from brittle to ductile failure. The elasto-plastic concrete model is based on a loading surface that is C^1-continuous, and that closes smoothly in equitriaxial compression, whereas the deviatoric trace expands from a triangular to a circular shape with increasing confinement. After exhausting the hardening regime, the failure surface opens in the cap region as concrete materials do not exhibit failure in equitriaxial compression. The overall formulation is based on the concept of isotropic hardening/softening plasticity, whereby the softening formulation introduces a characteristic length that is fracture energy based. In addition, a plastic potential function is introduced, which differs from the yield function to reduce excessive dilatation of associated plastic flow.

For structural analysis, a reinforced concrete column that is subject to axial loading and increasing shearing is analyzed. The column is a scale model of a bridge pier, which was tested at the University of California, San Diego [1] under quasi-static loads. The failure mode of reinforced concrete bridge columns depends on a great number of factors, including the amount/spacing of transverse reinforcement, their slenderness, and boundary conditions. While slender piers tend to fail in flexure, short piers designed in accordance with older code provisions with too little transverse reinforcement tend to fail in a brittle shear manner. In view of the dramatic failure of several bridge piers during the 1995 Kobe

* Corresponding author.

Reprinted from *Computers and Structures* **77 (5)**, 423-440 (2000)

earthquake, the ability to predict the failure mechanism and, thereby, the ductility of bridge piers under the earthquake loading is of great importance for the performance and seismic retrofit of existing bridge structures.

2. Triaxial concrete formulation

For the analysis of concrete failure under various load scenarios, the constitutive model presented recently by Kang [2] captures most features, which are representative for concrete except for the degradation of the elastic stiffness in tension. The triaxial formulation exhibits pressure sensitivity, inelastic dilatancy, growth of deviatoric strength, brittle–ductile transition, strain softening and limitation of hardening in equitriaxial compression. The concrete model is based on loading surfaces that are smooth in all load regimes except for the vertex in equitriaxial tension. The smooth cap in high triaxial compression opens up eventually when the hardening surface reaches the triaxial failure envelope, which is fixed in stress space. On the tension side, at hydrostatic stress levels higher than those at the transition point of brittle–ductile failure, the conical failure envelope contracts to a no-tension cut-off condition in the limit. At an intermediate stage of hardening and softening, the loading surface is comprised of three components:

$$F(\xi,\rho,\theta) = F(\xi,\rho,\theta)_{\text{fail}} + F(\xi,\rho,k(q_{\text{h}}))_{\text{hardg}} + F(\xi,\rho,c(q_{\text{s}}))_{\text{softg}} = 0. \quad (1)$$

The curvilinear shape of the failure envelope is a function of the three stress invariants $I_1 = \text{tr}\,\boldsymbol{\sigma}$, $J_2 = 1/2\mathbf{s}:\mathbf{s}$, $J_3 = \det(\mathbf{s})$ that are here expressed in terms of the Haigh–Westergaard coordinates, $\xi = (1/\sqrt{3})I_1$, $\rho = \sqrt{2J_2}$, and $\theta = (1/3)\cos^{-1}((3\sqrt{3}/2)(J_3/J_2^{1.5}))$. The failure envelope determines the triaxial strength in stress space in the form of a curvilinear triple-symmetric cone, which is depicted in Fig. 1(a) in terms of the meridional section, and in Fig. 1(b), where deviatoric sections are shown at different levels of hydrostatic stress. The C^1-continuous failure envelope is described by a power function as follows:

$$F(\xi,\rho,\theta)_{\text{fail}} = \frac{\rho r(\theta,e)}{f'_{\text{c}}} - \frac{\rho_1}{f'_{\text{c}}}\left(\frac{\xi-\xi_0}{\xi_1-\xi_0}\right)^{\alpha} = 0. \quad (2)$$

The order of the power function, $\alpha = 0.77$, and thus the shape of the meridian was determined by refining the failure envelope to fit the triaxial concrete test data of Launay and Gachon [3]. Solving $F(\xi_{\text{uc}},\rho_{\text{uc}},\theta_{\text{uc}},\rho_1) = 0$, the shear strength yields $\rho_1 = \rho_{\text{uc}}/[(\xi_{\text{uc}}-\xi_0)/(\xi_1-\xi_0)]^{\alpha}$. This assures that the compression meridian passes through the point of uniaxial compressive strength, $F_{\text{uc}} = F(\xi_{\text{uc}},\rho_{\text{uc}},\theta = 60°)$, where the subscript uc indicates uniaxial compression. In Eq. (2), the triaxial concrete model is delimited by the equitriaxial tensile strength $\xi_0 = \sqrt{3}f'_{\text{t}}$ and by the equitriaxial compressive strength $\xi_1 = -\sqrt{3}Sf'_{\text{c}}k$, which locates the cap surface in terms of the cap limiter S and the hardening parameter $0 \leqslant k \leqslant 1$.

The shape of the deviatoric trace is described by the radial distance from the hydrostatic axis [4]

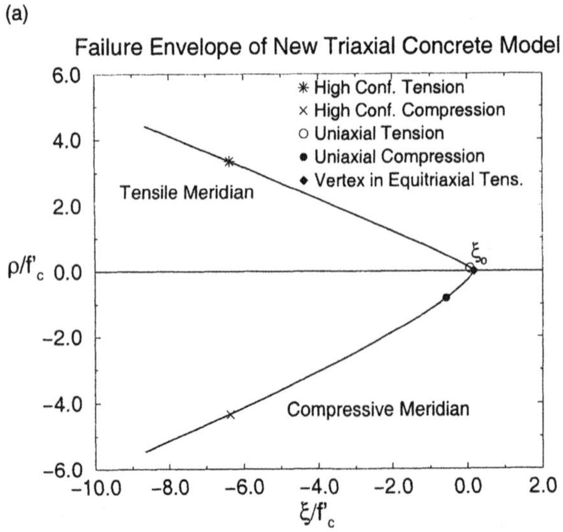

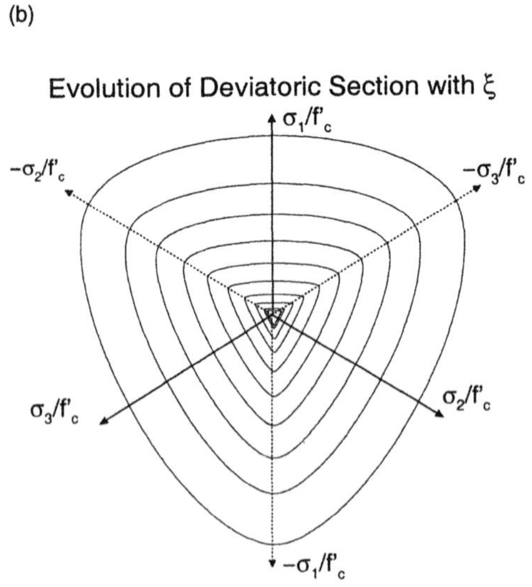

Fig. 1. (a) Meridional section, and (b) deviatoric tracings.

$$r(\theta, e) = \frac{4(1-e^2)\cos^2\theta + (2e-1)^2}{2(1-e^2)\cos\theta + (2e-1)\sqrt{4(1-e^2)\cos^2\theta + 5e^2 - 4e}}, \tag{3}$$

where the eccentricity $0.5 \leqslant e \leqslant 1.0$ describes the out-of-roundedness of the deviatoric trace and is given by the following function of hydrostatic stress $e(\xi) = 1 - 0.5\left((\xi_0 - 5.5f'_c)/(\xi - 5.5f'_c)\right)$. The dependence of the radial distance on the third invariant θ allows the deviatoric trace to expand from triangular to circular shapes with increasing hydrostatic pressure as shown in Fig. 1(b).

The behavior of concrete can be best described by the uniaxial stress–strain relation. For normal strength, concrete generally exhibits a linearly elastic behavior as long as the uniaxial compressive stress does not exceed 45–50% of the compressive strength, f'_c, and it exhibits a nonlinear strain-hardening behavior when the compressive stress varies between $0.50f'_c$ and f'_c. Concrete exhibits softening when it is loaded beyond its peak resistance due to the occurrence of macrocracks and shear bands that jeopardize the integrity of the material. Concrete in tension follows similar behavior, except that the strength is much lower in tension, the change of stiffness in the hardening regime is very small, and softening occurs in a very abrupt and drastic manner due to formation of highly localized tensile cracks.

2.1. *Elasto-plastic formulation*

Assuming infinitesimal deformations, the strain rate decomposes additively into an elastic component and a plastic component, $\dot{\epsilon} = \dot{\epsilon}_e + \dot{\epsilon}_p$. The elastic strain rate is related to the stress rate by Hooke's law:

$$\dot{\boldsymbol{\sigma}} = \mathscr{E} : \dot{\epsilon}_e = \mathscr{E} : (\dot{\epsilon} - \dot{\epsilon}_p). \tag{4}$$

$\mathscr{E} = \Lambda\boldsymbol{\delta} \otimes \boldsymbol{\delta} + 2G\mathscr{I}$ denotes the isotropic elasticity tensor where Λ, G are the Lame moduli, $\boldsymbol{\delta}$ the unit second-order tensor, and $\mathscr{I}$ the unit fourth-order tensor. Assuming no damage, the elastic properties remain constant during the entire plastic deformation process.

According to classical flow theory of plasticity, the elastic range is delimited by the plastic yield condition,

$$F(\boldsymbol{\sigma}, q_h, q_s) = 0. \tag{5}$$

The size and shape of the surface depend on two internal variables q_h, and q_s. One describes the increase of the strength due to hardening and the other the degradation of strength due to softening. In the case of plastic loading, the direction of the plastic strain rate is governed by the non-associated flow rule,

$$\dot{\epsilon}_p = \dot{\lambda}\frac{\partial Q}{\partial \boldsymbol{\sigma}} = \dot{\lambda}\boldsymbol{m}, \tag{6}$$

where Q denotes the plastic potential and $\dot{\lambda}$ the plastic multiplier. The latter is determined with the help of the incremental consistency condition, $F_{n+1}(\Delta\lambda) = 0$, which assures that the constitutive behavior under persistent plastic deformations satisfies the yield condition at the end of the finite time interval $\Delta t = t_{n+1} - t_n$.

2.2. *Isotropic hardening*

The nonlinear behavior of concrete in the pre-peak regime is described with the help of isotropic hardening plasticity. Intermediate stages of the hardening and softening surfaces are shown in Fig. 2(a) and (b).

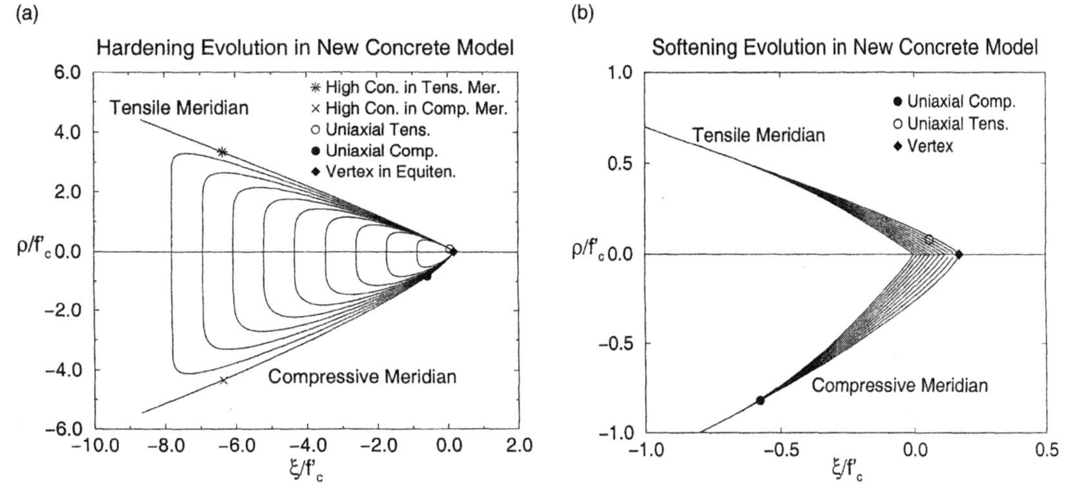

Fig. 2. Construction of triaxial failure envelope with (a) hardening, and (b) softening.

The evolution is expressed by an additional function, which varies with the hardening parameter k. The hardening component F_{hardg} of the yield function in Eq. (1) introduces a compression cap [5] to the conical loading surface, which expands as $0 \leqslant k_0 \leqslant k < 1$ increases, and which opens up entirely when the hardening surface reaches the failure envelope, $k = 1$. The mathematical description of the hardening function involves

$$F(\xi, \rho, k(q_h))_{\text{hardg}} = -\frac{\rho_1}{f_c'}\left[\left(\frac{\xi - \xi_1}{\xi_0 - \xi_1}\right)^{\beta} - 1\right]. \tag{7}$$

The exponent $\beta = 0.25[(1 - k^2)/(1 - k_0^2)]$ is a function of the hardening parameter $0 \leqslant k(q_h) \leqslant 1$, which in turn is a monotonically increasing scalar function of the plastic strain: $q_h = \int_t \sqrt{\dot{\epsilon}_p : \dot{\epsilon}_p} = \int_t \dot{\lambda}\|\boldsymbol{m}\|$, where $\boldsymbol{m}$ denotes the gradient of the plastic potential. In order to assure smooth transition from the initial elastic response to elasto-plastic hardening and subsequent softening, the *Hognestad* parabola is used for describing hardening in terms of $k = k(q_h, \xi)$, where the hardening parameter, k, varies also with hydrostatic stress in order to account for the increase of ductility under increasing confinement. Note that the softening contribution remains inactive during the entire hardening process.

2.3. Isotropic softening

The failure envelope contracts due to tensile decohesion after the stress path has reached the failure envelope, $k = 1$, at hydrostatic stress levels higher than the transition point of brittle–ductile failure, $\xi_c = -T\xi_0$. The mathematical description of the softening function involves

$$F(\xi, \rho, c(q_s))_{\text{softg}} = \frac{\rho_1}{f_c'}(1 - c)\left(\frac{\xi_0}{\xi_0 - \xi_1}\right)^{\alpha}\left(\frac{\xi_c - \xi}{\xi_c}\right)^2. \tag{8}$$

The softening parameter $1 \geqslant c(q_s) \geqslant 0$ is a monotonically decreasing scalar function of the tensile fracture strain $\epsilon_f = \int_t \dot{\lambda}\|\boldsymbol{m}_t\|$ or rather the crack opening displacement $q_s = \epsilon_f \ell_c$. Note, ℓ_c denotes an internal characteristic length, which accounts for the fracture energy release [6], while $\boldsymbol{m}_t$ extracts the tensile components from the gradient of the plastic potential. In order to assure smooth transition from hardening to softening, the vertex of the *Hognestad* parabola is matched to the *Gaussian* decay function, which describes softening in terms of $c = c(q_s, \xi)$. In the case of the fracture energy based embedded crack formulation, the softening modulus E_f is enhanced by a *Dirac Delta* function [7,8] such that $E_f = \partial c/\partial \epsilon_f = (\partial c/\partial q_s)(\partial q_s/\partial \epsilon_f) = \bar{E}_f \ell_c$, where $\bar{E}_f$ denotes the plastic modulus of the Gaussian c–q_s relation, while the characteristic length ℓ_c regularizes the softening after the onset of localization. Note that the hardening contribution remains inactive during the entire softening process.

2.4. Non-associated plastic flow rule

Concrete materials exhibit microcracking at an early stage of the deformation process, which introduces inelastic strains in addition to elastic deformations. Strain softening causes dissipation of plastic energy that is transformed into fracture energy when major cracks develop. These inelastic phenomena are described by plastic flow, i.e. by an evolution law that defines the plastic strain rate $\dot{\epsilon}_p = \dot{\lambda}(\partial Q/\partial \boldsymbol{\sigma})$ in a tensorial sense. The plastic potential determines the direction of plastic flow in terms of the gradient $\boldsymbol{m} = (\partial Q/\partial \boldsymbol{\sigma})$, which controls inelastic dilatation. In order to avoid excessive inelastic dilatation, a non-associated plastic potential, $Q \neq F$, is adopted in its volumetric contribution, so that $\boldsymbol{m}_v \neq \boldsymbol{n}_v$, where $\boldsymbol{n}_v = (\partial F/\partial \xi)(\partial \xi/\partial \boldsymbol{\sigma})$ and $\boldsymbol{m}_v = (\partial Q/\partial \xi)(\partial \xi/\partial \boldsymbol{\sigma})$. In this case, the exponent $\alpha = 0.77$ in the failure envelope in Eq. (2), which determines the slope of the conical failure envelope, is reduced to $\bar{\alpha} = 0.23$ in the expression of the plastic potential Q_{fail} in order to control inelastic dilatation, see Fig. 3(a) and (b).

2.5. Model calibration and performance

The constitutive model presented in this article was calibrated with the aid of conventional triaxial compression tests by Hurlbut [9] with increasing confinement pressures of $p = 0$, 0.69, 3.45 MPa (0, 100, 500 psi), respectively. The parameters and functions that describe the loading surface at an intermediate stage of hardening and softening are summarized in Table 1. For performance studies of the concrete model, we need to add the elastic properties $E = 19\,320$ MPa (2800 ksi) and $\nu = 0.2$ to the description of the inelastic material properties. Plots of the axial compressive stress versus the axial (z-component) and lateral strains (x, y-components) obtained from the experiments and the constitutive study are compared in Fig. 4(b). Strength, softening, and lateral dilatation are well captured for $p = 0.69$ MPa (100 psi). However, the strength under $p = 3.45$ MPa (500 psi) is underestimated by the concrete model. The behavior under uniaxial tension in Fig. 4(a) compares well with the direct tension experiments by Hurlbut [9], it replicates the stiffness and strength but overestimates the measured fracture energy release.

2.6. Incremental format in constitutive driver

In the finite time step $\Delta t = t_{n+1} - t_n$, the constitutive driver integrates the rate form of the elasto-plastic constitutive relations. Given the initial conditions at stage $t = t_n$ in terms of $\boldsymbol{\sigma} = \boldsymbol{\sigma}_n$, $q = q_n$, and $\epsilon_p = \epsilon_n^p$, the incremental integration scheme detects first, whether

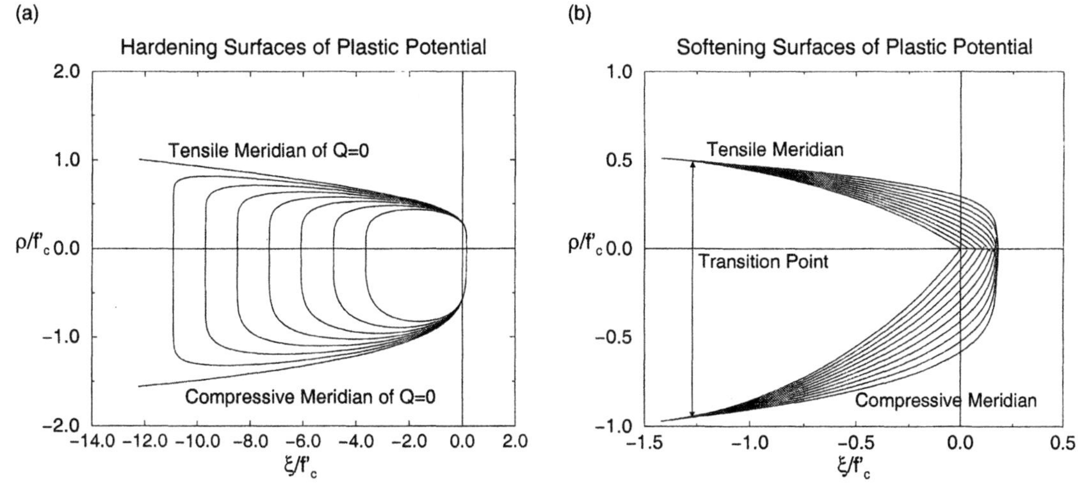

Fig. 3. Construction of plastic potential envelope with (a) hardening, and (b) softening.

Table 1
Six parameters and three functions of evolution for the eccentricity and hardening/softening

$F_{\text{fail}} = 0$	$F_{\text{hardg}} = 0$	$F_{\text{softg}} = 0$	$Q_{\text{fail}} = 0$
$e = e(\xi)$	$\beta = \beta(k(\epsilon_{\text{p}}))$	$c = c(u_{\text{f}})$	$\bar{\alpha} = 0.23$
$f_{\text{c}}' = 22.08,\ f_{\text{t}}' = 2.48$ MPa	$\dot{\epsilon}_{\text{p}} = \dot{\lambda}\|\boldsymbol{m}\|$	$\dot{u}_{\text{f}} = \dot{\epsilon}_{\text{f}} l_{\text{c}}$	
$\alpha = 0.77$	$\boldsymbol{m} = \partial Q/\partial \boldsymbol{\sigma}$	$\dot{\epsilon}_{\text{f}} = \dot{\lambda}\|\boldsymbol{m}_{\text{t}}\|$	
$S = 7.0$		$T = 8.2$	

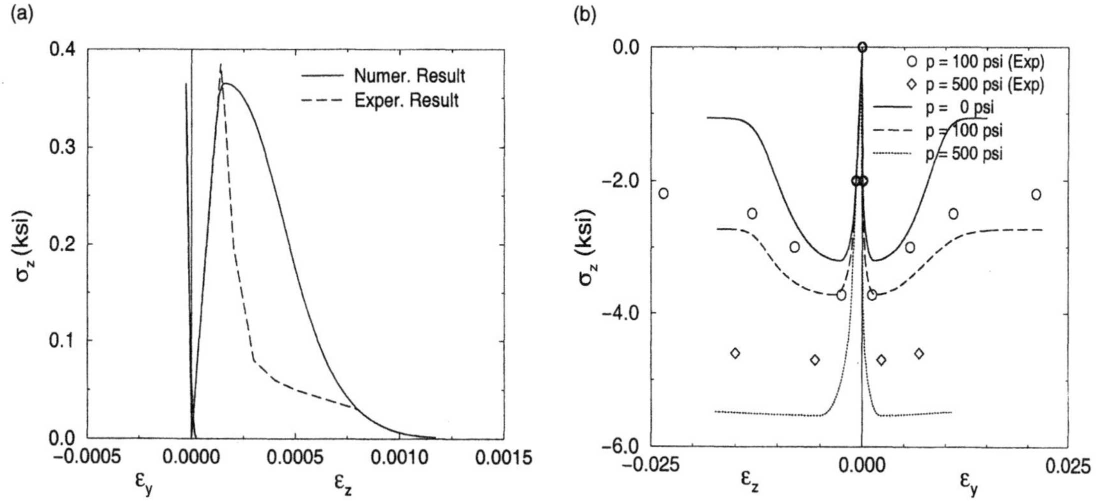

Fig. 4. Calibration of the model in (a) uniaxial tension, and (b) confined compression.

loading is in the plastic regime by evaluating the yield function at the trial stress state, to see if $F_{\text{trial}} = F(\boldsymbol{\sigma}_{\text{T}}, k, c) > 0$, where $\boldsymbol{\sigma}_{\text{T}} = \boldsymbol{\sigma}_n + \mathscr{E} : \Delta\boldsymbol{\epsilon}$. The flow direction $\boldsymbol{m} = \partial Q/\partial \boldsymbol{\sigma}$ for the plastic corrector step and the magnitude of the incremental plastic multiplier $\Delta\lambda$ are subsequently determined iteratively by enforcing incremental consistency $F_{n+1}(\Delta\lambda) = 0$ in the spirit of the *Backward Euler Method*:

$$\Delta\epsilon_p = \Delta\lambda \boldsymbol{m}_{n+1}. \tag{9}$$

Then, the elastic stress increment after a cycle of elastic prediction and plastic correction is

$$\Delta\boldsymbol{\sigma} = \mathscr{E} : \Delta\epsilon - \Delta\lambda\mathscr{E} : \boldsymbol{m}_{n+1}. \tag{10}$$

The return mapping algorithm projects the elastic trial stress back to the final yield surface $F_{n+1} = 0$, in order to satisfy the *Kuhn–Tucker* condition under persistent plastic deformation. With the above stress increment, the final state of stress at $t = t_{n+1}$ is evaluated as

$$\boldsymbol{\sigma}_{n+1} = \underbrace{\boldsymbol{\sigma}_n + \mathscr{E} : \Delta\epsilon}_{\boldsymbol{\sigma}_T} - \Delta\lambda\mathscr{E} : \boldsymbol{m}_{n+1}, \tag{11}$$

where $\boldsymbol{\sigma}_T$ denotes the trial stress state. For a given direction $\boldsymbol{m} = \boldsymbol{m}^k_{n+1}$ of the kth iterative implementation of the Backward Euler Method, the magnitude $\Delta\lambda$ of the plastic corrector in the return mapping algorithm is initially estimated from the linearized consistency condition $\dot{F}^k_{n+1} = 0$.

$$\begin{aligned} \dot{F}^k_{n+1} &= \frac{\partial F^k_{n+1}}{\partial \boldsymbol{\sigma}} : \dot{\boldsymbol{\sigma}} + \frac{\partial F^k_{n+1}}{\partial q}\dot{q} = 0 \\ \Delta\lambda^{\text{lin}} &= \frac{\boldsymbol{n}^k_{n+1} : \mathscr{E} : \Delta\epsilon}{\boldsymbol{n}^k_{n+1} : \mathscr{E} : \boldsymbol{m}^k_{n+1} + H^p_{n+1}}. \end{aligned} \tag{12}$$

The incremental plastic multiplier $\Delta\lambda$ is determined by enforcing full consistency, $F^k_{n+1}(\Delta\lambda) = 0$. During hardening, the variables, ξ_1, ρ_1, and β in the loading function $F_h(\boldsymbol{\sigma}, k) = 0$ vary with the hardening parameter k such that the effective hardening modulus, H^{hardg}_p is given as

$$\begin{aligned} H^{\text{hardg}}_p &= -\frac{\partial F}{\partial k}\frac{\partial k}{\partial q_h}\|\boldsymbol{m}^k\| \\ &= \left(\frac{\partial F}{\partial \xi_1}\frac{\partial \xi_1}{\partial k} + \frac{\partial F}{\partial \rho_1}\frac{\partial \rho_1}{\partial k} + \frac{\partial F}{\partial \beta}\frac{\partial \beta}{\partial k}\right)\frac{\partial k}{\partial q_h}\|\boldsymbol{m}^k\|, \end{aligned} \tag{13}$$

$$\text{where} \quad \frac{\partial k}{\partial q_h} = E_p = \frac{2}{h_D}(1 - k_0)\left(\frac{h_D}{\sqrt{2h_D q_h}} - 1\right). \tag{14}$$

Here, h_D denotes the hardening ductility function. During softening, $F_s(\boldsymbol{\sigma}, c) = 0$, the effective softening modulus H^{softg}_p becomes

$$H^{\text{softg}}_p = -\frac{\partial F}{\partial c}\frac{\partial c}{\partial q_s}\frac{\partial q_s}{\partial \epsilon_f}\|\boldsymbol{m}^k_t\|, \tag{15}$$

$$\frac{\partial c}{\partial q_s} = \bar{E}_f = -2g\left(\frac{q_s}{s_D}\right)^2 \exp\left(-\delta\left(\frac{q_s}{s_D}\right)^2\right), \tag{16}$$

where $\boldsymbol{m}_t$ denotes $\langle \boldsymbol{m} \rangle$, whereby the McCauley bracket $\langle\ \rangle$ extracts the tensile (*positive*) component of the gradient tensor $\boldsymbol{m}$. The partial derivative $\partial q_s/\partial\epsilon_f = \ell_c$ introduces the characteristic length, $\delta = 10^{4.3}$ is a parameter, which describes the rate of change of softening, and s_D denotes the softening ductility function.

In the current time step, the gradient tensors $\boldsymbol{n}$ and $\boldsymbol{m}$ are used to evaluate the plastic multiplier in Eq. (12). The gradient tensors $\boldsymbol{n}$ and $\boldsymbol{m}$ of the loading surface and the plastic potential are evaluated with the help of the chain-rule

$$\begin{aligned} \frac{\partial F(\xi,\rho,\theta)}{\partial \boldsymbol{\sigma}} &= \frac{\partial F}{\partial \xi}\frac{\partial \xi}{\partial \boldsymbol{\sigma}} + \frac{\partial F}{\partial \rho}\frac{\partial \rho}{\partial \boldsymbol{\sigma}} + \frac{\partial F}{\partial \theta}\frac{\partial \theta}{\partial \boldsymbol{\sigma}}, \\ \frac{\partial Q(\xi,\rho,\theta)}{\partial \boldsymbol{\sigma}} &= \frac{\partial Q}{\partial \xi}\frac{\partial \xi}{\partial \boldsymbol{\sigma}} + \frac{\partial Q}{\partial \rho}\frac{\partial \rho}{\partial \boldsymbol{\sigma}} + \frac{\partial Q}{\partial \theta}\frac{\partial \theta}{\partial \boldsymbol{\sigma}}. \end{aligned} \tag{17}$$

2.6.1. *Incremental consistency condition*

As the linearized plastic multiplier $\Delta\lambda^{\text{lin}}$ yields only an approximate magnitude of the plastic corrector, incremental consistency $F_{n+1}(\Delta\lambda) = 0$ is enforced in the context of the backward euler method. The incremental plastic multiplier $\Delta\lambda^{k+1}$ is the smallest real-value root of $F(\Delta\lambda) = 0$, which is evaluated here via Newton–Raphson iteration. The incremental plastic multiplier is obtained from

$$\Delta\lambda^{k+1} = \Delta\lambda^k - \frac{F^k_{n+1}}{\text{d}F^k_{n+1}/\text{d}\Delta\lambda}, \tag{18}$$

where $F^k_{n+1} = F(\boldsymbol{\sigma}^k_{n+1}, q^k_{n+1})$ denotes the yield residual, and

$$\frac{\text{d}F^k_{n+1}}{\text{d}\Delta\lambda} = \frac{\partial F^k_{n+1}}{\partial \boldsymbol{\sigma}_{n+1}}\frac{\partial \boldsymbol{\sigma}_{n+1}}{\partial \Delta\lambda} + \frac{\partial F^k_{n+1}}{\partial q_{n+1}}\frac{\partial q_{n+1}}{\partial \Delta q}\frac{\partial \Delta q}{\partial \Delta\epsilon_p}\frac{\partial \Delta\epsilon_p}{\partial \Delta\lambda}, \tag{19}$$

the tangential Jacobian of the Newton–Raphson solver to locate the root in the incremental consistency condition.

2.7. *Numerical solution of non-linear finite element problem*

At the structural level the modified Newton–Raphson scheme is used in the current study to solve the nonlinear equations governing global equilibrium. It reduces the overall computing time when the tangent stiffness $\mathbf{K}^*_{\text{tan}}$ is updated at the second iteration of each increment at time $t = t^{i=2}_n$, and is used $\mathbf{K}^{i=2}_n$ throughout the iterative cycle of the increment.

2.7.1. *Algorithmic tangent operator*

In order to set up the tangent stiffness $\mathbf{K}^*_{\text{tan}}$ at the second iteration of the current time step, the algorithmic consistent tangent operator $\mathscr{E}^*_{\text{tan}}$ needs to be first determined from the d$\boldsymbol{\sigma}$–dϵ relation. To this end, the incremental format of the stress–strain relation in Eq. (10) is differentiated, which yields

$$\begin{aligned} \Delta\boldsymbol{\sigma} &= \mathscr{E} : (\Delta\epsilon - \Delta\lambda\boldsymbol{m}) \\ \text{d}\Delta\boldsymbol{\sigma} &= \mathscr{E} : (\text{d}\Delta\epsilon - \text{d}\Delta\lambda\boldsymbol{m} - \Delta\lambda\text{d}\boldsymbol{m}). \end{aligned} \tag{20}$$

The fourth-order Hessian tensor appears in $\mathrm{d}\boldsymbol{m} = (\partial m/\partial\boldsymbol{\sigma}) : \mathrm{d}\Delta\boldsymbol{\sigma} = \mathcal{M} : \mathrm{d}\Delta\boldsymbol{\sigma}$, where $\mathcal{M} = \partial^2 Q/\partial\boldsymbol{\sigma}\otimes\partial\boldsymbol{\sigma}$. Rearranging Eq. (20), the linearized tangential format of the stress–strain relation becomes

$$\mathrm{d}\Delta\boldsymbol{\sigma} = \underbrace{[\mathcal{E}^{-1} + \Delta\lambda\mathcal{M}_{n+1}]^{-1}}_{\mathcal{E}^*} : (\mathrm{d}\Delta\boldsymbol{\epsilon} - \mathrm{d}\Delta\lambda\boldsymbol{m}_{n+1}), \tag{21}$$

where $\mathcal{E}^*$ denotes the modified fourth-order algorithmic elastic stiffness tensor [10], and $\mathrm{d}\Delta\lambda$ the plastic multiplier of the tangential linearization, which is evaluated at $t = t_{n+1}$. Taking the second derivatives of the plastic potential Q with respect to the stress tensor $\boldsymbol{\sigma}$ yields the fourth-order Hessian tensor $\mathcal{M}$ as

$$\begin{aligned}\mathcal{M} &= \frac{\partial^2 Q}{\partial\boldsymbol{\sigma}\otimes\partial\boldsymbol{\sigma}}\\ &= \left(\frac{\partial^2 Q}{\partial\xi^2}\frac{\partial\xi}{\partial\boldsymbol{\sigma}} + \frac{\partial^2 Q}{\partial\xi\partial\rho}\frac{\partial\rho}{\partial\boldsymbol{\sigma}} + \frac{\partial^2 Q}{\partial\xi\partial\theta}\frac{\partial\theta}{\partial\boldsymbol{\sigma}}\right)\otimes\frac{\partial\xi}{\partial\boldsymbol{\sigma}} + \frac{\partial Q}{\partial\xi}\\ &\quad\times\frac{\partial^2\xi}{\partial\boldsymbol{\sigma}\otimes\partial\boldsymbol{\sigma}} + \left(\frac{\partial^2 Q}{\partial\rho\partial\xi}\frac{\partial\xi}{\partial\boldsymbol{\sigma}} + \frac{\partial^2 Q}{\partial\rho^2}\frac{\partial\rho}{\partial\boldsymbol{\sigma}}\right.\\ &\quad\left. + \frac{\partial^2 Q}{\partial\rho\partial\theta}\frac{\partial\theta}{\partial\boldsymbol{\sigma}}\right)\otimes\frac{\partial\rho}{\partial\boldsymbol{\sigma}} + \frac{\partial Q}{\partial\rho}\frac{\partial^2\rho}{\partial\boldsymbol{\sigma}\otimes\partial\boldsymbol{\sigma}}\\ &\quad + \left(\frac{\partial^2 Q}{\partial\theta\partial\xi}\frac{\partial\xi}{\partial\boldsymbol{\sigma}} + \frac{\partial^2 Q}{\partial\theta\partial\rho}\frac{\partial\rho}{\partial\boldsymbol{\sigma}} + \frac{\partial^2 Q}{\partial\theta^2}\frac{\partial\theta}{\partial\boldsymbol{\sigma}}\right)\otimes\frac{\partial\theta}{\partial\boldsymbol{\sigma}}\\ &\quad + \frac{\partial Q}{\partial\theta}\frac{\partial^2\theta}{\partial\boldsymbol{\sigma}\otimes\partial\boldsymbol{\sigma}}.\end{aligned} \tag{22}$$

The linearized tangential form of the plastic multiplier $\mathrm{d}\Delta\lambda$ is obtained from the first-order differential format of the consistency condition.

$$\begin{aligned}\mathrm{d}F_{n+1} &= \frac{\partial F_{n+1}}{\partial\boldsymbol{\sigma}} : \mathrm{d}\Delta\boldsymbol{\sigma} + \frac{\partial F_{n+1}}{\partial q}\mathrm{d}\Delta q = 0\\ &= \boldsymbol{n} : \mathcal{E}^* : (\mathrm{d}\Delta\boldsymbol{\epsilon}_{n+1} - \mathrm{d}\Delta\lambda : \boldsymbol{m}_{n+1})\\ &\quad + \frac{\partial F}{\partial\kappa}\frac{\partial\kappa}{\partial q}\mathrm{d}\Delta q_{n+1} = 0,\end{aligned} \tag{23}$$

where κ denotes the hardening/softening parameter, (i.e. $\kappa = k(\epsilon_{\mathrm{p}})$ in hardening, and $\kappa = c(\epsilon_{\mathrm{f}})$ in softening), and $\mathrm{d}\Delta q_{n+1}$ involves the differential form of the inelastic internal variables as

$$\begin{aligned}\mathrm{d}\Delta q_{\mathrm{h}}^{n+1} &= \mathrm{d}\Delta\lambda\|\boldsymbol{m}_{n+1}\| + \Delta\lambda\frac{\partial\|\boldsymbol{m}_{n+1}\|}{\partial\boldsymbol{m}}\frac{\partial\boldsymbol{m}}{\partial\boldsymbol{\sigma}}\mathrm{d}\Delta\boldsymbol{\sigma}_{n+1},\\ \mathrm{d}\Delta\epsilon_{\mathrm{f}}^{n+1} &= \mathrm{d}\Delta\lambda\|\boldsymbol{m}_{n+1}^t\| + \Delta\lambda\frac{\partial\|\boldsymbol{m}_{n+1}^t\|}{\partial\boldsymbol{m}}\frac{\partial\boldsymbol{m}}{\partial\boldsymbol{\sigma}}\mathrm{d}\Delta\boldsymbol{\sigma}_{n+1},\end{aligned} \tag{24}$$

where $\|\boldsymbol{m}_{n+1}\| = \sqrt{\boldsymbol{m}_{n+1} : \boldsymbol{m}_{n+1}}$. Substituting $\mathrm{d}\Delta q_{n+1}$ into Eq. (23), and solving for $\mathrm{d}\Delta\lambda$ yields the linearized tangential form of the plastic multiplier,

$$\mathrm{d}\Delta\lambda = \frac{(\boldsymbol{n} : \mathcal{E}^* + \frac{2(\partial F/\partial\kappa)E_{\mathrm{p}}\Delta\lambda}{3\|\boldsymbol{m}\|}\boldsymbol{m} : \mathcal{M} : \mathcal{E}^*) : \mathrm{d}\Delta\boldsymbol{\epsilon}}{\boldsymbol{n} : \mathcal{E}^* : \boldsymbol{m} - \partial F/\partial\kappa E_{\mathrm{p}}\|\boldsymbol{m}\| + \frac{2(\partial F/\partial\kappa)E_{\mathrm{p}}\Delta\lambda}{3\|\boldsymbol{m}\|}\boldsymbol{m} : \mathcal{M} : \mathcal{E}^* : \boldsymbol{m}}, \tag{25}$$

where the subscript $n+1$ has been omitted for the sake of simplicity. Substituting $\mathrm{d}\Delta\lambda$ into Eq. (21), the consistent tangent operator $\mathcal{E}^*_{\mathrm{tan}}$ is obtained in the form of $\mathrm{d}\Delta\boldsymbol{\sigma}_{n+1} = \mathcal{E}^*_{\mathrm{tan}} : \mathrm{d}\Delta\boldsymbol{\epsilon}_{n+1}$.

$$\mathcal{E}^*_{\mathrm{tan}} = \mathcal{E}^* - \frac{\mathcal{E}^* : \boldsymbol{m}\otimes\boldsymbol{n} : \mathcal{E}^* + \frac{2(\partial F/\partial\kappa)E_{\mathrm{p}}\Delta\lambda}{3\|\boldsymbol{m}\|}\mathcal{E}^* : \boldsymbol{m}\otimes\boldsymbol{m} : (\mathcal{M} : \mathcal{E}^*)}{\boldsymbol{n} : \mathcal{E}^* : \boldsymbol{m} - (\partial F/\partial\kappa)E_{\mathrm{p}}\|\boldsymbol{m}\| + \frac{2(\partial F/\partial\kappa)E_{\mathrm{p}}\Delta\lambda}{3\|\boldsymbol{m}\|}\boldsymbol{m} : \mathcal{M} : \mathcal{E}^* : \boldsymbol{m}}. \tag{26}$$

The final form of the tangent element stiffness reads $\mathbf{k}^*_{\mathrm{tan}} = \int_V \boldsymbol{B}^t\boldsymbol{E}^*_{\mathrm{tan}}\boldsymbol{B}\,\mathrm{d}V$, whereby $\boldsymbol{E}^*_{\mathrm{tan}}$ is the matrix form of the fourth-order tensor $\mathcal{E}^*_{\mathrm{tan}}$. Fig. 5 illustrates the sequence of calculation steps and iteration cycles at the constitutive as well as the structural levels in the form of a flow chart. Finally, the Boolean assembly operator **a** is used to connect the global system and local element matrices. Thus, the global tangent stiffness and the global internal force vector are assembled respectively as

$$\mathbf{K}^*_{\mathrm{tan}} = \sum_{e=1}^{nel}\mathbf{a}^t\mathbf{k}^*_{\mathrm{tan}}\mathbf{a}, \tag{27}$$

$$\mathbf{S}_{n+1} = \sum_{e=1}^{nel}\mathbf{a}^t\left(\int_{V_{\mathrm{el}}}\boldsymbol{B}^t\boldsymbol{\sigma}_{n+1}\,\mathrm{d}V\right). \tag{28}$$

2.8. Localization study of concrete model

In order to illustrate the potential failure pattern of the concrete structure, the results of uniaxial tension, uniaxial compression, and shear tests are briefly illustrated with the aid of a 3-D eight-node brick element, which is loaded in displacement control (except for the case of pure shear) as indicated in Fig. 6. It is understood that the prescribed displacements introduce a uniform state of deformation in the cubical element, and thus all the examples constitute material rather than structural test problems.

2.8.1. Uniaxial tension

The input parameters for the numerical tests are given in Table 1 with $f'_{\mathrm{c}} = 22.08$ MPa (3.2 ksi), $f'_{\mathrm{t}} = 2.484$ MPa (0.36 ksi), and positive quantities represent tension and negative ones indicate compression. The result of the uniaxial tension test in Fig. 7 indicates mode I brittle fracture due to formation of discontinuous

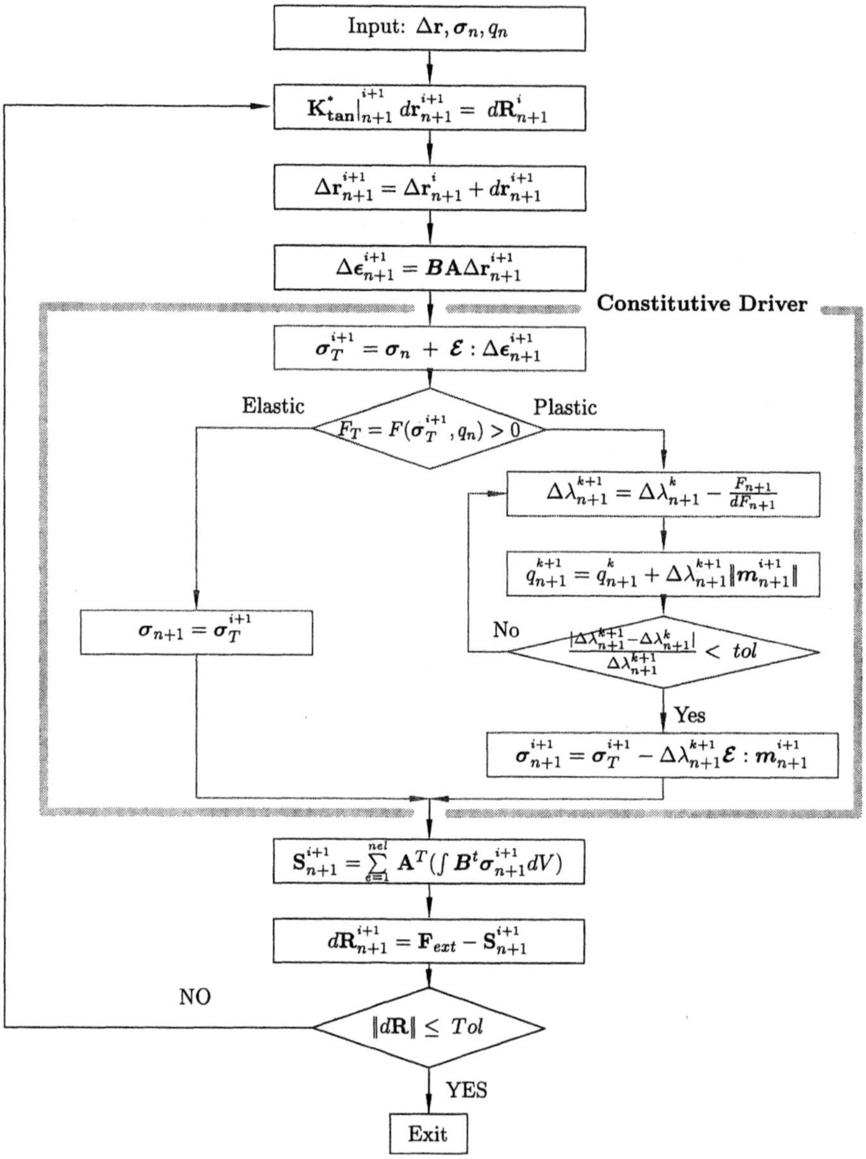

Fig. 5. Flow chart of incremental finite element procedure.

cracks at the localization point, [11]. This brittle failure mode results in sudden degradation of peak strength at the structural level that is accompanied by the formation of spatial discontinuities in the strain and displacement fields perpendicular to the major principal stress according to the Rankine criterion of maximum normal stress. Compared to the Rankine model, there is a slight difference in the value of the eccentricity, $e = 0.56$, but it still agrees well with the accepted cracking mode of failure orthogonal to the direction of major principal stress.

2.8.2. Axial compression

In compression, the lack of associativity reduces the lateral deformations when compared to the axial deformations. Localization analysis in uniaxial compression shows ductile failure, [12] which remains continuous with a tendency for axial splitting as shown in Fig. 7. This phenomenon has been observed recently in a series of uniaxial concrete compression tests that minimize interface friction [13,14]. This failure pattern is confirmed in Fig. 8 in the form of phase velocity plot, and in Fig. 9 in the form of degradation of the local-

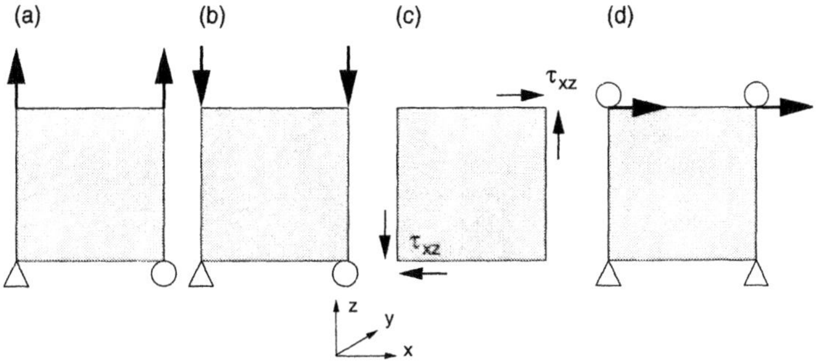

Fig. 6. Failure predictions of non-associated flow rule: (a) uniaxial tension, (b) uniaxial compression, (c) pure shear, and (d) simple shear.

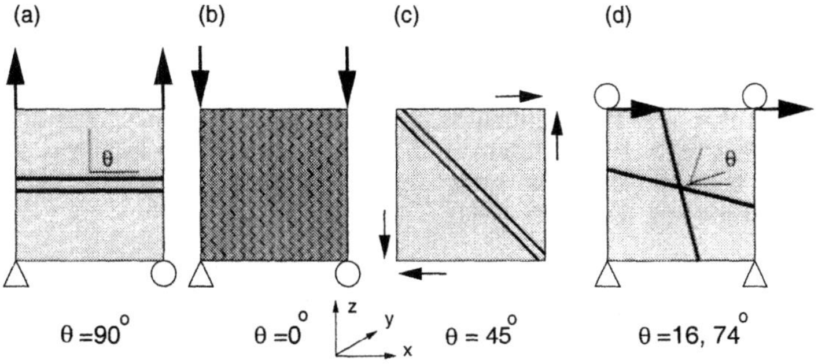

Fig. 7. Failure predictions of non-associated flow rule: (a) uniaxial tension, (b) uniaxial compression, (c) pure shear, and (d) simple shear.

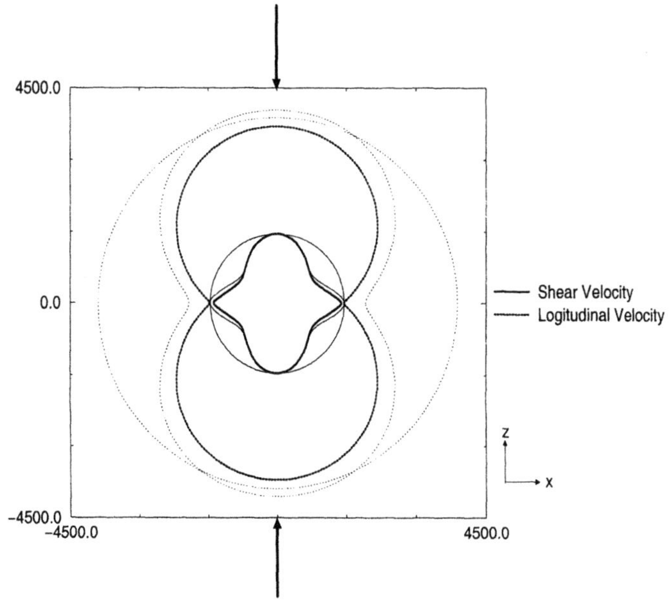

Fig. 8. Polar plots of phase velocities in acoustic analysis under uniaxial compression.

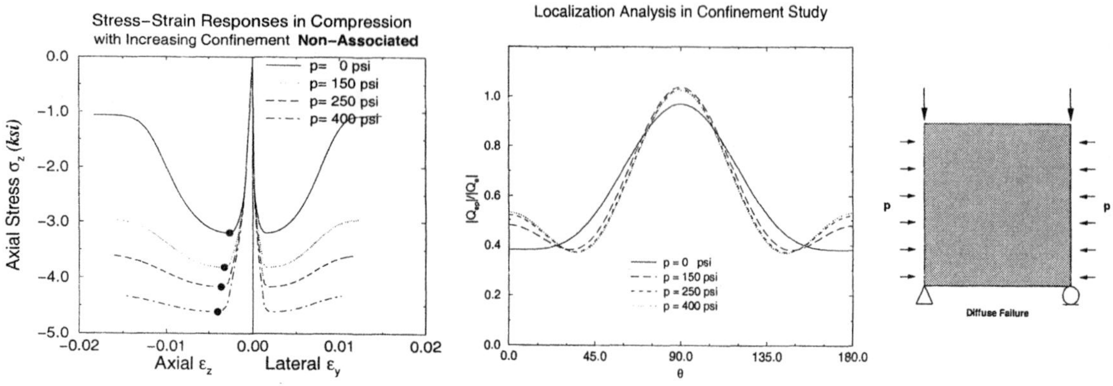

Fig. 9. Deformation and failure of compression with increasing confinement.

ization tensor, $\boldsymbol{Q}_{\tan} = \boldsymbol{N} \cdot \mathscr{E}_{\tan} \cdot \boldsymbol{N}$. This ductile failure mode results in a gradual reduction of strength at the structural level that is accompanied by the formation of diffuse cracks rather than spatial discontinuities in the strain and displacement fields. Non-associativity plays a significant role in controlling the excessive inelastic dilatancy by enforcing the plastic return direction.

In the case of confined compression, the localization failure mode indicates mixed shear as depicted in Fig. 9, which agrees well to the diffuse mixed mode failure pattern of triaxial compression tests [9,15]. With increasing hydrostatic pressure, the non-associated flow rule reduces the lateral dilatation and thus the lateral deformation behavior of concrete.

Unlike the uniaxial compression with or without confinement, the out-of-plane strain constraint in plane strain analysis breaks the symmetry of the cylindrical stress state in axial compression, and enhances the axial strength by 10% above the uniaxial compressive strength as depicted in Fig. 10. The singularity of the localization tensor, $\det(\boldsymbol{Q}_{\tan}) = 0$, which appears due to loss of axial symmetry of the stress state in plane strain, indicates a localized shear mode. This phenomenon explains the criticality of loss of symmetry in analogy to buckling of cylindrical shells. In reinforced concrete structures, imperfections of the constraint introduced by stirrups or transverse reinforcement may induce similar types of failure modes.

2.8.3. Shear

The performance of the concrete model under pure shear and simple shear was studied recently with particular focus on the difference of shear failure under different levels of confinement [16,11]. Apparent shear banding is at best a mixed mode failure phenomenon permitting a variety of interpretations. In consequence, we need to distinguish very carefully between stress controlled experiments with only $\tau_{xz} \neq 0$ and deformation controlled experiments with only $\gamma_{xz} \neq 0$. The former is called pure shear and the latter simple shear. In pure shear, the stress driven format of elasto-plastic tangent compliance becomes

$$\dot{F} = \frac{\partial F}{\partial \boldsymbol{\sigma}} : \dot{\boldsymbol{\sigma}} + \frac{\partial F}{\partial q}\dot{q} = 0 \rightarrow \mathscr{C}_{\tan} = \mathscr{C}_{\mathrm{e}} + \frac{\boldsymbol{m} \otimes \boldsymbol{n}}{H_{\mathrm{p}}}, \tag{29}$$

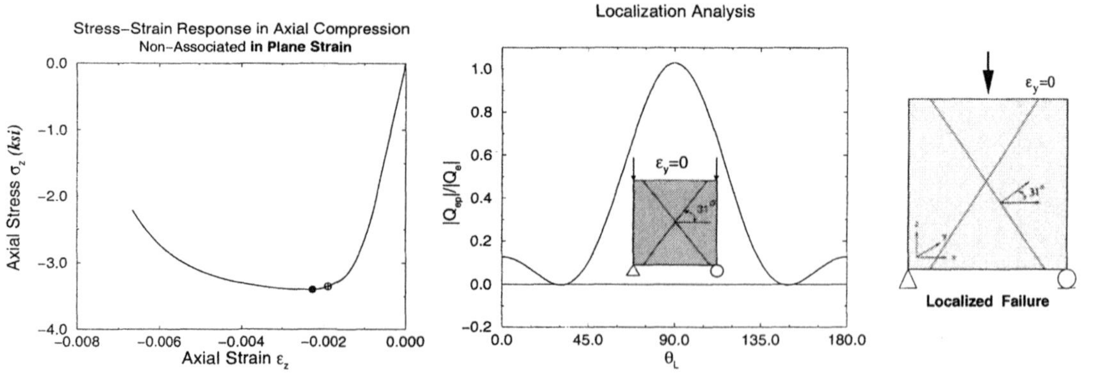

Fig. 10. Deformation and failure of axial compression under plane strain.

where the denominator $H_p = (\partial F/\partial q)\|\boldsymbol{m}\|$ denotes the plastic modulus. Henceforth, when the plastic modulus H_p becomes zero at the peak strength, the elasto-plastic compliance tensor $\mathscr{C}_{tan}$ is infinite as the plastic contribution is divided by zero. In other words, as the shear strength of concrete is very limited, any stress increase beyond the limit point cannot be stabilized by stress control, see Fig. 11(a). Consequently, the stress–strain curve in pure shear shows no softening beyond peak due to the stress driven failure. Thereby, discontinuous tensile cracking takes place at $\Theta_L = 45°$ being the angle of the localization direction, according to the maximum stress hypothesis of Rankine.

The simple shear test is an experiment where the shear strain $\gamma_{xz} > 0$ monotonically increases whereas all other strain components remain zero. The shear capacity of the concrete model under deformation control is shown in Fig. 11(b). The boundary conditions shown in Fig. 6 enforce isochoric deformations due to the kinematic constraints. The so-called Reynolds effect of frictional materials introduces normal stresses under increasing shear deformations which remain negligible in the case of linear isotropic elastic behavior under infinitesimal deformations. This phenomenon occurs since the pressure-sensitive concrete formulation introduces large inelastic dilatation which in turn induce direct stresses and confinement when the dilatational deformations are suppressed. Thus, the stress–strain curve τ_{xz} vs. γ_{xz} in the simple shear test exhibits apparent hardening without ever reaching the limit point, det $(\mathscr{E}_{tan}) = 0$ [17] along the load path. The suppression of inelastic dilatation induces compression due to confinement, which leads to shear localization dominated by boundary conditions rather than the material.

Failure analysis with the non-associated flow rule in simple shear indicates in-plane (i.e. x–z plane) localized failure with $\Theta_L = 16°, 74°$ that initiates before reaching the limit point. At this stage, the single brick element is unable to capture the formation of discontinuous bifurcation det $\boldsymbol{Q}_{tan} = 0$ and continues the numerical solution along the hardening branch as depicted in Fig. 11(b). We note that discontinuous bifurcation develops in the ascending branch of simple shearing because of loss of symmetry, $\boldsymbol{Q}_{tan} \neq \boldsymbol{Q}^t_{tan}$, due to non-associative flow. We should be aware that the stress–strain diagram exhibits apparent hardening though $k = 1$ as the stress path moves up the conical failure envelope and moves along the deviatoric trace from $\theta = 30° \rightarrow 45°$ due to the normality in the deviatoric section.

In sum, shear failure of concrete materials has been a long-term issue from the fundamental standpoint of mechanics of materials that exhibit large differences between the tensile and compressive strength values. Pressure-sensitive concrete materials introduce large dilatational effects which in turn induce normal stresses and confinement when the dilatational deformations are suppressed. Thus, simple shear produces strength and ductility values which are far higher than those under pure shear, due to the confinement effect induced by the kinematic constraint of zero normal strain. The very large discrepancy of strength and ductility in pure and simple shear reflects the large difference in shear behavior of reinforced concrete columns due to the different amount of transverse strirrups.

3. Numerical results on R/C column

In sequel to the constitutive study, a finite element analysis is performed to examine the behavior of a rectangular reinforced concrete column which was tested at the University of California, San Diego [1] under axial

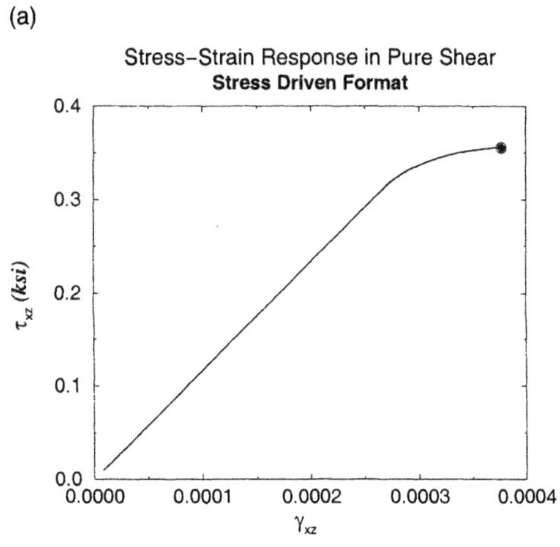

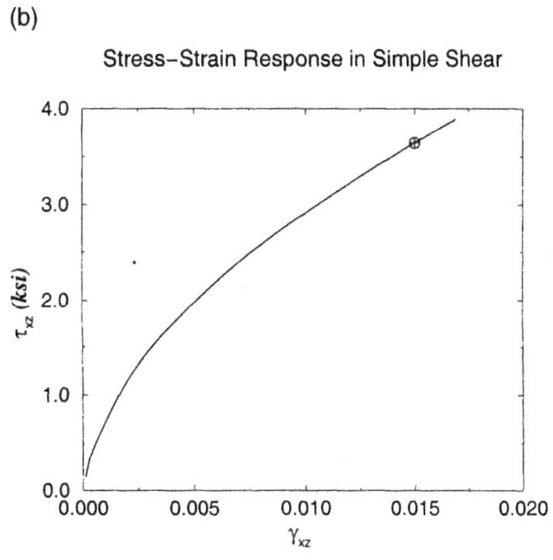

Fig. 11. Responses of (a) pure shear and (b) simple shear.

loads and lateral displacements. The finite element analysis involves several factors such as spatial discretization, boundary conditions, loading histories, and material assumptions, which dominate the limit state performance of the column. First, the structure needs to be discretized properly in order to capture regions of high strain gradients and the most likely places, where spatial discontinuities develop in the finite element mesh due to localization; otherwise, spatial failure bands cannot be formed. Second, the boundary conditions of the numerical simulation have to reproduce faithfully the boundary conditions, which the real structure does experience. In fact, the boundary conditions play a significant role in failure analysis, because they provide the static and kinematic constraints. Finally, the loading needs to be applied in a realistic manner in the numerical simulation in order to capture the proper history effects in the inelastic path-dependent failure process. The numerical simulation needs robust algorithms, which are able to handle severe non-linearities such as tensile cracking and shear softening in the reinforced concrete column. For the study of the concrete column, the finite element program FEAP [18] is used.

3.1. Material properties

The average concrete strength of $f_c' = 37.95$ MPa (5.5 ksi) reported by Xiao et al. [1] is used to analyze the reinforced concrete column. Grade 40 ($f_y = 317.4$ MPa (4.6 ksi)) steel is used for the longitudinal as well as the transverse reinforcement. Young's moduli of concrete and steel are $E_c = 24150$ MPa (3500 ksi) and $E_s = 200100$ MPa (29000 ksi), respectively. In the absence of extensive material test data, representative values for Poisson's ratio, the tensile strength, and the fracture energy are used for the concrete material. The geometry of the column in the test setup and in the finite element mesh is shown in Figs. 12 and 13. As the nonlinear 3-D finite element analysis is taxing computer resources, the finite element mesh lumps the longitudinal reinforcement in the y-direction at the corner of the cross-section (i.e. in the plane orthogonal to the x-load direction), and increases the spacing of the transverse hoop reinforcement in the z-direction. Thereby, the same amount of steel area for axial reinforcement and transverse hoops are used in the finite element analysis as in the experiment. For simplicity, axial bar elements with elastic–perfectly plastic behavior are used to idealize the longitudinal as well as the transverse reinforcing bars. The concrete is described by eight-node solid brick elements using full $2 \times 2 \times 2$ Gauss integration. For the sake of computer time, the mesh of eight-node solid elements was not subdivided in the y-direction in analogy to a layered representation of the cross-section. As all bar and brick elements are sharing the same

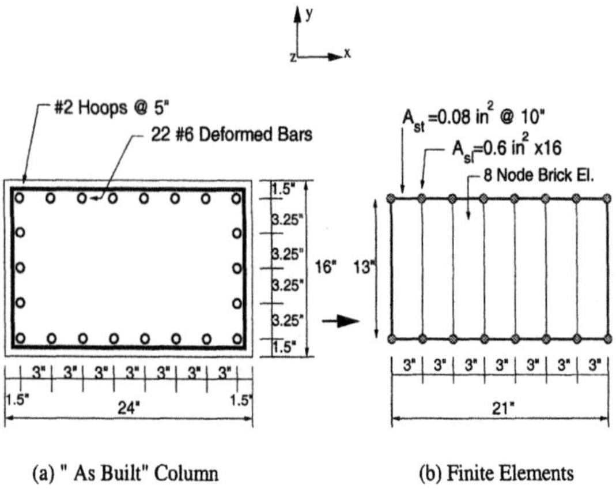

Fig. 12. (a) Section of column in test, and (b) section of column in finite element mesh.

nodes, concrete and reinforcement of the column are assumed to be fully bonded permitting no slip.

3.2. Axial compression

Concrete columns in centric compression induce lateral dilatation. Thus, reinforcing the column with transverse stirrups confines the concrete and increases the strength and ductility of the concrete column. Before considering the combined axial and transverse load history, the axial compression capacity of the reinforced concrete column is examined.

3.2.1. Plain concrete column

The unreinforced concrete column is numerically tested first to estimate the capacity of plain concrete. Figs. 14(a) and 15(a) illustrate the overall behavior of plain concrete in terms of the global force–displacement curve and of the deformed shape. In analogy to the failure pattern in Fig. 7 when a single plain concrete element was subjected to uniaxial compression the column under axial compression does not form a spatial discontinuity. In fact, the localization diagrams in Figs. 8 and 9 suggest a tendency towards diffuse axial splitting.

3.2.2. Reinforced concrete

Similar to the plain concrete specimen under confined compression, the transverse reinforcement generates passive confinement which increases the strength as well as the ductility of the concrete column. For comparison, Fig. 14(a) super imposes the increase of stiffness, strength, and ductility of the column onto the response of the unreinforced column. The distribution of lateral normal strains (i.e. ϵ_{xx}) in Fig. 15(b) indicates some loss of uniform lateral dilatation of the column. In other terms, the reinforced concrete column develops a triaxial

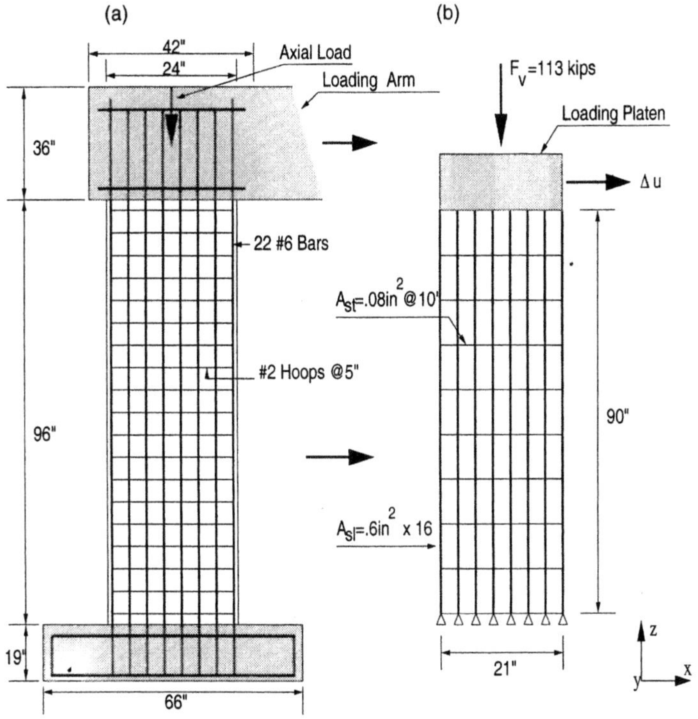

Fig. 13. (a) Column in test setup, and (b) finite element mesh of column.

state of confined compression, which is no longer uniform due to the transverse reinforcement.

3.3. *Axial tension*

In order to revisit the tensile behavior of plain concrete, the unreinforced concrete column is subjected to axial tension first. Similar to the single element test shown in Fig. 7, the finite element mesh of the plain concrete column experiences a transverse crack in the form of a limit point when det $\mathscr{E}_{\tan} = 0$ at the same time as localization det $\boldsymbol{Q}_{\tan} = 0$ [19,20] takes place at the peak of the stress–strain response. The structural behavior of the plain concrete column is illustrated in the

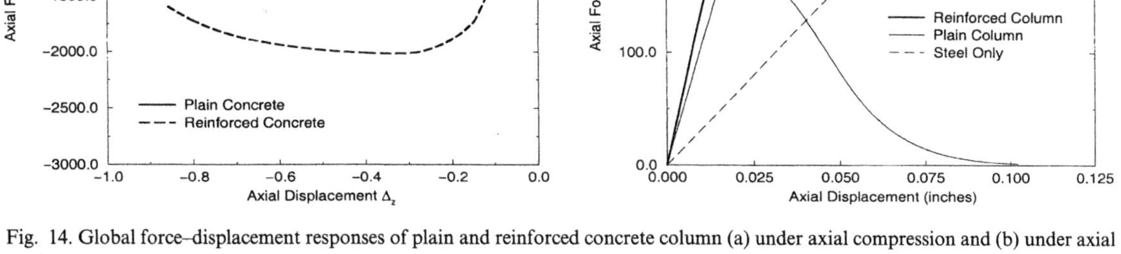

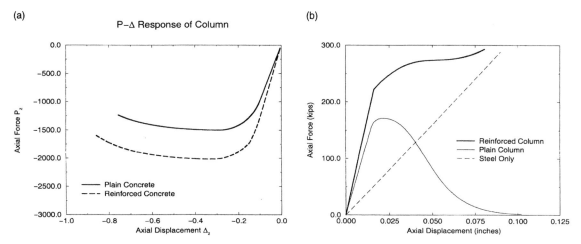

Fig. 14. Global force–displacement responses of plain and reinforced concrete column (a) under axial compression and (b) under axial tension.

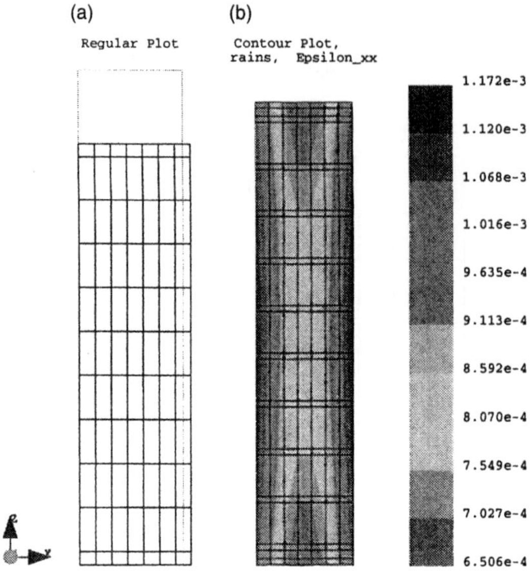

Fig. 15. (a) Deformed shape of plain concrete column subject to axial compression. (b) Deformed shape and strain distribution of reinforced concrete column subject to axial compression.

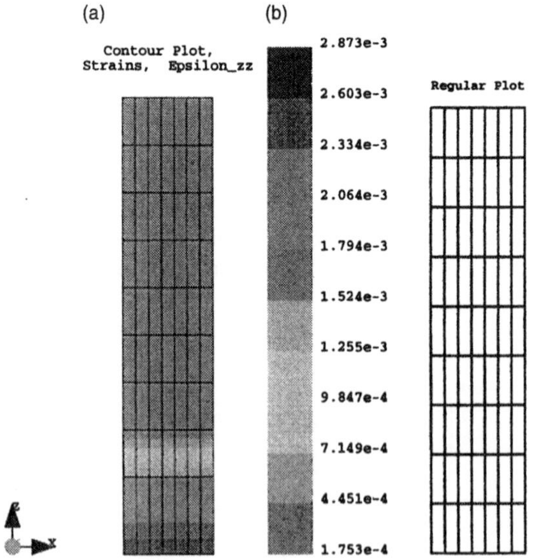

Fig. 16. (a) Deformed shape and normal strain ϵ_{zz} distribution of plain concrete column subject to uniaxial tension. (b) Deformed shape reinforced concrete column subject to axial tension.

form of the load–displacement response in Fig. 14(b) and the deformed shape in Fig. 16(a). In the absence of axial reinforcement, the tensile strain ϵ_{zz} is trapped in the bottom layer of finite elements of the plain concrete column. For comparison, the force–displacement response of the reinforced concrete column is super imposed in Fig. 14(b) exhibiting tension stiffening due to concrete softening. The deformed shape is presented in Fig. 16(b), which shows no localization in the presence of axial reinforcement. Thereby, the reinforced concrete column when subjected to axial tension exhibits a non-uniform state of triaxial tension due to the lateral restraint of the stirrups.

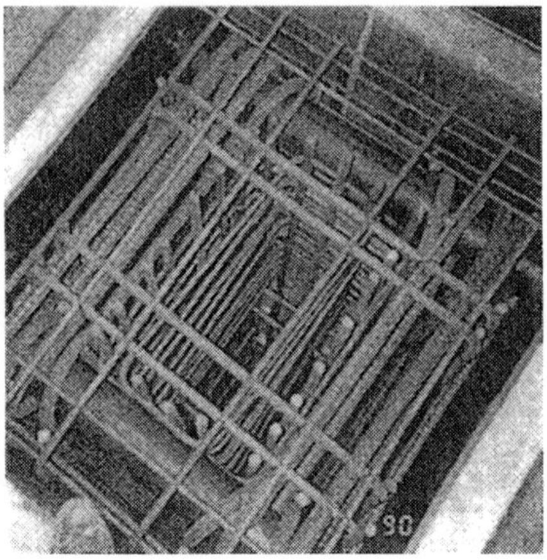

Fig. 17. Reinforcement in column top stub.

3.4. Combined axial load and lateral displacement

In general, reinforced concrete columns act primarily as compression members. However, lateral loads due to wind and earthquake introduce combinations of axial, flexural and transverse shear loading. To this end, the computational simulation of the 1/3 scale model of a prototype bridge pier is presented as a final example, which was tested at the University of California, San Diego [1]. As shown in Fig. 13, cyclic lateral displacements with a constant axial load of $P = -113$ kips were applied to the reinforced concrete pier in the experiment in order to estimate the shear capacity of the column under the earthquake motion. The experimental results of column R1 (out of three columns tested with different reinforcement) do exhibit axial, flexural and shear deformations. The longitudinal reinforcement was anchored to the footing and top slab of the column, (Figs. 13 and 17). Fig. 18 illustrates the experimental results of the load–displacement response behavior under the cyclic test, whereby the peak of each experimental load cycle is marked by an ∘ symbol. In order to verify the flexural as well as the shear capacities of the reinforced concrete column, the finite element mesh depicted in Figs. 12 and 13 is used for the numerical simulation under a monotonically increasing lateral displacement with a constant axial load. The top and

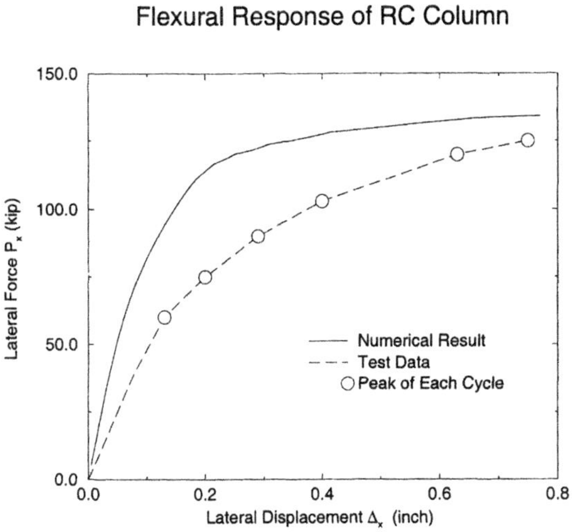

Fig. 18. Lateral force–displacement response of reinforced concrete column subject to axial and lateral loading.

bottom of the column are fixed from rotation. To this end, uniform vertical deflections are prescribed at the top of the column. In fact, as the stress in each element varies, the convergence of the iterative stress redistribution in the numerical simulation is very sensitive to the size of the load increment.

3.4.1. *Numerical results*

Due to the fixed end conditions at the top and bottom, the column deforms anti-symmetrically with respect to the mid-height under combined axial and lateral loading. As delineated in the constitutive study, the shear response of concrete with kinematic constraints demonstrates an enhanced shear strength because of dilatancy-induced confinement and stress redistribution. This may reduce the shear deformation of a reinforced concrete column subject to axial and lateral loads. Figs. 19(a) and (b) illustrate the shear deformations of the column, which indicate' smaller shear deformations in the numerical result when compared to the experiment.

The lateral force of the numerically tested column reads approximately 28.1 kN (125 kips) at the peak, which agrees reasonably well with the experimental test data. In the report of the University of California at San Diego, the experimental value of the ultimate lateral force is about 27 kN (120 kips). However, the finite element model, which comprises eight-node brick elements and two-node bar elements, is stiffer than the experimentally tested column illustrated in Fig. 18. Moreover, the overall numerical results exhibit less ductility than the actual column. This discrepancy may result from the softening behavior of the tensile part of the column, where the fracture energy release rate in direct tension was assumed to be $G_f \sim 1.505$ N/mm (0.17 kips/in.), which corresponds to the area under the σ–Δ curve. Furthermore, since the cut-offs of the longitudinal reinforcement bars were not hooked to the horizontal bars in the end slabs as depicted in Fig. 17, bond slip might develop at the end slabs due to bending. This might have caused the difference in the behaviors illustrated in Fig. 18, in addition to the influence of the different load histories (cyclic vs. monotonic).

The stress–strain responses of the longitudinal reinforcement in the tension part of the column first reach yielding when the lateral force $P_x = 21.8$ kN (97 kips) and the lateral displacement $\Delta_x = 5.08$ mm (0.2 in.), whereas in the experiment, the longitudinal reinforcement first yield when the lateral force is 20.2 kN (90 kips) and the lateral displacement is about $\Delta_x =$ 7.62 mm (0.3 in.). In order to examine the behavior

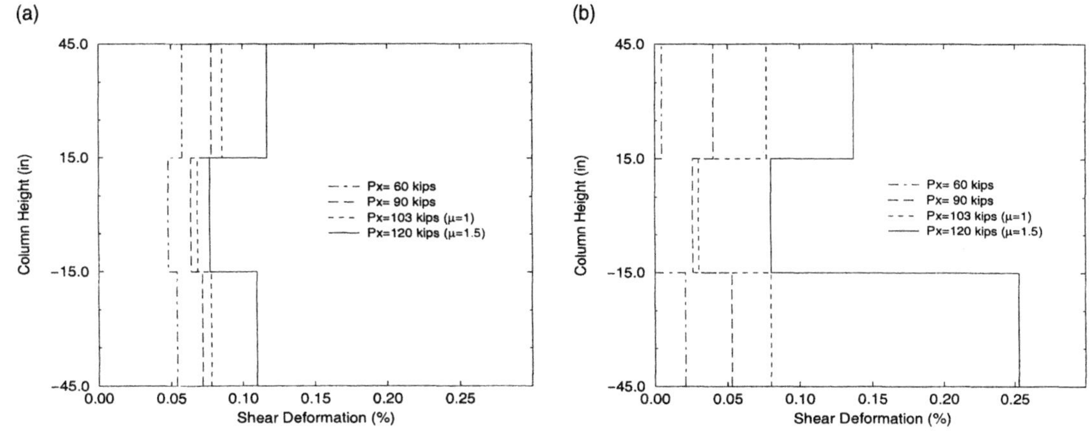

Fig. 19. (a) Numerical and (b) experimental results of shear deformation distributions.

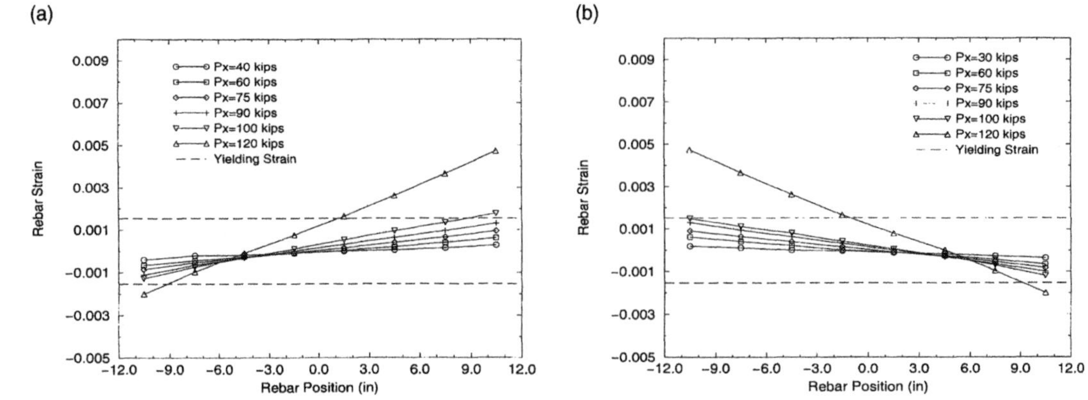

Fig. 20. Strain distributions of longitudinal reinforcement (a) at the top, and (b) at the bottom critical sections.

of the longitudinal reinforcement in detail, the strain distributions at the top and bottom sections of the column are plotted in Fig. 20(a) and (b), respectively. The strain values of the longitudinal reinforcement at the critical top and bottom sections reach the yield strength of Grade 40 steel as depicted in Fig. 20. The strain distributions over the cross-section show nearly linear distribution, which confirms that '*plane sections remain plane*', at least up to 80% of the peak resistance.

Fig. 21 presents the deformed mesh, the distribution of normal strain ϵ_{zz}, and stress σ_{zz} of the reinforced concrete column subject to axial and lateral loads, where some of the elements experience shear deformations. The distribution of normal strain ϵ_{zz} implies that tensile cracks develop at the right-hand corner of the top, and left-hand corner of the bottom of the column. The corresponding tensile stress distribution explains that the tension part of the column is already exhausted from softening. However the tensile crack pattern demonstrates distributed cracks because of the longitudinal reinforcement in the tension zones. In other terms, a macrocrack pattern is not present in the tension zone of the reinforced concrete column. This effect is also illus-

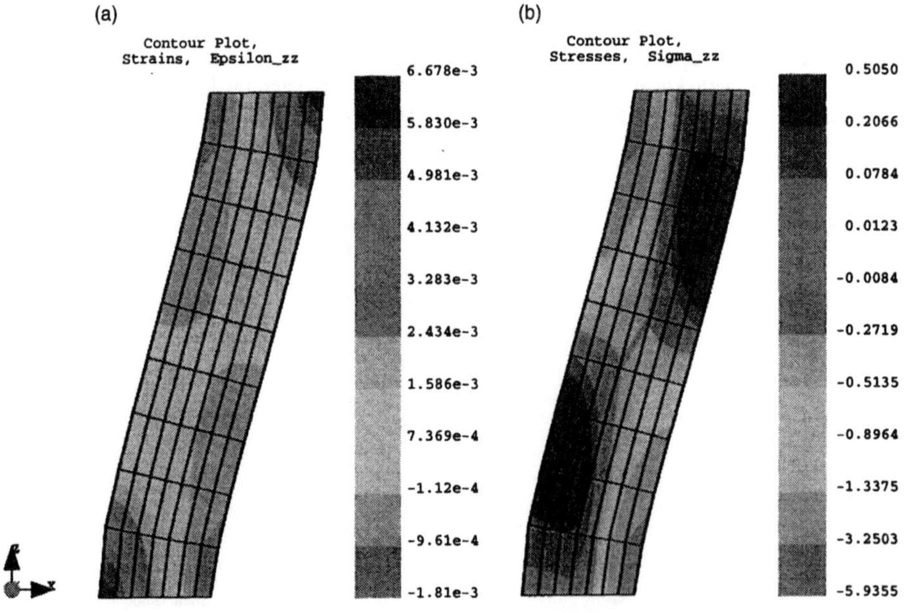

Fig. 21. (a) Normal strain ϵ_{zz} and (b) normal stress σ_{zz} distributions of reinforced concrete column subject to axial and lateral loading.

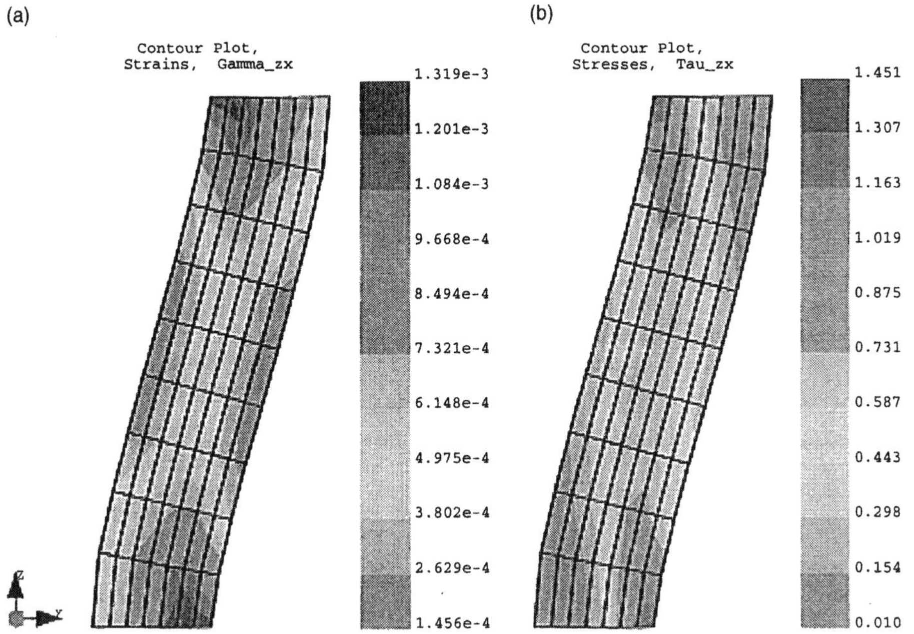

Fig. 22. (a) Shear strain γ_{xz}, and (b) shear stress τ_{xz} distributions of reinforced concrete column subject to axial and lateral loading.

trated in the deformed shape of the reinforced concrete column subject to axial tension as shown in Fig. 16(b). The shear strain and stress distributions shown in Fig. 22 demonstrate the potential development of shear cracks which are normally developed in bridge columns subject to lateral earthquake loading.

In summary, we observe fairly good agreement between experiments and numerical simulations conducted on the reinforced concrete column subject to axial and lateral loading even though the proposed model does not capture all the details of the experimental results. The discrepancies of the results, in general, are related to the boundary conditions, load histories and the mesh discretization as well as possible bond slip in the experiment.

4. Conclusions

Localization analysis at the material level indicates that tensile failure takes place in the form of discontinuous bifurcation and thus brittle cracking (mode I) as predicted by the Rankine concept of maximum stress. In contrast, compression failure does not localize and remains continuous with a tendency to fail in a diffuse manner in the form of axial splitting. In reality imperfections break cylindrical symmetry and lead to localized failure in mixed mode shear-compression as illustrated for axial compression in plane strain.

On the contrary, shear failure depends critically on the level of confinement and ranges from tensile cracking in pure shear to mixed mode shear-compression failure in simple shear. Thereby, deformation control leads to large normal stress redistribution, where the failure mode resembles compressive shear failure rather than tensile cracking. In all cases, shear failure develops during apparent hardening in the form of localization because of non-associated plastic flow. Consequently, shear failure necessitates special mesh design to capture emerging discontinuities in the strain and eventually in the displacement field irrespective of the boundary conditions at hand.

The finite element study of the reinforced concrete column captures the flexural and shear response under the combination of a constant axial load and a monotonically increasing lateral shearing. The finite element prediction of the column behavior is too stiff in comparison to the experimental results, which may be partly attributed to bond slip introduced by anchorage condition of the axial rebars. However, the localized shear failure at the material level and the structural level requires special mesh provisions to capture inclined shear failure realistically.

In sum, the constitutive behavior of the triaxial concrete model does reproduce the main deformation characteristics of concrete material under a variety of loading scenarios. Localization analysis at the material level exhibits realistic failure phenomena in terms of the failure mode and failure pattern. In addition, the triaxial concrete model contributes added insight into the behavior of the reinforced concrete column under severe load conditions.

Acknowledgements

The authors wish to acknowledge the support of this work under National Science Foundation grant CMS 9622940 *Performance of reinforced concrete bridge piers during the 1995 Hyogoken-Nanbu earthquake* to the University of Colorado at Boulder. However, opinions expressed in this paper do not necessarily represent those of the sponsor.

References

[1] Xiao Y, Priestley MJN, Seible F. Steel jacket retrofit for enhancing shear strength of short rectangular reinforced concrete columns. Report No. SSRP-92/07, University of California, San Diego, July 1993.
[2] Kang H. A triaxial constitutive model for plain and reinforced concrete behavior. Ph.D. Dissertation, CEAE Department, University of Colorado, Boulder, 1997.
[3] Launay P, Gachon H. Strain and ultimate strength of concrete under triaxial stress. Am Concrete Inst Spec Publ 1971;34:13–43.
[4] Willam KJ, Warnke E. Constitutive model for triaxial behavior of concrete. Proc concrete structures subjected to triaxial stresses. Int Assoc Bridge Struc Engng, Zurich, 1975;19:1–30.
[5] Desai CS, Somasundaram S, Frantziskonis G. A hierarchical approach for constitutive modelling of geologic materials. Int J Num Anal Meth Geomech 1986;10:225–57.
[6] Willam KJ, Montgomery K. Fracture energy-based softening plasticity model for shear failure. Proc Int Symp Interaction Conv Munition Protective Struct Mannheim 1987;2:679–91.
[7] Simo JC, Oliver J, Armero F. An analysis of strong discontinuities induced by strain-softening in rate-independent inelastic solids. Comput Mech 1993;12:277–96.
[8] Kang H, Willam K. An analysis of strong discontinuities in concrete panels. In: Choi C-K, Yun C-B, Lee D-G, editors. Proc. 3rd Asian-Pacific Conference on Computational Mechanics, vol. 2. Seoul: Seoul Techno-Press, 1996, p. 1529–34.
[9] Hurlbut BJ. Experimental and computational investigation of strain-softening in concrete. MS Thesis, University of Colorado, Boulder, 1985.
[10] Simo JC, Taylor RL. Consistent tangent operators for rate-independent elastoplasticity. Comp Methods Appl Mech Engng 1985;48:101–18.
[11] Kang H, Willam K. Localization characterizations of triaxial concrete model. ASCE J Engng Mech 1999;125: 941–50.
[12] Ortiz M, Leroy Y, Needleman A. A finite element method for localized failure analysis. Comp Meth Appl Mech Engng 1987;61:189–214.
[13] Vonk RA. Softening of concrete loaded in compression. PhD Thesis, Eindhoven University of Technology, Netherlands, 1992.
[14] Lee YH, Willam K. Mechanical properties of concrete in uniaxial compression. ACI Mater J 1997;94(6):457–71.
[15] Smith SS. On fundamental aspects of concrete behavior. MS Thesis, CEAE Deparment, University of Colorado, Boulder, 1987.
[16] Willam K, Kang H, Shing B, Spacone E. Analysis of shear failure in concrete materials. In: de Borst R, van der Giessen E, editors. Proc IUTAM Symposium on Material Instabilities in Solids, Delft University of Technology, UK: Wiley, June 1997. p. 27–39 [chapter 3].
[17] Hill R. A general theory of uniqueness and stability in elastic–plastics solids. Mech Phys Solids 1958;6:236–49.
[18] Taylor RL, FEAP, Finite element analysis program, University of California, Berkeley, 1997.
[19] Runesson K, Ottosen NS, Peric D. Discontinuous bifurcations of elastic–plastic solutions at plane stress and plane strain. Int J Plasticity 1991;7:99–121.
[20] Neilsen MK, Schreyer HL. Bifurcations in elastic–plastic materials. Int J Solids Struct 1993;30:521–44.

Geometrically nonlinear finite element reliability analysis of structural systems. I: theory

Kiyohiro Imai [a], Dan M. Frangopol [b,*]

[a] *Planning Division, Maintenance Department, Honshu Shikoku Bridge Authority, First Operation Bureau, 4-115 Higashi-Maiko-cho, Tarumi-ku, Kobe, 655-0047, Japan*

[b] *Department of Civil, Environmental, and Architectural Engineering, University of Colorado, Boulder, CO 80309-0428, USA*

Received 8 January 1999; accepted 1 December 1999

Abstract

This article reviews the theory of finite element reliability analysis of geometrically nonlinear elastic structures (GNS) based on the total lagrangian formulation. It also provides computer implementation developments and establishes the basis of understanding of the applications presented in the second part of this investigation. Because of the slenderness of GNS, the structural responses are nonlinear even if the strains are within the elastic range. For this reason, the nonlinear relationships between strains and displacements should be considered. Since the failure surface is nonlinear, this study reviews the evaluation of structural reliability of GNS by using both first-order and second-order reliability methods. To evaluate the structural reliability, the linkage of system reliability analysis program RELSYS with the finite element analysis program (FEAP) is presented. The computer code RELSYS–FEAP is readily applicable to the evaluation of system reliability of GNS. © 2000 Elsevier Science Ltd. All rights reserved.

Keywords: Finite element; Geometrically nonlinear analysis; Structural reliability; First-order reliability method; Second-order reliability method; System reliability

1. Introduction

This article reviews the theory of finite element reliability analysis of geometrically nonlinear elastic structures (GNS) based on the total lagrangian formulation. It also provides computer implementation developments and establishes the basis of understanding of the applications presented in the second part of this investigation.

Because of the slenderness of GNS, the structural responses are nonlinear even if the strains are within the elastic range. For this reason, the nonlinear relationships between strains and displacements should be considered. To determine the responses of this type of structures, geometric nonlinearity has to be considered [1–3].

The determination of the reliability index of structural components and systems is an optimization problem in the standard normal space [4–6]. There are two basic methods to estimate the structural reliability: the first-order and the second-order reliability method. These methods approximate the limit state by a linear surface (i.e., hyperplane) and a paraboloid, respectively. If the limit state is nonlinear, the second-order approximation will produce more accurate results [7]. However, if the failure surface is nearly flat, the two methods will produce almost the same results.

If the structural response can be described analytically, the reliability can be evaluated without the aid of the finite element method. However, for most structural systems, it is impossible to obtain the response without the aid of the finite element method. Therefore, the finite element reliability analysis has been developed. The formulation of the finite element reliability for GNS has

* Corresponding author. Tel.: +1-303-492-7165; fax: +1-303-492-7317.

E-mail address: frangopo@spot.colorado.edu (D.M. Frangopol).

Reprinted from *Computers and Structures* **77 (6)**, 677-691 (2000)

been introduced by Liu and Der Kiureghian [8,9]. They were the first to create a general purpose reliability code to perform finite element reliability analysis of GNS. The applications of the finite element reliability analysis to geometrically and/or material nonlinear structures are found in Refs. [8–13], among others. As indicated in Ref. [10], two main efforts are involved in the finite element reliability analysis. The first is in the implementation of a first-order or a second-order reliability method. This effort consists in finding the minimum distance from the origin to the failure surface in the standard normal space. The second effort consists in the finite element implementation for the reliability analysis by computing the gradients of structural response in terms of load, material properties, and geometry variables. These efforts are reviewed in this study.

To implement the finite element reliability analysis at the University of Colorado, the computer code RELSYS–FEAP was created by interfacing the system reliability analysis program RELSYS developed by Estes and Frangopol [14] with the finite element analysis program (FEAP) developed by Taylor [15].

The linkage between RELSYS and FEAP is based on the work of Liu and Der Kiureghian [8,9] who linked CALREL [16] with FEAP. However, RELSYS–FEAP uses the new version of FEAP (version 6.0), to which some useful macro commands had been added. The reliability analysis program RELSYS can deal with the system reliability analysis without using simulation by modeling the structure as a series of parallel subsystems.

For the reliability analysis, one needs response gradients as well as structural responses. Ryu et al. [17] and Wu and Arora [18] derived an analytical formula for the response gradients of GNS. Liu and Der Kiureghian [8,9] applied the analytical formula of the response gradients to the reliability analysis of GNS.

This study reviews the analytical formulas of the response gradients from the equilibrium equation of internal and external forces. Because of the geometric nonlinearity, a first-order approximated limit state surface does not exactly fit the limit state surface of GNS. To evaluate the first-order approximation of the failure surface, the first-order reliability indices of GNS have to be compared with the second-order reliability indices.

This study addresses the system reliability analysis of GNS. Liu and Der Kiureghian [8] and Dagher et al. [19] considered the system reliability of GNS. However, most of the applications are limited to series systems of components (weakest-link system), where the structure has no redundancy. In the real world, most structural systems are redundant. Therefore, system failure has to be modeled by the failure of a series of parallel subsystems. The computation of the reliability indices of parallel (i.e., redundant) systems needs the reduced stiffness matrix and equivalent internal forces [20]. The internal forces in the geometrically nonlinear formulation are obtained from both an elastic stiffness matrix and a geometric stiffness matrix. Generally, it is difficult to derive the reduced stiffness matrix and equivalent internal forces by the static condensation method. In this study, the reduced stiffness matrix and equivalent internal forces of truss elements are derived in the geometrically nonlinear formulation. However, for frame elements, the system reliability analysis is implemented by inserting a hinge element with the aid of the macro command of FEAP. By using these procedures, system reliability indices of geometrically nonlinear redundant structures can be evaluated.

2. Review of finite element formulation of geometrically nonlinear elastic structures

Small displacement theory is no longer applicable to geometrically nonlinear structures [21]. In the linear analysis, a linear relation is assumed between strains and displacements. However, if there are large displacements and rotations, the nonlinear relation between strains and displacements cannot be ignored. Also, the equilibrium equation of internal and external forces should be considered in the deformed configuration.

The geometrically nonlinear analysis may be described by using the total or the updated lagrangian formulations. The total lagrangian formulation is derived with respect to the initial configuration. The updated lagrangian formulation is derived with respect to the current configuration. Bathe and Bolourchi [22], Washizu [21], Bathe [1], and Crisfield [3] described the comparison between these two formulations. Practical applications of the total and updated lagrangian formulations are found in Refs. [23–27]. To analyze GNS, the tangent stiffness matrix and internal forces play important roles. The comparison between the two formulations considered is described in Table 1, where $\mathbf{K}_E$ is the elastic stiffness matrix, $\mathbf{K}_G$, the geometric stiffness matrix, and $\mathbf{K}_U$, the initial displacement matrix [2,3], $\mathbf{T}_0$ and $\mathbf{T}$ are the transformation matrices from global co-

Table 1
Comparison of the total lagrangian formulation and the updated lagrangian formulation

Main descriptors	Total lagrangian formulation	Updated lagrangian formulation
Tangent stiffness matrix	$\mathbf{T}_0^T(\mathbf{K}_E + \mathbf{K}_G + \mathbf{K}_U)\mathbf{T}_0$	$\mathbf{T}^T(\mathbf{K}_E + \mathbf{K}_G)\mathbf{T}$
Internal force vector	$\mathbf{T}_0^T(\mathbf{K}_E + \mathbf{K}_G)\mathbf{T}_0 \cdot \mathbf{u}$	$\mathbf{T}^T(\mathbf{K}_E)\mathbf{T} \cdot \mathbf{u}$

ordinates to local coordinates of the initial and current configurations, respectively, and $\mathbf{u}$, the vector of displacements.

The total lagrangian formulation constructs the tangent stiffness matrix with respect to the initial configuration. On the other hand, the updated lagrangian formulation constructs the tangent stiffness matrix with respect to the current configuration. The updated lagrangian formulation is computationally effective [22] because it does not include the initial displacement matrix $\mathbf{K}_U$. In a reliability analysis, however, we need response gradients of structures [8,9]. Since the elastic stiffness matrix $\mathbf{K}_E$, geometric stiffness matrix $\mathbf{K}_G$, and transformation matrix $\mathbf{T}$ are functions of displacements in the updated lagrangian formulation, these matrices make the computation of response gradients complicated. Therefore, the total lagrangian formulation was selected for the finite element discretization because response gradients are easier to compute [8]. In the total lagrangian formulation, the initial configuration remains constant. This simplifies the computation [23].

According to Refs. [2,3,28,29], the total lagrangian formulation is as follows:

Displacements of structures $\mathbf{u}$ are given by the product of shape functions and nodal displacements:

$$\mathbf{u} = \sum \mathbf{N} \cdot \mathbf{U}, \tag{1}$$

where $\mathbf{N}$ is the vector of shape functions and $\mathbf{U}$ is the vector of nodal displacements. Because of large displacements and rotations, Green's strain is adopted for the nonlinear relationships between strains and displacements. The Green's strain ε_G includes linear and nonlinear terms:

$$\varepsilon_G = (\mathbf{B}_L + \mathbf{B}_{NL}(\mathbf{U}))\mathbf{U}, \tag{2}$$

where $\mathbf{B}_L$ is the vector which relates the linear strain term to the nodal displacements, and $\mathbf{B}_{NL}(\mathbf{U})$, the vector which relates the nonlinear strain term to the nodal displacements. From Eq. (2), the incremental form of the strain–displacement relationship is

$$\delta\varepsilon_G = (\mathbf{B}_L + \mathbf{B}_{NL}(\mathbf{U}))\,\delta\mathbf{U}. \tag{3}$$

Using the principle of virtual displacement, the virtual work δW is given as

$$\delta W = \int_{V_0} (\sigma_S \delta\varepsilon_G)\,\mathrm{d}V_0 - \mathbf{q}_{ext}\,\delta\mathbf{U}, \tag{4}$$

where σ_S is the second Piola–Kirchhoff stress, V_0, the volume of initial configuration, and $\mathbf{q}_{ext}$, the vector of external loads. Since Green's strain is based on a small strain, stress can be given by Hook's law as

$$\sigma_S = E\varepsilon_G = E(\mathbf{B}_L + \mathbf{B}_{NL}(\mathbf{U}))\mathbf{U}, \tag{5}$$

where E is the modulus of elasticity. Substituting $\delta\varepsilon_G$ from Eq. (3) into Eq. (4) results in

$$\begin{aligned}\delta W &= \int_{V_0} (\sigma_S(\mathbf{B}_L + \mathbf{B}_{NL}(\mathbf{U}))\,\delta\mathbf{U})\,\mathrm{d}V_0 - \mathbf{q}_{ext}\,\delta\mathbf{U} \\ &= \int_{V_0} (\sigma_S(\mathbf{B}_L + \mathbf{B}_{NL}(\mathbf{U})))\,\mathrm{d}V_0\,\delta\mathbf{U} - \mathbf{q}_{ext}\,\delta\mathbf{U} \\ &= \left[\int_{V_0} (\sigma_S(\mathbf{B}_L + \mathbf{B}_{NL}(\mathbf{U})))\,\mathrm{d}V_0 - \mathbf{q}_{ext}\right]\delta\mathbf{U}.\end{aligned} \tag{6}$$

Since $\delta\mathbf{U}$ is arbitrary, the vector of internal forces $\mathbf{q}_{int}$ is

$$\mathbf{q}_{int} = \int_{V_0} (\sigma_S(\mathbf{B}_L + \mathbf{B}_{NL}(\mathbf{U})))\,\mathrm{d}V_0. \tag{7}$$

Taking the derivative of $\mathbf{q}_{int}$ with respect to the nodal displacements $\mathbf{U}$ gives the tangent stiffness matrix $\mathbf{K}_T$ as

$$\begin{aligned}\mathbf{K}_T &= \frac{\partial \mathbf{q}_{int}}{\partial \mathbf{U}} \\ &= \int_{V_0} \left(\frac{\partial \sigma_S}{\partial \mathbf{U}}(\mathbf{B}_L + \mathbf{B}_{NL}(\mathbf{U})) + \sigma_S \frac{\partial \mathbf{B}_{NL}(\mathbf{U})}{\partial \mathbf{U}}\right)\mathrm{d}V_0.\end{aligned} \tag{8}$$

In addition, this expression can be written by substituting Eq. (5) into Eq. (8) as

$$\begin{aligned}\mathbf{K}_T &= \int_{V_0} E\mathbf{B}_L\mathbf{B}_L\,\mathrm{d}V_0 \\ &\quad + \int_{V_0} \sigma_S \frac{\partial \mathbf{B}_{NL}(\mathbf{U})}{\partial \mathbf{U}}\,\mathrm{d}V_0 + \int_{V_0} E(\mathbf{B}_L\mathbf{B}_{NL}(\mathbf{U}) \\ &\quad + \mathbf{B}_{NL}(\mathbf{U})\mathbf{B}_L + \mathbf{B}_{NL}(\mathbf{U})\mathbf{B}_{NL}(\mathbf{U}))\,\mathrm{d}V_0.\end{aligned} \tag{9}$$

In Eq. (9), the first term is the elastic stiffness matrix, $\mathbf{K}_E$, the second term is the geometric stiffness matrix, $\mathbf{K}_G$, and the third term is the initial displacement stiffness matrix, $\mathbf{K}_U$ [2,3]:

$$\mathbf{K}_E = \int_{V_0} E\mathbf{B}_L\mathbf{B}_L\,\mathrm{d}V_0, \tag{10}$$

$$\mathbf{K}_G = \int_{V_0} \sigma_S \frac{\partial \mathbf{B}_{NL}(\mathbf{U})}{\partial \mathbf{U}}\,\mathrm{d}V_0, \tag{11}$$

$$\begin{aligned}\mathbf{K}_U = \int_{V_0} & E(\mathbf{B}_L\mathbf{B}_{NL}(\mathbf{U}) + \mathbf{B}_{NL}(\mathbf{U})\mathbf{B}_L \\ & + \mathbf{B}_{NL}(\mathbf{U})\mathbf{B}_{NL}(\mathbf{U}))\,\mathrm{d}V_0.\end{aligned} \tag{12}$$

The vector of internal forces (7) and the tangent stiffness matrix (9) are used to analyze geometric nonlinear structures. Appendices A and B present for two-dimensional truss and frame elements, respectively, the expressions of components of the vector of internal forces (7) and matrices (10)–(12).

A typical solution procedure for this type of nonlinear analysis is obtained by using the Newton–Raphson iterative procedure [1–3,29]

$$\mathbf{K}_T^i \Delta\mathbf{U} = \mathbf{q}_{ext} - \mathbf{q}_{int}^i, \tag{13}$$

$$\mathbf{U}^{i+1} = \mathbf{U}^i + \Delta\mathbf{U}, \tag{14}$$

where i and $i+1$ are the iteration numbers at which the equations are computed, $\mathbf{K}_T^i$ is the tangent stiffness matrix based on the ith displacements, $\mathbf{U}$, the vector of nodal displacements, $\Delta\mathbf{U}$, the vector of correction nodal displacements, and $\mathbf{q}_{int}^i$, the vector of internal forces based on the ith displacements. By using Eqs. (13) and (14) iteratively, a converged solution for GNS is obtained.

3. Review of reliability analysis methods: first-order reliability and second-order reliability methods

3.1. First-order reliability method

The design of any structure requires that its resistance, R, is greater than the load effect, Q. This requirement (i.e., $R > Q$) is described by using the limit state function

$$g(\mathbf{X}) = R - Q = 0, \tag{15}$$

where $\mathbf{X} = \{X_1, X_2, \ldots, X_n\}^T$ is the vector of random variables, $g(\mathbf{X}) = 0$ is the limit state, and $g(\mathbf{X}) < 0$ is the failure state. The reliability index is given as [4–6]

$$\beta = \frac{\mu_R - \mu_Q}{\sqrt{\sigma_R^2 + \sigma_Q^2}}, \tag{16}$$

where μ_R and μ_Q are the mean resistance and mean load effect, respectively, and σ_R and σ_Q are the standard deviation of the resistance and the standard deviation of the load effect, respectively. If the resistance R and the load effect Q are normally distributed, then the probability of failure can be determined as

$$P_f = \Phi(-\beta), \tag{17}$$

where Φ is the standard normal probability function. Calculation of the reliability index β is a constrained optimization problem consisting in finding the nearest point on the limit state surface in the standard normal space [4–6]:

$$\begin{aligned} &\text{minimize} && |\mathbf{X}'| \\ &\text{subject to} && g(\mathbf{X}) = 0, \end{aligned} \tag{18}$$

where $\mathbf{X}' = \{X_1', X_2', \ldots, X_n'\}^T$ is the vector of standard normal variables:

$$X_i' = \frac{X_i - \mu_{X_i}}{\sigma_{X_i}}, \tag{19}$$

where μ_{X_i} is the mean of X_i, and σ_{X_i}, the standard deviation of X_i. Standard algorithms for solving this optimization problem are available [5,6,30,31].

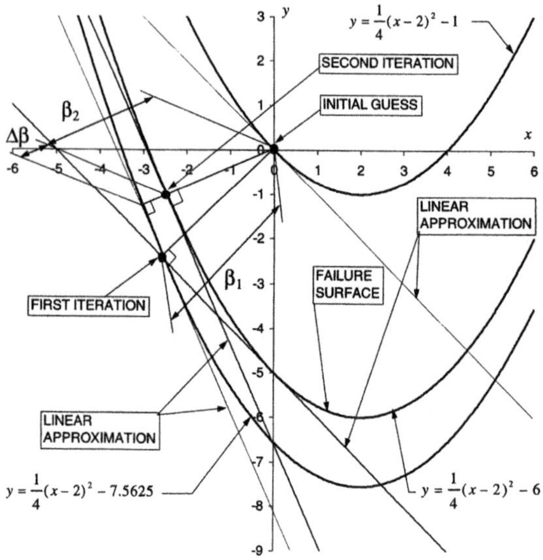

Fig. 1. Illustration of iterations in standard normal space.

The procedure to determine the first-order reliability index which is the minimum distance from the origin to the limit state surface in the standard normal space is illustrated in Fig. 1 for the limit state function

$$g(X,Y) = Y - \left\{\tfrac{1}{4}(X-4)^2 - 3\right\} = 0, \tag{20}$$

where X and Y are the normal random variables with parameters [2.0; 1.0] and [3.0; 1.0], respectively. In the notation [μ, σ], μ and σ are the mean and standard deviation, respectively. Transformation from the original space to the standard normal space is given as

$$x = X - 2, \tag{21a}$$

$$y = Y - 3. \tag{21b}$$

Substituting Eqs. (21a) and (21b) into Eq. (20) gives the limit state function in the standard normal space as

$$g(x,y) = y - \left\{\tfrac{1}{4}(x-2)^2 - 6\right\} = 0. \tag{22}$$

The initial guess of the failure point is the origin of the (x, y) space. Successive approximations of the limit state surface produce the minimum distance from the origin as shown in Fig. 1.

Gradient information is essential to execute first-order reliability analysis (FORM). However, the evaluation of derivatives of the limit state function is not always an easy task.

3.2. Second-order reliability method: second-order reliability method

In the second-order reliability method (SORM), the limit state surface is approximated by a paraboloid. Several approximations of the probability of failure based on a second-order approximation have been proposed [7]. The probability of failure is provided by Breitung [32] as

$$P_f = \Phi(-\beta)\prod_{i=1}^{n-1}\frac{1}{\sqrt{1+\kappa_i\beta}}, \tag{23}$$

where β is the reliability index based on the first-order approximation, $\Phi(-\beta)$, the probability of failure based on the first-order approximation, κ_i, the principal curvature with respect to ith random variable, and n, the number of random variables. Two cases are presented in Fig. 2. Let us assume that β is positive. When the principal curvature is positive, $1/\sqrt{1+\kappa_i\beta}$ is less than 1. Therefore, the probability of failure P_f in Eq. (23) is smaller than $\Phi(-\beta)$. On the other hand, when the principal curvature is negative, $1/\sqrt{1+\kappa_i\beta}$ is larger than 1 and, consequently, P_f in Eq. (23) is larger than $\Phi(-\beta)$. Therefore, the probability of failure associated with a nonlinear performance function estimated on the basis of FORM will be conservative or unconservative depending on the sign (i.e., positive or negative) of the principal curvature of the performance function in the standard normal space.

The second-order reliability method was implemented by Liu et al. [16] in the program code CALREL. This code employs various approximations of the probability of failure. One is an improved Breitung's formula given by Hohenbichler and Rackwitz [33] and the other is given by Tvedt [34] (see also Refs. [7,8,35]).

The computation of the principal curvatures with respect to the random variables is an extra task in SORM. Two types of paraboloid approximations for the limit state surface were examined by Der Kiureghian et al. [7]. One is a curvature fitted paraboloid proposed by Tvedt (see the last two references in Ref. [34]), and the other is a point fitted paraboloid proposed by Der Kiureghian et al. [7].

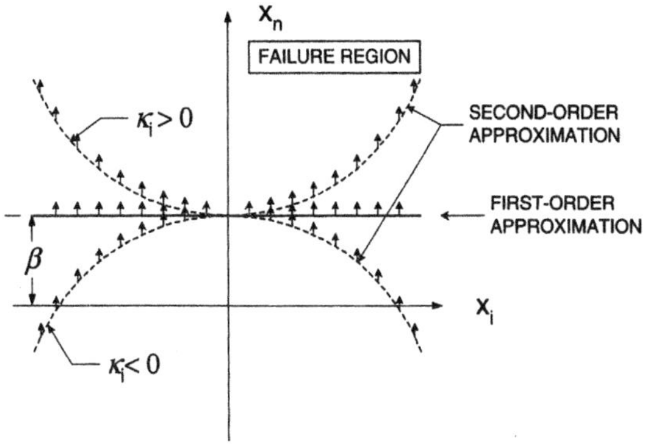

Fig. 2. Justification of curvature correction term in Breitung's formula.

The curvature fitting method needs the second derivative of the limit state function and the eigen solution of the second derivative matrix of the limit state function. The second derivatives of the limit state function are provided by the finite difference method in CALREL [16]. Therefore, the user need not provide the second derivatives of the limit state function. On the other hand, the point fitting method needs neither the second derivatives of the limit state function nor the eigen solution. Therefore, the first derivative of the limit state function is enough to execute the point fitting method. This method, however, needs an iterative search to determine the fitting points. For an in-depth discussion of second-order reliability approximations the reader is referred to Ref. [7].

4. System reliability analysis using RELSYS

System reliability can be computed by modeling a structure as a series of parallel systems. Estes and Frangopol [14] developed the system reliability analysis program (RELSYS) (RELiability of SYStems). In this program, the failure surface of series, parallel, or series–parallel systems is approximated by a hyperplane. The system reliability is computed by successively reducing the series and parallel systems until the system has been simplified to a single equivalent component. Equivalent alpha vectors are used to account for the correlation between failure modes during the system reduction process. This study uses RELSYS for the evaluation of system reliability.

4.1. Series system

The probability of failure of a series system can be written as the union of events [20]

$$P_f = P\left(\bigcup_{i=1}^{N}\left\{g_i(\mathbf{X}) \leqslant 0\right\}\right) = P\left(\bigcup_{i=1}^{N}E_i\right), \tag{24}$$

where g_i is the ith limit state function, N, the number of components, and E_i, the event $g_i(\mathbf{X}) \leqslant 0$. Computation of P_f needs numerical integration of the multi-normal distribution function. This numerical integration, however, is extremely time consuming or even impossible for values of N greater than, say 4 [20]. Therefore, approximations are almost always necessary. In this regard, lower and upper bounds of P_f are useful [6]. Two kinds of bounds, uni-modal bounds [36] and bi-modal bounds [37] have been suggested. RELSYS adopts the bi-modal bounds. The upper bound provided by Ditlvesen [37] is

$$P_{\mathrm{f\,upper}} = \sum_{i=1}^{N} P(g_i(\mathbf{X}) \leqslant 0) - \sum_{i=2,j<i}^{N} \max\left[P\{(g_i(\mathbf{X}) \leqslant 0) \cap (g_j(\mathbf{X}) \leqslant 0)\}\right] \quad (25a)$$

and the lower bound is

$$P_{\mathrm{f\,lower}} = \max_{i=1}^{N} P(g_i(\mathbf{X}) \leqslant 0) + \sum_{i=2}^{N} \max\left[P(g_i(\mathbf{X}) \leqslant 0) - \sum_{j=1}^{i-1} P\{(g_i(\mathbf{X}) \leqslant 0) \cap (g_j(\mathbf{X}) \leqslant 0)\}; 0\right]. \quad (25b)$$

The probability of the series system P_f is between these two bounds

$$P_{\mathrm{f\,lower}} \leqslant P_f \leqslant P_{\mathrm{f\,upper}}. \quad (26)$$

RELSYS treats P_f as the average of upper and lower bounds [14]:

$$P_f = \left(P_{\mathrm{f\,lower}} + P_{\mathrm{f\,upper}}\right)/2. \quad (27)$$

Computation of Eqs. (25a) and (25b) is based on both the individual probability of failure $P(g_i(\mathbf{X}) \leqslant 0)$ and the joint probability of failure $P\{(g_i(\mathbf{X}) \leqslant 0) \cap (g_j(\mathbf{X}) \leqslant 0)\}$. The individual probability of failure is calculated from first-order reliability method. The joint probability of failure is calculated by numerical integration. This integration is

$$P\{(g_i(\mathbf{X}) \leqslant 0) \cap (g_j(\mathbf{X}) \leqslant 0)\} = \int_{\beta_i}^{\infty}\int_{\beta_j}^{\infty} \frac{1}{2\pi\sqrt{1-\rho_{\mathrm{sys}\,ij}^2}} \exp\left[-\frac{1}{2}(1-\rho_{\mathrm{sys}\,ij}^2) \times (\beta_i^2 + \beta_j^2 - 2\rho_{\mathrm{sys}\,ij}\beta_i\beta_j)\right] \mathrm{d}\beta_i\,\mathrm{d}\beta_j, \quad (28)$$

where β_i and β_j are the reliability indices associated with the ith and jth limit state functions and $\rho_{\mathrm{sys}\,ij}$ is the correlation coefficient between these functions. This coefficient is calculated from the vector of direction cosines [6] as follows:

$$\rho_{\mathrm{sys}\,ij} = \sum_{k=1}^{n} \alpha_{ik}^{*}\alpha_{jk}^{*}, \quad (29)$$

where n is the number of random variables, α_{ik}^* and α_{jk}^* are the direction cosines of the performance functions $g_i(\mathbf{X})$ and $g_j(\mathbf{X})$, respectively, at the most probable failure points. α_{ik}^* and α_{jk}^* are computed as indicated in Ref. [6]

$$\alpha_{ik}^{*} = \frac{\left(\partial g_i/\partial X_k'^{*}\right)}{\sqrt{\left(\mathrm{d}g_i/\mathrm{d}\mathbf{X}'^{*}\right)^{\mathrm{T}}\left(\mathrm{d}g_i/\mathrm{d}\mathbf{X}'^{*}\right)}}, \quad (30a)$$

$$\alpha_{jk}^{*} = \frac{\left(\partial g_j/\partial X_k'^{*}\right)}{\sqrt{\left(\mathrm{d}g_j/\mathrm{d}\mathbf{X}'^{*}\right)^{\mathrm{T}}\left(\mathrm{d}g_j/\mathrm{d}\mathbf{X}'^{*}\right)}}. \quad (30b)$$

4.2. Parallel system

The probability of failure for a parallel system can be written as the probability of intersection of events [20] as

$$P_f = P\left(\bigcap_{i=1}^{N}\{g_i(\mathbf{X}) \leqslant 0\}\right) = P\left(\bigcap_{i=1}^{N} E_i\right), \quad (31)$$

where N is the number of components. In general, numerical integration of the multi-normal distribution is extremely time consuming. Therefore, approximate or bounding techniques are used. Simple (first-order) bounds for Eq. (31) are [6]

$$\prod_{i=1}^{N} P\{g_i(\mathbf{X}) \leqslant 0\} \leqslant P_f \leqslant \min_{i=1}^{N} P\{g_i(\mathbf{X}) \leqslant 0\}. \quad (32)$$

However, these bounds will in most cases be so wide that they are of very little use [20]. Therefore, numerical integration of Eq. (31) must be evaluated. The probability (31) is calculated from the integration:

$$P\{(g_1(\mathbf{X}) \leqslant 0) \cap (g_2(\mathbf{X}) \leqslant 0) \cap \cdots \cap (g_N(\mathbf{X}) \leqslant 0)\} = \int_{\beta_1}^{\infty}\int_{\beta_2}^{\infty}\cdots\int_{\beta_N}^{\infty} \frac{1}{(2\pi)^{N/2}\sqrt{\det\left[\rho_{\mathrm{sys}}\right]}} \times \exp\left[-\frac{1}{2}\{\beta\}\left[\rho_{\mathrm{sys}}\right]^{-1}\{\beta\}^{\mathrm{T}}\right]\mathrm{d}\{\beta\}, \quad (33)$$

where $\{\beta\} = \{\beta_1, \beta_2, \ldots, \beta_N\}$ is the reliability index vector and $\left[\rho_{\mathrm{sys}}\right]$, the correlation matrix described by:

$$\left[\rho_{\mathrm{sys}}\right] = \begin{bmatrix} 1 & \rho_{21} & \cdots & \rho_{1N} \\ \rho_{21} & 1 & \cdots & \vdots \\ \vdots & \vdots & \ddots & \vdots \\ \rho_{N1} & \cdots & \cdots & 1 \end{bmatrix}. \quad (34)$$

To compute this integration numerically, $\{\beta\}$ and $\left[\rho_{\mathrm{sys}}\right]$ are both required. The reliability of each component is calculated from the first-order reliability method. The components of the correlation matrix are calculated based on Eq. (29).

Estes [38] compared various numerical integration methods with the exact values. For a two-member parallel system, RELSYS [14] numerically integrates the two-variable normal distribution by using the composite Simpson's rule [39]. For a three-member parallel system, RELSYS numerically integrates the three-variable normal distribution by using Gaussian quadrature with 11 points if $\rho_{sys} \leqslant 0.9$ and 25 points if $\rho_{sys} > 0.9$. For parallel systems with four members or more, the Hohenbichler–Rackwitz approximation [40] is used.

4.3. Combined series–parallel system

The probability of failure of a series of N parallel systems (i.e., general system) is

$$P_f = P\left(\bigcup_{i=1}^{N}\bigcap_{j=1}^{m_i}\{g_{ij}(\mathbf{X}) \leqslant 0\}\right), \tag{35}$$

where N is the number of parallel systems and m_i, the number of components in ith parallel system. The reliability of this general system can be calculated by reducing each parallel system to one component of a series system [14].

The computation of the reliability of a series system needs the correlation between failure modes of components. As shown in Eq. (29), the correlation is computed by the direction cosines. Therefore, to compute the system reliability of the series system of N components, direction cosines of the performance functions which represent each of the N parallel subsystems are required. For this purpose, the sensitivities of the system reliability of each parallel subsystem with respect to each random variable are required to compute the equivalent direction cosines [14].

Let us consider the ith parallel system that contains m_i components. The reliability of this system is denoted as β_{sys}, and the reliability of jth component is denoted as $\beta_j (j = 1, 2, \ldots, m_i)$. If we consider a small change φ of the kth random variable X_k, the revised reliabilities of each component become [14]

$$\begin{aligned} \beta_{1k} &= \beta_1 + \varphi\alpha^*_{1k}, \\ \beta_{2k} &= \beta_2 + \varphi\alpha^*_{2k}, \\ &\vdots \\ \beta_{m_ik} &= \beta_{m_i} + \varphi\alpha^*_{m_ik}, \end{aligned} \tag{36}$$

where β_{jk} is the revised reliability index, and α^*_{jk}, the direction cosine of the jth component with respect to the random variable X_k. Using these revised reliability indices and the correlation matrix, integration of the multi-normal distribution (33) can be computed. Then, the revised system reliability $\beta_{sys(new)}$ is obtained. The equivalent direction cosines can be computed as [14]:

$$\frac{\partial(\beta_{sys})}{\partial(X_k)} = \frac{\beta_{sys(new)} - \beta_{sys}}{\varphi}, \tag{37}$$

where φ is an arbitrary small constant. Using these equivalent direction cosines, the reliability of a general system can be found as follows:

Step 1: Evaluate the probability of failure of each parallel subsystem according to Eq. (33);

Step 2: Evaluate the equivalent direction cosines for each parallel subsystem according to Eq. (37);

Step 3: Evaluate the probability of failure of the series system according to Eq. (27).

5. Review of response gradient formulation

5.1. Component reliability for linear analysis

The response gradients must be derived to execute FORM. These gradients are derived by using the direct stiffness method [8,41–43]. The equilibrium equation of the internal forces $\mathbf{q}^{(g)}_{int}$ and the external forces $\mathbf{q}^{(g)}_{ext}$ in the global coordinates is

$$\mathbf{q}^{(g)}_{int} = \mathbf{q}^{(g)}_{ext}. \tag{38}$$

In this section, superscripts (g) and (e) denote the global and element coordinates, respectively. In the direct stiffness method, the internal forces in the global coordinates are described by the stiffness matrix $\mathbf{K}^{(g)}$ and the nodal displacements $\mathbf{U}^{(g)}$. Therefore,

$$\mathbf{q}^{(g)}_{int} = \mathbf{K}^{(g)}\mathbf{U}^{(g)}. \tag{39}$$

Consequently, the equilibrium equation is

$$\mathbf{q}^{(g)}_{int} = \mathbf{K}^{(g)}\mathbf{U}^{(g)} = \mathbf{q}^{(g)}_{ext}. \tag{40}$$

In general, limit states restrain displacements and/or member forces. Member forces are derived from displacements in the direct stiffness method. Therefore, if we can evaluate the gradients of the displacements with respect to random variables, the member forces can be evaluated from these gradients. Differentiating Eq. (39) results in

$$\begin{aligned} \frac{d\mathbf{q}^{(g)}_{int}}{d\mathbf{v}} &= \left.\frac{\partial\mathbf{q}^{(g)}_{int}}{\partial\mathbf{U}^{(g)}}\right|_{\mathbf{v}} \frac{d\mathbf{U}^{(g)}}{d\mathbf{v}} + \left.\frac{\partial\mathbf{q}^{(g)}_{int}}{\partial\mathbf{v}}\right|_{\mathbf{U}^{(g)}} \\ &= \mathbf{K}^{(g)}\frac{d\mathbf{U}^{(g)}}{d\mathbf{v}} + \left.\frac{\partial\mathbf{q}^{(g)}_{int}}{\partial\mathbf{v}}\right|_{\mathbf{U}^{(g)}} = \frac{d\mathbf{q}^{(g)}_{ext}}{d\mathbf{v}}, \end{aligned} \tag{41}$$

where $\mathbf{v}$ is the vector of random variables. In this equation, $(\partial\mathbf{q}^{(g)}_{int}/\partial\mathbf{U}^{(g)})|_{\mathbf{v}}$ represents the partial derivative of $\mathbf{q}^{(g)}_{int}$ with respect to $\mathbf{U}^{(g)}$ calculated at $\mathbf{v}$, and $d\mathbf{q}^{(g)}_{int}/d\mathbf{v}$, $d\mathbf{U}^{(g)}/d\mathbf{v}$, and $d\mathbf{q}^{(g)}_{ext}/d\mathbf{v}$ represent total derivatives. Solving Eq. (41), $d\mathbf{U}^{(g)}/d\mathbf{v}$ is obtained as follows:

$$\frac{d\mathbf{U}^{(g)}}{d\mathbf{v}} = \mathbf{K}^{(g)^{-1}}\left(\frac{d\mathbf{q}^{(g)}_{ext}}{d\mathbf{v}} - \left.\frac{\partial\mathbf{q}^{(g)}_{int}}{\partial\mathbf{v}}\right|_{\mathbf{U}^{(g)}}\right). \tag{42}$$

There are three categories of random variables, which are loads, $\mathbf{v}_l$, material properties, $\mathbf{v}_m$, and geometry, $\mathbf{v}_g$ [8]. The notation used here for these vectors is identical to that used by Der Kiureghian and coworkers [8]. The gradients of the displacements with respect to random variables are calculated as follows:

- gradients with respect to external load:

$$\frac{d\mathbf{U}^{(g)}}{d\mathbf{v}_l} = \mathbf{K}^{(g)^{-1}}\left(\frac{d\mathbf{q}^{(g)}_{ext}}{d\mathbf{v}_l}\right), \tag{43}$$

- gradients with respect to material property:

$$\frac{d\mathbf{U}^{(g)}}{d\mathbf{v}_m} = \mathbf{K}^{(g)^{-1}}\left(-\left.\frac{\partial\mathbf{q}^{(g)}_{int}}{\partial\mathbf{v}_m}\right|_{\mathbf{U}^{(g)}}\right), \tag{44}$$

- gradients with respect to geometry

$$\frac{d\mathbf{U}^{(g)}}{d\mathbf{v}_g} = \mathbf{K}^{(g)^{-1}}\left(\frac{d\mathbf{q}^{(g)}_{ext}}{d\mathbf{v}_g} - \left.\frac{\partial\mathbf{q}^{(g)}_{int}}{\partial\mathbf{v}_g}\right|_{\mathbf{U}^{(g)}}\right). \tag{45}$$

In the reliability analysis, calculations of gradients are iterated due to the change of the failure point $\mathbf{X}'^{*}$.

5.2. *System reliability for linear analysis*

To calculate the system reliability, conditional probabilities are needed. These conditional probabilities represent the probabilities of failure given that one (or more) member(s) has (have) failed. To compute these probabilities, Thoft-Christensen and Murotsu [20] formulated a reduced stiffness matrix $\mathbf{k}^{(e)}_R$ and equivalent nodal forces $\mathbf{q}^{(e)}_{R\,int}$ by applying the static condensation method.

Using the reduced stiffness matrix and the equivalent nodal forces, the equilibrium equation of the structure given the failure of one (or more) element(s) is

$$\mathbf{q}^{(g)}_{int} = \left(\mathbf{K}^{(g)} + \mathbf{K}^{(g)}_R\right)\mathbf{U}^{(g)} + \mathbf{q}^{(g)}_{R\,int} = \mathbf{q}^{(g)}_{ext}, \tag{46}$$

where $\mathbf{K}^{(g)}$ is the stiffness matrix assembled from the intact elements, $\mathbf{K}^{(g)}_R$, the reduced stiffness matrix assembled form the failed elements, and $\mathbf{q}^{(g)}_{R\,int}$, the equivalent internal forces in the global coordinates. To perform FORM, it is necessary to compute the derivative of Eq. (46) with respect to random variables of this equation as

$$\begin{aligned}\frac{d\mathbf{q}^{(g)}_{int}}{d\mathbf{v}} &= \left.\frac{\partial\mathbf{q}^{(g)}_{int}}{\partial\mathbf{U}^{(g)}}\right|_{\mathbf{v}}\frac{d\mathbf{U}^{(g)}}{d\mathbf{v}} + \left.\frac{\partial\mathbf{q}^{(g)}_{int}}{\partial\mathbf{v}}\right|_{\mathbf{U}^{(g)}} + \frac{d\mathbf{q}^{(g)}_{R\,int}}{d\mathbf{v}} \\ &= \left(\mathbf{K}^{(g)} + \mathbf{K}^{(g)}_R\right)\frac{d\mathbf{U}^{(g)}}{d\mathbf{v}} + \left.\frac{\partial\mathbf{q}^{(g)}_{int}}{\partial\mathbf{v}}\right|_{\mathbf{U}^{(g)}} + \frac{d\mathbf{q}^{(g)}_{R\,int}}{d\mathbf{v}} \\ &= \frac{d\mathbf{q}^{(g)}_{ext}}{d\mathbf{v}}.\end{aligned} \tag{47}$$

Therefore, the gradient of the displacements with respect to $\mathbf{v}$ is

$$\frac{d\mathbf{U}^{(g)}}{d\mathbf{v}} = \left(\mathbf{K}^{(g)} + \mathbf{K}^{(g)}_R\right)^{-1}\left\{\frac{d\mathbf{q}^{(g)}_{ext}}{d\mathbf{v}} - \left.\frac{\partial\mathbf{q}^{(g)}_{int}}{\partial\mathbf{v}}\right|_{\mathbf{U}^{(g)}} - \frac{d\mathbf{q}^{(g)}_{R\,int}}{d\mathbf{v}}\right\}. \tag{48}$$

The difference between Eqs. (48) and (42) is that $(\mathbf{K}^{(g)} + \mathbf{K}^{(g)}_R)$ replaces the stiffness matrix $\mathbf{K}^{(g)}$, and another term $d\mathbf{q}^{(g)}_{R\,int}/d\mathbf{v}$ is added. $(\mathbf{K}^{(g)} + \mathbf{K}^{(g)}_R)$ is available from structural analysis. Therefore, we need to evaluate the term $d\mathbf{q}^{(g)}_{R\,int}/d\mathbf{v}$. At the element level, $d\mathbf{q}^{(e)}_{R\,int}/d\mathbf{v}$ is easily computed. The vectors $d\mathbf{q}^{(e)}_{R\,int}/d\mathbf{v}$ from failed elements are assembled into $d\mathbf{q}^{(g)}_{R\,int}/d\mathbf{v}$. By using this procedure, the gradients of the displacement can be obtained.

5.3. *Response gradient formulation for geometrically nonlinear structures*

The response gradients of GNS are derived as those of linear structures from the equilibrium equation of internal and external forces [8,9,17]. In the geometrically nonlinear analysis, the equilibrium equation of internal and external forces $\mathbf{q}^{(g)}_{int} = \mathbf{q}^{(g)}_{ext}$ coincides with a converged solution of the Newton–Raphson method.

The response gradients are computed by taking the derivative of the internal forces with respect to the random variables as follows [8,9,17]:

$$\begin{aligned}\frac{d\mathbf{q}^{(g)}_{int}}{d\mathbf{v}} &= \left.\frac{\partial\mathbf{q}^{(g)}_{int}}{\partial\mathbf{U}^{(g)}}\right|_{\mathbf{v}}\frac{d\mathbf{U}^{(g)}}{d\mathbf{v}} + \left.\frac{\partial\mathbf{q}^{(g)}_{int}}{\partial\mathbf{v}}\right|_{\mathbf{U}^{(g)}} \\ &= \mathbf{K}^{(g)}_T\frac{d\mathbf{U}^{(g)}}{d\mathbf{v}} + \left.\frac{\partial\mathbf{q}^{(g)}_{int}}{\partial\mathbf{v}}\right|_{\mathbf{U}^{(g)}} = \frac{d\mathbf{q}^{(g)}_{ext}}{d\mathbf{v}},\end{aligned} \tag{49}$$

where $\mathbf{K}^{(g)}_T$ is the tangent stiffness matrix in the global coordinates. Solving Eq. (49) for $d\mathbf{U}^{(g)}/d\mathbf{v}$ gives the gradients of the displacements with respect to the random variables as

$$\frac{d\mathbf{U}^{(g)}}{d\mathbf{v}} = \mathbf{K}^{(g)^{-1}}_T\left(\frac{d\mathbf{q}^{(g)}_{ext}}{d\mathbf{v}} - \left.\frac{\partial\mathbf{q}^{(g)}_{int}}{\partial\mathbf{v}}\right|_{\mathbf{U}^{(g)}}\right). \tag{50}$$

After the convergence of the Newton–Raphson method, no iterations to calculate the response gradients are necessary [9] because the derivation of Eq. (50) starts from a converged equilibrium equation. For this reason, the tangent stiffness approach is found to be appropriate for design sensitivity analysis [17].

The gradients of the displacements with respect to the various random variables are as follows:

- gradient with respect to external load:

$$\frac{d\mathbf{U}^{(g)}}{d\mathbf{v}_l} = \mathbf{K}^{(g)^{-1}}_T\left(\frac{d\mathbf{q}^{(g)}_{ext}}{d\mathbf{v}_l}\right), \tag{51}$$

- gradient with respect to material property:

$$\frac{d\mathbf{U}^{(g)}}{d\mathbf{v}_m} = \mathbf{K}_T^{(g)^{-1}} \left(- \frac{\partial \mathbf{q}_{int}^{(g)}}{\partial \mathbf{v}_m} \bigg|_{\mathbf{U}^{(g)}} \right), \tag{52}$$

- gradient with respect to geometry:

$$\frac{d\mathbf{U}^{(g)}}{d\mathbf{v}_g} = \mathbf{K}_T^{(g)^{-1}} \left(\frac{d\mathbf{q}_{ext}^{(g)}}{d\mathbf{v}_g} - \frac{\partial \mathbf{q}_{int}^{(g)}}{\partial \mathbf{v}_g} \bigg|_{\mathbf{U}^{(g)}} \right). \tag{53}$$

In the reliability analysis, the computation of the gradients is iterated due to the changes in failure points. For GNS, converged solutions from the Newton–Raphson method are used for the failure points.

6. Geometrically nonlinear truss and frame elements

6.1. Geometrically nonlinear truss element

A two-dimensional truss element which fails by yielding can be formulated in the geometrically nonlinear case by using a zero modulus of elasticity and a constant axial force. Appendix A provides the vector of internal forces and the elastic, geometric, and initial displacement stiffness matrices of a truss element which fails due to yielding.

According to Appendix A, the vector of internal forces is

$$\mathbf{q}_{int} = T \begin{Bmatrix} -1 \\ -\varphi \\ 1 \\ \varphi \end{Bmatrix}, \tag{54}$$

where T is the axial force and $\varphi = (v_2 - v_1)/L$. Since the truss element fails by yielding, the axial force is

$$T = A\sigma_y, \tag{55}$$

where A is the cross-section area and σ_y, the yield stress.

The tangent stiffness matrix is

$$\mathbf{K}_T = \int_L EA\left(\mathbf{B}_u^T + \mathbf{B}_v\mathbf{U}\mathbf{B}_v^T\right)\left(\mathbf{B}_u + \mathbf{B}_v\mathbf{U}\mathbf{B}_v\right) dx + \int_L T\mathbf{B}_v^T\mathbf{B}_v\, dx. \tag{56}$$

Substituting a zero modulus of elasticity into Eq. (56) results in

$$\mathbf{K}_T = \int_L T\mathbf{B}_v^T\mathbf{B}_v\, dx. \tag{57}$$

Therefore, the reduced tangent stiffness matrix is given by

$$\mathbf{K}_T = \int_L T\mathbf{B}_v^T\mathbf{B}_v dx = \frac{T}{L}\begin{bmatrix} 0 & 0 & 0 & 0 \\ 0 & 1 & 0 & -1 \\ 0 & 0 & 0 & 0 \\ 0 & -1 & 0 & 1 \end{bmatrix}. \tag{58}$$

By using the above formulations for the yielded truss elements, the system reliability analysis can be performed.

6.2. Geometrically nonlinear frame element

The vector of internal forces and the elastic, geometric, and initial displacement stiffness matrices for a two-dimensional geometrically nonlinear frame element are provided in Appendix B. It is not easy to introduce a plastic hinge into a geometrically nonlinear frame element. This is due to the fact that internal forces are obtained by using elastic and geometric stiffness matrices. The static condensation method is complicated and, furthermore, the tangent stiffness matrix cannot be derived straightforward. Because of these reasons, a hinge element with no length is inserted into the section which fails due to a plastic moment (Fig. 3).

To implement this procedure, the command LINK provided in FEAP controls displacements of the hinge element according to the user's definition. By using this command, the translations of both nodes of the hinge element are restrained to be the same in the x and y directions, and rotations of both ends of the hinge element can take different values. In addition, the plastic moments are applied to both ends as equivalent internal forces. This produces the same x, y displacements and different rotations with constant plastic moments on the hinge element.

The stiffness matrix and equivalent internal forces used in FEAP are described in Fig. 4. The notations 1–4 are nodal numbers, (1)–(3) are element numbers, where (2) denotes the hinge element. By using the LINK command, the displacements u_3 and v_3 are reduced in the assembled stiffness matrix and equivalent internal forces.

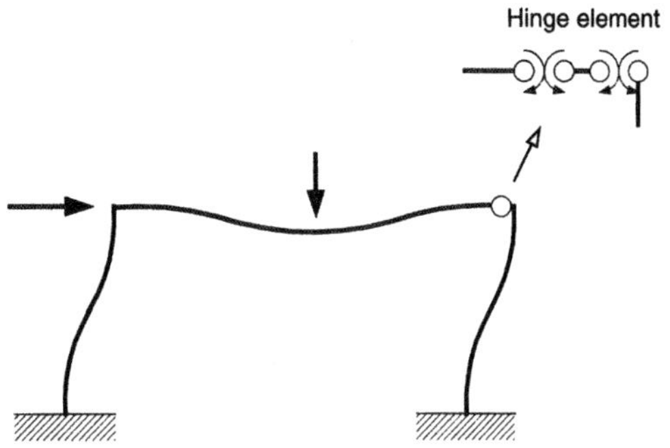

Fig. 3. Inserting a hinge element into a frame structure.

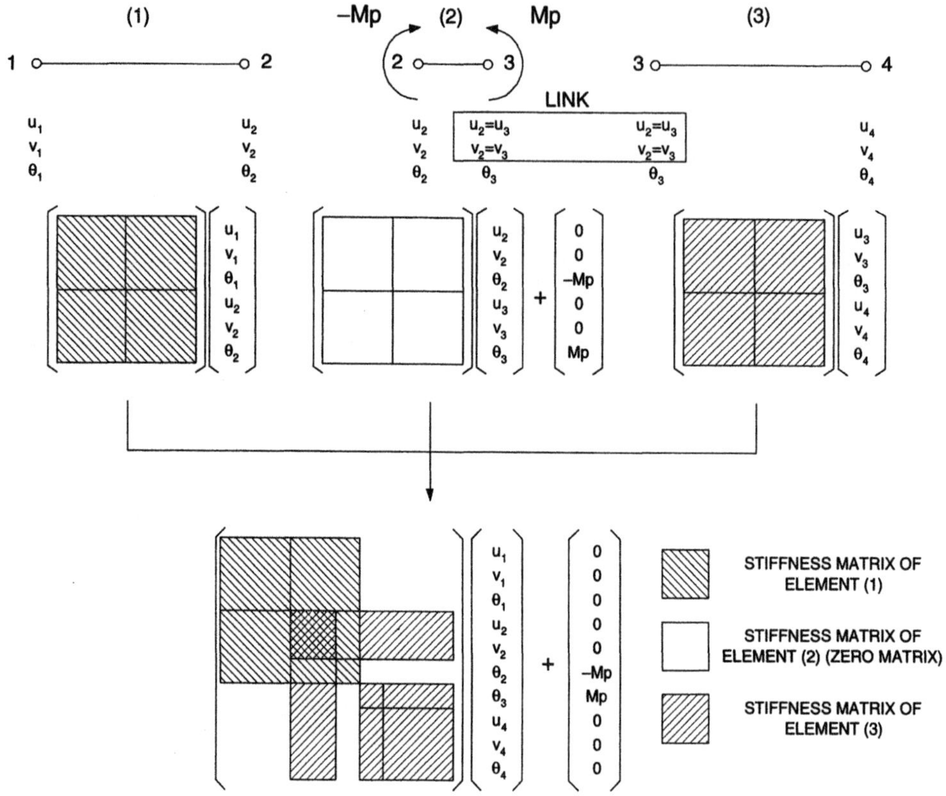

Fig. 4. Computer implementation by using LINK command.

Different rotations θ_2 and θ_3 with the constant plastic moment can be obtained by using the assembled stiffness matrix.

7. Computer implementation

To implement the above procedures at the University of Colorado, the linkage code RELSYS–FEAP was developed (Fig. 5). This code consists of three parts, RELSYS, INTERFACE, and FEAP. The main program is RELSYS. INTERFACE and FEAP are treated as subroutines of RELSYS. In the reliability analysis, the subroutine FEAP is executed from RELSYS until RELSYS finds the converged reliability index.

The linkage between RELSYS and FEAP has been achieved without modifying these programs. The reason is that both programs are long documented and well organized and modifications may cause unnecessary disturbances.

7.1. RELiability of SYStems

Component reliability in RELSYS is computed by the direct derivation approach developed by Lee [44] that was originally incorporated into the program RELTRAN [45]. System reliability is calculated by modeling the structure as a series–parallel combination of its failure modes [14].

The subroutine limitfg defines the limit state function and computes its value. The subroutine gradfg provides the gradient of the limit state function. These two subroutines play an important role to link RELSYS with FEAP. The main program calls these subroutines.

7.2. Finite element analysis program

FEAP is a general-purpose finite element analysis program developed by Taylor [15], which can analyze nonlinear as well as linear elastic structures. FEAP has an element library and accepts a user-defined ele-

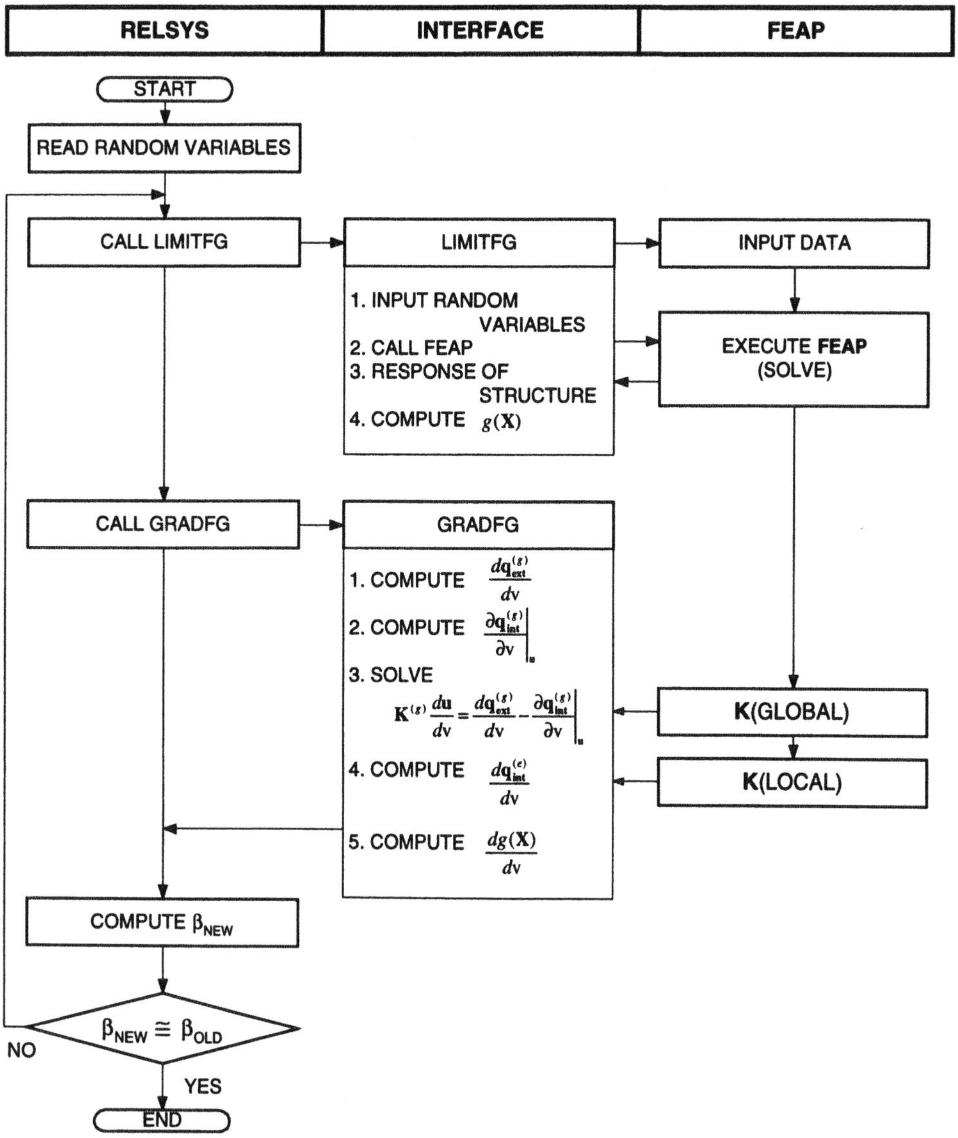

Fig. 5. RELSYS–FEAP linkage.

ment. The user can choose the element from this library and add the user-defined element. This user-defined element is useful for the development of the finite element reliability analysis, because the user can add the gradient computation routine to the element subroutine.

To execute FEAP, the user must provide the input-file, which defines the problem. In reliability analysis, RELSYS provides the random variables in the input file of FEAP and then FEAP solves the problem. The macro command INCLude is used for this procedure. The detailed input file and macro commands are available in FEAP user's manual [15].

7.3. Linkage of RELSYS with finite element analysis program

To link RELSYS with FEAP, new subroutines are created from subroutines of FEAP to compute the gradients of the limit state functions. The main functions of subroutines are as follows:

- Execute FEAP and return the response displacement and member forces.
- Calculate various gradients as indicated in the subroutine gradfg in Fig. 5.

As mentioned above, the linkage of RELSYS with FEAP is based on the work of Liu and Der Kiureghian [8, 9] who linked CALREL [16] with an earlier version of FEAP.

8. Conclusions

This article reviewed the theory of finite element reliability analysis of geometrically nonlinear elastic structures based on the total lagrangian formulation. To implement the finite element system reliability analysis of geometrically nonlinear elastic structures, the linkage of the system reliability analysis program RELSYS with the deterministic finite element analysis program FEAP was created. The computer code RELSYS–FEAP should find applicability in the system reliability analysis of slender elastic structures with geometric nonlinearities. This is shown in the companion paper [46].

Acknowledgements

The work described in this article was conducted by the first writer as part of his Ph.D. studies at the University of Colorado under the supervision of the second writer. The financial support from the Honshu-Shikoku Bridge Authority is gratefully acknowledged. Also, the partial financial support of the US National Science Foundation through grants CMS-9506435 and CMS-9522166 is gratefully acknowledged.

Appendix A

Fig. 6 represents the large displacements of a truss element. By using the notations in this figure, the internal forces and the components of the tangent stiffness matrix are as follows:

The internal forces are

$$\mathbf{q}_{\text{int}} = T\begin{Bmatrix} -1 \\ 0 \\ 1 \\ 0 \end{Bmatrix} + T\begin{Bmatrix} 0 \\ -\varphi \\ 0 \\ \varphi \end{Bmatrix}. \tag{A.1}$$

The elastic stiffness matrix is

$$\mathbf{K}_{\text{E}} = \frac{EA}{L}\begin{bmatrix} 1 & 0 & -1 & 0 \\ 0 & 0 & 0 & 0 \\ -1 & 0 & 1 & 0 \\ 0 & 0 & 0 & 0 \end{bmatrix}. \tag{A.2}$$

The geometric stiffness matrix is

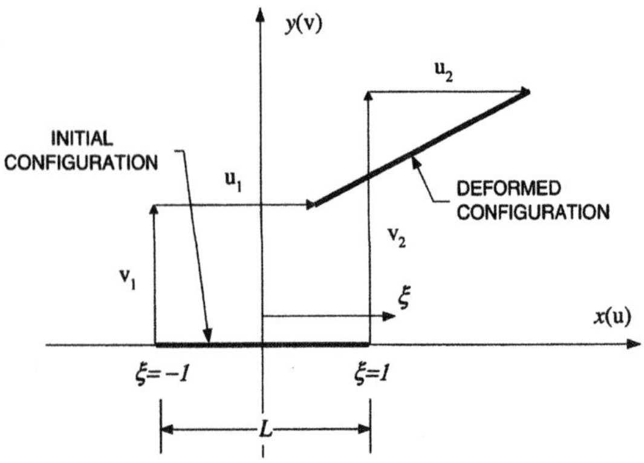

Fig. 6. Large displacements of a two-dimensional truss element.

$$\mathbf{K}_{\text{G}} = \frac{T}{L}\begin{bmatrix} 0 & 0 & 0 & 0 \\ 0 & 1 & 0 & -1 \\ 0 & 0 & 0 & 0 \\ 0 & -1 & 0 & 1 \end{bmatrix}. \tag{A.3}$$

The initial displacement stiffness matrix is

$$\mathbf{K}_{\text{U}} = \frac{EA}{L}\begin{bmatrix} 0 & \varphi & 0 & -\varphi \\ \varphi & \varphi^2 & -\varphi & -\varphi^2 \\ 0 & -\varphi & 0 & \varphi \\ -\varphi & -\varphi^2 & \varphi & \varphi^2 \end{bmatrix}, \tag{A.4}$$

where $\varphi = v_2 - v_1/L$ and

$$T = EA\varepsilon_{\text{G}} = EA\left\{\frac{u_2 - u_1}{L} + \frac{1}{2}\left(\frac{v_2 - v_1}{L}\right)^2\right\}.$$

Appendix B

Fig. 7 represents the large deformations and rotations of a two-dimensional frame element. Fig. 8 represents the deformation of the cross-section. By using the notations in Figs. 7 and 8, the internal forces and the components of the tangent stiffness matrix are as follows:

The internal force vector is:

$$\mathbf{q}_{\text{int}} = \begin{Bmatrix} -T \\ 0 \\ 0 \\ T \\ 0 \\ 0 \end{Bmatrix} + [\mathbf{K}_{\text{E}} + \mathbf{K}_{\text{G}}]\begin{Bmatrix} u_1 \\ v_1 \\ \theta_1 \\ u_2 \\ v_2 \\ \theta_2 \end{Bmatrix}, \tag{B.1}$$

where

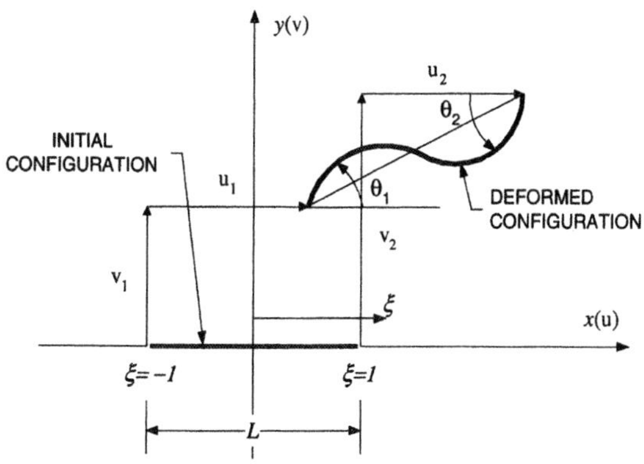

Fig. 7. Large displacements of a two-dimensional frame element.

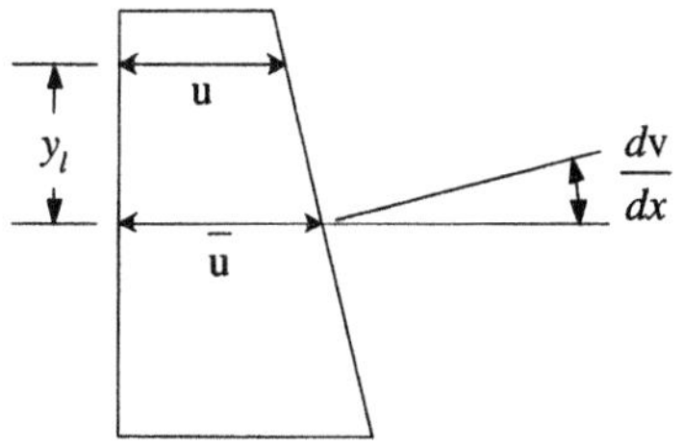

Fig. 8. Deformation of the cross-section of a two-dimensional frame element.

$$T = EA\varepsilon_G = EA\left\{\frac{u_2-u_1}{L}+\frac{1}{2}\left(\frac{v_2-v_1}{L}\right)^2\right\}.$$

The elastic stiffness matrix is

$$\mathbf{K}_E = \begin{bmatrix} \frac{EA}{L} & 0 & 0 & -\frac{EA}{L} & 0 & 0 \\ & \frac{12EI}{L^3} & \frac{6EI}{L^2} & 0 & -\frac{12EI}{L^3} & \frac{6EI}{L^2} \\ & & \frac{4EI}{L} & 0 & -\frac{6EI}{L^2} & \frac{2EI}{L} \\ & & & \frac{EA}{L} & 0 & 0 \\ & \text{sym.} & & & \frac{12EI}{L^3} & -\frac{6EI}{L^2} \\ & & & & & \frac{4EI}{L} \end{bmatrix}. \tag{B.2}$$

The geometric stiffness matrix is

$$\mathbf{K}_G = \frac{T}{L}\begin{bmatrix} 0 & 0 & 0 & 0 & 0 & 0 \\ & \frac{6}{5} & \frac{L}{10} & 0 & -\frac{6}{5} & \frac{L}{10} \\ & & \frac{2L^2}{15} & 0 & -\frac{L}{10} & -\frac{L^2}{30} \\ & & & 0 & 0 & 0 \\ & \text{sym.} & & & \frac{6}{5} & -\frac{L}{10} \\ & & & & & \frac{2L^2}{15} \end{bmatrix}. \tag{B.3}$$

The initial displacement stiffness matrix is

$$\mathbf{K}_U = \frac{EA}{L} \times \int_L \begin{bmatrix} 0 & -A_1 & -A_2 & 0 & A_1 & -A_3 \\ -A_1 & 0 & 0 & A_1 & 0 & 0 \\ -A_2 & 0 & 0 & A_2 & 0 & 0 \\ 0 & A_1 & A_2 & 0 & -A_1 & A_3 \\ A_1 & 0 & 0 & -A_1 & 0 & 0 \\ -A_3 & 0 & 0 & A_3 & 0 & 0 \end{bmatrix} dx$$
$$+EA \times \int_L \begin{bmatrix} 0 & 0 & 0 & 0 & 0 & 0 \\ 0 & A_1^2 & A_1A_2 & 0 & -A_1^2 & A_1A_3 \\ 0 & A_1A_2 & A_2^2 & 0 & -A_1A_2 & A_2A_3 \\ 0 & 0 & 0 & 0 & 0 & 0 \\ 0 & -A_1^2 & -A_2A_2 & 0 & A_1^2 & -A_1A_3 \\ 0 & A_1A_3 & A_2A_3 & 0 & -A_1A_3 & A_3^2 \end{bmatrix} dx. \tag{B.4}$$

The integrals in Eq. (B.4) are as follows:

$$\int_L A_1\,dx = -\frac{6}{5L}(v_2-v_1)+\frac{1}{10}(\theta_1+\theta_2),$$

$$\int_L A_2\,dx = -\frac{1}{10}(v_2-v_1)+\frac{L}{30}(4\theta_1-\theta_2),$$

$$\int_L A_3\,dx = -\frac{1}{10}(v_2-v_1)-\frac{L}{30}(\theta_1-4\theta_2),$$

$$\int_L A_1^2\,dx = \frac{72}{35L^3}(v_2-v_1)^2+\frac{3}{35L}(\theta_1^2+\theta_2^2) - \frac{18}{35L^2}(v_2-v_1)(\theta_1+\theta_2),$$

$$\int_L A_2^2\,dx = \frac{3}{35L}(v_2-v_1)^2+\frac{L}{210}(12\theta_1^2+\theta_2^2-3\theta_1\theta_2) + \frac{1}{70}(v_2-v_1)(\theta_1-\theta_2),$$

$$\int_L A_3^2\,dx = \frac{3}{35L}(v_2-v_1)^2+\frac{L}{210}(\theta_1^2+12\theta_2^2-3\theta_1\theta_2) - \frac{1}{70}(v_2-v_1)(\theta_1-\theta_2),$$

$$\int_L A_1A_2\,dx = \frac{9}{35L^2}(v_2-v_1)^2-\frac{1}{140}(\theta_1^2-\theta_2^2-2\theta_1\theta_2) - \frac{6}{35L}(v_2-v_1)\theta_1,$$

$$\int_L A_1A_3\,dx = \frac{9}{35L^2}(v_2-v_1)^2+\frac{1}{140}(\theta_1^2-\theta_2^2+2\theta_1\theta_2) - \frac{6}{35L}(v_2-v_1)\theta_2,$$

$$\int_L A_2A_3\,dx = -\frac{L}{420}(3\theta_1^2+3\theta_2^2-4\theta_1\theta_2) - \frac{1}{70}(v_2-v_1)(\theta_1+\theta_2).$$

References

[1] Bathe KJ. Finite element procedures in engineering analysis. Englewood Cliffs, NJ: Prentice-Hall, 1982.

[2] Zienkiewicz OC, Taylor RL. The finite element method, 4th ed, vol. 2. Berkshire (UK): McGraw-Hill, 1991.

[3] Crisfield MA. Non-linear finite element analysis of solid and structures. Chichester (UK): Wiley, 1991.

[4] Hasofer M, Lind C. Exact and invariant second-moment code format. J Engng Mech 1974;100(EM1):111–21.

[5] Shinozuka M. Basic analysis of structural safety. J Struct Engng 1983;109(3):721–40.

[6] Ang AH-S, Tang WH. Probability concepts in engineering planning and design. New York: Wiley, 1984.

[7] Der Kiureghian A, Lin H-Z, Hwang S-J. Second-order reliability approximations. J Engng Mech 1987;113(8): 1208–25.

[8] Liu P-L, Der Kiureghian A. Finite element reliability methods for geometrically nonlinear structures. Report No. UCB/SEMM-89/05. Department of Civil Engineering, University of California, Berkeley, CA, 1989.

[9] Liu P-L, Der Kiureghian A. Finite element reliability of geometrically nonlinear uncertain structures. J Engng Mech 1991;117(8):1806–25.

[10] Der Kiureghian A, Ke J-B. The stochastic finite element method in structural reliability. Probab Engng Mech 1988;3(2):83–91.

[11] Teigen JG, Frangopol DM, Sture S, Felippa CA. Probabilistic FEM for nonlinear concrete structures. I: theory. J Struct Engng 1991;117(9):2674–89.

[12] Teigen JG, Frangopol DM, Sture S, Felippa CA. Probabilistic FEM for nonlinear concrete structures. II: applications. J Struct Engng 1991;117(9):2690–707.

[13] Frangopol DM, Lee Y-H, William K. Nonlinear finite element reliability analysis of concrete. J Engng Mech 1996;122(12):1174–82.

[14] Estes AC, Frangopol DM. RELSYS: a computer program for structural system reliability. Struct Engng Mech 1998;6(8):901–19.

[15] Taylor RL. FEAP – a finite element analysis program, version 6.0. Department of Civil Engineering, University of California, Berkeley, CA, 1996.

[16] Liu P-L, Lin H-Z, Der Kiureghian A. CALREL user manual. Report No. UCB/SEMM-89/18. Department of Civil Engineering, University of California, Berkeley, CA, 1989.

[17] Ryu YS, Haririan M, Wu CC, Arora JS. Structural design sensitivity analysis of nonlinear response. Comput Struct 1985;21(1–2):245–55.

[18] Wu CC, Arora JS. Design sensitivity analysis and optimization of nonlinear structural response using incremental procedure. AIAA J 1987;25(8):1118–25.

[19] Dagher HJ, Lu Q, Peyrot AH. Reliability of transmission structures including nonlinear effects. J Struct Engng 1998;124(8):966–73.

[20] Thoft-Christensen P, Murotsu Y. Application of structural systems reliability theory. Berlin (Germany): Springer, 1986.

[21] Washizu K. Variational methods in elasticity and plasticity, 3rd ed. Oxford: Pergamon, 1982.

[22] Bathe KJ, Bolourchi S. Large displacement analysis of three-dimensional beam structures. Int J Num Meth Engng 1979;14:961–86.

[23] Wood RD, Zienkiewicz OC. Geometrically nonlinear finite element analysis of beams, frames, arches and axisymmetric shells. Comput Struct 1977;7(7):725–35.

[24] Adeli H, Zhang J. Fully nonlinear analysis of composite girder cable-stayed bridges. Comput Struct 1995;54(2): 267–77.

[25] Fleming JF. Nonlinear static analysis of cable-stayed bridge structures. Comput Struct 1979;10:621–35.

[26] Nazmy AS, Abdel-Ghaffar AM. Three-dimensional nonlinear static analysis of cable-stayed bridges. Comput Struct 1990;34(2):257–71.

[27] Chen ZQ, Agar TJA. Geometric nonlinear analysis of flexible spatial beam structures. Comput Struct 1993;49(6): 1083–94.

[28] Kanchi MB. Matrix methods of structural analysis. New Delhi (India): Wiley-Eastern, 1993.

[29] Felippa CA. Lecture notes in nonlinear finite element methods. Center for Aerospace Structures, University of Colorado, Boulder, CO, 1996.

[30] Rackwitz R. Practical probabilistic approach to design, Bulletin 112. Comite European du Beton, Paris, France, 1976.

[31] Rackwitz R, Fiessler B. Structural reliability under combined random load sequences. Comput Struct 1978;9: 489–94.

[32] Breitung K. Asymptotic approximations for multinormal integrals. J Engng Mech 1984;110(3):357–66.

[33] Hohenbichler M, Rackwitz R. Improvement of second-order reliability estimates by importance sampling. J Engng Mech 1988;114(12):2195–9.

[34] Tvedt L. Distribution of quadratic forms in normal space – application to structural reliability. J Engng Mech 1990; 116(6):1183–97.

[35] Lin H-Z. Methods for structural system reliability analysis, PhD Thesis. Structural Engineering, Mechanics, and Materials. Department of Civil Engineering, University of California, Berkeley, CA, 1990.

[36] Cornell CA. Bounds on the reliability of structural systems. J Struct Engng 1967;93(2):171–200.

[37] Ditlevsen O. Narrow reliability bounds for structural systems. J Struct Mech 1979;7(4):453–72.

[38] Estes AC. A system reliability approach to the lifetime optimization of inspection and repair of highway bridges, PhD Thesis. Department of Civil, Environmental, and Architectural Engineering, University of Colorado, Boulder, CO, 1997.

[39] Burden RL, Faires JD. Numerical analysis, 5th ed. Boston: PWS Publishing, 1993.

[40] Hohenbichler M, Rackwitz R. First-order concepts in system reliability. Struct Safety 1983;1:177–88.

[41] Ishii K, Suzuki M. Stochastic finite element method for slope stability analysis. Struct Safety 1987;4:111–29.

[42] Lee J-S. Application of stochastic finite element method to system reliability analysis of offshore structures, vol. 1. In: Schuëller GI, Shinozuka M, Yao JTP, editors. Structural safety and reliability. Rotterdam: Balkema. p. 355–60.

[43] Yamazaki K. Design sensitivity analysis in structural optimization. In: Kodiyalam S, Saxena M, editors. Geometry and optimization techniques for structural design. Southampton: Computational Mechanics Publications [chapter 8].
[44] Lee Y-H. Stochastic finite element analysis of structural plain concrete, PhD Thesis. Department of Civil, Environmental, and Architectural Engineering, University of Colorado, Boulder, CO, 1994.
[45] Lee Y-H, Hendawi S, Frangopol DM. RELTRAN: a structural reliability analysis program, version 2.0. Report No. CU SR-93/6. Structural Engineering and Structural Mechanics Series. Department of Civil, Environmental, and Architectural Engineering, University of Colorado, Boulder, CO, 1993.
[46] Frangopol DM, Imai K. Geometrically nonlinear finite element reliability analysis of structural systems. II: applications. Comput Struct 2000;77(6):693–709.

Papers from

Construction and Building Materials

Performance characteristics of surface coatings applied to concrete for control of reinforcement corrosion

A.M.G. Seneviratne, G. Sergi, C.L. Page*

Department of Civil Engineering, Aston University, Aston Triangle, Birmingham B4 7ET, UK

Received 30 April 1999; received in revised form 15 June 1999; accepted 6 January 2000

Abstract

Three elastomeric surface coatings were applied to naturally carbonated concrete components obtained from buildings that were suffering from reinforcement corrosion. Monitoring of the internal relative humidity of the concrete revealed that all three coating systems were able to exclude water from the carbonated components for a period of 2 years but only one of them was able to sustain its performance for a period of up to 5 years in an urban UK outdoor environment. Dynamic mechanical thermal analysis suggested that the most successful coating was able to maintain its elastomeric properties over the required period of exposure and over a wide range of operational temperatures. It also had a relatively low but uniform bond strength to the concrete and this appeared to have a beneficial effect on its ability to accommodate movements of the substrate. Such coatings are considered capable of extending the service-lifetimes of carbonated reinforced concrete structures in cases where significant chloride contamination does not exist and where the only substantive route for moisture ingress is via the coating. © 2000 Elsevier Science Ltd. All rights reserved.

Keywords: Carbonation; Reinforcement corrosion; Surface coatings

1. Introduction

The high inherent alkalinity of concrete provides protection to steel reinforcement allowing it to form a thin compact passive oxide film. Penetration of the concrete by deleterious agents such as chlorides and neutralization of the alkalis in the cover concrete by carbonation can lead to the breakdown of this protective film and to corrosion of the steel. The fact that the corrosion rate of reinforcement in carbonated concrete is controlled by the electrolytic resistance of the material [1,2] allows it to be reduced to an insignificant level if the concrete can be maintained in a dry condition [2,3]. Surface coatings, capable of preventing water ingress, are thus prime candidates for this type of protection and three elastomeric coating systems, with good crack-bridging capabilities, have been shown to be successful over a period of 2 years in controlling reinforcement corrosion in carbonated concrete which was free from significant chloride contamination [4]. Over the longer term, one of the coatings was found to maintain an adequate level of protection but the other two were only partially effective in retaining their waterproofing character after weathering [5,6]. This paper presents the results of an investigation to determine the effect of weathering on properties considered likely to influence the ability of the three surface coatings to exclude water from the underlying concrete. The coatings were applied to naturally carbonated concrete components and were exposed outdoors for a period of up to 5 years. The components were initially exposed for 2 years at a south facing site as part of a building in

*Corresponding author. Tel.: +44-121-359-3611; fax: +44-121-333-3389.

Wexham Springs but were relocated to a north facing site in Birmingham for a further 3 years when the building was demolished.

The work described in this paper formed part of a collaborative research programme involving Aston University and several industrial and science-based partners. The primary aim of the project was to evaluate the effectiveness of a number of commercially available surface treatment systems for extending the service-lives of concrete structures where reinforcement corrosion had been initiated as a consequence of carbonation and to determine whether their performance was affected by natural weathering.

2. Experimental

The selection of the structures, the choice of surface coatings, the application of the coatings and monitoring of the corrosion behaviour of the steel reinforcement have all been described in other publications [4–6]. In this paper, only those aspects of the work relating to the performance of the coatings are reported.

The three specialist materials manufacturers, who were involved in the research project, applied their own coating systems, according to their recommended procedures to selected concrete components which had suffered from varying degrees of carbonation. Application procedures included cleaning of the dry concrete surface with a wire brush, filling of any large voids and cracks on the surface with repair mortar and application of the primer, followed by the coating using a paintbrush. Details of the coating systems are listed in Table 1. The following techniques were used to assess the properties of the coatings and their changes on weathering:

- dynamic mechanical thermal analysis (DMTA); and
- tensile bond strength measurement (pull-off test).

The former of the above methods required small samples, which were carefully removed from the treated components at the appropriate times. The pull-off tests were performed in-situ but, again, only small areas of the coatings were required. It was considered essential that the damage to the coatings be kept to a minimum so that their long-term performance could continue to be monitored. In every case, the area from which the coating was removed was carefully covered with a thin layer of waterproof epoxy resin.

The results obtained by the analyses were examined in relation to the ability of the coatings to exclude moisture from the underlying concrete components. These investigations were conducted over a period of 5 years by applying cycles of wetting and drying to the components and measuring the internal relative humidity (RH) of the concrete through stainless steel inserts at fixed locations [7].

2.1. Dynamic mechanical thermal analysis

DMTA is a sensitive method of examining the thermo-mechanical properties of polymer test specimens. It involves the application of a sinusoidal time-dependent force to the specimen and measurement of both the amplitude and phase angle of the displacement. From these data, the elastic (energy storage) and viscous (energy dissipation) properties of the material may be evaluated as a function of time and temperature. The technique allows the glass transition temperature of a polymer (T_g) and other secondary effects to be characterized.

2.2. Tensile bond strength measurement

The bond strength test was performed in-situ by means of an 'Elcometer' adhesion tester (Model 106). A 20 mm diameter aluminium disc-shaped dolly was first fixed on the coating with rapid-hardening 'Araldite' adhesive. After 48 h a sharp cut was made around the dolly through the coating to separate the test area from the surrounding material. An increasing tensile force was then applied to the dolly until de-bonding occurred and the bond strength was recorded. Tests were carried out in triplicate on 18 separate specimens (six each per coating type). The specimens were carbonated to varying degrees and were either unweathered or weathered for a period of approximately 5 years.

3. Results and discussion

The ability of the three elastomeric coatings to resist

Table 1
Details of the surface treatment systems used in the investigation

ID	Description
A	Acrylic modified siloxane primer/water based acrylic elastomeric coating
B	Low-viscosity water based acrylic primer/water based acrylic elastomeric coating
C	Water based epoxy primer/water based acrylic elastomeric coating

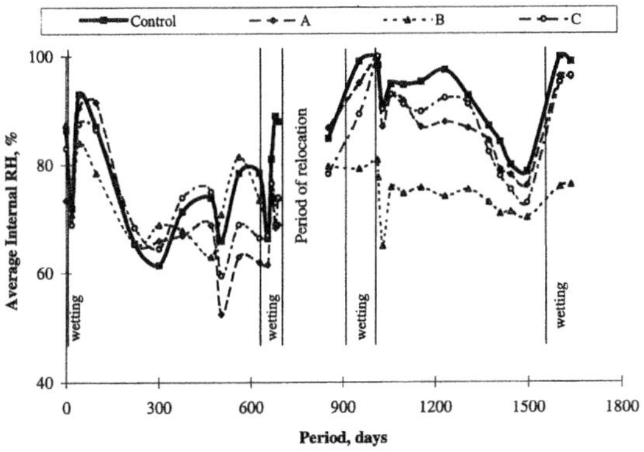

Fig. 1. Average internal RH of carbonated concrete columns.

ingress of water to the underlying concrete is illustrated by the results obtained on highly carbonated concrete columns which were part of a building at the former BCA site at Wexham Springs. Three of the reinforced concrete columns had the whole of their exposed surfaces treated with one of the three coating systems. One other was left untreated for comparison. Stainless steel RH inserts were fixed at two positions on each column, which enabled the monitoring of the internal RH of the concrete to be carried out. Fig. 1 shows the variation of the RH of each column with time. As the columns were in a fairly sheltered location, some artificial watering was employed at intervals. These periods of artificial watering are indicated on the same figure. It is evident from the results that the RH of the control column mirrored that of the external environment and increased during wetting cycles. That of the coated columns, however, remained relatively constant for a period of approximately 2 years and rarely exceeded 80% RH, except at the beginning of exposure when all the columns were pre-wetted.

An RH of approximately 80% was shown in previous laboratory experiments to be an approximate threshold to sustain significant rates of corrosion of steel reinforcement embedded in carbonated concrete [2]. To confirm this, four standard 100-mm OPC concrete cubes were produced with a 0.6 water/cement ratio. They contained four 6-mm diameter steel bars cast at 4, 8, 12 and 20 mm below the only exposed surface. The other faces were masked with a combination of copper leaf and epoxy resin in order to ensure that moisture and gas transport could occur only through the exposed face. After an initial curing period of 7 days under water, the cubes were exposed to a constant 60% RH environment at 20°C for approximately 18 months. Carbonation depth measurements on control cubes were found to be at least 10 mm which meant that the steel bars at the 4 mm cover depth were lying in totally carbonated concrete. All the cubes were then exposed at a range of constant RH values from 60 to 100% until the corrosion rates of the bars at the 4 mm cover depth, obtained by linear polarization, had reached a steady value. Fig. 2 shows how the mean corrosion rate of the four samples varied with external RH. When the RH was maintained at or below 80%, the concrete had a high enough resistivity to restrict the corrosion rate of the reinforcement to levels below 0.1 $\mu A/cm^2$. Such low values are within the range typically recorded for corrosion rates of passive steel in non-carbonated concrete and are therefore not considered to be significantly detrimental [8].

The concrete columns had to be relocated at Aston University in Birmingham as a consequence of the sale of the Wexham Springs site. Following some minor repairs to the coatings, monitoring of the RH continued for another 2 years. During this further period of exposure, coatings A and C started to fail and allowed some moisture to penetrate the concrete. Coating B, however, continued to exclude external water and maintained the internal RH of the underlying concrete below the critical level of 80% (Fig. 1). The same sustained performance of coating B was observed on other treated concrete components exposed externally in a similar way. In order to elucidate the features that enabled coating B to maintain its integrity for a longer period than the other two elastomeric coatings, a number of tests were performed on each coating as explained earlier.

The three coatings were found to be of a similar thickness which normally exceeded 400 μm. Typical examples of the coatings are shown in cross-section in Fig. 3. DMTA measurements, carried out on the three coatings at four different periods of weathering, identified some differences in T_g between coating A and the other two coatings (Table 2). T_g is a characteristic temperature below which polymers become likely to

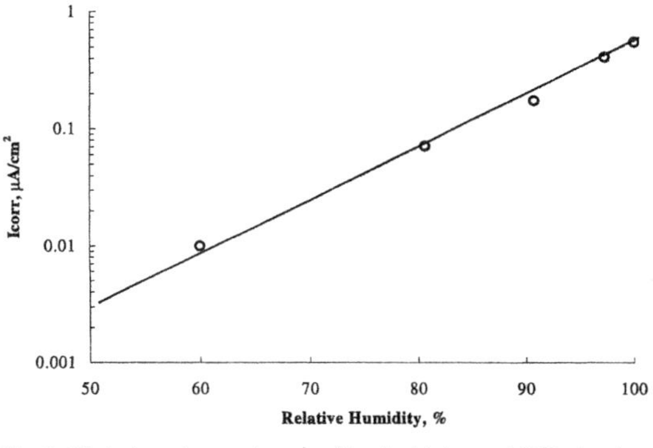

Fig. 2. Variation of corrosion rate (I_{corr}) with internal RH of carbonated concrete.

Table 2
Glass transition temperature (°C) for different exposure periods

Exposure period	DMTA: glass transition temperature (°C)		
	Coating A	Coating B	Coating C
Initial	−1.0	−11.0	< −20
18 months	−2.5	−12.6	−12.2
2 years	−3.5	−11.0	−11.9
4 years	−2.2	−11.0	−10.0

fail in a brittle manner when stressed. Coatings B and C throughout the 4 years of weathering maintained a T_g which was lower than −10°C, a temperature that is reached only on rare occasions in the UK. Coating A, on the other hand, had a T_g of only −1 to −3.5°C, a range of temperatures that occurs quite frequently during the winter months. This could explain, at least partly, the small cracks that appeared on some specimens treated with coating A. It does not, however, explain the better overall performance of coating B which appeared to tolerate movement of the underlying concrete better than coatings A and C. The bond strength values obtained offer a possible explanation for this.

Fig. 4 shows all the bond strength results obtained for coatings A, B and C. The results indicate very clearly that the bond strength of coating B, throughout 5 years of weathering, remained relatively low and uniform. The bond strength of coating A and, in particular, coating C were, on the other hand, more variable and were, in some cases, two to three times higher than the maximum bond strength obtained for coating B. The observed difference in the behaviour of the coatings may have been caused by differing properties of the primer used in each case. It is interesting in this respect to note that the highest overall bond strength was obtained for coating C after 5 years of exposure. The primer used in this case was a water-based epoxy resin which, with time, may have undergone substantial cross-linking of the polymer chains.

The relatively modest uniform bond strength of coating B to the concrete would appear to have allowed this coating to remain intact and prevent water ingress to the substrate over the five year period. Movement or localized cracking of the underlying concrete is likely to introduce stresses to the coating. If the stresses are restrained and therefore concentrated over a limited area owing to the high bond strength, as may have been the case with coatings A and C, localized failure of the coating is likely to occur. If, on the other hand, the coating is weakly bonded to the substrate, as appears to be the case with coating B, stresses induced in the substrate are liable to be spread over a somewhat larger area by local disbondment and can, therefore, be relieved if the coating is sufficiently elastomeric.

4. Conclusions

It was shown in this project that the use of an elastomeric coating system with particular properties can protect carbonated concrete from water ingress. It can thus extend the service life of a reinforced concrete structure by controlling the rate of corrosion of embed-

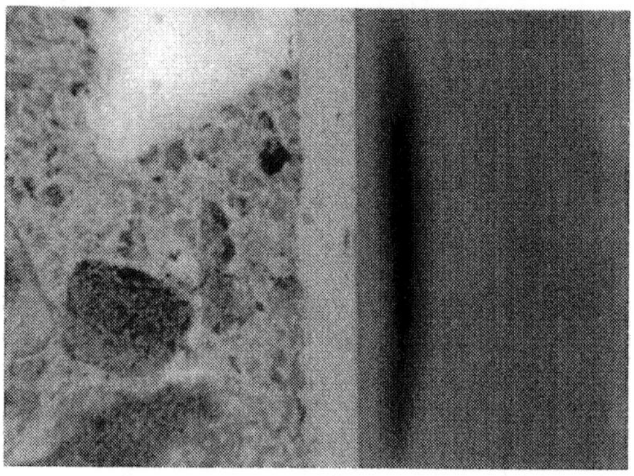

Coating A

Coating B

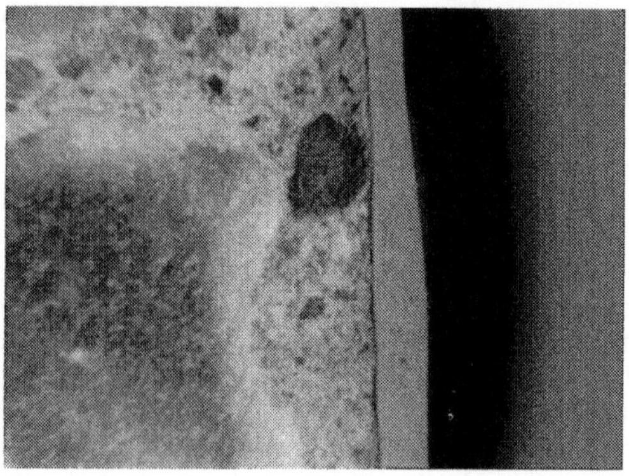

Coating C

Fig. 3. Cross-sectional view of the three coatings on concrete substrate (mag. ×10).

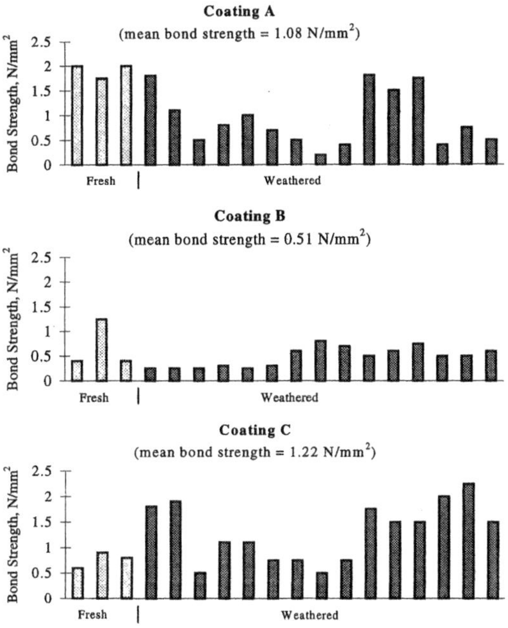

Fig. 4. Variation of the bond strength of coatings. *Note*: fresh (3 months after coating), weathered (4.5 years after coating).

ded steel in cases where significant chloride contamination does not exist and where the only substantive route for moisture ingress is via the coating.

It is suggested that, for a coating to be successful in this type of application, it must maintain its elastomeric properties over an appropriately wide range of temperatures throughout the required period of exposure. It appears also to be desirable for the coating to develop only a modest adhesional bond to the concrete substrate, so that it can tolerate stresses associated with movements in the concrete without cohesional failure. This may be achieved by the use of an appropriate primer.

Monitoring of the internal RH of surface-treated carbonated concrete is a convenient technique for assessing the performance of surface coatings applied for the prevention of water ingress and control of steel reinforcement corrosion.

Acknowledgements

The authors thank Mr Ray Owens of Loughborough Consultants Ltd. for his advice on interpretation of DMTA results. This work was carried out by Aston University and Aston Material Services Ltd. under contract to DETR as part of the LINK Programme on Construction Maintenance and Refurbishment, which is co-sponsored by EPSRC. Industry participants in the programme were: Aston Material Services Ltd., British Cement Association, Building Research Establishment, Fosroc International Ltd., MBE Consultants, Mouchel Consulting Ltd., Pullar–Strecker Consultancy Ltd., Weber & Broutin (UK) Ltd., Scott Wilson Kirkpatrick & Co. Ltd., Stirling Lloyd Polychem Ltd., Taywood Engineering and the Transport Research Laboratory.

References

[1] Gonzalez JA, Algaba S, Andrade C. Corrosion of reinforcing bars in carbonated concrete. Br Corros J 1980;15:135–9.

[2] Glass GK, Page CL, Short NR. Factors affecting the corrosion rate of steel in carbonated mortars. Corros Sci 1991; 32:1283–94.

[3] Sergi G, Lattey SE, Page CL. Influence of surface treatments on corrosion rates of steel in carbonated concrete. In: Page CL, Treadaway KWJ, Bamforth PB, editors. Corrosion of reinforcement in concrete. London: Elsevier, 1990:409–19.

[4] Seneviratne AMG, Sergi G, Maleki MT, Sadegzadeh M, Page CL. Effect of surface treatment of concrete on reinforcement corrosion. In: Page CL, Bamforth PB, Figg J, editors. Corrosion of reinforcement in concrete construction. London: Elsevier, 1996:567–76.

[5] Seneviratne AMG, Sergi G, Maleki MT, Sadegzadeh M, Page CL. Surface coatings for controlling reinforcement corrosion in carbonated concrete. In: Swamy RN, editor. Infrastructure regeneration and rehabilitation, improving the quality of life through better construction, a vision for the next millennium. Sheffield: Sheffield Academic Press, 1999:663–72.

[6] Sergi G, Seneviratne AMG, Maleki MT, Sadegzadeh M, Page CL. Control of reinforcement corrosion by surface treatment of concrete. Proc ICE Buildings Struct, in press.

[7] Parrott LJ. Moisture profiles in drying concrete. Adv Cement Res 1988;1(3):164–70.

[8] Andrade C, Alonso MC, Gonzalez JA. An initial effort to use the corrosion rate measurements for estimating rebar durability. In: Berke NS, Chaker V, Whiting D, editors. Corrosion rates of steel in concrete, ASTM STP 1065. Philadelphia: American Society for Testing and Materials, 1990:29–37.

Structural Engineering Compendium I

Investigation procedures for the diagnosis of historic masonries

L. Binda*, A. Saisi, C. Tiraboschi

Department of Structural Engineering (D.I.S.), Politecnico of Milan, Piazza Leonardo da Vinci 32, 20133 Milan, Italy

Received 15 March 1999; accepted 20 January 2000

Abstract

The Friuli, Irpinia and Umbria earthquakes, and subsequent experiences, have emphasized the need for adequate damage assessment prior to seismic rehabilitation. Furthermore, assessment can be enhanced by preventive studies under the guidance of those in charge of hazard mitigation (architects, engineers, etc.). Research procedures must be defined so that findings can be used for damage assessment and as input data for structural analysis and control models. This paper provides evaluations of in-situ and laboratory tests on materials for existing unreinforced masonry structures and, in particular, touches upon the difficulty of interpreting the results of NDE tests. Also, the need for a design of the investigation prepared by persons responsible for the rehabilitation is indicated. © 2000 Elsevier Science Ltd. All rights reserved.

Keywords: Masonry; Investigation; Non-destructive techniques

1. Introduction

Many experiences of restoration and rehabilitation of damaged masonry buildings have been accumulated in Europe for more than 50 years; in many cases the interventions were carried out after a long period of misuse and lack of maintenance following the Second World War. During this period also, major earthquakes have taken place in most of the Mediterranean countries.

Several unsuccessful results have emphasized the need for adequate assessment prior to any restoration or rehabilitation. In fact, when neither the real state of damage nor the effectiveness of repairs are known, the effectiveness of any intervention project is also unknown. Prevention and rehabilitation can be successfully accomplished only if diagnosis of the state of damage of the building has been carefully carried out.

In the past, the need for rehabilitation and repair of damaged masonry buildings allowed for the application and experimentation of both traditional and advanced techniques; the latter were used at first without previous control, due to a sense of urgency and lack of time.

The symptoms and their causes should first of all be determined. In some cases the correlation is clear (Figs. 1 and 2).

Frequently the correlation between effects and causes cannot be done without and experimental and analytical investigation (Figs. 3 and 4) [1,2].

Several investigation procedures have been implemented in recent years; the attempt is to use non-destructive (ND) techniques as much as possible. Nevertheless, there is still very little possibility at present

* Corresponding author. Tel.: +390-2-2399-4318; fax: +390-2-2399-4220.
E-mail address: binda@stru.polimi.it (L. Binda)

Reprinted from *Construction and Building Materials* **14 (4)**, 199-233 (2000)

Fig. 1. Damages after Umbria earthquake.

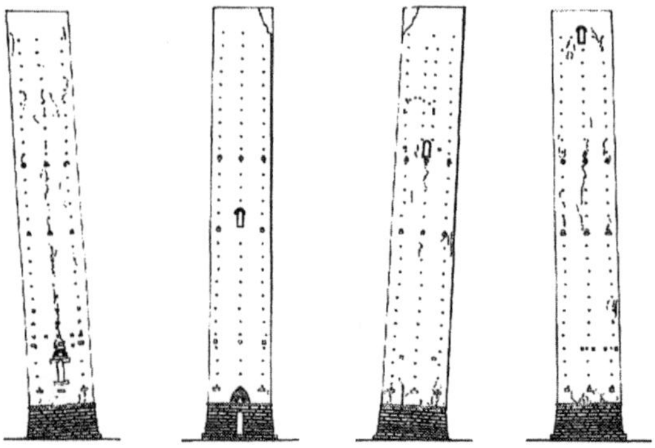

Fig. 2. The Garisenda leaning Tower in Bologna.

to correlate non-destructive evaluation (NDE) test data with masonry performance. Therefore, when the designer is not sufficiently skilled to interpret them, a great amount of data can be inappropriately used or even not used at all.

NDE can be helpful in finding hidden characteristics (internal voids and flaws and characteristics of a wall section) which cannot be determined other than through destructive tests. Sampling of masonry specimens is a costly operation, which also can lead to misunderstanding when the operation is not carried out in the appropriate manner. When an overall knowledge of the wall is needed, ND tests can be useful. For example, in order to qualify the state of preservation of stone or brick masonry walls and piers, the internal defects, the presence of multiple leaves and their type of connection, a knowledge of the inside is important. Two possible methods can be used to find out this information: (i) destructive survey through coring or local demolition, (ii) non-destructive survey carried out by NDT (non-destructive techniques) such as thermography, sonic and EM or X-ray tests.

Fig. 3. The Bell Tower of Pavia after collapse.

Fig. 4. The Cathedral of Noto (Siracusa).

The types of in-situ tests available at present are mainly based on the detection of the physical properties of the wall. The in situ mechanical tests available are flat-jack, hardness, penetration and pull-out tests. The flat-jack tests give local measurements and are slightly destructive: nevertheless, they can directly give the values of some mechanical parameters.

In the case of ND tests, a correlation between the measured parameters and the mechanical ones is usually difficult, but they can give an overall qualitative response of the masonry. At present, the most diffused ND techniques are the sonic (or ultrasonic), radar and thermography tests. Sonic and radar tests seem to be very promising (being widely used also in other fields like medicine or geotechniques), where a diagnosis has to be done on the state of damage present in inhomogeneous materials.

At present, most of the ND procedures can give only qualitative results; therefore the designer is asked to interpret the results and use them at least as comparative values between different parts of the same masonry structure or by using different ND techniques.

It must be clear that even if there is a need of consulting experts in the field, it is the designer, or a member of the design team, who must be responsible of the diagnosis and must: (i) set up the in-situ and laboratory survey project; (ii) constantly follow the survey; (iii) understand and verify the results; (iv) make technically acceptable use of the results including their use as input data for structural analyses; (v) choose appropriate models for the structural analysis; (vi) arrive at a diagnosis at the end of the study.

These operations can be accomplished with the help of experts in the field. Therefore, information is needed for architects and engineers on the availability and reliability of the investigation techniques.

This paper is an attempt to draw some guidelines for the diagnostic project based on the authors experience in the evaluation of masonry structures making use of case histories to illustrate examples. Following a brief description of the most important procedures for investigation an attempt is made to suggest the right choice of in-situ and laboratory testing according to the needs and taking into account the advantages and the limits of each procedure. The work tries to help avoiding the useless and expensive investigations frequently carried out by architects and engineers lacking a good knowledge of their problems.

2. Safety of masonry buildings

Historic masonry buildings, whatever use is made of them at present or in the future, have to show structural stability. From the point of view of the risk to human life, they may belong to the following categories, depending on the use of the building [3]: (i) isolated and non-accessible buildings; (ii) buildings belonging to the urban area; (iii) buildings open to the public; and (iv) buildings open to large assembles of people (cathedrals, theatres, etc.).

For each of the mentioned categories a certain amount of risk, as it is for new buildings, has to be accepted; the assessment of the structural state for several historic building has shown that for some of them the structural safety seems to be very low [3]. Certainly, some historic buildings exist whose stability is so precarious that they may collapse under slight earthquakes, strong winds, etc. (for example the Civic Tower of Pavia collapsed suddenly in 1989 without any apparent warning sign).

An appropriate and rational use of structural analysis can help in defining the eventual state of danger and in forecasting the future behaviour of the structure. To this aim, the definition of the mechanical properties of the materials, the implementation of constitutive laws for decayed materials and of methods of analysis for damaged structures and the improvement of reliability criteria are needed. Nevertheless, when the structure is a complex one, often linear elastic models are easily usable. Non-linear models or limit state design models are difficult to apply, also because the constitutive laws needed for the material are seldom available.

3. Materials and construction techniques

The structural performance of a masonry wall structure can be understood provided the following factors are known:

- the geometry;
- the characteristics of its masonry texture (single or multiple-leaf walls, connection between the leaves, joints empty or filled with mortar, physical, chemical and mechanical characteristics of the components (bricks, stones, mortar); and
- the characteristics of masonry as a composite material.

In the case of brickwork structures, and particularly in the case of new production masonry, the codes of practice can generally suggest some laws which allow the strength of masonry to be calculated as a function of that of the components.

In the case of stonework masonry, the problem is much more complex and some questions spontaneously rise. For instance, can the masonry texture, which strongly influences the bearing capacity of the wall, be easily identified? How can the characteristic strength of a highly non-homogeneous material be experimentally determined if sampling has to be numerous enough (for the results to be statistically representative) but at the same time be non-destructive? How can experimental tests be carried out and laboratory prototypes be built which are representative of real situations (think for instance to the great difficulty of building a wall made of river gravel in the laboratory)? Therefore, how can the physical–mechanical characteristics of stones and mortars be determined or be considered statistically reliable when sampling difficulties very often hinder the achievement of significant results? And again, can strength and deformability parameters (Young's modulus, Poisson's ratio) of the components, which indeed complete the mechanical characterization of the materials, be used to extrapolate the global strength and deformability of the masonry?

For these reasons, can the structural diagnosis of stonework masonry buildings be worked out from the knowledge of its components or can it be better performed through direct in-situ measurement of the mechanical characteristics of the masonry as a whole? How can the local and global behaviour of a stone masonry be determined under vertical and lateral loads? Provided good connections between the different structural elements in a stone masonry building have been introduced, how the behaviour of the wall is influenced by the construction technology under seismic in-plane or out of plane loads?

The actions which most compromise the stability of stonework masonry are the non-vertical ones. When some stones cannot fulfil the local equilibrium because of an insufficient mutual contact, a local mechanism can rise which suddenly evolves in a total collapse of the structure. This possibility is generally easier when the wall is subjected to non-vertical actions; in fact, with respect to the vertical ones, equilibrium had to be assured since the beginning of construction otherwise, without a sufficient horizontal constraint between the stones, the wall could have never been built [4].

Some answers to the above problems can be looked for in the literature on stonework masonry, although not much has been done up to now.

Giuffrè [4] proposes to first evaluate some aspects of the mechanical behaviour of a stone wall by visual inspection, looking for the presence of particular characteristics which permits the wall to be considered as built according to 'the rule of art'. There is in fact a complex of rules, which were precisely formulated in the treatises of the 19th century, although they were already known previously, which deal with the layout of the stones and are aimed to guarantee a good response of the wall toward external actions. A masonry wall built according to these rules should behave monolithically, and reach the collapse, for instance, under seismic loads through the formation of cylindrical springs, whereas the portions of the wall which do not crack maintain a 'rigid body'-like behaviour. Such a mechanism of collapse has the advantage to be predictable and possibly preventable through suitable interventions (Fig. 5).

The worst defect of a masonry wall is to be not monolithic in the lateral direction, and this can happen, for instance, when the wall is made by small pebbles or by two external layers well ordered but not mutually connected and containing a rubble infill. This makes the wall to become more brittle particularly when external forces act in the horizontal direction (Fig. 6). The same problem can happen under vertical loads if they act eccentrically.

Some indications on the mechanical behaviour of stonework masonry can also be given based on the research recently carried out by the authors. First of all, given the great number of existing cross-sections

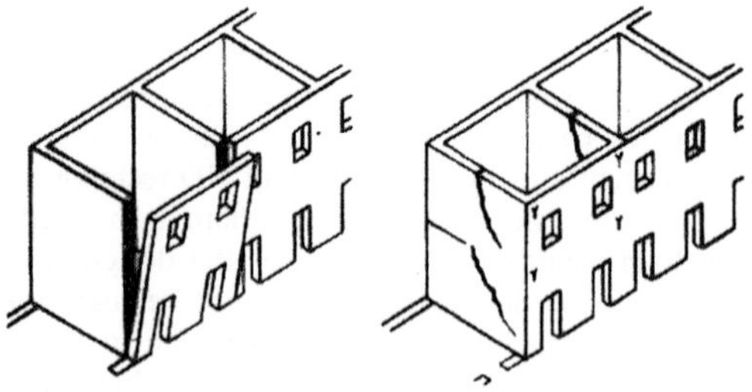

Fig. 5. Example of predictable mechanisms of collapse [4].

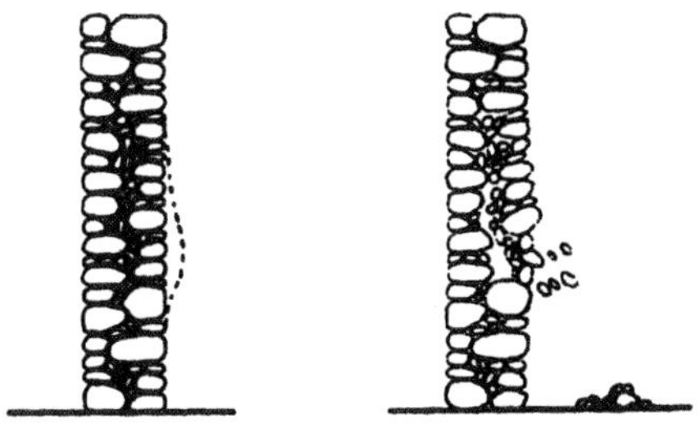

Fig. 6. Deformation and collapse of the external leaf of the right side of a stone masonry wall [4].

and the great influence of the building technique on the masonry behaviour, a systematic study on the mechanical behaviour of stonework masonry should begin from an extensive investigation of the different geometry and building techniques which takes into account the different layers constituting the wall and the kind of constrain3ts which may or may not be present between the layers themselves. In fact, the ancient building techniques and particularly those adopted in the poorer architecture still need to be carefully investigated.

In Fig. 7 the survey of an old single storey house (*casa terranea*) at Montescaglioso (MT) built in Gravina calcarenite is shown as an example of a correct analysis of the building. The survey is part of a final year thesis performed at the Department of Structural Engineering (D.I.S.) of Politecnico di Milano [5].

More research is in progress in different Italian regions by Abbaneo et al. [6], Baronio et al. [7], Binda et al. [8] studying stonework walls of buildings, the cross-sections of which can be inspected; the operation can be more easily conducted in those areas where the buildings were damaged by the earthquake and have not yet been repaired. The survey consists of a graphic and photographic procedure which includes taking a photograph with a camera with a 50-mm lens using a tripod which ensures the parallelism between the plane of the photograph and that of the wall; working out the metric survey of the walls first manually and then on a PC; creating a rich database organized in tables like that presented in Fig. 8 and included in a final year thesis [9].

The cataloguing has been carried out on 200 sections so far, belonging to different Italian regions (Lombardia, Friuli, Liguria, Basilicata, Trentino). The survey of the sections allowed to define some important parameters, e.g. the percentage distribution of stones, mortar, voids which allows to make comparisons between the different regions (Fig. 9); the ratio between the dimensions of the different layers and that between the dimension of each layer and the whole cross-section; the dimension and distribution of voids in the cross-section (Fig. 10). These parameters, together with the chemical, physical and mechanical properties of the materials allow better description of the masonry and constitute a fundamental basis of any conservative intervention.

A study like that presented above can be a good starting point for a classification of the different cross-sections to be carried out, particularly of multiple leaf ones. In fact, a correct structural analysis of these structures can only be performed provided some criteria are singled out in order to identify homogeneous groups of walls, not only on a geometric basis but also on a mechanical one. To this respect many investigations are in progress not only in Italy.

Giuffrè [4] for instance proposes a classification based on a parameter called δ which indicates the ratio of the distance d between two subsequent diatons to the thickness s of the masonry wall.

The study previously described, which led to an initial cataloguing of multiple leaf walls based on the percentage of mortar, stones and voids measured on the area of the cross-section, also allowed a subsequent classification based on the number of different layers and on the type of constraint between them (Fig. 11). This kind of classification, in particular, allows formulation of an important hypothesis on the static behaviour of the masonry.

To this respect some works [8,10] have to be mentioned which were aimed at studying, on an experimental and a numerical basis, the stress distribution between the different layers of a three-leaf masonry panel. During the research, tests have been performed to study the shear behaviour of the interface between the different layers and to interpret it numerically. Given the difficulties of building in the laboratory full-scale or small-scale models of masonry structures due to their high non-homogeneity and to the difficulty

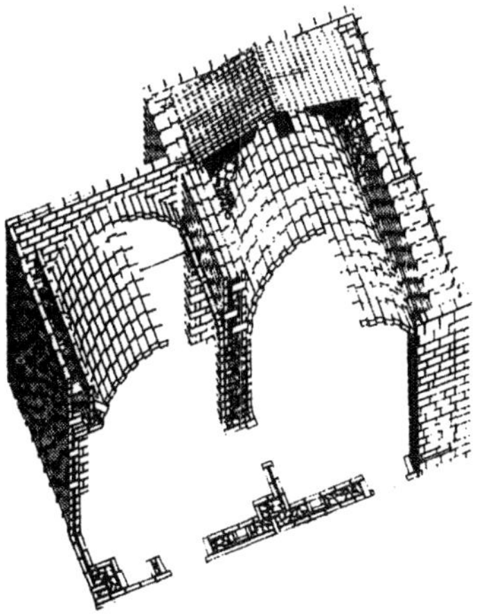

Fig. 7. Montescaglioso (Matera), a single-storey house (*casa terranea*).

PANEL *SAM 8.1*

village	**SANT'ANTONIO AI MONTI** (m. 710) Ramponio - Verna
in situ survey	PANEL *SAM 8.1:* - cross section - external front - internal front
date	Nov. 7th, 1993
plan of the building	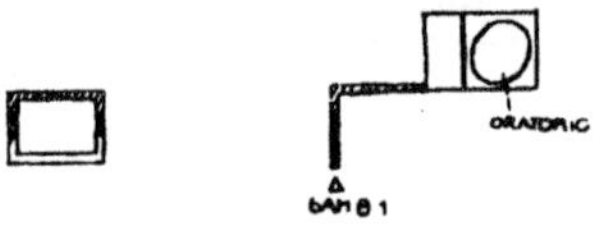

characteristics of the cross section

st	**type:** stone of Moltras	**color:** grey	**conservation:** good
83.	**source:** local emergences	**workmanship:** - chopped and squar-ed stones - scabbings	
mo	**nature:** probably origin	**color:** dark grey	**consistence:** friable
14.	**aggregate:** roundish	**color of the aggregate:** brownish	**function:** filling
v(			
2.2			
observations	- the panel is formed by a one-leaf section, with squared blocks (approximately 40 cm long), alternatively overleaved, with good scarf-joints. - the lateral free space is filled by little scabbings, bonded by the mortar		

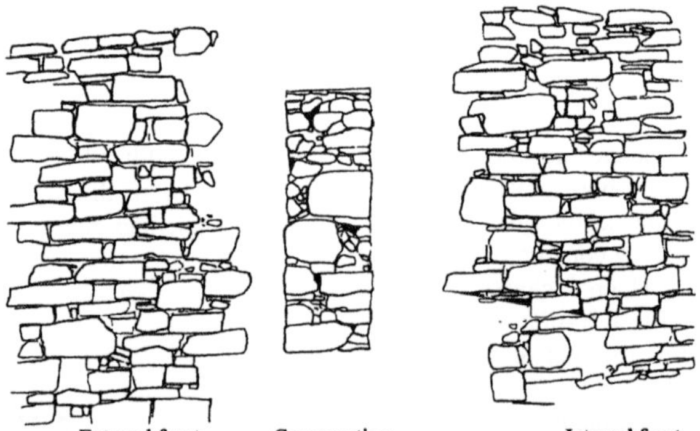

Fig. 8. S. Antonio ai Monti (Ramponio-Verna) oratory; example of a field sheet [9].

of reproducing the characteristics of ancient mortars, simplified models were built to analyse the mechanical behaviour at a local level. Four panels were built, two of which using a sandstone coming from Tuscany and two using a limestone coming from Matera, trying to reproduce the type of bonds between the layers (with and without offset between the stone blocks) which are shown in Fig. 11.

The samples (Fig. 12) were then subjected to shear tests obtaining interesting results (Fig. 13). Samples built with the weaker stone (limestone) showed a higher shear strength of the interface, probably because of the high porosity which gave a better adhesion with mortar. In the cases of panels having blocks offset, the vertical joint turned out to be stiffer than in absence of the offset, and the load was mainly transferred through the stone block. Looking at the crack pattern after failure it appeared that in the specimens made with the weaker stone, the upper block collapsed well before the mortar joint failed. This was probably due to the excessive

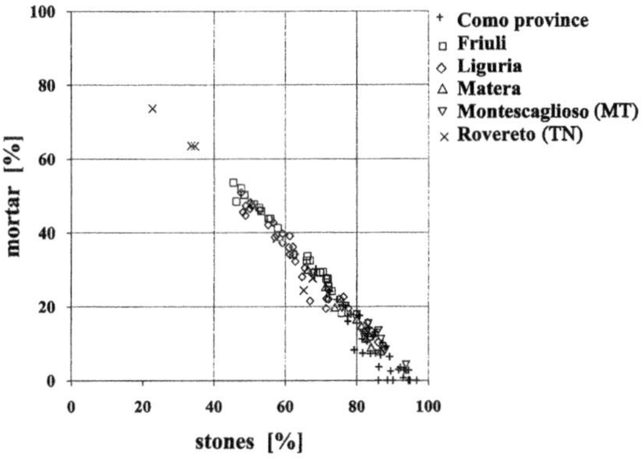

Fig. 9. Percentage of mortar vs. percentage of stones referred to the area of the cross-section of stonework walls in various Italian regions [11].

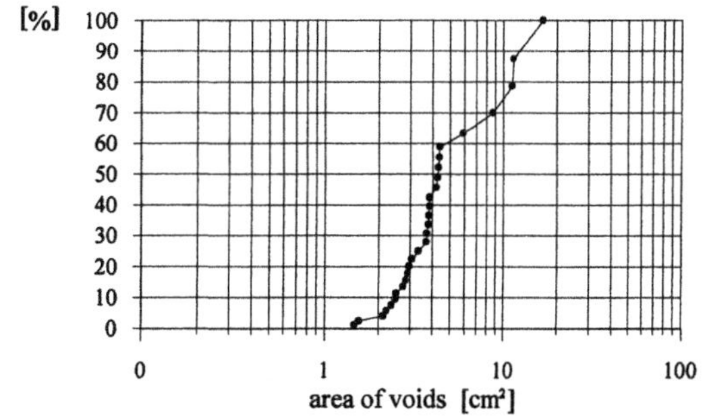

Fig. 10. Size and distribution of the voids within the section of the wall [11].

strength of the mortar used, compared with the strength of the stone (Table 1). The results show the importance of the bond between the wall leaves.

4. Investigation procedure for diagnosis

The necessity of establishing the building integrity or the load carrying capacity of a masonry building arises from several causes including: (i) assessment of the safety coefficient of the structure in a seismic area (before or after an earthquake, or following accidental events like hurricanes, fire, etc.); (ii) change of use or extension of the building; (iii) assessment of the effectiveness of repair techniques applied to structures or materials; and (iv) long-term monitoring of material and structural performance.

The flow chart of Fig. 14 schematically represents the needs to be fulfilled by the experimental investigation together with the techniques required.

4.1. Geometrical and crack pattern survey

A preliminary in-situ survey is useful in order to provide details on the geometry of the structure and in order to identify the points where more accurate observations have to be concentrated. Following this survey a more refined investigation has to be carried out, identifying irregularities (vertical deviations, rotations, etc.).

In the meantime the historical evolution of the structure has to be known in order to explain the signs of damage detected on the building.

Especially important is the survey and drawing of the crack patterns (Figs. 15 and 16). The interpretation of the crack pattern can be of great help in understanding the state of damage of the structure, its possible causes and the type of survey to be performed, provided that the development history of the building is already known. The crack pattern of Fig. 15 could be easily interpreted knowing that the large room A was inserted three centuries after the construction of the building and the large masonry vault had no rods or buttresses.

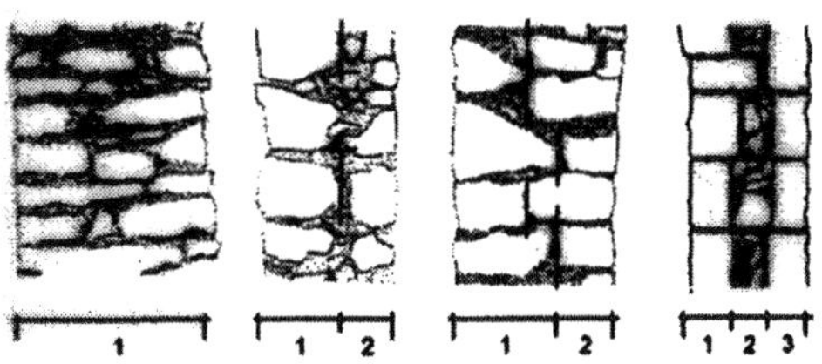

Fig. 11. Classification of the cross-sections of stonework walls: (a) single-leaf; (b) two leaves without connection; (c) two leaves with connection; (d) three leaves [11].

Table 1
Mechanical and physical properties of the material tested

	Composition	f_k [N/mm2]	E [N/mm2]	Water abs. (%)
Mortar	1:3 (lime/sand)	1.50	2400	–
S1	Sandstone	106.03	25 000	2
S2	Limestone	5.94	6147	21

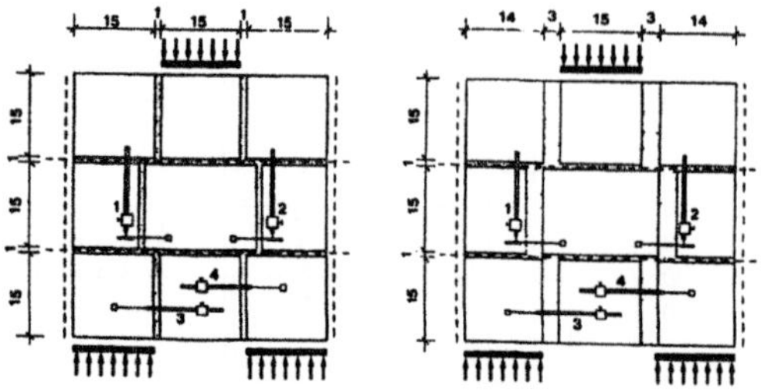

Fig. 12. Specimens subjected to shear tests: (a) with units offset; (b) without units offset [10].

Sometimes similar situations due to some causes can be understood through the study of the crack pattern (Fig. 17).

Often the geometrical details of the structure need a special refined survey if they are complex or of difficult interpretation with the usual procedures. Photogrammetry can be of great help in defining the geometry of vaults and arches. Fig. 18 presents the photogrammetric survey of the dome of S. Vitale in Ravenna (5th century AD). Knowledge of the exact geometrical shape is of fundamental importance for the stability assessment of thin masonry vaults. All the irregularities of the geometry are detected in detail and can be given as input data to a structural analysis model.

4.2. Structure control by monitoring

4.2.1. Static monitoring

Where an important crack pattern is detected and its progressive growth is suspected due to soil settlements, temperature variations or to excessive loads, the measure of displacements in the structure as function of time have to be collected. Monitoring systems can be installed on the structure in order to follow this evolution; in some cases the knowledge of the crack pattern evolution can help preventing the collapse of the structure [14].

This type of survey is frequently applied to important constructions, like bell towers (e.g. to the five remaining Pavia towers after the collapse of the Civic Tower in 1989, see P.P. Rossi lecture, to the Pisa leaning Tower, to the Dome of the Florence Cathedral in Italy, Fig. 19) or cathedrals and the system may stay in place for years before a decision can be taken for repair or strengthening.

Very simple monitoring systems can be also applied to some of the most important cracks in masonry walls, were the opening of the cracks along the time can be measured by removable extensometers with high resolution (see Section 5). This simple system can give very important information on the persistence of settlements, etc., to the designer. Also, in this case the monitoring should be long-term, not less than 1.5 years, in order to rule out the influence of temperature variation at every reading of the eventual displacements.

4.2.2. Dynamic monitoring

In-situ testing using dynamic methods can be considered a reliable non-destructive procedure to verify the structural behaviour and integrity of a building.

The principal objective of the dynamic tests is to control the behaviour of the structure to vibration. The first test carried out could be seen also as the starting one of a periodical survey using vibration monitoring inside a global preventive maintenance programme. Acceptance of vibration monitoring as an effective technique of diagnosis has been supported by different

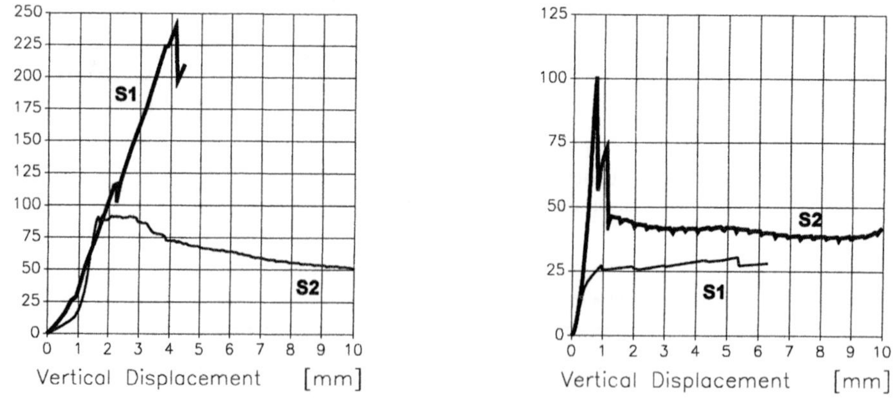

Fig. 13. Results obtained from shear tests on specimens (a) with units offset; (b) without units offset. S1, sandstone; S2, limestone; [10].

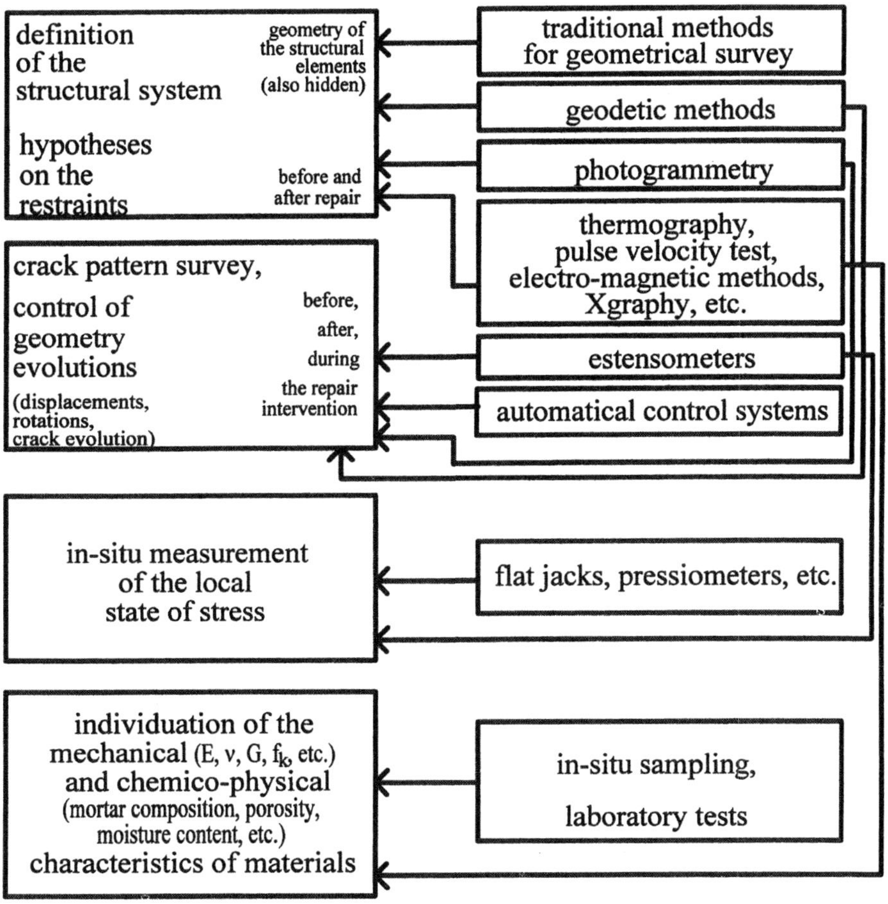

Fig. 14. Information required and correspondent investigation techniques [12].

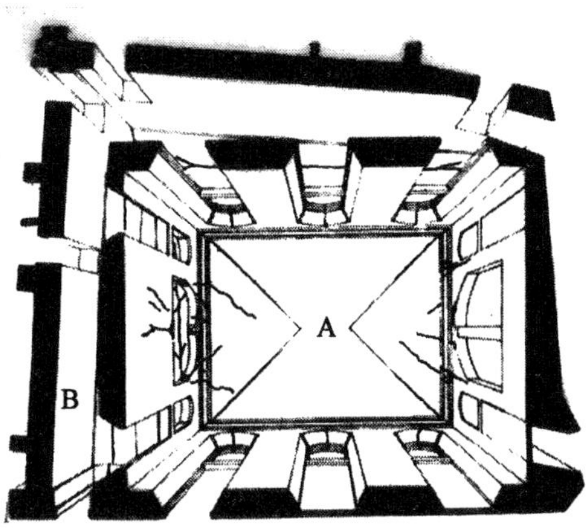

Fig. 15. Villa Crivelli, Inverigo, Italy. Drawing of the main cracks on the perspective of room A [12].

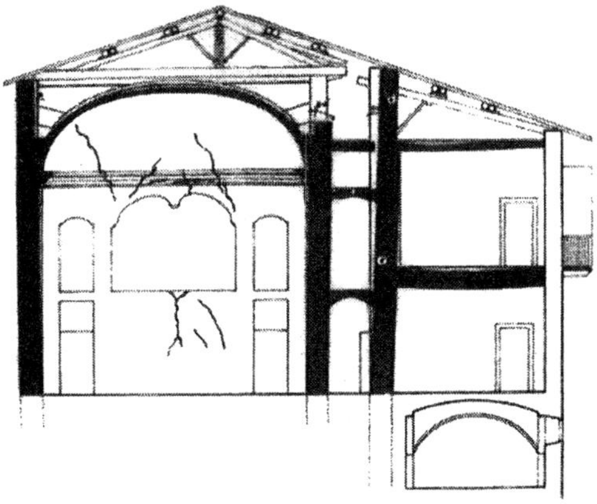

Fig. 16. Villa Crivelli, Inverigo, Italy. Crack distribution on wall B [12].

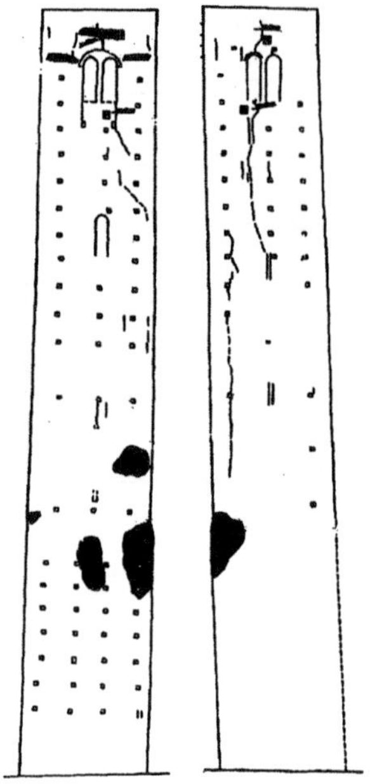

Fig. 17. Civic Tower of Ravenna. The crack pattern survey.

studies [17,18]. These tests are very important to detect eventual anomalies in the diagnosis phase and to calibrate efficient analytic models (FEM). In this manner it is possible to verify the effectiveness of the computational methods used in the analysis and control of the structure. The availability of an efficient numerical model allows for checking and predicting the structure behaviour to dynamic actions such as, e.g. wind effects and seismic actions.

The testing procedure could be direct to the updating of a preliminary analysis of the structure by FE modelling with the information gathered from the field tests. The result could be an FE model of the structure which is as close to reality as possible. This model could then be used to analytically determine:

- the system's response to static or dynamic loads within the limits of linearity;
- the consequences of structural changes like crack pattern propagation as continuous damage or strengthening and repair; and
- any changes to the dynamic properties which have occurred in the time elapsed between two measurements.

The key to this vibration analysis/preventive maintenance programme is a systematic, scheduled check of the structures, before, during and after the repair phases. The analysis should be able to assess the condition and the general trend of the structural behaviour and advise, for instance, whether there could be a change of the controlled parameters. Results are then compared with the original records from which any long-term change in the structure can be observed. Since only long-term trends are being monitored, subsidiary effects (e.g. temperature effects) are not considered to affect the results significantly.

The programme therefore could involve the use of environmental vibration (passive tests) or forced vibration (active tests) and include a systematic vibration recording and comparison of the analysed data with the model results.

The environmental excitation sources could be the wind, the traffic or the bell ringing in the particular case of towers [14]. The forced vibrations could be produced by local hammering systems or by the use of vibrodines. It has been known for several decades that the frequencies and mode shapes obtained from environment tests are generally in good agreement with results of forced vibration tests.

An accelerometer net is installed in significant chosen parts of the structure. Spectral analysis can be used to extract modal parameters from vibration data. The frequency-domain technique involve frequency analysis of the vibration signal and further processing of the resulting spectrum to obtain clearly defined information. With the modal analysis the vibration response consists of summing up the contribution of the infinite number of natural modes, each multiplied by a function of time; the normal modes, detected from the vibration tests analysis are functions of the system properties and the boundary conditions only.

The vibration tests allow detection of the frequencies, the modal shapes and the correspondent modal

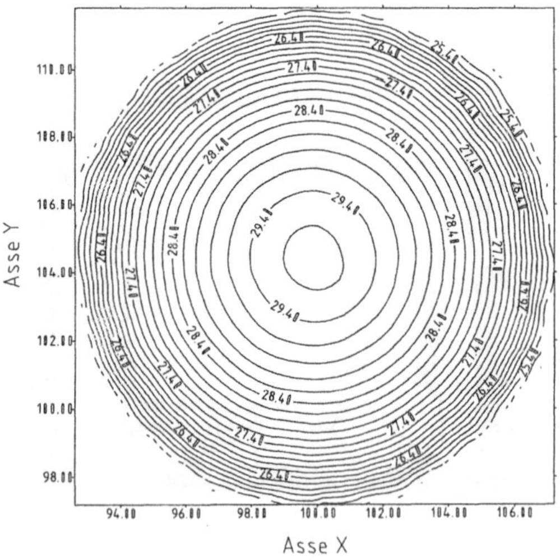

Fig. 18. Photogrammetric survey of the intradox of the dome of San Vitale at Ravenna [13].

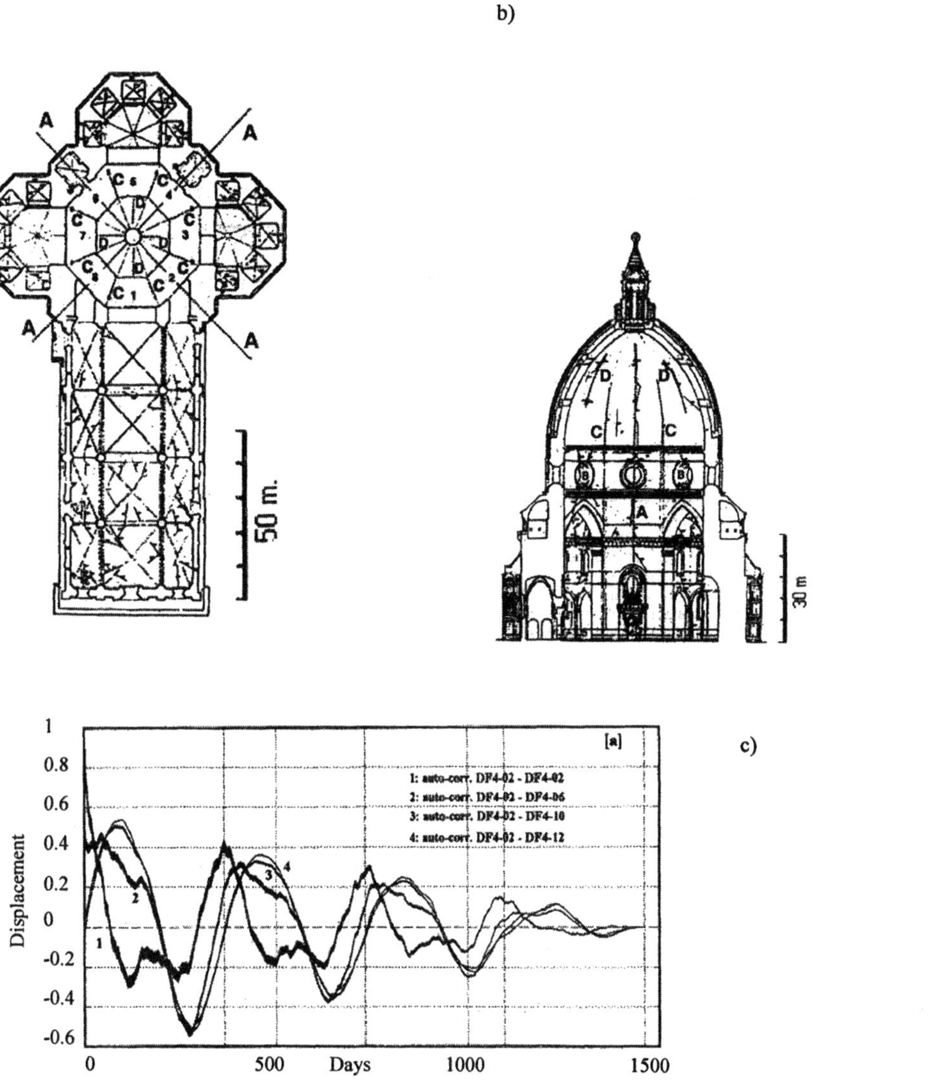

Fig. 19. S. Maria del Fiore, Florence Italy: (a) position of the major cracks denominated A, B, C, D on the dome plan; (b) position of the major cracks denominated A, B, C, D on the dome [15]; (c) curves displacement-time obtained as a correlation of the monitored [16].

damping of a structure. Spectral analysis provides a frequency-domain resolution of these component physical relationships. These parameters are characteristics of the local and global behaviour of a structure. They could be used both to verify the results of a theoretical or numerical model, or to monitor the behaviour with time.

The modal frequencies are parameters representative of the global behaviour of the system, while the modal shapes allow detection of the local performance. The analysis of the modal shapes could eventually localize damaged areas. The analysis of the sets of modal shapes deriving from different tests could be done following two global criteria:

1. 'modal assurance criterion' (MAC), which is a measure of the correspondence of two vectors of modal shapes. The values of MAC = 1 show a perfect correlation, while MAC = 2 shows an independence between the set of data; and
2. 'coordinate modal assurance criterion' (COMAC). The correspondence between the modes is evaluated in a defined position by the index COMAC. It is a measure of the correlation between all the

displacements in a point, concerning all the set of corresponding modes. Values of COMAC = 1 indicate a good agreement of the modal shapes in the chosen point.

There are two classes of spectral data: autospectrum (power spectrum) and cross-spectrum. The latter could be a measure in the frequency domain of the similarity of two functions. Modal frequencies are identified by the locations of peaks in the auto-spectra (ASD) and cross-spectra (CSD); mode shapes are estimated using the ratios of (square-root ASD) amplitudes and cross-spectral phases are used to determine directions of relative motion. For each natural frequency, mode shapes were normalized to the maximum modal value (amplitude 1, phase 0). Coherence values are computed to examine the influence of noise and non-linear response of the structure in each mode.

Magnitudes of the coherence function range from 0 to 1 and denoted on how well two signals at each frequency are related. Both the autospectrum and the coherence function are used to analyse vibrations and identify their possible sources by measuring the spectra at one point and its coherence value. The result is independent of the power levels at the two points and the transmission gain between them. The coherence function examines the phase difference between auto and cross-spectral values over several spectrum measurements. Thus, it can detect frequency differences so small to be only variation of the phase. A unit value of the coherence function means that the spectral line at the monitored point is completely coherent with the measured source. The use of this parameter allows detection of torsional modes.

The forced vibration technique using vibrodine and ambient vibration technique were used to analyse the structural condition of several towers. The results together with those supplied by other tests like flat-jack tests and NDT are intended to be used for detection of extended anomalies in the structural behaviour.

4.3. Foundation and soil survey

Small exploratory shafts or core drilling have to be done for the geometrical survey of the foundations. The material characteristics and the soil properties have also to be known.

In-situ and laboratory tests on the foundation soil are also needed in order to predict settlements or the state of stress and strain of the soil under new loading conditions after repair or to understand the causes of a collapse [19].

4.4. Laboratory tests

If samples of the materials are needed for destructive tests they must be cored from the walls inflicting the lowest possible damage. The technique of sampling is very important, since samples must be as undamaged as possible in order to be representative of the material in situ. The aims of these tests are the followings: (i) to characterize the material from a chemical, physical and mechanical point of view; (ii) to detect its origin in order to use similar materials for the repair; (iii) to know its composition and content; and (iv) to measure its decay and the durability to aggressive agents. Since it is very difficult to sample prisms representative of the walls, only single components or small assemblages are removed.

4.4.1. Sampling of bricks, stones and mortars

The method of sampling depends on the characteristics of the materials in situ. Some simple principles have to be applied:

1. sampling must be carried out respecting the existing building;
2. the quantity of sampled material must be consistent with the scope and the requirements of the test procedures;
3. if determination of the type and the extent of damage is involved, sampling must be carried out on different portions of the building in order to study all the types of degradation;
4. sampling has to be carried out dry in those portions of the building not subjected to the action of rain or by a previous repair, especially when mortar binder and aggregate characteristics are needed; and
5. the number of samples should be quite high, in order to represent statistically the situation of the existing masonry, but this is never possible.

4.4.2. Tests on mortars

At present there are no standardized tests to define the composition and the chemical–physical and mechanical characteristics of mortars sampled from an existing building. It is often very difficult to drill samples having the consistent dimensional tolerance needed to conduct mechanical tests; then the only useful information which can be obtained concerns the mortar composition and the state of decay. Chemical and mineralogical–petrographical analyses are useful (and less expensive than other more sophisticated tests) to determine: the type of binder and of aggregate, the binder/aggregate ratio, the extent of carbonation, the presence of chemical reactions which produce new formations (pozzolanic reactions, binder–aggregate reactions, alkali–aggregate reactions) [20].

The grain size and distribution of the aggregates can also be measured by separating the binder from aggregates through chemical or thermic treatments [21]. The above-mentioned tests permit the determination of the composition of the existing mortars and permit the reproduction of mortars and grouts for repairing the masonry.

Sheet A shows as an example the results of a mortar characterization reported on a special form set up by L. Binda and collaborators.

4.4.3. Tests on damaged and new bricks and stones

When masonry is damaged by aggressive agents the decay is never uniform; if maintenance is needed and only some bricks or stones or decorations are affected by the damage, the best remedy is frequently the substitution of the most decayed elements. In this case, laboratory tests can give useful information for the choice of the appropriate material for substitution.

When substitution is not possible and/or the decay is very extensive, a surface treatment may be required; again laboratory tests are needed for the right choice of the treatment. The tests have to be carried out on both deteriorated existing bricks or stones, and on undamaged and new ones. The following tests are suggested:

- *Mechanical tests*: compressive and indirect tensile tests, hardness tests on different points of the brick or stone sections in order to determine the depth of the decay.
- *Physical tests*: the volumetric mass, the water absorption by total immersion, the water absorption by capillary rise are important characteristics needed to determine the durability of the materials and the effects of surface treatments; the initial rate of suction of bricks and stones and the water retentivity of mortars can be useful when choosing mortars and grouts for repairs; the X-ray diffraction measurements can detect the type of salts found inside or on the surface of a decayed masonry; mercury porosimetry is a useful technique for evaluating durability and surface treatments; thermal and water expansion coefficients must also be measured on new bricks and stones.
- *Chemical tests*: tests for alkaline sulfate can be conducted on material samples taken at different depth of the masonry in order to detect the presence and quantity of these very aggressive salts.
- *Optical and mineralogical analysis*: optical observations (stereomicroscopy, SEM microscopy, define the deterioration, its causes and the presence of salts. Petrographic observations on thin sections determine the pore size distribution of the material, the size and distribution of the aggregates, the geographical origin of clays and stones, the firing temperature of bricks and the decay and its causes.
- *Durability tests*: freeze/thaw and salt crystallization tests are needed for new bricks and stones in order to determine their performance under aggressive agents.

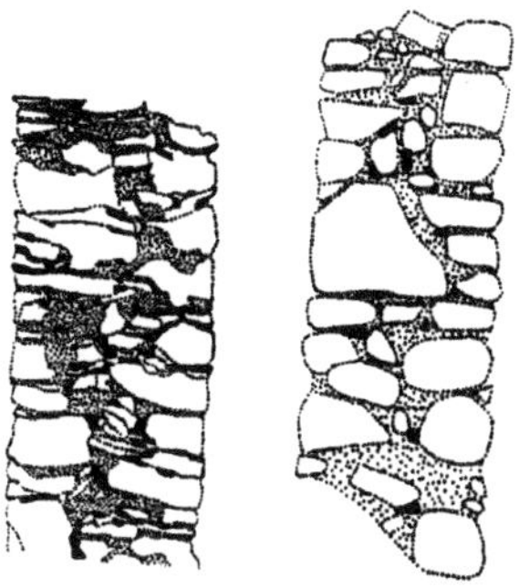

Fig. 20. Sections of the studied walls.

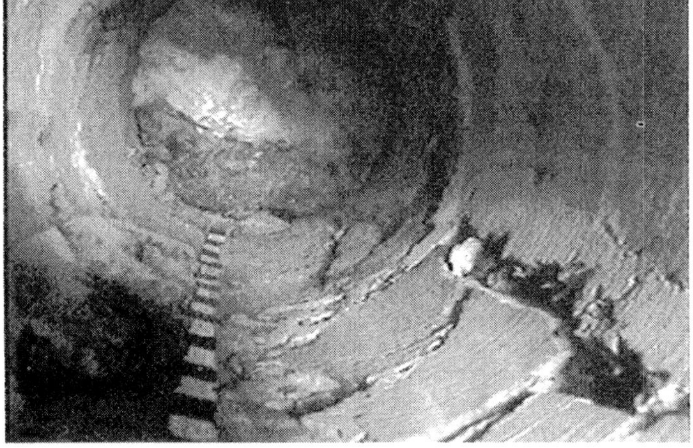

Fig. 21. Boroscopy view in a drilled core.

Sheet B and C show some characterization of a stone reported on the same type of form in Sheet A.

4.5. In-situ testing

The more tests can be carried out on site, the best information is collected, provided the designer is able to choose the most significant positions and areas and to interpret the data.

4.5.1. Slightly destructive testing

To understand the morphology of a masonry wall, direct inspection is important. Sometimes this can be performed by removing few bricks or stones and surveying photographically and drawing the section of the wall (Fig. 20).

In some cases it is possible to core boreholes in the most representative points of the walls. Coring should be done with a rotary driller using a diamond cutting edge. This operation is rather simple but has limits. The drilled core is usually very decohesioned so it is almost impossible to detect the quality of the original materials. Inside the boreholes additional investigations can be made by the use of borescopy. A small camera is inserted into the borehole allowing a detailed study of its surface and try a reconstruction of the wall section. The results of this study may be recorded in a video cassette for further analysis. The information obtained include the measurements of large cavities and a general view of the materials. Nevertheless, the interpretation of the photograms is a very difficult operation, sometimes hopeless; it should be remembered that boroscopy can only give a stratigraphy of the section (Fig. 21).

4.5.1.1. Flat-jack test. The method was originally applied to determine the in-situ stress level of the masonry and it has been extended to the detection of its deformability characteristics. The first applications of this technique on some historical monuments [22], clearly showed its great potential. It appeared to be the only way to achieve reliable information on the main mechanical characteristics of a masonry structure (deformability, strength, state of stress). The single flat-jack test is carried out by introducing a thin flat-jack into the mortar layer. The test is only slightly destructive. After the test is completed, the flat-jack can easily be removed and the mortar layer restored to its original condition [23].

The determination of the state of stress is based on the stress relaxation caused by a cut perpendicular to the wall surface; the stress release is determined by a partial closing of the cutting, i.e. the distance between two points after the cutting is lower than before. A thin flat-jack is placed inside the cut and the pressure is gradually increased to obtain the distance measured before the cut. The displacement caused by the slot and the ones subsequently induced by the flat-jack are measured by a removable extensometer before, after the slot and during the tests. P_f corresponds to the pressure of the hydraulic system driving the displacement equal to those read before the slot is executed. The equilibrium relationship is the fundamental requirement for all the applications where the flat-jack are currently used:

$$S_f = K_j K_a P_f$$

S_f = calculated stress value; K_j = jack calibration constant (<1); K_a = jack/slot area constant (<1); P_f = flat-jack pressure.

In brick masonry, the cut can be easily made in the horizontal joints. For this type of masonry a rectangular flat-jack is used. The cut can also be made by a steel disk, with a diamond cutting edge. The flat-jack has the same shape of the cut. In fact, the use of flat-jacks for stone masonries made with irregular stones is not so easy, due to the difficulty of finding regular joints; therefore the cut for the insertion of the jack is done directly in the stone courses. It must be pointed out that the flat-jack test in the case of multiple-leaf walls gives results concerning only the outer leaves.

The reliable determination of the equilibrium pressure is the fundamental requirement for the test, regardless of the details of the type of application. Conflicting information, due to the effect of the concentration of stresses and/or inelastic deformations or to the detection of very low stresses (e.g. one- or two-floor building) usually require a significant amount of subjective judgement, which may compromise the reliability of the entire procedure [24].

The test described can also be used to determine the deformability characteristics of a masonry. A second cutting is made, parallel to the first one and a second jack is inserted, at a distance of approximately 40–50 cm from the other. The two jacks delimit a masonry sample of appreciable size to which a uni-axial compression stress can be applied. Measurement bases for removable strain-gauge or LVDTs on the sample face provide information on vertical and lateral displacements. In this manner, a compression test is carried out on an undisturbed sample of large area. Several loading cycles can be performed at increasing stress levels in order to determine the deformability modulus of the masonry in its loading and unloading phases.

Fig. 22 shows the application of the double flat-jack test in the case of a brick masonry with thin and thick joints and of an irregular stone masonry. In the third example the slots were made with a special eccentric saw within the stones, being the mortar joint very irregular and weak.

Difficulties or impossibilities in applying this test can also be found in the case of low rise buildings (one or

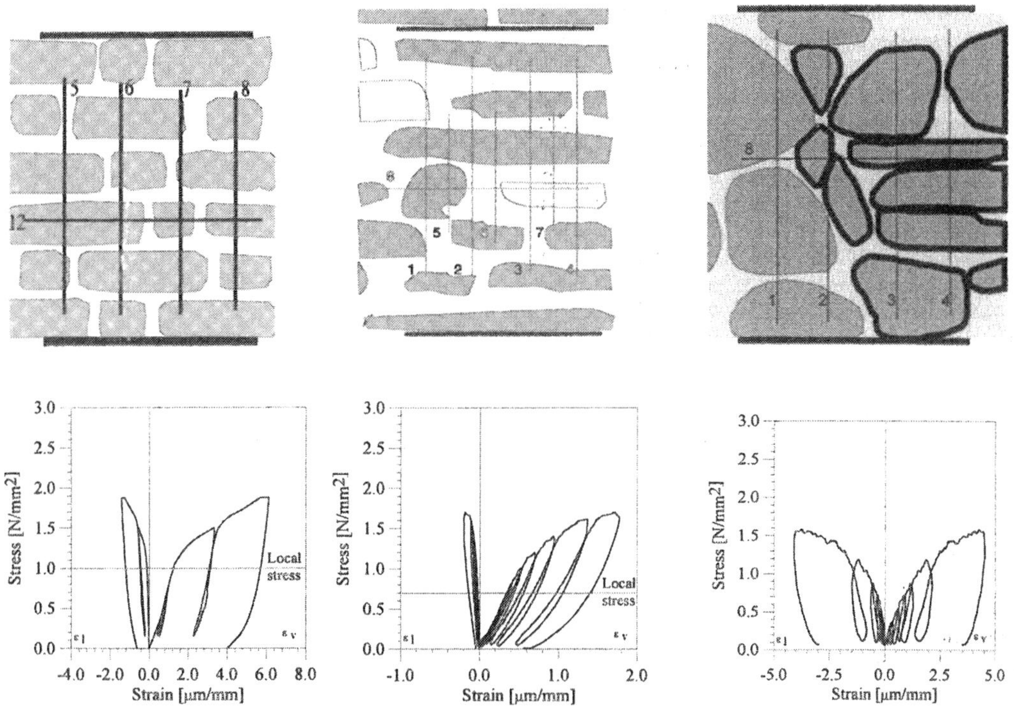

Fig. 22. Double flat-jack test on a Monza Tower (Lombardia), S. Vitale (Ravenna) and Castelletto (Toscana).

two stories high) due to the lack of stress contrast in the masonry caused by the low stresses found in it. In these cases the situation can be known after carrying out a single flat-jack test and the double jack test can be either avoided or carried out up to a low state of stress just in order to have information on the elastic parameters.

4.5.1.2. Hardness, penetration and pull-out tests. Other slightly destructive tests can be used to give more information on site about masonry and masonry components. They can be considered as surface or small penetration techniques, which can be used for a preliminary investigation. Some of them can are mentioned here:

1. the Schmidt hammer rebound test to detect the quality of mortar joints, has some limits in the present equipment which was set up to be used on cement mortar and can have too high energy for a lime mortar;
2. the penetration tests proposed in different ways, like probes, drillers, etc., correlate the depth of penetration to the material mechanical properties. Unfortunately a correlation is impossible to the real strength of ancient mortars; so the calibration of these tests is very difficult. Furthermore, the depth of penetration is low, so only the repointing mortars are usually detected; and
3. the pull-out tests can only be used on bricks and stones, very rarely on mortar joints, unless they are not very thick.

Other surface tests have been proposed so far; all of them can be useful to give an overall rough idea of the masonry condition on the surface and they can be meaningful for a preliminary survey of the structure, but they only give a qualitative interpretation of masonry condition. All these tests can be really useful for the quality control of a new masonry [25].

4.5.2. Non-destructive evaluation techniques

The importance of evaluating existing masonry buildings by non-destructive investigation carried out in situ has been mentioned by many authors. NDE techniques can be used for several purposes: (i) detection of hidden structural elements, like floor structures, arches, piers, etc.; (ii) qualification of masonry and of masonry materials, mapping of non-homogeneity of the materials used in the walls (e.g. use of different bricks in the history of the building); (iii) evaluation of the extent of mechanical damage in cracked structures; (iv) detection of the presence of voids and flaws; (v) evaluation of moisture content and capillary rise; (vi) detection of surface decay; and (vii) evaluation of mortar and brick or stone mechanical and physical properties.

4.5.2.1. Thermography. Thermovision is a NDT which has been applied since several years to works of art and monumental buildings. The thermographic survey has the advantage of being applicable to wide surfaces of walls; it is a telemetric method and has high thermal and spatial resolution.

The thermographic analysis is based on the thermal conductivity of a material and may be passive or active. The passive application analyses the radiation of a surface during thermal cycles due to natural phenomena (insulation and subsequent cooling). If the survey is active, forced heating to the surfaces analysed are applied.

The thermal radiation is collected by a camera sensitive to infrared radiation. In fact, each material emits energy (electromagnetic radiation) in this field of radiation; this radiation is characterized by a thermal conductivity, i.e. the capacity of the material itself of transmitting heat, and its own specific heat.

Each component of an inhomogeneous material like masonry shows different temperatures. The thermovision detects the infrared radiation emitted by the wall. The result is a thermographic image in a coloured or black and white scale. Each tone corresponds to a temperature range. Usually the differences of temperatures are fraction of a degree.

The total flux of energy E emitted by a surface, is the sum of the energy E_c emitted by the surface by thermal excitation and the flux E_r that is emitted by the surface around each point

$$E = E_c + E_r$$

The infrared camera, measures the energy flux E. The test is carried out at a certain distance without any physical contact with the surface.

Active thermovision can be also carried out for tests on depth. The surface of the tested wall should be heated for a certain time. In this way, the thermal conductivity of the internal part of the masonry is shown up to a certain depth. The infrared camera transforms the thermal radiation into electric signals, successively converted into images. These images can be visualized on a monitor and recorded on a computer. In the video camera, the infrared radiation that reaches the objective is transmitted by an optical system to a semiconductor element. The latter converts the radiation into a video signal, while the surveying unit signal processes the video camera signals and shows the thermographic image (Fig. 23) [26].

Thermovision can be very useful in diagnostic; in fact it is used to identify areas under renderings and plasters that can hide construction anomalies. It is particularly interesting for studies on frescoed walls, where it is not possible to take samples or use testing techniques that come in contact with the frescoed surfaces. Other applications can be: (i) survey of cavities; (ii) detection of inclusions of different materials; (iii) detection of water and heating systems; and (iv) moisture presence. In the presence of moisture, the camera will find the coldest surface areas, where there is continuous evaporation. The evaporation is due to the difference in relative humidity between the inside of the masonry and the environment outside and to natural air movements.

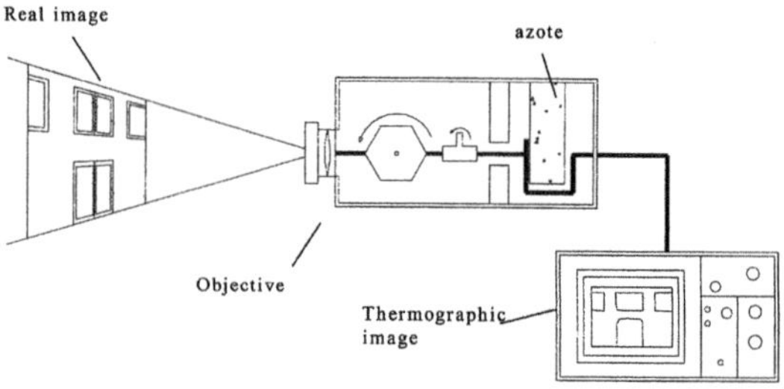

Fig. 23. The thermovision camera.

In the diagnosis of old masonries, thermovision allows the analysis of the more superficial leaves, in the absence of thermal irradiation. It is necessary to point out that the penetration depth of this technique is limited, so it is unable to locate anomalies which are hidden in the inner part of the masonry. The technique is often sensible to the boundary condition of the tests. Sometimes shapes are detected which are caused by different local emissions and not by effective variations.

The interpretation of thermographic data can be more accurate when an appropriate software is available. Nevertheless, it is necessary to focus the mathematical model on the specific problem to obtain specific indications on how infrared images are to be recorded [26]. This could be used as an effective mean for the analysis of heat transfer through building materials. By using this model it is possible to study the energy balance of the wall surface, and to measure the temperature of the surfaces where transpiration occurs. The amount of transpiration is affected by the air and wall temperature, the radiant heat, the wind speed and the relative humidity of the air inside the pores of the materials.

4.5.2.2. Sonic tests. Among the ND investigation methods, sonic methods are without doubt, the most widely used. The testing technique is based on the generation of sonic or ultrasonic impulses at a point of the structure. A signal is generated by a percussion or by an electrodynamics or pneumatic device (transmitter) and collected through a receiver which can be placed in various positions (Fig. 24).

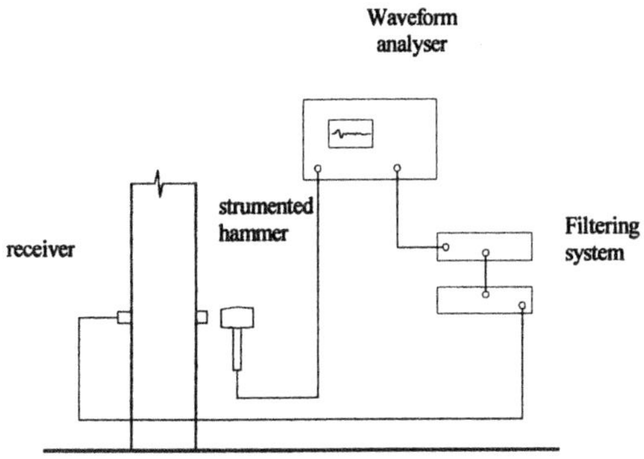

Fig. 24. The sonic test equipment.

The elaboration of the data consists in calculating the wave velocity by measuring the time the impulse takes to cover the distance between the transmitter and the receiver. The use of sonic tests for the evaluation of masonry structures has the following aims:

- to qualify masonry through the morphology of the wall section, to detect the presence of voids and flaws and to find crack and damage patterns; and
- to control the effectiveness of repair by injection technique in others which can change the physical characteristics of materials.

The first applications of sonic tests to the evaluation of masonry materials and structures have been carried out a long time ago in the 1960s [27]. The difficulty of interpretation of the results in the case of inhomogeneous materials like masonry was always known and the first results were clearly interpreted as qualifying rather than quantifying values. Several efforts have been put in the tentative of interpretation of the data from sonic and ultrasonic tests [28,29].

The limitation given by ultrasonic tests in the case of very inhomogeneous material made the sonic pulse velocity tests more appealing for masonry. Nevertheless in the case of low porosity units and mortar used in solid or cavity walls, ultrasonic tests can also be successfully used. Efforts have been made by the authors to correlate the sonic parameter to the mechanical characteristics of the material, but this correlation seems difficult [30]. Nevertheless, after the tests, a high number of data are collected which are certainly meaningful not only for the local situation but also for the overall state of the wall.

Limits of sonic tests to masonry can be defined as follows:

- cost of the operations due to the high number of measurements which has to be carried out;
- difficult elaboration of the results due to the difficulties created by the inhomogeneity of the material; and
- need for the calibration of the values to the different types of masonry.

The fundaments of wave propagation through solids allows to recognize the theoretical capabilities and limitations of the technique. The velocity of a stress wave passing through a solid material is proportional to the density ρ, dynamic modulus E, and Poisson's ratio ν of the material. Resolution in terms of the smallest recognizable features is related to the dominant wavelength (as determined by the frequency) of the incident wave and also to the size of the tested element.

The wavelength, λ, is determined by a simple relationship between velocity, v and frequency f:

$$\lambda = v/f$$

Hence, for a given velocity as the frequency increases the wavelength decreases, providing the possibility for greater resolution in the final velocity reconstruction. It is beneficial, therefore, to use a high frequency to provide for the highest possible resolution. However, there is also a relationship between frequency and attenuation of waveform energy. As frequency increases the rate of waveform attenuation also increases limiting the size of the wall section which can be investigated. The optimal frequency is chosen considering attenuation and resolution requirements to obtain a reasonable combination of the two limiting parameters.

In general, it is preferable to use a sonic pulse with an input of 3.5 kHz for masonry. Ultrasonics are not indicated in the case of highly inhomogeneous masonry such as multiple-leaf walls.

Mechanical pulse velocity equipment can be used to acquire pulse velocity data. The input signals are generated by a hammer, often instrumented, and the transmitted pulse is received by an accelerometer positioned on the masonry surface. Some other instruments generate impulses by means of a pendulum apparatus, which permits repeatable input waves. In the latter case the hammers provide a mass falling down from the same distances [30]. The frequency and energy content of the input pulse are governed by the characteristics of the hammer [31]. Small metal plates can be glued to the masonry surface with a rigid epoxy resin to act as points of application of the receiving accelerometer. Signals are stored by a waveform analyser coupled with a computer for further processing.

Three types of tests can be conducted: (1) direct (or through-wall) tests in which the hammer and accelerometers are placed in line on opposite sides of the masonry element; (2) semi-direct tests in which the hammer and accelerometers are placed at a certain angle to each other; and (3) indirect tests in which the hammer and accelerometer are both located on the same face of the wall in a vertical or horizontal line.

The velocity and waveform of stress waves generated by mechanical impacts can be affected by:

- input frequency generated by different types of instrumented hammers and transducers;
- number of mortar joints crossed from the source to the receiver location; the velocity tends to decrease with the number of joints.
- local and overall influence of cracks.
- input frequency changes with the characteristics of the superficial material (e.g. presence of thick plaster or cracks). The sonic test in this case shows a very important limit. Due to the wall structure or to the presence of a thick plaster (with fresco) the high frequency components could be filtered. The output signals have a rather low frequency content. Since the wavelength is equal to the ratio between wave velocity and frequency, this effect led to an output signal which contained only very long wavelengths.

The sonic test in this case does not have a resolution which can detect in detail the wall morphology, but it does give an overall description of the position of low velocity points.

Fig. 25 shows the results of sonic tests applied in the case of injection by grouting to the base of one of the collapsed piers of Noto Cathedral. The increase in velocity means successful injection.

4.5.2.3. Georadar. Among the techniques and procedures of investigation which have been proposed in these last years, georadar seems on the one hand to be most promising, but on the other hand needs a great deal more study and research. Since masonry is inhomogeneous material, all the techniques applicable to the homogeneous ones seem to fail unless an appropriate calibration is made.

The application of radar procedures are the following:

- to locate the position of large voids and inclusions of different materials, like steel, wood, etc. (Fig. 26);

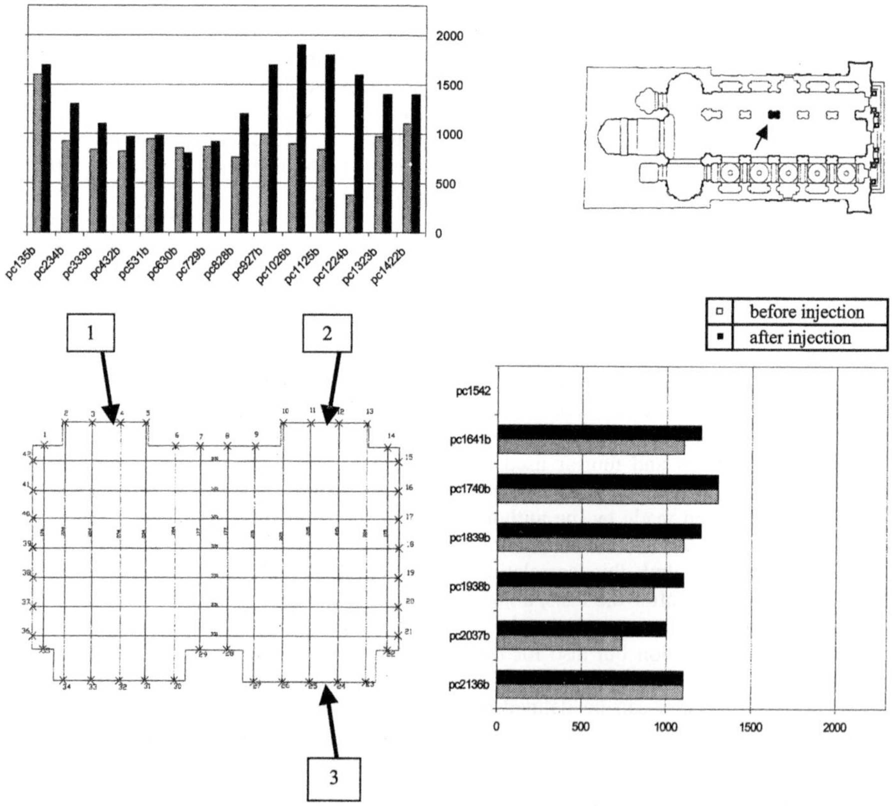

Fig. 25. Noto Cathedral: results of sonic tests applied in the case of injection by grouting to one of the collapsed piers.

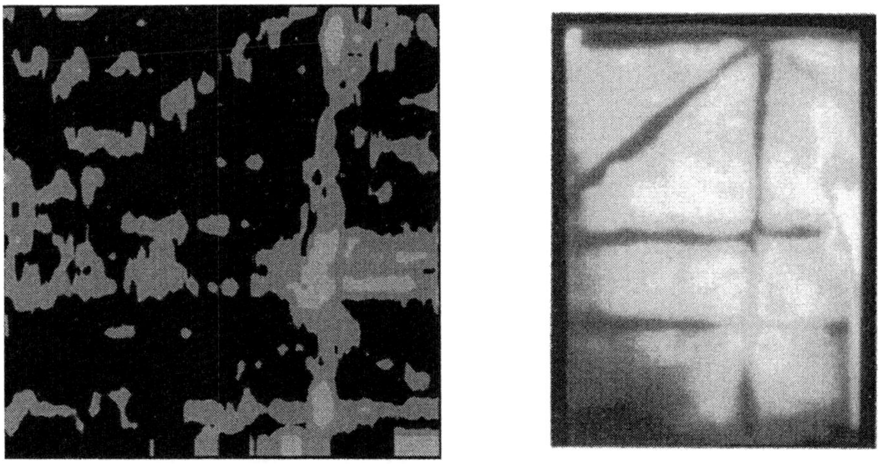

Fig. 26. Comparison between the results of the radar and thermography tests.

- to qualify the state of conservation or damage of the walls;
- to define the presence and the level of moisture;
- to control the effectiveness of repair by injection techniques; and
- to detect the morphology of the wall section in multiple-leaf stone and brick masonry structures (Fig. 27).

Georadar seems to be a powerful tool to detect the presence of voids and structural irregularities, the presence of moisture and hopefully the presence of multiple-leaves in stone masonry.

The method is based on the propagation of short electromagnetic impulses, which are transmitted into the building material using a dipole antenna. The impulses are reflected at interfaces between materials with different dielectric properties, i.e. at the surface and backside of walls, at detachments, voids, etc. (Fig. 28). When the transmitting and receiving antennas, which are often contained in the same housing, are moved along the surface of the object under investigation, radargrams (colour or grey scale intensity charts giving the position of the antenna against the travel time) are produced. Measuring the time range between the emission of the wave and the echo, and knowing the velocity of propagation in the media it would be possible to know the depth of the obstacle in the wall. In the real cases, the velocity is unknown because it changes from one material to the other or in the presence of voids. Furthermore, the velocity is higher in dry walls, and lower in wet walls.

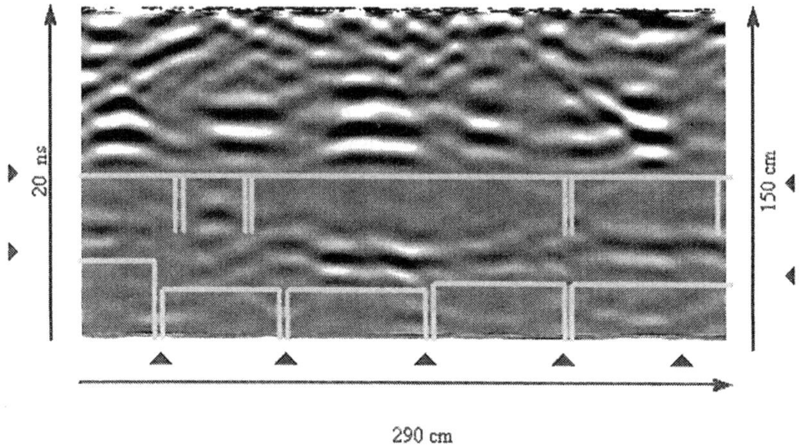

Fig. 27. Radargrams of the wall sections of Fontanella Church Bergamo, Italy.

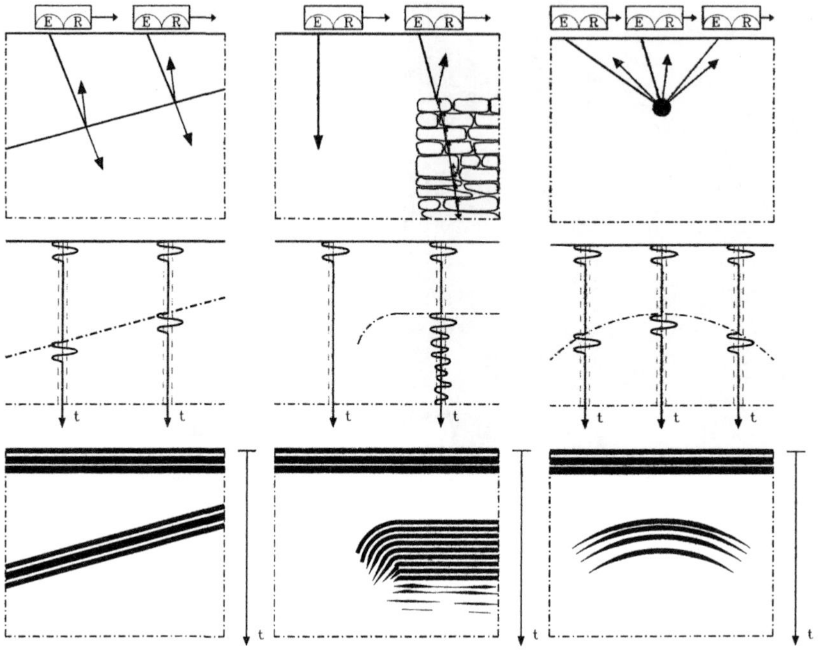

Fig. 28. Example of features in the radar testing of masonry [33].

The interpretation of the radar data involves the identification of significant anomalies. It should be a recognition process detecting features on the records that are characteristic of known signatures. Identifiable features on a radar record are continuous reflection from layers or reflection from discontinuities such as voids and local inhomogeneities in the masonry. A systematic approach was attempted by the authors [32] with the aim of giving guidelines to the designers for an easy interpretation of the results. Special procedures have been studied to eliminate the typical interferences and hopefully to produce easily readable plots after elaboration of the test results. The main problems in the data interpretation are caused by disturbances in the signals due to the following causes:

1. breakthrough effect. This effect is commonly visible on radar records and it hides partially the wall characteristics. It is caused by the fact that the antenna itself reacts to the electromagnetic wave;
2. multiple echoes due to the presence of layers and joints; and
3. superposition of the lateral echoes that create images of parallel reflectors. This effect could be enhanced for a wall by the lack of mortar joints and the regularity of the stones.

Data reduction is generally limited to simple filtering operations to remove unwanted noises due to the data acquisition and to the equipment interference. The single signals are processed by running average (high-cut filter), running median (low-cut filter), offset (zero-crossing trace alignment), etc.

Some trials have been done to remove the break through effects. First of all different antennas have been used. The use of higher frequency antennas could be the most simple method to solve the problem. For example in the tests carried out by the authors it was possible to observe that the 900-MHz antenna has a long break-through effect but it is able to detect the opposite side of the wall. On the contrary, the 1-GHz antenna has a short break through effect (so it is possible to see clearly the first layer of the models) but it produces waves which are too weak, unable to reach the opposite surface [32].

Experimental procedures and data processing. Radar tests always need a preliminary calibration to verify the characteristics of the antenna in relation to the aim of the research. The test lines should be selected to be representative of the masonry and of the problem to investigate. The horizontal or vertical surveying lines are usually located away from the wall edges in order to avoid boundary effects or away from other sources of surface noise (Fig. 29). There are different types of acquisition in the function of the aim of the tests. Usually echo modality is used, i.e. with the receiver and transmitter on the same side of the wall, moving the antenna along a surface (Fig. 30b). Other types, e.g. by transmission could be useful for tomographic acquisitions (Fig. 30).

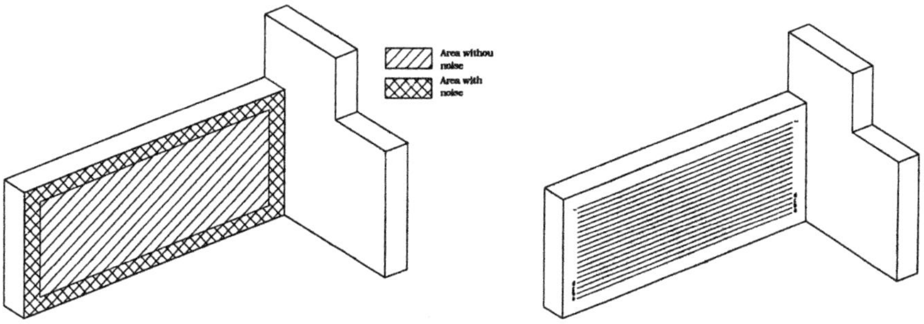

Fig. 29. Geometry of acquisition.

The first step is to carry out some measures in transmission, i.e. with the transmitter and receiver positioned on the two opposite sides of the wall (Fig. 30d). It is, therefore, possible to verify if the emitted signal is enough powerful to detect the opposite side of the wall and the wave speed. This step allows calibration of the relationship between the time and space scales, e.g. to localize anomalies like voids or layers in the depth of the wall. Some types of equipment give this transformation directly by setting up a value of the dielectric constant of the masonry. The value is an average of the characteristics of all the materials crossed by the wave.

A metal plate should be always positioned at the surface of the wall opposite to the emission surface. This procedure verifies that enough energy exists in the signal to allow the detection of the thickness of the wall (Fig. 31) and to control the data processing results. The tests are always carried out along four to five different parallel lines, every 5 cm moving the antenna regularly without time delay/advance or vertical fluctuations. Reasonable values for the velocity of the antenna along survey lines are 5 cm/s or 2.5 cm/s. The previous operation is repeated rotating the antenna in three directions: horizontal, vertical and at 45°. The antenna signal is strongly polarized. The different orientation of the antenna allows for enhancing horizontal or vertical characteristics. It has been observed that the records with a 45° orientation of the antenna contain less local reflections and show clearly the general structure of the wall. This effect is due to the masonry characteristics with vertical and horizontal joints and to the emitted signal that is highly polarized. Furthermore, the redundancy of the data is necessary for the data processing procedure.

The choice of the antenna must be made on a site basis. During the test it is important to control the radar potentialities in relation to the frequency used [34]. These characteristics are not constant; they change for every test. The phenomenon is connected to the wall depth and to the presence of humidity; in fact it is known that the humidity decreases the wave velocity. For some particular cases the repetition of the echo radar data acquisition on the other side of the wall along the opposite line could be necessary.

The following processing of the data is finalized to two main aims: (i) to clean the signals from general noise due to local echoes; (ii) to enhance the recognized features. The second aim tries to increase the utility and reliability of the radar data. In fact usually radar data are clearly readable only by experts. It is important to show results, as radargrams and graphics, which are significant to operators like architects and engineers.

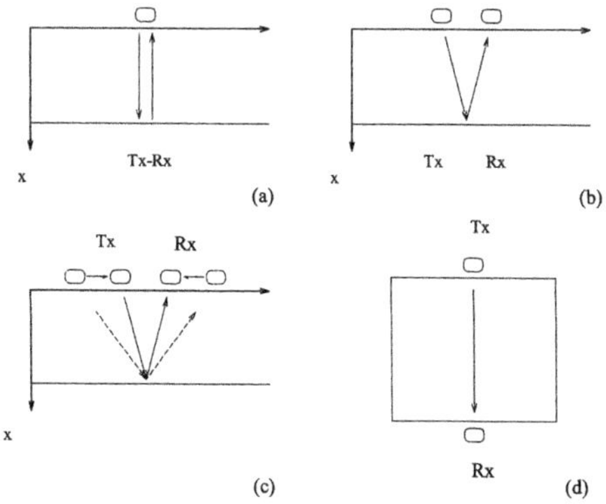

Fig. 30. Modality of acquisition.

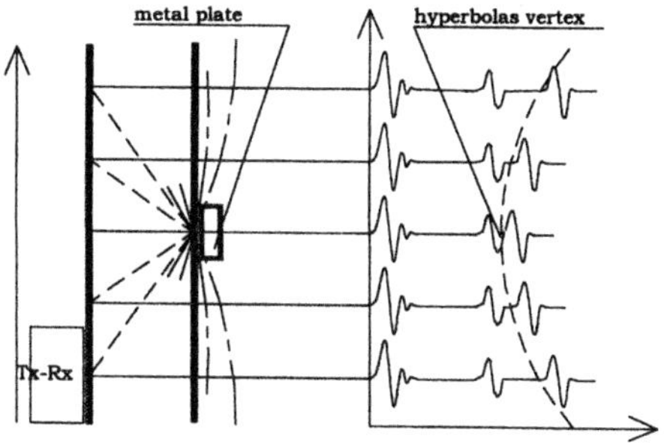

Fig. 31. Effect of a metal plate.

The first data processing on the single signal are filtering operations like running average and running median. The antenna signals decreases with the depth of the wall; these characteristics are usually less readable.

An exponential gain technique is often used to amplify the signals ends. The plot of the derivative of the signals give a better visualization of the wave. The noise on this type of radar records mainly comes from a variety of local irregularity in the masonry. The noise reduction is obtained by the sum or by the median of the signals of more radar traces recorded at a distance of 50 mm and with different antenna orientations. The differences in the radar traces could be considered as secondary reflections and in this way are reduced. The noises due to local reflections occur at different time positions. By summing each trace of the radar section, the common and global coherent characteristics should be enhanced with respect to the noises.

A technique could be applied to point out local defects. The technique can be very efficient when the aim of the radar surveying is to detect local voids, defects or irregular structural elements. A significant signal of the radar section is subtracted from each signal. The significant signal represents what is common in all the radar sections, including the break through effect. By subtracting it, the anomalies are shown clearly. Usually the average, the median or minimum value between all the traces are used.

4.5.2.4. Radar and sonic tomography. One of the most difficult application for sonic and EM tests concerns the detection of leaves and of their characteristics (connections, voids, etc.) in a multiple-leaf wall or pier. Sonic or radar measurement by transparency or echo are seldom successful due to the fact that: (i) the leaves are parallel to the surface where the antenna is applied (Fig. 32); (ii) the rubble and mortars used for the internal leaf are usually made with the same material as the external ones [35].

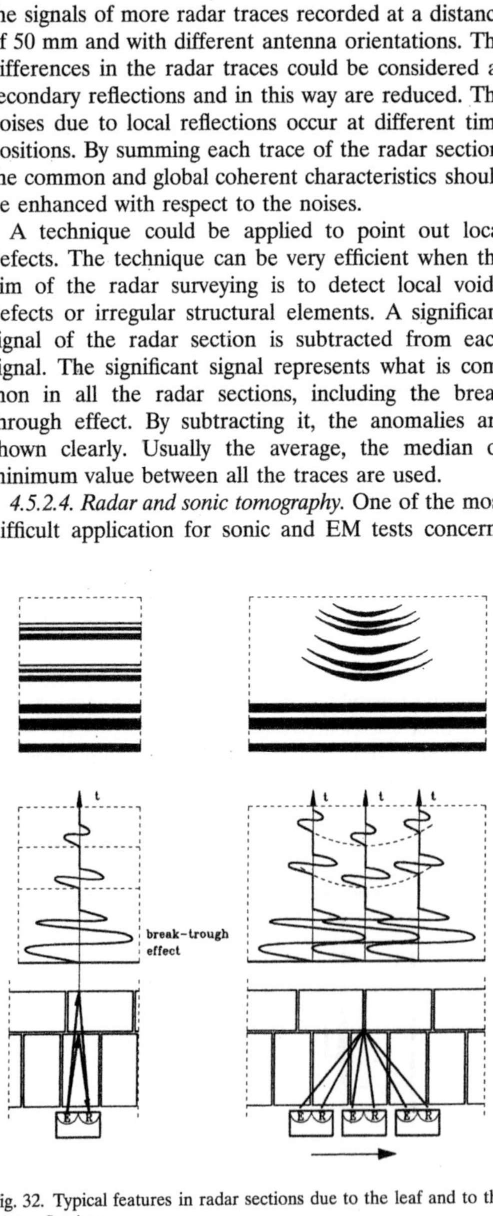

Fig. 32. Typical features in radar sections due to the leaf and to the joint reflections.

Among the ND applications the tomographic technique is quite attractive for the high resolution that can be obtained [36,37]. Tomography, developed in medicine and in several other fields, seems to be a valuable tool to give two- or and three-dimensional representation of the physical characteristics of a solid. Tomography, from Greek 'tomos' (slice), reproduces the internal structure of an object from measurements collected on its external surface. Tomographic imaging is a computational technique which utilizes an iterative method for processing a large quantity of data. Standard pulse velocity data or radar data could be used to reconstruct a velocity distribution within a solid material, thus providing an 'image' of the masonry interior. Tomography's principle is the Fourier Slice theorem [38], that shows how a complete slice of an object can be extracted from a proper set of measurements. The testing technique gives a map of the velocity distribution on a plane section of the structure under investigation. The method consists of obtaining numerically the time taken by a wave along several directions, which uniformly cover the section under investigation. The computation is made by using the inversion process which, starting from the time of signal propagation, reconstructs the field velocity. The section of the masonry is marked by a mesh grid whose dimension is related to the distance between two subsequent transmission or receiving points. The calculation is carried out in the case of sonic tests, under the hypothesis that (in a non-uniform velocity field) sonic impulses did not propagate in a straight line but following a curved line cause by refraction.

Because of the cost of a tomographic survey (acquisition time and processing complexity) a understanding of which results can be achieved and how is required. In fact, the accuracy of tomography depends on many parameters: the source (sonic or electromagnetic), the number and the position of measurements, the equipment settings and the reconstruction algorithms.

It is essential to stress that the resolution capabilities of tomography can be evaluated only by taking into account the measurements locations (i.e. the angular distribution of the observations and their spatial sampling) by means of the above theorem. Furthermore, the physical limits related to the wavelength should be considered. In fact, sonic and EM sources produce diffraction phenomena that limit the resolution. Thus, when the measurements are properly carried out (i.e.

the angular coverage and the spatial sampling honour the Fourier Slice theorem), the resolution has a physical limit strictly related to the wavelength involved in the survey. On the contrary, when the measurements do not satisfy the above conditions because of the environment or because of the structure geometry (i.e. in a wall survey), the resolution limits could even more unfavourable. All these concept are developed in [39]. The result of the tomographic inversion is a map of a property of the materials. In case of travel time tomography (TT) the measured quantity is the travel time of the signal and the map is the distribution of the propagation velocity within the object. In the case of amplitude tomography (AT) the measured quantity is the amplitude of the signal and the map is related to the distribution of the absorption coefficient.

Special care should be devoted to the: (i) choice of location of measurements; (ii) accuracy in their topological location; (iii) choice of optimal transducers/antennas (operating frequency); (iv) availability of auxiliary tools for a faster and easier acquisition procedure; and (v) settings of the acquisition electronic equipment (time range, filters, gains, time sampling, number of bits for A/D conversion). All of those items must contribute to obtain a reasonable number of good sonic/radar signals: the object section will be uniformly illuminated from any direction (provided that it is possible) and each single trace will contain clear information of its crossed path; hence, the signals will be clean with low noise, not clipped and long enough. In such a manner it will be easier to extract from the signal all the needed characteristics for inversion.

Point (iii) will be accomplished by a spectral analysis of the received signal; in fact, the choice of an operating frequency as high as possible suffers the limitation of absorption that is proportional to the frequency and to the object size. Hence the trade-off between resolution (that requires high frequency) and depth of penetration is a critical aspect in the design of a tomographic survey.

The determination of the absolute time scale is an important item. Usually the zero time does not correspond to the first sample of a trace. In the case of radar tests it can be obtained in different ways: by coupling the antennas and measuring the cable and electronics delay, or acquiring a sequential set of traces with antennas located at different distances. In this second case the regression of arrival times vs. distance allows computation of the zero time.

The specific characteristics of commercial radar equipment some different tools allow to be used and procedures to make the radar tomography faster and easier than the sonic one. TT and AT do no require huge pre-processing of the signals. The most critical step in TT is the first arrival picking. In addition AT (only for GPR) needs the knowledge of the antenna directivity pattern for a correct interpretation of amplitudes.

Both sonic and radar systems are suitable for travel time tomography (TT). Usually, materials present complementary behaviours with respect to sonic and EM velocity (i.e. slow sonic materials are generally fast materials for EM waves and vice versa). This indicates that one method may be more appropriate than the other depending on the material nature. In some cases, where it is not possible to apply the radar method (large presence of metals or water), sonic tomography can be the only solution; vice versa when the sonic fails fully (large presence of voids or chaotic inhomogeneities) radar can detect the main elements of the object section.

An advantage of sonic tomography is that elastic parameters are expected to be more correlated with the sonic velocity rather than with EM velocity. However, radar systems, provided the antennas are calibrated, are much more indicated than sonic tomography for the application of other powerful reconstruction techniques such as amplitude tomography, migration and diffraction tomography. Both the sonic and the radar systems suffer the difficulty of the source signature extraction.

Finally, cost considerations are also important and with respect to these it should be pointed out that the GPR equipment is much more expensive but the acquisition times for radar data may be an order of magnitude below the correspondent sonic acquisition times.

5. Design for investigation and evaluation of the results for the diagnosis

This brief description of available techniques and procedures illustrates the difficulty faced by the designer who must select the most technically and economically correct method to define the state of preservation or damage of the structure to be restored. Fig. 33 shows which information can be available from in situ and laboratory surveys and how they can constitute the input data for the structural analysis. When the design of the survey is previously available, then conclusions from the experimental and numerical investigation will bring to the diagnosis the real state of the structure. Fig. 34 shows the possibilities available to a designer when dealing with the structural analysis, from the elastic to the inelastic or limit analysis, provided that one can obtain enough information from the experimental survey; nevertheless, appropriate constitutive laws for the masonry materials are still not well developed.

Since every investigation described in the previous sections has its cost, it is evident that every single operation must be designed to obtain the results de-

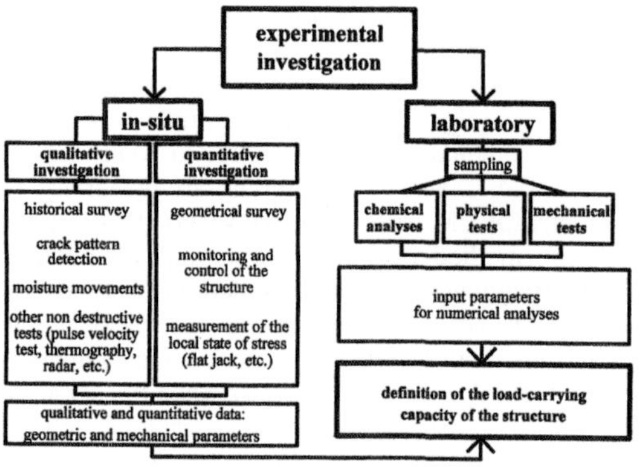

Fig. 33. Finalization of the experimental survey to the structural analysis [12].

sired by the designer; a survey can never be considered as a stream of required steps, but must be designed every time according to the knowledge requirements for the building under consideration.

6. Case history: the Bell Tower of Monza

The 16th century Bell Tower adjacent to the Cathedral of Monza, a town 30 km from Milan (Italy) has been investigated in order to assess its safety [40]. The western and eastern load-bearing walls of the tower, which is 70 m high bear two major passing-through cracks for more than 50 years. After the sudden collapse of the Civic Tower of Pavia occurred in 1989 and causing the death of four people, an extensive investigation on site and in laboratory together with analytical calculations have been carried out in order to assess the safety of the tower against a possible instability.

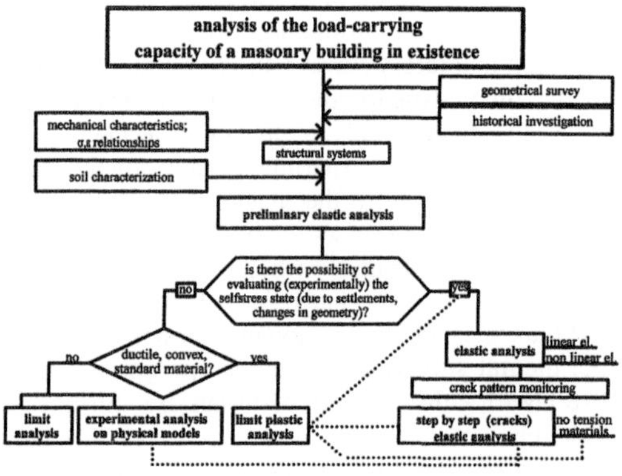

Fig. 34. Phases and alternatives of numerical analysis for an existing structure [12].

Fig. 35. Bell Tower of Monza.

6.1. *Historical notes and description of damage symptoms*

The study of historical documents concerning the construction and following events of the tower (Fig. 35), gave useful information for the diagnosis of the structure.

The tower construction started in 1592 perhaps following the design of Pellegrini and ended in 1605 [41]. The fact that the construction of the tower only required 14 years suggests that uniformity of construction techniques and materials were used, as in fact it was detected. This characteristic is very seldom found in other towers such as, e.g. the Civic Tower of Pavia, which were built over several centuries.

The only damage to the tower reported by the documents occurred in 1740 [42] and was due to the fire which developed in the bell tower and caused the collapse of the bells and of the supporting frame down to the vault of the first floor at 11 m. In 1755 a heavy clock was installed and some restoration works were carried out.

Other known events were: a lightning in 1816 [43] with no evident damages reported, a thunderstorm in 1842 [44] which caused limited damages to the roof and a serious thunderstorm in 1928 which caused the collapse of a pinnacle of the adjacent cathedral, but no damages were reported. Nevertheless, in 1927 some glass devices were applied across the main cracks to detect their movements. The observation of three photographs taken between 1916 and 1920, in 1932 and 1956 shows that the main crack already existed on the tower certainly in 1920.

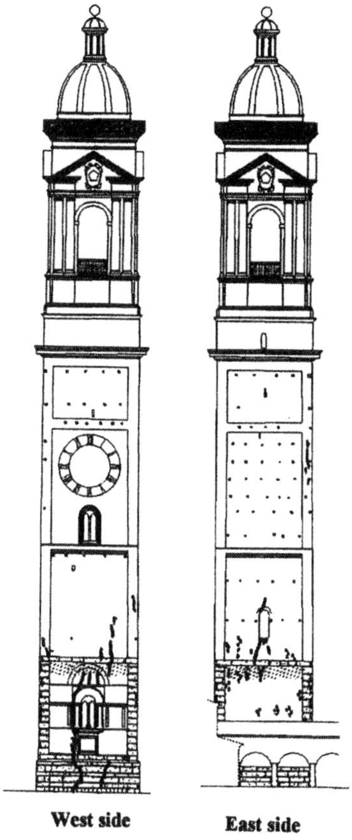

Fig. 36. Crack pattern.

Fig. 37. Crack on the west facade of the tower.

No repair works have been done to avoid the propagation of the cracks, the which were mainly developing vertically and passing through the wall on the west and east facade of the tower (Fig. 36). The major one can be seen from the square in front of the entrance of the cathedral (Fig. 37).

From 1978 the cracks have been surveyed with removable extensometers showing a slowly increasing of their opening along the time.

From 1988 the rate of opening seems to be increasing; this fact is causing a great concern among those responsible for the safety of the tower itself.

Fig. 38 shows the monitored opening of the main crack from 1978 to 1997 with a clear tendency toward a faster increase from 1988. Other cracks can be seen from the internal walls of the tower; they are very thin, vertical and diffused along the four sides of the tower and deeper at the sides of the entrance were the stresses are more concentrated.

The cracks mainly occur across the bricks and go 450 mm deep inside the masonry walls, as an accurate survey has shown. Since the crack pattern has developed slowly over the years a possible time-dependent behaviour of the material can be supposed due to the heavy dead load. This phenomenon, together with the effects of cyclic loads such as wind and temperature variations, can eventually cause the collapse of the structure.

The situation can be considered serious if other similar cases such as the Civic Tower in Pavia, collapsed in 1989 [55], the Tower of St. Magdalena in

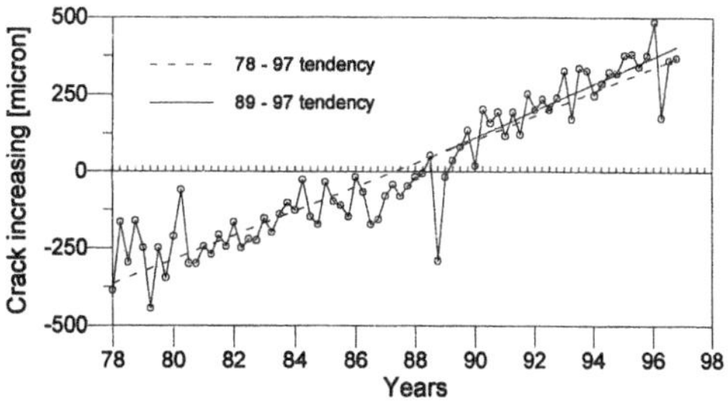

Fig. 38. Crack increasing of the base 2a.

Goch, collapsed in 1992 [45] and the Cathedral of Noto collapsed in 1994 [46].

They all were bearing vertical cracks for a rather long time and fell down suddenly without dramatic causes.

6.2. On site investigation

The first item of information needed by the structural engineer was a geometrical survey of the structure; this survey could be used as geometrical model for the structural analysis. Even if the main cracks had been monitored from 1978, no plans nor prospects of the tower were found.

The second item of information needed concerns the crack pattern and distribution, the depth of cracks, which have to be reported on plans and prospects, in order to give an initial interpretation of the damages occurred to the structure and as reference for a mathematical model eventually taking into account situations beyond the elastic limit or non-linear behaviour.

6.2.1. Geometrical survey

Shortage of time and money suggested choosing a photogrammetric survey of the external facades and a traditional survey in the interior. A geodetic net set up in the square for the cathedral in 1993 was used as support; based on some points of the net some significant points of the west facade were surveyed and used to straighten and return a series of photographic images [47,48]. A Rollei 6006 camera was used which allows a calibrated grid to overlap the image in order to have measurable references (Fig. 35). The bell tower was not surveyed since its precise dimensions were not needed for the structural analysis for which this part was assumed only as a dead load. The restitution of the images and dimensions was carried out with a computer program for simplified photogrammetry of the last generation (ELCO VISION 10). Two distinct products were obtained: (1) a detailed three-dimensional model from which the external and internal prospects and the sections for the internal prospects were obtained; (2) a simplified model for which only the essential aspects of the geometry were preserved for the structural analysis. No relevant leaning was measured due to the small subsidence which is taking place in the square.

6.2.2. Crack pattern study and inside survey of the wall section

As mentioned above, the presence of the main running cracks suggested carrying out a complete crack pattern survey on the walls of the tower. Cracks were surveyed visually and photographically and reported on plans prospects and sections. In the meantime the measurement of the main cracks continued while an automatic monitoring system was being set up. The tendency of opening of the three surveyed cracks was calculated as 30.6, 31.3 and 39.7 $\mu m/year$ from 1978 to 1995. Actually, if this tendency is considered from 1988 to 1997 the values change, respectively, into 41.2, 35.2 and 56.2. Special attention should be paid to the vertical cracks at the edges of the northern and southern facades of the tower; they unfortunately cannot be monitored without scaffolding from the exterior of the building.

The survey campaign revealed a distribution of thin vertical cracks crossing the bricks inside and outside starting from 12 m up to two-thirds of the tower. The situation can be made clear from the Fig. 36, showing

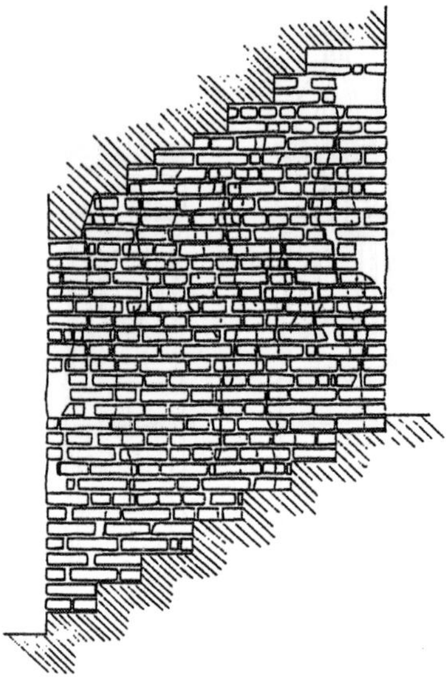

Fig. 39. Thin cracks on the entrance wall.

Fig. 40. Crack pattern inside the wall.

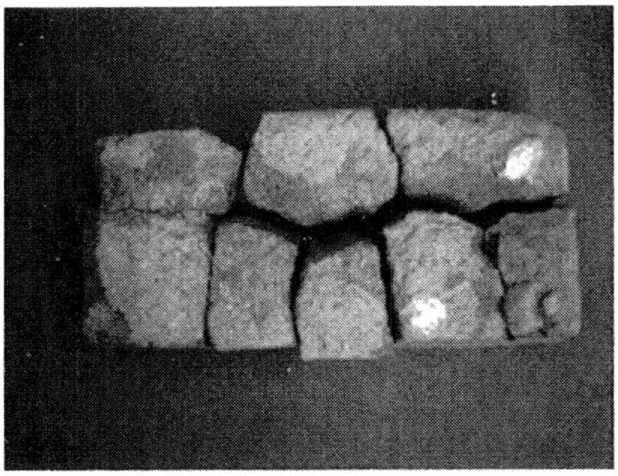

Fig. 41. Cracked brick inside the damaged wall.

the two most damaged walls. The distribution of the thin cracks is shown in Fig. 39, representing the wall corresponding to the entrance of the tower.

Since the tower appeared as being made from bricks, and it was not clear if the inside of the wall did not contain rubble instead, as in other cases of ancient towers, some bricks were gently sampled from the wall and later on replaced back, after photographic survey. Figs. 40 and 41 show, respectively, the depth of cracks and the real situation of a brick found inside one of the most damaged walls. Four points were inspected in the same way and confirmed what had been found in the first one. This inspection also allowed detection if the walls were made with solid bricks and to sample some representative parts of mortars for laboratory tests.

6.2.3. Drilling of cores

This operation is usually intended to be less destructive than sampling manually and therefore used in several more numerous points. The limit of drilling is due to the fact that it spoils the materials (especially mortars), it that it gives only a stratigraphy of the wall interior and, therefore, it does not allow reconstruction of the wall section. Furthermore, it can be destructive when the materials are very weak. In the case of the Monza Tower, a first core drilling attempt failed due to the softness of the external bricks which could not bear the weight of the attached driller.

Finally, it was useful to confirm that the wall was made with solid bricks, but no other information, apart from the brick cores could be obtained.

6.2.4. Flat-jack and dynamic tests

Flat-jack tests are carried out to measure the value of the vertical compressive stress and the stiffness of the material [49,50].

Seven single flat-jack tests were performed, respectively, at 5.4, 5.6 13.0, 14.0, 31.5, 38.0 m height of the tower and the detected stress values are reported in Fig. 42. It is possible to see that, as expected from analytical calculations, the stress values are increasing from the top to the bottom of the tower. These tests allowed also to take into account some criticism on the measurement of displacement [51,24]. The highest values seemed to be particularly dangerous, taking into account the strength values detected usually on these type of masonry.

Therefore, some double flat-jack tests were carried out in order to check the mechanical behaviour of the masonry under compression.

Fig. 43 shows, as an example, the results obtained in the case of the weakest point among the ones which were tested. Table 2 reports the values at which the masonry started cracking in the four cases, showing

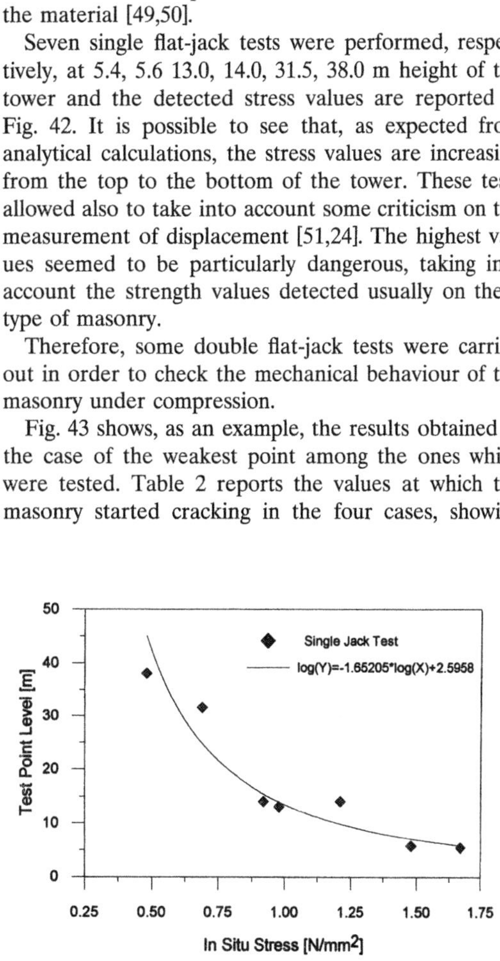

Fig. 42. Single flat-jack tests performed, respectively, at 5.4, 5.6, 13.0, 14.0, 31.5, 38 m.

Table 2
Flat-jack test results

Test number	Single test State of stress (N/mm^2)	Double test Onset of cracking (N/mm^2)	Elastic modulus E_s (20–50%) (N/mm^2)	Poisson coeff. ν_s (20–50%)
TMJ 1,2	0.98	1.87	985	0.13
TMJ 6,7	1.67	2.62	1380	0.19
TMJ 8,9	0.69	2.24	1372	0.07
TMJ 10,11	0.48	2.24	465	0.2

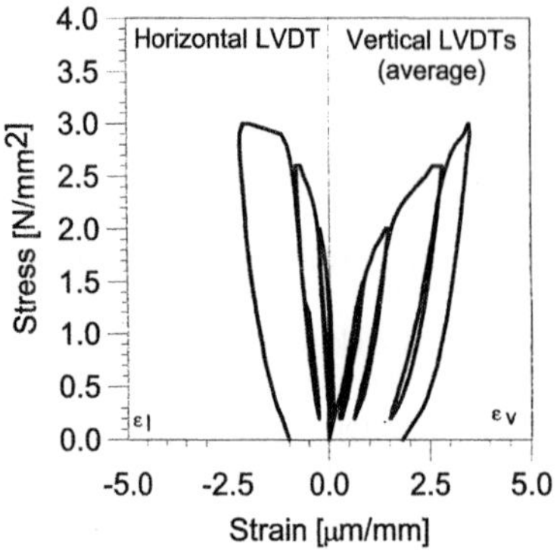

Fig. 43. Double flat-jack test on west side of tower.

that the state of stress seems to be not so far from unsafe, if the safety factors adopted by the codes for the new masonry are taken into account. The results also explain the diffused cracks in the walls of the tower.

In order to verify the response of the structure to dynamic loading and their effect in terms of stress variation, two dynamic tests were planned using the environmental excitation: the first one measured the effects of the bells ringing, the second one, not yet concluded, measured the effects of a strong wind. The response of six horizontal servo-accelerometers mounted in pairs at different height from the ground was monitored and elaborated, together with the readings of some transducers mounted across the major cracks near the front window at the base of the tower.

The sensitivity of the accelerometers was in the range of 0–700 Hz, that is from the low (0.5–5 Hz) frequencies expected for the tower movements to the frequencies associated with the sound propagation in the masonry.

Dynamic tests based on active excitation of the structure can inform about possible difference in stiffness along the height of the tower due also to material damage or to different construction techniques.

6.2.4.1. Sonic tests. As confirmation of the state of damage of the structure and also as a calibration of the procedure, sonic tests have been carried out on the most damaged part of the walls. The equipment used for the survey is composed of an instrumented hammer as transmitter, an accelerometer as receiver and an oscilloscope [28,30].

Two areas were detected at approximately 12 m height and with a depth of the masonry wall of 1800 mm:

- Zone 1 — at 2.28 m from the floor and 1.55 from the entrance to the tower, where the masonry was considered in good condition; no presence of cracks or damage was detected and the flat-jack test gave a compressive stress of 0.92 N/mm^2 sufficiently high. A surface of 800 × 800 mm was controlled with 16 measurements by transparency.
- Zone 2 — at 0.50 m from the floor and 0.22 m from the entrance where the wall presents a series of thin and large cracks (Fig. 36). Also in this area, a grid of 16 points was established; 16 and 32 measurements were taken by transparency and radially (Fig. 44).

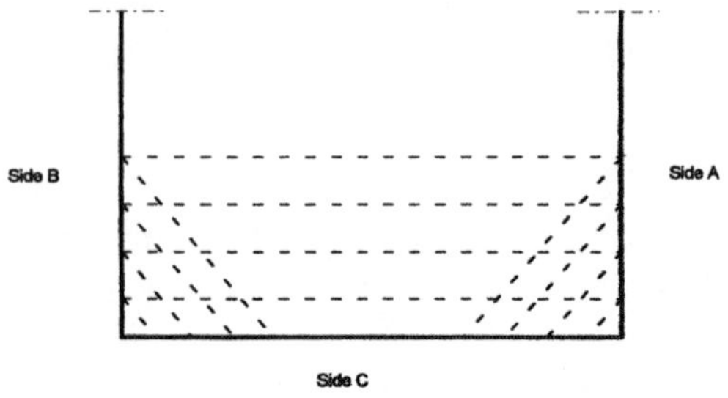

Fig. 44. Direction of measurements.

The velocity values were generally very low compared with other masonries, due to the fact that the materials are rather weak (see Section 4), and the presence of a large number of mortar joints in the wall which attenuate the sonic waves. For this reason, radial tests were also included where possible and the influence of this attenuation is shown in Fig. 45 reporting the results of zone 2 (Fig. 44). Nevertheless, a comparison between the two zones was possible and clearly zone 1 gave systematically much lower results than

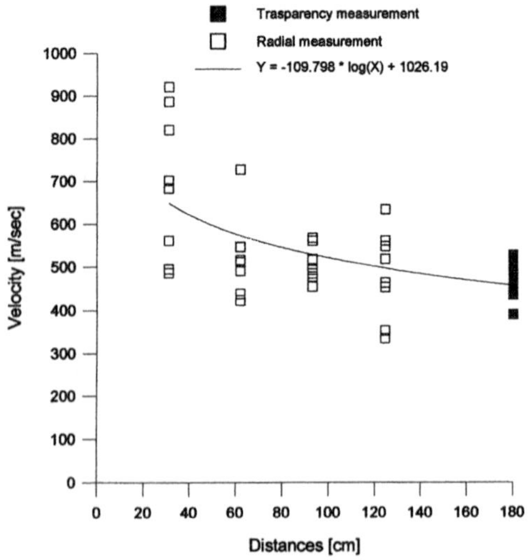

Fig. 45. Pulse velocity on zone 2.

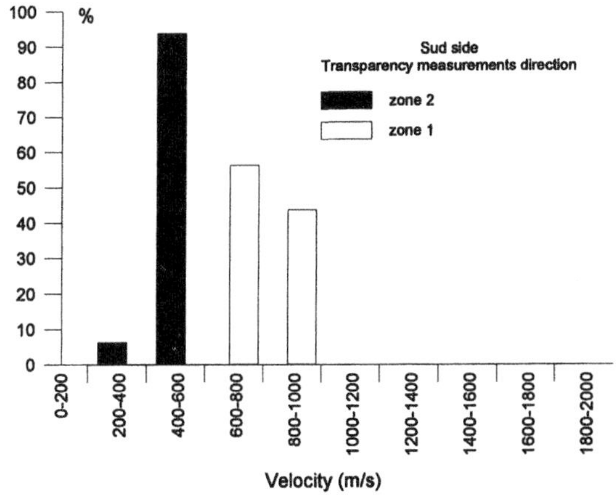

Fig. 46. Pulse velocity frequency histogram on zone 1 and zone 2.

zone 2. The difference can be clearly seen in the frequency histograms of Fig. 46. Fig. 47 shows the results of the sonic tests carried out by transparency on the two areas. The differences of the velocity ranges are clearly visible. In the damaged area (area 2), the velocity values are very low between 400 and 600 m/s. On the contrary, in the area not damaged, area 1, the values are almost double, approximately in the range of 700–900 m/s.

6.3. *Laboratory investigation*

The materials sampled from the walls of the tower were tested in laboratory to be characterized.

In the meantime, some masonry prisms recovered from the walls of the cathedral crypt which was built at the same time as the tower, as the historians have reported. The prism materials were also tested showing that the historic affirmation is reliable, i.e. the materials are of the same type.

Mortars and bricks sampled from the tower were subjected to chemical, mineralogical–petrographical analyses and physical and mechanical tests.

Mortars are mainly based on putty lime and siliceous aggregates (Fig. 48), coming from the near Ticino river. These mortars are very weak and rather decohesioned and they could not be tested mechanically. Nevertheless, the bond between mortars and bricks seems to be very good.

The bricks belong to two types different in colour: brown and light red. This two types show also great differences in the physical and mechanical characteristics. The brown brick is the less porous and absorbent (absorption by total immersion 13%), stronger (compressive strength between 28 and 33 N/mm^2 and elastic modulus between 2050 and 5300 N/mm^2). The red

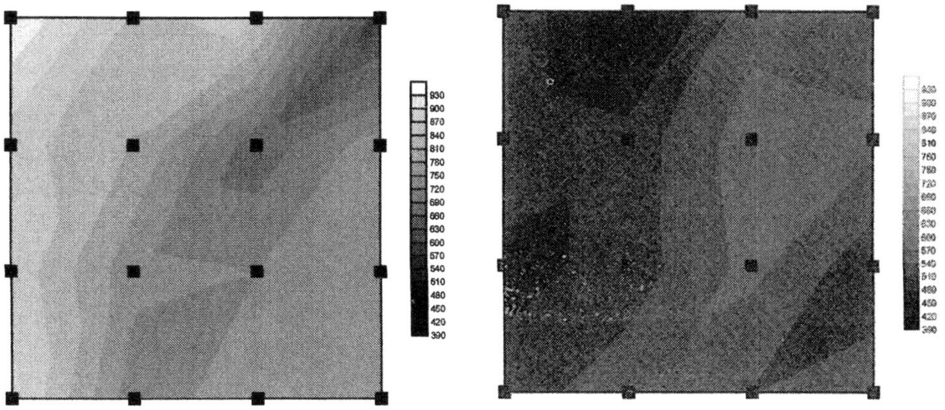

Fig. 47. Sonic velocity on zone 1 and zone 2 by transparency.

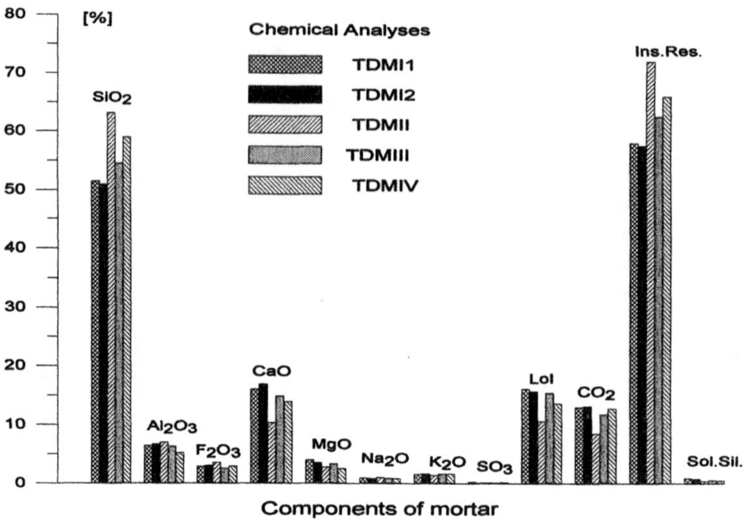

Fig. 48. Chemical analyses histogram of mortars.

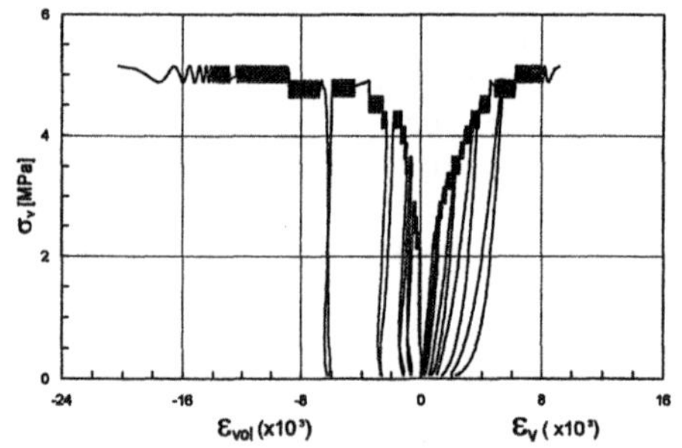

Fig. 49. Result of a fatigue test carried out on a prism (200 × 200 × 500) mm belonging to the crypt of the cathedral.

brick has a higher water absorption (18%) and lower strength (between 4 and 12 N/mm^2) and modulus of elasticity between 500 and 1330 N/mm^2.

Petrographic–mineralogical observation confirmed the red brick was produced at a low temperature of 800°C and hence had low characteristics [53]. Unfortunately, the red brick is the most diffused in the construction; this means that the masonry components in the tower are very weak and therefore also the masonry is weak.

Two large pieces of wall have been recovered during the opening of a door from the crypt of the Duomo. The crypt was apparently built during the same period and reasonably likely with the same technique as the tower.

Prisms of approximately 200 × 200 × 500 mm have been obtained from the wallets and have been subjected to three series of uniaxial compression tests of different type [52,54] after being characterized through sonic tests: (i) monotonic tests have been carried out initially to have a first indication on the compressive strength of the masonry; (ii) cyclic tests were carried out subsequently during which cycles of $\pm 0.15\ N/mm^2$ at 1 Hz were applied at increasing stress levels (Fig. 49); (iii) finally, compression tests were also carried out applying the loads in subsequent steps kept constant for a defined time interval of approximately 1.5 h (Fig. 50).

It can be observed that during the application of the cycles a deformation takes place the amount of which becomes higher when the stress level is higher. Moreover, considering the volumetric strain, it appears that dilation occurs since the beginning of the test.

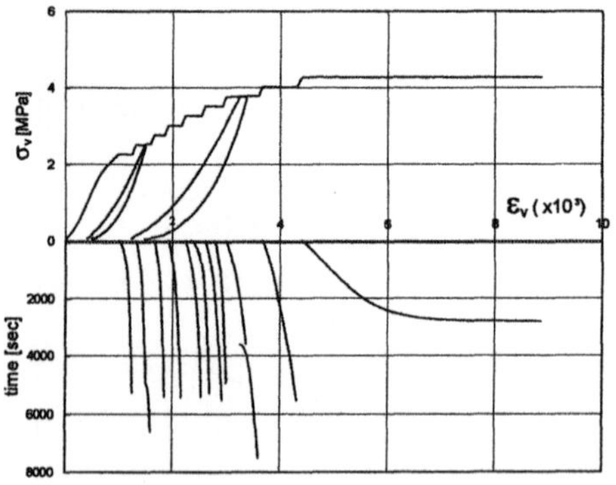

Fig. 50. Result of a test with subsequent load steps carried out on a prism (200 × 200 × 500) mm belonging to the crypt of the cathedral.

Side C

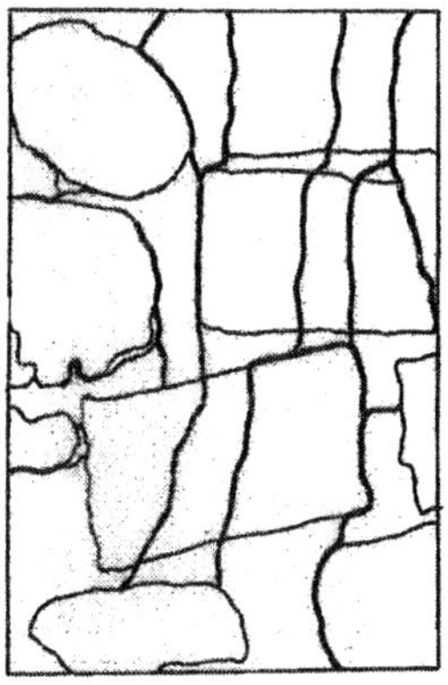

Fig. 51. Crack pattern of the prism.

Creep strain can be clearly observed while the load is kept constant, with the appearance of tertiary creep during the application of the last load step. The crack pattern at the end of the test shows vertical and diffused cracks indicating severe damage of the material (Fig. 51).

7. Conclusion

After the presentation of testing procedures and of their application some concluding remarks can be drawn.

- Masonry is a general term to define composite materials which can differ highly from each other

MORTAR		
BUILDING:	Castelletto CS	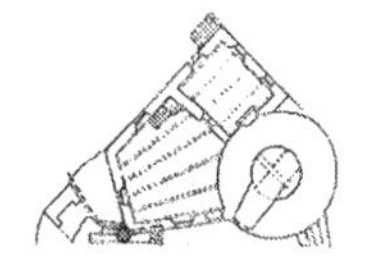
SAMPLE ORIGIN:	masonry CS1p	
SAMPLE DENOMINATION	CS 2m	
MACROSCOPIC CHARACTERISTICS: The mortar appears and grey crumbly.		

MINERALOGICAL PETROGRAPHIC ANALYSIS	
At the polarized light microscope the aggregate shows a very fine grain and is made of: mono- and polycrystalline quartz and quartzites; carbonatic rocks with generally fine grain; sandstones with fine grain; quartz-feldspar gneiss rocks; flints with fine grain; isolated individuals of feldspar, micas, calcites, granites and epidotes; rare small brick fragments.	

CHEMICAL ANALYSIS					
Loss on ignition	13,70	Fe_2O_3	2,77	Insoluble residue	64,06
CO_2	11,70	CaO	14,83	Na_2O	1,17
SiO_2	50,38	MgO	1,61	K_2O	1,30
$Al_2O_3 + Fe_2O_3$	15,48	SO_3	0,19	Sil. Sol.	0,38
Binder/aggr. ratio:				Carbonatation L..	

GRAN SIZE DISTRIBUTION OF THE AGGREGATE			
SAMPLE WEIGHT	SIEVE (mm)	PASSING (%)	GRANULOMETRIC CURVE
	16,00	100,00	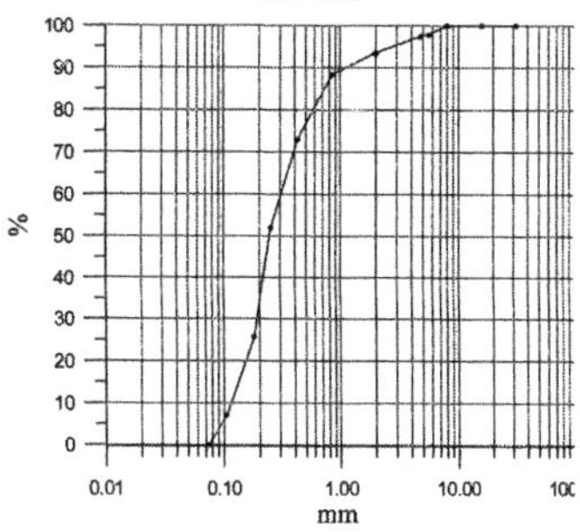
	8,000	100,00	
	5,600	97,92	
METHOD:	4,750	97,41	
	2,000	93,50	
	0,850	88,36	
Thermal attack followed by hand sieving	0,425	73,05	
	0,250	52,12	
	0,180	25,67	
		7,11	
		0,00	

Sheet A. Characterization of a mortar sample.

STONE

BUILDING:	Castelletto CS	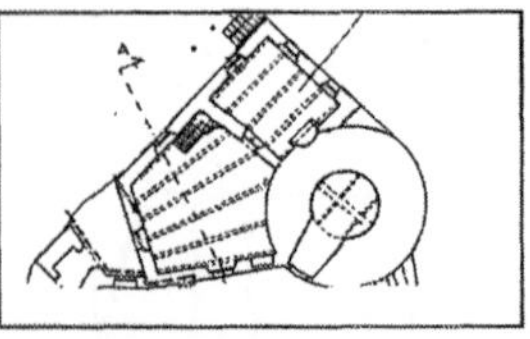
SAMPLE ORIGIN:	masonry CS 1p	
SAMPLE DENOMINATION	CS1.1.v CS1.2.v CS1.4.o	
MACROSCOPIC CHARACTERISTICS: The sample is made of a light grey ochre compact rock of medium grain.		

MINERALOGICAL PETROGRAPHICAL ANALYSIS

STONE TYPE	Macigno Sandstone	
At the polarized light microscope, in thin section, the sample shows the characters of a sandstone with rather small grains and clayey cement. Debris grains are made of: • mono and polycrystalline quartz e quartzites • carbonatic rocks with more or less fine grain volcanic rocks often with porfiric structure		

PHYSICAL ANALYSIS

Sample denomination	CS 1.1.v	CS 1.2.v	CS 1.4.o	Average
Apparent bulk density (Kg/m^3)	2567	2580	2590	2579
Porosity	-	-	-	-
Initial Rate of Suction (g/cm^2.)	0,05	0,04	-	0,045
Absorption Coefficient by Capillary Rise (g/cm^2 sec$^{0.5}$)	0,00028	0,00034	0,00025	0,00029
Absorption Coefficient by Immersion (%)	0,67	0,65	0,62	0,65

DIAGRAM OF CAPILLARY RISE

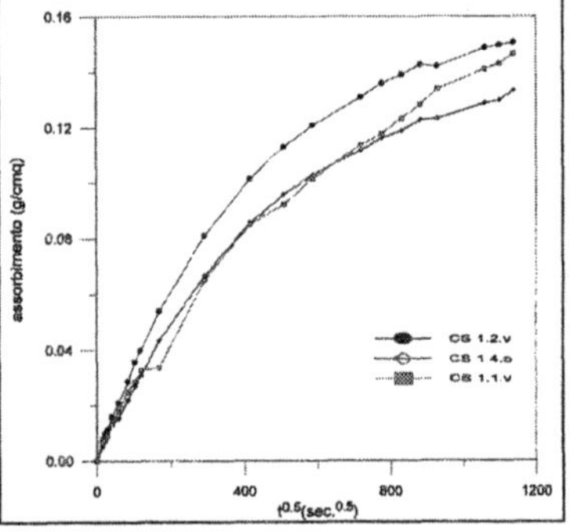

Sheet B. Characterization of a stone sample.

from the point of view of: (i) components; (ii) assemblage; (iii) technique of construction.

- The characteristic of masonry is inhomogeneity which influences the effectiveness of investigation and repair technique.
- Geometrical and crack pattern survey together with the structure monitoring and control can give fundamental information on the structure performance over time.
- In-situ mechanical testing (flat-jack) can give quantitative values of the mechanical parameters of masonry, while laboratory tests are important for the characterization of single components or small masonry specimens.
- NDT can give useful information extended to large wall areas once local specific information have been collected. Nevertheless they still need a great deal of research.

Due to all the problems mentioned it is clear that the main responsibility for choosing an adequate investigation procedure and the type of destructive and non-destructive tests has to be taken by the designer.

Therefore, it should be possible for him to draw a diagnosis research project eventually with the support of different experts.

Acknowledgements

The authors wish to thank A. Anzani, G. Lenzi, N.

MECHANICAL TESTS

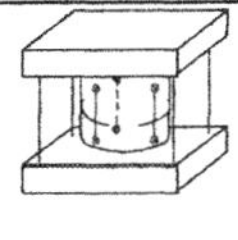

COMPRESSION TEST

Sample denomination	f_c N/mm²	Secant Young modulus (30-45%) E_{dd1} N/mm²	Secant Young modulus (30-45%) E_{lvdt} N/mm²	Poisson ratio (30-60%) $\frac{\Delta\varepsilon_o}{\Delta\varepsilon_v}$ dd1	Poisson ratio (30-60%) $\frac{\Delta\varepsilon_o}{\Delta\varepsilon_v}$ lvdt
CS 1.1.v	108		25500	0.43	0.25
CS.1.2.v	109		25900	0.25	0.24
Average	108,5		25700	0.34	0.245

STRESS VS. STRAIN DIAGRAM

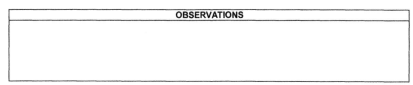

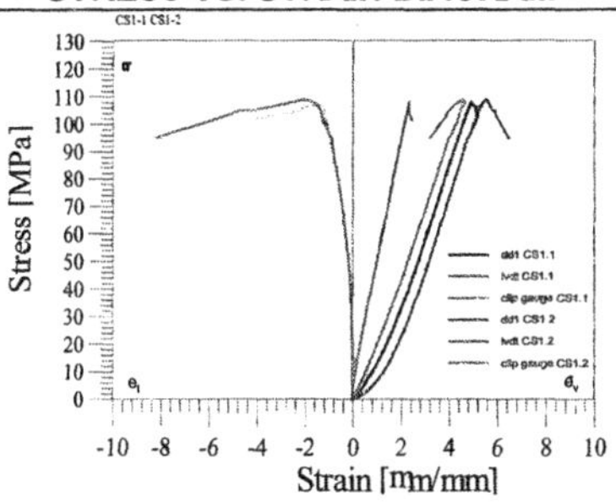

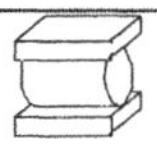

INDIRECT TENSILE TEST (splitting)

denomination	f_{ti} (N/mm²)
CS 1.4.o	5.730
average	5.730

OBSERVATIONS

Sheet C. Mechanical characterization of stone sample.

Ludwig, D. Penazzi, E. Rosina, R. Tongini Folli, S. Valle, L. Zanzi.

The research was carried out with the support of CNR-GNDT, MURST (Cofin 98) and of the European Community.

References

[1] Binda L, Anzani A, Mirabella Roberti G. The failure of ancient towers: problems for their safety assessment. International IABSE Conference on Composite Construction — Conventional and Innovative, Zurich, 1997:699–704.

[2] Rossi PP. Analysis of mechanical characteristics of brick masonry tested by means of in situ tests. 6th IBMaC, Rome, Italy, 1982.

[3] Macchi G. Monitoring and diagnosis of monumental structures. COMETT course. Pavia: Monitoraggio e Indagini Non Distruttive di Strutture Monumentali, 1992.

[4] Giuffrè A. Letture sulla meccanica delle murature antiche. Roma: Kappa, 1991.

[5] Carbone A, Di Palma R. Conservazione della casa terranea di Montescaglioso: tipologia, struttura e tecniche di riparazione. Milan: Tesi di Laurea della Facoltà di Architettura del Politecnico di Milano, 1995.

[6] Binda L., Tiraboschi C., Abbaneo S., Experimental research to characterize masonry materials, Convegno Stoke-on-Trent 1995; Masonry International, Vol. 10, N. 3, pp. 92–101, 1997.

[7] Baronio G, Binda L, Cantoni F, Rocca P. Durability of stone and brick-work protectives and consolidants: experimental research on full-scale models. 6th International Conference on Durability of Building Materials and Components Japan, 2, 1993:824–33.

[8] Binda L, Modena C, Baronio G, Abbaneo S. Repair and investigation techniques for stone masonry walls. Proceeedings of Structural Faults and Repair, 3, 1995:201–10.

[9] Lodigiani V, Manzoni V, Refaldi S. Murature tradizionali dell'area lariana; materiali, tecnologie e comportamento strut-

turale. Milan: Tesi di Laurea della Facoltà di Architettura del Politecnico di Milano, 1995.

[10] Binda L, Fontana A, Mirabella G. Mechanical behaviour and stress distribution in multiple-leaf walls. 10th International Brick/Block Masonry Conference, Calgary, 1, 1994:51–9.

[11] Binda L, Anzani A. Structural behaviour and durability of stone masonry, saving our architectural heritage: the conservation of historic stone structures. New York: Wiley, 1997: 112–48.

[12] Binda L, Mirabella G, Abbaneo S. The Diagnosis Research Project. Earthquake Spectra 1994;10(1):151–70.

[13] Binda L, Lombardini N, Guzzetti F. St Vitale in Ravenna: a survey on materials and structures. International Conference on Historical Buildings and Ensembles, Invited Lectures, 1996:113–24.

[14] Binda L, Anzani A, Mirabella Roberti G. The failure of ancient towers: problems for their safety assessment. Internatio-nal IABSE Conference on Composite Construction — Conventional and Innovative, Zurich, 1997:699–704.

[15] Chiarugi A, Fanelli M, Giuseppettti G. Diagnosis and strengthening of the Brunelleschi Dome, IABSE symposium, Rome. Zurich: IABSE-AIPC-IVBH, 1993:441–8.

[16] Bartoli G, Blasi C, Gusella V. Il sistema di monitoraggio della cupola del Brunelleschi: analisi dei dati rilevati (1988/1990), IV national congress ASSI.R.C.CO, Prato, Italy. Rome: Kappa, 1992:445–55.

[17] Niederwanger G. Structural repair of damaged old bell towers based on dynamic measurements. 5th International Conference STREMA, S. Sebastian, Spain, 1997:447–56.

[18] Rossi PP. Possibilities of the experimental techniques for the structural analysis of historical constructions. Barcelona: CIMNE, 1997:24–46.

[19] Binda L, Gatti G, Mangano G, Poggi C, Sacchi Landriani G. The collapse of the Civic Tower of Pavia: a survey of the materials and structure. Masonry Int 1992;6(1):11–20.

[20] Binda L, Baronio G. Survey of brick/binder adhesion in 'powdered brick' mortars and plasters. Masonry Int J 1988;2(3):87–92.

[21] Baronio G, Binda L. Experimental approach to a procedure for the investigation of historic mortars. 9th International Brick/Block Masonry Conference, Berlin, 1991:1397–464.

[22] Rossi PP. Analysis of mechanical characteristics of brick masonry tested by means of in situ tests. 6th IBMaC, Rome, 1982.

[23] ASTM. Standard test method for in situ compressive stress within solid unit masonry estimated using the flat-jack method, ASTM C 1196-91. Philadelphia: ASTM, 1991.

[24] Ronca P, Tiraboschi C, Binda L. In situ flat-jack tests matching new mechanical interpretations. 11th International Brick/Block Masonry Conference, Shanghai, China, 1, 1997:357–66.

[25] TC 127-MS. Tests for masonry materials and structures. Mater Struct J 1998;31(210):363–75.

[26] Lenzi G, Ludwig N, Rosina E, Saisi A, Binda L. Analisi di tessiture murarie mediante termografia e georadar. V congresso nazionale ASS.I.R.C.CO, Orvieto, 1997:142–6.

[27] Aerojet General Corporation. Investigation on sonic testing of masonry walls: final report. California: Deptartment of General Services of Architecture and Construction, 1967.

[28] Abbaneo S, Berra M, Binda L, Fatticcioni A. Non destructive evaluation of bricks–masonry structures: calibration of sonic wave propagation procedures. International Symposium on Non Destructive Testing in Civil Engineering (NDT-CE), Berlin, Germany, 1, 1995:253–60.

[29] Colla C, Das PC, McCann D, Ford MC. Sonic, electromagnetic and impulse radar investigation of stone masonry bridges. J Non-Destruc Testing Eval Int 1997;30(4):249–54.

[30] Abbaneo S, Berra M, Binda L. Pulse velocity test to qualify existing masonry walls: usefullness of waveform analyses. 3rd Conference on Non Destructive Evaluation of Civil Structures and Materials, Boulder CO, 1996:81–95.

[31] Suprenant B, Schuller M. Non destructive evaluation and testing of masonry structures, reviewed and recommended by TMS. USA: The Aberdeen Group, 1994.

[32] Binda L, Saisi A, Valle S, Zanzi L. Indagini soniche applicate alle murature in mattoni: calibrazione e individuazione di parametri significativi. V congresso nazionale ASS.I.R.C.CO, Orvieto, 1997:77–81.

[33] Kahle M, Illich B, Serexhe B. Erkundung del Untergrundes del Kathedrale von Autun mit dem Radarverfahren. Jahrbuch: Universität Karlsruhe, 1990.

[34] Padaratz IJ, Forde MC. Influence of antenna frequency on impulse radar surveys of concrete structures. Proceedings of The 6th International Conference on Structural Faults and Repair 1995;2:331–6.

[35] Binda L, Lenzi G, Saisi A. NDE of masonry structures: use of radar test for the characterisation of stone masonries. 7th International Conference and Exhibition, Structural Faults and Repair, Edinburgh, 3, 1997:505–14.

[36] Schuller MP, Berra M, Atkinson R, Binda L. Acoustic tomography for evaluation of unreinforced masonry. 6th Conference on Structural Faults and Repair, 3, 1995:195–200.

[37] Valle S, Zanzi L, Lenzi G, Bettolo G. Structure inspection with radar tomography. International College for Inspection and Monitoring of the Cultural Heritage, IABSE-ISMES, 1997.

[38] Kak AC, Slaney M. Principles of computerized tomographic imaging. IEEE Press, 1994.

[39] Valle S, Zanzi L. Resolution in radar tomography for wall or pillar inspection. 6th International Conference on Ground Penetrating Radar, Sendai, Japan, 1996.

[40] Binda L, Tiraboschi C, Tongini Folli R. On site and laboratory investigation on materials and structure of a bell-tower in Monza. 2nd International Conference RILEM on Rehabilitation of Structures, Melbourne, 1998:542–56.

[41] Scotti A. L'età dei Borromei in Monza. Il Duomo nella storia e nell'arte. Milan: Electa, 1989.

[42] Archivio Capitolare di Monza, Sezione Cronache 2, 1805–1838. Milan: Archivio di Stato di Milano, Fondo Religione, Cartella 2566.

[43] Biblioteca Capitolare di Monza, Burocco-Sirtori, Annuali di Monza, 1770–1850, vol. III.

[44] Biblioteca Capitolare di Monza, Burocco-Sirtori, Annuali di Monza, 1770–1850, vol. IV.

[45] Gantert Engineering Studio. Technical opinion about the collapse of the bell tower of St. Maria Magdalena in Goch (Germany), 1993.

[46] Iacono B. Noto, la Cattedrale: cenni storici ed architettonici dalle origini al crollo. la Cattedrale-Architettura ed Urbanistica del Centro Storico (1976–1995). Noto (Siracusa): Sicula Editrice Netum, 1996.

[47] Binda L, Mirabella Roberti G, Poggi C. Il Campanile del Duomo di Monza: valutazione delle condizioni statiche, no. 7/8. Editoriale L'Edilizia, 1996:44–53.

[48] Astori B, Bezoari G, Guzzetti F. Analogue and digital methods in architectural photogrammetry, XVII International Congress of Photogrammetry and Remote Sensing, Commission V. Washington, 1992.

[49] ASTM C 1196, 1197–991.

[50] RILEM LUM 90/2 D.2, D.3.

[51] Ronca P. The significance of the gauging system in the flat-jack in situ test. Masonry Soc J 1996;14(1):79–86.

[52] Lenczner D, Warren DJN. In situ measurement of long-term movements in a brick masonry tower block. Proceedings of the 6th IBMaC, Rome, 1982:1467–77.
[53] Binda L, Poggi C. Ricerca volta a stabilire le condizioni statiche del Campanile del Duomo di Monza mediante analisi chimiche, fisiche e meccaniche dei materiali. Milan: D.I.S. Politecnico di Milano, 1996.
[54] Binda L, Poggi C. Ricerca volta a determinare il comportamento meccanico della muratura del Campanile del Duomo di Monza mediante procedure sperimentali e simulazioni numeriche. Milan: D.I.S. Politecnico di Milano, 1997.
[55] Anzani A, Binda L, Mirabella Roberti G. A numerical interpretation of long-term behaviour of masonry materials under persistent loads. 4th STREMA, architectural studies, materials and analysis, 1. Computational Mechanics, 1995:179–86.

Papers from

Journal of Wind Engineering and Industrial Aerodynamics

Structural Engineering Compendium I

Turbulence closure schemes suitable for air pollution and wind engineering

Albert F. Kurbatskii[a,b,*], Sergey N. Yakovenko[a,b]

[a] *Institute of Theoretical and Applied Mechanics SD RAS, 630090 Novosibirsk, Russia*
[b] *Department of Physics, Novosibirsk State University, 630090 Novosibirsk, Russia*

Accepted 3 July 2000

Abstract

Computationally efficient turbulence closure schemes are formulated and evaluated. Fully explicit algebraic expressions for Reynolds stresses and turbulent scalar fluxes are derived in the weak-equilibrium assumptions from the transport equation for these functions. The two-equation Eulerian turbulent diffusion model is used for describing evolution of the concentration field of contaminant emitted from a ground source in the turbulent boundary layer in the absence of buoyancy forces. Results of modelling passive scalar propagation from a continuous line finite-size source located on the underlying surface of the boundary layer with using the non-local two-parameteric turbulence model and the transport equation of mean concentration are presented. In proposed diffusion model the turbulent diffusion coefficient changes not only with the vertical coordinate but also with the distance downstream from the source according to the laboratory experiments data. The results of modelling reproduce well the measurements data for structural features of the concentration field transformation. Using the differential transport equation model for turbulent matter flux and the low Reynolds number corrections for normal Reynolds stresses approximation gives the same results and somewhat more exact reproduction of fine features of diffusion field both near the source and near rigid surfaces. Diffusion of passive contaminant from a line source in a flow behind a backward-facing step is investigated. © 2000 Elsevier Science Ltd. All rights reserved.

Keywords: Contaminant; Turbulence; Boundary layer; Sharp-edged boundaries

*Correspondence address: Institute of Theoretical and Applied Mechanics SD RAS, Institutskaya Street, 4/1, 630090 Novosibirsk, Russia. Tel.: +7-383-2-30-34-30; fax : +7-383-2-34-22-68.
E-mail address: kurbat@itam.nsc.ru (A.F. Kurbatskii).

Reprinted from *Journal of Wind Engineering and Industrial Aerodynamics*
87 (2-3), 231-241 (2000)

1. Introduction

Turbulent diffusion of matter and heat is of primary importance in industrial, chemical and atmospheric studies. Since a source of such contaminants is in many cases close to solid boundaries, the study of diffusion in turbulent boundary layer flows is of special interest. The purpose of this work is to develop closure models for study of both transport and dispersion of pollutants in environmental shear flows.

Fackrell and Robins [1], Poreh and Cermak [2] have shown that the turbulent diffusivity depends on the distance from a pollutant source. As known, a shortcoming of the down-gradient diffusion model for the turbulent fluxes of matter is that it is a local model. However, because of its simplicity, the K-theory remains an attractive approximation due to the fact that it provides acceptably accurate results in a number of applications.

One of the aims of this paper is to reproduce passive contaminant concentration field characteristics in a turbulent flow with sharp-edged boundaries. This case models peculiarities of pollutant diffusion in a near-ground atmospheric layer with buildings. Complex turbulent flows around obstacles are characterized by substantial growth of turbulent intensities in a thin layer above an obstacle and behind it. In mixing layers formed above recirculation zones, large-scale (coherent) structures are formed and destroyed suddenly during reattachment. These formations interact with stationary eddies in a cascade of primary and secondary recirculation zones. Such complex structure of a flow with sharp-edged boundaries produces problems for its description by turbulence models. Moreover, flows with sharp-edged boundaries are a serious test for numerical procedure and experimental device because of sharp change of pressure, velocity and turbulent stresses near sharp edges.

The state of the art of utilizing turbulence models in wind engineering has been reviewed by Murakami [3,4]. In these overviews the revisions of the standard $k-\varepsilon$ model for adopting the bluff body aerodynamics are presented. The results of computations by the Reynolds stress model (RSM) applied to a cubic model are shown, and thus the shortcoming of this model is discussed. The two-layer model [5] as well as different versions of the $k-\varepsilon$ model (e.g., Refs. [6,7]) were also applied for calculation of flows around bluff bodies.

In this work, several models of turbulent diffusion are evaluated for describing an evolution of the concentration field of a pollutant emitted from a ground line source in the turbulent boundary layer in the absence of buoyancy forces. The present paper includes also the results of concentration field prediction using the standard two-parameter model of turbulent transport for a turbulent flow with a line source located in the recirculation zone behind a backward-facing step.

2. Prediction of the velocity and concentration field characteristics in the turbulent boundary layer with a near-ground line source

Applying the Eulerian approach can accurately solve diffusion problems in a real shear flow of the atmospheric boundary layer. The transport equation for the contaminant concentration C is

$$\frac{\partial C}{\partial t} + U_i \frac{\partial C}{\partial x_i} = \frac{\partial}{\partial x_i}\left[D \frac{\partial C}{\partial x_i} - \langle u_i c\rangle\right], \quad (1)$$

where D is the molecular diffusion coefficient. For deriving the closed form of Eq. (1) the turbulent contaminant flux vector $\langle u_i c\rangle$ is to be defined.

2.1. *Modified "K-closure" formulation and turbulent fluxes model*

To describe an evolution of the concentration field of a plume emitted from a ground line source in the turbulent boundary layer, the $k-\varepsilon$ model is applied. It is able to reproduce structural features of the concentration field transformation because the turbulent diffusion coefficient D_T obtained from parameters k and ε varies not only with the thickness of the boundary layer, but also with the distance from the source as in experiments [1,2]. The $k-\varepsilon$ turbulence model is a non-local theory with respect to k, ε and these parameters are determined by history.

The Euler diffusion model includes equations for the mean velocity, the turbulence kinetic energy k, the dissipation ε and the mean concentration of matter. An algebraic model for the turbulent mass-flux vector $\langle u_i c\rangle$ is derived from the closed differential transport equation for this correlation using the assumption of equilibrium turbulence [8]. This approximation can be written, assuming the linear Boussinesq hypothesis for the Reynolds stress tensor $\langle u_i u_j\rangle$, as the explicit algebraic mass–flux model

$$-\langle u_i c\rangle = D_T \frac{\partial C}{\partial x_i} - \frac{1}{C_{1C}} \frac{k}{\varepsilon} \{[2\nu_T + (1 - C_{2C})D_T]S_{ik} + (1 - C_{2C})D_T \Omega_{ik}\} \frac{\partial C}{\partial x_k}, \quad (2)$$

where $S_{ik} = \frac{1}{2}(\partial U_i/\partial x_k + \partial U_k/\partial x_i)$, $\Omega_{ik} = \frac{1}{2}(\partial U_i/\partial x_k - \partial U_k/\partial x_i)$, $D_T = \nu_T/\sigma_T$, $\nu_T = C_\mu f_\mu k^2/\varepsilon$, σ_T is the turbulent Schmidt number ($C_\mu = 0.09, C_{1C} = 3.0$, $C_{2C} = 0.4$). The damping near-wall function $f_\mu = \left(1 + 3.45/\sqrt{\mathrm{Re}_t}\right)[1 - \exp(-y^+/70)]$ is introduced into the turbulent viscosity coefficient ν_T to allow direct integration of the model to a wall in order to satisfy the asymptotic behavior of the turbulence statistics in the near-wall region.

The first term in Eq. (2) is a simple gradient transport approximation (K-model) while the second term gives a non-zero turbulent mass flux in the streamwise direction even when there is no streamwise concentration gradient. In other words, turbulent eddies generate the longitudinal flux $\langle uc\rangle$ due to their interaction with the mean concentration gradient normal to the streamwise direction. With the addition of the second term which relates the flux $\langle uc\rangle$ with the cross-stream concentration gradient, an improved estimate can be obtained.

Using the fully explicit algebraic expression (2), the closed form of Eq. (1) is obtained. Model (1), (2) can be used in conjunction with any two-equation or Reynolds-stress model for the velocity field, thus greatly simplifying a numerical solution of pollutant transfer problems. It could be easily extended to buoyant flows without having to solve additional equations.

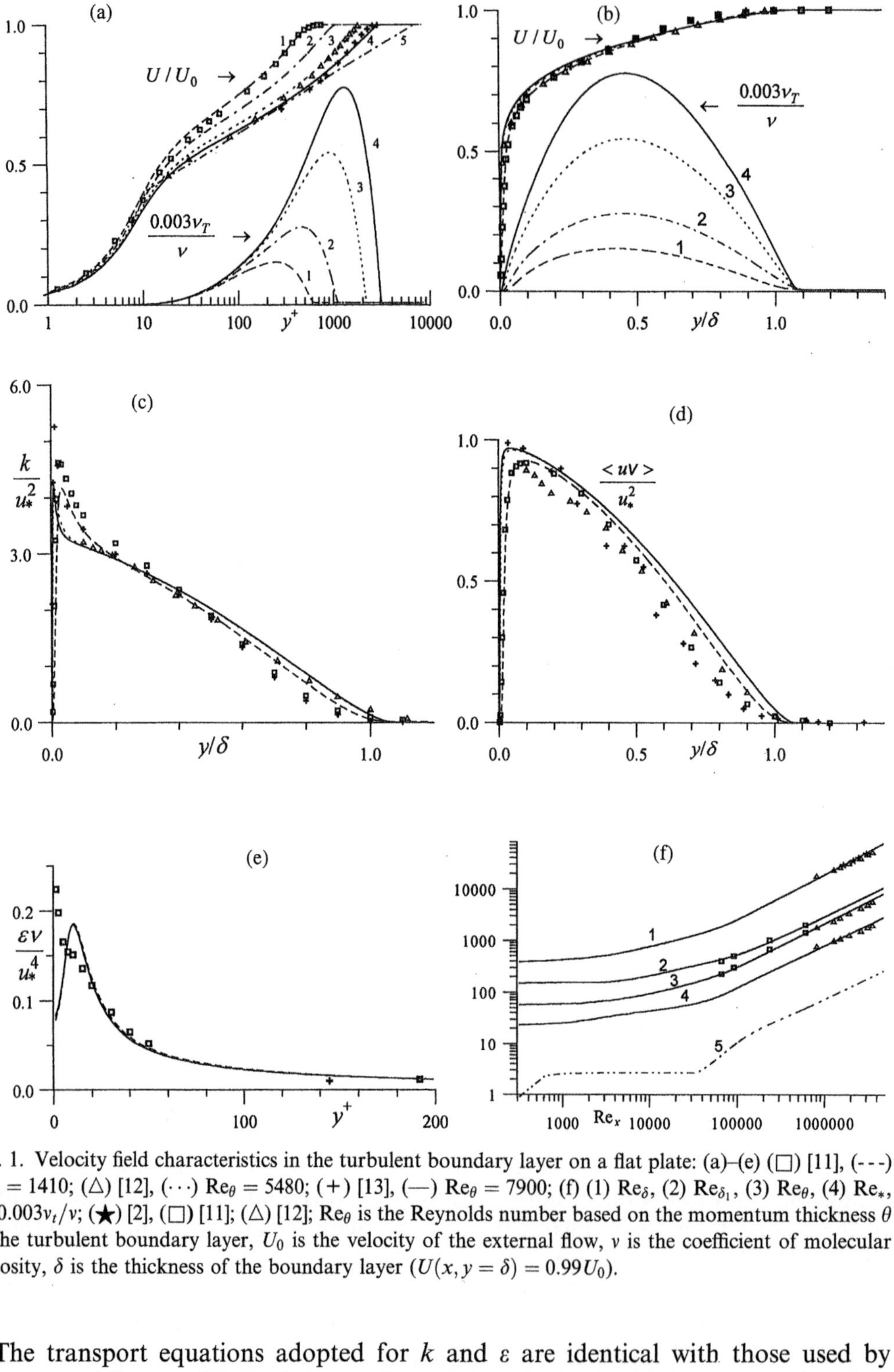

Fig. 1. Velocity field characteristics in the turbulent boundary layer on a flat plate: (a)–(e) (□) [11], (- - -) $Re_\theta = 1410$; (△) [12], (···) $Re_\theta = 5480$; (+) [13], (—) $Re_\theta = 7900$; (f) (1) Re_δ, (2) Re_{δ_1}, (3) Re_θ, (4) Re_*, (5) $0.003\nu_t/\nu$; (★) [2], (□) [11]; (△) [12]; Re_θ is the Reynolds number based on the momentum thickness θ of the turbulent boundary layer, U_0 is the velocity of the external flow, ν is the coefficient of molecular viscosity, δ is the thickness of the boundary layer ($U(x, y = \delta) = 0.99U_0$).

The transport equations adopted for k and ε are identical with those used by Nagano and Tagawa [9]. Turbulent values of the velocity field (Fig. 1) reproduced by this model and behavior of the concentration field (Fig. 2) are in good agreement

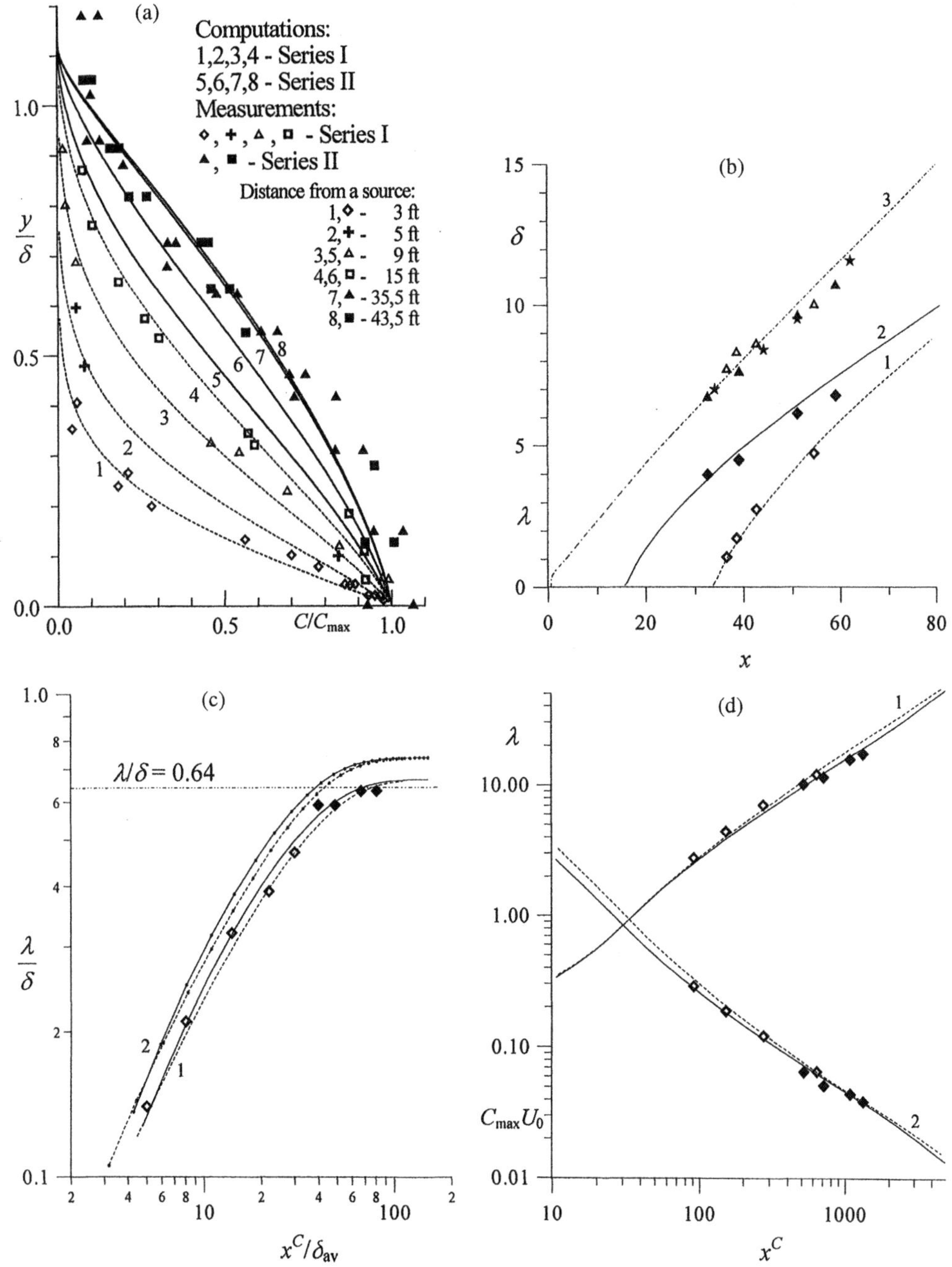

Fig. 2. Diffusion field characteristics in the turbulent boundary layer (lines 1 in (c) and all lines in other figures are obtained at $\sigma_T = 0.90$, line 2 in (c) – at $\sigma_T = 0.72$): (a) profiles of the mean concentration in different regions from a source; (b) thickness δ (lines 3, (★) are without the contaminant source, (△) S.I, (▲) S.II) and contaminant jet half-height λ (1, 2, ◇, ◆) in inches versus the distance x in feet from the boundary layer origin; (c) ratio δ/λ ($C(x, y = \lambda) = 0.5C_{\max}$); (d) half-height λ in cm (lines and markers 1) and the value of $C_{\max}U_0$ in mg/(cm^2 s) (lines and markers 2) versus the distance x in cm from the source. S.I (- - -) computation, (◇) measurement ; S.II (——) computation, (◆) measurement.

with experimental data. It should be pointed that the eddy diffusivity of pollutant $D_T \sim \nu_T$ is obtained in the present model to vary not only with distance from a wall, but also with downstream distance from a source (Fig. 1a and b). The model predictions of the power laws of growth of the vertical plume size and of decrease of the maximum surface concentration on downstream distance from the source are in complete agreement (Fig. 2b–d) with the experimental data [2].

2.2. Differential model for turbulent scalar flux

The closed differential transport equation for the turbulent mass-flux vector $\langle u_i c\rangle$ has been formulated in Ref. [8]. It can be modified allowing for wall effect and written for vertical turbulent concentration flux $\langle vc\rangle$ in the steady two-dimensional boundary layer flow as

$$U\frac{\partial\langle vc\rangle}{\partial x} + v\frac{\partial\langle vc\rangle}{\partial y} = \frac{\partial}{\partial y}\left[\left(\chi + 2C_S\langle v^2\rangle\frac{k}{\varepsilon}\right)\frac{\partial\langle vc\rangle}{\partial y}\right] - \langle v^2\rangle\frac{\partial C}{\partial y}$$
$$- \left[C_{1C} + C'_{1C}\frac{k^{3/2}}{\varepsilon y} + \left(1 + \frac{D}{\nu}\right) f_C\right]\frac{\varepsilon}{k}\langle vc\rangle, \qquad (3)$$

where $\chi = (D + 2\nu)/3$, $f_C = \exp\{-(\mathrm{Re}_T/80)^2\}$, $C_S = 0.11$, $C_{1C} = 3.0$, $C'_{1C} = 0.806$.

This equation can be reduced in the equilibrium [8] or non-equilibrium [10] approximations to the algebraic relations for the vertical turbulent scalar flux $\langle vc\rangle$ (A-model and AN-model, correspondingly).

It is evident that in the algebraic models and in the differential model (3) not only the velocity components U and V but also the mean-squared vertical velocity fluctuation $\langle v^2\rangle$ influence the $C(x, y)$ distribution. The linear Boussinesq model used in the $k-\varepsilon$ turbulence model gives $\langle v^2\rangle = 2k/3$, over-predicting measured values of $\langle v^2\rangle$. It does not describes significant anisotropy of normal components of the Reynolds stress tensor $\langle u_i u_j\rangle$ caused by mean shear effects and damping of fluctuation v by a wall. As a result, this approximation leads to change for the worse prediction of $C(x, y)$ near a surface-placed source in comparison with that of K-model. The latter gives satisfactory agreement with the experimental data of Poreh and Cermak [2] (Figs. 1 and 2) due to adequate reproducing of vertical profiles of $U, V, k, \varepsilon, \langle uv\rangle$ and the boundary layer thickness δ by the modified $k-\varepsilon$ model resolving the viscous sublayer and giving a good agreement with the data of measurements and DNS [11–13]. More exact value $\langle v^2\rangle \approx 0.46k$ estimated from RSM of Launder et al. [14] is not able to improve prediction of $C(x, y)$ quite satisfactorily.

For more accurate definition of $\langle v^2\rangle$ in the near-wall layer, the low Reynolds number model for the transport equations of the normal Reynolds stresses is applied. Suggested by Lai and So [15], the pressure–strain and dissipation terms models, reproducing correctly asymptotic behavior of these terms in the near-wall layer, are introduced into the equations. One can derive an algebraic model (near-wall correction of the algebraic RSM following from the differential one [14]) from these differential equations at the local balance approximation proposed by Rodi

[16]. It is written for the vertical normal Reynolds stress as

$$\frac{\langle v^2\rangle}{k}(P-\varepsilon) = -C_1\varepsilon\left(\frac{\langle v^2\rangle}{k}-\frac{2}{3}\right)+\frac{2}{3}\alpha P-\frac{4}{3}\beta P-\frac{2}{3}\varepsilon(1-f)$$
$$+\left[C_1\varepsilon\left(\frac{\langle v^2\rangle}{k}-\frac{2}{3}\right)-\frac{2}{3}\alpha^* P-2\frac{\langle v^2\rangle}{k}\varepsilon\right]f-\frac{4\langle v^2\rangle/k}{1+3\langle v^2\rangle/(2k)}\varepsilon f, \tag{4}$$

$$\alpha=\frac{8+C_2}{11},\quad \beta=\frac{8C_2-11}{11},\quad C_1=1.5,\quad C_2=0.4,\quad \alpha^*=0.45,$$

$$f=\exp\left[-(\mathrm{Re}_\mathrm{T}/150)^2\right].$$

Use of the Boussinesq model for the shear stress $\langle uv\rangle = -v_\mathrm{T}(\partial U/\partial y) = C_\mu f_\mu(k^2/\varepsilon)(\partial U/\partial y)$ allows to obtain an explicit algebraic model for the normal Reynolds stresses (near-wall correction of the normal stresses algebraic model). This model describes data of DNS [11] and experiments [12] quite good (Fig. 3). Differential equations for normal stresses (near-wall correction [15] of normal Reynolds stresses model of Launder et al. [14]) give similar results.

The differential flux model (1), (3) with correction (4) predicts diffusion characteristics, namely the half-height $\lambda(x)$ of pollutant jet and maximum value $C_{\max}(x)$ of pollutant concentration on a rigid surface, in the same good agreement as that for the K-model (Figs. 1 and 2). The differential model reproduces also the non-monotonic near-wall behavior of $\partial C/\partial y$ (Fig. 4) observed by Poreh and Cermak [2] at the first measuring section. The K-model does not describe accurately this fine feature (results of the A-model and AN-model are close to that of the K-model).

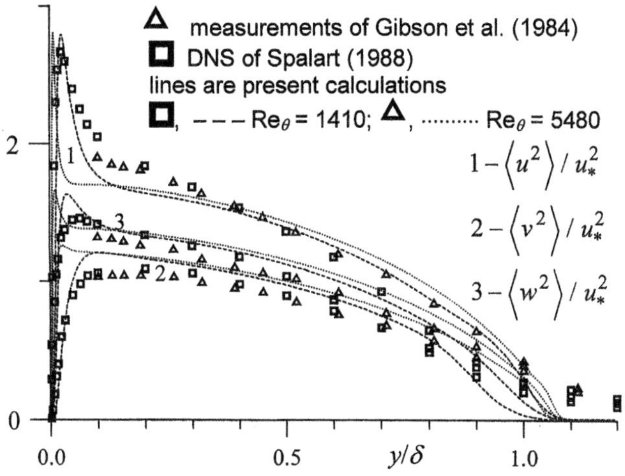

Fig. 3. Normal Reynolds stresses in the turbulent boundary layer.

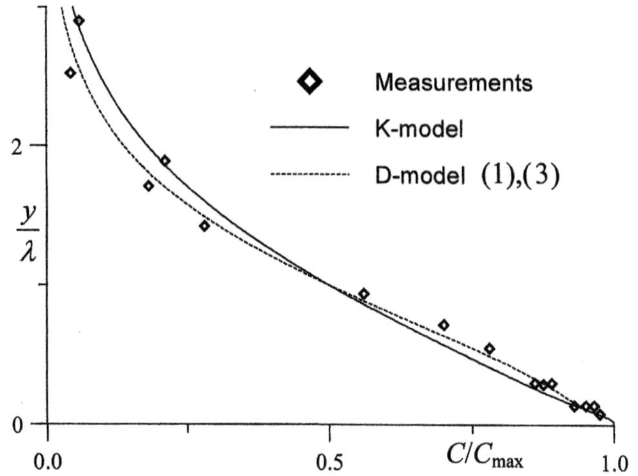

Fig. 4. Normalized mean concentration at distance $x^C = 3\,\mathrm{ft}$ from a source (Series I).

3. Flows with sharp-edged boundaries and scalar sources

Reynolds-averaged Navier–Stokes equations and standard $k-\varepsilon$ turbulence model are used to compute the mean velocity and pressure fields in flows with sharp-edged boundaries. Details of initial and boundary conditions, equations discretization, computation procedure for such flows and results of computations of flow structure with a quadratic obstacle are given by Kurbatskii and Yakovenko [17].

A turbulent flow behind a backward-facing step has been chosen due to available measurements data of Quante and Enteridge presented by Turfus [18]. In these experiments passive scalar diffused in the rear recirculation zone from sources at locations 1 and 2, namely $(x/H, y/H) = (0.08, 0.8)$ and (2.4, 0.4), where $x = 0$ is the step coordinate, $y = 0$ is the vertical coordinate of the bottom at outlet, expansion ratio is ER$=1.111$, $\mathrm{Re}_H = 8 \cdot 10^4$. The flow rate Q of contaminant through a thin porous tube and the diameter of this tube producing a line source were very small. Therefore, we can neglect [18] by influence of the source on the velocity fields characteristics which can be taken from separate computation without a source. The passive scalar source is realized then by the assignment of Q at locations of contaminant emission. It is the same way that has been used in calculation of the boundary layer on a flat plate.

Concentration distributions behind the back-step were obtained by the transport equation for the mean concentration written in the case of two-dimensional elliptic flow

$$U\frac{\partial C}{\partial x} + V\frac{\partial C}{\partial y} = \frac{\partial}{\partial x}\left[\left(\frac{v}{\sigma} + \frac{v_{\mathrm{T}}}{\sigma_{\mathrm{T}}}\right)\frac{\partial C}{\partial x}\right] + \frac{\partial}{\partial y}\left[\left(\frac{v}{\sigma} + \frac{v_{\mathrm{T}}}{\sigma_{\mathrm{T}}}\right)\frac{\partial C}{\partial y}\right]. \tag{5}$$

This equation has been closed with the help of the down-gradient K-model:

$$-\langle uc\rangle = D_T(\partial C/\partial x), \quad -\langle vc\rangle = D_T(\partial C/\partial y), \quad D_T = \nu_T/\sigma_T.$$

Use of more refined models for components of the turbulent scalar flux vector is restricted by the lack of experimental data for unknown terms of the transport $\langle u_i c\rangle$ equation as well as for the mass–flux vector itself.

There is no measured distributions of velocity field characteristics in Ref. [18] for a flow with scalar sources. Such comparison (see Fig. 5 for example) is carried out in one more run corresponding to close conditions of the experiments of Driver and Seegmiller [19] with ER = 1.125 and sufficiently high Reynolds number.

In Figs. 6 and 7 results of the scalar field computations of Turfus [18] by the discrete vortex model for defining trajectories of released particles (models 1 and 2 use different values of effective eddy diffusivity) are plotted together with the measurement and present computation results. One can see that in the whole K-model gives closer agreement with measurements data than the models 1 and 2 from Ref. [18] except near-step locations. Grid independence has been checked in

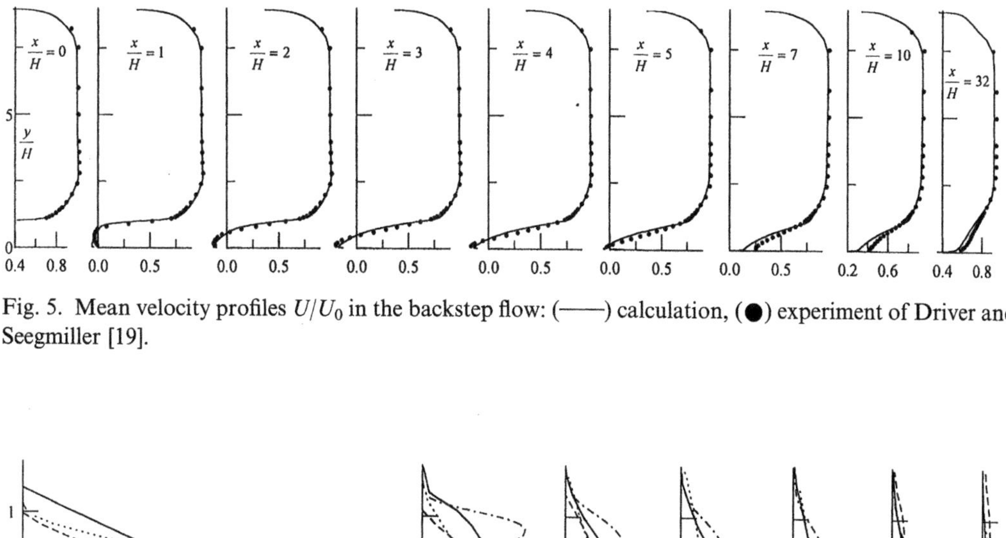

Fig. 5. Mean velocity profiles U/U_0 in the backstep flow: (——) calculation, (●) experiment of Driver and Seegmiller [19].

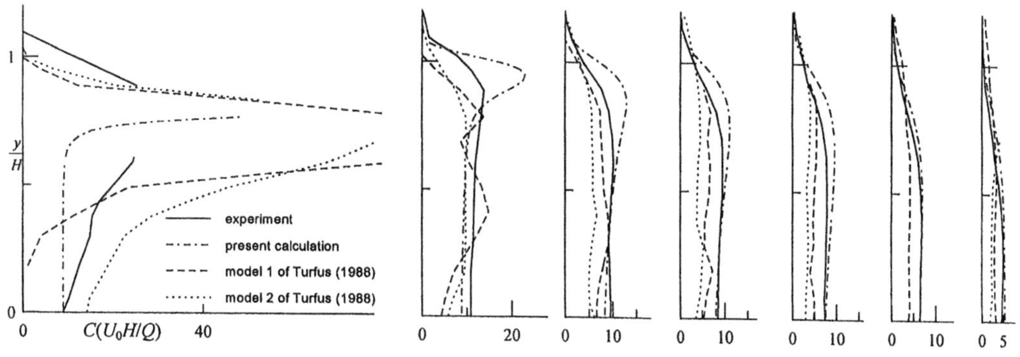

Fig. 6. Concentration profiles in the backstep flow at $x/H = 0$; 1; 2; 2.5; 3; 4; 5 (source location 1).

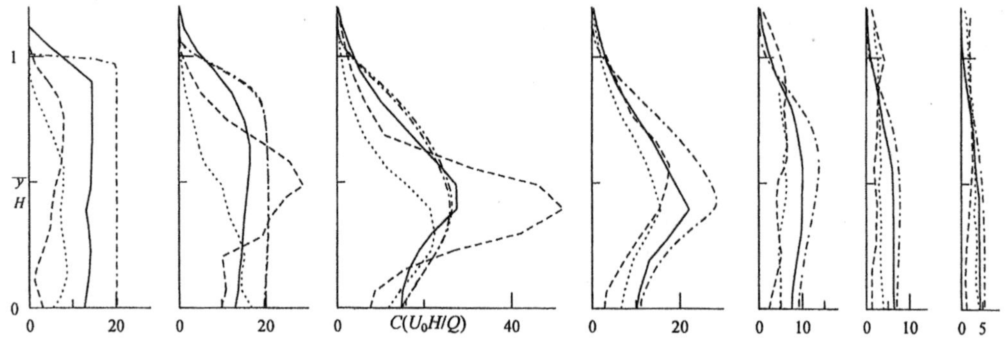

Fig. 7. Concentration profiles in the backstep flow at $y/H = 0$; 1; 2; 2.5; 3; 4; 5 (source location 2). Denotations are the same as in Fig. 6.

calculations at $H/\delta = 50$ (dash–dot lines) and 100 (dash–dot–dot lines in Fig. 7 at $x/H = 1$ and 2).

4. Conclusions

Modeling of the turbulent diffusion of passive contaminant in the boundary layer on a plane smooth underlying surface from a near-ground line source of the assigned strength has been carried out. The developed turbulent transport model includes the turbulent transport equations for the mean velocity, the turbulence kinetic energy, dissipation and mean concentration. The model reproduces characteristics of both the velocity field and the concentration field which are in good agreement with the experimental data and the results of the direct numerical simulations. The experimental observations about the dependence of diffusion field characteristics on the distance from the contaminant source have been confirmed. It is found that the turbulent diffusion coefficient is a function of both the vertical coordinate and the distance downstream from the source. Power laws of variation of the typical vertical jet size and the maximum concentration on the underlying surface are not the same in different regions of the contaminant jet, which is consistent with the measurements of Poreh and Cermak [2].

The differential model with the $\langle vc \rangle$ transport equation and near-wall corrections of RSM as well as the simplified down-gradient diffusion K-model with the low-Reynolds-number $k{-}\varepsilon$ model allow us to obtain fairly realistic predictions of structural features of the concentration field of contaminant emitted from a ground line source. The differential model for $\langle vc \rangle$ can give more exact reproduction of fine features of diffusion field both near the source and near rigid surfaces. However, formulation of an adequate second-order model of scalar transfer for flows with sharp-edged obstacles is restricted by lack of experimental data. K-model describes the concentration field in the most part of a complex separated flow and can be used as the first approximation for solving environmental problems and further development of more accurate models of turbulent heat and matter transfer.

Acknowledgements

The present work is supported by the Russian Fundamental Research Foundation Grant 99-05-64143.

The results of the paper were presented at the Tenth International Conference on Wind Engineering (Copenhagen, 21–24 June 1999), Eight International Symposium on Computational Fluid Dynamics (Bremen, 5–10 September 1999) and First International Symposium on Turbulence and Shear Flow Phenomena (Santa Barbara, 12–15 September 1999). Authors would like to thank Organizers of these forums for invitation and support making possible to disseminate the results of the research performed.

References

[1] J.E. Fackrell, A.G. Robins, Concentration fluctuations and fluxes in plumes from point sources in a turbulent boundary layer, J. Fluid Mech. 117 (1982) 1–26.

[2] M. Poreh, J.E. Cermak, Study of diffusion from a line source in a turbulent boundary layer, Int. J. Heat Mass Transfer 7 (1964) 1083–1095.

[3] S. Murakami, Overview of turbulence models applied in CWE-1997, In: G. Solari (Ed.), Proceedings of the Second European and African Conference on Wind Engineering (Genova, Italy), SGE, Padova, Vol. 1, 1997, pp. 3–24.

[4] S. Murakami, Current status and future trends in computational wind engineering, J. Wind Eng. Ind. Aerodyn. 67/68 (1997) 3–36.

[5] D. Lakehal, W. Rodi, Calculation of the flow past a surface-mounted cube with two-layer turbulence model, J. Wind Eng. Ind. Aerodyn. 67/68 (1997) 65–78.

[6] S. Lee, Unsteady aerodynamic force prediction on a square cylinder using k–ε turbulence model, J. Wind Eng. Ind. Aerodyn. 67/68 (1997) 79–90.

[7] M. Tsuchiya, S. Murakami, A. Mochida, K. Kondo, Y. Ishida, Development of a new k–ε model for flow and pressure fields around bluff body, J. Wind Eng. Ind. Aerodyn. 67/68 (1997) 169–182.

[8] B.E. Launder, Heat and mass transfer, in: P. Bradshaw (Ed.), Topics in Physics, Vol. 12, Turbulence, Springer, New York, 1978, pp. 231–287.

[9] Y. Nagano, M. Tagawa, An improved k–ε model for boundary layer flows, Trans. ASME I: J. Fluid Eng. 112 (1990) 33–39.

[10] M.M. Gibson, B.E. Launder, On the calculation of horizontal turbulent free shear flow under gravitational influence, J. Heat Transfer, Trans. ASME 99 C1 (1976) 81–87.

[11] P.S. Spalart, Direct simulation of turbulent boundary layer up to $R_\theta = 1410$, J. Fluid Mech. 187 (1988) 61–98.

[12] M.M. Gibson, C.A. Verriopoulos, N.S. Vlachos, Turbulent boundary layer on a mildly curved convex surface. Part 1: Mean flow and turbulence measurements, Exp. Fluids 2 (1984) 17–24.

[13] P.S. Klebanoff, Characteristics of turbulence in a boundary layer with zero pressure gradient, NACA Report 1247, 1955.

[14] B.E. Launder, G.J. Reece, W. Rodi, Progress in the development of a Reynolds–stress turbulence closure, J. Fluid Mech. 68 (1975) 537–566.

[15] Y.G. Lai, R.M. So, On near-wall turbulent flow modelling, J. Fluid Mech. 221 (1990) 641–673.

[16] W. Rodi, A new algebraic relation for calculating the Reynolds stresses, Z. Agnew. Math. Mech. 56 (1976) T219–T221.

[17] A.F. Kurbatskii, S.N. Yakovenko, Numerical investigation of turbulent flow around two-dimensional obstacle in the boundary layer, Thermophys. Aeromech. 3 (1996) 137–155.

[18] C. Turfus, Calculating mean concentrations for steady sources in recirculating wakes by a particle trajectory method, Atmos. Environ. 22 (1988) 1271–1290.

[19] D.M. Driver, H.L. Seegmiller, Features of a reattaching turbulent shear layer in divergent channel flow, AIAA J. 23 (1985) 163–171.

PALLAS: a novel optical measurement technique in air pollutant transport studies

István Goricsán[a,*], János Vad[a], Balázs Tóth[a], Pál Greguss[b]

[a] *Department of Fluid Mechanics, Budapest University of Technology and Economics, Budapest Bertalan L. u.4-6, 111 Budapest, Hungary*
[b] *Department of Manufacture Engineering, Budapest University of Technology and Economics, Budapest Bertalan L. u.4-6, 111 Budapest, Hungary*

Accepted 3 July 2000

Abstract

This paper presents the principles of a novel laser optical method in the measurement of concentration distribution of air pollutants. It is based on a concerted application of panoramic annular lens (PAL) optical detection system and a laser sheet illumination equipment (PALLAS). The PAL optic provides full 360° angle of panoramic view information. This is the most important property which makes the PALLAS especially suitable for extensive investigation of concentration field around models of buildings and features of terrain. This paper presents the advantages of this system in comparison to the conventional laser sheet technique. The first results are also presented and measurement technical improvement of PALLAS method is summarised. © 2000 Elsevier Science Ltd. All rights reserved.

Keywords: Panoramic annular lens (PAL); Laser sheet; Optical measurement technique

1. Introduction and objectives

The prediction of the transport of gaseous and dusty pollutants in the lower section of the atmospheric boundary layer is of growing importance in air quality management. In several cases, the transport processes occur in the presence of geographical (i.e., features of terrain) or artificial (e.g. building blocks) inhomogeniety of the flat ground surface. The influence of such objects on transport

*Corresponding author. Tel.: +36-1-463-3465; fax: +36-1-463-3464.
E-mail address: goricsan@simba.ara.bme.hu (I. Goricsán).

Reprinted from *Journal of Wind Engineering and Industrial Aerodynamics* **87 (2-3)**, 259-270 (2000)

processes is usually significant and not yet completely considered in mathematical pollutant transport models used in practice. For example, certain national standards ordering on the preparation method of air quality section of environmental impact assessments prescribe the use of the Gaussian transport model which is valid exclusively for pollutant transport over a flat surface. Thus, the reliability of emission values predicted according to the standards is doubtful in the cases when the pollutant source is located, e.g. in the neighbourhood of hills, valleys or a plateau. This fact emphasises an urgent need for the renovation of these standards, necessitating the improvement of pollutant transport models.

As a basis for the improvement of mathematical pollutant transport modelling, extensive wind tunnel experiments have to be carried out to simulate and study the pollutant transport phenomena via monitoring the flow properties and the pollutant concentration field. In the cases when the geometry of objects or object groups influencing the pollutant transport is very complex, such as industrial establishments, wind tunnel simulation appears at present as the only reliable method for the investigation of the transport process by means of physical modelling of the transport problem.

In the past years, the laser sheet (LS) flow visualisation technique, complemented with a digital image processing (DIP) procedure appeared as a powerful tool for high-resolution tomographic mapping of the pollutant concentration field in wind tunnel experiments. The LS technique proved its suitability in a number of research programs on pollutant transport studies, e.g. Refs. [1–3]. However, there are certain limitations in the use of the conventional LS + DIP technique (e.g. necessity for high-power laser source, shading effects, etc.), which are desirable to be surmounted.

The main objective of present paper is to deliver a brief introduction on a novel measurement technical methodology, termed by the authors as PALLAS concept, which offers a favourable potential for an extended application of the LS + DIP method in pollutant transport studies. The very first experiments based on the PALLAS concept are reported in this paper. Recent measurement technical improvement of PALLAS technique is also outlined.

2. The "PALLAS" concept in pollutant transport studies

The PALLAS method appears as a concerted application of a panoramic annular lens (PAL, e.g. Refs. [4–7] based optical detection system and a laser sheet illumination equipment. Before discussing briefly the PALLAS concept, the LS + DIP technique used conventionally in wind tunnel experiments is overviewed (see Fig. 1). In a conventional LS + DIP arrangement, the laser sheet generating unit is placed above the test section and the optical axis of the camera is adjusted more or less normal to the laser sheet. In this configuration, the sideward-scattered light coming from the tracer particles is detected by the camera. Since the sideward-scattered light of a spherical scatterer object has generally the lowest intensity [8], the conventionally used LS + DIP technique necessitates the use of a high-power illuminating laser source. In conventional arrangements shaded flow zones may

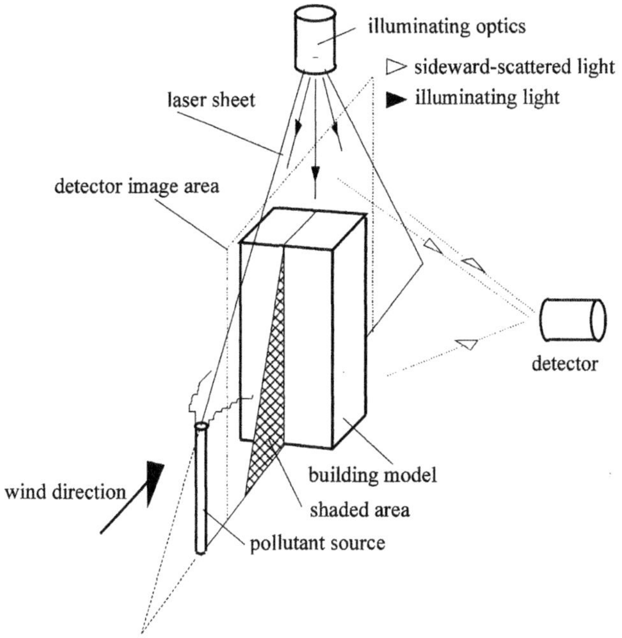

Fig. 1. A conventional laser optical configuration for concentration distribution measurements.

appear, making impossible the mapping of flow and concentration field just in the vicinity of the obstacles, which would be very important from the viewpoint of simulating the immission at solid boundaries (e.g. walls of building models). The mapped (imaged) flow zone is limited by the image area of the camera optics, compelling the user in some cases to compromise on the study of only a part of the concentration field.

The PALLAS method gives a possibility for overcoming some of these problems and also aims to introduce further favourable features in laser optical concentration field measurements. The main novel feature of the PALLAS method compared to the conventional LS + DIP is the way of detection: a PAL-based detection optics is involved, yielding a panoramic view information on the scattered image of full 360° polar object angle. The PAL has already proved its suitability in optical systems of robotics, satellite based research, astronautics, endoscopic techniques, biology, medicine, and other fields (e.g. Refs. [4,9]). However, no application of PAL has been reported in air pollutant transport studies and even in flow diagnostics in general.

The PAL has four surfaces: two concave reflective, one convex refracting and one flat refracting ones. The PAL optical system comprises one optical block; however, an auxiliary lens is also necessary for projecting the virtual PAL image to a CCD detector array. The longitudinal section of PAL is shown in Fig. 2, representing the PAL imaging scheme and the field of view [10,5–7].

If the relative position of the illuminating laser sheet and the PAL are known, the image points transmitted by the PAL can be corresponded to spatial points fitting to the laser sheet plane, by means of a related DIP software. With the knowledge of

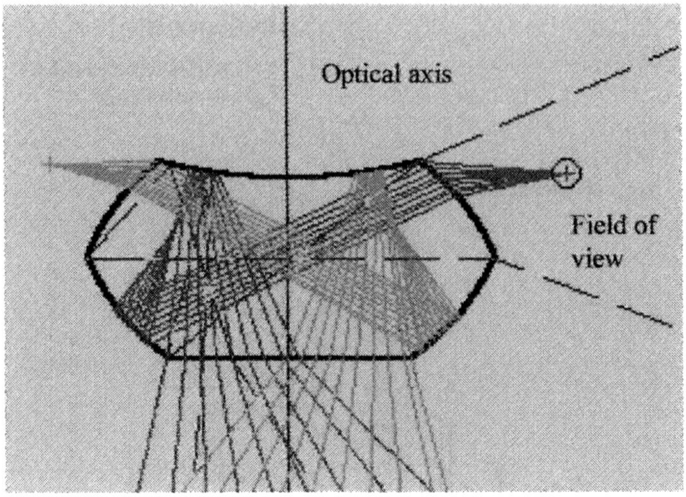

Fig. 2. The longitudinal section and ray tracing of PAL.

intensity distribution of illuminating light and transmission capability of the PAL and also considering other effects influencing the light intensity distribution of recorded image (e.g. transmission on opaque surfaces, variance in direction of detection, see below), the grey-level distribution of the image can be corrected via a DIP method such a way that it can be correlated to the concentration field, similarly to correlation methodology used in conventional LS + DIP experiments.

The fully developed PALLAS technique is expected to comprise an illuminating optics which forms a circularly shaped, plane laser sheet and thus, it acts virtually as a two-dimensional light point source. The relatively extensive azimuthal object angle of PAL allows the illuminating laser sheet to be placed (and thus, the "slice" of light scattering tracer smoke to be imaged) close to the PAL unit, normal to its axis, without an unfavourable reduction of the viewed flow area. If the virtual light point source fits to the PAL optical axis and the conditions mentioned above are fulfilled as well, mostly the backward-scattered light is utilised during the experiment. Since the backward-scattered light has a considerably higher intensity than the sideward-scattered light [8], use of a lower power laser source (e.g. a laser diode) is expected to be sufficient.

The scheme in Fig. 3 illustrates further potential benefits of PALLAS technique. In the configuration described above, the illumination and detection optics can be integrated into a small-scale, mobile, compact unit, make traversing easier (e.g. for tomographic mapping of the concentration field) and also offering a possibility for on-site concentration measurements. Since the PAL optical lens includes a dead zone not utilised in image formation, a miniaturised, diode-laser-based light probe generator can be integrated in the detector optics in itself.

If the model obstacle influencing the pollutant transport process – a building block model in Fig. 3 – is prepared from sheets of a transparent material and the compact PALLAS device is placed inside the model, the flow stays perfectly undisturbed by the measurement device. Furthermore, the surrounding flow field may be illuminated

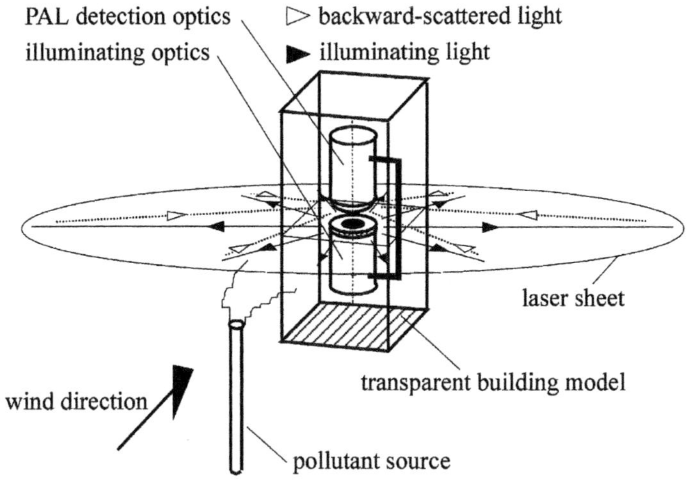

Fig. 3. Illustration of PALLAS concept applied for studies on pollutant concentration distribution around enclosures.

and also observed at a full 360° polar angle, thus avoiding shading effects (full 360° view around the model) and making possible a detailed investigation on the complex pollutant transport phenomena occurring in the boundary layers developing on the solid surfaces.

Based on the above, the potential advantages of PALLAS method versus the traditional LS + DIP methodology (Fig. 1) are summarised as follows:

(1) the probed (imaged) flow zone can be extended due to the panoramic imaging method,
(2) due to quasi-backscattering detection, a lower power laser source can be used,
(3) a small-scale, compact, mobile measurement device can be developed (integrated illuminating and detection functions),
(4) the disturbance of flow due to the measurement device can be fully eliminated,
(5) the shading effect can be avoided and, thus, complex flow and pollutant transport phenomena in the vicinity of solid boundaries as well as emission parameters can be extensively studied.

It is acknowledged that these advantages usually require a more complex preparation of the experiments (e.g. preparation of transparent test objects, elaboration of a more complex DIP and PAL image-retransformation software) and the application of PALLAS method has some optical limitations (total reflection, transmission problems at sharp edges, etc.). These problems must be thoroughly treated.

In the following section, the very first records supplied by the PALLAS method are reported and evaluated, which have been obtained in a preparatory phase of the technique described above. The PALLAS results are compared with the concentration data obtained by a conventional LS + DIP technique.

3. Experiments and discussion

The experimental investigation on pollutant concentration field have been carried out in the large-scale horizontal wind tunnel of the Department of Fluid Mechanics, Technical University of Budapest. The wind tunnel is a recirculating system of Goettingen type. Technical details of the wind tunnel have been summarised in Ref. [10]. Some basic dimensions are presented in Fig. 4, which also shows the experimental set-up. Upstream of the test section, a turbulence generating grating has been placed for the simulation of the atmospheric boundary layer. The experimental set-up presented here comprises a straight model stack mounted vertically 2.16 m downstream of the grating and a model plateau the leading edge of which has been placed 30 mm downstream of the stack. The internal diameter of the model stack is 6 mm and its height is 40 mm. The model plateau has a sloping plate of 1 m and a horizontal plate of 1 m. Its width is 0.6 m. In order to offer optical transparency for PALLAS experiments, the plateau has been prepared from two plexiglass plates of 8 mm thickness. For the experiments presented here, the steepness of the sloping plate was adjusted to 7°.

The axis of the stack is adjusted fitting to the symmetry plane of the plateau, which coincides with the midplane of the wind tunnel test section, which is parallel to the wind direction.

During the reported experiments, oil smoke was emitted by the model stack at an outlet velocity of 4.0 m/s in order to simulate the emission of a pollutant, which is transported downstream in a plume. The wind velocity at the edge of the boundary layer was 2.5 m/s, which was reduced to 1.75 m/s at the outlet height of the stack.

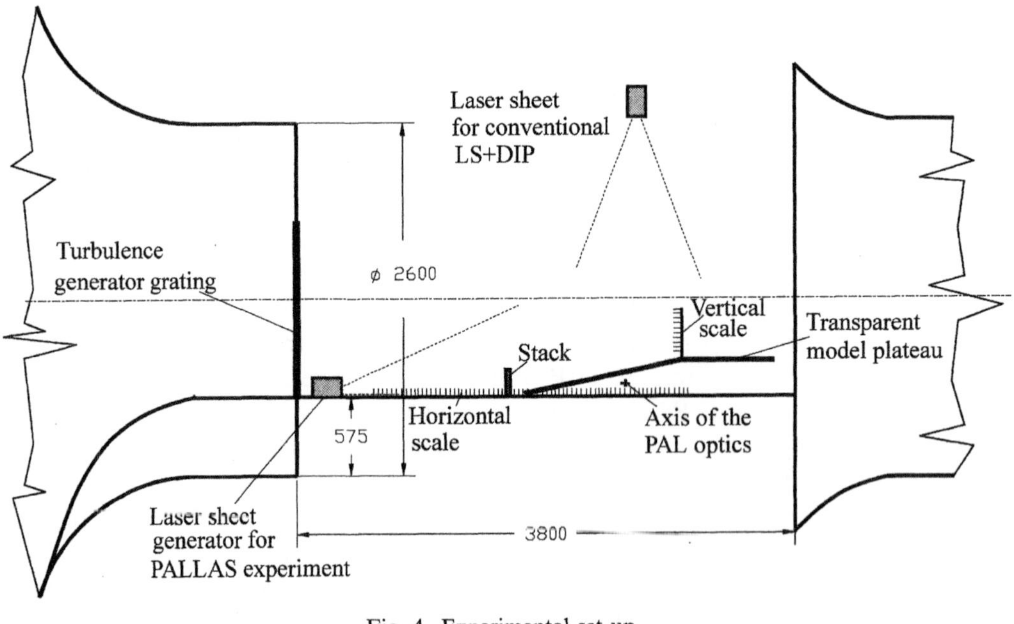

Fig. 4. Experimental set-up.

In the present, preparatory stage of developing a high capability PALLAS system described in Section 2, not a circular but a sharp sectorial illuminating laser sheet of 30° inclination angle was used, which was generated by a LAS-635S-15 1L30 laser sheet generator (LSG) diode modul of 15 mW. In order to study the capability of PALLAS method for pollutant transport studies, it was felt reasonable to compare the PALLAS results with the measurement data obtained by a conventional LS + DIP technique (see Fig. 7).

For both conventional and PALLAS techniques, the vertical laser sheet was adjusted fitting to the axis of the stack and also to the symmetry plane of the plateau.

For the conventional method, the outlet of LSG was placed to the height of 1.9 m above and 0.6 m in front of the breakline of the plateau, with its optical axis adjusted vertically. For Fig. 4, the optical axis of the common light detector unit (including a CCD detector) is to be imagined fitting to the breakline of the plateau for the conventional LS + DIP study.

For the present preparatory state of PALLAS test, the LSG was placed with its outlet 1 m upstream of the stack in such a way that the lower borderline of the laser sheet touched the "ground" (see Fig. 4). Hence, the concept of integration of illuminating and detection optics into a compact measurement equipment was temporarily given up. The PAL detector optics (including the same CCD detector) was placed below the transparent plateau, with its optical axis parallel to the plateau breakline, 0.5 m upstream and 90 mm below the breakline. The leading plane of the PAL was adjusted 3 mm off the laser sheet. In order to give a view on the imaging characteristics of PAL, the daylight PAL image of the experimental set-up is shown in Fig. 5, which illustrates the capability of PAL to supply a substantial panoramic view also including objects located on the periphery of the view field. The distorted horizontal scale at the bottom and vertical scale on the right-hand side of the

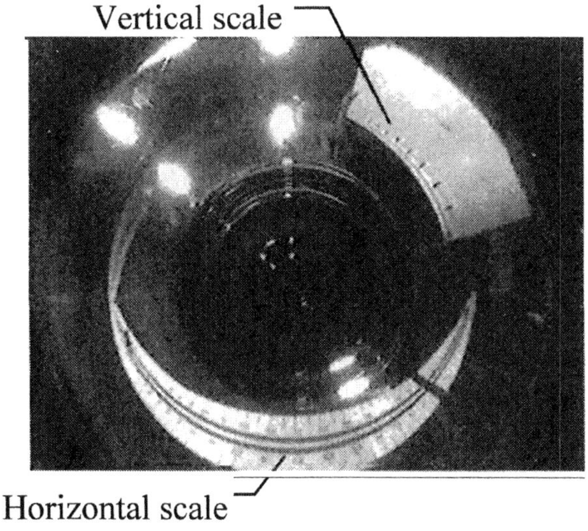

Fig. 5. Experimental set-up "from the viewpoint" of the PALLAS method.

panoramic image correspond to the horizontal and vertical scales indicated in Fig. 4. These scales are destined for presenting the image-distorting peculiarity of PAL to the reader and are also used as references when re-transforming certain spatial points from the PAL image (e.g. the stack and the edge of the plateau, see Fig. 6 later).

First, the concentration distribution within the central plane of the plume has been determined using the conventional LS + DIP method (Fig. 7).

The conventional method is described in detail in Ref. [11]. The percental concentration values refer to the stack outlet concentration as 100%. The grey levels represent concentration ranges of 10% width (although not well visible due to the small size of Fig. 7, the concentration field has been classified into 10 subranges). The influence of the plateau on the transport process is apparent in the figure: the centreline of the plume is deflected upwards and the vertical concentration distributions that would have approximately a Gaussian profile in the case of a transport process above a flat surface (according to the Gaussian diffusion model) show slight asymmetry.

With extensive, systematic series of experiments comprising variable flow and geometrical conditions, these effects can be investigated in the future and the results can be transformed in descriptive relationships which can form the basis of air pollutant transport model improvement mentioned in Section 1.

Following the tests based on the conventional LS technique, a test record has been taken with use of the PALLAS method. Fig. 8 shows the PAL image of the plume. With use of a PAL image-retransforming and DIP software, this plume image has been transformed to the one shown in Fig. 7, which is considered as a grey-level

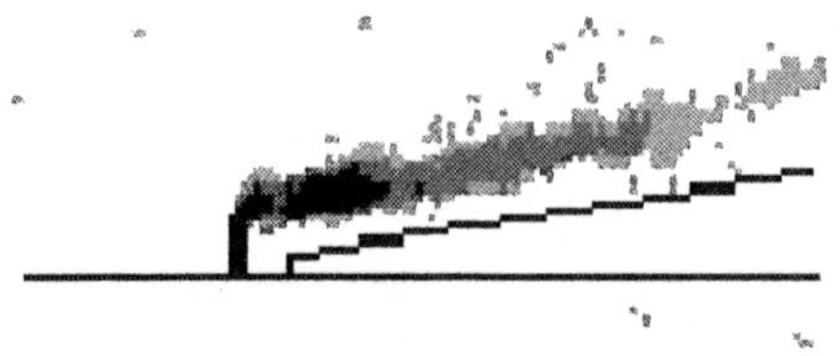

Fig. 6. Concentration distribution in a plume: PALLAS experiment.

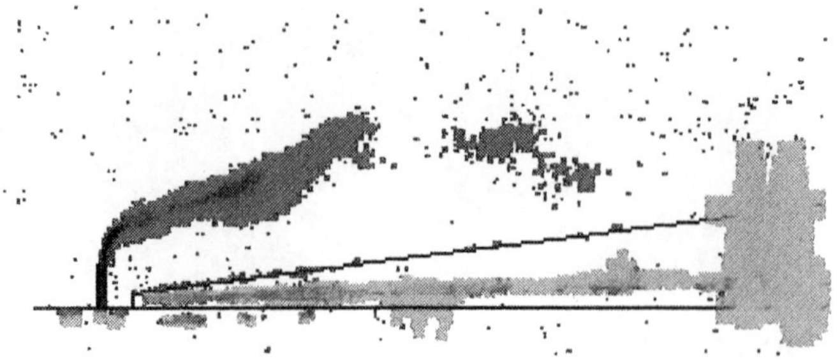

Fig. 7. Concentration distribution in a plume: conventional LS experiment.

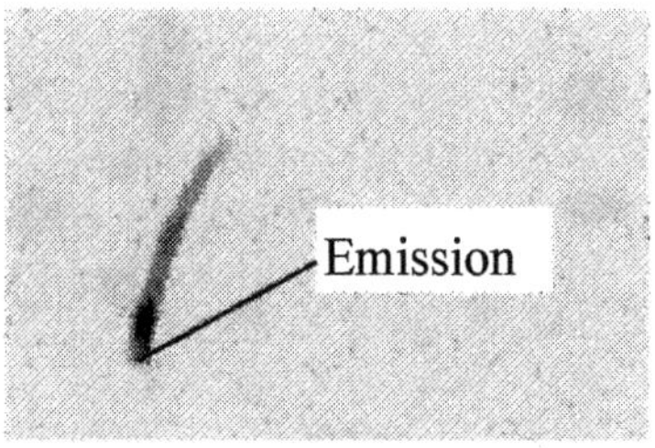

Fig. 8. PALLAS view of a plume.

distribution corresponding to a raw concentration map. The waving motion of the plume causes some difficulties when comparing Figs. 6 and 7.

The qualitative and, to a certain extent, also quantitative agreement between concentration maps in Figs. 6 and 7 encouraged the authors to state that with an appropriate improvement in the PALLAS hardware and the related DIP and grey-level correction software, the PALLAS method can be improved to a powerful tool of concentration field measurements [12]. A patent has been filed with regard to concepts of concerted application of PAL detection optics and a laser light probe in fluid mechanical measurements.

4. Measurement technical improvement

During evaluating the first PALLAS measurement results, needs for a considerable measurement technical improvement of PALLAS system have been pointed out. Such improvement concerns predominantly the PAL image-retransforming software, elaboration of a procedure for correction of detected grey levels, and increase of resolution of detected image.

The refinement in resolution of the image has been solved by obtaining a higher resolution CCD camera. The advanced method of grey-level correction will be reported in a future paper. In the following section, recent improvement in image retransformation is outlined.

In conventional "see-through-a-window" optical systems, e.g. in telescopes or camera optics, the life-like image of objects is developed with more or less distortion by the optical system in itself. In the case of centric minded imaging systems such as PAL detection optics [13], an additional, usually computer-based image retransforming procedure must be applied to the detected (annular, panoramic) image in order to obtain a true object view. As Fig. 9 illustrates, in general applications the PAL detection unit usually receives information from objects located arbitrarily in the extensive spatial zone determined by the full 360° polar object angle and the azimuthal angular range characterising the PAL unit. On the contrary, in PALLAS techniques the possible location of objects (i.e. light scattering particles or droplets) is restricted to the plane determined by the light probe positioned close to the PAL optical unit. According to such conditions, the scattered image has special distortion

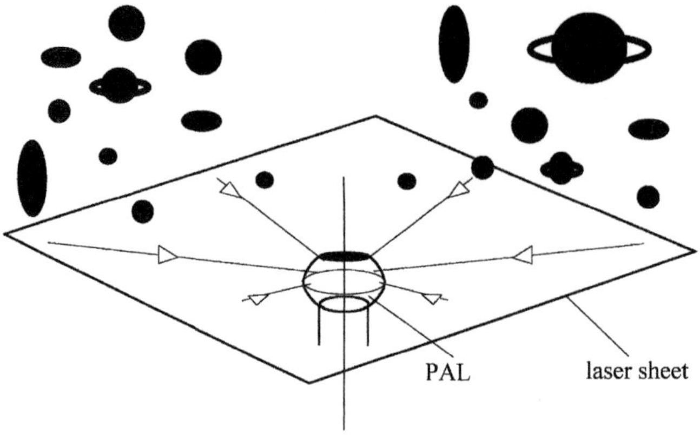

Fig. 9. Location of objects in different PAL applications: ▷ detected light scattered from objects passing the laser sheet in PALLAS applications; ● detected objects in general PAL applications.

characteristics, which are to be treated with use of a unique image retransformation method. The image of the plume in Fig. 7 has been derived by means of the original PAL image retransformation software used in PAL applications in general. The image of the plume seems to be rectified, indicating problems with use of original PAL image retransformation methodology in PALLAS experiments.

In order to surmount such problems, a special procedure has been established and is under improvement for appropriate retransformation of PAL images scattered from a plane such as a LS light probe. The retransformation procedure comprises semi-empiric and analytical models as well as interpolation techniques for treatment of PAL imaging peculiarities.

The appropriateness of the new retransformation technique is demonstrated in a case study. Fig. 10a shows a planar pattern of the plane which has been positioned normal to the PAL axis. Fig. 10b presents the annular image of the pattern formed by PAL. The image of the pattern obtained with the use of the new retransformation procedure is shown in Fig. 10c. (The central, circular dark part of the image corresponds to the blind zone of PAL.) The original pattern and the retransformed image in Fig. 10a and c show a fair agreement, which can be improved by further refinements in the retransformation method. A special graphic file format was used to store the PAL image and the retransformed image; therefore, the images were converted to an appropriate file format for word processor. The lines and letters in the pattern have been emphasised by hairline in Fig. 10b and c in order to get a better view of pattern.

5. Summary

The PALLAS method, a novel laser optical concentration measurement concept has been outlined in this paper. The use of this method has been illustrated for wind tunnel pollutant transport studies with presence of solid boundaries (features of

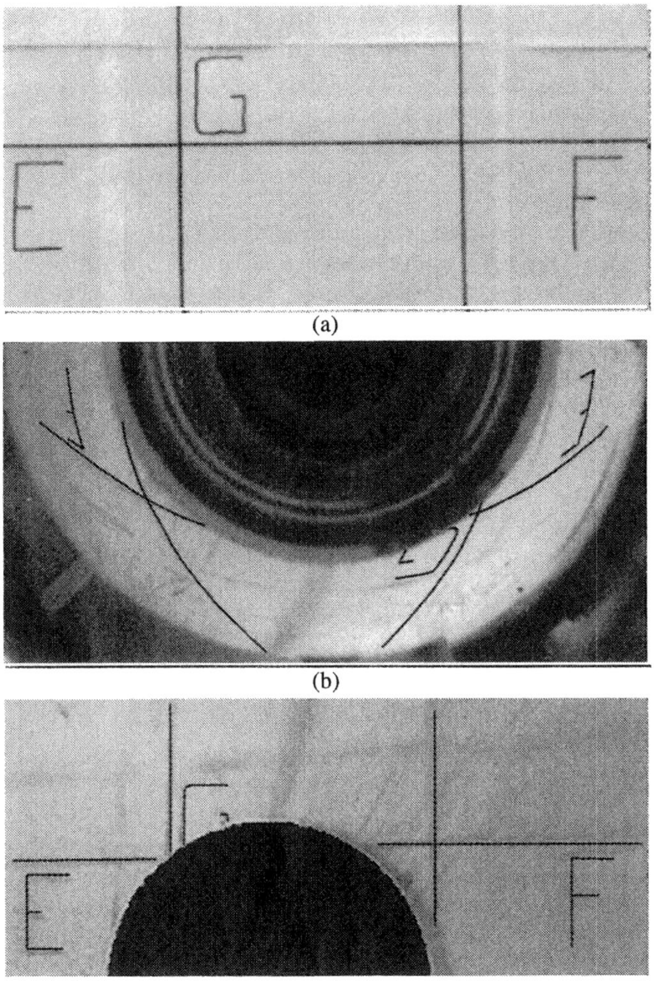

Fig. 10. A case study for PALLAS image retransformation: (a) original pattern, (b) annular PAL image of pattern, (c) retransformed image of pattern.

terrain, human objects). The suitability of PALLAS concept has been justified through a comparison between concentration measurements carried out using a conventional laser sheet technique and a PALLAS set-up. Demands for measurement technical improvement of PALLAS equipment have been pointed out and recent improvement results have been presented.

Acknowledgements

Authors acknowledge the contribution of OTKA (National Foundation for Scientific Research) supporting this research in the framework of OTKA T 030116 and M 27242 Project.

References

[1] D. Olivari, V. Babuska, Use of video camera recording and digital image processing for the analysis of pollutant dispersion in the near wake of a cube, J. Wind Eng. Ind. Aerodyn. 34, 291–301.

[2] B. Leitl, Entwicklung und Erprobung eines für den Einsatz in großen Windkanälen geeigneten laseroptischen Konzentrationsfeldmeßverfahrens, Dissertation der Fakultät Maschinenwesen der Technischen Universität Dresden, 1994.

[3] J. Brechling, H. Ihlenfeld, The modelling of the atmospheric boundary layer in a wind tunnel with an open test section and its application for investigations of the flow around buildings, Proceedings of the flow and dispersion through groups of obstacles, University of Cambridge, March 1994, The Institute of Mathematics and Its Applications, pp. 28–30.

[4] P. Greguss, Panoramic imaging in biology and medicine. Proceedings of the International Optical Computing Conference'83, IEEE Computer Society Conference Proceedings, 1983, pp. 67–72.

[5] Hungarian Patent 192125.

[6] USA Patent 4566763.

[7] Japanese Patent 1962784.

[8] C.F. Bohren, D. Huffmann, Absorption and scattering of light by small particles, Wiley, New York, 1983.

[9] D.R. Matthys, J.A. Gilbert, P. Greguss, Endoscoping measurement using radial metrology with digital correlation, Opt. Eng. 30 (1991) 1455–1460.

[10] T. Lajos, R. Szombati, Zs. Szepesi, Wind tunnel simulation of pollutant transport in the atmospheric boundary layer, Heat Eng. Environ. Protect. 22–25 (1995) 191–197.

[11] I. Goricsán, Zs. Szepesi, J. Vad, Development of a laser sheet technique for concentration distribution measurements, GEPESZET'98, Proceedings of the First Conference on Mechanical Engineering, Vol. 2 (1998) 820–825.

[12] I. Goricsán, J. Vad, T. Lajos, P. Greguss, Application of PALLAS method in air pollutant concentration measurements, in: A. Larsen, G.L. Larose, F.M. Livesey (Eds.), Wind Engineering into the 21st Century, Vol. 2, Balkema Publishers, Rotterdam, 1999, pp. 759–764 (Proceedings of the 10th International Conference on Wind Engineering, Copenhagen, Denmark 1999).

[13] M. Veres, P. Greguss, New method in space research: centric minded imaging. GÉPÉSZET'98, Proceedings of First Conference on Mechanical Engineering, Vol. 1, pp. 286–290.

Structural Engineering Compendium I

Probability distribution of dispersion from a model plume in turbulent wind

J.C.K. Cheung*, W.H. Melbourne

Department of Mechanical Engineering, Monash University, Clayton, Victoria 3168, Australia

Accepted 22 July 2000

Abstract

Wind tunnel experiments were performed to measure mean, standard deviation and peak mass concentrations of a discharge at vertical and horizontal traverses downstream of a model stack. Time-series data of concentration fluctuations measured from 10 fast response flame-ionization-detectors were collected for statistical analysis. The measurements were made for a single buoyant and non-buoyant model plume source discharged from a stack with two exit velocities in neutral, suburban terrain roughness turbulent wind flows. With the inclusion of a building placed between the stack and the measurement locations, the effects of the interaction with the building wake were also examined. Cumulative probability distributions presented in Weibull plots have shown the relatively higher intermittency characteristics of dispersion at the extremities of a plume close to the stack source. At distances further away from the stack, the intermittency of the plume has been shown to become lower. © 2000 Elsevier Science Ltd. All rights reserved.

Keywords: Model plume; Dispersion; Intermittency; Probability distribution

1. Introduction

There has been a considerable advance in recent years in the experimental technique for modelling the plume dispersion process in a wind tunnel. A lot of progress has also been made in understanding the fundamental side of the dispersion in wind. While the development of the theoretical work often provides guidance for the experimental studies, detailed wind tunnel measurements are also required to

*Corresponding author.
E-mail address: john.cheung@eng.monash.edu.au (J.C.K. Cheung).

Reprinted from *Journal of Wind Engineering and Industrial Aerodynamics*
87 (2-3), 271-285 (2000)

determine the necessary parameters in the analytical as well as computational predictions of the concentration levels downwind. A collaborative research project has commenced recently in the 1 MW environmental wind tunnel at Monash University to measure the stochastic properties of dispersion for a single plume in neutral, convective and stable stratified boundary layer wind flows. The effects of interaction of a building wake with its coherent structures are also examined. This paper mainly presents the probability distributions data for the plume in the neutral flow conditions only.

Previous studies of concentration fluctuations of a plume have shown that the dispersion process is highly intermittent with some occasional high peaks occurred over periods of zero concentration levels. The highest peak concentrations are particularly of interest as it is often necessary to estimate whether concentrations are likely to exceed certain limits, even though the time-averaged mean concentrations may be very low. In addition, it is equally important to quantify the plume behaviour with the probability of occurrence of these peak concentrations. One of the early studies [1] presented wind tunnel results for the variance, intermittency, peak values, probability density functions (p.d.f.) and spectra of concentration fluctuations for

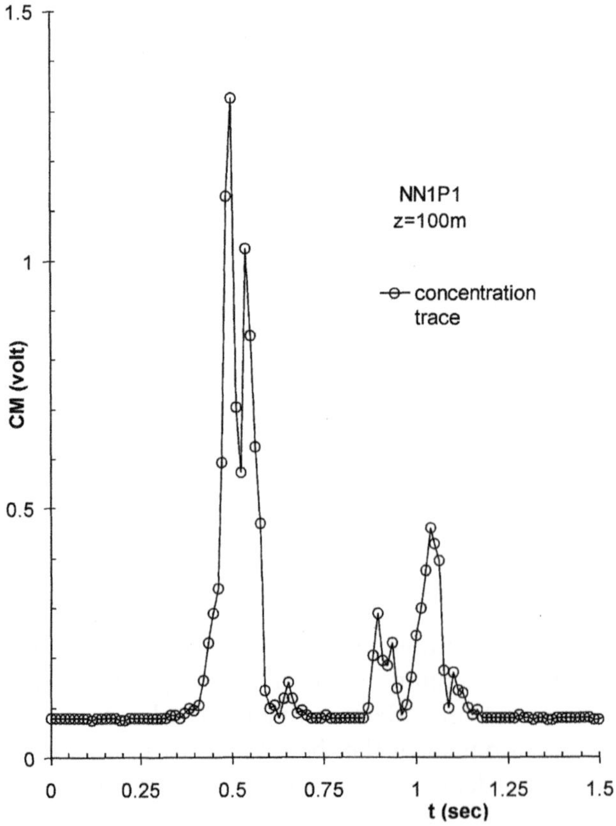

Fig. 1. An example of the concentration trace showing the digitized sample points as a function of time (non-buoyant plume exit velocity = 10 m/s at P1: one stack height high and one stack height downstream).

two passive plumes. The p.d.f. measured at an elevated level was shown to change from an exponential form towards a more Gaussian one near the ground. The review [2] on several concentration fluctuation models noted other forms of probability distributions with field and laboratory data sets. The log-normal distribution was believed to be more appropriate for urban or multiple point sources and the exponential distribution to be more appropriate for isolated point sources. If the data were dominated by zeros (i.e., the plume was not present at the measurement location most of the time), the log-normal distribution would be relevant only for the non-zero part of the plume concentration time series. From full-scale field data [3], a simple exponential distribution was found appropriate for intermittent concentrations measured far from the plume centreline. Near the plume centreline, the distribution was found closer to the Gaussian form. With conditional statistics having ignored all zero concentration data [4], the results were best represented by a clipped-normal frequency distribution or by a slightly over-estimated exponential distribution. A log-normal distribution was shown to overestimate the frequency of high concentrations. Further field data [5] confirmed the clipped normal distribution in better agreement with concentration measurements than the exponential

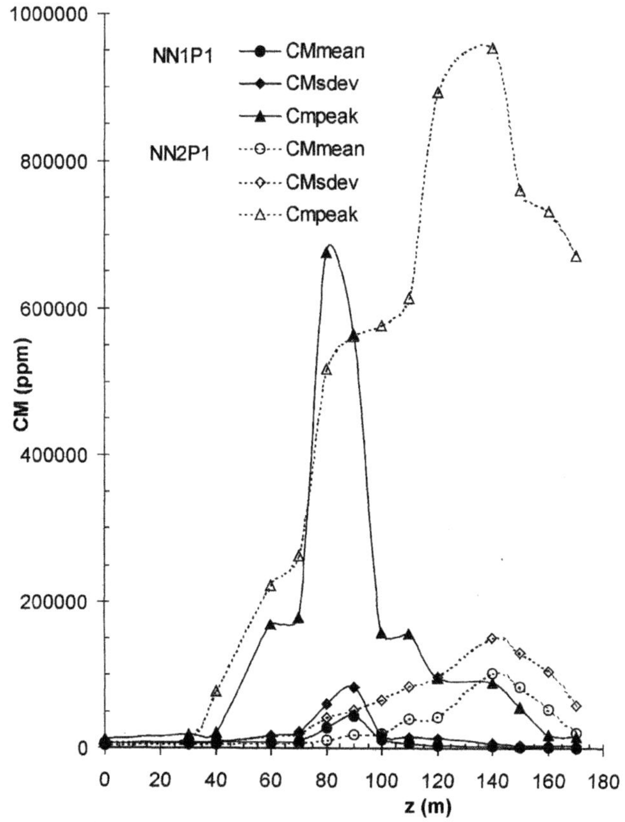

Fig. 2. Mean, standard deviation and peak concentration by mass as a function of height from ground for non-buoyant plume with exit velocities of 10 and 20 m/s.

distribution function. However, more recent field data [6], with some longer sampling periods, suggested different observations. The exponential distribution fits better close to the source for plumes dominated by meandering, while the clipped normal applies after the instantaneous plume has grown within the mean plume by mixing processes.

In summary, the p.d.f. for a dispersion process may take any of the forms, exponential, Gaussian, log-normal or clipped normal, depending on the range in distance and height or on different sites. With the mean, variance and the intermittency being fraction of non-zero readings in a concentration time series, theoretical derivations of these p.d.f. forms are given [4]. As intermittency tends to zero, the clipped normal distribution will asymptotically approach to the exponential

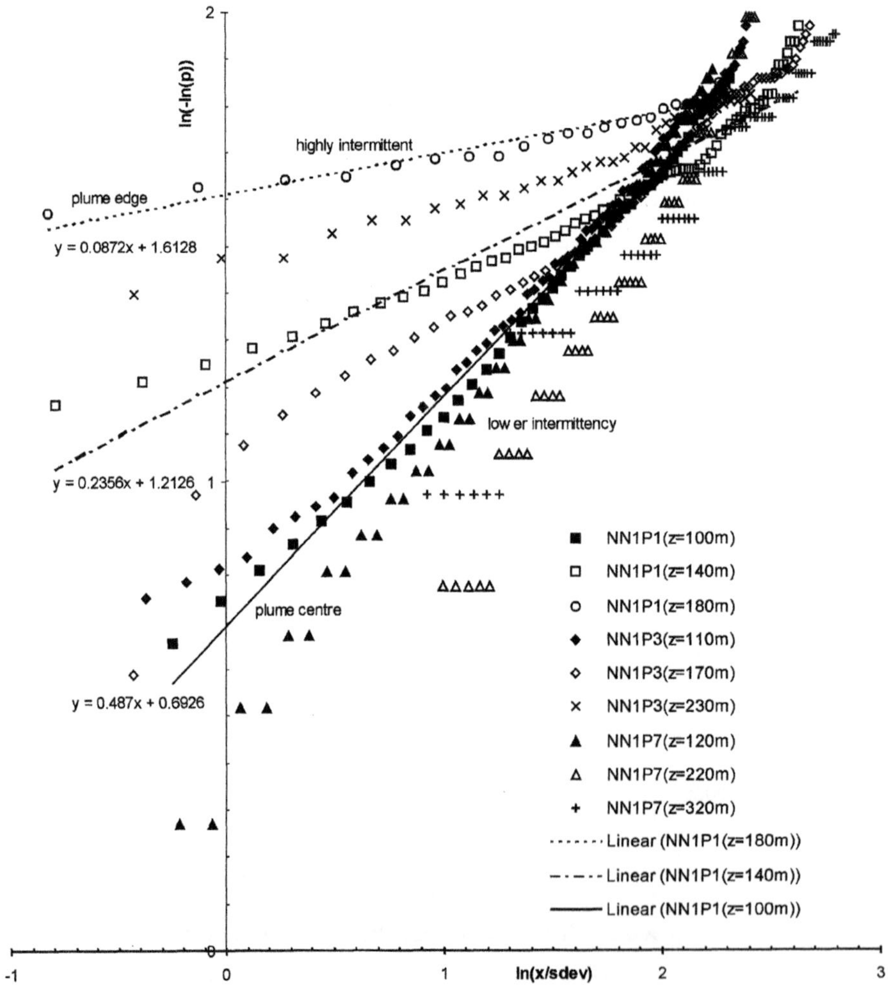

Fig. 3. A Weibull plot of the cumulative probability density function against the number of standard deviation of exceedance of concentration measured at near field and far field downstream vertically across the non-buoyant plume from the centre to the plume edge extremities (exit velocity = 10 m/s).

distribution. The log-normal distribution is the model for a variable whose logarithm follows a normal (Gaussian) distribution. In addition, the exponential and Gaussian distributions can be expressed in the Weibull format. This format is given by a log–log plot of the negative logarithm of the cumulative probability against the number of standard deviation of exceedance. To identify the relative effects of intermittency resulting from the plume extremities and building wake interference, the p.d.f. for all data in this paper are plotted in this Weibull form. Time-averaged data of dispersion with the interference of building wake can be found in previous studies [7–9].

2. Experimental set-up

The experiments were conducted in the 1 MW wind tunnel at the Department of Mechanical Engineering, Monash University. The detailed features and operational characteristics of the upper level 4 m high by 12 m wide by 40 m long working section are described [10]. A 1/200-scale natural wind boundary layer modelling was

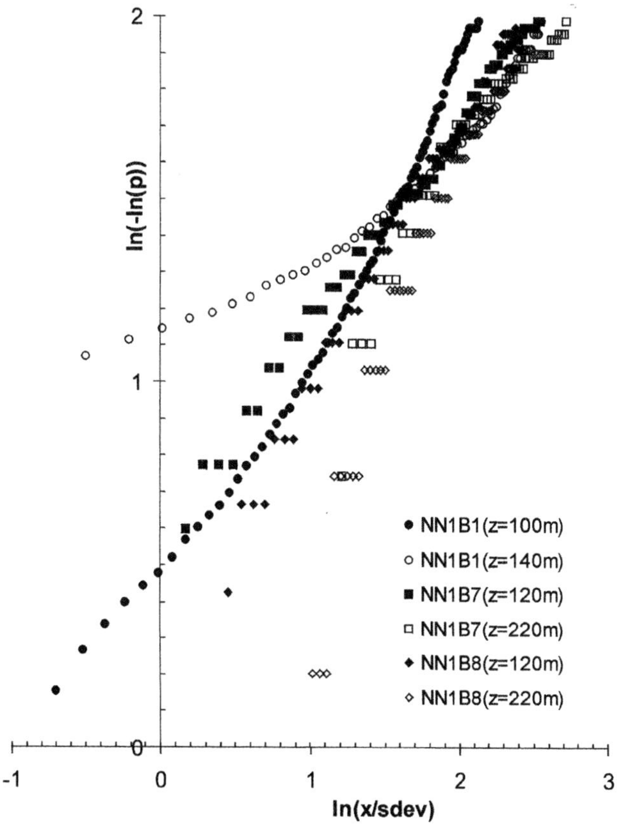

Fig. 4. Cumulative probability density function of near- and far-field concentrations measured at plume centre and edge with building wake interference (non-buoyant plume exit velocity = 10 m/s).

generated using floor roughness from the heating elements installed for modelling the convective flows. Vertical profiles for mean velocity and turbulence intensity were obtained with a cross hot-wire probe. Calibrations were performed in the free-stream with reference to measurements with a Pitot-static tube. The turbulent wind characteristics of a neutral flow conditions were simulated. The free-stream mean velocity as a function of height followed a power law with an exponent of 0.18 together with the longitudinal turbulence intensity of 0.19 at a height of 100 m. Also, the corresponding vertical component of turbulence intensity was 0.21. The flow uniformity was within 20% across 600 m from the source and up to 1600 m downstream for a velocity range of 4–40 m/s full scale.

A single stack, 5 m in diameter and 100 m high, was modelled to a geometric scale of 1/200. The full-scale stack exit velocities were 10 and 20 m/s. The stack exit temperatures of 0 and 500°C above ambient were modelled for the non-buoyant and buoyant plumes respectively. Strict Froude scaling was used such that the velocity scale was given by the square root of the geometric scale. The discharge gas from the stack in the measurements was made up of a mixture of helium, air and ethylene to

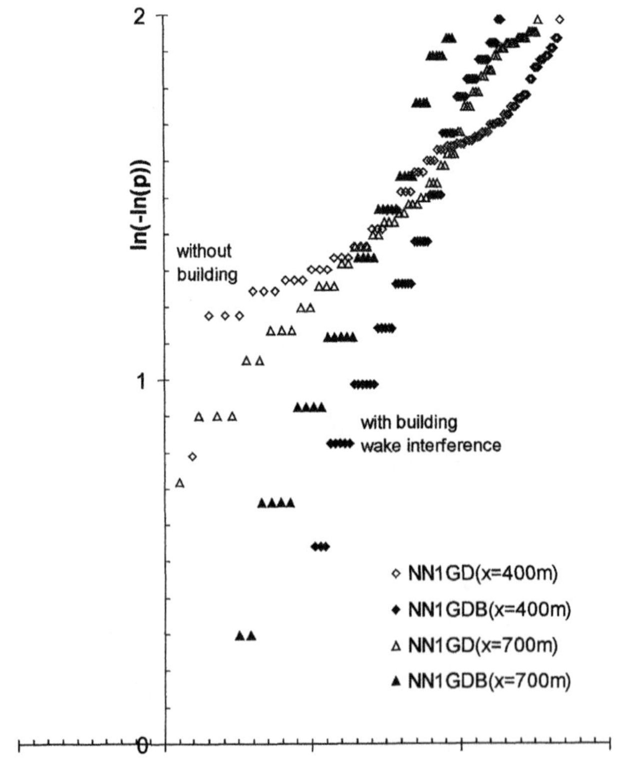

Fig. 5. Cumulative probability density function of near- and far-field ground level concentrations with and without building wake interference (non-buoyant plume exit velocity = 10 m/s).

model the buoyancy and momentum of the stack exhaust. The helium, air and ethylene flow rates were metered to a plenum from where the gas mixture was discharged through the stack. Ethylene was used as a tracer in the plume. The concentrations of ethylene, and hence the discharge gas from the stack, were measured by a set of flame ionization detectors (FID) located downstream. For testing with building interference effects, a building block 25 m by 100 m wide by 50 m high was used. Mean, standard deviation and maximum (peak) concentrations were measured for traverses on ground level behind the stack. Also, vertical and horizontal traverses were measured from the plume centre at various distances downstream. Time series for the concentrations as well as wind velocities from a hot wire for these configurations were also recorded. It is the objective of the present paper to examine the probability distributions of these time series data for dispersion.

The frequency response of the flame ionization detectors to a step-input in concentration [10] is 200 Hz. Eight channels of analogue voltage output were recorded at a frequency of 78 Hz each for periods over 2 min. The voltage signals for each channel of concentrations sampled were low pass filtered at 33 Hz. In all experiments, the gas mixture was discharged through the stack during the sampling period at a constant rate to ensure stationary measurements. Mass concentration

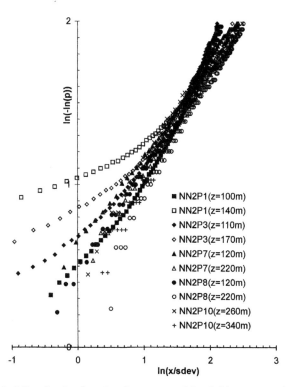

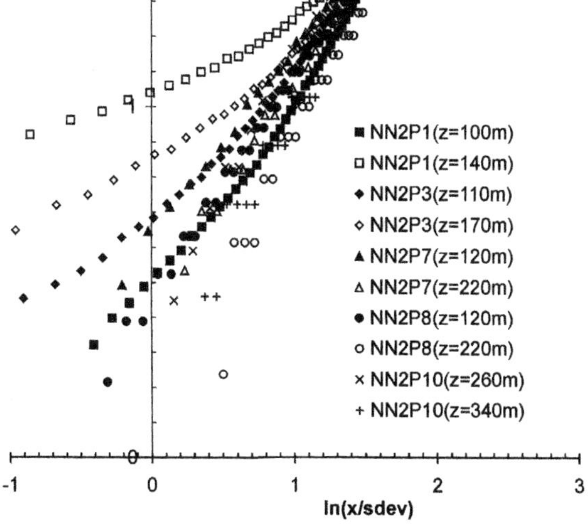

Fig. 6. Cumulative probability density function for near- and far-field concentrations measured at plume centre and edge without building wake interference (non-buoyant plume exit velocity = 20 m/s).

measurements have shown maximums occurring at wind speeds of about 15–20 m/s. All time series were recorded at about 17.5 m/s for 10 240 points in the spreadsheet format for analysis. The baseline or minimum value of each time series was subtracted from that channel. The histograms and the cumulative probabilities for each time series were calculated for different standard deviation levels of exceedance and then plotted in the Weibull format.

The first four characters of the name of a test run designate each concentration time series as follows:

1st character:	N = Neutral flow
2nd character:	N = Non-buoyant plume or
	B = Buoyant plume
3rd character:	1 = stack exit velocity 10 m/s or
	2 = stack exit velocity 20 m/s
4th character:	Measurement Position P
	with building B or on ground level G

This designation is then followed by some other descriptions such as distance of measurements downstream. For instance, NN1P1 denotes the measurement being

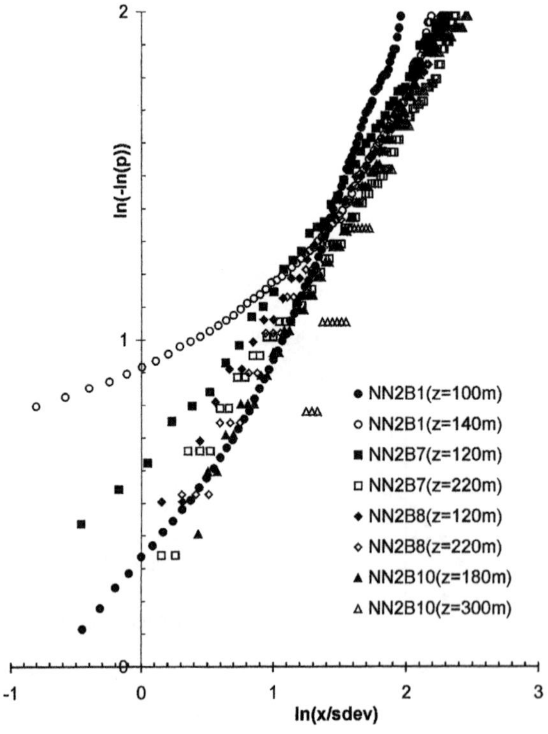

Fig. 7. Cumulative probability density function for near- and far-field concentrations measured at plume centre and edge with building wake interference (non-buoyant plume exit velocity = 20 m/s).

taken at a position of one stack height downstream from the stack. The plume is non-buoyant with an exit velocity of 10 m/s in a neutral flow condition.

3. Experimental results

3.1. Non-buoyant plumes

A typical section of the time series of the data trace (designated NNIPI) measured at 100 m above ground and one stack height downstream is shown in Fig. 1. The plume is non-buoyant with an exit velocity of 10 m/s in neutral wind conditions. The concentration appears to be intermittent with high-concentration peaks separated by intervals of zero concentration. The slope of the cumulative probability distribution plot in the Weibull format can represent the dispersion intermittency, which is defined as the fraction of time of non-zero concentration [5].

Mean, standard deviation and peak concentrations in parts per million (ppm) by mass of the discharge are evaluated for the various measurement configurations. An example plot of the concentrations from vertical traverse measurements at a

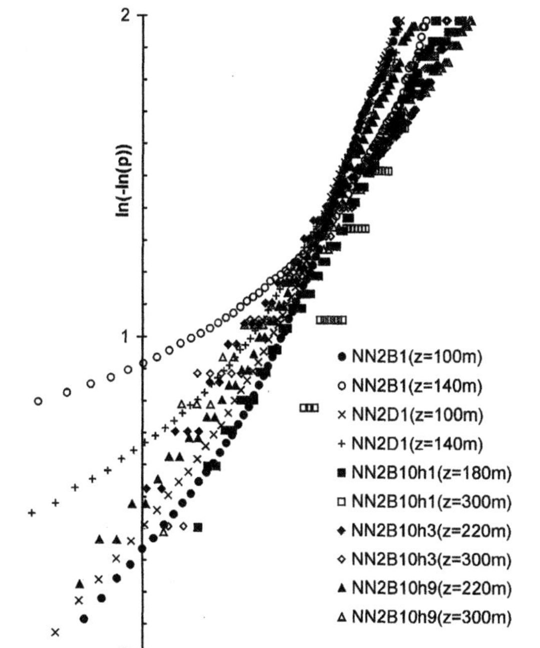

Fig. 8. Cumulative probability density function for near- and far-field concentrations measured at plume centre and edge with building wake interference from taller and closer buildings (non-buoyant plume exit velocity = 20 m/s).

free-stream mean wind speed of 17.5 m/s is given in Fig. 2. For a stack exit velocity of 10 m/s (run NNIPI), maximum mass concentration occurs at about 80–90 m above ground. For a higher efflux velocity of 20 m/s (run NN2PI), the maximum concentration is seen to be higher and to occur at a height of about 120–140 m above ground.

The concentration p.d.f. was calculated for each signal recorded. Fig. 3 represents the cumulative p.d.f of concentration measured at different height z at various distances of one, three and seven stack heights downstream. The plume is non-buoyant with an exit velocity of 10 m/s in neutral flow condition. From the vertical traverse one stack height 100 m downstream (run P1), the FID at height $z = 100$ m is close to the plume path centre. As z increases, the detector is moving away from the plume path. As can be seen in Fig. 3, the slope of the probability distribution of concentration near the plume centre is much steeper than that near the plume extremities at $z = 180$ m. This slope is seen to decrease from 0.487 to 0.087 as the measurement location moves away from the plume, indicating an increase in intermittency of the plume. As the downstream distance of the FID measurement location from the stack increases (run P3), the probability distribution of concentration near the plume centre remains approximately the same. However, the slope of the probability distribution of concentration near the plume edge

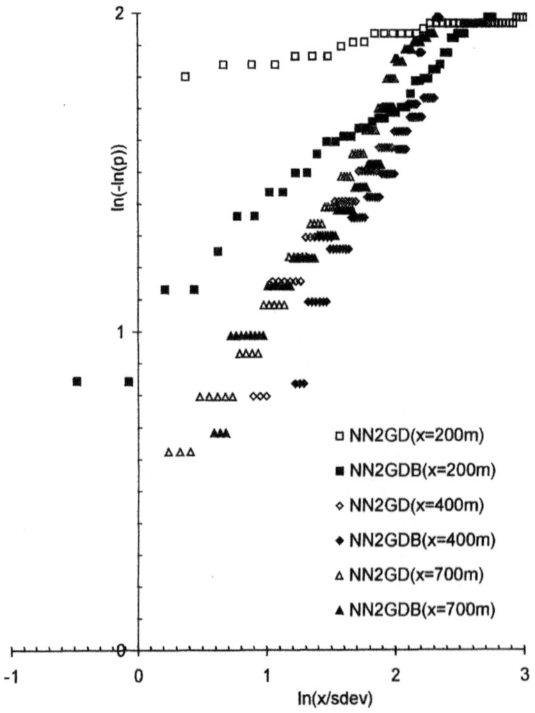

Fig. 9. Cumulative probability density function for near- and far-field ground level concentrations with and without building wake interference (non-buoyant plume exit velocity = 20 m/s).

becomes steeper. This agrees with previous observations [11] that when the plume size is smaller near the stack, the plume meanders back and forth more due to the advection of atmospheric eddies, resulting in higher intermittency. When the plume is farther away from the stack and becomes larger than the eddies, they do not meander the whole plume as much but merely cause minor fluctuations deep within the plume, resulting in lower intermittency. The steeper slope of the cumulative probability distribution plot as shown in Fig. 3 can describe this lower intermittency. For far-field measurements at a distance of seven-stack height downstream (run P7), the slopes for both the plume centre and plume edge configurations are all relatively steep, showing relatively lower intermittency across the plume.

The effects of building wake interference are shown in Figs. 4 and 5. The rectangular block building 25 m longitudinally, 100 m laterally and 50 m high was placed directly behind the stack. The p.d.f. with building interference have similar characteristics for near and far field, and plume centre and edge configurations as without the building interference. Nevertheless, with building coherent structures, the overall behaviour is less intermittent as shown by steeper slopes generally. This effect of coherent structures from the wake of the interference building is more

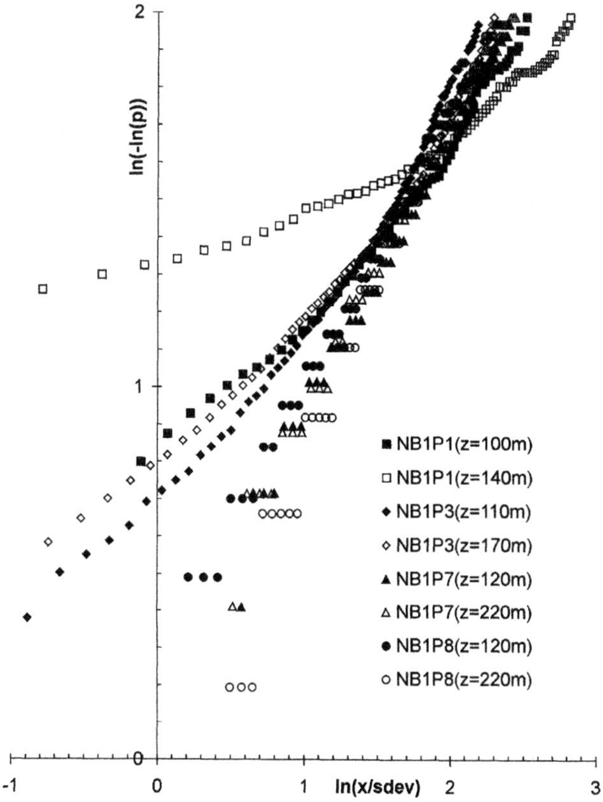

Fig. 10. Cumulative probability density function for near- and far-field concentrations measured at plume centre and edge without building wake interference (buoyant plume exit velocity = 10 m/s).

apparent, as shown in Fig. 5, from measurements of ground level concentrations directly behind the building. As the measurement location is farther away from the building, this effect of reducing intermittency becomes less significant.

Figs. 6–9 show similar results for the non-buoyant plume with a higher exit velocity of 20 m/s. The plume behaviour close to the stack is more intermittent near the edge extremities than its centre while the intermittency for far-field concentrations is slightly higher at the plume centre than its edge. The effect of building wake interference is seen to reduce the intermittency of the dispersion generally. In Fig. 8, the p.d.f. of concentrations from far-field measurements (ten stack height downstream) are shown for different locations of the interference building at one (NN2B10h1), three (NN2B10h3) and nine (NN2B10h9) stack height distance downstream from the stack. The effect of the location of the interference building on far-field concentration is seen to be small. In addition, a higher interference building 25 m long by 50 m wide by 100 m high was located behind the stack for the near-field measurements designated by runs NN2D1. The intermittency of concentration is seen to decrease outside the wake region but slightly increase inside the building wake. As can be seen in Fig. 9, the p.d.f. of the ground level concentrations with the

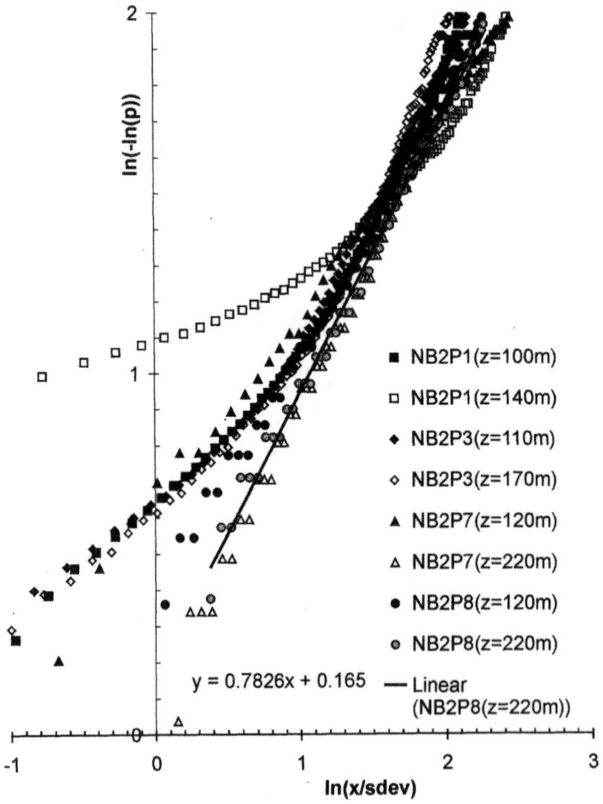

Fig. 11. Cumulative probability density function for near- and far-field concentrations measured at plume centre and edge without building wake interference (buoyant plume exit velocity = 20 m/s).

stack exit velocities of 10 and 20 m/s are similar and in general, the measured data show a small decrease in intermittency with a higher stack efflux velocity.

3.2. Buoyant plumes

The probabilistic behaviours of a buoyant plume in terms of near field, far field, in plume and plume edge concentrations are similar to those of a non-buoyant plume. Comparing Figs. 10 and 11 with Figs. 3 and 6, the effect of plume buoyancy is seen to lower the probability of the measured concentration and thus increase the intermittency of the plume slightly. As shown in Fig. 11, a straight-line fits quite well to the data NB2P8 ($z = 220$ m). Hence, the concentration data measured are shown to be represented by a Weibull distribution. This Weibull representation is generally valid for far-field location measurements. The cumulative p.d.f. for plume edge concentrations at near field (P1), medium field (P3) and far field (P8) locations are plotted against log–log and linear values of the number of standard deviation of exceedance in Figs. 12 and 13 to illustrate the appropriateness of data fit to log-normal and extreme value distribution, respectively. Fig. 12 shows that log-normal distribution is appropriate to the data only for the high standard deviation range.

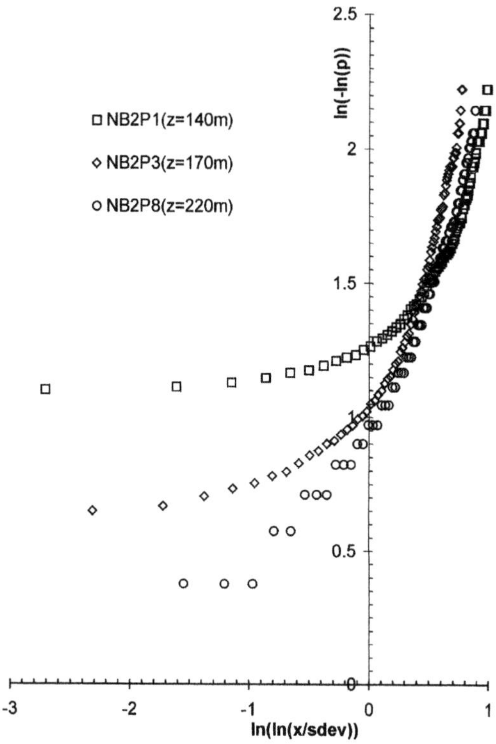

Fig. 12. Cumulative probability density function for near- and far-field concentrations measured at plume edge plotted against logarithm of logarithm of standard deviation (buoyant plume exit velocity = 20 m/s).

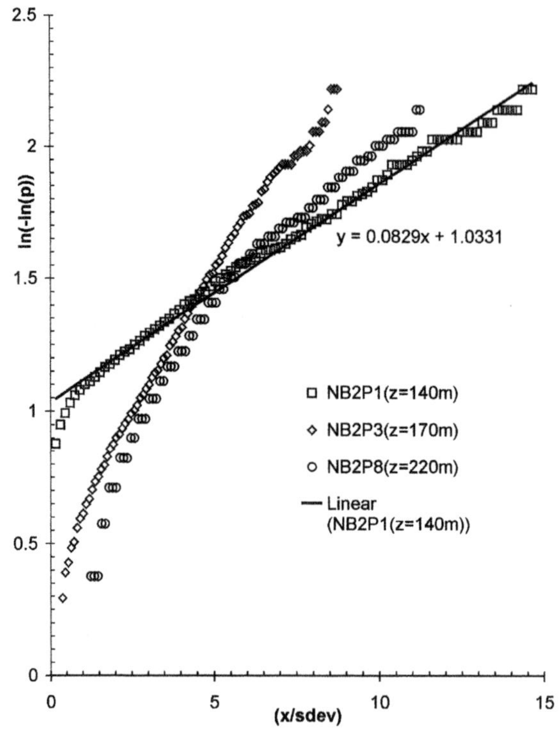

Fig. 13. Cumulative probability density function for near- and far-field concentrations measured at plume edge plotted against the number of standard deviation (buoyant plume exit velocity = 20 m/s).

This can be used in the analysis of conditionally sampled data for the dispersion process. Fig. 13 demonstrates that the highly intermittent data of the plume edge concentration measured at near field close to the stack (run NB2P1 (z = 140 m)) can be represented by an extreme value type of distribution.

4. Conclusion

Cumulative probability distributions of dispersion from a model plume have been analysed for cases with various efflux momentum and buoyancy in neutral boundary layer wind flow. The effect of building wake interference was also examined. The data were presented in Weibull format, from which the slope of the plot can be used to indicate the intermittency of the plume.

The intermittency of the plume was found to increase from the centre of the plume to the edge extremities near the stack, but decrease slightly for far field dispersion. The buoyancy of the plume would generally increase the intermittency while the effect of increasing stack exit velocity and building wake interference were found to reduce the intermittency of the concentration distribution.

References

[1] J.E. Fackrell, A.G. Robins, Concentration fluctuations and fluxes in plumes from point sources in a turbulent boundary layer, J. Fluid Mech. 117 (1982) 1–26.

[2] S.R. Hanna, Concentration fluctuations in a smoke plume, Atmos. Environ. 18 (6) (1984) 1091–1106.

[3] B.L. Sawford, C.C. Frost, T.C. Allan, Atmospheric boundary-layer measurements of concentration statistics from isolated and multiple sources, Boundary Layer Meteorol. 31 (1985) 249–268.

[4] B.L. Sawford, Conditional concentration statistics for surface plumes in the atmospheric boundary layer, Boundary Layer Meteorol. 38 (1987) 209–223.

[5] N. Dinar, H. Kaplan, M. Kleiman, Characterization of concentration fluctuations of a surface plume in a neutral boundary layer, Boundary Layer Meteorol. 45 (1988) 157–175.

[6] K.R. Mylne, P.J. Mason, Concentration fluctuation measurements in a dispersing plume at a range of up to 1000 m, Q. J. Roy. Meteorol. Soc. 117 (1991) 177–206.

[7] J.E. Fackrell, An examination of simple models for building influenced dispersion, Atmos. Environ. 18 (1984) 89–98.

[8] A.H. Huber, Wind tunnel and Gaussian plume modelling of building wake dispersion, Atmos. Environ. A 25 (7) (1991) 1237–1249.

[9] J.C.K. Cheung, W.H. Melbourne, Building downwash of plumes and plume interactions, J. Wind Engng. Ind. Aerodyn. 54/55 (1995) 543–548.

[10] T.J. Taylor, Wind tunnel modelling of atmospheric neutral and convective boundary layer flows and buoyant and non-buoyant plume dispersion, Ph.D. Thesis. Monash University, Clayton, Australia, 1997.

[11] S.R. Hanna, E.M. Insley, Time series analyses of concentration and wind fluctuations, Boundary Layer Meteorol. 47 (1989) 131–147.

Mean and fluctuating wind loads on rough and smooth parabolic domes

C.W. Letchford*, P.P. Sarkar

Wind Science and Engineering Research Center, Department of Civil Engineering, Texas Tech University, P.O. Box 4089-1023, Lubbock, TX 79409, USA

Accepted 5 June 2000

Abstract

Simultaneous pressure measurements have been obtained on rough and smooth parabolic domes in simulated atmospheric boundary layer flow. Mean and fluctuating pressure distributions compare favorably with earlier studies for similar shape and Reynolds number. The effect of surface roughness is to reduce suctions over the apex of the dome and increase suctions in the wake region on the leeward face. The consequence for mean and fluctuating overall loads is reduced uplift but increased drag for rougher surfaces. Correlation analysis of the fluctuating pressures reveals that the first two eigenvectors account for approximately 60% of the fluctuating pressure energy and follow the mean pressure coefficient and its gradient with respect to horizontal wind direction as predicted by the quasi-steady theory. Overall base shear and uplift forces on the domes can be well approximated by this theory.

1. Introduction

Many wind tunnel studies have been undertaken to determine wind loads on domes and hemispheres in boundary layer flow [1–8]. Maher's classical study was conducted on large models (600 mm diameter) but in more or less uniform flow with little turbulence and no attempt to simulate the earth's atmospheric boundary layer. Only the studies by Ogawa and Taylor present measurements of fluctuating pressures and only Taylor presents contour maps of maximum and minimum point pressures for hemispheres and truncated spheres. However, the correlation of fluctuating

*Corresponding author.
E-mail address: chris.letchford@coe.ttu.edu (C.W. Letchford).

Reprinted from *Journal of Wind Engineering and Industrial Aerodynamics*
88 (1), 101-117 (2000)

pressures was not obtained and thus effective overall loading distributions producing the worst eccentric or unsymmetrical loading are not available.

An additional complication with wind tunnel model studies of these types of structures is due to the curved surface, which leads to Reynolds number effects. Here, Reynolds number is defined as $\rho DU/\mu$, where ρ and μ are fluid properties, D is the base diameter and U is the mean velocity at top of model height. Maher examined these effects in the range 6×10^5–18×10^5, while Taylor dealt with this issue more systematically by evaluating loading under different turbulence intensities and Reynolds numbers ranging from 1×10^5 to 3×10^5. Ogawa [7,9] also investigated a range of turbulence intensities with Reynolds numbers ranging from 1.2×10^5 to 2.1×10^5. Taylor suggests that as long as the Reynolds numbers exceeds 2×10^5 and turbulence intensity exceeds 4% the surface pressure distributions are largely independent of Reynolds number. Maher suggests a critical Reynolds number of 1.4×10^6 but his tests were conducted in very low turbulence. Maher did study the influence of surface roughness, but in the absence of turbulence.

The present study sought to re-examine wind loading of domes and specifically to determine the effect of Reynolds number, surface roughness, and overall fluctuating load distributions to aid designers of these structures.

2. Experimental procedure

The wind tunnel tests were undertaken in the Colorado State University's Industrial Wind Tunnel. The wind tunnel is a closed circuit, 1.8 m wide with a ceiling adjustable to ~1.8 m. There is an upstream fetch of approximately 15 m for developing appropriate simulations of the earth's atmospheric boundary layers. A simulation of Exposure C at 1:300 was achieved by employing 4 spires, a 240 mm fence, and 18 mm chain at 200 mm spacing right up to the wind tunnel turntable. The mean velocity and turbulence intensity profiles are shown in Figs. 1 and 2, respectively, and compared with ASCE 7-98 [10] target values. The comparison with longitudinal velocity component spectrum at top of dome height is shown in Fig. 3, where the model-scale integral length scale of turbulence is ~700 mm. In all cases the simulation achieved compares very well with the Exposure C target. For the pressure measurements, the mean wind speed and turbulence intensity at the top of the dome was 18 m/s and 15%, respectively.

The nominal design gust wind velocity was 59 m/s at 10 m in flat open terrain. Using the approach of ASCE 7-98 [10], the design gust wind speed at the top of the simulated dome (45 m) is expected to be $K_z^{1/2} \times 59 = 1.17 \times 59 = 69$ m/s. The mean velocity at this height would be $0.82 \times 59 = 48$ m/s where the ratio of the 3 s gust to the 1 h mean velocity, $G = 1.17/0.82 = 1.43$. The model scale gust factor is expected to be very similar given the good agreement for mean and turbulence intensity profiles (Figs. 1 and 2). The maximum mean wind speed in the wind tunnel at 150 mm was measured to be ~18 m/s and thus the velocity scale was 18/48 ~ 1/3. Combining the length and velocity scales a time scale of $(1/3)/(1/300) = 1/100$ was obtained. Thus 36 s in the wind tunnel is equivalent to 1 h at full scale.

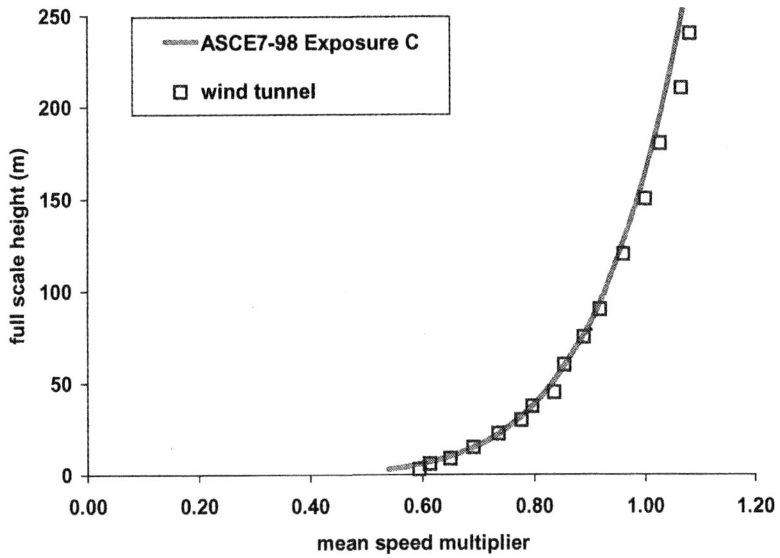

Fig. 1. Comparison of wind tunnel and ASCE 7-98 mean velocity profiles.

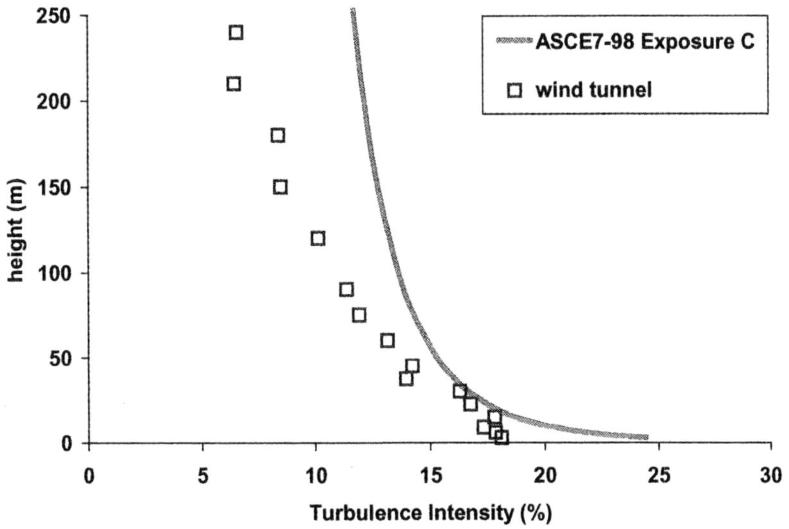

Fig. 2. Comparison of wind tunnel and ASCE 7-98 [10] turbulence intensity profiles.

A model dome was constructed with a base diameter (D) of 480 mm and height (h) of 150 mm. The nominal model scale was 1 : 300. The model was constructed from fiberglass to a parabolic profile that well approximates a sphere of radius 280 mm ($\pm$ 1%). The exterior surface was covered with 'gelcoat' to obtain a very smooth finish. To obtain a rougher surface the model was covered with a tailored fly screen mesh of 0.3 mm diameter fiber at 1.5 mm spacing and secured by tape around the base of the dome. This gave a nominal roughness coefficient of $\varepsilon/D = (0.3-1.5)/480 \sim 0.001$. Although wind tunnel blockage was not great ($\sim$1.5%) the wind

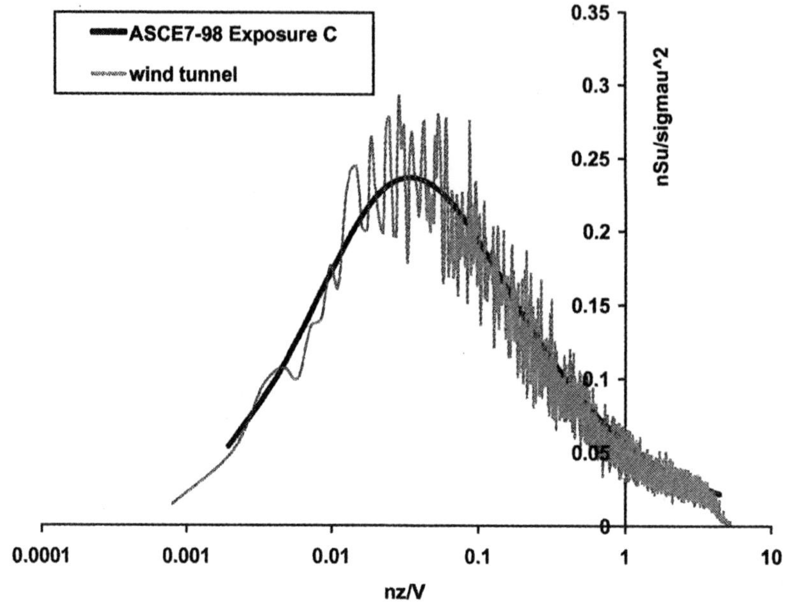

Fig. 3. Comparison of wind tunnel and ASCE 7-98 [10] turbulence spectra at top of dome height.

tunnel roof was positioned to minimize the static pressure gradient in the wind tunnel induced by the simulation and the model.

The dome was instrumented with 85 pressure taps, with one tap at the apex and seven taps non-uniformly distributed along 12 meridians at 30° increments, to achieve equal tributary area. Fig. 4 shows a general arrangement of the dome showing tapping grid and wind direction definition.

The taps were connected by tubing (240 mm long, 0.51 mm diameter), tubing to a multi-port PSI pressure transducer and sampled simultaneously at 480 Hz for a total of 96 s. The instrumentation frequency response was linear to 160 Hz, equivalent to 1.6 Hz at full scale. Mean and standard deviation (rms) pressures were obtained from the full 96-s record. To estimate mean extreme or peak pressures, each record was divided into 16 segments, equivalent to 10 min full scale, and a Fischer-Tippett type 1 extreme value distribution fitted. From this the 10-min mode and dispersion were obtained and the mean hourly extreme or peak pressure coefficient was estimated according to Eq. (1), after Cook [11]:

$$(\overline{\Delta\hat{p}} \text{ or } \overline{\Delta\check{p}}) = \text{mode} + (0.577 + \ln(6)) \times \text{dispersion}_{10\ \text{min}}. \tag{1}$$

All pressures were then non-dimensionalized by the mean dynamic pressure $(1/2\rho U^2)$ at the top of the dome. The definition of the coefficients is shown below with positive external pressure acting towards the surface and suction away. Δp is the instantaneous pressure difference between the surface pressure and a reference pressure in the wind tunnel.

$$\overline{Cp} = \frac{\overline{\Delta p}}{\frac{1}{2}\rho\bar{U}^2} \quad \text{mean pressure coefficient,} \tag{2}$$

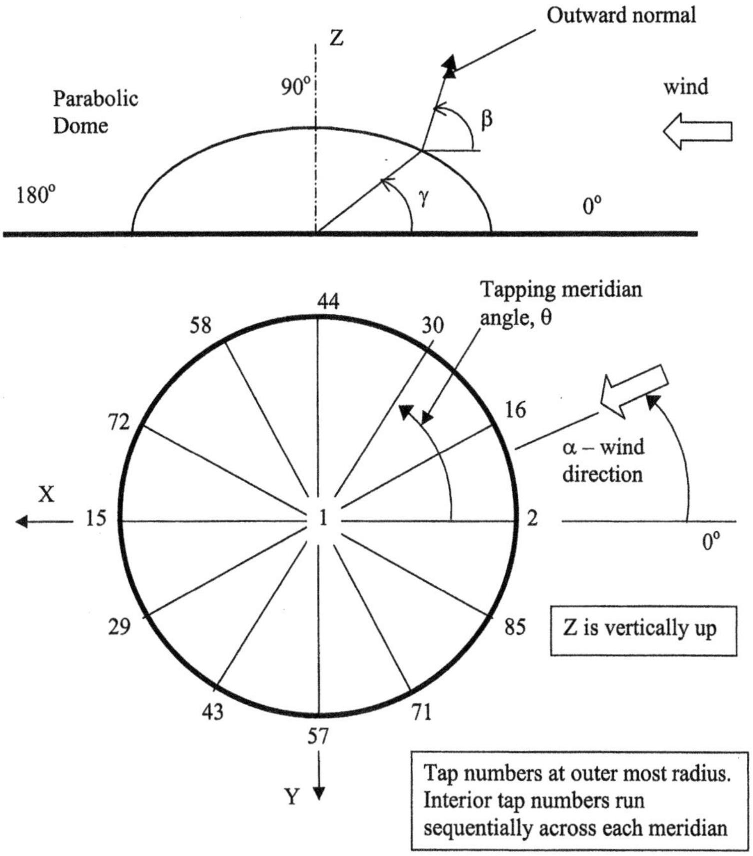

Fig. 4. Tapping arrangement and wind direction definition for single dome tests.

$$Cp_{\text{rms}} = \frac{\Delta p_{\text{rms}}}{\frac{1}{2}\rho \bar{U}^2} \quad \text{standard deviation pressure coefficient,} \tag{3}$$

$$\hat{C}p = \frac{\overline{\Delta \hat{p}}}{\frac{1}{2}\rho \bar{U}^2} \quad \text{mean peak maximum pressure coefficient,} \tag{4}$$

$$\check{C}p = \frac{\overline{\Delta \check{p}}}{\frac{1}{2}\rho \bar{U}^2} \quad \text{mean peak minimum pressure coefficient.} \tag{5}$$

Using the gust velocity for design purposes, it is often convenient to define pseudo-steady pressure coefficients which are the peak pressure coefficients defined above (Eqs. (4) and (5)) divided by the gust velocity/mean velocity ratio squared. In this way, it is possible to compare directly the mean pressure coefficient and the pseudo-steady coefficient with the latter being more realistic as it takes into account the fluctuating pressures over the structure:

$$G = \frac{\hat{U}_3}{\overline{U}_{1h}} \quad \text{gust velocity ratio.} \tag{6}$$

The corresponding coefficients are

$$\tilde{Cp}\hat{p} = \frac{\hat{C}p}{G^2} \quad \text{mean maximum pseudo-steady pressure coefficient,} \tag{7}$$

$$\tilde{Cp}\check{p} = \frac{\check{C}p}{G^2} \quad \text{mean minimum pseudo-steady pressure coefficient.} \tag{8}$$

If the quasi-steady theory of wind loading [12] which attributes the source of all pressure fluctuations to upwind velocity fluctuations, holds true, then the mean and pseudo-steady mean maxima (or mean minima depending on the sign of the mean) should be equal.

Overall fluctuating forces (or other load effects) on the dome are obtained by appropriate weighting of the pressures and integrating over the entire surface. This effect is shown in Eq. (9).

$$C_{F\phi} = \sum_{j=1}^{85} Cp_j \phi_j (\delta A/A), \tag{9}$$

where ϕ_j is the weighting coefficient or influence line for the force or load effect ($C_{F\phi}$) required, δA is the tributary area of the tapping (here $\frac{1}{85}$ of the surface area of the dome), and A has been set as the projected plan area of the dome.

Thus given the influence lines for any particular load effect (e.g., overall drag, overall uplift, top of dome deflection, or bending moment), the fluctuating time history of that effect can be calculated and the statistics, mean, rms, mean maxima and mean minima estimated.

Here the overall force coefficients in the X (parallel to wind direction $\alpha = 0°$), Y (perpendicular to X in the horizontal plane), and Z (vertical or uplift), were calculated because the influence coefficients can be determined without recourse to structural analysis of the dome. Here $\phi_X = \cos\beta\cos\theta$, $\phi_Y = \cos\beta\sin\theta$, and $\phi_Z = \sin\beta$. β is defined as the angle between horizontal and a line perpendicular to the surface and θ is the tapping angle; both are defined in Fig. 4.

3. Experimental results

A series of tests were conducted varying wind speeds, surface finishes, and wind direction. They are summarized in Table 1. The five wind directions were used to obtain a detailed picture of the mean pressure field and to provide a check on the results.

Table 1
Summary of testing regimes

Wind speed (m/s)	Surface finish	Wind directions, α (deg)
9 and 18	Smooth	0, 7.5, 15, 30, 180
18	Rough	0, 7.5, 15, 30, 180

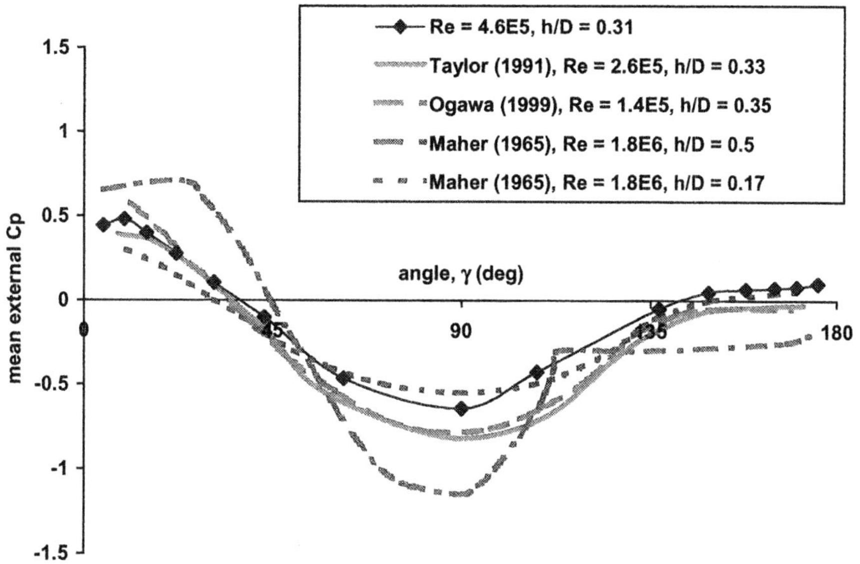

Fig. 5. Comparison of mean pressure coefficient along centerline of a smooth dome.

3.1. *Comparison with other studies*

The earlier studies of Taylor [8] represent the best work for comparison since similar geometries, in similar boundary layers were tested and mean and fluctuating pressures were reported. Ogawa [9] also provided data for comparison. Fig. 5 compares the mean pressure coefficient along the centerline of a smooth dome plotted against angle γ, defined in Fig. 4. The results for the domes tested by Maher [1], that bracket the current h/D ratio, are included. Generally, there is reasonable agreement for mean pressures even with interpolating (within h/D) Maher's tests, where no attempt was made to scale the atmospheric boundary layer. This is encouraging because Maher's tests were conducted at four times the Reynolds number of the present study and lend further support to Taylor's criteria for valid pressure measurements on domes.

Figs. 6 and 7 compare the fluctuating pressures measured along the centerline of the smooth dome with those of Taylor [8]. It is observed that there is excellent agreement for peak maximum and minimum pressure coefficient, with only a small discrepancy in distribution of rms pressure coefficient, which is in part due to the

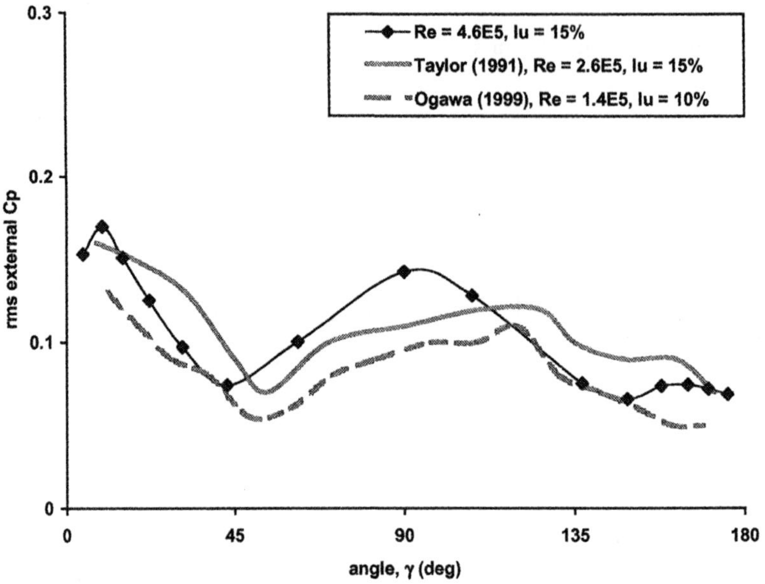

Fig. 6. Comparison of rms pressure coefficient along centerline of a smooth dome.

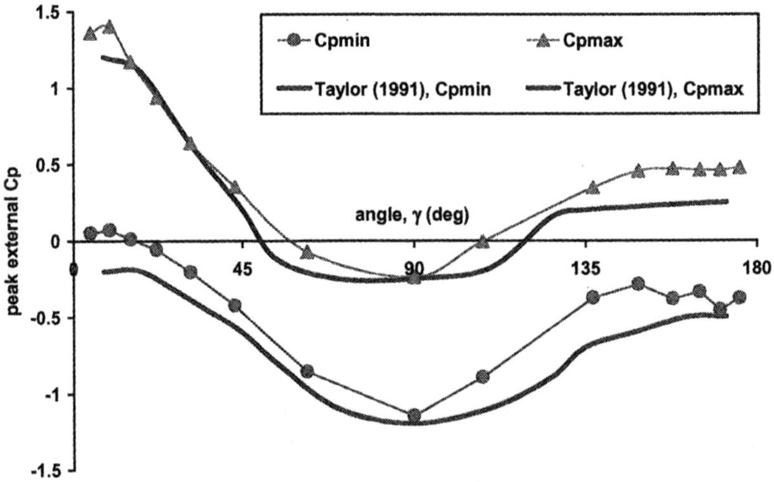

Fig. 7. Comparison of peak pressure coefficients along centerline of a smooth dome.

different dome shape. Ogawa et al. [7] described but did not present results for the truncated dome but has kindly provided them for this paper Ogawa [8]. His results are for a lower turbulence intensity at top of dome height (10%), which is reflected in the lower rms pressure fluctuations.

3.2. *Effects of surface finish on cladding pressures*

Maher [1] investigated the effect of surface finish on pressures over a hemisphere. As described above, his flow was basically uniform with only a shallow boundary layer on the wind tunnel floor. There was little turbulence in the flow. The flow Reynolds number was 1.8×10^6. Mean pressures were obtained on a smooth hemisphere and on one roughened by sand grains such that $\varepsilon/D = 0.001$, where ε is a measure of the roughness height and D the diameter of the hemisphere. His results indicate a reduction in maximum mean pressure coefficient near the top of the dome from 1.15 to 0.8, $\sim 30\%$, due to the addition of this roughness. This reduction is to be expected since surface roughness promotes a turbulent boundary layer over the dome surface, which will tend to separate earlier than that on the smooth surface. The consequences of earlier separation are reduced suctions at the point of separation but higher suction overall in the wake, leading to reduced uplift and increased drag. This result is born out in Maher's experiments.

In the present study, similar effects were observed. Figs. 8, 9 and 10 compare the mean, rms and peak pressure coefficients along the centerline of the smooth and rough single domes, respectively. Reduced suctions, both mean and fluctuating, over the apex of the dome are present but to a lesser extent than in Maher's study due to smaller h/D, here 0.31, cf. Maher's hemisphere with $h/D = 0.5$. In addition, the greater suctions in the wake lead to increased drag on the rougher dome. Overall forces on the dome are discussed in Section 3.5.

3.3. Effect of Reynolds number

The pressure distribution was measured on the smooth dome at a wind speed half that of the previous tests. This was the lowest wind speed that could be run and maintain reasonable sensitivity from the pressure measurement instrumentation. The Reynolds numbers of these tests was 2.3×10^5. The mean, rms and peak pressure coefficients along the centerline are compared with the earlier results at the higher Reynolds numbers in Figs. 11, 12 and 13, respectively. It is evident that these results

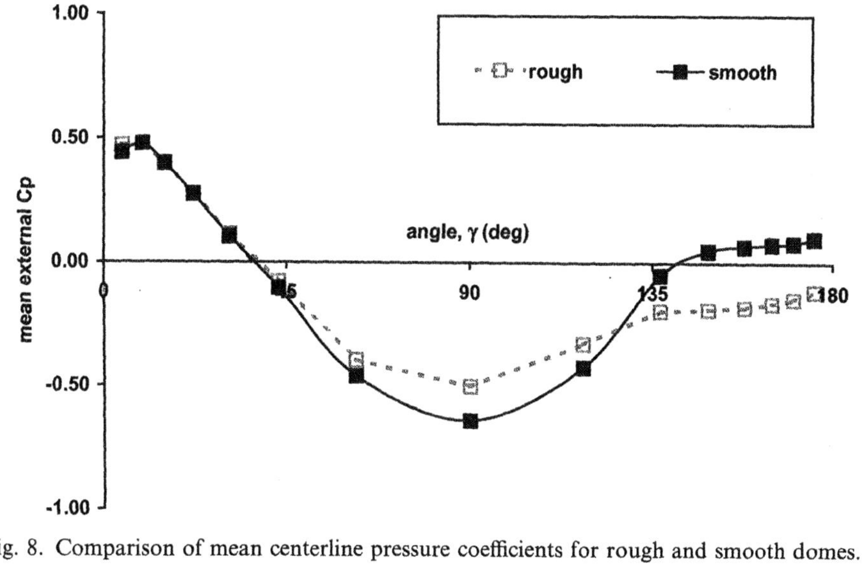

Fig. 8. Comparison of mean centerline pressure coefficients for rough and smooth domes.

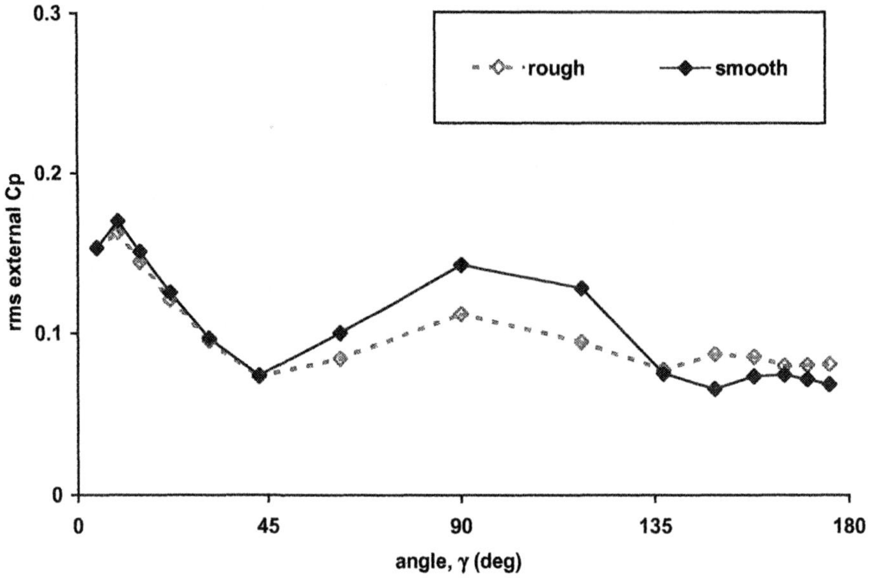

Fig. 9. Comparison of rms centerline pressure coefficients for rough and smooth domes.

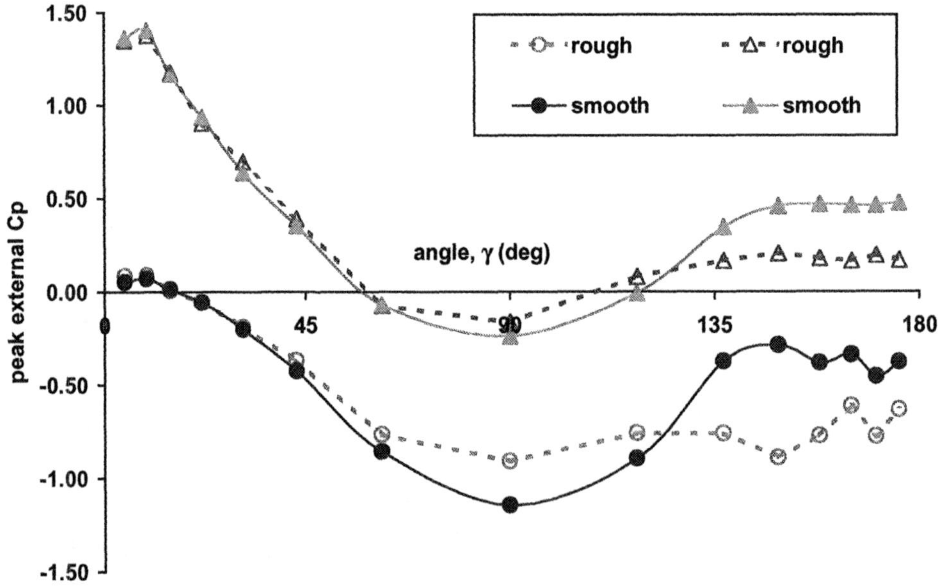

Fig. 10. Comparison of peak centerline pressure coefficients for rough and smooth domes.

indicate little sensitivity to Reynolds number over this small range for both mean and fluctuating pressures. On the graph of mean pressure coefficients, the result of Taylor [8] and Ogawa [9] are reproduced and the large difference between the present results and theirs may have been caused by static pressure gradients in their wind tunnels that remained uncorrected or the slightly different geometry of the dome.

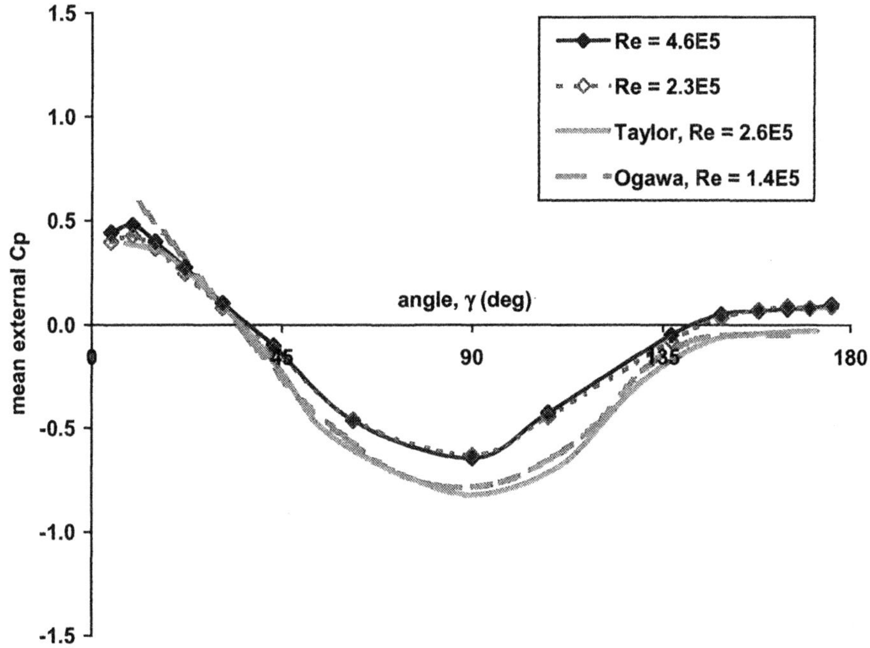

Fig. 11. Comparison of mean pressure coefficients along the centerline of a smooth dome for different Reynolds number.

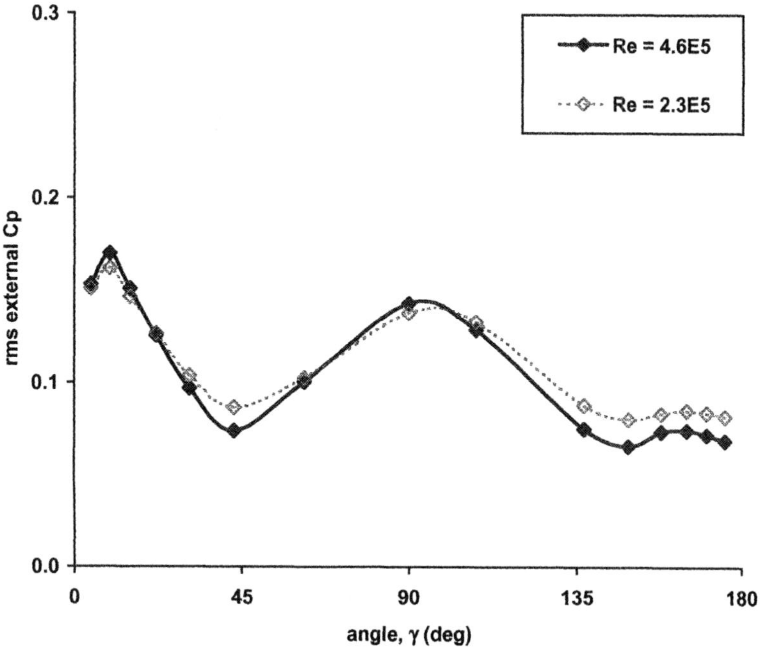

Fig. 12. Comparison of rms pressure coefficients along the centerline of a smooth dome for different Reynolds number.

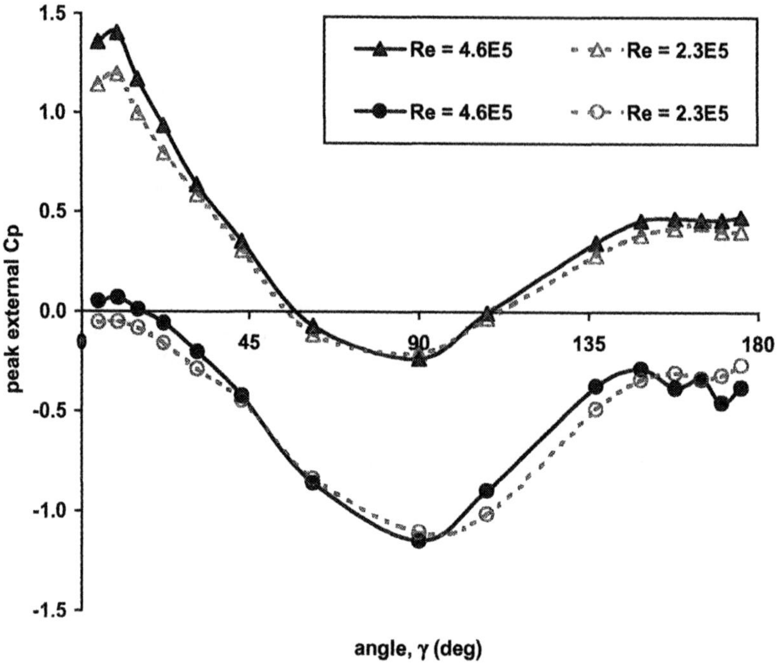

Fig. 13. Comparison of peak pressure coefficients along the centerline of a smooth dome for different Reynolds number.

Table 2
First 10 eigenvalues for the smooth and rough domes

Eigenvector number	Eigenvalue (%) smooth	Eigenvalue (%) rough
1	44	36
2	14	12
3	6.1	7.3
4	5.1	7.1
5	3.5	4.1
6	3.3	3.5
7	2.9	2.5
8	2.2	2.3
9	1.6	1.7
10	1.4	1.5
Sum of first 10 eigenvalues	83	77

3.4. Correlation of fluctuating surface pressures

A covariance matrix was formed from the 85 simultaneous time histories using MATLAB for both the smooth and rough domes for the 0° wind direction. An eigenvector/eigenvalue analysis was then undertaken to further interpret the fluctuating pressure field. The first 10 eigenvalues contribute approximately 80% of the total fluctuating pressure energy with the first eigenvalue accounting for some 40% in each case. Table 2 compares the first 10 eigenvalues for the rough and smooth domes. The first eigenvector is compared with the mean pressure distribution for the smooth dome in Fig. 14 and for the rough dome in Fig. 15. In each case, there is excellent correspondence as might be anticipated from the quasi-steady theory. Fig. 16 shows a contour plot, with full-scale dimensions, of the second eigenvector for the

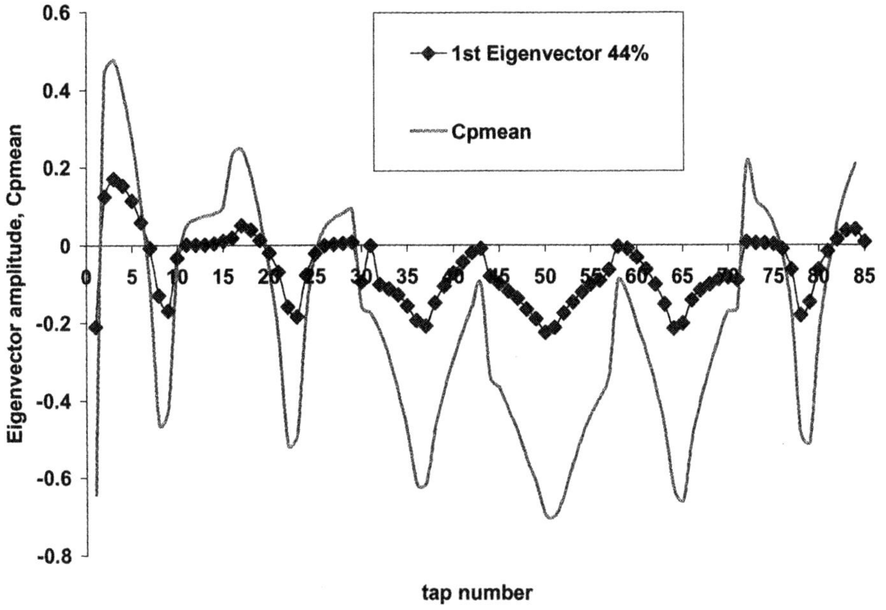

Fig. 14. Comparison of first eigenvector and variation of mean *Cp* with pressure tapping for the smooth dome.

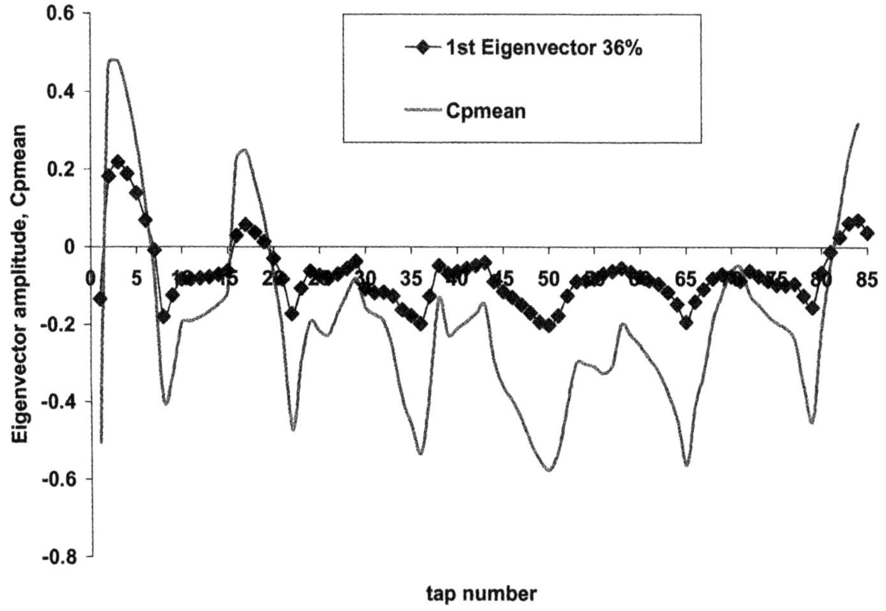

Fig. 15. Comparison of first eigenvector and variation of mean *Cp* with pressure tapping for the rough dome.

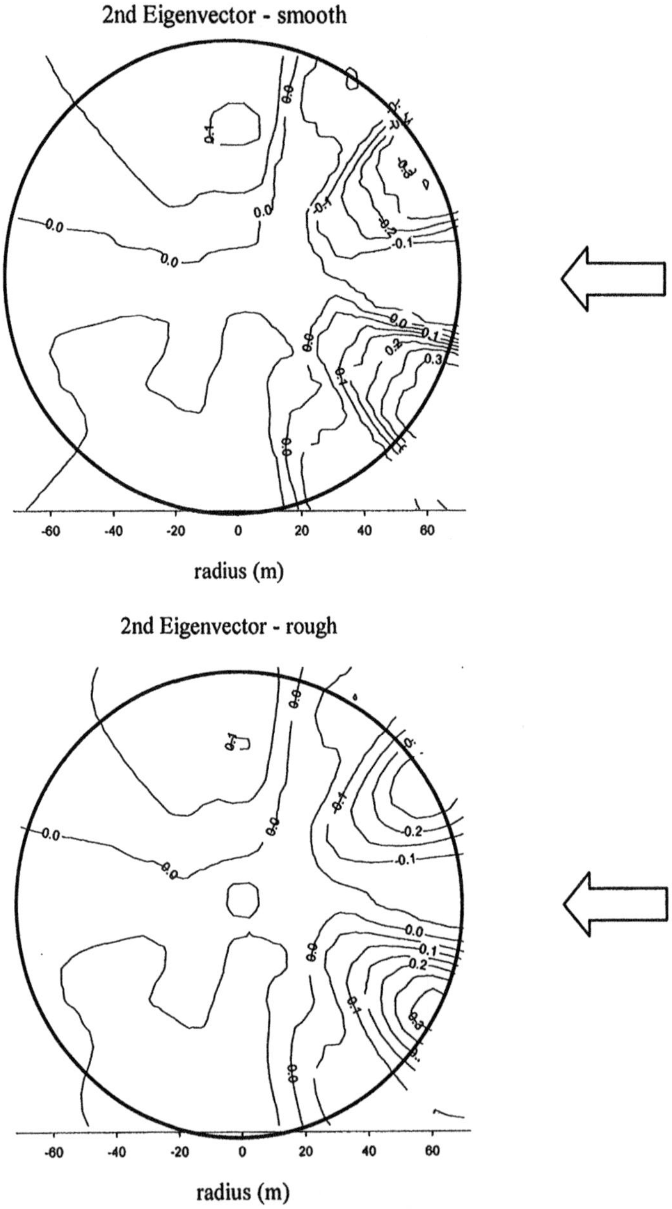

Fig. 16. Contour plot of second eigenvector for smooth and rough domes.

smooth and rough domes, respectively. The dominant feature is the asymmetry on the frontal portion, which closely resembles the gradient of mean pressure coefficient $\mathrm{d}Cp/\mathrm{d}\alpha$. Once again, this result would be anticipated from the quasi-steady theory.

3.5. Overall forces on a dome

As indicated in Section 2, by weighting the individual pressure taps accordingly, any particular load effect and its statistics can be calculated; all that is required are

Table 3
Comparison of overall force coefficients on smooth and rough domes

Dome	Force component	mean	rms	mean maxima	mean minina
Smooth	X	0.023	0.008	0.057	− 0.008
	Y	0.001	0.016	0.056	− 0.061
	Z	− 0.236	0.053	− 0.079	− 0.430
Rough	X	0.058	0.014	0.115	0.019
	Y	0.002	0.013	0.056	− 0.053
	Z	− 0.144	0.041	− 0.017	− 0.319

Table 4
Compares the pseudo-steady maxima and minima force coefficients with the mean force coefficients

Dome	Force component	mean	Pseudo-steady maxima	Pseudo-steady minima
Smooth	X	0.023	0.028	− 0.004
	Y	0.001	0.027	− 0.030
	Z	− 0.236	− 0.039	− 0.211
Rough	X	0.058	0.057	0.010
	Y	0.002	0.027	− 0.026
	Z	− 0.144	− 0.008	− 0.157

the influence coefficients for the load effect in question. For instance, the peak deflection at the top of the dome or the peak axial force in a chord member could be calculated.

As an example, the overall forces on a dome in the X, Y and Z directions (defined in Fig. 4) have been calculated according to Eq. (9). The results are in force coefficient form using the mean dynamic pressure at the top of the dome and the projected plan area of the dome base for non-dimensionalizing. Table 3 compares the X, Y and Z force coefficients for the smooth and rough domes. Both mean and fluctuating coefficients are presented.

It is evident from Table 3 that the smooth dome has nearly twice the uplift (Z) and less than half the drag (X) of the rough dome, and these differences are more pronounced in the mean force coefficients than the peak (maxima and minima) force coefficients.

Table 4 compares the mean force coefficients with the pseudo-steady force coefficients as defined by Eqs. (7) and (8). It is interesting to note that the quasi-steady theory predicts quite well the maximum drag on each dome, but would slightly overpredict the maximum uplift on the smooth dome while underpredicting the uplift on the rough dome.

4. Conclusions

Domes are commonly used to enclose large spaces because of their structural efficiency and consequent economic benefits. Such structures are used to cover sporting grounds, shopping centers and raw material stockpiles. These structures are sensitive to load distributions, and whereas they are excellent at resisting symmetric loading, asymmetric loading may cause structural distress [13,14]. In this study wind loading of domes has been revisited, and a comprehensive wind tunnel study produced, for the first time, simultaneous external fluctuating surface pressures over an entire dome. In a correctly scaled atmospheric boundary layer, this study has examined the influence of Reynolds number, and surface roughness as well as presented overall base shear loads on circular domes with height/base diameter aspect ratio of 1/3.

The specific conclusions of this study are:

- Mean and fluctuating pressure distributions compare favorably with earlier studies for similar shape, flow characteristics, and Reynolds number. The parabolic dome is well approximated by a spherical dome of the same height to diameter ratio.
- Pressure distributions were independent of Reynolds number in the range 2.3×10^5–4.6×10^5 defined by velocity at top of dome and base diameter.
- The effect of surface roughness is to reduce suctions over the apex of the dome and increase suctions in the wake region on the leeward face. The consequence for mean and fluctuating overall loads is reduced uplift but increased drag for rougher surfaces.
- Correlation analysis of the fluctuating pressures reveals that the first two eigenvectors account for approximately 60% of the fluctuating pressure energy and follow the mean pressure coefficient and its gradient with respect to horizontal wind direction as predicted by the quasi-steady theory. Overall base shear and uplift forces on the domes can be well approximated by this theory.

Acknowledgements

The authors wish to thank Dr. Bob Meroney and Dr. David Neff of Colorado State University for their assistance during the wind tunnel work. We also thank Dr. Toshiyuki Ogawa for kindly providing data.

References

[1] F.J. Maher, Wind loads on basic dome shapes, J. Struct. Div. ASCE ST3 (1965) 219–228.

[2] J. Blessmann, Pressures on domes with several wind profiles, Proceedings of the 3rd International Conference on Wind Effects and Building Structures, Tokyo, 1971, pp. 317–326.

[3] S. Taniguchi, H. Sakamoto, Time-averaged aerodynamic forces acting on a hemisphere immersed in a turbulent boundary, J. Wind Eng. Ind. Aerodyn. 9 (1981) 257–273.

[4] N. Toy, W.D. Moss, E. Savory, Wind tunnel studies on a dome in turbulent boundary layers, J. Wind Eng. Ind. Aerodyn. 1 (1983) 201–212.

[5] B.G. Newman, U. Ganguli, S.C. Shrivastava, Flow over spherical inflated buildings, J. Wind Eng. Ind. Aerodyn. 17 (1984) 305–327.
[6] E. Savory, N. Toy, Hemispheres and hemisphere-cylinders in turbulent boundary layers, J. Wind Eng. Ind. Aerodyn. 23 (1986) 345–364.
[7] T. Ogawa, M. Nakayama, S. Murayama, Y. Sasaki, Characteristics of wind pressures on basic structures with curved surfaces and their response in turbulent flow, J. Wind Eng. Ind. Aerodyn. 38 (1991) 427–438.
[8] T.J. Taylor, Wind pressures on a hemispherical dome, J. Wind Eng. Ind. Aerodyn. 40 (1991) 199–213.
[9] T. Ogawa, private communication, 1999.
[10] ASCE 7-98, ASCE Minimum Design Loads for Buildings and Other Structures, SEI, Reston, VA, 1999.
[11] N.J. Cook, The Designer's Guide to Wind Loading of Building Structures: Part 1, BRE/ Butterworths, London.
[12] C.W. Letchford, R.E. Iverson, J.R. McDonald, The application of the Quasi-steady theory to full scale measurements on the Texas tech building, J. Wind Eng. Ind. Aerodyn. 48 (1993) 111–132.
[13] J.D. Holmes, Determination of wind loads for an arch roof, Civil Eng. Trans. IEAust. CE26 (1984) 247–253.
[14] J.D. Holmes, Optimised peak load distributions, J. Wind Eng. Ind. Aerodyn. 41–44 (1992) 267–276.

Experimental measurements and computations of the wind-induced ventilation of a cubic structure

M.P. Straw[a], C.J. Baker[b,*], A.P. Robertson[c]

[a] *School of Civil Engineering, University of Nottingham, UK*
[b] *School of Civil Engineering, University of Birmingham Edgbaston, Birmingham B15 2TT, UK*
[c] *Silsoe Research Institute, Bedfordshire, UK*

Abstract

This paper presents the results of an experimental, theoretical and computational investigation of the wind-driven ventilation through a 6 m cube with openings on opposite faces. Measurements were made of the surface pressures coefficients and mean and total ventilation rates through the cube for the faces with the openings both normal and parallel to the wind. These measurements were then compared with a number of methods for the prediction of mean and fluctuating ventilation rates. For the normal configuration the mean component of ventilation was considerably greater than the fluctuating component, whilst for the parallel configuration the mean component was close to zero, and the ventilation was dominated by the fluctuating component. For the normal configuration the standard discharge coefficient method was shown to significantly underpredict the mean ventilation rate. A CFD calculation was however reasonably accurate in this regard. By contrast, for the parallel configuration the use of the standard discharge coefficient resulted in a small overprediction of the measured values of ventilation rate. The relative magnitudes of the ventilation produced by the various fluctuating flow mechanisms (broad banded, resonant and shear layer) were established, and methods of calculating the total ventilation rate from the mean and fluctuating components discussed. Finally, a simple method is presented for the estimation of shear layer ventilation.

Keywords: Wind-induced ventilation; Discharge coefficient; Full-scale experiments; CFD

*Corresponding author. Tel.: +44-121-414-5067; fax: +44-121-414-3675.
E-mail address: c.j.baker@bham.ac.uk (C.J. Baker).

Reprinted from *Journal of Wind Engineering and Industrial Aerodynamics*
88 (2-3), 213-230 (2000)

1. Introduction

The natural ventilation through a building consists of two components – the ventilation caused by thermal effects, and that caused by wind effects. It is with the fundamental nature of the second of these effects that this paper is concerned, and thus attention is restricted to relatively high wind speed conditions. The wind-driven ventilation itself is commonly considered to consist of two components – a mean component driven by the mean pressure field at the ventilation openings, and a fluctuating component driven by the fluctuating pressures and unsteady flows around the openings. The former is likely to be dominant when there are a number of openings around the buildings, in regions of different wind-induced pressures. The second component is likely to be dominant when there is only one major opening on the building, or where all openings are in regions of similar pressure. This fluctuating component of ventilation can be further considered to consist of a number of distinct phenomena [1]. The first mechanism is referred to rather loosely in [1] as "continuous" airflow, and seems to represent fluctuations in the ventilation flow caused by surface pressure fluctuations at the openings across a wide range of frequencies. In what follows we will refer to this mode of unsteady ventilation as broad banded ventilation. The second mechanism is pulsation flow, caused by a body of fluid being driven perpendicular to the opening by the difference between the external and internal pressures. Such ventilation flows are significantly affected by the geometry of the enclosure, and by air compressibility. The ventilation rate spectrum will have a peak at the Helmholz resonant frequency of the enclosure, and for this reason this mode will be referred to in what follows as resonant ventilation. The third is known as eddy penetration, and is caused by fluid transfer due to eddies in unstable shear layers that exist when the external flow is across the orifice. This will be referred to as shear layer ventilation in what follows.

In the past a number of experimental investigations have been carried out to investigate wind driven ventilation – for example, at wind tunnel scale the work of Refs. [2–5] and at full scale the work of Refs. [5–8]. A number of methods also exist for calculating the ventilation due to the components listed above. Mean ventilation rates can be calculated using simple zonal methods based on orifice flows [9]. The fundamental equation that is used to obtain the dimensionless mean discharge, through an orifice, $\bar{Q}$, is the simple discharge coefficient relationship

$$\bar{Q} = C_d \sqrt{\Delta C_p}, \tag{1}$$

where ΔC_p is the mean pressure difference across the orifice and C_d is the orifice discharge coefficient. Here, and in all that follows, the actual discharge is non dimensionalised with the opening area and the reference velocity. The discharge coefficient is conventionally taken as 0.61, which is the value for an orifice with flow parallel to its axis. However it should be noted that it is a weak function of Reynolds number and is a strong function of orifice shape and thickness. This equation arises from the use of the energy and continuity equations, and the discharge coefficient allows for real flow effects. As the flow is steady inertial effects are assumed to be negligible. The effect of flow across the orifice (i.e. in its plane) is not usually considered (but see Ref. [4]), although for many ventilation openings on the surface of buildings there may be a significant crossflow. In what follows it will be seen that we will be analysing flow through a structure with two openings of equal area. In this case Eq. (1) can be written as

$$\bar{Q} = C_{\mathrm{d}}\sqrt{\Delta C_{\mathrm{p}}/2}. \tag{2}$$

In this case the pressure coefficient difference is that measured across the two openings, and the factor of $\sqrt{2}$ is because the two openings are effectively in series.

In recent years, CFD packages have become a more popular tool for the prediction of mean ventilation flows [10], principally through the direct integration of flow across the ventilation openings. Using such methods it is not necessary to assume a value for discharge coefficient. Further if such calculations are based on a calculated flow field around the building, they will, in principle, take into account any cross flow effects that might exist.

Determination of the fluctuating components of ventilation is considerably more complex. Perhaps, the first thing to appreciate is that whilst flow through an individual ventilation opening can be either in or out (positive or negative), by definition ventilation is a measure of the total air exchange and both inflows and outflows result in positive ventilation. Having said this let us firstly consider broad banded ventilation. Broad banded fluctuations follow the fluctuations in the oncoming wind across a wide range of frequencies. Most of the energy in such fluctuations will be at relatively low frequencies (<0.1 Hz) and correspond to large-scale variations in wind direction, and consequent changes in the flow pattern around the structure. This ventilation mechanism can effectively be regarded as a modification of the mean ventilation mechanism. For a two opening enclosure such as will be considered here, the total ventilation due to the mean and broad banded fluctuating mechanisms together is given by the area underneath the ventilation time history divided by the length of record. If the ventilation time history does not change sign i.e. the magnitude of the fluctuations is less than the mean, this of course corresponds simply to the mean ventilation rate. If the value of the fluctuations is greater than the mean, then there will be some rectification of the ventilation time series around zero and the total ventilation due to these two mechanisms will be in excess of the mean value. If we assume that in such a case we have sinusoidal variations in ventilation with a true non-dimensional mean $\bar{Q}$ and a true non-dimensional r.m.s. value of σ_Q then it is straightforward to show that the total non-dimensional ventilation rate due to the mean and broad banded ventilation mechanisms is given by

$$Q_{\mathrm{B}} = \bar{Q} + \left(\frac{2\sqrt{2}}{\pi}\right)\sigma_Q\sqrt{1-\frac{1}{2}\left(\frac{\bar{Q}}{\sigma_Q}\right)^2}. \tag{3}$$

The true mean and r.m.s. values mentioned above would be such as could be calculated from velocity measurements in the ventilation outlets. Eq. (3) applies only for $\left(\bar{Q}/\sigma_Q<\sqrt{2}\right)$. Above this value, the total ventilation is equal to the mean ventilation.

When calculating the "resonant" ventilation, the usual approach is to derive the momentum equation for each opening, for a slug of fluid that is forced in and out of the orifice by the difference between the internal and external pressures – see Refs. [2,4,11,12] amongst others, and thus compressibility and inertia effects are taken into account. This thus represents a type of ventilation that is not allowed for in the discharge coefficient approach. These equations are then combined with the continuity equation to give a set of non-linear equations that are then either solved numerically or linearised using one of a number of approaches to produce equations that are analytically tractable. These are second-order differential equations relating the discharge through each opening (and the internal pressure coefficient) to the external pressure coefficient at the openings. Frequency-domain approaches are then used to calculate the relationship between the external pressure spectrum, the internal pressure spectrum and the ventilation rate spectrum. Ref. [13] shows that the use of the technique of proper orthogonal decomposition of the surface pressure field leads to an elegant solution for these parameters. The ventilation rate spectrum is given by

$$S_{QR} = \frac{\left(\omega\rho u_{\mathrm{R}}^2/2B\right)^2\left(\prod_1^2 S_{T1} + \prod_{21}^2 S_{T2}\ldots\ldots.\right)}{\left(\left(1-\omega^2/\omega_{\mathrm{n}}^2\right)^2+\left(2c\omega/\omega_{\mathrm{n}}\right)^2\right)} \tag{4}$$

where ω is an angular frequency, ω_{n} is the natural frequency of the system $\left(=\left(ABN/\rho\, L\right)^{0.5}\right)$ and c is the damping of the system $\left(=K/2(\rho LABN)^{0.5}\right)$; ρ is the density of air, u_{R} is a reference velocity, K is a coefficient of linearisation, A is the orifice area, L is the effective orifice length (actual orifice length $+\,0.89\sqrt{A}$) and N is the number of orifices; $B=\gamma P_{\mathrm{R}}/V$ where γ is the ratio of specific heat, P_{R} is a reference pressure and V is the volume of the enclosure; $\prod_i=\sum_j P_i(x_j)/N$ where $P_i(x_j)$ is the eigenvector of mode i at point j, the position of the openings on the surface; S_{Ti} is the spectrum of mode i.

In deriving the above equation the linearisation method of Ref. [11] has been used, although other approaches would be equally valid. It can be seen from the form of the above equations that the ventilation spectrum will peak at the resonant (Helmholz) frequency of the system. For most buildings this frequency will be quite high (>1 Hz). Further note that the spectrum will tend to zero at low frequencies. Physically, this corresponds to the filtering out of long period fluctuations, due to the finite size of the enclosure. It is thus likely that the broad banded and resonant ventilation spectra will be separated in the frequency domain. From such spectra the r.m.s. values of non-dimensional ventilation rate σ_{QR} can thus be found. Then, assuming to a first approximation that the variation is sinusoidal, the ventilation due to this mechanism can be shown to be given as

$$Q_{\mathrm{R}} = \sigma_{QR}\frac{\sqrt{2}}{\pi}. \tag{5}$$

The particular utility of this method is that, when applied to the case with more than one opening, it obviates the need for the calculation of the cross spectra of pressures between orifices that would otherwise be required. This becomes particularly significant where the number of openings becomes large. This being said, in what follows we will consider only the two opening case, for which the main use of Eqs. (4) and (5) lies in their relative simplicity.

With regard to the other unsteady ventilation mechanism due to shear layer unsteadiness across orifices, to the author's knowledge, no methods exist for the prediction of non-dimensional ventilation due to this mechanism Q_S. Note however that as this type of ventilation is driven by the momentum of the fluid parallel to the opening, any ventilation caused by this mechanism will be in addition to that produced by the other mechanisms.

Thus it is possible to calculate both steady and unsteady ventilation through a number of different approaches. The question then arises as to how the total ventilation can be predicted. Full scale ventilation measurements are usually made using tracer gas experiments (see below) which effectively give the total ventilation of the enclosure and it is this quantity that needs to be predicted. If the mean ventilation rate is much greater than the fluctuating ventilation, then the total non dimensional ventilation Q should be given by $(\bar{Q} + Q_R + Q_S)$. If the mean ventilation is close to zero then the total ventilation will be given by $(Q_B + Q_R + Q_S)$.

This paper presents the results of a large scale experiment of wind driven ventilation and uses the experimental data to calculate ventilation rates by a variety of methods. A simple geometric arrangement has been chosen that, it is hoped, will allow a fundamental understanding of the ventilation flow mechanisms to be gained. The experiments will also provide a simple test case for the different methods of calculating mean and unsteady ventilation rates. The nature of the experiments is described in Section 2. Section 3 then presents the experimental results, together with the results of calculations of the mean ventilation rates (using direct velocity measurements, a calculation based on Eq. (2) and CFD calculations), the broad banded ventilation rate (based both on the use of a time varying form of Eq. (2) and on Eq. (3)), the resonant ventilation rate (based on Eqs. (4) and (5)), and the total ventilation, using various combinations of the above methods. Section 4 goes on to present a simple method for the calculation of shear layer ventilation, and its adequacy is discussed. Finally, conclusions are drawn in Section 5.

2. The experiments

The experiments were carried out on a 6 m cube constructed on an exposed site at Silsoe Research Institute. The cube was built on a turntable such that it could be rotated to any angle relative to the approaching wind direction. Two 1 m square openings were cut with their axes on the vertical centreline 0.5 m above the centre on opposite sides of the cube (i.e. the bottom of the openings were at a height of 3.0 m above the ground). Further details of the cube are given in Ref. [14]. Two sets of experimental results are presented as follows.

(a) The normal case with the faces containing the openings positioned normal to the wind i.e. with the openings on the windward and the leeward sides.
(b) The parallel case with the faces containing the openings positioned parallel to the wind i.e. with the openings on the side faces.

These two cases represent distinctly different ventilation conditions. One would expect the former to be dominated by mean flow effects, and the second by fluctuating flow effects. Reference wind conditions were measured using a sonic anemometer mounted at cube height (6 m), 2 m to the side of the cube and 18 m upstream. The atmospheric boundary layer at the site has been measured in the past and shown to be a typical rural boundary layer with a surface roughness length of approximately 0.01 m.

The following measurements were carried out.

(a) Pressure measurements on the external surfaces of the cube (with the openings closed – the sealed case) and within the cube (with the openings exposed – the open case), using pressure tappings and probes connected to pressure transducers sampled at 5 Hz.
(b) Three-dimensional velocity measurements within the cube using a sonic anemometer, sampled at 20 Hz. Measurements were made within the cube along the cube centreline, around the openings and also across a number of planes perpendicular to the plane of the opening.
(c) Tracer gas measurements using carbon monoxide (CO) sampled with a GFC Ambient CO analyser at a frequency of 0.1 Hz. A constant injection method was utilised with the tracer gas being released at nine equally spaced points within the structure. The sampling point was varied in order to ensure that adequate mixing was taking place. The tracer gas measurements provided the total effective wind driven ventilation. Leakage tests were performed for the sealed case.

For the purposes of the present investigation the results were analysed to determine the mean and unsteady pressure and velocity characteristics at the orifice positions and within the cube. Table 1 shows the values of the mean and r.m.s. reference velocities for both geometric configurations and for the sealed and open cases, together with the values of the mean and standard deviations of external and internal pressure coefficients and the mean and standard deviations of the dimensionless flow rates at the openings, as calculated from the velocity measurements. The actual measured values of the flow rates were non-dimensionalised with the reference velocity and the orifice area. The total dimensionless flow rate out of the cube, as measured by the tracer gas experiments, is also shown. Leakage from the cube has been allowed for in arriving at this figure. Note that, for the normal configuration, the flow rate values are given as measured (which was at a point 0.2 m into the cube

Table 1
Flow characteristics

	Normal case		Parallel case	
	Mean	Standard deviation	Mean	Standard deviation
Reference velocity – sealed case (m/s)	11.78	2.47	11.37	2.50
Reference velocity – open case (m/s)	6.70	1.44	7.20	1.40
External pressure coefficient – opening 1	0.87	0.439	−0.649	0.380
External pressure coefficient – opening 2	−0.390	0.186	−0.698	0.347
Internal pressure coefficient – open case	0.081	0.203	Not measured	Not measured
Opening 1 – non-dimensional discharge $\bar{Q}$ and σ_Q	0.884 (measured) 0.67 (corrected)	0.136 (measured) 0.103 (corrected)	0.113 (measured)	0.101 (measured)
Opening 2 – non-dimensional discharge $\bar{Q}$ and σ_Q	0.681 (measured) 0.71 (corrected)	0.103 (measured) 0.107 (corrected)	0.115 (measured)	0.105 (measured)
Total non-dimensional outflow – tracer gas Q	0.787	—	0.380	-

from the centre of the plane of the orifice) and also corrected to the mean value at the plane of the orifice. This correction was carried out using the results of the CFD calculations (see below) by multiplying by the ratio of the calculated mean velocity at the orifice plane to the calculated velocity at the measurement point. When this process has been carried out it can be seen that the discharges at the two orifices are similar to each other as would be expected.

Firstly, consider the normal case. The pressure coefficient on the front face is positive, and that on the rear face is negative as expected. The different ventilation results obtained from the velocity measurements can be seen to be reasonably consistent, with a value of $\bar{Q}$ of 0.69 ($\pm$0.02) and a value of σ_Q of 0.105 ($\pm$0.002). The overall measured ventilation rate Q is 0.787, suggesting that around 10% of the ventilation is accounted for by resonant ventilation and shear layer ventilation. For the parallel case the situation is rather different. The pressure coefficients at each of the two openings are very similar (but not identical, suggesting that the sides of the cube were not completely parallel to the mean flow direction). $\bar{Q}$ is around 0.1 whilst the measured Q is around 0.4, suggesting a relatively greater fluctuating component than in the normal configuration.

The reference velocity spectra and external pressure coefficient spectra are shown in Fig. 1 for the two configurations. The velocity spectra for all three velocity components are broadly as expected with a slope close to the value of $-\frac{5}{3}$ at high frequencies. The spectrum of the vertical component is flatter than the others as would be expected in such near ground conditions. The pressure spectra are

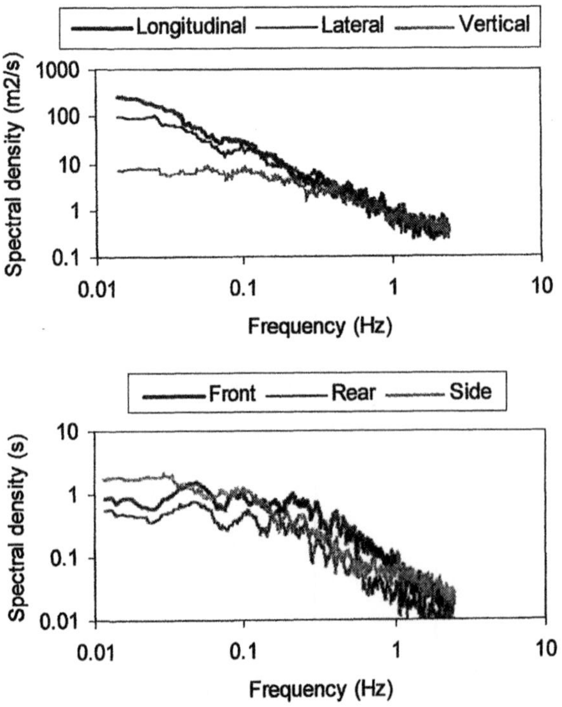

Fig. 1. Velocity and pressure coefficient spectra.

broadly similar in form to the velocity spectra. Fig. 2 shows the spectra of the internal pressure coefficient plotted in the form of spectral density × frequency/variance. There is an indication of a small high-frequency peak in the internal pressure spectrum at around 2 Hz. This will be seen to be of some significance in what follows.

The measured longitudinal velocity distribution through the cube for the normal configuration, on the centre-line directly between openings, is shown in Fig. 3 together with CFD predictions. In this figure the actual velocity is normalised with a reference velocity. The CFD calculations will be discussed further below. At this point it is sufficient to state that the curves are similar in form, although the experimental values vary rather more across the cube than the predicted values. It can be seen that the velocity at the windward opening reduces rapidly with a recovery towards the leeward opening. It would be expected that the velocity at the openings should be equal for mass conservation. However, as mentioned above it was not possible to measure velocities in the plane of the openings, with the nearest point being 0.2 m from the plane (hence the use of corrected values as outlined above). The results of Table 1 suggest that the average values at the openings should

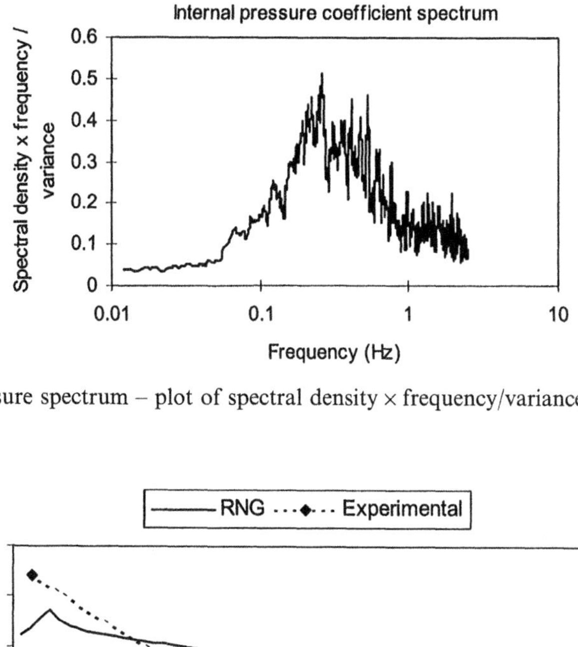

Fig. 2. Internal pressure spectrum – plot of spectral density × frequency/variance against frequency.

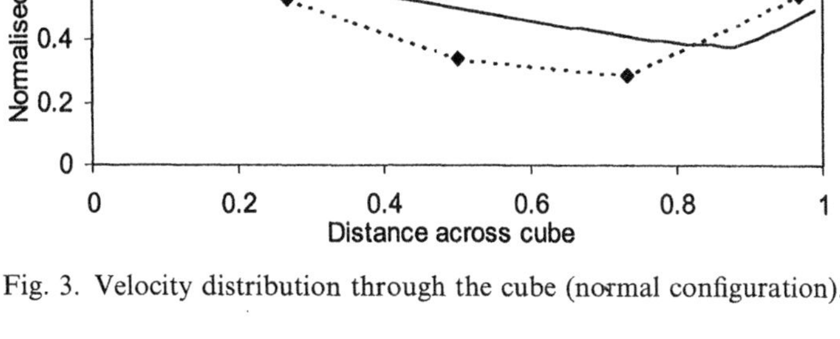

Fig. 3. Velocity distribution through the cube (normal configuration).

be 0.69. This implies that the measurements near the windward opening were made in a strongly accelerated "vena contracta" region.

To enable the method of Ref. [13], to be used to calculate the unsteady ventilation rate, a proper orthogonal decomposition was carried out on the surface pressure coefficients around the centre line of the cube for both configurations. Fig. 4 shows the distribution of the first three eigenvectors that are obtained in the analysis and the proportion of fluctuating energy associated with each mode for the normal and

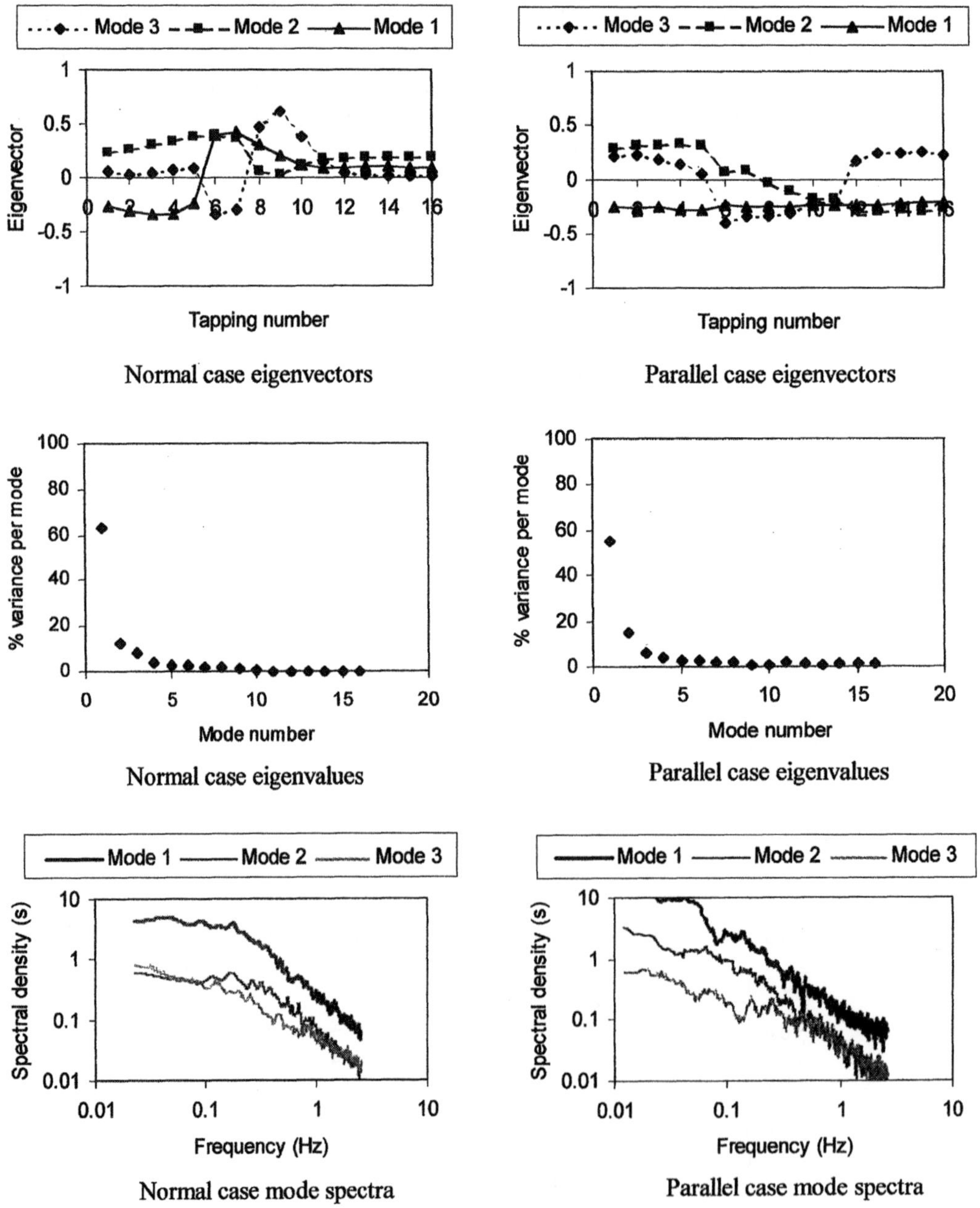

Fig. 4. Results of POD analysis for cube.

parallel configurations. The modal spectra for the first three modes for each configuration are also shown. Note that in the plots for the normal case tappings, 1–5 are on the windward face of the cube, tappings 6–11 on the cube roof and tappings 12–16 on the leeside of the cube. For the parallel case tappings 1–5 and 12–16 are on the side faces of the cube, and tappings 6–11 are on the roof. The centre of the orifices corresponds to the positions of tappings 3 and 14. For the normal case the mode shapes are broadly as expected, with the shape of the first mode mirroring that of the external pressure distribution, suggesting that this mode is due to longitudinal turbulence fluctuations (see Ref. [13]). The physical cause of the second mode is not obvious, but in Ref. [13] it is suggested that it might be related to the unsteadiness induced by the distortion of longitudinal turbulence as it passes around the cube. Ref. [13] further relates the third mode to vertical turbulence fluctuations. For the parallel case, the first mode is flat, and similar in form to the mean pressure distribution, again suggesting that this is caused by longitudinal turbulence fluctuations. The second mode is highly asymmetric, and may well be caused by large-scale vortex shedding from the cube. The third mode is symmetric, and may again be related to vertical turbulence fluctuations. As is usual in such analysis, nearly all of the energy is contained within the first few modes, suggesting that the series in Eqs. (2) and (3) can be truncated after a few terms with little loss of precision. The spectra are similar in form to the reference velocity spectra, with the mode 3 spectra showing relatively more energy at high frequencies, which would be expected if this mode were related to vertical turbulence fluctuations.

3. Calculation of ventilation rates

3.1. Mean ventilation rates – normal case

The non-dimensional mean ventilation rates $\bar{Q}$ were calculated for the normal case using two methods as follows:

(a) Using the discharge coefficient formula (Eq. (2)) for the measured mean pressure coefficients, and a standard value of the discharge coefficient of 0.61. The pressure coefficient increment (ΔC_p) was taken as that measured between tappings 3 and 14 for the sealed cube measurements.
(b) Using a CFD solution of the flow through the cube (using CFX-F3D). Simulations utilised the RNG $k-\varepsilon$ turbulence model with CCCT differencing. A fully independent grid utilising 2×10^5 cells produced a fully converged solution of the external flow field and the consequent internal airflow pattern. Both external and internal flow fields were simulated simultaneously. This required a domain of sufficient size that would not affect the external flow field around the cube. The domain size utilised was 5 cube heights upstream, above and to the side of the cube, and 10 cube heights downstream. Data from the site of the cube was used to develop a mean boundary layer profile in terms of both velocity and turbulent kinetic energy profiles with a roughness height of 0.01 m. This

calculation was only carried out for the normal configuration, as the symmetry of the cube suggests a zero value should be predicted if the cube sides were perfectly aligned with the flow.

The results of these calculations are shown in Table 2. It can be seen that the discharge coefficient method significantly underpredicts the mean discharge through the cube. This could be for a number of reasons – inaccuracies in the methods of measuring the experimental discharges (but this is unlikely given the consistency of the various results); the assumed (ideal) value of the discharge coefficient being too low or the pressure coefficients measured in the sealed cube case being significantly different when the openings were present. Detailed pressure measurements around the orifice would be required to determine whether this was the case. The CFD prediction provides a result which is far closer to the measured mean ventilation rate.

3.2. Combined mean and broad banded unsteady ventilation calculations – the parallel case

For the parallel case the combined mean and broad banded ventilation rate Q_B was calculated in three ways as follows:

(a) Using Eq. (3) with the true mean and standard deviation $\bar{Q}$ and σ_Q as given by the velocity measurements in the plane of the openings – i.e. assuming the adequacy of Eq. (3) that was derived for a sinusoidal fluctuation.
(b) As in (a) but with $\bar{Q}$ and σ_Q calculated from the time series of obtained by using Eq. (2) with the time series of the pressure coefficient difference, with the flow direction taken into account i.e. allowing for positive and negative values of the flow through any one opening.
(c) Directly from the integration of the time series produced in (b), but with the absolute (rectified) values of the ventilation rectified about zero.

The results shown in Table 3 are reasonably consistent. The two values obtained using the discharge coefficient assumption ((b) and (c)) are close to each other, which gives some confidence in the use of Eq. (3). These are both above the value obtained using the velocity measurements, suggesting that in this case the standard value of the discharge coefficient is somewhat too high. The absolute differences are however small. It is of interest to note at this point that the value of ventilation rate calculated using the mean values of the pressure coefficients in Table 1 to form the pressure coefficient difference, results in a value of non-dimensional ventilation of 0.095 which

Table 2
Mean ventilation rates $\bar{Q}$ – normal case

Values from velocity measurements	0.67/0.71
Discharge coefficient method	0.483
CFD calculations	0.648

Table 3
Mean and broad banded fluctuating ventilation rates Q_B – parallel case

Values from velocity measurements of mean and r.m.s. values and Eq. (3)	0.185
Values from discharge coefficient calculations of mean and r.m.s. values and Eq. (3)	0.207
Values from discharge coefficient calculations of ventilation time histories	0.220

is close to the mean value from the velocity measurements, but significantly less than the calculated values of Q_B.

3.3. Resonant ventilation calculations

The resonant ventilation rates were also calculated using the frequency domain method [13] and the technique of proper orthogonal decomposition (Eqs. 4 and 5). The following values were assumed for the parameters in Eqs. (4) and (5) – $u_R = 6.7\,\text{m/s}$, $A = 1\,\text{m}^2$, $L = 1.09\,\text{m}$ (based on an actual orifice length of 0.2 m), $V = 216\,\text{m}^3$, $\gamma = 1.4$, $P_R = 100\,000\,\text{Pa}$ These give values of ω_n and c of 30.8 r/s and 0.142, respectively.

The results of the calculation give a natural frequency for the ventilated cube system of 4.9 Hz. It can be seen from Fig. 2 that the internal pressure spectrum shows a peak at about 2 Hz which may correspond to this natural frequency. This difference is likely to be caused by leakage from the cube and the flexibility of the side of the cube, causing a change in the effective bulk modulus of the flow. It can be shown that this is equal to the product $P_R\gamma$ in Eqs. (4) and (5). Ref. [4] points out that the effect of building flexibility can reduce the effective value of this parameter to as low as 20% of the normal value. This is consistent with the observed shift in the natural frequency from its predicted value. To allow for this effect the measured natural frequency will be used in what follows. With the natural frequency at this value, it is likely that some of the unsteady ventilation will take place at frequencies of up to, say, 10 Hz i.e. higher than the sampling frequency. To enable the calculations to be made up to this frequency the following procedure was adopted.

(a) The modal spectra shown in Fig. 2 were fitted with a power-law curve for frequencies between 0.25 and 2.5 Hz and were extended to higher frequencies using this curve fit.
(b) The measured power spectra were used in the calculations using Eqs. (1) and (2) at frequencies below 2.5 Hz, with the extrapolated values being used at the higher frequencies.

Calculations were carried out using the first three POD modes only i.e. assuming only three terms in Eqs. (4) and (5) are of significance. The results of this procedure are shown in Table 4 where the values for the non-dimensional resonant ventilation rates Q_R are presented. The immediate thing to notice about these results is that the ventilation rates predicted for the two cases are, compared to the mean ventilation rates, relatively small. The latter point will be taken up further below. The distribution of the ventilation between modes (i.e. the different terms in Eqs. (4) and (5)) is, however, very different. For the normal case the percentage of the discharge that can be attributed to modes 1, 2 and 3, respectively, is 51.6%, 47.7% and 1.6%, whilst for the parallel case the corresponding figures are 86.4%, 0% and 13.6%. The ventilation power spectra for each mode shown in Fig. 5 also illustrate this. For both cases therefore, the resonant ventilation seems to be largely due to the modes that reflect oncoming longitudinal turbulence fluctuations. Note also that most of the ventilation due to this effect occurs at frequencies around the resonant frequency (>0.5 Hz). At such frequencies there is little energy in the oncoming wind. The ventilation due to broad banded ventilation can be expected to occur mainly at frequencies significantly lower than this.

3.4. Total ventilation calculations

For the normal case the total ventilation rate measured in the experiments is the sum of the mean ventilation rate (which effectively incorporates the broad banded unsteady ventilation), and the resonant and shear layer ventilation. For the parallel case the total ventilation is approximately given by the sum of the mean and broad banded (Eq. (3)), the resonant and shear layer mechanisms (see Section 1). Table 5 compares the measured total discharge with the sum of the mean values and resonant values calculated in a number of ways as follows.

(a) From the values of the mean and broad banded mechanisms calculated from the velocity measurements in the opening plus the calculated resonant ventilation

Table 4
Resonant fluctuating ventilation rates Q_R

	Normal configuration	Parallel configuration
Resonant ventilation rates	0.008	0.014

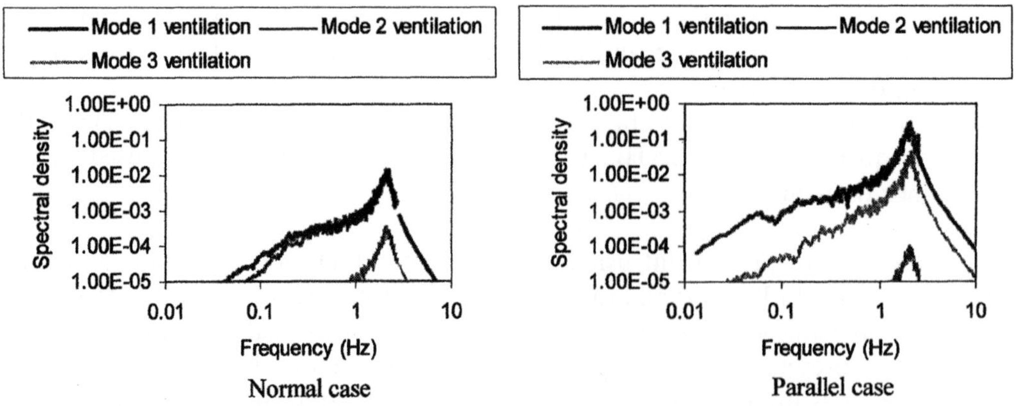

Fig. 5. Results of POD resonant ventilation analysis for cube.

Table 5
Total ventilation rates

	Normal configuration ($\bar{Q}+Q_R$)	Parallel configuration (Q_B+Q_R)
Measured total ventilation Q	0.787	0.380
From velocity measurements and Eq. (3) + calculated resonant ventilation	0.69 + 0.008 = 0.698	0.185 + 0.014 = 0.199
Discharge coefficient calculations + calculated resonant ventilation	0.483 + 0.008 = 0.491	0.207 + 0.014 = 0.221, or 0.220 + 0.014 = 0.234
CFD calculated mean + calculated resonant ventilation	0.648 + 0.008 = 0.656	—

(b) From the values of the mean and broad banded rates calculated using a standard discharge coefficient, plus the calculated resonant fluctuating ventilation.
(c) From the value of the mean ventilation calculated using CFD and the calculated resonant ventilation (normal configuration only).

The first point to note is the relative magnitudes of the fluctuations caused by the different mechanisms – with the resonant ventilation component being relatively small for both configurations. However, for different geometries this will not always be the case – for example, for single opening enclosures, one might expect this type of ventilation to play a much greater role. However the major point that arises is that there is still a significant discrepancy between the measured total ventilation rates, and the various calculated values – of about 0.1 for the normal case, and 0.2 for the parallel case. It is likely that this discrepancy is caused by the ventilation mechanism that has so far not been considered – shear layer ventilation, to which we turn our attention next.

4. Shear layer ventilation

It is apparent from what has been said above that the effect of a cross flow across a ventilation opening can be of considerable importance. In terms of the mean ventilation the experimental results for the normal configuration suggest that a higher than expected discharge coefficient is required for the ventilation rate predicted by Eq. (2) to be consistent with the measured values. One reason for this that was suggested was that the ventilation might be increased by a cross flow across the orifice. Some unpublished experimental data obtained recently by staff at Silsoe Research Institute suggests that the discharge coefficient might reach values as high as 0.75 with a large cross flow velocity. This is presumably caused by the directing of some external flow directly into the opening (see Ref. [14]). Whether or not it is sensible to allow for this by an increased discharge coefficient is debatable, although it is undoubtedly convenient.

Now let us consider unsteady shear layer ventilation – the so-called eddy penetration mechanism of Ref. [1]. Here the ventilation is caused by vortices in the shear layer across the opening transporting flow into and out of the opening. Simple physical reasoning suggests that the dimensional ventilation rate due to this mechanism, q_S will be a function of opening length l, opening width w, velocity parallel to the opening u_o, and shear layer thickness δ. A simple dimensional analysis leads to the functional expression

$$\frac{q_S}{u_o l w} = fn\left(\frac{l}{w}, \frac{l}{\delta}\right). \tag{6}$$

Assuming that the opening geometry is fixed and that the dependence upon the shear layer thickness is small, this reduces to

$$\frac{q_S}{u_o l w} = k, \tag{7}$$

where k is a constant.

Now writing down the energy equation along a streamline, and making the (very approximate) assumption that energy is conserved for a streamline around a building

$$0.5\rho u_R^2 + p_R = 0.5\rho u_o^2 + p_o, \tag{8}$$

where subscript R indicates reference upstream values, and subscript o indicates values at the opening. This leads directly to

$$\frac{u_o}{u_R} = \left(1 - C_{po}\right)^{1/2} \tag{9}$$

and thus

$$\frac{q_S}{u_R l w (1 - C_{po})^{1/2}} = \frac{Q_S}{(1 - C_{po})^{1/2}} = k \tag{10}$$

Using this expression the experimental data allow some estimates for the parameter k to be obtained. We assume that for each configuration the ventilation that has so far not been accounted for is all due to this mechanism (in dimensionless terms Q_S is about 0.1 for the normal case, and 0.2 for the parallel case – see Table 5). Eq. (10) is written down for each opening, together with a continuity equation relating Q_S to the sum of the discharges at the two openings. The values of pressure coefficient at each opening are given by those in Table 1. One can thus calculate the ventilation rate for each opening and a value of k for each configuration. For the normal case it was thus calculated that the front opening has a non dimensional ventilation rate of 0.023, and the rear opening of 0.077, with a value of k of 0.065. For the parallel case, the ventilation is more or less evenly partitioned between the two openings as would be expected (0.101 and 0.099) with a value of k of 0.077. Such values of k are consistent and seem physically reasonably – one would expect an effective ventilation velocity to be an order of magnitude smaller than the velocity outside the shear layer as is implied by these results. This suggests that the use of Eq. (10) with a value of $k = 0.1$ should give a conservative estimate of shear layer ventilation.

5. Concluding remarks

From the above results and calculations the following conclusions can be drawn.

(1) For the normal configuration the mean component of ventilation is significantly greater than the fluctuating component. For the parallel configuration the mean component is close to zero, and the ventilation is dominated by fluctuating effects.
(2) For the normal configuration the total ventilation is the sum of the mean ventilation, the resonant ventilation and the shear layer ventilation. For the parallel configuration the total ventilation is given by the combined mean and broad banded ventilation together with the resonant ventilation and the shear layer ventilation.
(3) For the normal configuration the mean ventilation rates are not well predicted by the discharge coefficient method using the standard value of discharge coefficient. The most likely reason for this is that there is a component of mean ventilation due to the flow across the openings. To allow for this a value of discharge coefficient significantly higher than the standard value is required. This ventilation was however, well predicted by a routine CFD calculation of the combined internal and external flow fields. For this configuration the resonant ventilation component was small, and the majority of the fluctuating ventilation was due to the shear layer ventilation mechanism.
(4) For the parallel configuration the use of the standard discharge coefficient resulted in a small overprediction of the mean and broad banded ventilation rate. The resonant ventilation was again small, but represented a larger proportion of the total fluctuating ventilation. The remainder could be attributed to shear layer ventilation.
(5) A simple formula has been derived for the prediction of fluctuating shear layer ventilation, but this needs further verification and calibration before it can be widely used.

Acknowledgements

During the course of the study the first author was supported by a Silose Research Institute/University of Nottingham studentship.

References

[1] H.K. Malinowski, Wind effect on the air movement inside buildings, Proceedings of the Third International Conference on wind on buildings and structures, Tokyo, 1971, pp. 125–134.
[2] J.D. Holmes, Mean and fluctuating pressure induced by wind, Proceedings of the Fifth International Conference on Wind Engineering, Fort Collins, CO, 1979, pp. 435–450.

[3] H. Liu, K.H. Rhee, Helmholz oscillation in building models, J. Wind Eng. Ind. Aerodyn. 24 (1986) 95–115.
[4] B.J. Vickery, C. Bloxham, Internal pressure dynamics with a dominant opening, J. Wind Eng. Ind. Aerodyn. 41 (1992) 193–204.
[5] J. Rao, Assessment of the effect of mean and fluctuating wind induced pressures on air infiltration and ventilation in buildings – a system theoretic approach, Ph.D. Thesis, Concordia University, Montreal, Canada, 1993.
[6] D. Bienfait, H. Phaff, L. Vandaele, J. van der Maas, R. Walker, Single sided ventilation, Proceedings of the 12th AIVC Conference, Ottawa, Canada, 1991, pp. 24–27.
[7] F. Haghighat, J. Rao, J. Riberon, Modelling fluctuating airflow through large openings – ventilation for energy efficiency and optimum indoor air quality, Proceedings of the 13th AIVC Conference, Nice, France, 1992, pp. 77–85.
[8] J. Furbringer, J. van der Maas, Suitable algorithms for calculating air renewal rate by pulsating air flow through a large single opening, Building Environ. 30 (1995) 493–503.
[9] BSI, Code of practice for ventilation – principles and designing for natural ventilation, British Standard, 5925, 1991.
[10] H.B. Awbi, Application of computational fluid dynamics in room ventilation, Building Environ. 24 (1) (1989) 73–84.
[11] F. Haghighat, J. Rao, P. Fazio, The influence of turbulent wind on air change rates – a modelling approach, Building Environ. 26 (2) (1991) 95–109.
[12] G. Chaplin, C.J. Baker, J. Randall, The turbulent ventilation of a single opening enclosure, J. Wind Eng. Ind. Aerodyn. 85 (2000) 145–161.
[13] C.J. Baker, Aspects of the use of the technique of orthogonal decomposition of surface pressure fields, Proceedings of the 10th International Conference on Wind Engineering, Copenhagen, 2000.
[14] M.P. Straw, Computation and measurement of wind induced ventilation, Ph.D. Thesis, University of Nottingham, 2000.

Aeroelastic complex mode analysis for coupled gust response of the Akashi Kaikyo bridge model

Nguyen Nguyen Minh*, Hitoshi Yamada, Toshio Miyata, Hiroshi Katsuchi

Department of Civil Engineering, Yokohama National University, 79-5 Tokiwadai, Hodogaya-ku, Yokohama 240, Japan

Abstract

An analytical approach for the coupled gust response of long-span bridges is presented. The calculation scheme is based on the direct complex eigenanalysis of the integrated system of three-dimensional FEM model of a long-span bridge and the aeroelastic forces. Actual vibration modes of the bridge in wind flow, which are called aeroelastic complex modes, are appropriately determined at each mean wind speed. A complex modal analysis scheme is then formulated for gust response analysis in either time or frequency domains. By using the aeroelastic complex modes, the aeroelastic coupled gust responses are effectively captured and interpreted. A numerical example is given for the Akashi Kaikyo Bridge model. The calculated results agree very well with the experimental ones. Parameter studies on the effect turbulence properties and the advantages of using the aeroelastic complex modes are then addressed © 2000 Elsevier Science Ltd. All rights reserved.

Keywords: Coupled gust response; Long-span bridges; Complex eigenanalysis; Aeroelastic complex modes; Turbulence properties; Spatial coherence; Time domain analysis; Frequency domain analysis

1. Introduction

The Akashi Kaikyo Bridge, the longest suspension bridge in the world, has been successfully realized. In the design of the bridge, a full-model test in a large boundary layer wind tunnel has been conducted from which many interesting findings on the

*Corresponding author; Present address: Structure Department, Abiko Research Laboratory, Central Research Institute of Electric Power Industry (CRIEPI), 1646 Abiko, Abiko city, Chiba 270-1194, Japan. Tel.: +81-471-82-1181; fax: +81-471-83-2962.

E-mail address: mstunbi@yahoo.com (N.N. Minh).

Reprinted from *Journal of Wind Engineering and Industrial Aerodynamics* **88 (2-3)**, 307-324 (2000)

dynamic behaviors of the bridge were obtained [1,2]. Under the action of turbulent flows, the full model of the bridge exhibited a strongly coupled three-dimensional vibration with remarkable signs of aeroelastic phenomenon. Prediction of such gust responses, therefore, emerges as a major consideration in the design of all long-span bridges.

In the past, a couple of methods for predicting coupled gust response of long-span bridges have been proposed and proved to be effective [3–6]. However, these methods, mostly following the modal analysis technique, have performed the analysis using the set of *mechanical modes*, namely the modes of the bridge system determined at no-wind condition. It is obvious that due to the aeroelastic effects, the mechanical modes are not the actual vibration modes of the bridge vibrating in wind flows. Such analyses are therefore rendered complicated by mathematical formulations or transformation techniques to incorporate and to express the aeroelastic effects.

In this regard, the Direct FEM Flutter Analysis Method by Miyata and Yamada [7,8], and its developments – the Mode Tracing Method by Dung et al. [9,10] for flutter prediction could provide a better representation of the dynamic behaviors of long-span bridges in terms of *complex flutter modes*. Complex eigenanalysis is directly made for an integrated system consisting of the 3-D FEM model of a bridge and the aeroelastic forces caused by the wind flow, resulting in the complex modes, which are apparently the actual vibration modes of the bridge in wind flow. The effectiveness and importance of the complex modes in predicting the flutter wind speed and flutter mode shape of long-span bridges have been well confirmed and reported [8,10].

In this study, an approach for gust response analysis of long-span bridges using the complex modes is developed. The complex mode herein is renamed as *aeroelastic complex mode* to indicate the fact that the aeroelastic effect is incorporated in the mode. By employing the mode tracing method [9], the aeroelastic complex modes at a certain wind speed can be appropriately obtained. The modal decomposition then can be made straightforward by a complex modal analysis scheme to facilitate the gust response analysis. The outline of the method in time domain has been introduced in Ref. [11]. In this paper, the approach will be consistently presented in detail, and the formulation in frequency domain is also provided.

Coupled gust response of the Akashi Kaikyo Bridge model is analyzed and interpreted. Very good agreements between analytical and experimental results are obtained. Parameter studies to address the effect of turbulence properties and the effectiveness of using the aeroelastic complex modes are presented.

2. Aeroelastic complex mode

The equation of motion of a full-model bridge in the presence of aeroelastic phenomena can be expressed as

$$\boldsymbol{M}\ddot{\boldsymbol{u}} + \boldsymbol{C}\dot{\boldsymbol{u}} + \boldsymbol{K}\boldsymbol{u} = \boldsymbol{F}_{\mathrm{ae}} + \boldsymbol{F}_{\mathrm{b}}, \tag{1}$$

where $\boldsymbol{M}$, $\boldsymbol{C}$ and $\boldsymbol{K}$ are the mass, damping and stiffness matrices formed by the finite element method (FEM), $\boldsymbol{u}$ is the displacement vector. A typical FEM model of a long suspension bridge is depicted in Fig. 1. The wind field is normally considered as a superposition of a mean wind speed U perpendicular to the deck and three turbulent velocities of horizontal $u(t)$, lateral $v(t)$ and vertical $w(t)$. The applying forces and response of the deck can be demonstrated on its cross-section in Fig. 2. The applying forces consist of drag, lift and moment, which cause the deck response in horizontal, vertical and rotational directions. In terms of mechanism, the wind force is considered as a superposition of two kinds of forces: aeroelastic (or self-excited) force $\boldsymbol{F}_{ae}$ caused by aeroelastic phenomena, and buffeting force $\boldsymbol{F}_b$ caused by the turbulence.

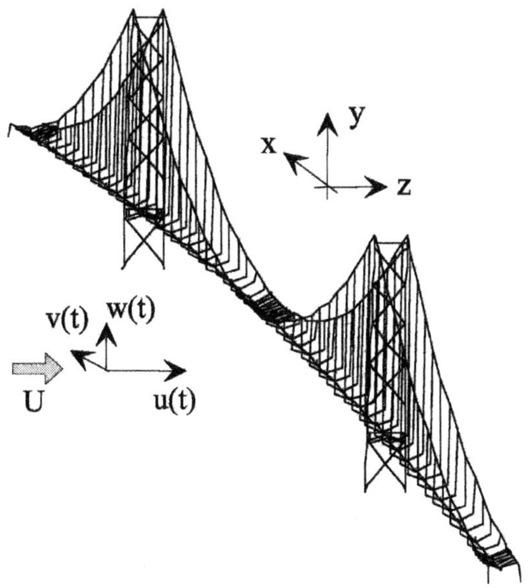

Fig. 1. 3-D FEM mesh of a bridge.

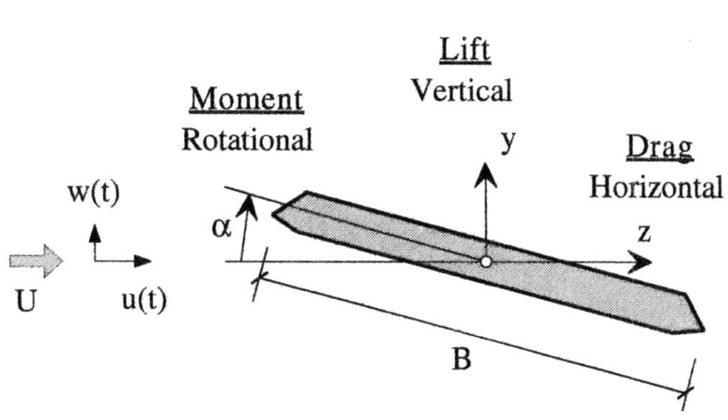

Fig. 2. Cross-section of a bridge deck.

The direct FEM flutter analysis [7] is employed for formulating $\boldsymbol{F}_{\mathrm{ae}}$ and establishing equation of complex eigen problem. Assuming harmonic oscillation, the self-excited force $\boldsymbol{f}_{\mathrm{ae}}$ on a unit length of bridge deck in local coordinates can be expressed as

$$\boldsymbol{f}_{\mathrm{ae}}\begin{Bmatrix} L_{\mathrm{ae}} \\ D_{\mathrm{ae}} \\ M_{\mathrm{ae}}/B \end{Bmatrix} = -(\rho\pi B^2)\begin{bmatrix} L_{y\mathrm{R}}+\mathrm{i}L_{y\mathrm{I}} & L_{z\mathrm{R}}+\mathrm{i}L_{z\mathrm{I}} & L_{\alpha\mathrm{R}}+\mathrm{i}L_{\alpha\mathrm{I}} \\ D_{y\mathrm{R}}+\mathrm{i}D_{y\mathrm{I}} & D_{z\mathrm{R}}+\mathrm{i}D_{z\mathrm{I}} & D_{\alpha\mathrm{R}}+\mathrm{i}D_{\alpha\mathrm{I}} \\ M_{y\mathrm{R}}+\mathrm{i}M_{y\mathrm{I}} & M_{z\mathrm{R}}+\mathrm{i}M_{z\mathrm{I}} & M_{\alpha\mathrm{R}}+\mathrm{i}M_{\alpha\mathrm{I}} \end{bmatrix}$$

$$\times\begin{Bmatrix} \ddot{y} \\ \ddot{z} \\ \ddot{\alpha}B \end{Bmatrix} = -(\rho\pi B^2)\,\boldsymbol{F}_{\mathrm{w}}\boldsymbol{w}, \tag{2}$$

where $\mathrm{i}=\sqrt{-1}$, $\boldsymbol{w}$ is the local displacement vector, in which y, z, and α are the vertical, horizontal displacements and the rotational angle as depicted in Fig. 2, L_{ae}, D_{ae}, M_{ae} are the aeroelastic lift, drag and moment, respectively, ρ is the air density. The matrix $\boldsymbol{F}_{\mathrm{w}}$ contains a full set of 9-complex unsteady coefficients, which exclusively depend on reduced frequency K,

$$K=\omega B/U, \tag{3}$$

where ω is the circular frequency, other notations are depicted in Fig. 2. The equivalence between the unsteady coefficients with the well-known flutter derivatives (see Ref. [5]) is as follows:

$$\begin{aligned}
&L_{y\mathrm{R}}=H_4^*/2\pi, && D_{y\mathrm{R}}=-P_6^*/2\pi, && M_{y\mathrm{R}}=-A_4^*/2\pi,\\
&L_{y\mathrm{I}}=H_1^*/2\pi, && D_{y\mathrm{I}}=-P_5^*/2\pi, && M_{y\mathrm{I}}=-A_1^*/2\pi,\\
&L_{\alpha\mathrm{R}}=-H_3^*/2\pi, && D_{\alpha\mathrm{R}}=P_3^*/2\pi, && M_{\alpha\mathrm{R}}=A_3^*/2\pi,\\
&L_{\alpha\mathrm{I}}=-H_2^*/2\pi, && D_{\alpha\mathrm{I}}=P_2^*/2\pi, && M_{\alpha\mathrm{I}}=A_2^*/2\pi,\\
&L_{z\mathrm{R}}=-H_6^*/2\pi, && D_{z\mathrm{R}}=P_4^*/2\pi, && M_{z\mathrm{R}}=A_6^*/2\pi,\\
&L_{z\mathrm{I}}=-H_5^*/2\pi, && D_{z\mathrm{I}}=P_1^*/2\pi, && M_{z\mathrm{I}}=A_5^*/2\pi,
\end{aligned} \tag{4}$$

The local $\boldsymbol{f}_{\mathrm{ae}}$ distributing along the bridge deck is lumped to apply to each shear node, as depicted in Fig. 3. For each shear node of the beam element to model the

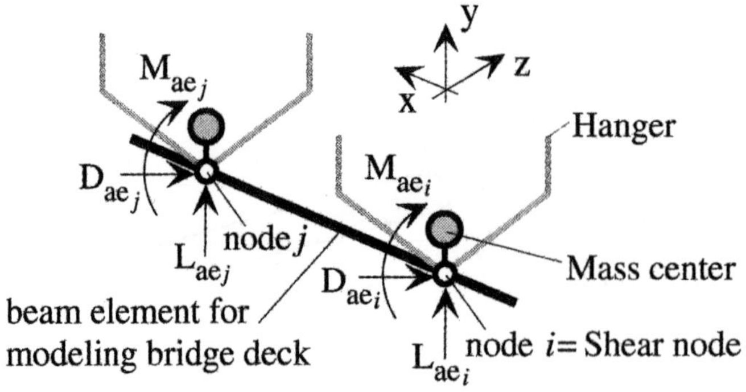

Fig. 3. Application of self-excited force on bridge deck.

bridge deck, there are six degrees of freedom (DOF), as illustrated in Fig. 4. With the local self-excited force by Eq. (2), the corresponding force vector at each shear node in local coordinate is

$$\begin{Bmatrix} f_x \\ f_y \\ f_z \\ m_x/B \\ m_y/B \\ m_z/B \end{Bmatrix} = \begin{Bmatrix} \cdots \\ L_{\mathrm{ae}} \\ D_{\mathrm{ae}} \\ M_{\mathrm{ae}}/B \\ \cdots \\ \cdots \end{Bmatrix}$$

$$= -\rho\pi l_{\mathrm{e}} B^2 \begin{bmatrix} 0 & 0 & 0 & 0 & 0 & 0 \\ 0 & L_{y\mathrm{R}} + \mathrm{i}L_{y\mathrm{I}} & L_{z\mathrm{R}} + \mathrm{i}L_{z\mathrm{I}} & L_{\alpha\mathrm{R}} + \mathrm{i}L_{\alpha\mathrm{I}} & 0 & 0 \\ 0 & D_{y\mathrm{R}} + \mathrm{i}D_{y\mathrm{I}} & D_{z\mathrm{R}} + \mathrm{i}D_{z\mathrm{I}} & D_{\alpha\mathrm{R}} + \mathrm{i}D_{\alpha\mathrm{I}} & 0 & 0 \\ 0 & M_{y\mathrm{R}} + \mathrm{i}M_{y\mathrm{I}} & M_{z\mathrm{R}} + \mathrm{i}M_{z\mathrm{I}} & M_{\alpha\mathrm{R}} + \mathrm{i}M_{\alpha\mathrm{I}} & 0 & 0 \\ 0 & 0 & 0 & 0 & 0 & 0 \\ 0 & 0 & 0 & 0 & 0 & 0 \end{bmatrix} \begin{Bmatrix} \ddot{u}_x \\ \ddot{u}_y \\ \ddot{u}_z \\ \ddot{\phi}_x B \\ \ddot{\phi}_y B \\ \ddot{\phi}_z B \end{Bmatrix}, \tag{5}$$

where l_{e} is the lumped length of the node. The vector on the right-hand side includes the nodal accelerations corresponding to the nodal forces on the left-hand-side vector depicted in Fig. 4. The global self-excited vector $\boldsymbol{F}_{\mathrm{ae}}$ then can be expressed as

$$\boldsymbol{F}_{\mathrm{ae}} = \sum \kappa(-\rho\pi B^2 \boldsymbol{F}_w \ddot{\boldsymbol{w}}) = \boldsymbol{F}_w^{\mathrm{G}} \ddot{\boldsymbol{u}}. \tag{6}$$

The operator $\kappa(\cdot)$ indicates the transformation from local to global coordinate, and the sigma sign expresses the assembly process in FEM.

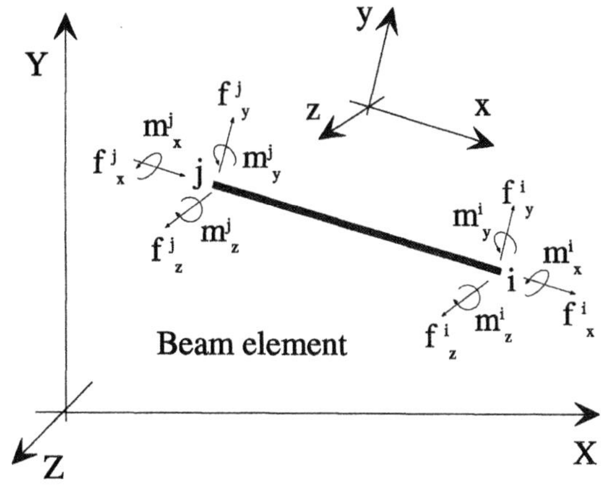

Fig. 4. Beam element's nodal forces.

Neglecting the structural damping matrix $\boldsymbol{C}$, Eq. (6) can be rewritten as

$$\boldsymbol{M}_{\mathrm{F}}\ddot{\boldsymbol{u}} + \boldsymbol{K}\boldsymbol{u} = \boldsymbol{F}_{\mathrm{b}}, \tag{7}$$

where, $\boldsymbol{M}_{\mathrm{F}} = \boldsymbol{M} - \boldsymbol{F}_{w}^{\mathrm{G}}$. Assume harmonic response for the free vibration of $\boldsymbol{u}$ with complex frequency λ as in Eq. (8), the equation of complex eigenproblem, Eq. (9), is formed:

$$\boldsymbol{u} = A\mathrm{e}^{\mathrm{i}\lambda t}, \tag{8}$$

$$\det\left|\boldsymbol{K} - \boldsymbol{M}_{\mathrm{F}}(K)\times\lambda^2\right| = 0. \tag{9}$$

$\boldsymbol{M}_{\mathrm{F}}$ is the complex function of the reduced frequency K, which depends on the wind speed U and unknown modal frequencies ω. To solve this, the mode tracing method [9,10] is employed. This method is an iteration scheme to trace the evolution of each aeroelastic complex mode with step-by-step increment of wind speed. The set of aeroelastic complex modes at a specified wind speed therefore can be determined to facilitate the gust response analysis. The method can be made fast by only one-time tracing mode-by-mode from $U = 0$ to the maximum interested wind speed. The wind speed steps are chosen coinciding with the interested wind speeds so that their complex modal properties can be evaluated and collected in the process.

3. Complex modal analysis

Since $\boldsymbol{M}_{\mathrm{F}}$ is complex and not symmetric, there are two orthogonal sets of complex eigenvectors, left $\boldsymbol{v}_{\mathrm{L}}$ and right $\boldsymbol{v}_{\mathrm{R}}$, which result from

$$\boldsymbol{K}\boldsymbol{v}_{\mathrm{R}n} = \lambda_n^2\boldsymbol{M}_{\mathrm{F}}\boldsymbol{v}_{\mathrm{R}n}, \tag{10}$$

$$\boldsymbol{K}\boldsymbol{v}_{\mathrm{L}m} = \lambda_m^2\boldsymbol{M}_{\mathrm{F}}^{\mathrm{T}}\boldsymbol{v}_{\mathrm{L}m}. \tag{11}$$

By the mode tracing method, complex eigenvalues λ^2 and right eigenvectors $\boldsymbol{v}_{\mathrm{R}}$ are obtained from Eq. (10), which is the equivalent form of Eq. (9). It is obvious that the eigenvalues from Eqs. (10) and (11) are the same for a mode. The left eigenvector therefore can be easily obtained together with the right along the mode tracing process. The orthogonality of the left and right eigenvectors with respect to $\boldsymbol{M}_{\mathrm{F}}$ and $\boldsymbol{K}$ can be proved. That is,

$$\left.\begin{array}{l}\boldsymbol{v}_{\mathrm{L}m}^{\mathrm{T}}\boldsymbol{K}\boldsymbol{v}_{\mathrm{R}n} = 0\\ \boldsymbol{v}_{\mathrm{L}m}^{\mathrm{T}}\boldsymbol{M}_{\mathrm{F}}\boldsymbol{v}_{\mathrm{R}n} = 0\end{array}\right. \quad \text{if } m \neq n \tag{12}$$

The uncouple equation of motion for Eq. (7) in generalized coordinate $\boldsymbol{r}$ then can be obtained

$$(\boldsymbol{v}_{\mathrm{L}}^{\mathrm{T}}\boldsymbol{M}_{\mathrm{F}}\boldsymbol{v}_{\mathrm{R}})\ddot{\boldsymbol{r}} + (\boldsymbol{v}_{\mathrm{L}}^{\mathrm{T}}\boldsymbol{K}\boldsymbol{v}_{\mathrm{R}})\boldsymbol{r} = \boldsymbol{v}_{\mathrm{L}}^{\mathrm{T}}\boldsymbol{F}_{\mathrm{b}} \tag{13}$$

or

$$\ddot{r}_n + \lambda_n^2 r_n = \frac{f_{\mathrm{b}n}}{m_n}, \tag{14}$$

where

$$\boldsymbol{u} = \boldsymbol{v}_{\mathrm{R}}\boldsymbol{r}, \tag{15}$$

$$m_n = \boldsymbol{v}_{\mathrm{L}n}^{\mathrm{T}}\boldsymbol{M}_{\mathrm{F}}\boldsymbol{v}_{\mathrm{R}n}, \tag{16}$$

$$k_n = m_n\lambda_n^2 = \boldsymbol{v}_{\mathrm{L}n}^{\mathrm{T}}\boldsymbol{K}\boldsymbol{v}_{\mathrm{R}n}, \tag{17}$$

$$f_{\mathrm{b}n} = \boldsymbol{v}_{\mathrm{L}n}^{\mathrm{T}}\boldsymbol{F}_{\mathrm{b}}, \tag{18}$$

in which m_n, k_n, $f_{\mathrm{b}n}$ are the complex modal mass, modal stiffness and modal buffeting force, respectively. The physical meaning of the left and right eigenvectors can be understood from Eqs. (13) and (15). The left eigenvectors decide the contribution of external forces to each mode, whereas the right eigenvectors express the vibration mode shapes.

The stiffness or frequency in Eq. (14) is complex, which implies the existence of a frequency phase lag representing the *aerodynamic damping* in the vibration. For ease of solving Eq. (14), further derivations to split the complex modal frequency λ into the modal aerodynamic damping and the modal frequency explicitly in real values are done as follows. Write the complex frequency (square root of the complex eigenvalue) into its real and imaginary parts, $\lambda_n = \lambda_{\mathrm{R}n} + \mathrm{i}\lambda_{\mathrm{I}n}$, and consider the second term of Eq. (14),

$$\lambda_n^2 r_n = (\lambda_{\mathrm{R}n} + \mathrm{i}\lambda_{\mathrm{I}n})^2\, r_n = \left[(\lambda_{\mathrm{R}n}^2 + \lambda_{\mathrm{I}n}^2) + 2\lambda_{\mathrm{I}n}\mathrm{i}(\lambda_{\mathrm{R}n} + \mathrm{i}\lambda_{\mathrm{I}n})\right] r_n. \tag{19}$$

From Eq. (15) and the harmonic motion assumption, Eq. (8), the following relation holds:

$$\dot{r}_n = \mathrm{i}\lambda_n r_n = \mathrm{i}(\lambda_{\mathrm{R}n} + \mathrm{i}\lambda_{\mathrm{I}n}) r_n. \tag{20}$$

Substituting Eq. (20) into Eq. (19) yields

$$\lambda_n^2 r_n = (\lambda_{\mathrm{R}n}^2 + \lambda_{\mathrm{I}n}^2) r_n + 2\lambda_{\mathrm{I}n}\dot{r}_n. \tag{21}$$

With this new expression of the second term, Eq. (14) can be rearranged in a more convenient and explicit form as

$$\ddot{r}_n + 2\xi_{\mathrm{ae}n}\omega_n\dot{r}_n + \omega_n^2 r_n = Q_{\mathrm{b}n}. \tag{22}$$

The modal aerodynamic damping ratio $\xi_{\mathrm{ae}n}$ and frequency ω_n here are *real* values as follows:

$$\omega_n = \sqrt{\lambda_{\mathrm{R}n}^2 + \lambda_{\mathrm{I}n}^2}, \qquad \xi_{\mathrm{ae}n} = \lambda_{\mathrm{I}n}\Big/\sqrt{\lambda_{\mathrm{R}n}^2 + \lambda_{\mathrm{I}n}^2}, \tag{23}$$

where $\lambda_n = \lambda_{\mathrm{R}n} + \mathrm{i}\lambda_{\mathrm{I}n}$; $Q_{\mathrm{b}n} = f_{\mathrm{b}n}/m_n$ is the normalized buffeting force. At this stage, under the assumption of proportional damping, the structural damping can be introduced in the form of modal structural damping ratio ξ_{sn}. The uncouple equation of motion becomes

$$\ddot{r}_n + 2\xi_n\omega_n\dot{r}_n + \omega_n^2 r_n = Q_{\mathrm{b}n} \tag{24}$$

in which $\xi_n = (\xi_{aen} + \xi_{sn})$ is the total modal damping ratio. The most important advantage of this equation of motion when compared to those of other existing methods is that it is completely uncoupled at a mean wind speed, so that it can then be easily solved in either time or frequency domain. This advantage is due to the use of the aeroelastic complex modes, in which the aeroelastic effects have been incorporated.

4. Gust response in time domain

The gust response using aeroelastic complex modes in time domain has been presented in Ref. [11]. A brief description is given here for the context. By the quasi-steady assumption, the buffeting force $\boldsymbol{f}_b$ on a unit length of bridge deck is computed from horizontal $u(t)$ and vertical $w(t)$ turbulent velocities, which are numerically generated by a digital filter scheme. The buffeting forces applying to the shear nodes along the deck at each time step are then determined and transformed into global coordinate to form the buffeting force vector $\boldsymbol{F}_b$ to incorporate into Eq. (13). The gust response analysis is then carried out by the Newmark β direct integration method for Eq. (24). The time history of the generalized displacement $\boldsymbol{r}$ is then transformed into physical displacement $\boldsymbol{u}$ by Eq. (15) at any node along the deck. The response spectra, RMS and ensemble average maximum amplitude then can be evaluated directly from the response time-histories.

5. Gust response in frequency domain

The equation of motion in frequency domain can be expressed as,

$$(-\omega^2 + 2\xi_n\omega_n\omega + \omega_n^2)\bar{r}_n = \bar{Q}_{bn} \tag{25}$$

or in matrix form as

$$\boldsymbol{E}\bar{\boldsymbol{r}} = \bar{\boldsymbol{Q}}_b, \tag{26}$$

where the overhead bar denotes the Fourier transform of the corresponding variable. Apparently, the impedance matrix $\boldsymbol{E}$ is diagonal. Using standard random vibration analysis for an MDOF system to develop the equation of motion, Eq. (26), into spectral form, the power spectral density (PSD) matrix for the generalized coordinate $\boldsymbol{r}$ can be expressed by

$$\boldsymbol{S}_{rr}(\omega) = \boldsymbol{E}^{-1}\boldsymbol{S}_{Q_bQ_b}\left[\boldsymbol{E}^{H}\right]^{-1}, \tag{27}$$

where $\boldsymbol{S}_{Q_bQ_b}$ is the PSD matrix of modal buffeting force. $\boldsymbol{E}^H$ is the Hermitian transpose of $\boldsymbol{E}$. There existed a developed form of $\boldsymbol{S}_{Q_bQ_b}$ by the mechanical modes in real number and K-domain for bridge analysis in Ref. [5]. However, since the aeroelastic complex modes here are complex and separated into left and right eigenvectors, necessary derivations in ω-domain consistent with complex modal analysis must be done [12]. Using analogous notation in Ref. [5], the general term of

$\boldsymbol{S}_{Q_b Q_b}$ then can be derived as

$$S_{Q_{bn}Q^{H}_{bm}}(\omega) = \left(\frac{\rho UB}{2}\right)^2 \frac{1}{m_n m_m^*} \int_0^L \int_0^L \Psi(x_A, x_B, \omega)\, dx_A\, dx_B, \tag{28}$$

where

$$\begin{aligned}\Psi(x_A, x_B, \omega) =& \{\tilde{q}_n(x_A)\, \tilde{q}_m^*(x_B)\, S_{uu}\,(x_A, x_B, \omega) \\ &+ \tilde{s}_n(x_A)\, \tilde{s}_m^*(x_B)\, S_{ww}\,(x_A, x_B, \omega) + [\tilde{q}_n(x_A)\, \tilde{s}_m^*(x_B) \\ &+ \tilde{s}_n(x_A)\, \tilde{q}_m^*(x_B)]\, \mathrm{Co}_{uw}\,(x_A, x_B, \omega) + \mathrm{i}[\tilde{q}_n(x_A)\, \tilde{s}_m^*(x_B) \\ &- \tilde{s}_n(x_A)\, \tilde{q}_m^*(x_B)]\, Q_{uw}\,(x_A, x_B, \omega)\}\end{aligned} \tag{29}$$

in which L is the bridge's length, x_A, x_B are the span locations, S_{uu}, S_{ww} are, respectively, the uu-cross-spectrum and ww-cross-spectrum, Co_{uw} is the cospectrum; Q_{uw} is the quadrature spectrum between two points x_A and x_B, the asterisk denotes the complex conjugate operation, and

$$\begin{aligned}\tilde{q}_m(x) &= 2\,[C_L y_{Lm}(x) + C_D z_{Lm}(x) + C_M \alpha_{Lm}(x)], \\ \tilde{s}_n(x) &= (C_L' + C_D)\, y_{Ln}(x) + C_D' z_{Ln}\,(x) + C_M' \alpha_{Ln}(x),\end{aligned} \tag{30}$$

where $y_{Ln}(x)$, $z_{Ln}(x)$ and $\alpha_{Ln}(x)$ are, respectively, the vertical, horizontal and rotational components of the left eigenvector of the nth aeroelastic complex mode at a span location x. C_L, C_D, C_M are the respective static coefficients. The prime (′) denotes the derivatives with respect to the angle of attack α. The PSD matrix for generalized response $\boldsymbol{S}_{rr}$ can be solved by Eq. (27). Then the PSD for physical displacements at a span location x are

$$\begin{aligned}S_{yy}(x, \omega) &= \sum_n \sum_m y_{Rn}(x) y_{Rm}^*(x)\, S_{rnrm}(\omega), \\ S_{zz}(x, \omega) &= \sum_n \sum_m z_{Rn}(x) z_{Rm}^*(x)\, S_{rnrm}(\omega), \\ S_{\alpha\alpha}(x, \omega) &= \sum_n \sum_m B^{-2} \alpha_{Rn}(x)\, \alpha_{Rm}^*(x)\, S_{rnrm}(\omega),\end{aligned} \tag{31}$$

where $y_{Rn}(x)$, $z_{Rn}(x)$ and $\alpha_{Rn}(x)$ are, respectively, the vertical, horizontal and rotational components of the right eigenvector of the nth aeroelastic complex mode at a span location x. The mean-square values of the gust response then can be evaluated by taking the integration of the physical displacement PSD with respect to frequency $f = \omega/2\pi$,

$$\begin{aligned}&\sigma_y^2(x) = \int_0^\infty S_{yy}(x, f)\, df, \quad \sigma_z^2(x) = \int_0^\infty S_{zz}(x, f)\, df, \\ &\sigma_\alpha^2(x) = \int_0^\infty S_{\alpha\alpha}(x, f)\, df.\end{aligned} \tag{32}$$

The expected maximum vibration response occurring in the time interval T is [13]

$$p_{\max}(x) = k_p(x)\,\sigma_p(x), \tag{33}$$

$$k_p(x) = [2\ln \nu(x)T]^{1/2} + \frac{0.577}{[2\ln \nu(x)T]^{1/2}},$$

$$\nu(x) = \left[\int_0^\infty f^2 S_{pp}(x,f)\,\mathrm{d}f \Big/ \int_0^\infty S_{pp}(x,f)\,\mathrm{d}f\right]^{1/2}, \tag{34}$$

where p stands for y, z or α, σ_p is the root-mean-square, k_p is the respective peak factor, $S_{pp}(x,f)$ is from Eq. (31) with ω replaced by f. This expected maximum vibration amplitude can be treated equivalent to the ensemble average maximum amplitude in the time domain analysis for comparison.

There are two assumptions in the above buffeting force formulation. First, the aerodynamic admittance functions equal unity (quasi-static assumption). Second, the strip assumption [13] holds for the spatial coherence functions of the buffeting forces along the deck. These assumptions are justifiable due to the fact that frequencies of the significant modes of a very long-span bridge are very small. However, these effects can be easily incorporated into the present approach if the information is available.

6. Numerical example: Akashi Kaikyo bridge model

A numerical example of the present method in both time and frequency domains is considered for the 3-D FEM model of the Akashi Kaikyo bridge. The gust response is calculated at eight levels of mean wind speeds: $U = 30$, 36, 44, 54, 63, 71, 76 and 80 m/s. Turbulence intensity $I_u = 9.6\%$, $I_w = 6\%$ for turbulent components $u(t)$ and $w(t)$, respectively. The modal structural damping ratio is $(0.03/2\pi)$. The first 32 modes of the bridge are considered. The frequency step is $\Delta f = 0.001$ Hz, being fine enough to tune to the modal frequencies. The time step for the turbulence simulation [11] and the Newmark β direct integration is 0.1 s. The duration to obtain response is 150 min, which is equivalent to the 15-min recorded response of the full-model test. The sets of static coefficients and unsteady coefficients of the modified cross-section of the bridge at zero angle of attack [2] are used. This is an assumption for simplicity in the analysis. Descriptions of this set in terms of flutter derivatives can be found in Ref. [6]. A more sophisticated investigation taking account of the change of the angle of attack along the deck due to mean wind-induced displacement was presented in Ref. [10].

The set of *direct fitting* turbulence properties [11] is used for the best match with the condition of the full-model test. This set was obtained from the direct analysis of the time–history records of the turbulence generated in the wind tunnel of the full-model test [11]. The scaled (from model to prototype) spectra and point cospectrum, and the coherence functions are expressed as follows:

$$\frac{fS_u(f)}{\overline{u^2}} = \frac{5.11 f_r}{(1+7.05 f_r)^{5/3}}, \quad \frac{fS_w(f)}{\overline{u^2}} = \frac{6.15 f_r}{1+3.7 f_r^{5/3}},$$
$$\frac{f\,\mathrm{Co}_{uw}(f)}{\overline{u^2}} = \frac{-1.5 f_r}{(1+2.94 f_r)^{2.4}}, \tag{35}$$

$$\mathrm{Coh}_{u_A u_B}(f) = (1 - 0.001\,\mathrm{d}x - 0.0003\,\mathrm{d}x^2)\exp(-f\,C_x^u\,\mathrm{d}x/U),$$

$$\mathrm{Coh}_{w_A w_B}(f) = (1 - 0.03\mathrm{d}x + 0.0002\,\mathrm{d}x^2)\exp(-f\,C_x^w\,\mathrm{d}x/U),$$

$$\mathrm{Coh}_{u_A w_B}(f) = \mathrm{Coh}_{w_A u_B}(f)$$
$$= 0.5\frac{\mathrm{Co}_{uw}^2(f)}{S_u(f)\,S_w(f)}[\mathrm{Coh}_{u_A u_B}(f) + \mathrm{Coh}_{w_A w_B}(f)], \tag{36}$$

where $\overline{u^2}$ is mean square of $u(t)$, $f_r = fz/U$, z is the deck's height, $C_x^u = 12$ and $C_x^w = 8$. The coherence functions are defined in square forms, and the quadrature spectra are neglected since they appeared very small in the analysis.

6.1. Coupled gust response of the bridge model

Table 1 lists the description of some typical aeroelastic complex modes for the main span of the bridge. In general, an aeroelastic complex mode may consist of 1, 2 or even 3 motion components, depending on the mode's type and the wind speed. Such evolution of the aeroelastic complex modes from $U = 0$ to 76 m/s can be seen in Table 1 from the decrease of modal frequencies and development of motions in mode shapes, such as for modes #3, #9 and #10. One of the most interesting developments is shown in Fig. 5 for mode #10, which is the typical mode to observe the aeroelastic effect. The aeroelastic complex mode #10 at $U = 0$ m/s, being the mechanical mode

Table 1
Descriptions of typical mode shapes of the main span of the Akashi Kaikyo bridge model[a]

Mode no.	Aeroelastic mode at $U = 0$ m/s = mechanical mode		Aeroelastic mode at $U = 76$ m/s	
	Description	Frequency (Hz)	Description	Frequency (Hz)
#1	H1	0.03936	H1	0.03916
#2	V1	0.06503	V1	0.06289
#3	H2	0.07908	H2 + V2	0.07875
#4	V2	0.08427	V2	0.08250
#7	V2	0.11234	V2	0.11065
#8	V3	0.12283	V3	0.12090
#9	R1 + H3	0.12983	R1 + H3 + V3	0.12869
#10	R1 + H3	0.14779	R1 + H3 + V3	0.13713

[a] 'H': horizontal, 'V': vertical, 'R': rotational motions; 1, 2, 3..., are the orders of the mode.

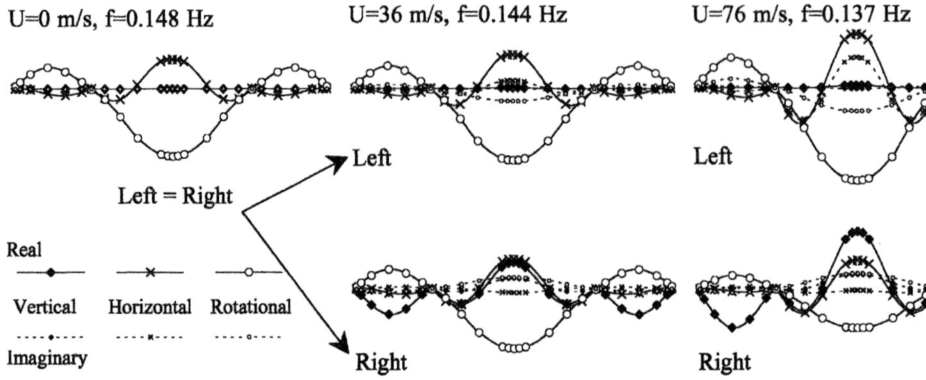

Fig. 5. Evolution of the aeroelastic complex mode #10 with wind speed.

with identical left and right eigenvectors, is only a 2-motion coupling mode of horizontal 'H3' and rotational 'R1' components. As the wind speed increases, e.g. to 36 m/s and then to 76 m/s, the mode evolves considerably due to the aeroelastic effect. The modal frequency decreases from 0.148 to 0.144 Hz and then to 0.137 Hz. The left and the right eigenvectors gradually become quite different from the mechanical mode, and also different from each other. The most interesting point is the appearance and development of the vertical component 'V3' to make the mode become more and more 3-motion coupling at higher wind speeds. A similar evolution is obtained for mode #9. Other modes also evolve to a lesser extent [12]. A more detailed interpretation of the coupled response can be found in Ref. [14].

In accordance with the evolution of the aeroelastic complex modes, the spectacular evolution of the vertical response spectrum with wind speeds is shown in Fig. 6. At low wind speeds, the vertical response is dominated by mode #2. As wind speed increases, modes #10 and #9 develop, and finally dominate the response. As a result, the response spectra in Fig. 7 of all three motions – vertical, horizontal and rotational – have peaks of almost the same order of magnitude at the same frequency of mode #10, and to a smaller extent at mode #9. This coupled gust response – response in all 3 motions at the same frequencies – therefore can be effectively characterized and predicted by the aeroelastic complex modes.

Fig. 7 shows a comparison of response spectra at 76 m/s. Very good similarities between those from time and frequency domain analyses are obtained. These calculated response spectra and their developments with wind speed are very analogous to the experimental results presented in HSBA's report [2]. More details of this comparison can be found in Ref. [12]. Finally, Fig. 8 presents the comparison of gust response in term of root-mean-square (RMS) and average maximum amplitude (AMA). Very good agreements between calculated results in both domains with the experimental results are achieved. The peak factors (AMA divided by RMS) by Eq. (34) in the frequency domain, and by direct calculation from results in time domain agree well with each other and also with the experimental results. The average peak factors are found to be 3.25, 2.95 and 3.2 for vertical, horizontal and

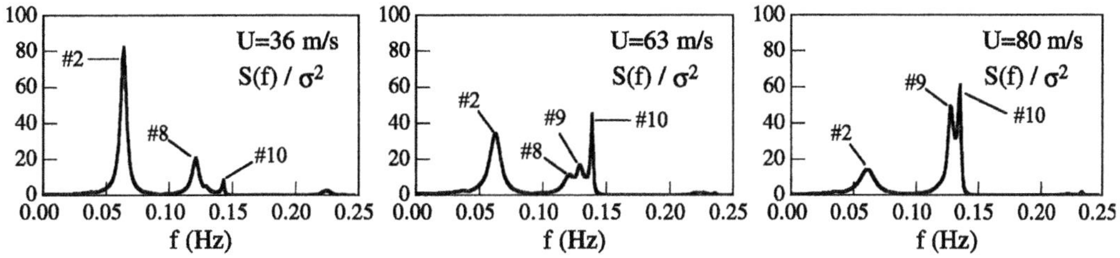

Fig. 6. Evolution of the vertical response spectrum with wind speed.

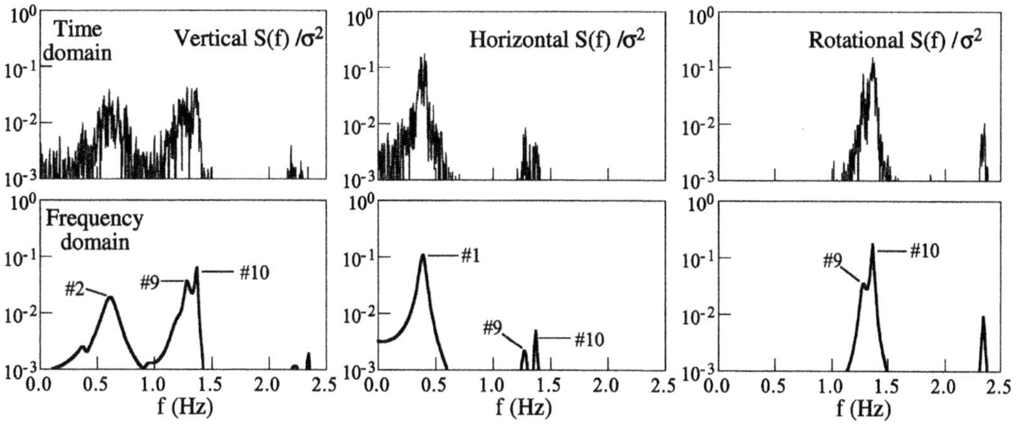

Fig. 7. Response spectra of the mid-span of the bridge at $U = 76\,\mathrm{m/s}$ in time domain (upper) and frequency domain (lower) analysis.

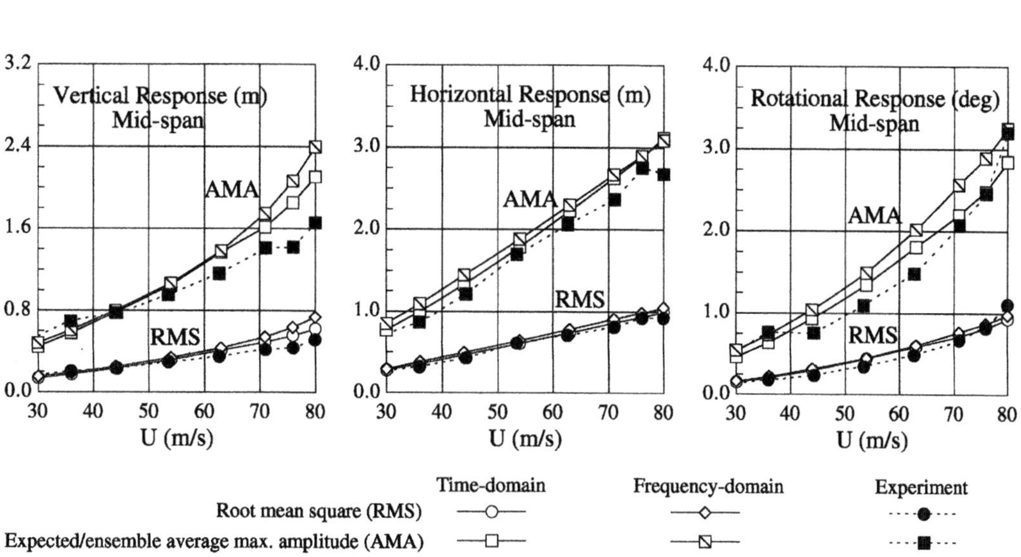

Fig. 8. Comparison of calculated and experimental responses at the mid-span of the bridge.

rotational responses, respectively. These peak factors are found to be not so much changed with the wind speed.

6.2. A parameter study on turbulence properties

The following turbulence properties have been reported as the targets of the wind-tunnel turbulence simulation of the full-model test (see Ref. [6]). An analysis by the present approach in time domain is made here using this set of turbulence properties (hereafter referred to as *reported target* set) as input. The Hino spectrum and Busch & Panofski spectrum hold for S_u and S_w, respectively, as follows:

$$\frac{f\,S_u(f)}{\overline{u^2}} = 0.4751\frac{f/\beta}{\left\{1+(f/\beta)^2\right\}^{5/6}}, \quad \beta = 0.01718\frac{\alpha K_r U_{10}}{I_u^3}\left(\frac{z}{10}\right)^{(2m-3)\alpha-1}, \tag{37}$$

$$\frac{f\,S_w(f)}{\overline{w^2}} = 0.6320\frac{f_r/f_{\max}}{1+1.5(f_r/f_{\max})^{5/3}}, \tag{38}$$

where $\alpha = 1/8$; $K_r = 0.0025$, I_u is the turbulence intensity, $m = 3$ is an empirical factor, U_{10} is the mean wind speed at 10 m height, $f_{\max} = 0.4$ is an empirical factor. Since there was no reported information for cospectra Co_{uw}, Kaimal's empirical form is taken [6]:

$$\frac{f\,\mathrm{Co}_{uw}(f)}{u^{*2}} = \frac{-14f_r}{(1+9.6f_r)^{2.4}} \tag{39}$$

in which u^* is the friction velocity. A modified form of the Von Karman spatial coherence was reported to agree with the measured one (see Ref. [6]),

$$\mathrm{Coh}(f) = \exp\left\{-\frac{cB_1\,\mathrm{d}x}{\pi L}\sqrt{1+(2\pi/B_1)^2(fL/U)^2}\right\}, \tag{40}$$

where $L = L_u^x = 70\,\mathrm{cm}$ for $\mathrm{Coh}_{u\mathrm{A}u\mathrm{B}}$; $L = L_w^x = 40\,\mathrm{cm}$ for $\mathrm{Coh}_{w\mathrm{A}w\mathrm{B}}$; $B_1 = 0.747$; decay factor $c = 8$.

Figs. 9 and 10 show the comparison of the *reported target* set with the *direct fitting* set [11] used in the previous section. All spectra and cospectrum have been

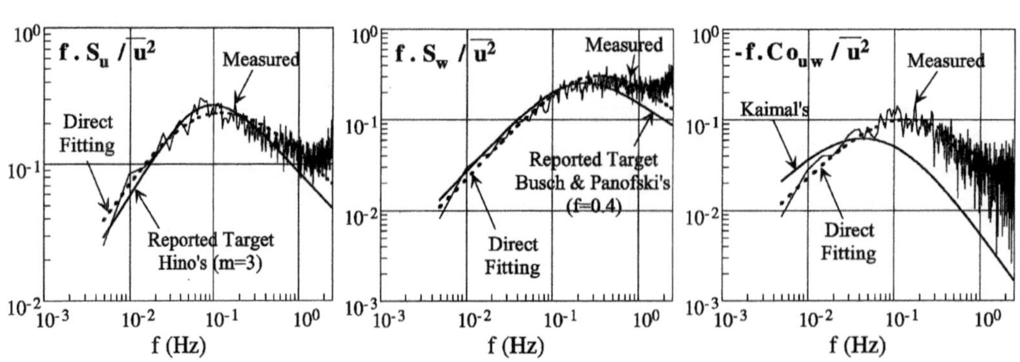

Fig. 9. Comparison of auto-spectra and cospectrum.

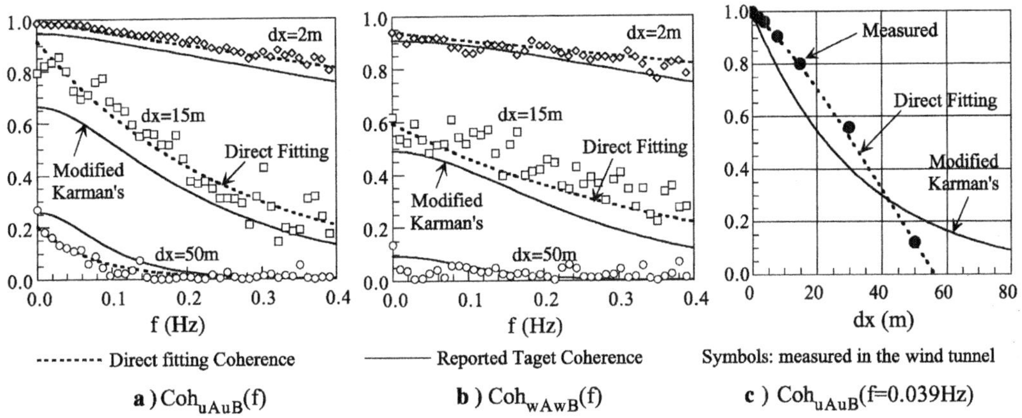

Fig. 10. Comparison of coherence functions.

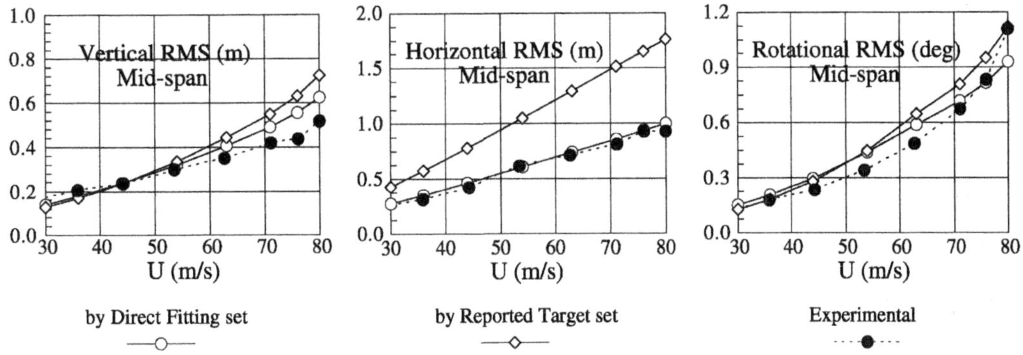

Fig. 11. RMS response comparison between analyses by different sets of turbulence properties.

transferred to the same normalizing value for comparison. The RMS responses are shown in Fig. 11. Compared to the experimental results, the calculated responses by both sets of turbulence agree very well with experimental ones, except the horizontal response, which is highly overestimated when using the reported target set. Other, previous works using the reported target set of turbulence also encountered similar discrepancies (Ref. [6], among others).

The reasons for the discrepancies are found as follows. From the response spectra (Fig. 7), the horizontal response is governed mainly by mode #1, which is the first symmetric horizontal mode 'H1' (see Table 1), and has a very low frequency of around 0.039 Hz. In the reported target set, the modified Von Karman coherence appears higher than the measured coherence of $u(t)$ at the far distance of $\mathrm{d}x = 50\,\mathrm{m}$, as in Fig. 10a. This can be seen more clearly in Fig. 10c, which presents the comparison of coherence versus separate distance at the frequency of mode #1 (0.039 Hz), which influences directly the horizontal response. The comparison indicates clearly that the modified Von Karman coherence is higher than the measured coherence from $dx = 43$ m and continuing to very far distances following exponential rule. On the contrary, the measured coherence decreases and appears to vanish quickly. For a very long-span bridge, the coherence at far distances, even small, still has an important effect on the response dominated by the first mode like 'H1' in this case. As a result, using the reported target set yields around 1.8 times overestimation of horizontal response. Almost 90% of the error is caused by the coherence of $u(t)$. Other properties, though also showing discrepancies with measured results as seen in Figs. 9 and 10b, contribute to the error to a much lesser extent.

6.3. *Effectiveness of using aeroelastic complex mode*

By using the aeroelastic complex-mode, in which the aeroelastic effect has been incorporated, the equation of motion is completely uncoupled at a mean wind speed. The present approach therefore inherently possesses the advantages of the normal modal analysis technique, including the following two issues:

(a) The response of the bridge can be well predicted by considering only a few significant modes. For example, the response at the bridge's mid-span as found in Ref. [11] was mainly due to only five aeroelastic complex modes: #1, #2, #8, #9 and #10 (see Table 1). The results by these modes are smaller than the results by the 32 modes, with maximum errors of only 2.0%, 0.12% and 4.9% for vertical, horizontal and rotational response, respectively.

(b) The effect of the cross terms (or off-diagonal terms) of the PSD matrix of modal buffeting force $\boldsymbol{S}_{Q_b Q_b}$ in Eq. (28) appears not insignificant. In the formulation of frequency domain, Eqs. (28) and (31) have been derived with multi-mode consideration, so that $\boldsymbol{S}_{Q_b Q_b}$ is a full matrix. Neglecting the off-diagonal terms in $\boldsymbol{S}_{Q_b Q_b}$ makes the procedure become the so-called single-mode analysis, which is much simpler and requires much lesser computational effort. The procedure for single-mode spectral formulation is then similar to the multi-mode spectral formulation with the following two revisions:

- The PSD matrix of modal buffeting force is diagonal, then Eq. (28) becomes the general term of the diagonal elements as follows:

$$S_{Q_{bn}Q_{bn}^{H}}(\omega) = \left(\frac{\rho UB}{2}\right)^2 \frac{1}{m_n m_n^*} \int_0^L \int_0^L \Psi(x_A, x_B, \omega)\, dx_A\, dx_B, \tag{41}$$

where

$$\begin{aligned}\Psi(x_A, x_B, \omega) = \{&\tilde{q}_n(x_A)\,\tilde{q}_n^*(x_B)\, S_{uu}(x_A, x_B, \omega) \\ &+ \tilde{s}_n(x_A)\,\tilde{s}_n^*(x_B)\, S_{ww}(x_A, x_B, \omega) + [\tilde{q}_n(x_A)\,\tilde{s}_n^*(x_B) \\ &+ \tilde{s}_n^*(x_A)\,\tilde{q}_n^*(x_B)]\, C_{uw}(x_A, x_B, \omega) + i[\tilde{q}_n(x_A)\,\tilde{s}_n^*(x_B) \\ &- \tilde{s}_n(x_A)\,\tilde{q}_n^*(x_B)]\, Q_{uw}(x_A, x_B, \omega)\}.\end{aligned}$$

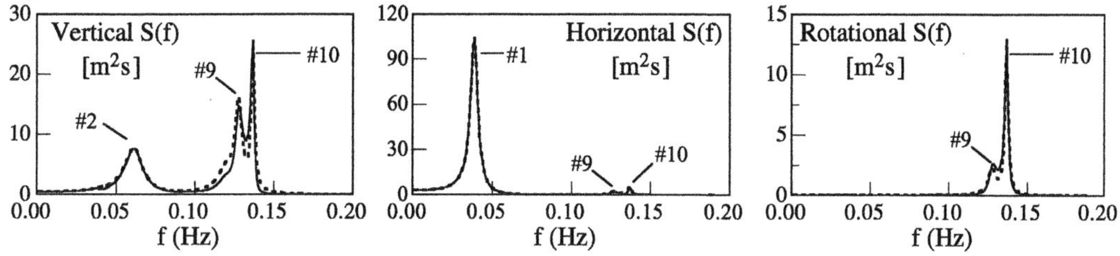

Fig. 12. Effects of cross-terms on response spectra at $U = 76\,\text{m/s}$. The continuous lines are results from analysis with the cross-terms (multi-mode); the dotted lines are results from analysis neglecting the cross-terms (single-mode) in frequency domain.

- The PSD for physical displacemements at span location x in Eq. (31) are

$$\begin{aligned} S_{yy}(x,\,\omega) &= \sum_n y_{\mathrm{R}n}(x)\, y^*_{\mathrm{R}n}(x)\, S_{rnrn}(\omega), \\ S_{zz}(x,\,\omega) &= \sum_n z_{\mathrm{R}n}(x)\, z^*_{\mathrm{R}n}(x)\, S_{rnrn}(\omega), \\ S_{\alpha\alpha}(x,\,\omega) &= \sum_n B^{-2}\alpha_{\mathrm{R}n}\,(x)\alpha^*_{\mathrm{R}n}\,(x) S_{rnrn}(\omega). \end{aligned} \tag{42}$$

Fig. 12 shows the comparison of the response spectra at a wind speed of 76 m/s between two cases: (a) analysis considering the cross-terms (multi-mode) and (b) analysis neglecting the cross-terms (single-mode). Notable differences between results from the two cases can be seen especially for closely spaced dominant modes, such as modes #9 and #10. The vertical spectrum is therefore most affected. Less effect is found for rotational spectrum, and almost no effect is seen for the horizontal spectrum. Comparisons at other wind speeds [12] show that the effect is more pronounced at higher wind speeds. However, the maximum errors, which are found at $U = 80\,\text{m/s}$, are small at 7.8%, 0.7% and 1.2% for vertical, horizontal and rotational RMS response, respectively.

7. Conclusion

A newly developed method for coupled gust response has been presented. Using the aeroelastic complex modes, a better understanding on the coupled gust response of long-span bridges due to aeroelastic effects has been achieved, which effectively predicts and explains the observed dynamic behaviors of the full-model test of the Akashi Kaikyo Bridge. From the numerical example and parameter studies, the following remarks are made:

(a) By using aeroelastic complex modes, the coupled gust response due to aeroelastic effects could be accurately captured. The appearance and development of the vertical motion to couple with other motions can be clearly represented by the aeroelastic complex modes.
(b) Due to very low vibration frequencies, the gust response of a long-span bridge is very sensitive to the turbulence properties, especially the spatial coherence. Some of their effects on the gust response of long-span bridges have been pointed out.

(c) Advantages of using aeroelastic complex mode have been addressed. Good prediction of response can be made from a few significant modes. The single-mode spectral formulation, which is simple and requires much lesser computational effort, has been shown to the effective enough.

The investigations and results have proved the effectiveness and accuracy of the presented method, which could be a reliable approach for the gust response analysis of long-span bridges in future.

References

[1] T. Miyata, H. Sato, R. Toriumi, M. Kitagawa, H. Katsuchi, Full-model wind tunnel study on the Akashi Kaikyo Bridge, Proceedings of the Ninth International Conference on Wind Engineering, New Delhi, India, 1995, pp. 793–802.

[2] The Honshu-Shikoku Bridge Authority (HSBA), Report on full model test of the Akashi Kaikyo Bridge, Japan, 1992 (in Japanese).

[3] V. Boonyapinyo, T. Miyata, H. Yamada, Combined flutter and buffeting response of suspension bridges in time domain, Proceedings of the Fourth Asia–Pacific Symposium on Wind Engineering, Australia, 1997, pp. 99–102.

[4] Y. Fujino, K. Wilde, J. Masukawa, B. Bhartia, Rational function approximation of aerodynamic forces on bridge deck and its application to active control of flutter, Proceedings of the Ninth International Conference on Wind Engineering, New Delhi, India, 1995, pp. 994–1005.

[5] A. Jain, N.P. Jones, R.H. Scanlan, Coupled flutter and buffeting analysis of long-span bridges, J. Struct. Eng. ASCE 122 (7) (1996) 716–725.

[6] H. Katsuchi, N.P. Jones, R.H. Scanlan, Multimode coupled flutter and buffeting analysis of the Akashi Kaikyo bridge, J. Struct. Eng. ASCE 125 (1) (1999) 60–70.

[7] T. Miyata, H. Yamada, Coupled flutter estimate of a suspension bridge, J. Wind Eng. 37 (1988) 485–492.

[8] T. Miyata, H. Yamada, K. Kazama, On application of direct flutter FEM analysis for long span bridges. Proceedings of the Ninth International Conference on Wind Engineering, New Delhi, India, 1995.

[9] N.N. Dung, T. Miyata, H. Yamada, The mode tracing method for flutter of long span bridges, Proceedings of the 14th National Symposium on Wind Engineering, Kyoto, Japan, 1996.

[10] N.N. Dung, T. Miyata, H. Yamada, N.N. Minh, Flutter responses in long span bridges with wind induced displacement by the mode tracing method, J. Wind Eng. Ind. Aerodyn. 77–78 (0) (1998) 367–379.

[11] N.N. Minh, T. Miyata, H. Yamada, Y. Sanada, Numerical simulation of wind turbulence and buffeting analysis of long span bridges, J. Wind Eng. Ind. Aerodyn. 83 (1–3) (1999) 301–315.

[12] N.N. Minh, Aeroelastic complex-mode method for coupled buffeting of long-span bridges, Dr. Eng. Dissertation, Yokohama National University, Japan, 1998.

[13] A.G. Davenport, Buffeting of a suspension bridge by storm winds, J. Struct. Division ASCE 88 (ST3) (1962) 233–268.

[14] N.N. Minh, H. Yamada, T. Miyata, An interpretation of aeroelastic phenomenon for bridge decks in wind flows, Proceedings of the Seventh East Asia–Pacific Conference on Structural Engineering & Construction, Kochi, Japan, Vol. 1, 1999, pp. 663–668.

Papers From

Marine Structures

Structural Engineering Compendium I

Experimental study of slam-induced stresses in a containership

J. Ramos[a], A. Incecik[b], C. Guedes Soares[a,*]

[a]*Unit of Marine Technology Engineering, Instituto Superior Técnico, Universidade Técnica de Lisboa, Av. Rovisco Pais, 1049-001 Lisboa, Portugal*
[b]*University of Newcastle upon Tyne, Department of Marine Technology, Armstrong Building, NE 7RU Newcastle upon Tyne, UK*

Received 31 December 1998; received in revised form 19 January 2000; accepted 28 February 2000

Abstract

Experiments for the ship motions and sea loads were carried out on a segmented model of a container ship in ballast condition. Comparisons between the measurements and the theoretical results were carried out for the vertical motions and bending moments. For the evaluation of the primary stresses it is assumed that the total vertical bending moment induced by waves is divided into one component obtained by the linear theory and another one is due to the slamming loads. Several formulations for the determination of the slamming loads are compared with experimental results. The vibratory response of the model is calculated by modelling the hull with rotational springs and rigid links. Linear finite elements with a consistent mass formulation are adopted for the structural model and the response is obtained by modal superimposing and direct integration methods. © 2000 Elsevier Science Ltd. All rights reserved.

1. Introduction

When a ship travels in a seaway its hull is vulnerable to unsteady wave, wind and current loading. Among those external loads the dynamic wave forces present the most significant design problem for ship owners, shipbuilders and classification societies. In general, the dynamic wave loads consist of two components: the global and primary loads. The global loads are induced by the unsteady hydrodynamic

* Corresponding author. Tel.: + 351-21-841-7607; fax: + 351-21-847-4015.
E-mail address: guedess@alfa.ist.utl.pt (C. Guedes Soares).

Reprinted from *Marine Structures* **13 (1)**, 25-51 (2000)

pressure due to the fluid oscillatory motions surrounding the hull while the local or secondary loads, such as slamming and whipping are due to wave impacts. In this paper both global and local loads are considered through both theoretical and experimental investigations.

Ochi and Motter [1] presented a complete description of the slamming problem. Due to a large number of unknowns required for the determination of the whipping stresses they suggested some simple formulae for the calculation of the slamming loads, for practical purposes. These formulae were based on experiments with frigate models. They stressed the importance at the design stage of the combined effect of wave-induced and whipping stresses, i.e. the total bending moment induced by the waves.

Kawakami et al. [2], whose study was based on experimental work for a tanker, proposed an expression for the time history of the slamming loads. They found that the Ochi and Motter [1] formulation tends to underpredict the maximum slamming pressure when compared with the experimental measurements.

Belik et al. [3] assumed that the bottom slamming could be divided into two different components: impact and momentum slamming. Based on this assumption they used the Ochi and Motter method for the determination of the maximum slamming pressure and the Kawakami et al. expression for the determination of the time history of the slamming impact force. After that, they carried out calculations for the vertical bending moments and shear forces in regular head seas.

Using the same approach for the calculation of the slamming loads, Belik and Price [4] made comparisons for two different slamming theories using time simulation of ship responses in irregular seas. They found that the magnitudes of the responses after a slam depend very much on the mathematical model adopted in the calculation of the slamming loads.

Yamamoto et al. [5] calculated the non-linear ship motions based on the equations given by the linear theory but with time-varying coefficients dependent on the instantaneous sectional draft. They also included the hydrodynamic impact component given by the rate of change of the sectional added mass, assuming that this force only acts on the vessel when the section is penetrating the water. After that they carried out experiments and calculations on a bulk carrier model for head seas. They found that the accuracy of the calculation of the hydrodynamic coefficients has a significant influence on the results of the slamming forces, and the computation with accurate coefficients results in better agreement with experiments. A study of the accuracy of the hydrodynamic coefficients for several ship sections, using the common methods, were made by Ramos and Guedes Soares [6].

Guedes Soares [7] developed a method in which the theory of Salvesen et al. [8] was used to solve the linear motions and the Frank close fit method was used for the evaluation of the sectional added mass. Slams were identified from a time domain simulation and slamming loads were calculated using the momentum theory.

Molin et al. [9] compared the results of impact force calculations obtained from conservation of momentum and conservation of kinetic energy. They concluded that both methods give identical results.

Tao and Incecik [10] investigated the large-amplitude motions and bow flare slamming pressures in regular waves. The non-linear restoring, damping and fluid momentum forces were considered in predicting ship motions in the time domain. The momentum slamming theory and Wagner theory were used to predict the bow flare slamming pressures. The bow flare slamming pressures were calculated by separating the pressure into the water immersion impact pressure and the wave striking impact pressure. A satisfactory correlation between the results of predictions and model test measurements was obtained.

Sames et al. [11] applied a finite-volume method to predict impact coefficients around the bow region of a ship during slamming. Ship motions in regular waves were predicted by a linear panel method which takes into account incident, diffracted and radiated waves. The impact pressures were calculated by processing the results of the computed pressure coefficients and the transfer functions of ship motions in the time domain. No comparisons with the measurements were given.

Comparisons between the full-scale measurements and theoretical predictions were carried out by Aksu et al. [12] for a fast patrol boat travelling in rough seas. Due to the uncertainty of the wave measurements in a real sea state, the experimental results of the vertical bending moments were compared with calculations for two different sea states in a histogram form and satisfactory results were found.

It was found by Ramos and Guedes Soares [13] that the several slamming load formulations can produce large differences in the slamming pressures, loads and also in primary stresses. The Ochi and Motter method tends to underpredict the pressures, loads and also bending moments when compared with the other methods.

It is apparent that despite the various formulations that have been developed, some open questions remain in the slam-induced response of ships, and the available experimental data are not very large. Therefore, it was decided to conduct an experimental program in order to obtain data to compare with the prediction of the computational approach developed. The experiments were carried out on a segmented model of a container ship in the Hydrodynamics Laboratory of the University of Glasgow.

2. Model details

2.1. Main particulars of the model

The model used to validate the motions and loads was the S175 containership, which was used earlier by the Seakeeping and Manoeuvring ITTC Committees in comparative studies. The scale ratio of the model is 1/70 and the experiments were carried out in the ballast condition. Table 1 shows the model's main particulars for the ballast condition.

In order to measure the vertical shear forces and bending moments, the model was cut in three longitudinal positions and the segments were connected using flexible bars. In each bar three strain gauges were mounted, two of which were used to measure the normal stresses, i.e. bending moments and the third one to measure the shear forces.

Table 1
Model's main particulars

L_{pp} (m)	2.50
B (m)	0.363
T (m)	0.10
Δ (kg)	48.8
C_b	0.54
LCG (aft of midship)	$0.8\% L_{pp}$
Pitch radius of gyration	0.61

Table 2
Channels description used in the experimental measurements

Channel	Description
1	Strain gauge for measuring VBM cut no. 1 (bottom of the bar)
2	Strain gauge for measuring VBM cut no. 1 (top of the bar)
3	Strain gauge for measuring shear force cut no. 1
4	Strain gauge for measuring VBM cut no. 2 (bottom of the bar)
5	Strain gauge for measuring VBM cut no. 2 (top of the bar)
6	Strain gauge for measuring shear force cut no. 2
7	Strain gauge for measuring VBM cut no. 3 (bottom of the bar)
8	Strain gauge for measuring VBM cut no. 3 (top of the bar)
9	Strain gauge for measuring shear force cut no. 3
10	Forward diode
11	Aft diode
12	Wave probe

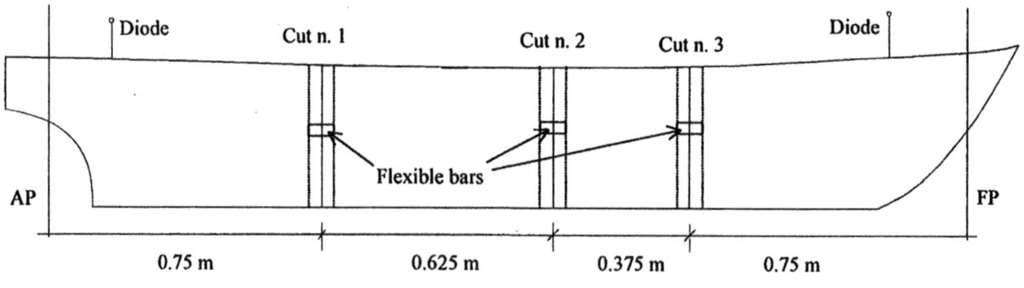

Fig. 1. Longitudinal position of the strain gauges and diodes.

Near the forward and aft perpendiculars, two light-emitting diodes (LED) were mounted on the model. Using the signals from the LEDs and taking into account their positions, the heave and pitch motions were obtained. In order to measure the incident wave height, one wave probe was fixed in the main carriage near the forward perpendicular of the model. This probe induced an electrical signal whose intensity depends on its wetted height. Fig. 1 shows the longitudinal positions of the strain gauges and the diodes.

Table 3
Measured bending moments

	M(kg m)
Strain gauge 1	1.45
Strain gauge 2	4.75
Strain gauge 3	2.40

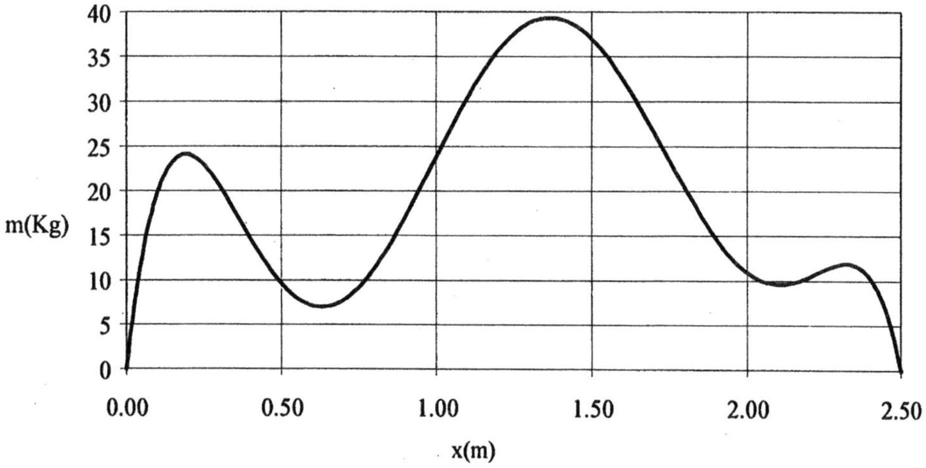

Fig. 2. Model mass distribution using the 8-coefficient polynomial.

All these signals were amplified and recorded in a microcomputer. After each experimental run the measurements were displayed in a graphical form on the microcomputer.

2.2. Weight distribution of the model

The weight distribution $w(x)$ of the S-175 container model is unknown because there were no available data about longitudinal distribution of the model hull weight. In order to solve this problem the weight distribution was assumed as an Nth-order polynomial function with $N + 1$ unknown coefficients,

$$w(x) = \sum_{n-0}^{N} C_n x^n. \tag{1}$$

The coefficients are determined using some properties obtained from measurements on the hull model such as the hull weight, the longitudinal centre of gravity, the radius of gyration and the still water bending moments in three longitudinal positions. Since the number of measurement quantities is eight, this is also the maximum number of coefficients that can be in the polynomial. The static bending moments of the model were measured and the results are shown in Table 3. The mass distribution obtained is presented in Fig. 2.

The still water bending moment distribution for this condition is illustrated in Fig. 3.

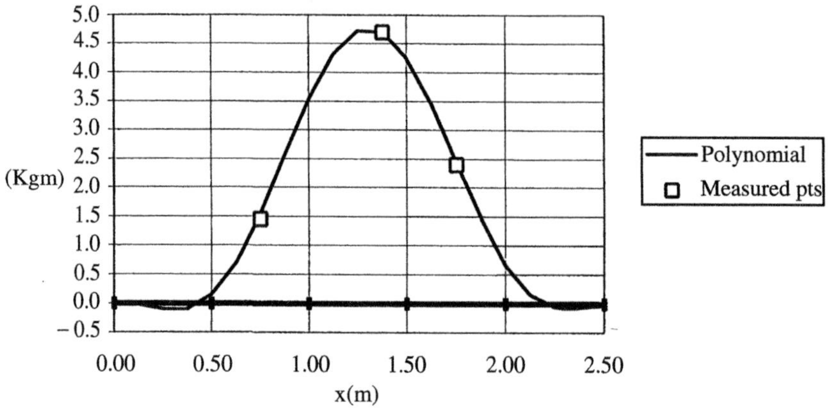

Fig. 3. Still water bending moments using the 8-coefficient polynomial.

3. Assessment of natural frequencies and mode shapes

In order to measure the vertical bending moment and shear forces, the model was segmented into three parts. From a structural point of view the model can be considered as three nodes with rotational stiffness and four rigid segments that are linked to the nodes.

In this system the unknown variables are the rotations. For the evaluation of the mass matrix and the force vector, the finite element approximation is used. Assuming that each element has only two degrees of freedom, linear shape functions are adopted.

Applying the principle of virtual displacement, the element mass matrix can be obtained using the following expression:

$$m_{ij} = \int_0^L m(x)\Psi_i\Psi_j \, dx, \tag{2}$$

where $m(x)$ is the total mass distribution, i.e. including the added mass contribution and Ψ represents the shape functions. The applied force vector can be evaluated in a similar form,

$$F_i(x, t) = \int_0^L q(x, t)\Psi_i \, dx. \tag{3}$$

The structural model requires the equivalent values of the rotational stiffener springs K and the associated structural damping. For this purpose three free vibration experimental tests were conducted. Fig. 4 shows the measured vertical bending moments (VBM) for one of the three strain gauges.

Using the measured bending moments from the free decay tests, the first natural frequency of the structure can easily be obtained.

Table 4 shows the mean first natural frequency obtained using the data from the three experimental tests.

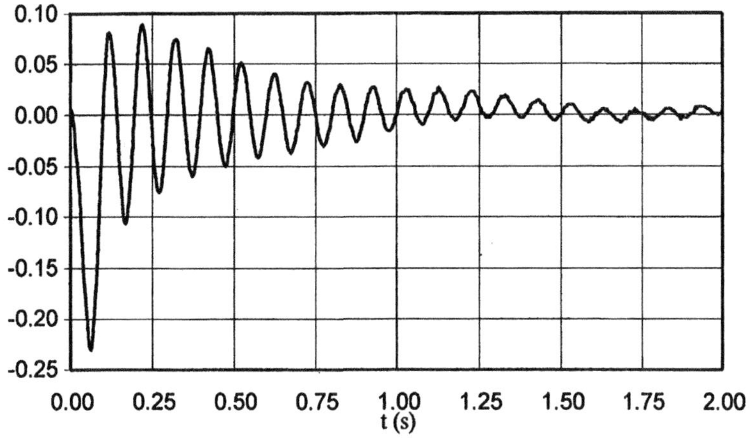

Fig. 4. Vertical bending moment for a free vibration test: Cut number 2.

Table 4
First natural frequency of the model structure

	Natural frequency (rad/s)
Strain gauge 1	61.37
Strain gauge 2	61.48
Strain gauge 3	61.29
Mean	*61.38*

The higher frequencies of the structural model cannot be obtained because the measured data are filtered. The logarithmic decrement of the dynamic system was obtained using the following expression:

$$\delta = \frac{1}{n} \ln \frac{u_1}{u_{n+1}}, \tag{4}$$

where n is the number of positive or negative peaks and u_1 is the first positive or negative peak and u_{n+1} the $n+1$ peak. This approximation is valid if the bending moments are oscillating about the zero value. Fig. 4 shows that there is a small oscillation in the mean position, so it is desirable to use an equivalent amplitude in Eq. (4) which is given by

$$\bar{u} = \frac{u_{\max} - u_{\min}}{2}, \tag{5}$$

where $u_{\max}$ is the maximum peak of the cycle n and $u_{\min}$ the minimum for the same cycle. Fig. 5 shows the amplitude decay for the three strain gauges.

Using the information contained in Fig. 5 in association with Eq. (5) the logarithmic decrement was calculated in the three sections during 10 cycles and is shown in Fig. 7.

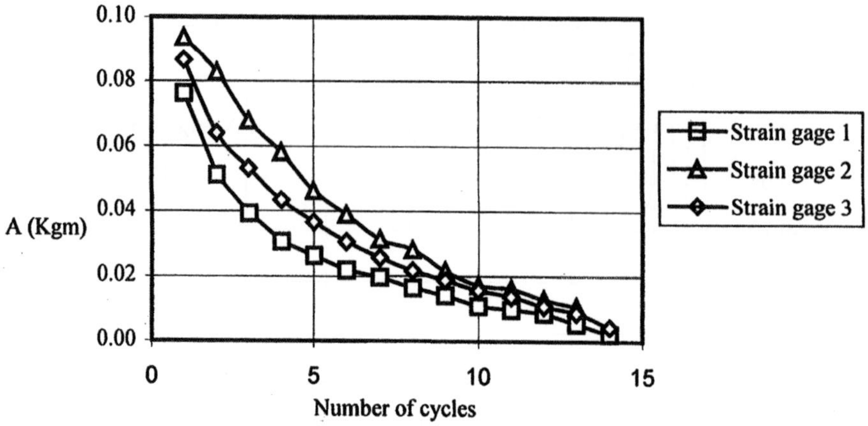

Fig. 5. VBM amplitude decay observed in the three strain gauges.

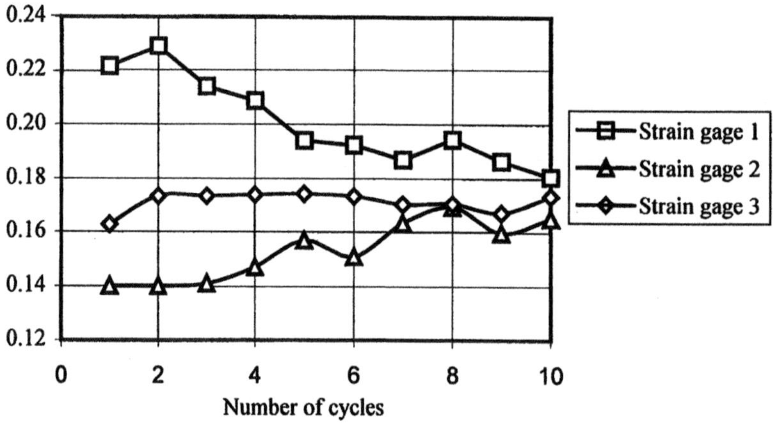

Fig. 6. Logarithmic decrement for the three strain gauges.

Fig. 6 shows that the logarithmic decrement for the three strain gauges converges to the same value. This means that the coupled terms of the damping matrix can be neglected. The mean value obtained for the logarithmic decrement is equal to

$$\delta_1 = 0.174.$$

Based on the value obtained for the logarithmic decrement, the damping ratio was obtained using the linear relation

$$\zeta_1 = \frac{\delta}{2\pi} = 2.77\%.$$

The rotational stiffness of the springs was found to be

$$K = 9.44KN.$$

Table 5
Natural frequencies for the wet mode

	Natural frequency (rad/s)
Mode 1	61.4
Mode 2	165.9
Mode 3	338.5

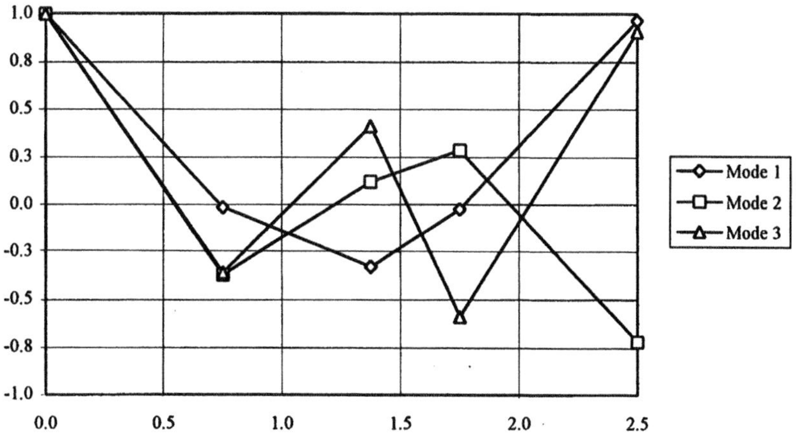

Fig. 7. Natural shapes for the wet mode.

Table 6
Coefficients of the decoupled dynamic system

Mode	Mass	Damping	Stiffness
1	7.76	26.56	3.04E4
2	2.86	55.49	7.87E4
3	9.14	616.91	1.04E6

Using this value, the three first natural frequencies and mode shapes were obtained and represented in Table 5 and Fig. 7.

Kumai [14] suggests a formula to evaluate the logarithmic decrement for the higher frequencies

$$\delta_N = \delta_1 \left(\frac{\omega_N}{\omega_1} \right)^{0.75}. \tag{6}$$

The mass and stiffness coefficients were obtained for the first three modes. The corresponding values are presented in Table 6.

Table 7
Comparison of the still water bending moments

	Real	Calculated	Dif. (%)
Spring 1	1.450	1.377	– 5.006
Spring 2	4.850	4.895	0.936
Spring 3	2.400	2.599	8.312

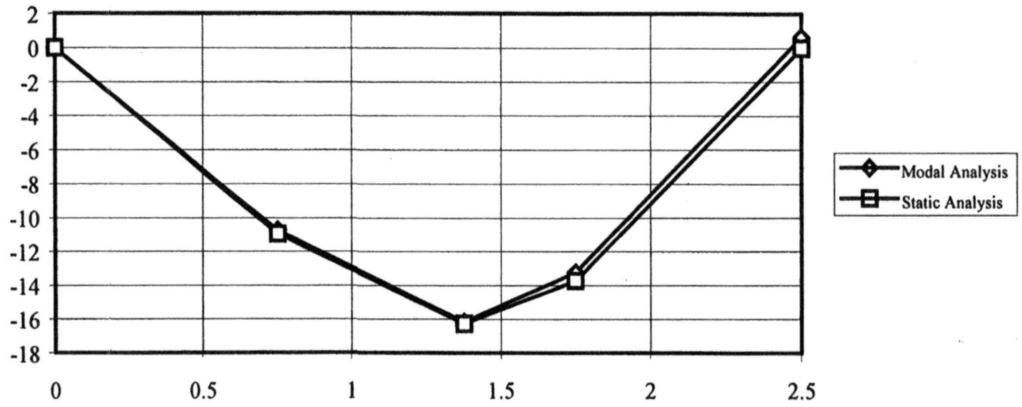

Fig. 8. Still water structural displacement using direct and modal calculations.

In order to check the derivation of the stiffness matrix, the node displacements were evaluated for the still water condition and afterwards the total rotations of the nodes were also calculated. Using these values the bending moments were obtained using the following relation:

$$M_i = K_i \theta_i. \quad (7)$$

Based on this approximation the bending moment values obtained are given in Table 7.

It can be concluded that this approximation gives reasonable results despite the small number of elements used in the model. Also, in order to check the modal analysis method, the structural nodal displacements were evaluated and compared with those obtained using a static analysis. Fig. 8 shows that the results for the two approximations agree well with each other.

4. Model tests

Twelve different runs were carried out with a Froude number equal to 0.2. Table 8 represents the main particulars of the tests.

Table 8
Model speed and wave data for the different runs

Run	Speed (m/s)	Wave frequency (Hz)	Wave amplitude (cm)	Wave amplitude/ship length
1	0.989	0.48	1.71	
2	0.989	0.56	1.84	2.0
3	0.999	0.64	1.77	1.5
4	0.985	0.79	1.58	1.0
5	0.990	0.96	1.31	
6	0.992	1.11	1.40	
7	0.992	0.56	3.62	2.0
8	0.991	0.64	3.51	1.5
9	0.996	0.79	3.56	1.0
10	0.988	0.56	5.37	2.0
11	0.993	0.64	5.40	1.5
12	0.994	0.79	4.35	1.0

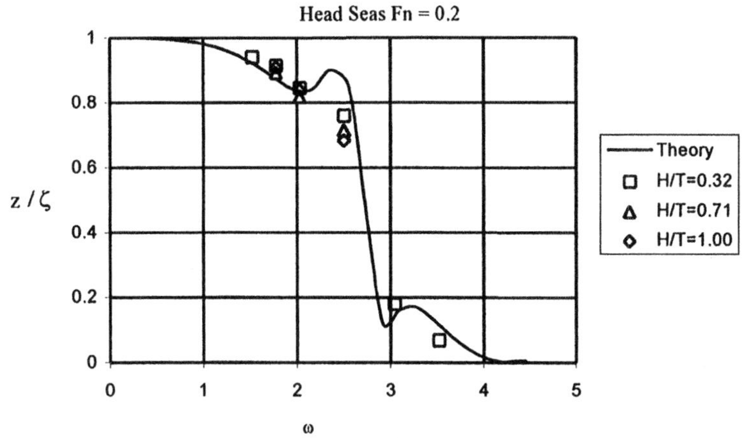

$\omega\sqrt{\frac{L}{g}}$

Fig. 9. Heave transfer function.

Since the aim of the measurements was to validate the theoretical predictions, comparisons between the measurements and calculations were included for all the tests.

The heave and pitch transfer functions were calculated using the method presented in Ramos and Guedes Soares [13] which is based on the formulation given by Gerritsma and Beukelman [14]. Figs. 9 and 10 show the experimental and theoretical results using the linear strip theory for the model with a Froude number of 0.2. The experimental results are shown for the different ratios of wave height (H) to the model draft (T). This is a non-common wave height ratio, which was adopted as it can

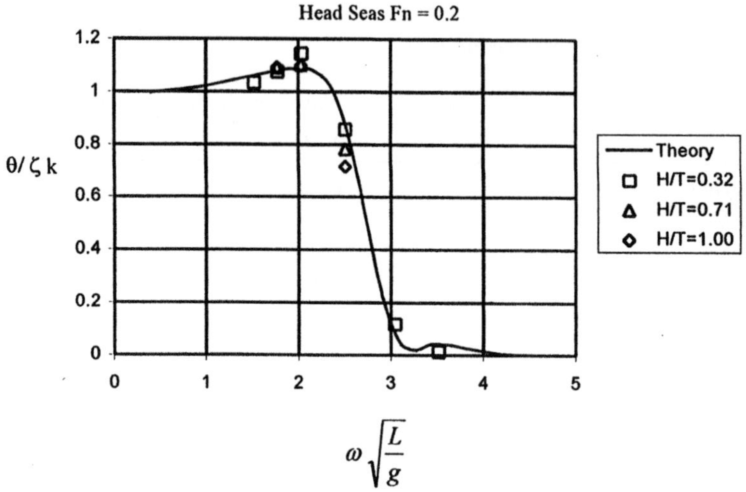

Fig. 10. Pitch transfer function.

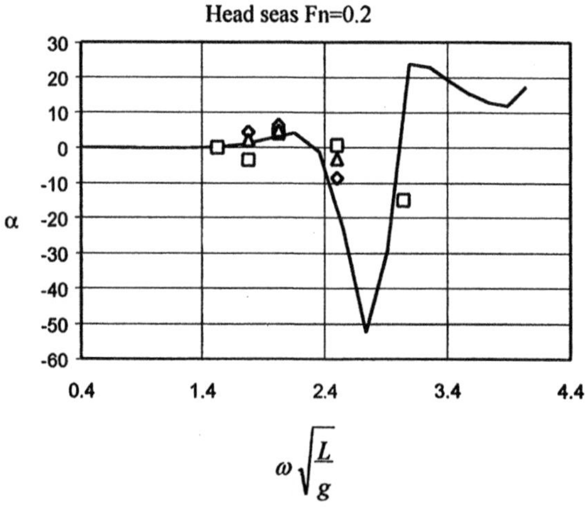

Fig. 11. Heave lag — results comparison.

provide indication about when slamming can occur. From these figures it can be concluded that the linear theory can predict quite well the heave and pitch motions even for severe seas.

The lag between the wave and the heave and pitch motions, which is shown in Figs. 11 and 12, was evaluated using the signals of the wave probe that was mounted on the forward perpendicular of the model.

Using the experimental data from the heave and pitch transfer functions and phase angles, the relative motion transfer functions for any longitudinal position were

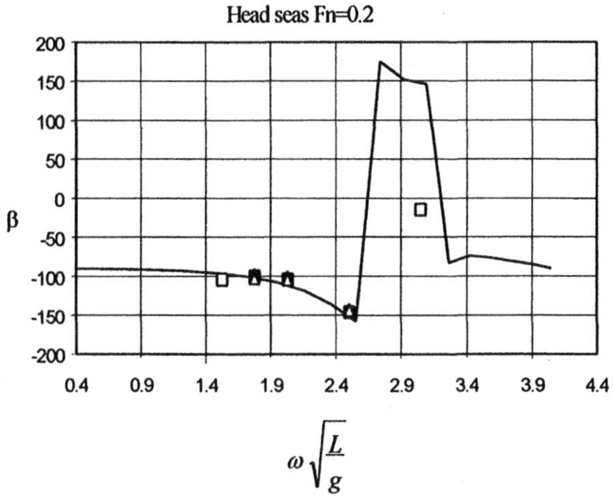

Fig. 12. Pitch lag — results comparison.

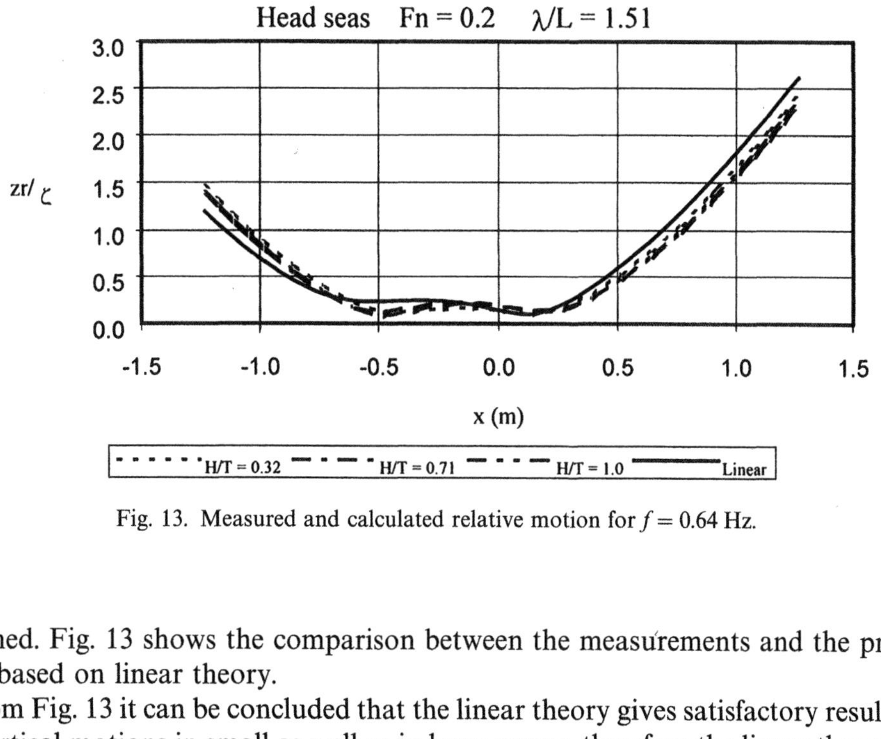

Fig. 13. Measured and calculated relative motion for $f = 0.64$ Hz.

obtained. Fig. 13 shows the comparison between the measurements and the predictions based on linear theory.

From Fig. 13 it can be concluded that the linear theory gives satisfactory results for the vertical motions in small as well as in large waves, therefore the linear theory can be applied with some confidence. Several other authors also stated the same conclusion. For the bow locations in which larger relative motion amplitudes occur the linear theory tends to overpredict the relative motion amplitudes.

The calculated low-frequency bending moments are compared with measurements in Figs. 14 and 15 for the first six runs in which the wave amplitude is small compared with the ship length. The vertical bending moments were non-dimensionalized using

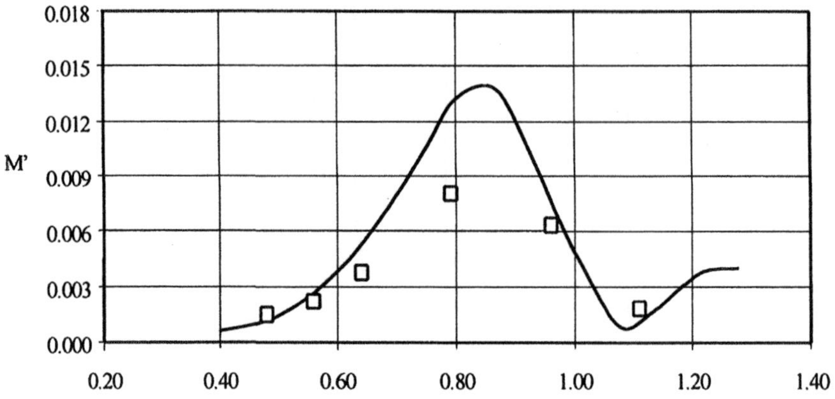

Fig. 14. Nondimensional VBM at strain gauge 1.

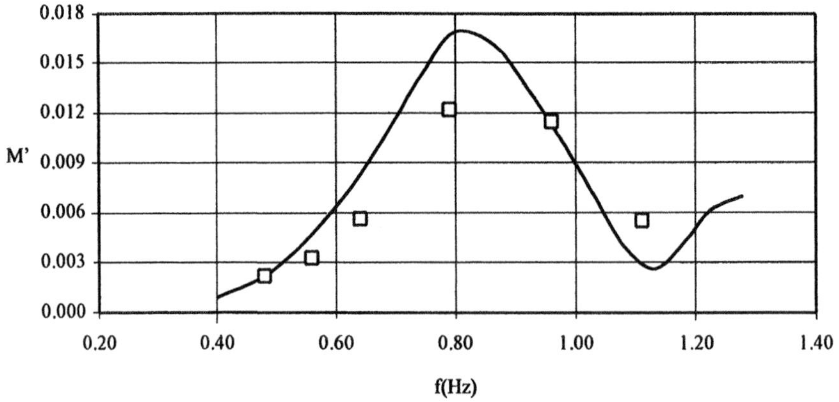

Fig. 15. Nondimensional VBM at strain gauge 2.

the following relation:

$$M' = \frac{M}{\rho g L^2 B}. \tag{8}$$

From Figs. 14–16 it can be concluded that satisfactory agreement between measured and predicted values for the wave bending moments was found, considering all possible inaccuracies due to the unknown weight distribution and measurement errors.

In the experiments it was observed that:

- More severe slamming occurs when the relative motion attains the maximum values and this occurs for tests 9 and 12 when λ ($f = 0.79$ Hz).
- For λ/L ($f = 0.64$ Hz) only a mild slamming occurs for the highest wave amplitude which corresponds to test number 11.
- For λ/L ($f = 0.56$ Hz) no slamming occurs.

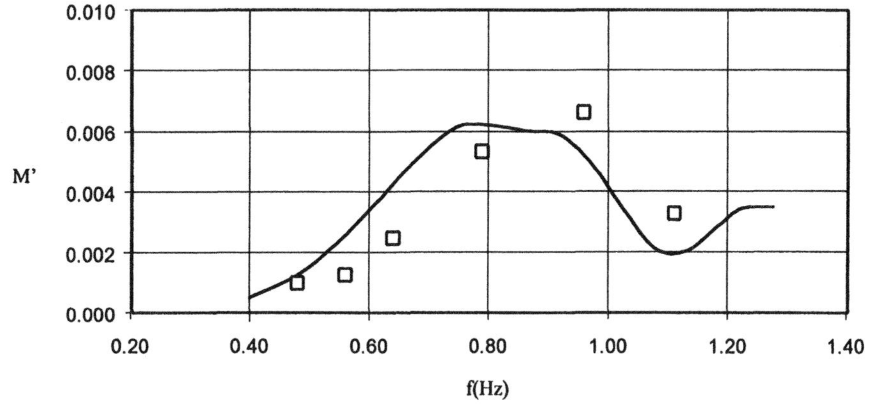

Fig. 16. Nondimensional vertical bending moment at strain gauge 3.

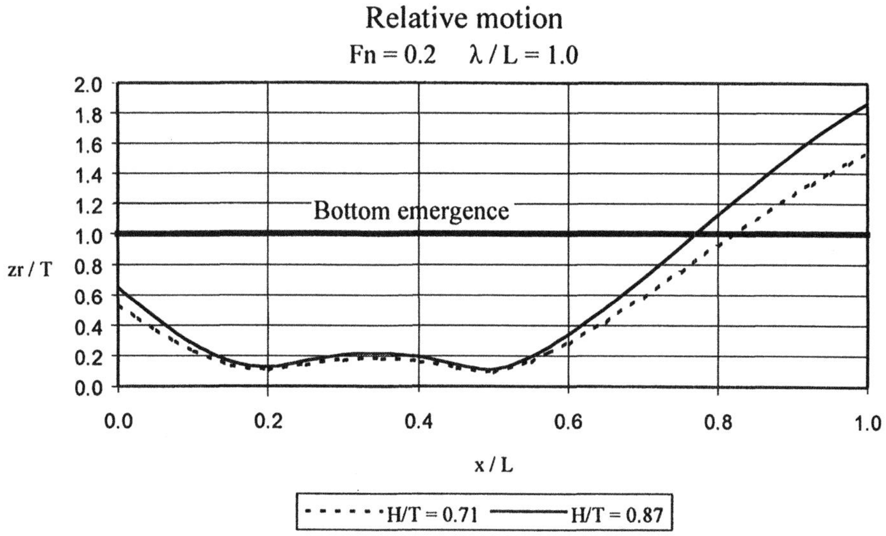

Fig. 17. Longitudinal variation of the relative motion for $\lambda/L = 1.0$.

Three experiments were carried out for $\lambda/L = 1.0$ with different wave amplitudes. In order to check the validity of the linear theory for the ship motions and wave loads, the first set of tests was carried out with small amplitudes. In the last two tests, the wave amplitude was chosen such that slamming occurs.

Fig. 17 shows the longitudinal variation of the relative motion amplitude divided by the ship draft, obtained from runs 9 and 12. Fig. 17 indicates that when the relative motion is greater than one the bottom will emerge. From the information contained in this figure it can be seen that the extent of slamming is about 20% of the ship length during run 9 and it is equal to 25% during the last run. These are typical

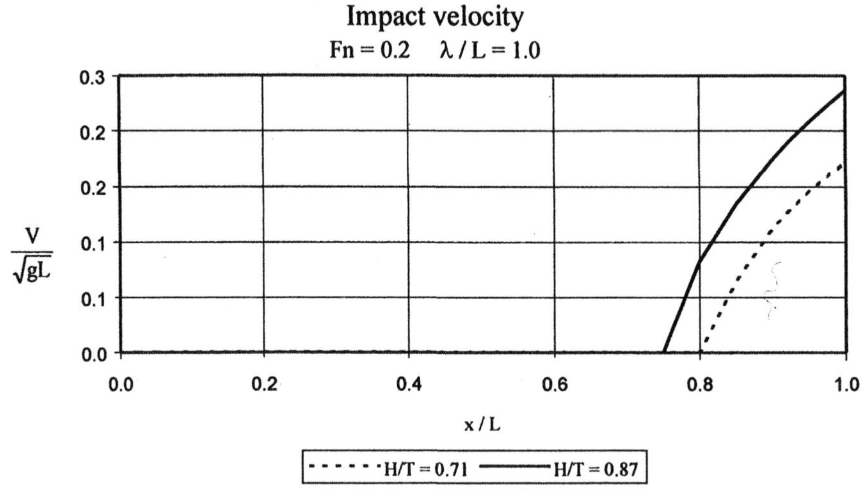

Fig. 18. Impact velocity for runs 9 and 12.

values that can be considered to represent severe slamming, according to Ochi and Motter [15].

The calculated impact velocities for runs 9 and 12 are shown in Fig. 18. The velocities increased, as expected, in the forward direction with the increase in wave amplitudes.

5. Combined low-frequency and whipping bending moments

In this section comparisons between the experimental results and some of the prediction methods for the determination of slamming loads will be carried out. Due to the large number of methods considered they were divided into two groups.

The first one uses the linear theory for the calculation of the wave bending moments and the empirical formulations for the calculation of the slamming loads and the resulting whipping stresses. Two different methods are compared in this group, one using the formulation proposed by Ochi and Motter [15] and the other is the combination of the Stavovy and Chuang [16] method for calculation of the maximum slamming force and the Kawakami et al. [2] formula for the evaluation of the time history of the slamming force (Figs. 19–24).

The second group uses the linear theory in association with the vertical derivative of the added mass for the evaluation of the slamming-induced bending moments.

5.1. Slamming loads from empirical methods

In this section the experimental measurements will be compared with two theoretical methods based on different approaches. The first one uses the linear theory

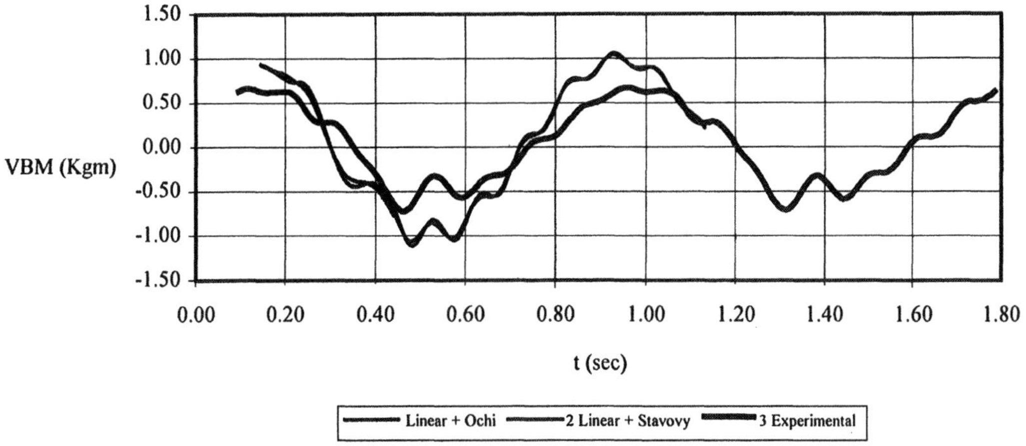

Fig. 19. Measured and predicted bending moments on SG 1 using empirical formulations for the whipping stresses $H/T = 0.71$, $\lambda/L = 1.0$.

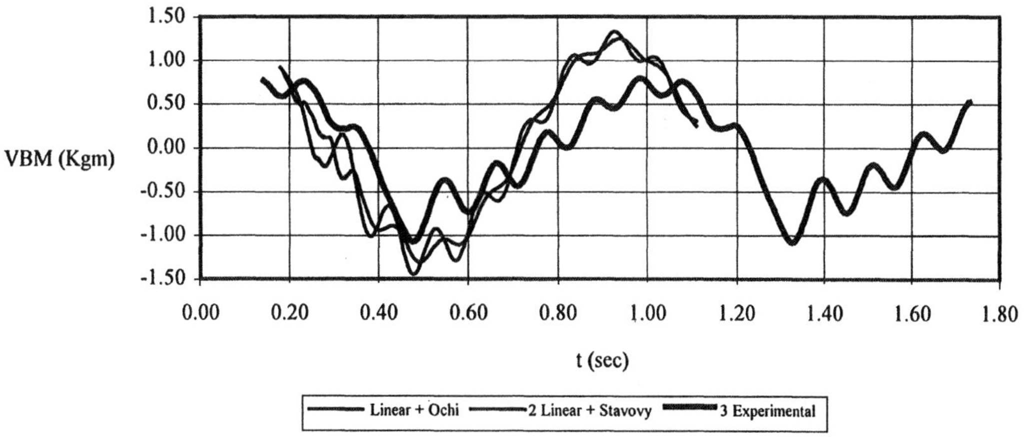

Fig. 20. Measured and predicted bending moments on SG 1 using empirical formulations for the whipping stresses $H/T = 0.87$, $\lambda/L = 1.0$.

combined with the slamming methods proposed by Ochi and Motter, and the second combines the linear theory with the slamming methods of Stavovy and Chuang and of Kawakami et al.

The bending moment values are measured and calculated at three longitudinal positions shown in Fig. 1. From Table 2 it can be seen that two strain gauges were used at each section, so the bending moments represented in the next figures with SG1, SG2 and SG3 labels are in fact the mean of the two strain gauges at each position as shown in Table 9.

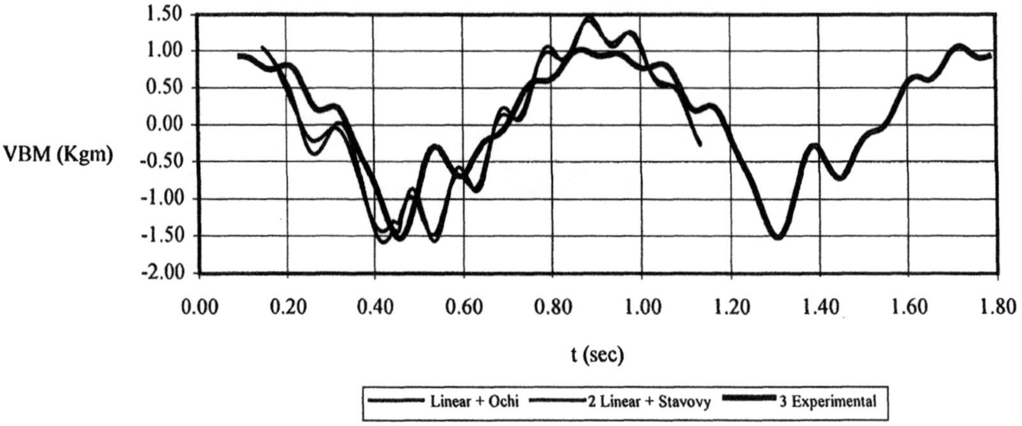

Fig. 21. Measured and predicted bending moments on SG 2 using empirical formulations for the whipping stresses $H/T = 0.71$, $\lambda/L = 1.0$.

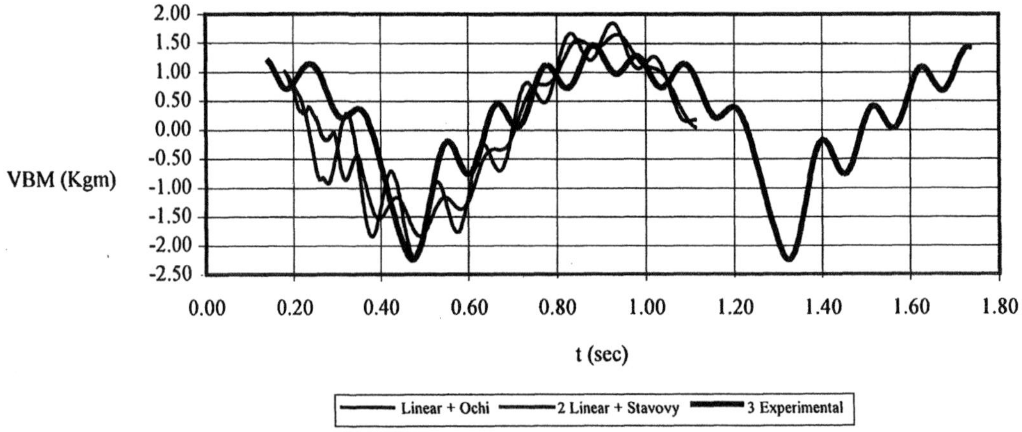

Fig. 22. Measured and predicted bending moments on SG 2 using empirical formulations for the whipping stresses $H/T = 0.87$, $\lambda/L = 1.0$.

5.2. Slamming loads from momentum theory

For a ship advancing in regular head waves, the vertical force according to the momentum theory can be calculated as follows:

$$F = \frac{\mathrm{D}}{\mathrm{D}t}(m'v) = \frac{\mathrm{d}m'}{\mathrm{d}t}v + m'\dot{v} - U\frac{\mathrm{d}m'}{\mathrm{d}x}v - U(\theta + k\xi)m'. \tag{9}$$

The first two terms of Eq. (9) are the same as those given in the Wagner theory. Eq. (9) can be rewritten as

$$F = F_1 + F_2 + F_3 + F_4, \tag{10}$$

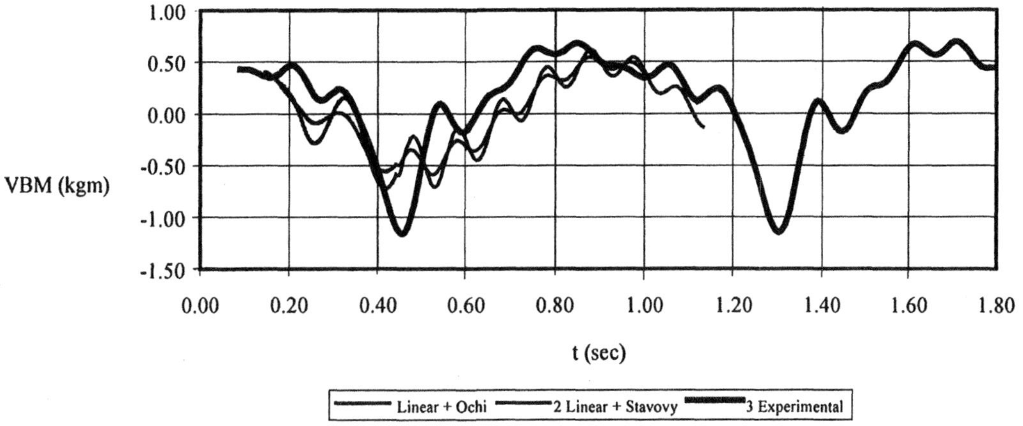

Fig. 23. Measured and predicted bending moments on SG 3 using empirical formulations for the whipping stresses $H/T = 0.71$, $\lambda/L = 1.0$.

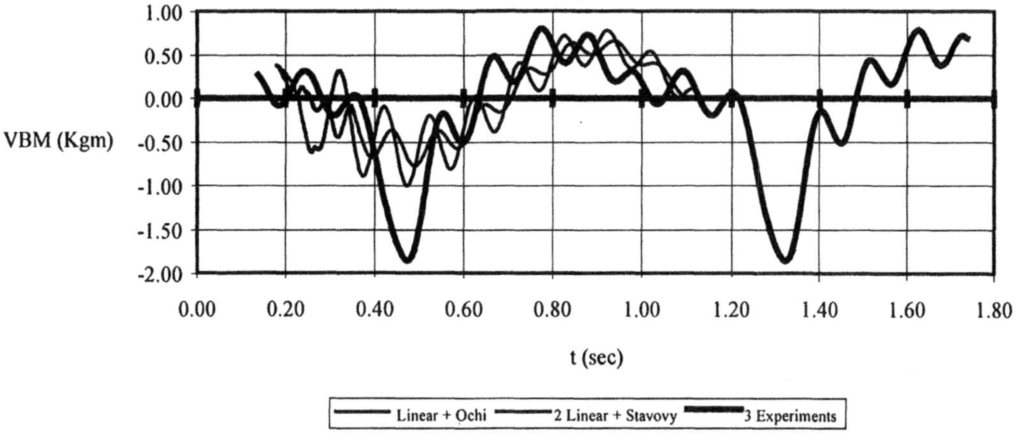

Fig. 24. Measured and predicted bending moments on SG 3 using empirical formulations for the whipping stresses $H/T = 0.87$, $\lambda/L = 1.0$.

where

$$F_1 = \frac{dm'}{dz} A_r^2 \omega_e^2 \sin^2(\omega_e(t+\tau)+\beta), \tag{11}$$

$$F_2 = -m' A_r \omega_e^2 \cos(\omega_e(t+\tau)+\beta), \tag{12}$$

$$F_3 = U \frac{dm'}{dx} A_r \omega_e \sin(\omega_e(t+\tau)+\beta), \tag{13}$$

$$F_4 = -Um'(A_p \cos(\omega_e(t+\tau)+\theta_p) + \xi k \cos(kx + \omega_e(t+\tau))). \tag{14}$$

Table 9
Equivalent strain gauges used in the comparisons

Denomination	Cut number (Fig. 1)	Channels
Strain gauge 1	1	1 and 2
Strain gauge 2	2	4 and 5
Strain gauge 3	3	7 and 8

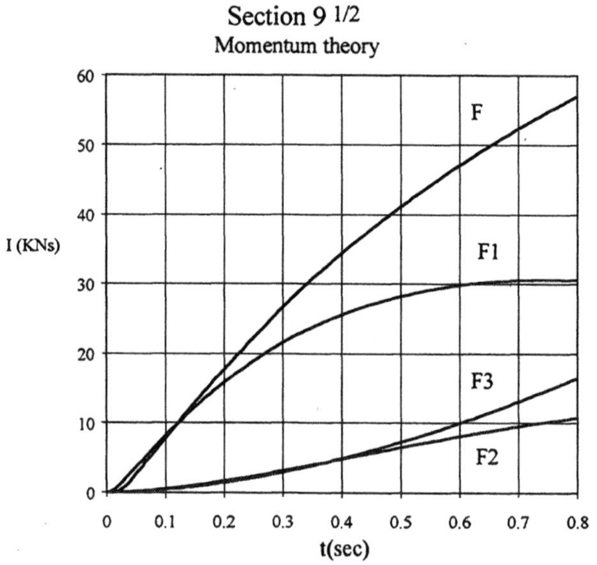

Fig. 25. Contributions of some of the components of the slamming force.

The problem with this method when combined with the linear theory for the prediction of the ship motions and the wave bending moments is that the last three terms of Eq. (9) are already included in the linear theory but with constant values of the coefficients. Fig. 25 shows that the last three terms have a small contribution in the initial stage of the slamming process. The term F_4 is not shown because it is close to zero. For the time equal to 0.3 s which corresponds to a draft equal to 1.9 m (0.2 T), this contribution is equal to 25% and for the time equal to 0.7 s (immersion equal to 0.63 T) the contribution is equal to 55% but for this immersion the linear theory will produce similar results for the last three contributions.

Fig. 25 represents the slamming forces for the foremost station. For this station where there is no flat bottom, the contribution of F_1 is smaller than the other components. The same calculations were carried out for other stations and for station 8 the first component contribution was equal to 84% at 0.2 T. So it seems appropriate not to include these last three components in the slamming forces and to use only the first term of Eq. (9) with a constant velocity term. The first curve in Fig. 26 represents

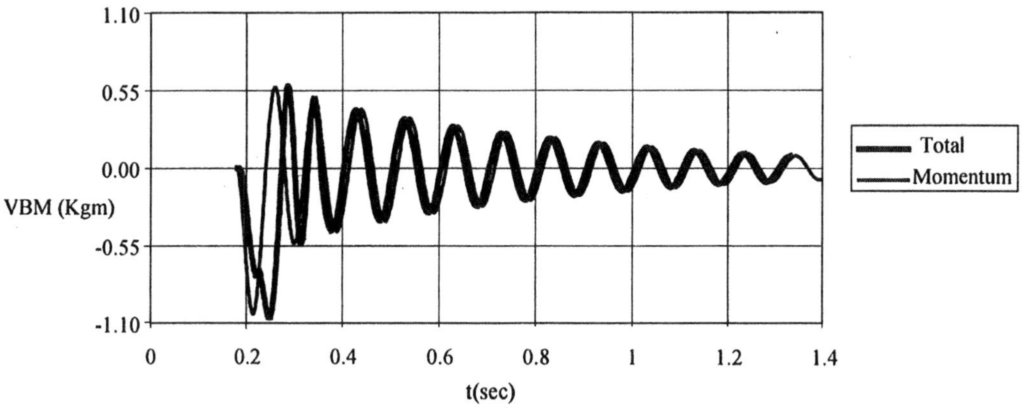

Fig. 26. Whipping stresses for the momentum method varying the hull domain of the slamming load $H/T = 0.87$, $\lambda/L = 1.0$.

the whipping stresses calculated using only the momentum term and the second using all the components of Eq. (9). From this figure one can conclude that the other components of the force will not have significant influence in determining the whipping stresses.

Another problem related to this method is the vertical extension of the hull domain where the impact force produces significant whipping stresses, i.e. resorting to Eq. (11),

$$F(t) = \frac{\mathrm{d}m}{\mathrm{d}z}(t)\dot{z}_r^2. \tag{15}$$

Knowing that this force will have a large value at the beginning of the impact and that the force will sharply decrease afterwards, the question is, how long and what is the vertical hull position where these forces will not have an influence on the whipping stresses. This question has a difficult answer which depends on several factors like, the impact velocities, the longitudinal extension of the bottom emergence and the natural frequencies of the ship hull. Fig. 27 shows that if the slamming loads are evaluated between 0 and 0.3 T the stresses are close to the ones when the loads are calculated between 0 and 0.6 T.

Using this information the total stresses were evaluated by superimposing the linear terms with the whipping stresses produced by the momentum forces which were evaluated from the bottom to 0.3 of the section draft. Afterwards, the results obtained from predictions were compared with those obtained from the mreasurements (Figs. 28–30).

6. Discussion of results

One of the main assumptions in this study is that the total bending moment acting on a ship can be divided into linear low-frequency and whipping components. In other

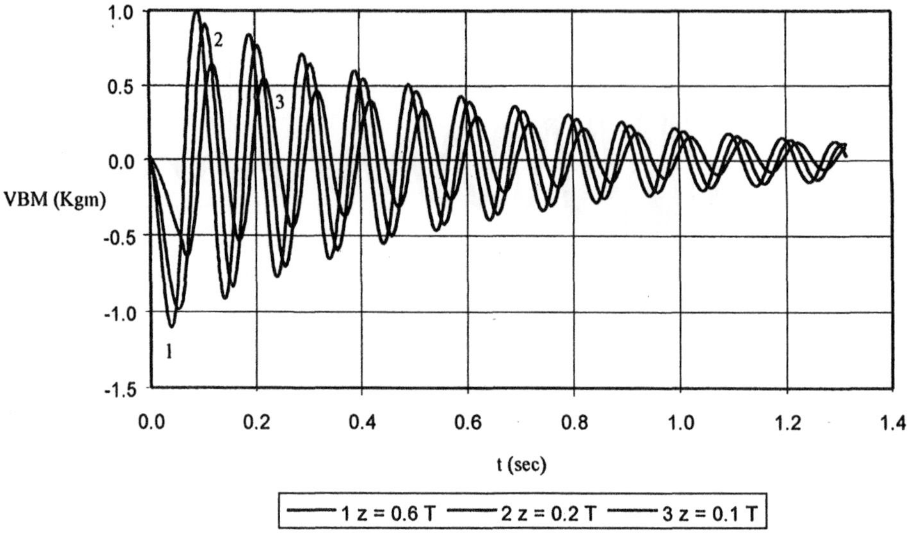

Fig. 27. Whipping stresses for the momentum method varying the maximum draft in which the slamming loads are taken into account $H/T = 0.87$, $\lambda/L = 1.0$.

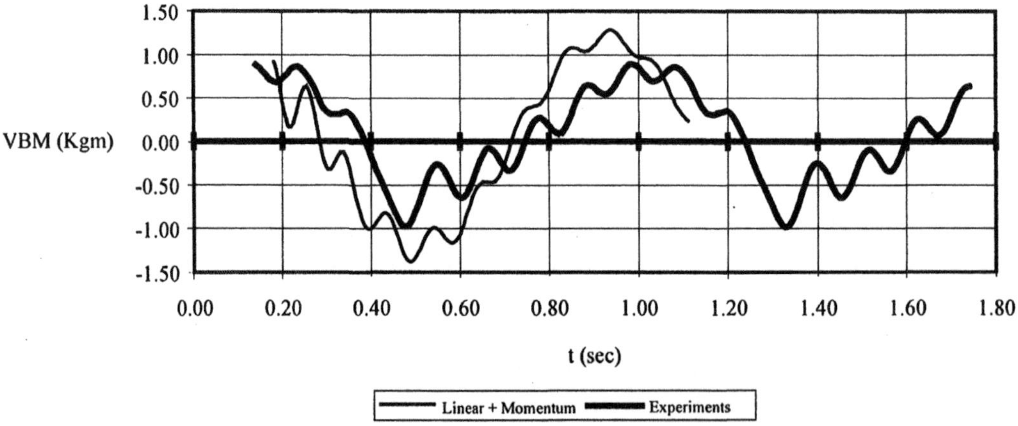

Fig. 28. Measured and predicted bending moments on SG 1 using the vertical derivative of the added mass for the whipping stresses $H/T = 0.87$, $\lambda/L = 1.0$.

words, linear strip theory is used for the calculation of the low-frequency stresses and the whipping stresses are superposed when slamming occurs. Figs. 31 and 32 compare the measured vertical bending moments for two different wave heights. As can be seen from Fig. 32 the large wave height causes non-linear bending moments. From these figures one can conclude that the non-linearity is related to the structural vibration due to the slamming loads and the superposition principle appears to be a reasonable assumption.

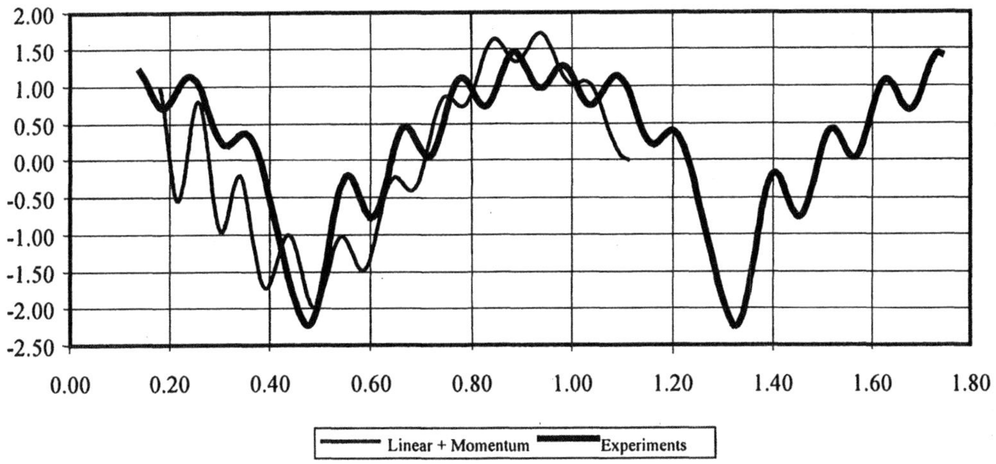

Fig. 29. Measured and predicted bending moments on SG 2 using the vertical derivative of the added mass for the whipping stresses $H/T = 0.87$, $\lambda/L = 1.0$.

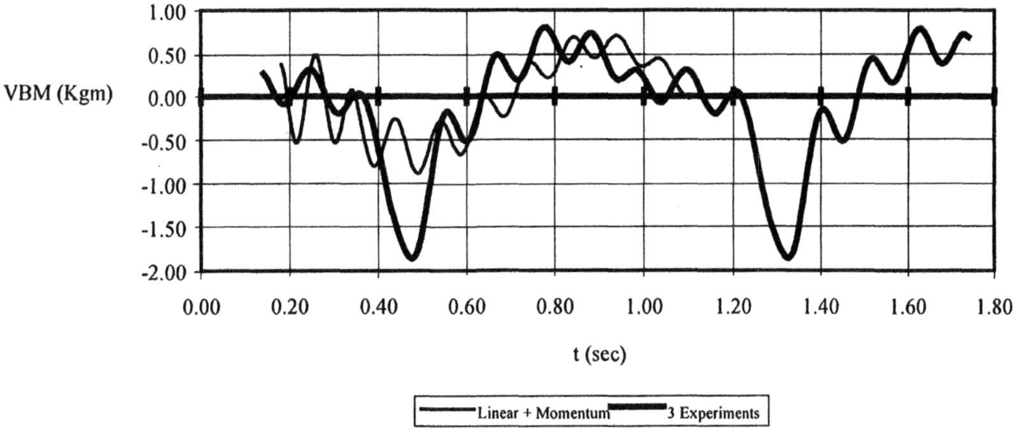

Fig. 30. Measured and predicted bending moments on SG 3 using the vertical derivative of the added mass for the whipping stresses $H/T = 0.87$, $\lambda/L = 1.0$.

The effect of slamming in amplifying the stress amplitudes is described by a coefficient K_s given by the following relation:

$$K_s = \frac{\text{Max}(M(t)) - \text{Min}(M(t))}{2M_a}, \tag{16}$$

where $M(t)$ is the bending moment time history and M_a is the amplitude of the vertical bending moment obtained using the linear theory. Figs. 32 and 33 show the slamming contribution K_s obtained from the experiments and the three different prediction methods (Figs. 34 and 35).

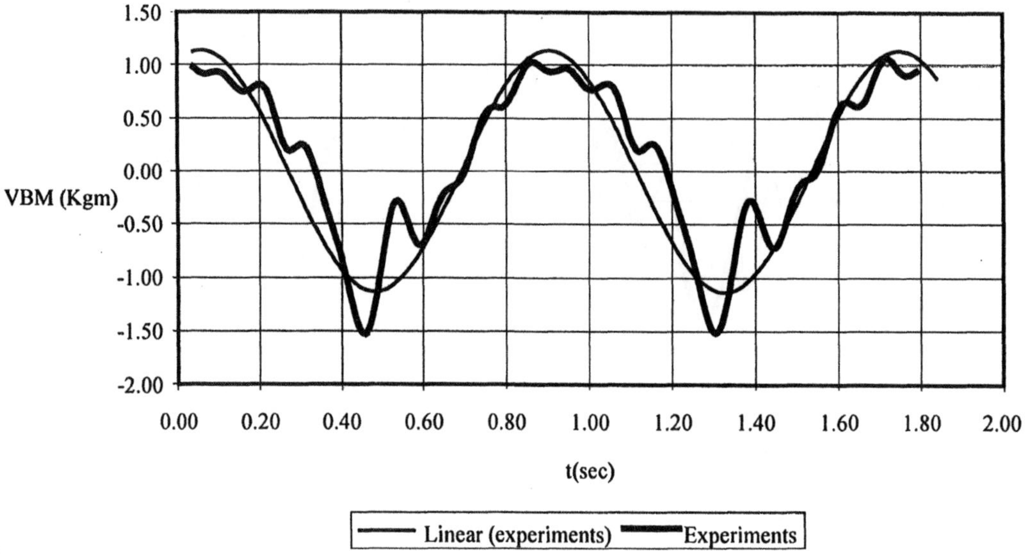

Fig. 31. Vertical bending moment assumed to be linear (obtained from the first experiment) and measured in the SG 2, $H/T = 0.71$, $\lambda/L = 1.0$.

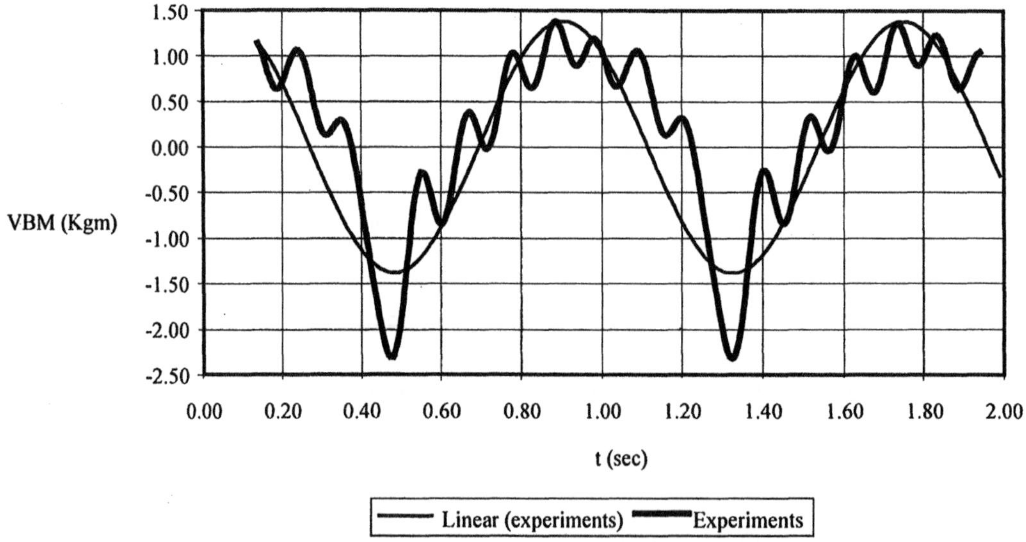

Fig. 32. Vertical bending moment assumed to be linear (obtained from the first experiment) and measured in the SG 2, $H/T = 0.87$, $\lambda/L = 1.0$.

Good agreement was found between the results of the measurements obtained from the first two strain gauges and the Stavovy and Chuang method and the momentum theory for coefficient K_s. However, the Ochi and Motter method tends to produce lower values for the coefficient.

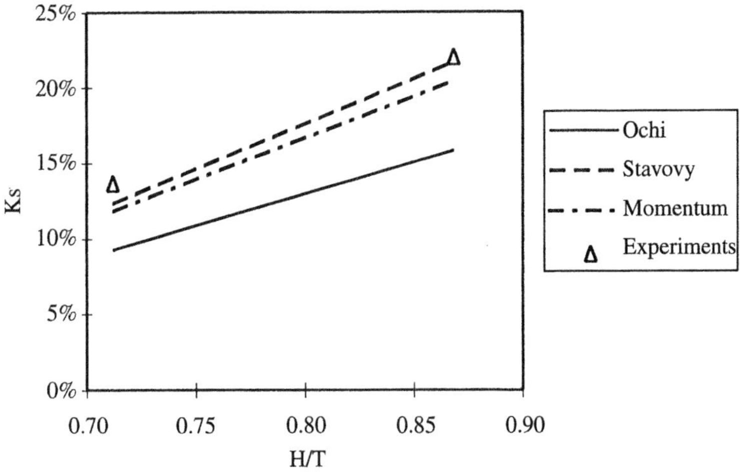

Fig. 33. Comparison of the coefficient K_s with the experiments and the several methods, SG 1, $\lambda/L = 1.0$.

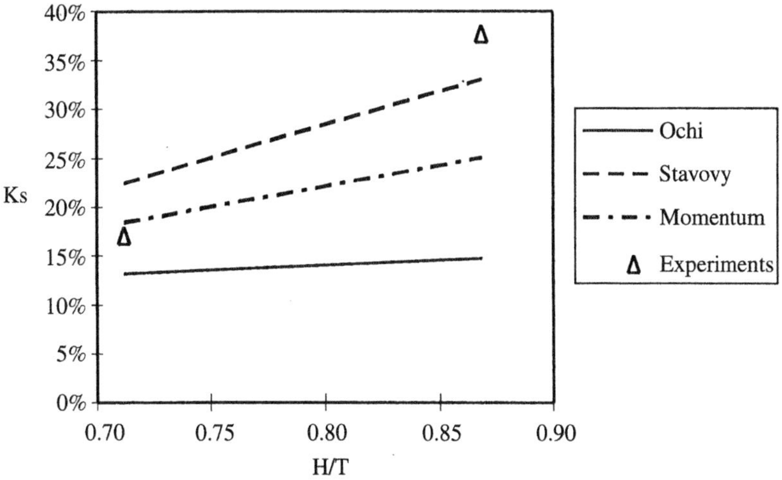

Fig. 34. Comparison of the coefficient K_s with the experiments and the several methods, SG 2, $\lambda/L = 1.0$.

The results of the measurements for the last strain gauge show large deviations between the experiments and the results of the predictions. No plausible explanation was found for the discrepancies.

Several factors, which are directly related to the slamming loading mechanism, could have produced these deviations, including a possible fault in the instrumentation. The higher vibration modes were not taken into account, but since the frequency is much higher this would not have affected those results. The non-linear low-frequency terms expressed in Eqs. (11)–(14) and illustrated in Fig. 26 can also affect the stress results, especially in the forward part of the model.

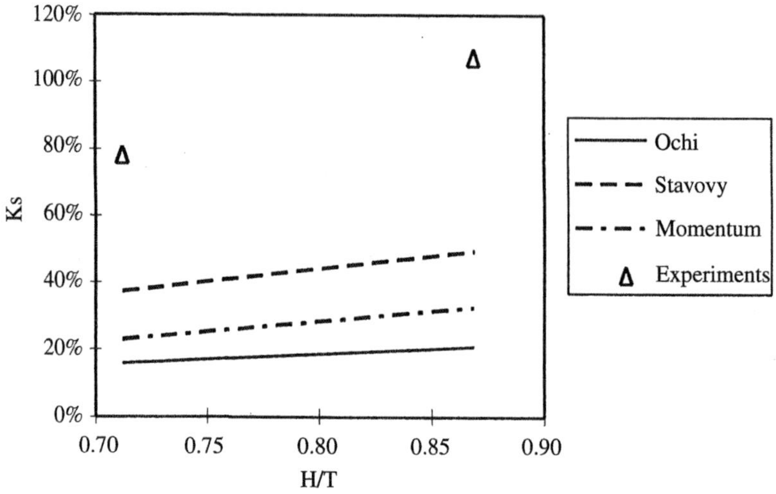

Fig. 35. Comparison of the coefficient K_s with the experiments and the several methods, SG 3, $\lambda/L = 1.0$.

Due to all these unknowns it seems appropriate to carry out more experiments with different models (with more cuts and with light bars in the extremes) in order to establish more consistent conclusions about the loads on those ship sections.

Considering all these possible error sources and the relative simplicity of the theoretical model, satisfactory and encouraging results were obtained for the momentum theory and for the Stavovy and Chuang method combined with the Kawakami et al. method for the prediction of the time history of the slamming forces.

7. Conclusions

The overall conclusions attained from this work can be summarized as follows:

- The linear strip theory used for the determination of the ship motions agrees in a very satisfactory way with the experimental results for all the wave frequencies and for the different wave heights.
- Good agreement was also found for the vertical bending moment for small waves by using the linear theory.
- From the experimental results it can be concluded that the total bending moment which can be divided into two different components as linear low-frequency and whipping, seems to be a reasonable approximation from the practical point of view.
- Good agreement was found between the measurements obtained from the first strain gauges and the Stavovy and Chuang method and momentum theory for the coefficient K_s. However there are some deviations between the measurements and the Ochi method, where the letter tends to produce lower values for the coefficient.
- For the last strain gauge large unexplained deviations were observed between the experiments and the calculations suggesting that further tests should be carried out.

- It has been established that segmented models, which have been used to determine wave induced load effects, can also be used to determine the vibratory response to slamming loads.

Acknowledgements

The first author has been financed by the European Project "*Reliability Methods for Ship Structural Design* (*SHIPREL*)", which has been partially financed by the European Commission under the Contract BREU-CT91-0501.

His stay at Glasgow University, where the experimental work was carried out, was partially financed by FUNDENAV.

References

[1] Ochi MK, Motter LE. A method to estimate the slamming characteristics for ship design. Mar. Technol 1971;8:219–32.
[2] Kawakami M, Michimoto J, Kobayashi K. Prediction of long term whipping vibration stress due to slamming of large full ships in rough seas. Int. Shipbuild Prog 1977;24:83–110.
[3] Belik O, Bishop RED, Price WG. Influence of bottom and flare slamming on structural responses. Trans RINA 1988;130:325–37.
[4] Belik O, Price WG. Comparison of slamming theories in the time simulation of ship responses in irregular waves. Int Shipbuild Prog 1982;29:173–87.
[5] Yamamoto Y, Sugai K, Inoue H, Yoshida K, Fujino M, Ohtsubo H. Wave loads and response of ships and offshore structures from the viewpoint of hydroelasticity. Advances in marine structures. Amsterdam: Elsevier, 1986, pp. 26–40.
[6] Ramos J, Guedes Soares C. On the assessment of hydrodynamic coefficients in heaving. Ocean Engng 1997;24(8):743–64.
[7] Guedes Soares C. Transient response of ship hulls to wave impact. Int Shipbuild Prog 1989;36:137–56.
[8] Salvesen N, Tuck E, Faltiseu O. Ship motions and sea loads. Trans SNAME 1970;78:250–87.
[9] Molin B, Cointe R, Fontaine E. On energy arguments applied to slamming force. Proceedings of the 11th International Workshop on Water Waves and Floating Bodies, Hamburg, Germany, 1996.
[10] Tao Z, Incecik A. Large amplitude ship motions and bow flare slamming pressures. Proceedings of the 15th International Conference on Offshore Mechanics and Arctic Engineering, OMAE'96, Florence, Italy, 1996.
[11] Sames PC, Schellin TE, Muzaferija S, Peric M. Application of a two-fluid finite volume method to ship slamming. Proceedings of the 17th International Conference on Offshore Mechanics and Arctic Engineering, OMAE'98, Lisboa, Portugal, 1998.
[12] Aksu S, Price WG, Temarel P. A comparative study of the dynamic behaviour of a fast patrol boat in rough seas. Mar Struct 1993;6:421–41.
[13] Ramos J, Guedes Soares C. Vibratory response of ship hulls to wave imapact loads. Int Shipbuild Prog 1998;45(441):71–87.
[14] Gerritsma J, Beukelman W. Analysis of the modified strip theory for the calculation of ship motions and wave bending moments. NSRC, TNO Report, No. 96S, 1967.
[15] Ochi MK, Motter LE. Prediction of slamming characteristics and hull responses for ship design. Trans SNAME 1973;81:144–90.
[16] Stavovy AB, Chuang SL. Analytical determination of slamming pressures for high speed vessels in waves. J Ship Res 1976;20:190–8.

Determination of structural stress for fatigue assessment of welded aluminum ship details

Bård Wathne Tveiten[a], Torgeir Moan[b,*]

[a]*Halliburton AS, Structural and Marine Department, P.O. Box 200, N-4065 Stavanger, Norway*
[b]*Department of Marine Structures, Faculty of Marine Technology, Norwegian University of Science and Technology, N-7491 Trondheim, Norway*

Received 19 January 2000; received in revised form 5 June 2000; accepted 15 July 2000

Abstract

The main objective of this paper has been to review and to verify already published hot-spot stress extrapolation procedures for plate structures, and to develop and verify a new and general method for the structural stress extrapolation to be used together with a hot-spot design S-N curve for aluminum ship structures. The proposed extrapolation method has been based on the asymptotic behavior of the stresses adjacent to an idealized notch ('singularity'). On basis of the fatigue test S-N data obtained in this study, relevant S-N curves to be used together with a proposed extrapolation procedure and with already published extrapolation methods have been suggested. A hot-spot design S-N curve with a characteristic strength of 32 was suggested as a suitable choice for the fatigue assessment of profile ground fillet welded stiffener/bracket connections while a design hot-spot design S-N curve with a characteristic strength of 25 was suggested for as-welded stiffener/bracket connections. © 2000 Elsevier Science Ltd. All rights reserved.

1. Introduction

Aluminum high-speed vessels are increasingly applied as large fast ferries in service around the world and catamarans of more than 120 m length have already been built and operated for some time. Ships are subjected to significant cyclic loading and are therefore vulnerable to fatigue failure. Until recently fatigue has not been an explicit design consideration for aluminum ships [1]. It has been claimed that the fatigue

* Corresponding author. Tel.: + 47-7659-5541; fax: + 47-7359-5528.
E-mail address: tormo@marin.ntnu.no (T. Moan).

Reprinted from *Marine Structures* **13 (3)**, 189-212 (2000)

strength of aluminum ship structures is sufficient when certain 'allowable' stresses given in the ship rules are complied with. However, in the development of direct calculation methods for ships, explicit account of different limit states is crucial. A more rational approach would then be to have explicit fatigue requirements.

The main objective of this paper is to establish and verify methods for the assessment of the fatigue strength of welded aluminum ship details based on a structural stress range approach. The structural stress approach using S-N curves have been applied to numerous test S-N data obtained from constant amplitude fatigue tests of aluminum stiffener/bracket connections.

2. S-N curve approach

Fatigue is based on experimentally determined S-N curves combined with the hypothesis of linear cumulative damage, Miner–Palmgren hypothesis [2]. For welded joints where large tensile residual stresses are an inherent feature caused by the welding and the assembly process, the fatigue life is assumed to be solely dependent on the stress range, S, the 'stress-range philosophy'. Adopting the S-N curve approach together with the Miner–Palmgren hypothesis, the criterion for fatigue failure can be expressed as

$$D = \sum \frac{n_i}{N_i} \leqslant \Delta, \tag{1}$$

$$N_i S_i^{m_1} = \bar{a}_1 \quad \text{if } S_i \geqslant S_k, \tag{2}$$

$$N_j S_j^{m_2} = \bar{a}_2 \quad \text{if } S_j < S_k, \tag{3}$$

where n_i is the number of cycles in stress range block (i) and n_j is the number of cycles in stress range block (j), S_i is the stress range in stress range block (i) and N_i is the corresponding number of cycles to failure and S_j is the stress range in stress range block (j) and N_j is the corresponding number of cycles to failure, S_k is the stress range at some reference fatigue life, N_k, and $\bar{a}_1$, $\bar{a}_2$, m_1 and m_2 are geometrical/material parameters obtained from constant amplitude S-N curve tests. D is the cumulative damage index and in a constant amplitude stress history, $D = 1$, at failure. Δ is the allowable damage which would include a safety margin in addition to that implied by using characteristic S-N curves corresponding to mean minus 2 standard deviations.

3. Calculation of the structural stress in welded structures

3.1. Introduction

The structural stress range is the stress at the weld toe including the stress-raising effects due to structural geometry, excluding the stress concentrations due to the presence of the weld geometry. Excluding the weld notch effects can be achieved by carrying out an extrapolation procedure of the structural stress from outside the region close to the weld which is influenced by the local notch. The extrapolation points must be located such that the nonlinear stress peak caused by the local notch effects at the weld is not included in the stress results. At the same time, the points should be sufficiently close to catch the trend in the stress caused by global geometrical effects. The stress at the extrapolation points can be obtained by strain gage measurement, or by means of numerical stress analysis. However, to calculate the local stress distribution which captures the stress-raising effects due to the structural discontinuities, a local finite element model with a sufficient mesh is required. In contrast to the calculation of the nominal stresses where only frame models or coarse global finite element models are sufficient, the calculation of the structural stress requires more complex finite element models and consequently higher computational efforts. The mesh refinement of large, global finite element models at the hot-spot locations is not feasible. However, the structural geometry at the hot-spot locations can be reanalyzed by means of the submodeling technique. The displacements on the cut boundary of the global model are then specified as boundary conditions for the local finite element model (the submodel).

The analyst has several possibilities of modeling the structural geometry at the hot-spot location, ranging from finite solid elements (three-dimensional elements), thick or thin finite shell elements, or a combination of these. Finite shell element models as well as coarse finite solid element models (one finite element layer over the thickness) are characterized by a linear stress distribution over the plate thickness. Therefore, both types of finite element modeling are suitable for the calculation of the structural stress since the nonlinear stress distribution due to the presence of the weld toe in the thickness directions is excluded. The main problem which arises when finite shell elements are used is that the finite shell element formulation only provides a model for the mid-plane of the plates (the element thickness is given as an element property) and thus the local change of the stiffness associated with the weld shape at the hot-spot location cannot be modeled. It is therefore necessary with some sort of finite element model idealization to represent the actual stiffness of the weld seam. There have been attempts to solve the problem by joining the shell elements with some sort of rigid elements, thicker elements at the weld area, or by inclined shell elements to account for the 'missing' weld (e.g. [3–5]), or by an extrapolation to the intersection line which lies 'inside' the material and hence an extrapolation which potentially predicts the structural stress at a fictitious position (e.g. [6]). In a series of studies conducted in Finland [7] using finite shell element analysis, finite solid element analysis, and strain gage measurements it was concluded that both finite shell elements and finite solid elements were suitable for estimating the structural stress. However, when finite shell elements were used the influence of the welds has to be included (by e.g. rigid links, thicker elements at the weld area, or inclined finite shell elements). Note that most of the studies referred to above have been on typical longitudinal stiffener/bracket connections and care should be taken when applying the same principles to completely different structural geometries. Niemi [8] and Tveiten [9] present comprehensive recommendations and studies concerning the application of the finite element method for the determination of stresses in structural details.

3.2. *Methods for structural stress extrapolation*

Traditionally, most extrapolation methods assume that the local notch effects (local geometrical effects) are localized within a small distance close to the weld expressed as a fraction of the main plate thickness ($0.4t$ to $0.5t$) and that the structural stress can be obtained by an extrapolation of stresses from outside this region. An other criterion suggested is to present the local stress field on a double logarithm form, recognizing that the local stress field close to an idealized notch with zero radius becomes log-linear while any other stress components become nonlinear [10]. This makes it possible to separate the two components and perform some sort of extrapolation of the stresses not influenced by the notch to obtain the structural stress. Over the last several years several extrapolation procedures for the determination of the local structural stress at the weld toe of welded joints have been suggested. Most research has been focused on tubular steel joints used in offshore structures and a general method, the European Convention for Structural Steelwork (ECCS) procedure [11] has been well established and is currently used in fatigue design of tubular joints. For plated structures numerous procedures to obtain the structural stress have been proposed, but there is still no universally accepted method.

Niemi [12] has suggested two extrapolation methods for obtaining the structural stress in the vicinity of single-sided edge gussets welded to a stressed member. The structural stress includes the effects of structural discontinuities, but the local geometrical effects of the notch at the weld toe are not included. In this case there is no self-evident indication of the location of the extrapolation points, such as plate thickness, which is normally taken as basis. Niemi [12] proposes two alternative methods:

Method 1: Three extrapolation points at fixed distances at the plate edge are defined and quadratic extrapolation to the weld toe is performed to obtain the structural stress, σ_s.

Method 2: Two extrapolation points at the plate edge are defined relative to the apparent size, t_{app}, as: $0.15t_{app}$ and $0.30t_{app}$. The apparent size is determined as $t_{app} = min\{B, 1.5L, 15H\}$ where B is the height of the stressed member, L is the length of the attached gusset plate, and H is the height of the attached gusset plate.

The International Institute of Welding (IIW) [13] has suggested both a linear and a quadratic extrapolation method for the determination of the structural stress in welded plate and tubular joints based on strain gage measurements or finite element analysis. The two-point (linear) extrapolation method is to be used in cases of mainly membrane stress, while the three-point (quadratic) extrapolation is used in cases of shell bending caused by, e.g. eccentric attachments in large diameter tubes or at plane plates. The distance to the leading extrapolation point should be $0.4t$, which stems from the assumption that the local notch effects are localized within that distance. It is noted, however, that the methods do not apply to edge attachments.

3.3. *A proposed extrapolation method for obtaining the structural stress*

Existing extrapolation methods have traditionally been derived from experience and systematic finite element analysis of specific welded joint configurations. Instead of an extrapolation procedure dependent on some structural dimension to separate the local and global stress, the separation of the global and local geometrical effects close to a weld toe (modeled as a singularity) can be based on the asymptotic behavior of stresses adjacent to an idealized notch ('singularity'). The log-linear property of the local stress field is due to the fact that the stress field close to a singularity in a continuous elastic medium has been shown to be proportional to the distance from the singularity powered to an exponent, $\beta(\alpha)$, as a function of the vertex angle, α [14]

$$\sigma(r) = C_1 f(\theta) r^{\beta(\alpha)} + O(\theta), \tag{4}$$

where r and θ are polar coordinates, C_1 is a constant, $f(\theta)$ is a dimensionless function of θ, and $O(\theta)$ are higher order terms of the stress field solution. As $r \to 0$, it is shown that the leading term of the series expansion approaches infinity but other terms remain finite or approach zero. Thus the stress near the singularity varies with $r^{\beta(\alpha)}$.

On the basis of the asymptotic behavior of stresses adjacent to a singularity, the following structural stress extrapolation procedure has been proposed (Fig. 1):

- To separate the zone solely influenced by the structural geometry as the distance from the end of the influencing zone of the singularity to the point where the structural stress and the nominal stress coincide.
- To fit a curve to the stress field influenced by the structural geometry by means of a polynomial of second order and extrapolate to the weld toe location.

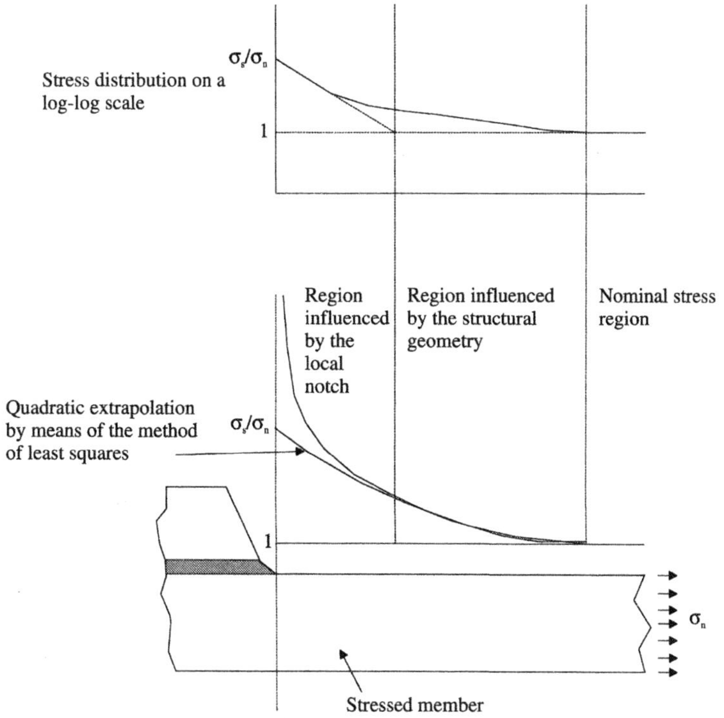

Fig. 1. Proposed extrapolation method based on a 'singularity check'.

The calculation of the local stress field by means of a finite element analysis requires a rather detailed finite element model with a very fine element mesh at and close to the notch in order to ensure that the stress singularity is accounted for. Preferably, 20-node finite solid elements should be used, however, 4/8-node shell elements can in many cases be used for some simple detail configurations. The weld shall be modeled with no radius ('singularity') and a weld toe angle, $\phi = 45°$, and the stress results used in the analysis should be evaluated at the finite element Gauss points. The weld toe angle, $\phi = 45°$, was assumed to represent an average value of the maximum weld toe angle found in practice for as-welded aluminum structures [15]. The curve fitting of the structural stress field by means of a polynomial of second order is easily done with the method of least squares. Even if higher order polynomial seems to provide an excellent fit of the curve within the data points, an extrapolation of the curve outside the data points may give rather poor results due to the highly oscillatory behavior towards the ends of the interpolation interval.

The proposed extrapolation method eludes the weakness of the extrapolation methods suggested by, e.g. Niemi [12] and IIW [13] in that the extrapolation method is independent of any structural dimension to determine the region influenced solely by the structural geometry. Sensitivity studies have shown that the structural stress and thus the predicted fatigue life is limited as variations of the rear end of the region influenced by the structural stress is concerned [9]. A drawback with the proposed extrapolation method is that it requires a rather detailed (time consuming) finite element analysis to ensure that the local singularity at the notch has been properly accounted for. However, it may be possible, on basis of systematic studies using the presented method, to come up with simple guidance for extrapolation points or simplified parametric equations for structural stress concentration factors of different welded joint configurations.

4. Numerical stress analysis

4.1. Introduction

The general-purpose finite element program system SESAM (Super Element Structural Analysis Modules) was used in the numerical stress analysis [16]. The purpose of the finite element analysis was to calculate the stress distribution close to the weld toe to obtain the structural stress concentration factor, K_g, using both 8-node finite shell elements and 20-node finite solid elements. The structural stress concentration factor in this study was determined by means of the extrapolation methods suggested by Niemi [12], IIW [13] and the proposed extrapolation procedure based on a separation of the structural and the local stress field by means of a 'singularity check' (Section 3.3). Two different load cases have been analyzed:

- Applied longitudinal membrane tensile stress.
- Three point bending by a concentrated load applied at the mid span (web frame).

4.2. Geometrical properties of test specimens

A total of 33 test specimens were included in the test program, 12 as-welded flat bar/bracket connections (Fig. 2) and 21 profile ground fillet welded bulb stiffener/ bracket connections (Fig. 3). The flat bar, the brackets, and the transverse plate were made of aluminum alloy 5083, while the extruded longitudinal stiffener with plate flange (shaded area at Fig. 3) was made of aluminum alloy 6082. In order to determine the 'weld quality' of the test specimens, the weld geometry was measured. A plastic cast which was sectioned and photographed through a microscope replicated the weld geometry at the hot-spot locations for all test specimens. For the flat bar/bracket

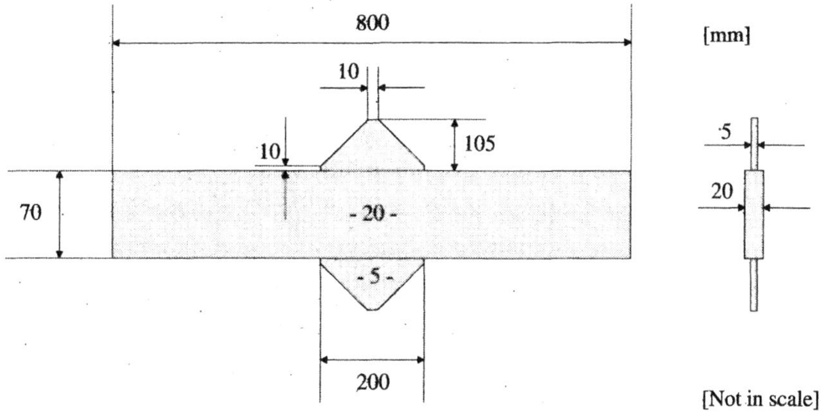

Fig. 2. Geometrical dimensions of flat bar/bracket connection.

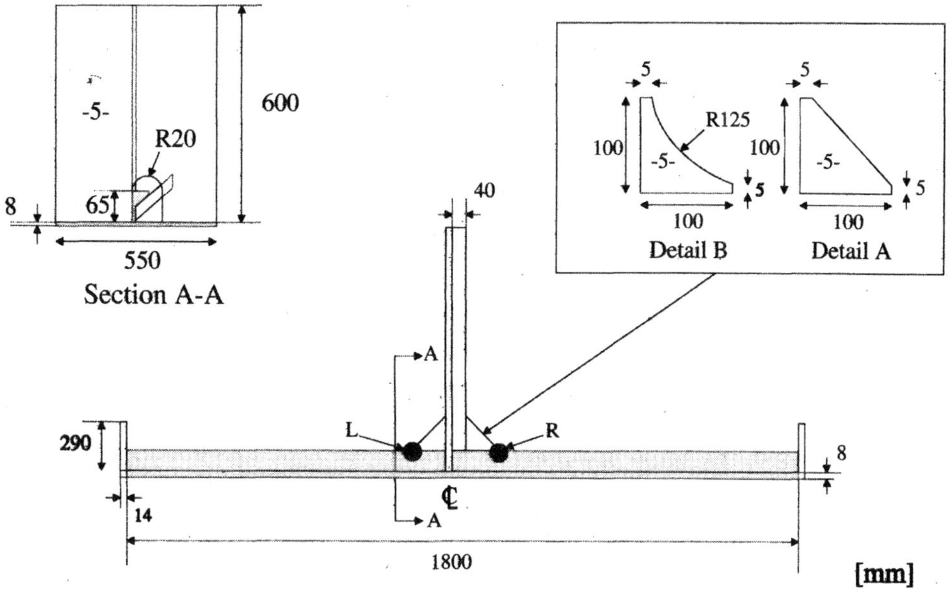

Fig. 3. Geometrical dimensions of bulb stiffener/bracket connection.

Table 1
Weld toe characteristics for test specimens

	ρ (mm)	ϕ (degrees)
Flat bar/bracket connection		
Smallest value	0.1	10
Largest value	12.0	85
Mean value	1.2	55
Standard deviation	1.8	17
Bulb stiffener/bracket connection (detail A)		
Smallest value	3.5	17
Largest value	14.6	65
Mean value	8.2	41
Standard deviation	3.1	16
Bulb stiffener/bracket connection (detail B)		
Smallest value	6.0	15
Largest value	13.5	55
Mean value	8.9	33
Standard deviation	2.1	11

connections and the longitudinal bulb stiffener/bracket connections (named detail A and B) the weld toe radius, ρ, and the weld toe angle, ϕ, were measured for each casting at one saw cut section. This two-parameter description of the weld toe has been shown to provide an accurate description of the local conditions at the weld toe location [17]. The characteristic values of the weld geometry of the flat bar/bracket connection and the bulb stiffener/bracket connection are presented in Table 1.

4.3. Structural stress concentration factors obtained from finite element analysis

The shell element formulation provides a model for the mid-plane of the plates, and the actual material thickness is given as an element property. Due to this two-dimensionality of the shell element formulation, the unsymmetrical bulb section had to be modeled with an equivalent built-up flange section which could produce approximately the same local nominal stress field as in the actual bulb section. The equivalent bulb section (L-stiffener) used in the finite shell element analysis consisted of an equivalent web thickness of 4.5 mm, an equivalent web height of 62.0 mm, an equivalent flange thickness of 13.0 mm, and an equivalent flange width of 15.5 mm. In this study, the effect of the fillet welds in the shell finite element model of the flat bar/bracket connection was introduced by means of inclined shell elements at the end of the brackets with the same thickness as the flat bar. For the bulb stiffener/bracket connection, longitudinal inclined shell elements along the bracket in the longitudinal direction, and at the bracket ends with thickness twice the bracket thickness were used for modeling the weld stiffness (Fig. 4). The weld that was included in the finite solid element model was modeled with an idealized weld profile with no radius and a weld toe angle, $\phi = 45°$. The element size at weld toe for the flat bar/bracket connection

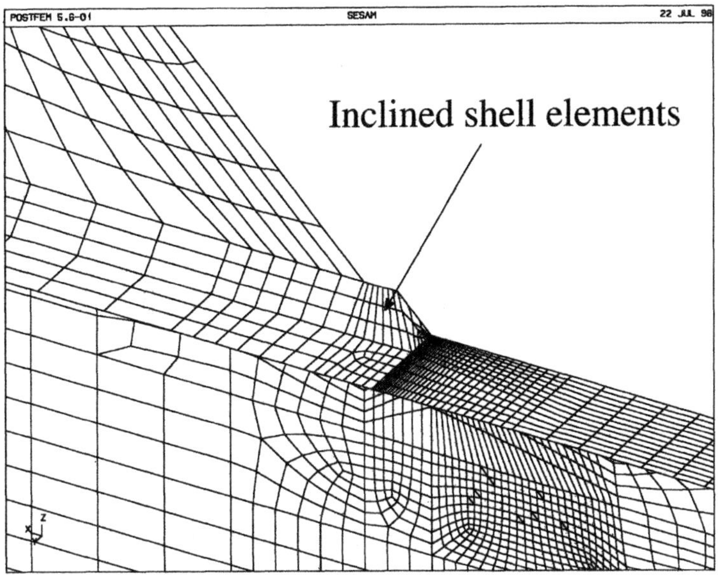

Fig. 4. Inclined shell element model of the bulb stiffener/bracket connection.

was 0.038 mm with a gradually increasing element size to 2.14 mm at the location where structural stress and nominal stress coincided. For the bulb stiffener/bracket connection, the element size was gradually increased from 0.032 to 6 mm.

For the flat bar/bracket connection, comparison of finite shell element results and finite solid element results did not show any significant difference between the stress results. For the longitudinal bulb stiffener/bracket connection, the inclined finite shell element solutions used in this study seems to give a good representation of the weld stiffness, that is the finite element stress results showed good correspondence in the region influenced solely by the structural geometry. Similar promising results were also reported by Fricke and Petershagen [3] on a longitudinal stiffener/bracket connection (T-stiffener) using inclined finite shell elements to model the weld stiffness. However, the stress results obtained from finite shell element models are highly dependent on the analyst's skills and previous experience. Thus, other models of the weld (e.g. by means of rigid elements) could have given equal or even better agreement with those obtained from the finite solid element model. It is in the authors' opinion that finite solid element models preferably should be used in the calculation of the structural stress, at least in cases where little or no previous experiences on the use of finite shell element models are available, to calibrate the finite shell element models.

For the flat bar/bracket connection, the extrapolation methods suggested by Niemi [12] for edge gussets were used for the structural stress extrapolation (Section 3.2). Niemi [12] has suggested two extrapolation methods for obtaining the structural stress in the vicinity of single-sided edge gussets welded to a stressed member. Niemi [12] does not present any element type or element size requirements but the extrapolation methods were originally derived on basis of a very fine mesh (element size from 0.02 mm to 0.1 mm) at the hot-spot location using parabolic plane stress elements. This implies the use of mesh refinement that ensures convergence for practical purposes. The extrapolation methods proposed by Niemi [12] were originally derived for single-sided gussets but in this study a two-sided gusset was used. Therefore, the plate strip breadth, *B*, was taken as half the breadth of the flat bar, $B = 35$ mm. With respect to the finite element analysis from which the extrapolation methods were derived from, this was interpreted to be the most appropriate choice for the particular detail tested in this study. Niemi [12] claims that any 'size effect' is automatically accounted for by the quadratic extrapolation procedure since the shape variation of the stress distribution in the crack plane due to change in the dimensions is accounted for in the local structural stress. While for the second method the 'size effect' has to be considered by multiplying the hot-spot fatigue strength by a conventional size factor, $f(t) = (t_{ref}/t_{app})^{0.25}$ where $t_{ref} = 25$ mm. Including the 'size effect' in the structural stress concentration factor, K_g, the structural stress concentration factor obtained from the extrapolation method 2 can be written as, $K_{g,corr} = K_g/f(t)$. Note that the plate strip breadth, *B*, in this study was taken as half the breadth of the flat bar, $B = 35$ mm, and thus giving an apparent size of, $t_{app} = 35$ mm and a size factor of, $f(t) = 0.92$.

For the longitudinal bulb stiffener/bracket connection, the quadratic extrapolation was performed by means of least-squares method with a polynomial of second order using the three specified extrapolation points given by the IIW [13]. Figs. 5–8 show the stress distribution for the flat bar/bracket connection and the bulb stiffener/bracket connection (detail A) obtained from the finite solid element model with no radius and a weld toe angle, $\phi = 45°$. For the flat bar/bracket connection it is

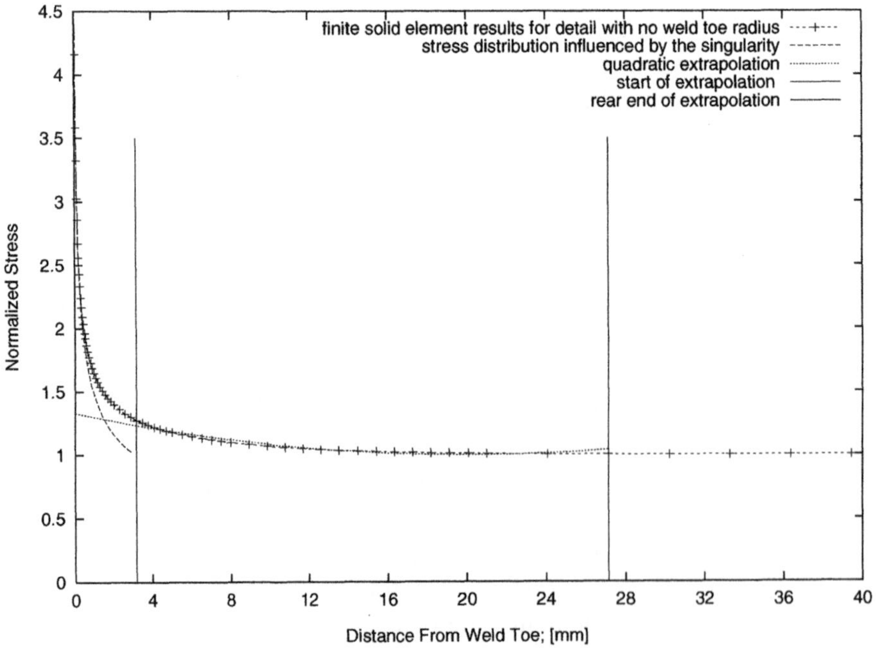

Fig. 5. Stress distribution, flat bar/bracket connection.

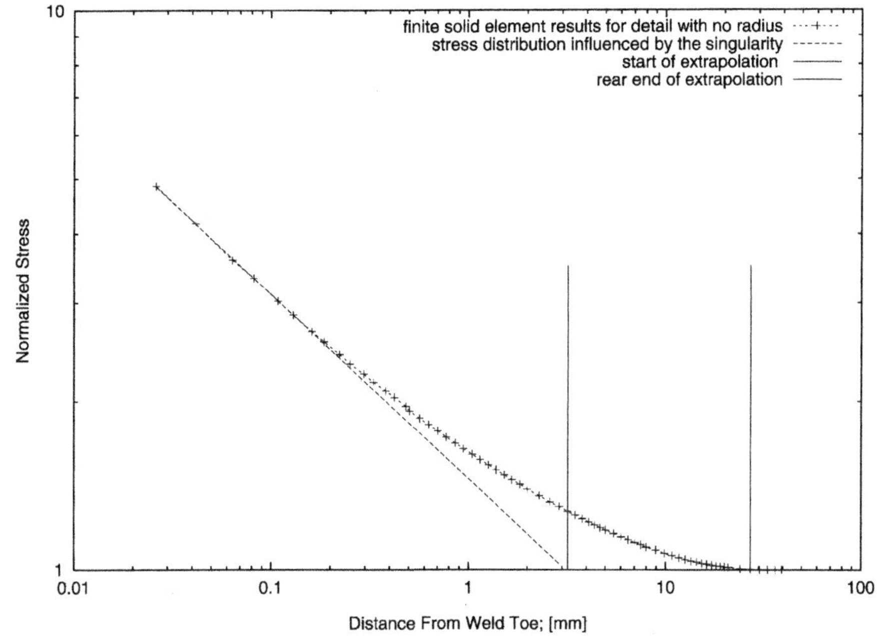

Fig. 6. Stress distribution on a double logarithm form, flat bar/bracket connection.

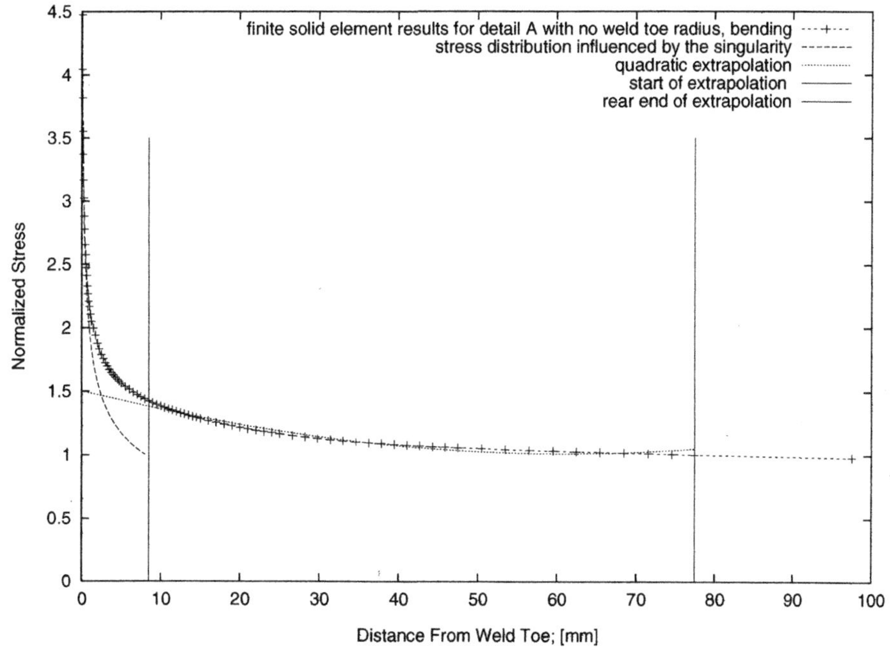

Fig. 7. Stress distribution, detail A.

seen that the notch effects were localized within a distance of approximately 3.2 mm. For the longitudinal bulb stiffener/bracket connections, the notch effects were localized within a distance of 8.1 mm (bending) and 5.9 (tensile) for detail A, and 5.7 mm

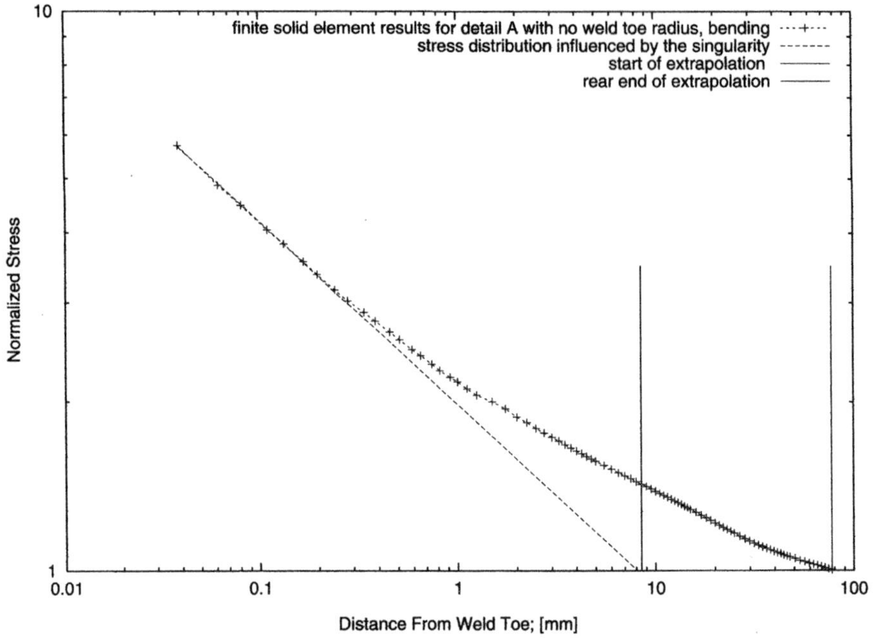

Fig. 8. Stress distribution on a double logarithm form, detail A.

(bending) and 4.7 mm (tensile) for detail B. The stress analysis results in the region that was assumed to be solely governed by the structural stress field (extending to 27.0 mm for the flat bar/bracket connection and 70.0–75.0 mm for the bulb stiffener/bracket connection in tensile and bending loading, respectively) were fitted by means of least-squares method with a polynomial of second order and extrapolated to the weld toe (Section 3.3).

The structural stress concentration factors obtained from the finite element analysis are presented in Tables 2–4. The structural stress concentration factor was defined as the extrapolated stress obtained from the finite element analysis divided by the nominal stress obtained from elastic beam theory. For the stiffener/bracket connection subjected to tensile loading, the nominal stress was simply calculated as the applied force divided by the cross-section area of the stiffener, F/A. For the stiffener/bracket connection subjected to bending, the local nominal stresses were calculated by means of elastic beam theory for unsymmetrical cross sections. The effect of shear lag was accounted for by a constant effective breadth, $b_{eff} = 0.54b$ where b is the breadth of the test specimen [18].

The accuracy of the structural stress obtained from the extrapolation methods will clearly depend on the location of the extrapolation points. In this study it is seen that for the extrapolation methods suggested by Niemi [12] all the extrapolation points are located outside the region affected by the local notch (singularity) and inside the region solely influenced by the structural geometry. This explains the good correspondence between the structural stress concentrations obtained from the proposed extrapolation method and the stress concentration factors obtained from the

Table 2
Structural stress concentration factor at the weld toe location, K_g, obtained from finite shell element analysis and finite solid element analysis with stress extrapolation for the flat bar/bracket connection

	K_g	$K_{g,corr}$
Extrapolation method suggested by Niemi [12]		
Quadratic extrapolation, method 1		
Shell element analysis	1.38	
Solid element analysis	1.43	
Linear extrapolation, method 2		
Shell element analysis	1.26	1.37
Solid element analysis	1.28	1.39
Proposed extrapolation method based on a 'singularity check'		
Shell element analysis	1.35	
Solid element analysis	1.34	

Table 3
Structural stress concentration factor at the weld toe location, K_g, obtained from finite shell element analysis or finite solid element analysis with stress extrapolation for detail A

	$K_{g,bending}$	$K_{g,axial}$	$\frac{K_{g,bending}}{K_{g,axial}}$
Extrapolation method suggested by the IIW [13]			
Shell element analysis			
Linear extrapolation	1.694	1.592	1.064
Quadratic extrapolation	1.767	1.667	1.066
Solid element analysis			
Linear extrapolation	1.708	1.558	1.096
Quadratic extrapolation	1.791	1.622	1.104
Proposed extrapolation method based on a 'singularity check'			
Shell element analysis, model 3	1.558	1.523	1.043
Solid element analysis	1.505	1.464	1.028

extrapolation methods suggested by Niemi [12]. From the results presented in Tables 3 and 4 it is noticed that the structural stress concentration factors obtained from the linear and the quadratic extrapolation procedure suggested by the IIW [13] were generally higher than the structural stress concentration factors obtained from the proposed extrapolation method (by 5–19%). Both the proposed extrapolation method and the quadratic extrapolation method suggested by the IIW [13] were performed by means of the method of least squares and the higher structural stress concentration factor obtained from the quadratic extrapolation method could be explained by the fact that the quadratic extrapolation was only performed in the leading part of the stress field influenced by the structural geometry and thus in a

Table 4
Structural stress concentration factor at the weld toe location, K_g, obtained from finite shell element analysis or finite solid element analysis with stress extrapolation for detail B

	$K_{g,bending}$	$K_{g,axial}$	$\frac{K_{g,bending}}{K_{g,axial}}$
Extrapolation method suggested by the IIW [13]			
Shell element analysis			
Linear extrapolation	1.571	1.578	0.996
Quadratic extrapolation	1.636	1.635	1.001
Solid element analysis			
Linear extrapolation	1.568	1.493	1.050
Quadratic extrapolation	1.634	1.549	1.051
Proposed extrapolation method based on a 'singularity check'			
Shell element analysis, model 3	1.488	1.522	0.978
Solid element analysis	1.442	1.407	1.025

region with an overall steeper stress gradient. The leading extrapolation point used in the linear and the quadratic extrapolation suggested by the IIW [13] applied on the stress results obtained from the finite finite solid element analysis of detail A was also within the region influenced by the singularity (notch) and this could explain the relatively higher stress concentration factor (13.0 and 19.0%) compared to the proposed extrapolation method.

5. Static test program

In order to assess the validity of the finite element analysis (Section 4) and to verify the actual stress state experienced in the test rig (Section 6), the stress results obtained from the finite element analysis were compared to measured strains obtained from static tests. The finite solid element analysis showed that stress in the transverse direction was less than 3% of the stress in the longitudinal direction at the location of the leading strain gage, thus the measured strains were converted to stresses using Hooke's law assuming an uni-axial stress state.

Figs. 9 and 10 show that the difference between the finite solid element stress results and the strain gage measurements in the region solely influenced by the structural stress was less than 6% for the flat bar/bracket connection (F2-2) and less than 8% for the bulb stiffener/bracket connection (A1-L). In the nominal stress region, the difference was less than 8% for the flat bar/bracket connection (F2-1) and less than 3% for the bulb stiffener/bracket connection (A2-L). The larger difference experienced by the flat bar/bracket connection in the nominal stress region was identified as secondary bending effects mainly caused by misalignment of test specimen in test rig and specimen bending distortions. Thus, all the test specimens included in the fatigue test program were strain gaged and tested statically prior to each test run. The test

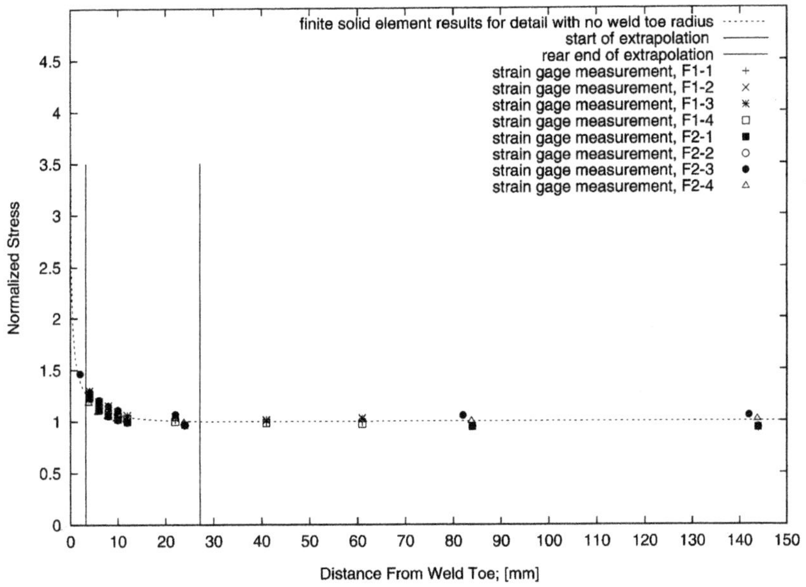

Fig. 9. Normalized stress obtained from finite solid element results (weld no radius) and strain gage measurements, flat bar/bracket connection. Two test specimens (F1 and F2) were tested statically with strain gages at all four hot-spot locations.

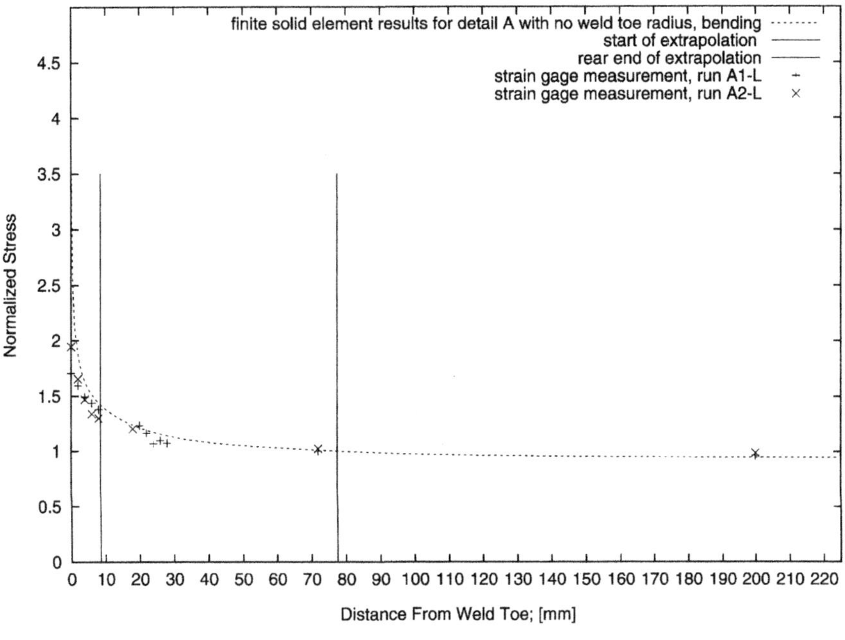

Fig. 10. Normalized stress obtained from finite solid element results (weld no radius) and strain gage measurements, detail A. Two test specimens (A1 and A2) were tested statically with strain gages at hot-spot location L (see Fig. 3).

specimens were then aligned in the test rig, followed by re-testing statically. This was done in order to minimize the secondary bending effects. The large difference seen for detail A in the region influenced by the local notch can be explained by the profile grinding of the weld surfaces of detail A giving the details a rather smooth and regular weld profile with a moderate notch stress concentration factor, whereas the finite element model was modeled with no radius and thus 'infinite' stress at the weld toe.

6. Fatigue test program

6.1. Fatigue test rig arrangement

Two different test rig arrangements were used in the fatigue test program. The as-welded flat bar/bracket connection was tested in tensile loading (Fig. 11), while the profile ground fillet welded longitudinal bulb stiffener/bracket connection was tested under three-point bending (Fig. 12).

The axial loading was provided by a servo hydraulic actuator capable of producing a maximum dynamic load of $\pm$ 500 kN. The fatigue loading was applied in load control at a *R*-ratio equal to 0.44. The actuator was mounted in a steel frame and the test specimens were locked by hydraulic clamps. The testing was performed in laboratory air at ambient temperature with a loading frequency of 10 HZ.

The test rig used for the longitudinal bulb stiffener/bracket connection was arranged in order to simulate the effect of lateral load transferred from longitudinal stiffeners into transverse girder web. To maintain a well-controlled testing environment a pinned boundary solution for the test specimen was employed. At each model end a 14 mm flat aluminum plate was attached and equipped with a fixed shaft. One of the end plates also had an additional ball bearing which eliminated all friction at the support and secured an ideal, pinned end support. The loading was provided by

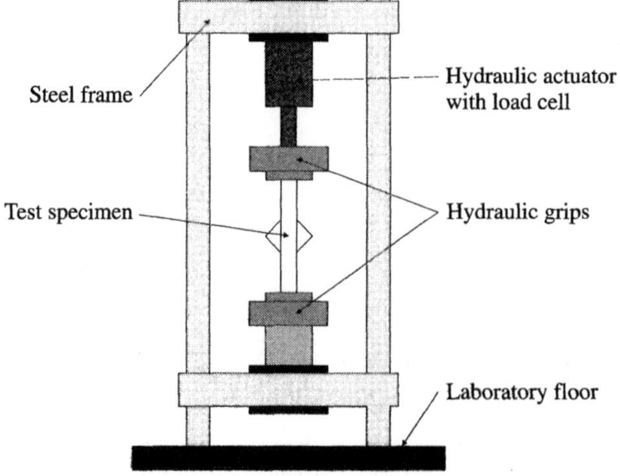

Fig. 11. Test rig arrangement for the bracket flat/bar connection, tensile loading.

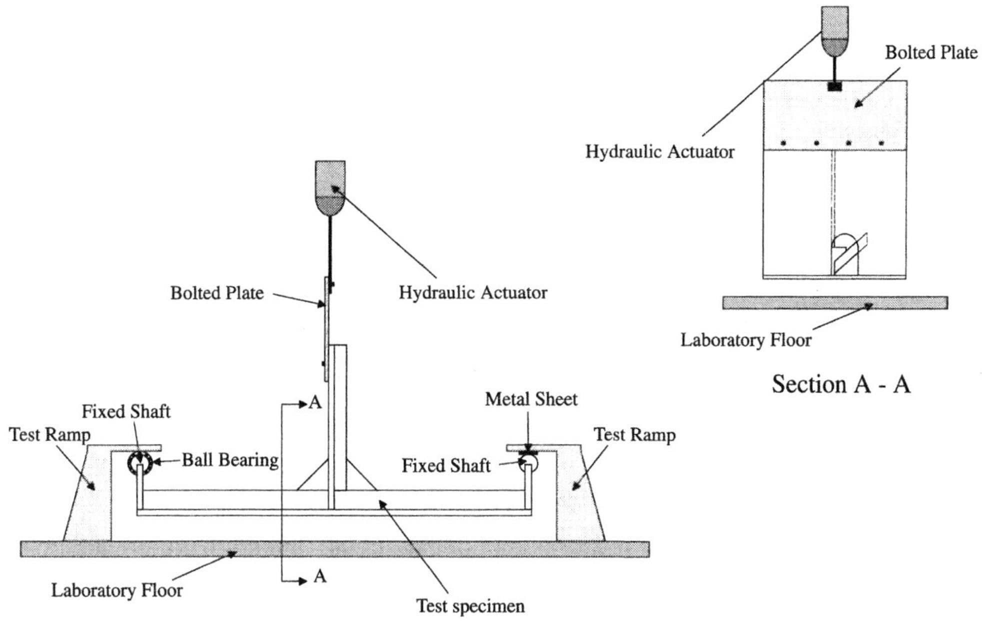

Fig. 12. Test rig arrangement for the bracket flat/bulb stiffener connection, bending loading.

a servo hydraulic actuator capable of producing a maximum dynamic load of $\pm$ 25 kN. The fatigue loading was applied in tension with load control at a R-ratio equal to 0.44. The testing was performed in laboratory air at ambient temperature with a loading frequency of approximately 4.5 HZ.

6.2. Constant amplitude fatigue test results

6.2.1. Flat bar/bracket connection

The S-N data were analyzed assuming a linear S-N curve on a log–log scale, using statistical regression analysis (Table 5). Due to the rather limited number of test specimens (12) the slope of the regression line, m, was fixed according to the slope of the appropriate design S-N curve. The details tested in this study had four possible locations of fatigue crack failure (hot-spot locations) with different weld geometry subjected to the same nominal stress state and environment. The fatigue failure was defined as a complete loss of the load bearing capacity of the cross-sectional area and no repair of the fatigue cracks was performed during the fatigue test program. Consequently, the S-N data derived from the fatigue tests in this study will be at the lower end of the scatter band since the fatigue life has been taken as the fatigue failure at the 'weakest link' location. The test S-N data together with their mean regression curve and the approximated parallel 95% lower limit confidence regression line using a structural stress range obtained from the proposed extrapolation method are presented in Figs. 13 and 14.

It has been shown that small-scale fatigue test specimens may be of insufficient size to retain the full amount of residual stresses experienced in full-scale structures [19],

Table 5
The logarithm of the standard deviations, $STDV_{\log N}$, obtained from the regression analysis using a fixed slope exponent, $m = -3.00$

Detail	Number of samples, n	$STDV_{\log N}$
Flat bar/bracket connection	12	0.094
Bulb stiffener/bracket connection		
Detail A	11	0.209
Detail B	10	0.156

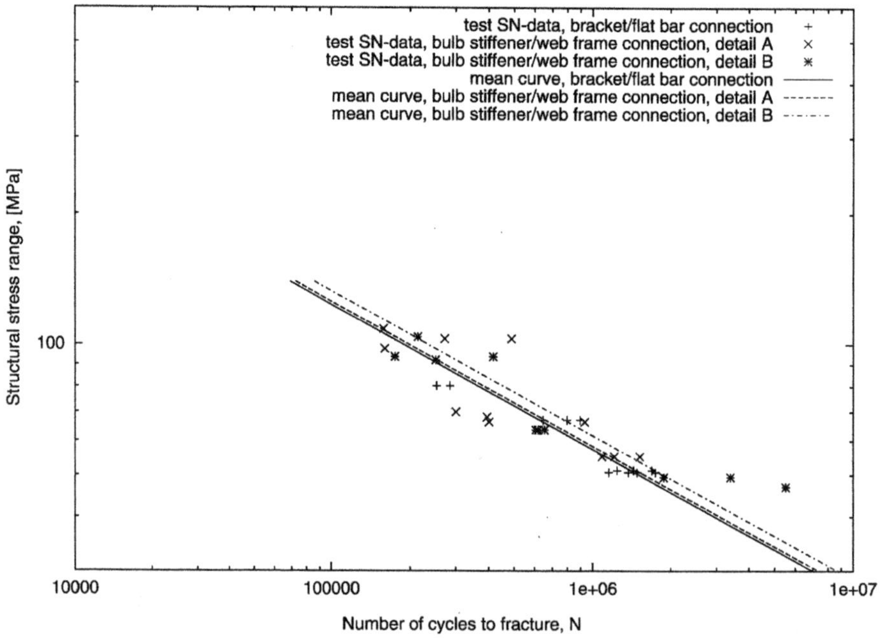

Fig. 13. Constant amplitude test S-N data obtained for bracket/flat bar connection and longitudinal stiffener/bracket connection using a structural stress range obtained from the proposed extrapolation method, mean regression S-N curve with fixed $m = -3.0$.

implying longer fatigue lives for small-scale fatigue test specimens, particularly in the long life region. It has, however, been reported that the lack of residual stresses in laboratory fatigue test specimens can be eliminated by testing the specimens on a high R-ratio [19]. Residual stress measurements on an unbroken test specimen selected randomly from the whole test sample gave a surface residual stress level of 118 MPa (error margin 19 MPa) at a distance 2 mm from the weld toe. This was approximately 77% of the yield strength of the base material. The residual stress level was suspected to be even higher closer to the weld toe due to the high residual stress gradients normally found close to the weld toe for welded aluminum structures [20]. The rather high residual stress level found in the test specimen confirmed that the fatigue test specimens in this study were able to retain high residual stresses in the magnitude of

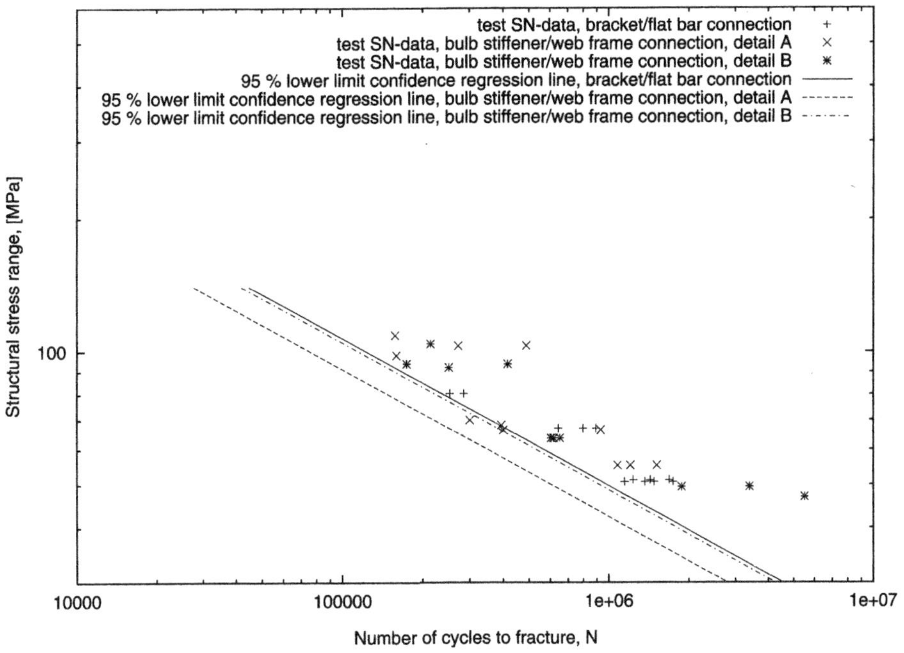

Fig. 14. Constant amplitude test S-N data obtained for bracket/flat bar connection and longitudinal stiffener/bracket connection using a structural stress range obtained from the proposed extrapolation method, approximated parallel 95% lower limit confidence regression line with fixed $m = -3.0$.

the yield stress (in the HAZ). Additionally, the constant amplitude fatigue testing was performed on a high R-ratio ($R = 0.44$) and it was therefore assumed that any positive effects on the fatigue life due to a possible lack of residual stresses in the test specimen could be ruled out.

In order to consider the present fatigue test results with regard to design implications, a factor of conservatism of design, ζ, was defined as the stress obtained at the approximated parallel 95% lower limit confidence regression line at a reference fatigue life, $N_{ref} = 2 \times 10^6$, obtained by test results, divided by the characteristic design stress range given by fatigue design S-N curve at the same reference fatigue life. Thus a $\zeta \geqslant 1.0$ would indicate a safe fatigue life prediction by means of the design S-N curve and a $\zeta < 1.0$ would indicate a possibility of an unsafe fatigue life prediction.

In this study the structural stress concentration factor was found by two extrapolation methods suggested by Niemi [12] and a proposed extrapolation method based on a 'singularity check' (Section 3.3). At the present time, no recommendations are given in codes on design S-N curves for the fatigue design of aluminum structures by means of a structural stress range approach. Since the effect of the local weld geometry is included in the design hot-spot S-N curve, it is necessary that the local conditions at the weld toe of the details are reflected in the design S-N curve. That means that different hot-spot S-N curves should be used for, e.g. as-welded, profile ground welds, or toe ground welds. The results obtained from the numerical stress analyses (Section 4.3) and the fatigue tests were used together with the design S-N curve 25 taken from the IIW fatigue design recommendation [13]. The design S-N curve 25 is commonly assigned to transversely loaded butt welds with weld toe angles $\geqslant 50°$. The design S-N curve 25 was chosen on basis of the measured weld geometry characteristics presented in Table 1 where an average value of the weld toe angle can be found to be more than 55°. Table 6 shows that the design hot-spot S-N curve 25 together with the extrapolation procedures to obtain the structural stress concentration factor, K_g, provide safe life predictions for the flat bar/bracket connection tested in this study.

Design S-N curves are determined by means of statistical regression analysis on fatigue test data [21] and include the effects of several parameters such as, e.g. weld geometry, weld imperfections, stress directions, residual stresses, metallurgical conditions, and welding process. The regression constants obtained from the fatigue testing are identified with an uncertainty, leading to a total uncertainty which increases from the center of gravity of the experimental data. The confidence level usually applied in fatigue design codes is 95% [21]. At high number of test specimens the 95% lower confidence limit can be approximated as the mean minus two standard deviation limit and the commonly used design curve in fatigue design is then exceeded by a 97.7% probability by test values. Normally, design codes are also derived from several different test batches which implies that joints with different mean values and standard deviations are included in the same detail class and this will further increase the scatter of the joint category. Table 6 shows that by using design S-N curve 25 to assess the fatigue strength of the flat bar/bracket connection provides rather high factors of conservatism (between 1.5 and 1.7). The logarithm of the standard deviation of the single batch of test specimens tested in this study, using a fixed slope exponent, was $STDV_{\log N} = 0.094$ (Table 5). This indicates a relatively narrow scatter of the fatigue test results and thus much higher lower bound fatigue strength than generally seen for

Table 6
Factor of conservatism of design, ζ, for the flat bar/bracket connection at a reference fatigue life of $N_{ref} = 2.0 \times 10^6$

	Reference stress (σ_s) based on design codes	Reference stress ($K_g\sigma_n$) based on test data	Factor of conservatism of design; ζ
Extrapolation method suggested by Niemi [12]			
Quadratic extrapolation, method 1			
Shell element model	25	39.80	1.58
Solid element model	25	41.24	1.68
Linear extrapolation, method 2			
Shell element model	25	39.51	1.59
Solid element model	25	40.09	1.60
Proposed extrapolation method based on a 'singularity check'			
Shell element model	25	38.93	1.55
Solid element model	25	38.64	1.54

the same type detail configurations in design codes and design recommendations (which normally has a standard deviation approximately 2–2.5 times larger). The rather conservative fatigue lives obtained in this study could therefore be explained by a rather narrow scatter of the test results and thus higher lower bound fatigue strength than normally seen for design S-N curves assigned to similar joint category.

6.2.2. Longitudinal bulb stiffener/bracket connection

The S-N data were analyzed assuming a linear S-N curve on a log–log scale, using statistical regression analysis (Table 5 and Figs. 13 and 14). Due to the rather limited number of as-welded test specimens (11 test specimens of detail A and 10 test specimens of detail B) the slope of the regression line, *m*, was fixed according to the slope of the appropriate design S-N curve [13]. All the welds had been profile ground at the yard before the delivery, that is the entire weld face has been machined and given a favorable shape to reduce the stress concentration factor. The welds had not been fully profile ground, that is toe grinding to a depth of 0.5 mm below the bottom of any visible undercut [22] to remove possible harmful defects at the weld toe.

In this study the results obtained from the numerical stress analyses (Section 4.3) and the fatigue tests were used together with the design hot-spot S-N curve 32 taken from the IIW fatigue design recommendation [13]. The design S-N curve 32 is commonly assigned to transversely loaded butt welds with weld toe angles $\leqslant 50°$. The design hot spot S-N curve 32 was chosen on the basis of the measured weld geometry characteristics presented in Table 1. Tables 7 and 8 show that the design hot-spot S-N curve 32 resulted in safe fatigue life predictions for both detail A and detail B.

It was expected that the constant amplitude fatigue test results of the bulb stiffener/bracket connection would show a larger variability than the test results of the flat bar/bracket connection in that the production and the shipment of the fatigue test specimens had been subjected to a larger extent of unknown and uncontrolled parameters. This was confirmed by the regression analysis where the logarithm of the

Table 7
Factor of conservatism of design, ζ, for the detail A at a reference fatigue life of $N_{ref} = 2.0 \times 10^6$

	Reference stress (σ_s) based on design codes	Reference stress ($K_g \sigma_n$) based on test data	Factor of conservatism of design; ζ
Extrapolation method suggested by the IIW [13]			
Linear extrapolation			
Shell element	32	35.87	1.13
Solid element model	32	36.16	1.13
Quadratic extrapolation			
Shell element	32	37.41	1.18
Solid element model	32	37.92	1.19
Proposed extrapolation method based on a 'singularity check'			
Shell element model 3	32	32.99	1.03
Solid element model	32	31.87	1.00

Table 8
Factor of conservatism of design, ζ, for the detail B at a reference fatigue life of $N_{ref} = 2.0 \times 10^6$

	Reference stress (σ_s) based on design codes	Reference stress ($K_g\sigma_n$) based on test data	Factor of conservatism of design; ζ
Extrapolation method suggested by the IIW [13]			
Linear extrapolation			
Shell element model 3	32	40.23	1.26
Solid element model	32	40.15	1.25
Quadratic extrapolation			
Shell element model 3	32	41.89	1.31
Solid element model	32	41.84	1.31
Proposed extrapolation method based on a 'singularity check'			
Shell element model 3	32	38.10	1.19
Solid element model	32	36.93	1.15

standard deviation of the longitudinal bulb stiffener/web frame connection was approximately 2 times the logarithm of the standard deviation of the flat bar/bracket connection (Table 5). Figs. 13 and 14 show that even if the mean regression curve of the flat bar/bracket connection was below the mean regression curves of detail A and B, the larger variability of the constant amplitude test S-N data of detail A and B resulted in a lower factor of conservatism of design, ζ, for detail A and B than for the flat bar/bracket connection. The logarithm of the standard deviations of detail A and detail B are approximately in the same order as normally seen for the same type of detail configurations in design codes and design recommendations. Any possible effect on the fatigue strength due to the loading mode (tensile or bending) was assumed to be limited since this effect was mostly covered by a change in the structural stress concentration factor (Tables 3 and 4).

In view of the results obtained in this study, it is recommended that a design S-N curve of fatigue class 32 can be used safely together with the proposed extrapolation procedure to determine the structural stress, or in cases with a well-defined main plate thickness together with the linear or quadratic extrapolation procedure suggested by the IIW [13] for the fatigue assessment of aluminum ship details with profile ground fillet welds.

7. Conclusions

The purpose of this paper has been to improve the design analysis methodology based on a consistent calculation of the structural stress range and the selection of appropriate design S-N curves. An S-N curve approach based on a structural stress range has been applied to constant amplitude data of aluminum stiffener/girder connections. A total of 12 small-scale test specimens and 21 full-scale test specimens have been included in the test program. The test results have been closely linked to finite shell and finite solid element analysis, and strain gage measurements of the tested details in order to accurately define the structural stress range which was used in the fatigue assessment procedure.

Existing methods for extrapolating stresses to hot-spots in welded plate structures have been assessed and a new and more general method for the structural stress extrapolation to be used together with a hot-spot design S-N curve has been established and validated. The method for determining the optimal location of points for stress extrapolation has been based on the asymptotic behavior of the stresses adjacent to an idealized notch ('singularity'). It was shown that the influence of the weld notch stress can be separated from the structural stress on different structural ship details by means of this 'singularity check'. Finite shell element and solid element analysis showed that the extrapolated structural stress obtained from existing extrapolation methods was rather sensitive to how the weld stiffness had been accounted for in the finite shell element model. It was therefore recommended that finite solid element models should be used in the calculation of the hot-spot stress, or at least to verify methods based on finite shell elements.

On the basis of the constant amplitude data presented in this paper, the most consistent device of S-N curves in combinations with the proposed extrapolation method to determine structural stress is an S-N curve with a characteristic strength of 32 corresponding to a fatigue life of 2×10^6 cycles, for the fatigue assessment of profile ground fillet welded stiffener/bracket connections and a characteristic strength of 25 for the fatigue assessment of as-welded stiffener/bracket connections.

References

[1] Fredriksen A. Fatigue aspects of high speed craft. In: Proceedings of the Fourth International Conference on Fast Sea Transportation, FAST'97, Sydney, Australia, vol. 1, Baird Publications, Hong Kong, July 1997, p. 217–24.

[2] Miner MA. Cumulative damage in fatigue. J Appl Mech 1945;12(3):A-159–64.

[3] Fricke W, Petershagen H. Detail design of welded ship structures based on hot-spot stresses. In: Caldwell JB, Ward G, editors. Practical design of ships and mobile units. Amsterdam: Elsevier, 1992, p. 21,087–21,100.

[4] Susuma Machida, Masaaki Matoba, Hitoshi Yoshinari, Ryuichi Nishimura. Definition of hot spot stress in welded plate type structure for fatigue assessment. Part 2 — derivation of hot spot stresses by finite element analysis. Technical Report IIW-XIII-1448-92, IIW — The International Institute of Welding, Cambridge, United Kingdom, 1992.

[5] Mikkola TPJ. Comments: modeling of hot spot area using rigid links. In: Moan T, Berge S, editors. Proceedings of the 13th International Conference of Ship and Offshore Structures Congress, ISSC'97, Trondheim, Norway, vol. 3, August 1997.

[6] Cramer E, Gran S, Holtsmark G, Lotsberg I, Løseth R, Olaisen K, Valsgård S. Class note: Fatigue assessment of ship structures. Technical Report 93-0432/Rev. 6, Det Norske Veritas Classification A.S., Høvik, Norway, September 1996.

[7] Lehtonen MJ. On the evaluation of hot-spot stresses using FEM. In: Blom AF, editor. Proceedings of the First North European Engineering and Science Conference, NESCO I, Stockholm, Sweden, EMAS Publishing, Cradley Heath, UK, October 1997.

[8] Erkki Niemi. Stress determination for fatigue analysis of welded components. Technical Report IIS/IIW-1221-93, IIW — The International Institute of Welding, Cambridge, United Kingdom, 1995.

[9] Bård Wathne Tveiten. Fatigue assessment of welded aluminium ship details. Ph.D. thesis, Department of Marine Technology, The Norwegian University of Science and Technology, NTNU, Trondheim, Norway, 1999.

[10] Meneghetti G, Tovo R. Stress gradient effects on fatigue strength of light alloy welds. Proceedings of the International Conference of Aluminium Structures, INALCO'98, United Kingdom, April 1998.
[11] Radenkovic D. Stress analysis in tubular joints. International Conference on Steel in Marine Structures, Paris, France, 1981.
[12] Niemi E. On the determination of hot spot stresses in the vicinity of edge gussets. Technical Report IIW-XIII-1555-94, IIW — The International Institute of Welding, Cambridge, United Kingdom, 1994.
[13] Hobbacher A. Recommendations for fatigue design of welded joints and components. ISO standard proposal. Technical Report IIW doc XIII-1539-96/XV-845-96, IIW — The International Institute of Welding, May 1996.
[14] Williams ML. Stress singularities resulting from various boundary conditions in angular corners of plates in extensions. J Appl Mech 1952;19:526–8.
[15] Paauw AJ, Aabø S, Engh B, Solli O. Fatigue of welded aluminium joints. Technical Report STF34 A85024, SINTEF, Trondheim, Norway, 1985 (in Norwegian).
[16] Veritas SESAM System A.S., Høvik, Norway. SESAM Technical Description, 96-7006/rev.0 edition, April 1996.
[17] Engesvik KM. Analysis of uncertainties in the fatigue capacity of welded joints. Ph.D. thesis, Department of Marine Technology, The Norwegian Institute of Technology, NTH, Trondheim, Norway, 1981.
[18] Shade HA. The effective breadth concept in ship-structure design. Transactions of the Society of Naval Architects and Marine Engineers, vol. 61, The Society of Naval Architects and Marine Engineers, 1953, p. 410–30.
[19] Aabø S, Paauw AJ, Engh B. The effect of the R-ratio on fatigue life. Third International Conference on Aluminium Weldments, April 1985, p. III.7.1–III.7.7.
[20] Mazzolani FM. Aluminium alloy structures, 2nd ed. London, UK: E and FN SPON, 1995.
[21] ASTM. Statistical analysis of linear or linearized stress-life and strain-life fatigue data. Technical Report E 739-91, American Society for Testing and Materials, Philadelphia, USA, 1991.
[22] Department of Energy, London. Offshore installations: guidance on design, construction and certification, 4 ed., 1990.

AUTHOR INDEX

Printed and bound by CPI Group (UK) Ltd, Croydon, CR0 4YY
08/05/2025
01864849-0002